MAMMALS OF AFRICA

VOLUME I

INTRODUCTORY CHAPTERS AND AFROTHERIA

Series Editors

Jonathan Kingdon

Department of Zoology, University of Oxford

David C. D. Happold

Research School of Biology, Australian National University

Thomas M. Butynski

Zoological Society of London / King Khalid Wildlife Research Centre

Michael Hoffmann

International Union for Conservation of Nature – Species Survival Commission

Meredith Happold

Research School of Biology, Australian National University

Jan Kalina

Soita Nyiro Conservancy, Kenya

MAMMALS OF AFRICA

VOLUME I

INTRODUCTORY CHAPTERS AND AFROTHERIA

EDITED BY JONATHAN KINGDON, DAVID HAPPOLD, MICHAEL HOFFMANN, THOMAS BUTYNSKI, MEREDITH HAPPOLD AND JAN KALINA

ILLUSTRATED BY JONATHAN KINGDON
PEN AND INK ILLUSTRATIONS OF SMALL AFROTHERIA BY MEREDITH HAPPOLD

BLOOMSBURY WILDLIFE
LONDON • OXFORD • NEW YORK • NEW DELHI • SYDNEY

BLOOMSBURY WILDLIFE
Bloomsbury Publishing Plc
50 Bedford Square, London, WC1B 3DP, UK
29 Earlsfort Terrace, Dublin 2, Ireland

BLOOMSBURY, BLOOMSBURY WILDLIFE and the Diana logo are trademarks of Bloomsbury Publishing Plc

First published in the United Kingdom 2013
This edition published 2024

A catalogue record for this book is available from the British Library

Library of Congress Cataloguing-in-Publication data has been applied for

ISBN: HB: 978-1-3994-2031-0; ePDF: 978-1-4081-8990-0

2 4 6 8 10 9 7 5 3 1

Commissioning editor: Nigel Redman
Design and project management: D & N Publishing, Baydon, Wiltshire
Printed and bound in India by Replika Press Pvt. Ltd.

FSC www.fsc.org
MIX
Paper | Supporting responsible forestry
FSC™ C016779

To find out more about our authors and books visit www.bloomsbury.com and sign up for our newsletters

Recommended citations:

Series: Kingdon, J., Happold, D., Butynski, T., Hoffmann, M., Happold, M. & Kalina, J. (eds) 2013. *Mammals of Africa* (6 vols). Bloomsbury Publishing, London.

Volume: Kingdon, J., Happold, D., Hoffmann, M., Butynski, T., Happold, M. & Kalina, J. (eds) 2013. *Mammals of Africa. Volume I: Introductory Chapters and Afrotheria.* Bloomsbury Publishing, London.

Chapter/species profile: e.g. Taylor, A. 2013. *Orycteropus afer* Aardvark; pp 210–215 in Kingdon, J. *et al.* (eds) 2013. *Mammals of Africa. Volume I: Introductory Chapters and Afrotheria.* Bloomsbury Publishing, London.

Donors and Patrons

T. R. B. Davenport, D. De Luca and the Wildlife Conservation Society, Tanzania

R. Dawkins

R. Farrand & L. Snook

R. Heyworth, S. Pullen and the Cotswold Wildlife Park

G. Ohrstrom

Viscount Ridley & M. Ridley

L. Scott and the Smithsonian UK Charitable Trust

M. & L. Ward

R. & M. Ward

Contents

Series Acknowledgements

Jonathan Kingdon, David Happold, Thomas Butynski,

Michael Hoffmann, Meredith Happold and Jan Kalina

The editors wish to record their thanks to all the authors who have contributed to *Mammals of Africa* for their expert work and for their patience over the very protracted period that these volumes have taken to materialize. We also thank the numerous reviewers who have read and commented on earlier drafts of this work.

We are also grateful for the generosity of our sponsoring patrons, whose names are recorded on our title pages, who have made the publication of these volumes possible.

Special thanks are due to Andy Richford, the Publishing Editor at Academic Press, who initiated and supported our work on *Mammals of Africa*, from its inception up to the point where Bloomsbury Publishing assumed responsibility, and to Nigel Redman (Head of Natural History at Bloomsbury), David and Namrita Price-Goodfellow at D & N Publishing, and the whole production team who have brought this work to fruition. We also acknowledge, with thanks, Elaine Leek who copy-edited every volume. We are grateful to Chuck Crumly, formerly of Academic Press and now the University of California Press, for being our active advocate during difficult times.

We have benefited from the knowledge and assistance of scholars and staff at numerous museums, universities and other institutions all over the world. More detailed and personal acknowledgements follow from the editors of each volume.

The editors are also grateful to the coordinating team of the Global Mammal Assessment, an initiative of the International Union for Conservation of Nature (IUCN), which organized a series of workshops to review the taxonomy and current distribution maps for many species of African mammals. These workshops were hosted by the Zoological Society of London, Disney's Animal Kingdom, the Owston's Palm Civet Conservation Programme, and the Wildlife Conservation Research Unit at the University of Oxford; additionally, IUCN conducted a review of the maps for the large mammals by the Specialist Groups of the Species Survival Commission. We owe a particular word of thanks to all the staff and personnel who made these workshops possible, and to the participants who attended and provided their time and expertise to this important initiative. We also thank IUCN for permission to use data from the *IUCN Red List of Threatened Species*.

ABOVE LEFT: Jan Kalina.
ABOVE: *(from left to right)* Jonathan Kingdon, Thomas Butynski, Meredith Happold, David Happold and Andrew Richford.
LEFT: Jonathan Kingdon (*left*) and Mike Hoffmann.

Acknowledgements for Volume I

Jonathan Kingdon

When Andrew Richford at Academic Press first took up my suggestion of a Handbook of African Mammals (partly modelled on *The Birds of Africa*, which he had edited), neither of us anticipated that it might take as long to reach print as it has. Even so the journey has been deeply rewarding in terms of scientific and intellectual stimulus and my gratitude to the authors and editors that have joined me in this enterprise is immense. Above all, my fellow-editors, Mike Hoffmann and Andy Richford, have offered a quality of fellowship that has been unprecedented in a lifetime of many shared endeavours. Tom Butynski, Jan Kalina, and David and Meredith Happold have also been splendid collaborators, and Tom a true companion and stimulating friend on the field trips we have shared. In assuming the task of drawing the many line drawings for the small mammal skulls, Meredith Happold relieved me of much labour, for which I am most grateful. Likewise, David and Meredith Happold, in taking exclusive charge of editing all small mammal profiles, significantly reduced the burden on other editors. This has freed us all to concentrate on our chosen areas of interest. Nigel Redman, David and Namrita Price-Goodfellow, and Elaine Leek have been wonderful collaborators in bringing these volumes to print. I am deeply grateful for their enthusiasm and hard work.

Many scientists have shared their time and expertise with us, some freely giving many hours, days and weeks, over several years to referee, review and edit texts. In this regard special mention and profound gratitude is due to Colin Groves, John Harris, Erik Seiffert, Clifford Jolly, Simon Bearder, Matt Cartmill, Richard Dawkins, Fritz Vollrath, Mike Lock, Bob Morley and the late Peter Grubb: as author, referee, reviewer and friend, Peter's words and influence on these volumes serve as a fitting memorial to an excellent scientist and a fine man.

I owe a personal debt to my many colleagues at Oxford University from the support of John Pringle, Peter Miller, Niko Tinbergen, Malcolm Coe and others, at my induction into the Zoology Department, well over 30 years ago, to my contemporaries, Marion Dawkins, Richard Dawkins, Paul Harvey, David Macdonald, Bob May, Mark Stanley-Price, Fritz Vollrath, Dawn Burnham, Steve Cobb and many others. Dick Southwood and Ryk Ward (the latter my colleague in the Institute of Biological Anthropology) are remembered with respect and affection. With the creation of WildCRU, Oxford University's conservation hub under David Macdonald, I found myself gratefully joining a community of conservation scientists all dedicated to the survival of plant and animal ecosystems worldwide.

In the long chain of events that eventually led to the decision to embark on an inventory and handbook of the mammals of Africa Julian Huxley was an early but profoundly formative influence. On retirement as the founding Director of UNESCO, he revisited many African countries in 1960 to update a previous landmark report on progress in Science and Education on our continent. During his visit to the Makerere University Zoology Field Course at Mweya, in western Uganda I was asked to guide him and the students. In the evening, as we listened to the choruses of hippos as they left the water to go grazing, Julian quizzed me about my local background and biological interests, remarking 'you should do something with the knowledge you have acquired'. Subsequently he helped shepherd me through the first two volumes of my *East African Mammals: An Atlas of Evolution in Africa*. His passionate interest in that enterprise was a legacy that helped drive my own determination to inform others about the mammals of our continent, their evolution and their relevance in the unfolding quest to discover the historical and prehistoric roots of our own humanity.

What and who inspires a youngster's interests is often arbitrary. I remember with affection Saidi Batale, who took me hunting hyraxes and cane-rats when I was six years old. Crawling over the rocks of Mwanza Bay with him, watching otters, mongooses and tomb bats laid foundations for my lifetime interest in the natural history of my home continent. As a schoolboy, banished to England, it was mutual loneliness and longing for the tropics that brought me and Percy Willoughby Lowe together. I became an eager sounding board for his reminiscences of scientific collecting expeditions, some of them made at the beginning of the twentieth century (several African mammals are named after him, notably a guenon from West Africa, gerbils and a bat from the Jebel Marra in Sudan and a genet from Tanzania). He put me in touch with C.H.B Grant and R.W. Hayman at the British Museum (Natural History). Soon thereafter, at Percy's and their behest, I, too, was collecting specimens and I still have a certificate of thanks, issued by the British Museum: I was 16. My friendship with Percy made discoveries and events that happened long before I was born tangible. Whenever I handle the bird and mammal specimens that he so carefully preserved and read his handwritten labels I am moved by more than memories of an endearing old man. I give thanks for the spirit of scientific enquiry and intellectual adventure that sent Percy (and before him, William Bates, Alfred Wallace and Charles Darwin) to remote places, to all our benefit. Their spiritual descendants are now countless but all the contributors and most readers of this work will be quick to acknowledge our communal debt to these pioneers and to the many workers in universities, laboratories, museums and conservation societies who strive to maintain the momentum of biological science in an age of ecological and economic barbarism.

I had the good fortune to have known 'Iodine' Ionides, an avid naturalist and collector, for many years a colleague of my father and Chief Game Warden in the Southern Province of Tanganyika (my essay on polymorphism in Giant Sengis in this volume owes much to our discussions and to the carefully transcribed field notes that he sent me in the early 1960s and to the specimens that he collected). Many aspects of our science still depend upon past and present observers and collectors of specimens and records. They are often inadequately acknowledged and in addition to Percy Lowe I should especially mention Tony Archer, Richard Gicheru, Robert Glen, David Harrison, Paul Bates, Peter Mwangi, Andy Williams and Raphael Abdullah (and his father, Ali) all of whom have made exceptionally important contributions in this regard.

In my teenage years I walked on foot over extensive areas of central Tanganyika with Bill Moore-Gilbert and Hamisi Sikana, tough-as-boots companions who were alert to the countless traces of passage that web the Nyika, from sengi pathways and broken cobwebs to honeyguide cries, caracal-scarred tree trunks, hedgehog tracks, Acacia rat 'corridors' or Ratel spoor. Thanks, too, to Mtemi Senge Masembe, in 1953 boyish chief of Kititima, Turu, Tanganyika: our dawn expeditions around Singida played an essential part in my development as a naturalist. I recently retrieved a poem in Kiswahili that I wrote to express the spirit of our long conversations. 'Ku safiri katika nyika ni kukutana maisha mpya' or 'to travel in the nyika is to encounter life anew'. For companionship in the field and for much practical help over the years I owe very special debts to all my immediate family, especially my mother, Dorothy, my wife, Elena, my sons and daughters, Margaret Ward and family, and among many generous friends, Tony Archer, Eric Balson, Bandusya, Isabirye Basuta, Leonard Beadle, Francois Bourliere, Tom Butynski, Malcolm Coe, Gerard and Ahn Galat, Jean-Pierre and Annie Gautier, Robert Glen, Margaret Kinnaird, Lysa Leland, Dan Livingstone, Reginald Moreau, Ian Parker, Galen Rathburn, Alan Root, Laura Snook, Tom Struhsaker, Desmond Vesey FitzGerald, Alan Walker and many others listed below.

It has been very reassuring that so many institutions, worldwide, have been sufficiently interested in the mammals of Africa to display some of the drawings and paintings that illustrate these volumes. I am grateful to the following for hosting these works, many of them major exhibitions: the Smithsonian Institution in Washington, USA, the Los Angeles County Museum, California, Duke University, North Carolina, the Senckenberg Museum in Frankfurt, Germany, Justus Leibig University, Geissen, Germany, the Natural History Museum in Geneva, Switzerland, the Zoological Society of London, The Barbican Centre, London, the Royal Society (1982 Darwin Centenary), the Royal Geographic Society, the Commonwealth Centre and the Wellcome Trust HQ all in London; the Ashmolean Museum, Oxford, University of Oxford Zoology Dept, the University of Oxford Natural History Museum, the Cambridge University Zoological Museum, exhibits in Bristol, Bath, Bury St Edmunds, Rolle College, Exmouth, Port Lympne, Pangolin Gallery, Chalford and the Welshpool gallery, Wales, Brunel University, Edinburgh, the Uganda Museum and Makerere University, Kampala, the Sorsbie and Watatu galleries and Kenyatta University, Kenya, the Windhoek Museum, Namibia, the South African Wildlife Society, Johannesburg, the British Council in Dar-es-Salaam, Tanzania, the British Council in Kyoto, Japan and the CSIRO Forest Research Centre in Atherton, Australia. The commissioning and response to these exhibits have demonstrated a widespread enthusiasm for the mammals of Africa and for their study. Apart from helping fund my work, the responses to these exhibitions have reinforced my confidence in the worth of such endeavours. Thanks also to the librarians in many institutional libraries, including Maria Garruccio of Bioversity International.

Many years ago Peter Medawar reviewed a proposal in which I sought to study some of the principles involved in visual signalling systems through a study of guenon monkeys. He said 'make sure you have a real biological point to make; don't just make a parish register'. That advice has been incompletely followed in *MoA*: we *have*, unashamedly, made parish registers but Peter's advice was taken to heart and a great many biological points have been made too.

Among the many people who have contributed data, ideas, practical help and facilitated the growth and realization of this project over the years the following are remembered with much gratitude: M. J. Adams, I. Aggundey, P. Agland, B. Agwanda, T. E. S. Ahmed, K. Albrooke, P. Andrews, W. F. H. Ansell, P. Arman, C. Bahal, C. Baker, E. W. Balson, A. Barili, W. Baumgartel, L. Beadle, S. Bearder, J. Bindernagel, I. Bishop, W. Bishop, L. Boitani, N. Bolwig, F. Bourliere, S. Braatz, R. Brain, A. C. Brooks, G. du Boulay, E. Bunengo, J. Bushara, R. Carr-Hartley, M. Cartmill, N. Chalmers, G. S. Child, V. Clausnitzer, M. Colyn, H. B. S. Cooke, B. Cooper, Y. Coppens, G. B. Corbet, M. Coryndon, H. Cronin, D. Cumming, T. R. B. Davenport, C. de Haes, M. Delany, D. de Luca, E. Delson, J. Dorst, I. Douglas-Hamilton, G. Dubost, P. Duncan, G. Durrell, B. Dutrillaux, K. Eltringham, R. D. Estes, Y. Evans, R. Farrand, T. Fison, T. Flannery, A. Forbes-Watson, B. Foster, U. Funaioli, A. K. Garewal, S. Gartlan, A. Gautier-Hion, J.-P. Gautier, A. Gentry, W. Gewalt, L. Goodwin, M. Gosling, P. J. Greenway, P. Grubb, A. Guillet, M. D. Gwynne, A. Haddow, A. Hamilton, W. D. Hamilton, M. Hammer, R. Hargreaves, G. Harrington, J. Harris, R. W. Hayman, B. Hedges, H. Heim de Balzac, M. van Heist, P. Hemmingway, J. Heyworth, J. E. Hill, C. Hillman, R. Hofmann, P. Hoppe, F. C. Howell, K. Howell, R. Hughes, E. Huxley, J. Huxley, J. Ingles, J. Itani, C. Jackson, C. Janis, J. Jarvis, P. Jewell, D. Jones, A. Jolly, C. Jolly, J. Kagoro, J. Kasenene, F. X. Katete, M. Kawai, J. Kielland, M. Kinnaird, T. A. Kindy, A. Kingdon, D. B. Kingdon, E. Kingdon, P. A. Kingdon, R. Kingdon, Z. A. Kingdon, Z. E. Kingdon, H. Klingel, P. Klopfer, J. Knowles, D. Kock, C. Koenig, S. Kondo, A. Kortlandt, H. Kruuk, T. Kuroda, H. Lamprey, R. Lamprey, R. M. Laws, M. Leakey, R. Leakey, L. S. B. Leakey, P. Leedal, L. Leland, W. Leuthold, P. Leyhausen, I. Linn, J. M. Lock, B. Loka-Arga, J. Lubonde, Q. Luke, A. D. MacKay, V. Maglio, G. Maloba, G. Maloiy, R. Malpas, R. D. Martin, D. Martins, J. Maynard-Smith, D. M. McCabe, A. McCrae, J. Meester, A. J. Mence, J. Mercer, G. Mills, X. Misonne, G. Moller, N. Monfort, R. E. Moreau, D. Morris, P. Morris, C. Moss, B. Mugerwa, P. Napier, J. Nel, H. Ngweno, T. Nishida, M. North, M. Norton-Griffiths, G. Ntenga, T. Nuti, F. Nyahoza, J. Obondio-Odur, T. O'Brien, R. O'Hanlon, B. O'Shea and M. O'Shea (for help with maps), W. K. Otim, N. Owen-Smith, M. Pagel, I. Parker, F. Petter, G. Petter, B. Pezzotta, J. Phillipson, J. Sabater Pi, S. Price, S. Pullen, P. Pullman, A. Pye, D. Pye, M. Rae, U. Rahm, M. Rainey, M. Ridley, S. Roach, W. A. Rodgers, J. P. Rood, A. Root, J. Root, D. R. Rosevear, I. Ross, L. Roth, F. Rovero, A. Ruweza, R. Rwazagiri, J. Sale, R. G. Savage, G. Schaller, K. Schmidt-Nielsen, L. Scott, I. Seki, C. Sekintu, D. Sheil, P. Shipman, C. Sillero-Zubiri, A. M. Simonetta, E. Simons, F. Simons, J. Skinner, R. Smithers, R. Southwood, C. Spinage, P. Ssali, Y. Ssenkebugye, M. R. Stanley-Price, A. Start, M. Stewart, C. Stringer, T. Struhsaker, C. Stuart, M. Stuart, Y. Sugiyama, A. Suzuki, T. Synnott, I. Tattersall, M. Taylor, L. Tennant, H. Thomas, C. Thouless, S. Tomkins, H. Tripp, M. Turner, C. Tutin, C. Turnbull, C. van Schaik, H. van Lawick, D. F. Vesey-FitzGerald, S. Wainwright, A. Walker, R. & M. Ward, M. & L. Ward, D. Webb, R. Wheater, J. White, J. G. Williams, P. Williams, N. Winser, J. Woodall, P. Zuckerman and S. Zuckerman.

CHAPTER ONE

Mammals of Africa: *An Introduction and Guide*

David Happold, Michael Hoffmann, Thomas Butynski and Jonathan Kingdon

Mammals of Africa is a series of six volumes that describes, in detail, every extant species of African land mammal that was recognized at the time the profiles were written (Table 1). This is the first time that such an extensive coverage has been attempted; all previous books and field guides have either been regional in coverage, or have described a selection of mammal species – usually the larger species. These volumes demonstrate the diversity of Africa's mammals, summarize what is known about the distribution, ecology, behaviour and conservation status of each species, and serve as a guide to identification.

Africa has changed greatly in recent decades because of increases in human populations, exploitation of natural resources, agricultural development and urban expansion. Throughout the continent, extensive areas of forest have been destroyed and much of the forest that remains is degraded and fragmented. Savanna habitats have been altered by felling of trees and development for agriculture. Many of the drier areas are threatened with desertification. As a result, the abundance and geographic ranges of many species of mammals have declined – some marginally, some catastrophically, some to extinction. Hence, it seems appropriate that our knowledge of each species is recorded now, on a pan-African basis, because the next few decades will see even more human-induced changes. How such changes will affect each mammalian species is uncertain, but this series of volumes will act as a baseline for assessing future change.

Table 1. The mammals of Africa.

Order	Number of families	Number of genera	Number of species
Hyracoidea	1	3	5
Proboscidea	1	1	2
Sirenia	2	2	2
Afrosoricida	2	11	24
Macroscelidea	1	4	15
Tubulidentata	1	1	1
Primates	4	25	93
Rodentia	15	98	395[a]
Lagomorpha	1	5	13
Erinaceomorpha	1	3	6
Soricomorpha	1	9	150
Chiroptera	9	49	224
Carnivora	9	38	83
Pholidota	1	3	4
Perissodactyla	2	3	6
Cetartiodactyla	6	41	93
16	**57**	**296**	**1116[b]**

[a] Including five introduced species. [b] Species profiles in *Mammals of Africa*.

The study of African mammals has taken several stages. During the era of European exploration and colonization, the scientific study of African mammals was largely descriptive. Specimens that were sent to museums were described and named. As more specimens became available, and from different parts of the continent, there was increasing interest in distribution and abundance, and in the ecological and behavioural attributes of species and communities. At first, it was the largest and most easily observed species that were the focus of most studies, but as new methodologies and equipment became available, the smaller and more cryptic and secretive species became better known. Many species were studied because of their suspected role in diseases of humans and livestock, and because they were proven or potential 'pests' in agricultural systems. During the past decade or so, there has been greater emphasis on the genetic and molecular characteristics of species. All these studies have produced a wealth of information, especially during the past 40 years or so. These volumes are not only a distillation of the huge literature that now exists on African mammals, but also of much unpublished information.

Readers will notice that there is a huge discrepancy among species in the amount of information available. Some species have been studied extensively for many years, especially the so-called 'game species', some species of primates, and a few species that are widespread and/or easily observed. In contrast, other species are known only by one or a few specimens, and almost nothing is known about them. Likewise, some areas and countries have been well studied, while other areas and countries have been neglected. During the preparation of these

Table 2. The six volumes of *Mammals of Africa*.

Volume	Contents	Number of species	Editors
I	Introductory chapters. Afrotheria (Hyraxes, Elephants, Dugong, Manatee, Otter-shrews, Golden-moles, Sengis and Aardvark)	49	Jonathan Kingdon, David C. D. Happold, Michael Hoffmann, Thomas M. Butynski, Meredith Happold and Jan Kalina
II	Primates	93	Thomas M. Butynski, Jonathan Kingdon and Jan Kalina
III	Rodents, Hares and Rabbits	408	David C. D. Happold
IV	Hedgehogs, Shrews and Bats	380	Meredith Happold and David C. D. Happold
V	Carnivores, Pangolins, Equids and Rhinoceroses	93	Jonathan Kingdon and Michael Hoffmann
VI	Pigs, Hippopotamuses, Chevrotain, Giraffes, Deer and Bovids	93	Jonathan Kingdon and Michael Hoffmann

volumes, the editors have often been surprised by the wealth of information about some species when little was anticipated, and by the paucity of information about others, some of which were assumed to be 'well known'. In addition to presenting information that is based on sound scientific evidence, the aims of these volumes are to point out where there are gaps in knowledge and to correct inaccurate information that has become embedded in the literature. For most taxa, the detail provided in the species profiles allows accurate identification.

Mammals of Africa comprises six volumes (Table 2). The volumes consist mainly of species profiles – each profile being a detailed account of the species. They have been edited by six editors who distributed their work according to the orders with which they were most familiar. Each editor chose authors who had extensive knowledge of the species (or higher taxon) and, preferably, had experience with the species in the field. Each volume follows the same general format with respect to arrangement, subheadings and contents. Because *Mammals of Africa* has contributions from 356 authors (each with a different background and speciality) and each volume was edited by one or more editors (each with a different perspective), it has not been possible or even desirable to ensure exact consistency throughout. Species profiles are not intended to be exhaustive literature reviews, partly for reasons of space. None the less, they are written and edited to be as comprehensive as possible, and to lead the reader to the most important literature for each species. Inevitably, not all information available could be accommodated for the better-known species, and so such profiles are a précis of available knowledge. Extensive references in the text alert the reader to more detailed information.

In addition to the species profiles, there are profiles for the higher taxa (genera, families, orders, etc.). Thus, there is a profile for each order, for each family within the order, for each genus within the family and for each species within the genus. For some orders there are additional taxonomic levels, for example, tribes (e.g. in Bovidae), subgenera (e.g. in *Procolobus*) and species-groups, or 'superspecies' (e.g. in *Cercopithecus*). The taxonomy used in these volumes mostly follows that presented in the third edition of *Mammal Species of the World: A Geographic and Taxonomic Reference* (Wilson & Reeder 2005), although authors have employed alternative taxonomies when there were good reasons for doing so. Volume I differs from the other volumes in that it contains a number of introductory chapters about Africa and its environment, and about African mammals in general.

The continent of Africa

For the purposes of this work, 'Africa' is defined as the continent of Africa (bounded by the Mediterranean Sea, the Atlantic Ocean, the Indian Ocean, the Red Sea and the Suez Canal) and the islands on the continental shelf, which, at some time in their history, have been joined to the African continent. The largest of the 'continental islands' are Zanzibar (Unguja), Mafia and Bioko (Fernando Po). All 'oceanic islands', e.g. São Tomé, Principe, Annobón (Pagulu), Madagascar, Comoros, Seychelles, Mauritius, Socotra, Canaries, Madeira and Cape Verde are excluded, with the exception of Pemba. Pemba is included because of its close proximity (ca. 50 km) to the mainland.

The names of the countries of Africa are taken from the *Times Atlas* (2005). The Republic of Congo is referred to as 'Congo' and the Democratic Republic of Congo (formerly Zaire) as 'DR Congo'. Smaller geographical or administrative areas within countries are rarely referred to except for Provinces in South Africa, which are used extensively in the literature. A political map of Africa, and of the Provinces of South Africa, is given (Figure 1), as well as a list of the 47 countries together with their previous names, which are used in the older literature on African mammals (Table 3).

Africa is the second largest continent in the world (after Asia), but it differs from other continents (except Australia and Antarctica) in being essentially an island. At various times in the past, Africa has been joined to other continents – a situation that has had a strong influence on the fauna and flora of the continent. Africa is a vast continent (29,000,000 km², 11,200,000 mi²) that straddles the Equator, with about two-thirds of its area in the northern hemisphere and one-third in the southern hemisphere. As a result, Africa has many varied climates (with seasons in each hemisphere being 6 months out of phase), many habitats (including deserts, savannas, woodlands, swamps, rivers, lakes, moist forests, monsoon forests, mountains and glaciers), and altitudes ranging from 155 m (509 ft) below sea level at L. Assal, Djibouti, in the Danakil (Afar) Depression, to 5895 m (19,341 ft) on Mt Kilimanjaro, Tanzania. Africa is comprised of 47 countries, some of which are very large (e.g. Sudan [2,506,000 km²; 967,000 mi²], Algeria (2,382,000 km², 920,000 mi²] and Congo [2,345,000 km², 905,000 mi²]), and others that are relatively small (e.g. Djibouti [23,200 km², 9000 mi²], Swaziland [17,400 km², 6700 mi²] and The Gambia [11,300 km², 4400 mi²]). The human population of each country also varies greatly, from about 346/km² in Rwanda to only about 2.5/km² in Namibia. With its great size and varied habitats, Africa supports a high biodiversity, including a large number of species of mammals. Likewise, most countries have a high diversity of mammals (especially when compared with temperate countries).

Figure 1. (a) Political map of Africa; (b) provinces of South Africa; (c) altitudes and major rivers of Africa.
South Sudan and Somaliland are not identified as separate countries in the text.

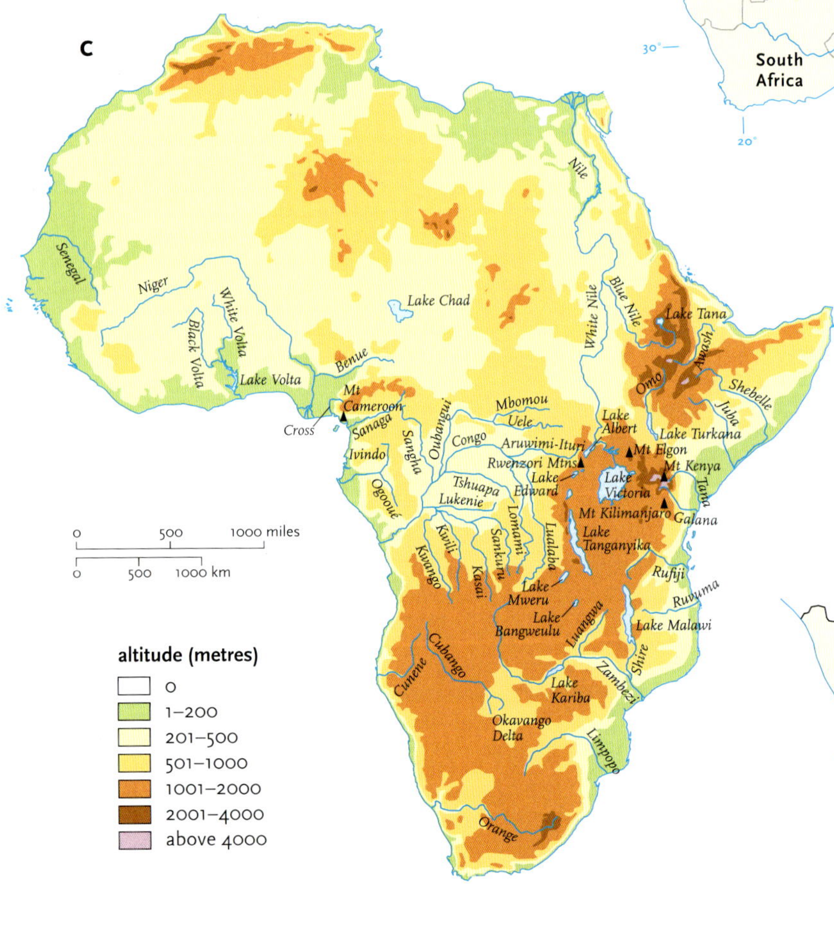

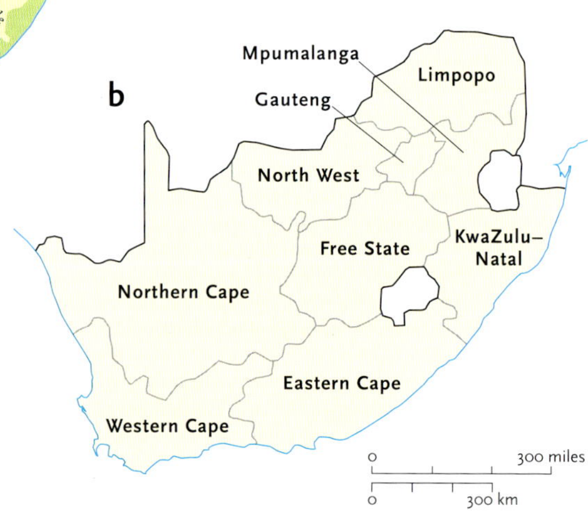

Table 3. The countries of Africa: names, areas and human population density.

Country name	Area (km^2) '000	Area (miles^2) '000	Human population '000 (2006)	People per km^2
Algeria	2,382	920.0	33,500	14.1
Angola (includes Cabinda)	1,247	481.0	15,800	12.7
Benin * [Dahomey]	113	43.0	8,700	77.0
Botswana [Bechuanaland]	582	225.0	1,800	3.1
Burkina Faso * [Upper Volta; Burkina]	274	106.0	13,600	49.6
Burundi [part of Ruanda-Urundi (= part of Belgian Congo)]	27.8	10.7	7,800	280.5
Cameroon [includes former French Cameroon, German Cameroon and part of Eastern Nigeria]	475	184.0	17,300	36.2
Central African Republic #	623	241.0	4,300	6.9
Chad [Tchad]	1,284	496.0	10,000	5.8
Congo [Republic of Congo]	342	132.0	3,700	10.8
Côte d'Ivoire * [Ivory Coast]	322	125.0	19,700	61.2
Democratic Republic of Congo [Belgian Congo; Congo (Kinshasha); Zaire]	2,345	905.0	62,700	26.7
Djibouti [French Somaliland]	23.2	9.0	800	34.5
Egypt	1,001	387.0	75,400	75.3
Equatorial Guinea # (includes Rio Muni [Spanish Guinea] and Bioko I. [Fernando Po])	28.1	10.8	500	17.8
Eritrea (formerly part of Ethiopia)	94	36.0	4,600	48.9
Ethiopia [Abyssinia]	1,128	436.0	74,800	66.3
Gabon #	268	103.0	1,400	5.2
The Gambia	11.3	4.4	1,500	132.7
Ghana [Gold Coast]	239	92.0	22,600	94.6
Guinea *	246	95.0	9,800	39.8
Guinea-Bissau [Portuguese Guinea]	36	13.9	1,400	38.9
Kenya	580	224.0	34,700	59.8
Lesotho [Basutoland]	30.4	11.7	1,800	59.2
Liberia	111	43.0	3,400	30.6
Libya	1,760	679.0	5,900	3.6
Malawi [Nyasaland]	118	46.0	12,800	108.5
Mali *	1,240	479.0	13,900	11.2
Mauritania *	1,030	412.0	3,200	3.1
Morocco [includes former Spanish Morocco and French Morocco]; (now also includes Western Sahara = former Spanish Sahara)	447	172.0	32,100	71.8
Mozambique [Portuguese East Africa]	802	309.0	19,900	24.8
Namibia [South-west Africa]	825	318.0	2,100	2.5
Niger *	1,267	489.0	14,400	11.3
Nigeria	924	357.0	134,500	145.6
Rwanda [part of Ruanda-Urundi (= part of Belgian Congo)]	26.3	10.2	9,100	346.0
Senegal *	197	76.0	11,900	60.4
Sierra Leone	71.7	27.7	5,700	79.5
Somalia ¥ [British Somaliland and Italian Somaliland; Somali Republic]	638	246.0	8,900	13.9
South Africa	1,220	471.0	47,300	38.7
Sudan § [Anglo-Egyptian Sudan]	2,506	967.0	41,200	16.4
Swaziland	17.4	6.7	1,100	63.2
Tanzania [German East Africa; Tanganyika] (now includes Zanzibar I., Mafia I. and Pemba I.)	945	365.0	37,900	40.1
Togo [Togoland]	56.8	21.9	6,300	110.9
Tunisia	164	63.0	10,100	61.6
Uganda	236	91.0	27,700	117.4
Zambia [Northern Rhodesia]	753	291.0	11,900	15.8
Zimbabwe [Southern Rhodesia]	391	151.0	13,100	33.5
Totals/mean density	**29,448**	**11,383**	**902,600**	**56.8**

Former names are listed in chronological order in square brackets, with the oldest name listed first. Obsolete names are listed because much of the older literature refers to past colonial entities. * = formerly part of French West Africa. # = formerly part of French Equatorial Africa. § At the time of going to press, the country of Sudan had been divided into two: the Republic of Sudan in the north, and the Republic of South Sudan in the south. ¥ The former British Somaliland is now a self-declared state under the name of the Republic of Somaliland, but remains internationally unrecognized.

Africa may also be categorized into Biotic Zones (see Chapter 5, Figure 3, p. 62). A Biotic Zone is defined as an area within which there is a similar environment (primarily rainfall and temperature) and vegetation, and which differs in these respects from other Biotic Zones. Africa can be divided into 13 Biotic Zones, two of which may be divided into smaller categories. The Biotic Zones concept provides a general assessment of the environmental conditions in which a species lives, as well as providing an assessment of the geographic distribution of the species. In a similar way, the Rainforest Biotic Zone (see Chapter 5, Figure 5, p. 69) and the South-West Arid Biotic Zone may be divided into regions and sub-regions that reflect the different biogeographical distributions of species, each region/sub-region having a community of mammals and other animals that is different to any other. Details of the Biotic Zones of Africa, and the Regions and Sub-regions of the Rainforest and South-West Arid Biotic Zones, are given in Chapter 5. The Biotic Zones map is reprinted in the Introductions to volumes II–VI of this series.

The Afrotheria of Africa

This volume, Volume I, is devoted to the orders Hyracoidea (hyraxes), Proboscidea (elephants), Sirenia (Dugong and Manatee), Afrosoricida (otter-shrews and golden-moles), Macroscelidea (sengis or elephant-shrews) and Tubulidentata (Aardvark), which collectively comprise the Afrotheria. In Africa, these orders contain 49 species (i.e. about 4% of all African mammals). Of these, 10 species are large to very large and 39 species (Afrosoricida and Macroscelidea) are small to very small. Some species described in this volume have been studied in detail and are well known; others, particularly the smaller species, have not been well studied, often because of their rarity and small geographic ranges.

Since the texts for this volume were prepared, two new species have been described within the order Macroscelidea:

Rhynchocyon udzungwensis Rathbun & Rovero, 2008, *J. Zool., Lond.* 274: 127. Distribution: Vikongwa River Valley, Ndundulu Forest, West Kilombero Scarp Forest Reserve, Udzungwa Mountains, Iringa Region, Tanzania [7°48.269' S, 36°30.355' E (Arc 1960 datum)], at 1350 m a.s.l. Remarks: location is *c.* 15 km south-east of Udekwa Village, Iringa Region, Tanzania. (Reference: Rovero *et al.* 2008).

Elephantulus pilicaudus Smit, 2008. *J. Mammal.* 89: 1263. Distribution: Vondelingsfontein Farm, Calvinia, Northern Cape Province, South Africa (31°48' S, 19°49' E), at 1449 m a.s.l.

The editors of the species profiles in this volume are Jonathan Kingdon and Michael Hoffmann (Hyracoidea, Proboscidea, Sirenia and Tubulidentata; 10 species), and David and Meredith Happold (Afrosoricida and Macroscelidea; 39 species).

Species profiles

Information about each species is given under a series of subheadings. The amount of information under each of these subheadings varies greatly between species; where no information is available, this is recorded as 'No information available' or words to this effect. The sequence of subheadings is as follows:

Scientific Name (genus and species) The currently accepted name of the species.

Vernacular Names English, French and German names are given, as available. The first given English name is the preferred vernacular name for the species; alternative names are given in parentheses for some species. Wilson & Cole (2000) list proposed vernacular names for all the world's mammals; most of these names were also given

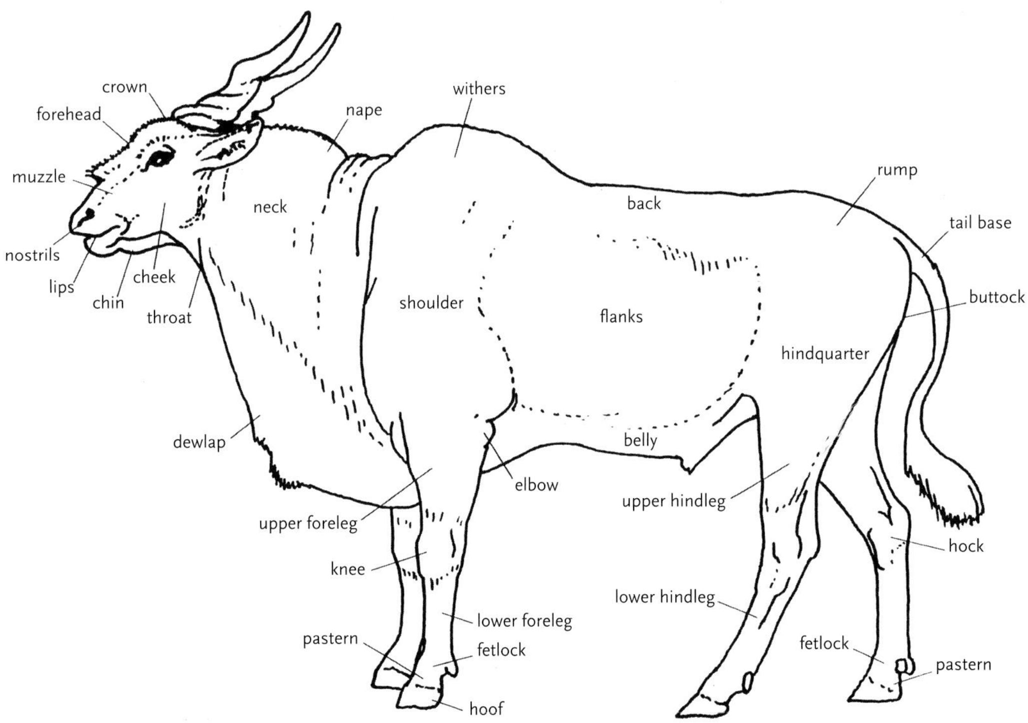

Figure 2. External features of a mammal: Common Eland *Tragelaphus oryx.*

Figure 3. External features of a mammal: *Genetta* sp. (side-view and face frontal).

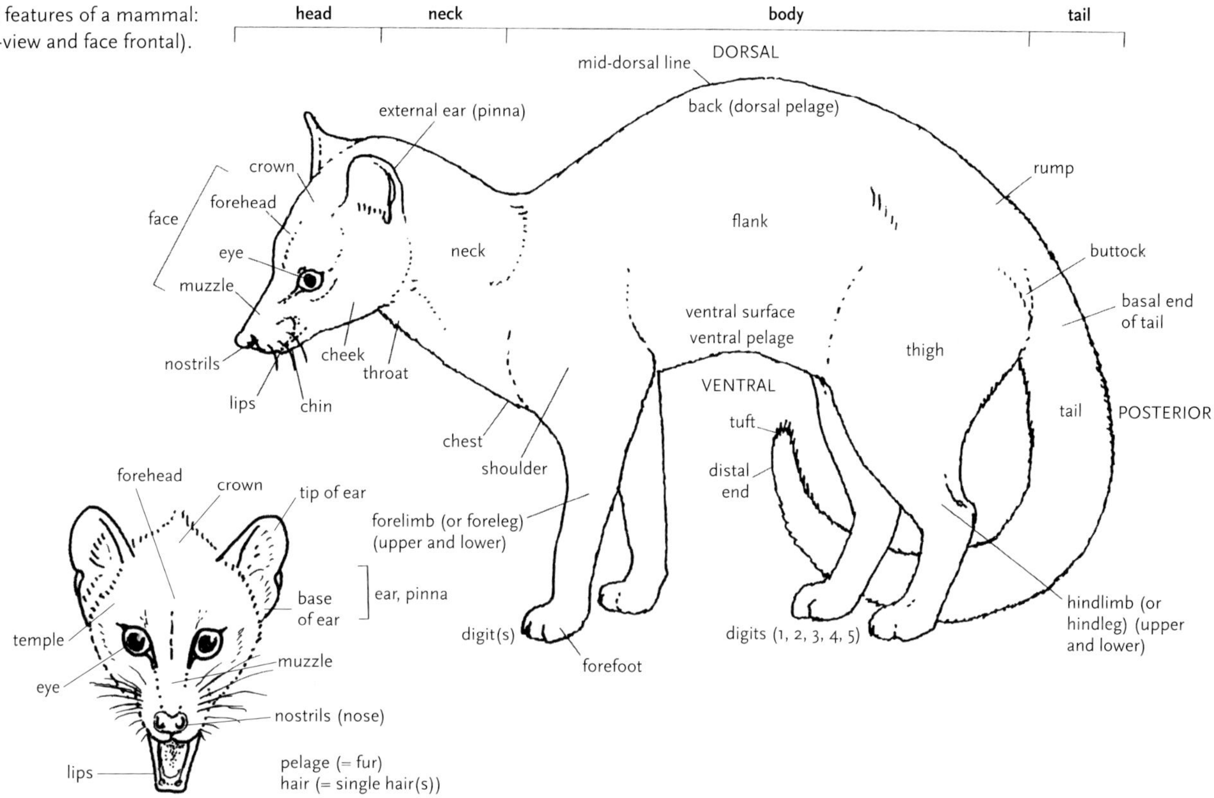

in the third edition of *Mammal Species of the World* (Wilson & Reeder 2005). Although these works have been consulted, the names used have not always been adopted in *Mammals of Africa*. French names were either provided by authors, or taken from Gunther (2003).

Scientific Citation This provides the full scientific name of the species, i.e. genus name, species name, authority name and date of authority. Parentheses around the authority's name and date indicate that the species was originally named in a different genus to its present generic allocation. The scientific name is followed by the publication where the species was described, and the location where the type specimen (or type series) was obtained. Most of this information is taken from Wilson & Reeder (2005).

Taxonomy This section contains information on taxonomic problems, if any, associated with the species, and its relationship with other species in the genus. For some species, there is considerable information about these topics; for others, there may be nothing. A list of synonyms (without the taxonomic authority for each) and the number of subspecies (if any) is presented, mostly taken from Wilson & Reeder (2005). The chromosome number is given if available, and in some cases this is followed by other information relevant to the chromosomes.

Description This section, together with the illustrations, provides the reader with adequate information to identify the species. The section begins with a brief overall description of the species, including an indication of size. This is followed by a detailed description of the external features of the species' head (and parts of the head), dorsal pelage, legs, feet, ventral pelage, and tail (in this order), as well as any special characteristics unique to the species. For some species, diagnostic characteristics of the skull are given. The characters described in this section are common to all subspecies of this species (unless otherwise noted). The mammary formula, i.e. the number and arrangement of nipples in adult females, is noted wherever this feature varies between the taxa being discussed.

In the profiles of the Afrosoricida and Macroscelidea, the word 'comparatively' is used in the context of describing the size of one character compared with the size of the same character in a *different* species. The word 'relatively' is used in the context of describing the size of one character relative to the size of a different character in the *same* species. This is usually expressed as a percentage, e.g. Tail 80% of HB. In contrast, authors of profiles in other orders may have given these words more generalized meanings.

Geographic Variation Variation within the species may be of two sorts: (a) clinal variation without subspecies, or (b) subspecific variation. If (a), there is a description of the character(s) that alter clinally across the geographic range of the species. If (b), each of the subspecies is listed with its geographic range and the characters that distinguish it from other subspecies of the species. For some species, subspecies have been described that are no longer considered to be valid; in some cases, such names may be listed but without further comment.

Similar Species Species that are sympatric or parapatric with the species under consideration, and with which it may be confused, are listed along with diagnostic characteristics (additionally, readers may refer to profiles of the similar species in question). In some instances, species that are allopatric in distribution are also included.

In the profiles of the Afrosoricida and Macroscelidea, when two similar species are distinguished on the basis of their size, one is only said to be 'larger' or 'smaller' than the other if there is no

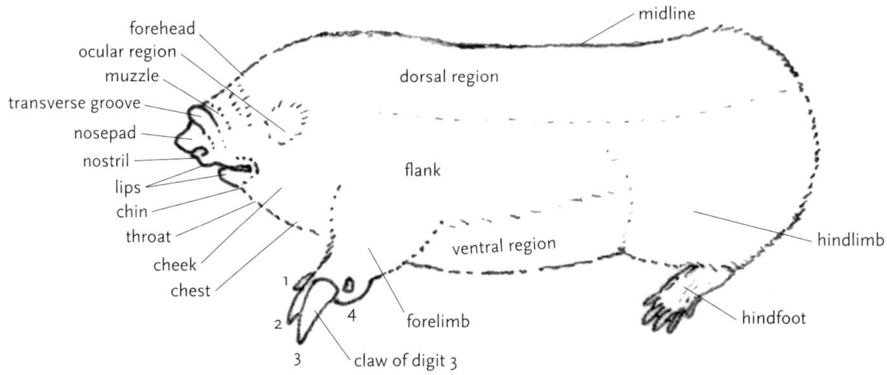

Figure 4. External features of a hypothetical golden-mole.

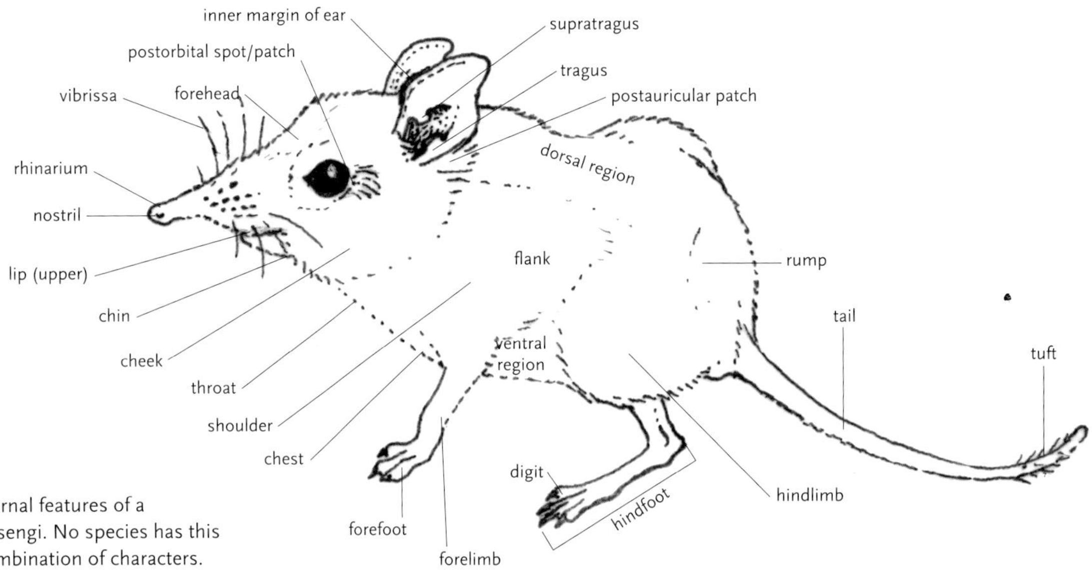

Figure 5. External features of a hypothetical sengi. No species has this particular combination of characters.

overlap in the size ranges of the two species. If there is overlap but the means are different, it is stated that one is 'larger on average' or 'smaller on average' than the other. In contrast, authors of profiles in other orders may have used 'larger' or 'smaller' in both of the above situations.

Distribution The first sentence 'Endemic to Africa' informs the reader that this is an African species and does not occur on any other continent; if a species also occurs outside Africa, this is noted at the end of this section. The next sentence usually gives the Biotic Zone (or Zones) where the species has been recorded; this provides the reader with a general impression of where the species occurs in Africa and the sort of habitat where the species occurs. Finally, the countries (or parts of countries) where the species has been recorded are generally listed; sometimes other data (such as altitude range and habitat) are also included. As a general rule, descriptions of the range for species with very restricted distributions are more precise in terms of information given (including, for example, geographic coordinates) than for more widespread species, where a more generalized range statement is adequate. A distribution map (see below) augments the information given here.

Habitat This section provides a description of the range of habitats where the species lives. Details of plant communities, plant species, vegetation structure, soil type and/or structure and water availability, etc. (if available) are also recorded. Other information may include average annual rainfall, altitudinal limits and seasonal variation in habitat characteristics.

Abundance A general indication of abundance in the habitat. This may be unquantified, such as abundant, common, uncommon, rare, or phrases such as 'rarely seen but frequently heard', etc. For better-known species, abundance may be expressed as estimates of density (e.g. number/ha or number/km^2), or relative abundance within the community (e.g. 'comprised 40% of small mammals captured', 'the second most numerous species'); for the better-known rare species, actual numbers of individuals for the species may be given. Other information may include seasonal changes in density, frequency of observations, or the relative abundance of specimens in collections.

Adaptations This section describes morphological, physiological and behavioural characteristics, which show how the species uniquely interacts with its environment, conspecifics and other animals. This section may also describe species-specific adaptations for feeding, locomotion, burrowing, mechanisms for orientation, production of sound, sensory mechanisms and activity patterns. In some instances comparison with related or convergent species allows the unique adaptations of the species under discussion to be detailed or highlighted.

Foraging and Food The first sentence briefly describes the food habits of the species (e.g. insectivorous, carnivorous, granivorous, etc.). This may be followed by the method of collecting food (foraging), size of home range and daily distance moved. The diet is then described either by a list of the taxa of animals or plants consumed, and/or as a quantitative measure based on direct observations or of examination of the contents of the stomach or the faeces.

Social and Reproductive Behaviour Topics in this section may include group structure (whether solitary, social, or colonial), group size and composition; agonistic and amicable behaviour, comfort behaviour, etc.; home-range (including quantitative data), territorial behaviour, courtship and mating behaviour, behaviour of young, parental–young interactions; presence of helpers, vocalizations, and interactions with other species (mammals, birds, etc.).

Reproduction and Population Structure This section begins with an assessment of reproductive strategy (if known) and the times/seasons of the year when individuals are reproductively active (pregnancy and lactation in females, active spermatogenesis in males). Other information may include length of gestation, times/seasons of births, including peaks of births, litter-size, birth-weight and size, spacing of litters, growth and time to weaning, maturity, longevity and mortality rates. Reproductive strategies, if known, are described with respect to locality, food availability and population density. Population structure (sex ratio, adult/young ratio, abundance of different cohorts in the population at different times of the year) may be described, and related to seasonal variations in reproduction and environmental variables.

Predators, Parasites and Diseases The known predators, parasites and diseases are listed. Additional information is given if the species is a host to diseases that affect humans and domestic stock, and if it is utilized as food for humans ('bushmeat').

Remarks This subheading subsumes five of the above subheadings (Adaptations, Foraging and Food, Social and Reproductive Behaviour, Reproduction and Population Structure, and Predators, Parasites and Diseases) in those instances where there is little or no information available.

Conservation The conservation status of the species is stated, as given by IUCN – International Union for Conservation of Nature Red List of Threatened Species (version 2011.2). The IUCN Red List categories follow the definitions given in the *IUCN Red List Categories and Criteria Version 3.1* (Table 4); for those species classified as threatened (Vulnerable, Endangered and Critically Endangered), readers may obtain detailed reasons for the classification on the IUCN Red List website. If a species is listed on Appendix I or Appendix II under CITES (Convention on International Trade in Endangered Species; www.cites.org), this is also indicated. For some species, additional information, such as presence in protected areas, major threats and current or recommended conservation measures are provided.

Measurements A series of morphological measurements is provided. For each species there is a standard set of measurements. The abbreviations for each measurement are given in the Glossary. A measurement is cited as the mean value (with minimum value

Table 4. Definitions for the IUCN Red List categories (from IUCN – Red List Categories, Version 3.1).

Category	Description
Extinct (EX)	A taxon is **Extinct** when there is no reasonable doubt that the last individual has died. A taxon is presumed Extinct when exhaustive surveys in known and/or expected habitat, at appropriate times (diurnal, seasonal, annual), throughout its historic range have failed to record an individual. Surveys should be over a time-frame appropriate to the taxon's life-cycles and life form.
Extinct in the Wild (EW)	A taxon is **Extinct in the Wild** when it is known only to survive in cultivation, in captivity or as a naturalized population (or populations) well outside the past range. A taxon is presumed Extinct in the Wild when exhaustive surveys in known and/or expected habitat, at appropriate times (diurnal, seasonal, annual), throughout its historic range have failed to record an individual. Surveys should be over a time-frame appropriate to the taxon's life-cycle and life form.
Critically Endangered (CR)	A taxon is **Critically Endangered** when the best available evidence indicates that it meets any of the criteria A to E for Critically Endangered, and it is therefore considered to be facing an extremely high risk of extinction in the wild.
Endangered (EN)	A taxon is **Endangered** when the best available evidence indicates that it meets any of the criteria A to E for Endangered, and it is therefore considered to be facing a very high risk of extinction in the wild.
Vulnerable (VU)	A taxon is **Vulnerable** when the best available evidence indicates that it meets any of the criteria A to E for Vulnerable, and it is therefore considered to be facing a high risk of extinction in the wild.
Near Threatened (NT)	A taxon is **Near Threatened** when it has been evaluated against the criteria but does not qualify for Critically Endangered, Endangered or Vulnerable now, but is close to qualifying for (or is likely to qualify for) a threatened category in the near future.
Least Concern (LC)	A taxon is **Least Concern** when it has been evaluated against the criteria and does not qualify for the Critically Endangered, Endangered, Vulnerable or Near Threatened categories. Widespread and abundant taxa are included in this category.
Data Deficient (DD)	A taxon is **Data Deficient** when there is inadequate information to make a direct, or indirect, assessment of its risk of extinction based on its distribution and/or population status. Data Deficient is not a category of threat. Listing of taxa in this category indicates that more information is required and acknowledges the possibility that future research will show that a threatened classification is appropriate.
Not Evaluated (NE)	A taxon is **Not Evaluated** when it has not yet been evaluated against the criteria.

to maximum value in parentheses) and sample size. For some, the standard deviation (mean ± 1 S.D.) is given instead of the range. For most measurements, data for males and females are combined but where there is sexual dimorphism, measurements for males and females are given separately. Where possible, measurements also detail the localities where the specimens were obtained, and the source of the data. Sources are either cited publications, or specimens in museums, or unpublished information from authors or others. The acronym BMNH corresponds to 'Natural History Museum, London, UK [formerly British Museum (Natural History)]'. Most museum records are provided by the author of the profile; others – when an author did not have the measurements or did not have the opportunity to visit museums – were provided by the editor(s).

Key References A select list of references, which provides more general information on the species. Each reference is given in full in the Bibliography.

Author The name of the author, or authors, is given at the end of each profile. All profiles should be cited using the author name(s).

Tables For selected taxa (mainly families and genera) tables provide details of the main characteristics of these taxa and can be used as an aid to identification.

Higher taxon profiles

The profiles for orders, families and genera are less structured than for the species profiles. Each profile usually begins with a listing of the taxa in the next lower taxon; for example, each family profile lists the genera in that family. An exception to this arrangement is where a taxon has only one lower taxon. Higher taxa profiles provide the characteristics common to all members of that taxon. Some of these characteristics may not be repeated in lower taxon profiles (unless essential for identification).

Distribution maps

Each species profile, with a very few exceptions, contains a pan-African map showing the geographic range of the species. Most maps were provided by the author of the profile and were compiled from literature records and museum specimens; some maps were provided by the editor(s) when it was not possible for the author to do so. We are grateful to the IUCN SSC African Elephant Specialist Group for permission to reproduce the map of the Savanna and Forest Elephants from the African Elephant database 2007. Each map shows the boundaries of the 47 countries of Africa, some of the major rivers (Nile, Niger–Benue, Congo [with the tributaries Ubangi, Lualaba and Lomani], Zambezi and Orange), and Lakes Chad, Tana, Turkana (formerly Rudolf), Albert, Edward, Victoria, Kyoga, Kivu, Tanganyika, Malawi, Mweru, Bangwuela and Kariba. The geographic distribution of a species is indicated as:

- red shading = current range (or ranges)
- × = isolated locations considered to be separate from the main geographic range(s); some locations indicated by × may include two or more closely spaced locations
- red arrow = recorded from the island indicated by the arrow

Editors of *Mammals of Africa*

Jonathan Kingdon, Department of Zoology, University of Oxford, WildCRU, Tubney House, Abingdon Road, Tubney OX13 5QL, UK. (Vols I, II, V & VI)
David Happold, Research School of Biology, Australian National University, Canberra, ACT 0200, Australia. (Vols I, III & IV)
Thomas Butynski, Eastern Africa Primate Diversity and Conservation Program, PO Box 149, Nanyuki 10400, Kenya, and Zoological Society of London, King Khalid Wildlife Research Centre, Saudi Wildlife Authority, PO Box 61681, Riyadh 11575, Kingdom of Saudi Arabia. (Vols I & II)
Michael Hoffmann, International Union for Conservation of Nature – Species Survival Commission, 219c Huntingdon Road, Cambridge CB3 0DL, UK. (Vols I, V & VI)
Meredith Happold, Research School of Biology, Australian National University, Canberra, ACT 0200, Australia. (Vols I & IV)
Jan Kalina, Soita Nyiro Conservancy, PO Box 149, Nanyuki 10400, Kenya. (Vols I & II)

CHAPTER TWO

Thinking Mammals: An Introduction to African Mammals in Science, Natural History and Culture

Jonathan Kingdon

Darwin saw clearly that the succession of life on this planet was not a formal pattern imposed from without, or moving exclusively in one direction. Whatever else life might be, it was adjustable and not fixed. It worked its way through difficult environments. It modified and then, if necessary, it modified again, along roads which would never be retraced.

As the only thinking mammals on the planet – perhaps the only thinking animals in the entire sidereal universe – the burden of consciousness has grown heavy upon us. Nevertheless, in the nature of life and in the principles of evolution we have had our answer. Of men elsewhere, and beyond, there will be none forever.

Loren Eiseley, *The Immense Journey*, 1946

Because we are thinking animals and because we depend so much upon animals for our subsistence and economies, it is impossible for us *not* to think about the rest of animal life (although, of course, many urban people actively avoid such thought and do not come into contact with non-human animals). That said, *how* people think about, behave towards, talk and write about animals has always said more about the upbringing, personality, culture and history of the humans concerned than it can possibly tell us about an evolved being as complex and interesting as a mammal. A work such as this recognizes the many potentialities for distortion that come with that legacy, but the scientific principles and purposes it serves require that they be minimized. We have brought scientific methods to bear but, none the less, our work is still an artefact, a collaborative product by many authors. We, in turn, are products of our time, place and culture, all humble students of beings whose depths we are still ill-equipped to plumb.

When those mammals number many hundreds of species from every part of a vast continent, anyone interested in them, individually or severally, must depend upon all those naturalists and scientists who have provided us with records. What form those records have taken ranges from systematic details in pocket books or computers, tables of measurement, maps, drawings, film and other images, to dead or living specimens, and surely anecdotes, scribbles on napkins or scraps of paper etc. as well. In any effort to synthesize such diverse sources of knowledge authors, editors and readers of a work such as this must extract some sort of truth-seeking portrait from a rich mix of reportage embedded in very varied cultural histories.

We also depend upon the serendipity that sometimes puts a passionate naturalist in a particular corner of Africa where she gets to tell us more about one sort of animal than had ever been learned in all of history. Names may seem invidious among a long line of illustrious naturalists, but it was Jane Goodall who introduced wild-living chimpanzees to the world, George Schaller who transformed our view of gorillas while David Macdonald created both interest and empathy for meerkats and foxes, but there are many others to whom we all owe deep debts.

While we have no option but to copy, assemble and integrate what other people have reported about the animals, many of us, especially those with first-hand experience of particular species, such as Jane, George and David, will be able to summon up, in the mind's eye, still richer internal visions of that animal's physical existence. But no matter how rich, how wide-ranging, deep and inspiring any one person's knowledge and writing is, it will belong to a time and place as will their experience. The fast growth of science itself ensures that all of us are in some sense pioneers that will very soon be overtaken, but that same growth will ensure that much of what goes on now, as before, will eventually become redundant. For knowledge to grow and deepen, the animals and their environments must continue to exist across the generations. African mammals incorporate many realities that are still hidden

Head of Blue Buck *Hippotragus leucophaeus* (from skins and contemporaneous documents).

from us, but for future humans, better equipped than we are, they may reveal much that is now unknown simply by *being* the animals they are.

In the effort to match verbal and other images to living, behaving mammals we exercise a very human faculty, but it is essential that we remember the ephemerality of our observations. Three centuries ago a hunter described a quality that he thought was unique to his quarry, he wrote, 'The coat looked like blue velvet in life but faded to lead colour in death'. Another hunter, who was also an amateur artist, tried to render its living appearance in water-colour, tinting his paper with cerulean blue. Appropriately this South African mammal was called 'Blue Buck', but by 1779 it was extinct. Something blue was gone and even the lead colouring of a few remaining skins now looks more like dust. The words of someone who had witnessed a quality that belonged only to life are more eloquent than the remaining dusty skins and skulls.

In the growth of science, skins and skulls, ideally deposited in an accessible museum, provided the first records upon which names and descriptions could be based. As science has progressed the range of recordings has vastly increased, but has it become any less cadaverous? Had that hunter had a tape-recorder would he merely have preserved the last cry of the Blue Buck? With an electronic tag would he have preserved an oscillogram of its heart's last beat? Of Africa's mammals many more will leave just such eccentric records of their existence in museums, recorders and in published words. For many species that is all that those with a curiosity about the mammals of Africa will find in the future.

While we have no choice but to be children of our time, attempting to place our work in some sort of historical perspective helps us define what is new about our own time. If we are self-conscious enough we may, perhaps, be less content to be passive witnesses to blue tints turning to leaden grey.

So where do texts and images of African mammals begin? Drawings and paintings, finer, more authoritative and better observed than almost anything depicted today, ornament many cave and rock-shelter walls across Africa, testifying to the intensity of observation and importance of large mammals for our ancestors. It was images, but mainly words, woven into tales about Africa's most dramatic mammals that made a historical progression (but also a verbal chain that resembled 'Chinese whispers') from campfires in Africa itself, to markets around the Mediterranean, to be translated eventually into books, mainly illuminated or printed in Europe and Asia. A few species of North African mammals were physically transported to the circuses and menageries of ancient Rome where they became 'curiosities' just as they did later in northern European menageries. In China and India royal courts marvelled at giraffes and zebras. That legacy long ago made some African mammals familiar to the rest of the world. What child has not heard of lions, zebras and elephants? Their likenesses fill storybooks and plaster school-room walls, just as they ornamented the very first maps of Africa, but most people are still more familiar with images and stories about African mammals than they are with any sort of reality.

The curiosity and awe that fuelled gossip in tenth-century souks, that marketed medieval bestiaries, created a hunger for zoos in nineteenth-century Europe and went global with twentieth-century wildlife films, is not entirely separable from the scientific curiosity that finds expression in this book. Certainly the history of books about African mammals faithfully records the dominant preoccupations of their authors and of the times in which they were published. From our descendants we can expect a cold-eyed examination of today's technological state and the position our current culture took on our relationship with other animals. Even our own profile accounts in these volumes will not escape such scrutiny.

Before the invention of books, stories about animals were as ephemeral as the memories of those who heard them. Books changed that, but because they were invented far from Africa it need surprise no one that whatever facts might have been there to begin with became mostly myth and fable. Beast-based fables, such as those told by the slave Aesop about 2600 years ago, contained moral messages that eventually found their way into books. One of the most interesting examples of myth-creation or, perhaps, myth-reinforcement, concerned one-eyed giants, Cyclopes, that were reputed to have inhabited certain caves in Sicily. Bolstering, or, perhaps, originating, belief in these Titanic monsters were elephant skulls that were found in the same caves, with the nasal orifice interpreted as a single eye-socket in the forehead of large, but otherwise somewhat humanoid skulls! The rhinoceros, familiar enough to many Africans and South Asians, became transformed into the mythological unicorn in ancient China, Vietnam and northern Europe and acquired various attributes, including that of a horned symbol of fertility. Such ideas were still in circulation in Rennaissance Italy, where Leonardo da Vinci imagined unicorns falling asleep in the laps of virgins! Likewise, two giraffes carried to Beijing in 1433 were acclaimed as magical 'Qilins', their capture and harnessing being paraded as emblematic of the Chinese Emperor's supreme power. By the time stories, body-parts or actual animals had found their way to the primitive cultures of Europe and Asia they were transformed by many similarly grotesque expectations or perversions.

The most pervasive and damaging of all cultural artefacts was a belief in several Mediterranean monotheistic cultures that animals were explicitly placed on earth for man's exploitation and that exotic animals were manifestations of 'thoughts' in the mind of an anthropomorphic creator god. The main 'purpose' of animals was their moral and symbolic role in a mental universe, they were essentially avatars, embodiments of concepts: thus monkeys, leopards, foxes, goats and rats were all manifestations of evil or 'the devil' in early Christian iconography, while purity found expression in a stoat! It was even believed that an animal went extinct when its 'creator' ceased to think about it! This anthropocentric (almost 'post-modern') tradition is still predominant in the thinking of large numbers of people, and it is still being reinforced by a variety of atavistic cultural thought-police.

In a history of the development of books and ideas about African mammals such ideas found expression in medieval 'bestiaries'. These emerged, from much earlier traditions, as the first of some four or five major phases. Intensely interesting as products of their feudal theocratic times, they now have little or no scientific interest. Likewise, traditional parables, typified by Aesop's fables, were among the first books to be published in Europe. Descendants of this tradition, in which animal characters speak for human stereotypes, continue to flourish, often in the form of cartoons, sometimes brilliantly funny but more often silly stories with cutesy names. Film, TV and DVD have, to a large extent, overtaken the book in this tradition.

Two rather different sorts of books and records emerged as European empires expanded into Africa, but the boundary between them was sometimes blurred. The year 1758 saw the publication of Linnaeus' tenth edition of Systema Naturae, where many African mammals were named and described systematically for the first time. This open-ended system wherein every organism could be placed, depended partly on written descriptions, partly on preserved specimens. Linnaeus, his contemporaries and his followers begged travellers to send them specimens to describe and their collection became a major *raison d'être* for many expeditions and one of the expressions of what has been called the scientific 'Enlightenment'.

Collecting specimens and measurements were among the primary objectives of the voyage of *The Beagle*, a small British navy survey vessel that circumnavigated the world between 1831 and 1836. Although it only docked in Africa for two weeks and was but one of innumerable exploratory expeditions sent out by European imperial powers, one passenger on *The Beagle*, Charles Darwin, ultimately changed humanity's perception of its relationship with the rest of nature. His ideas and writing continue to be relevant to how African mammals are viewed and valued.

The science of biology has many historical roots but Darwin's *The Origin of Species*, published in 1859, put the study of plants and animals, including humans, on an entirely new footing. While Darwin provided numerous examples from nature to exemplify the many dimensions of evolution by natural selection, most of his illustrations were drawn from Europe, Asia and the Americas, Africa being little-known to the outside world at that time. On his visit to Cape Town he learned that more species of plants were crowded together in the Cape of Good Hope than in any other quarter of the world and he took as serious, and in my view a more perceptive, an interest in African zebras than any contemporary scientist. He collected abundant evidence that mules and other inter-specific equid hybrids all tended to show striping on legs and shoulders, even when both parents showed none, and he witnessed, at first hand, barring on the legs and shoulders of individual horses. He wrote: 'I venture confidently to look back thousands of generations, and I see an animal striped like a zebra, but perhaps otherwise very differently constructed, the common parent of our domestic horse' (Darwin 1859). It has taken well over a century for comparable sensitivity and deep prehistorical insight to be brought to bear on the study, not just of zebras, but of African mammals as a whole.

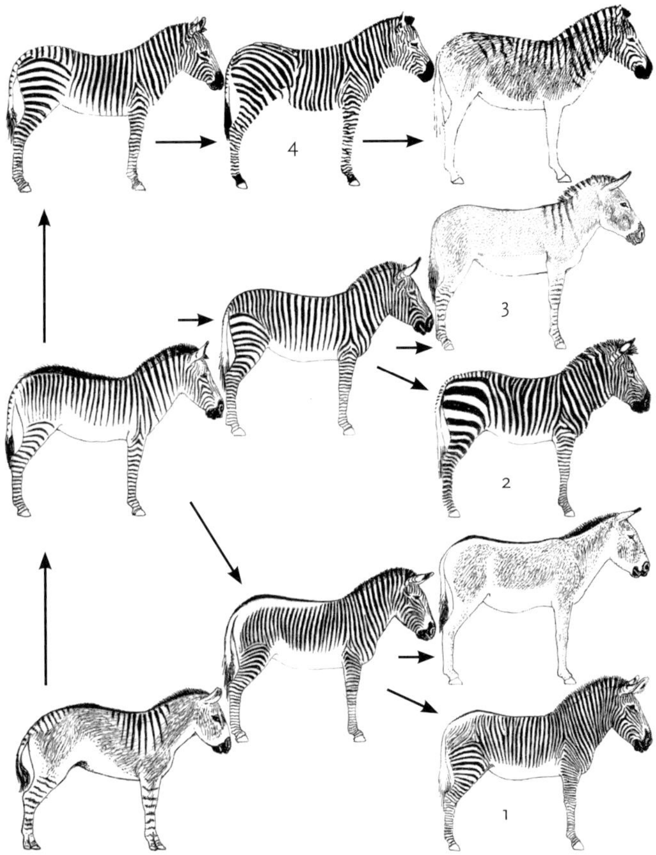

Darwin's prediction corroborated? Tentative relationships between five equine lineages (after Orlando *et al.* 2009). This chart assumes (as Darwin did) that extant equines (right column and top middle) had striped ancestors. Reconstruction of common ancestor for all lineages centre left column.

Hypothesis: vertical stripes originated in ancient equid ancestor as modest 'grooming targets' on croup and shoulder. Stripes then spread to the rest of the body, neck and head (lower left). Increased contrast (black and white) made broadside views of the entire animal highly visible to conspecifics.

Separate banding of the lower legs extended upwards. When enlarged, horizontal stripes on hindquarters enhance the signal value of overall striping because the optical spacing of such stripes is visible at greater distances and is maintained regardless of viewing angle. This improvement in visual clarity appears to have been a later development in zebroid patterning. Ancestors of the three zebra species probably entered Africa in sequence (as indicated by numbers). Extant equine species, right column, from top: *Equus quagga*, *E. africanus*, *E. zebra*, *E. hemionus* and *E. grevyi*.

Later, with the rise of imperialism, trophies as well as specimens became the lure. Collecting trophies imitated and extended feudal hunting practices as a display of power and privilege. Books from this period were sometimes titled 'Records of Big Game'; others were essentially memoirs of shooting trips, but several, notably Frederick Courtenay Selous' reminiscences (Selous 1881, 1893, 1896) and President Teddy Roosevelt's two books, included some useful scientific information within a mindscape that envisaged eastern Africa as a land teeming with 'beasts of the chase'. The latter, best-sellers in their time, are particularly interesting for Roosevelt's ability to combine an almost medieval imagination with factual records. More than 150 forms of mammals were collected and named and the Roosevelt collection in the American Museum of Natural History remains one of the major reference collections of African mammals in the world. Yet Roosevelt introduced 'African Game Trails' (Roosevelt 1909), recalling his visceral:

> joy of hunting the mighty and terrible lords of the wilderness, the cunning, the wary and the grim. ... there are dread brutes that feed on the flesh of man; and among the lower things that crawl, and fly, and sting, and bite, he finds swarming foes far more evil and deadly than any beast or reptile. ... there are creatures which are the embodiments of grace; and others whose huge ungainliness is like that of a shape in a nightmare.
>
> (Roosevelt 1909)

Contemporary mockery of such conceits needs to be tempered by the realization that many contemporary perceptions are no less inconsistent and primitive.

A symbolic change in vocabulary took place in the mid-twentieth century when 'game' became 'wildlife'. This cultural phase coincided with the independence of African countries, the spread of television and the development of mass tourism bringing large numbers of visitors to the newly formed National Parks of Africa. Parks, field guides, research stations and very large numbers of tourists provide the contexts, rationales and economic support in which much of contemporary mammalogy takes place.

Now new perspectives are developing and the sort of seriousness that Darwin and before him Lucretius had brought to bear on their view of nature has begun to re-emerge from the shadows. The most recent revolution in the way African mammals are perceived can be quite as anthropocentric as any earlier manifestation. It is strongly reinforced by contemporary genetics and palaeontology, which have demonstrated, without ambiguity, the genetic affinity of humans with other mammals, indeed with all of organic life.

In this view of life, animals and humans are infinitely complex genetic accretions that reflect countless adaptations to the environments and biotic communities in which they evolved over huge stretches of time. Of course there is a downside to contemporary preoccupation with molecular science. It is still a closeted subculture, its safe benches very far from the world inhabited by the animals themselves. (Academic leadership in leading universities and research centres often remains as short-sighted as ever, more often than not favouring molecules over ecological, behavioural and whole-organism biology.) However, this phase must eventually give way to matching genomes with the behaviour and ecological adaptedness that the genes actually code for.

There are important philosophical dimensions to the huge expansion in knowledge that has been led by molecular science. We desire and pursue knowledge but we know that the enquiring mind is but one expression of being, one faculty of the processes that have given us life, one small detail of the process that generated meerkats, gorillas, chimpanzees and the people who tell us about them. Greater is the desire for life itself. We all desire existence, as it is, and for its own sake, whether we are one in a rare band of foragers or members of a race numbering seven billion today or ten billion in 50 years time. In that we differ not at all from any other mammal, but in coming to terms with our exploding numbers and our cultures' exclusive preoccupation with our own rights and indifference to the survival of other mammals we face new ethical dilemmas.

It is in such perspectives that African mammals have acquired a special interest in the twenty-first century. We now know that embedded within the various communities of African mammals there were, in recent prehistory, a diversity of bipedal primates that included our own ancestors among their numbers. Unveiling the details of their evolutionary progression is a science still in its infancy but it is attracting some of the best minds of our time. We hope that *Mammals of Africa* will be a useful tool in the study of human origins, but most mammalogists would assert that their subjects need no human dimension to be of all-absorbing interest in their own right.

Beyond the present lie still deeper dimensions to the study of mammals. Because every one of us is a mammal we should, theoretically, be able to bring a large measure of self-knowledge to bear on the task of thinking about other mammals. When a mammal eats, drinks, excretes, copulates, gives birth, socializes, sniffs the air, trembles, dreams or sleeps, our personal experience of those activities is there to give us some insight as to what is going on in that other mammal – or is it? In fact that apparently straight-forward comparison is subverted by human culture. Eating, drinking, socializing, giving birth and all the rest are activities that have been so appropriated and ritualized by culture that most people perceive their expression in other animals as funny, embarrassing or, in many instances, offensive, even obscene and horrible. That horror, well illustrated by Queen Victoria's response to seeing an orang-utan ('frightful and painfully and disagreeably human') will continue to subvert the development of comparative psychology. It will slow the quest to explore how our fellow mammals think, and it will delay the changes in ethics that must be integral to our exploration of other minds.

When a culture persuades its adherents that other animals are obscene or funny when they perform the same functions as ourselves, that indoctrination effectively blocks both knowledge and empathy. Many cultures go still further, categorizing particular mammals as 'unclean', demonic or comical and have dogmas that enforce automatic responses of revulsion, fear or laughter when confronted by such animals or their images. It is just such indoctrination that blocks entire cultures from any appreciation of mammals as worthy subject-matter for adult intellect. Many, perhaps most, cultures permit children to indulge their natural interest in animals but then enforce coming-of-age customs that 'put away childish things', including fascination with animals. It is just such cultural denigration that has made an adult concern with animals into very much of a minority interest and our intellectual life is the poorer for it.

The biographies of several eminent scientists reveal how biological science and natural history at school allowed them to hang on to their 'childish' fascination with the natural world and later helped emancipate them from cultural brain-washing, while offering them a way of making a living as professional scientists (Wilson 1994, Hamilton 1996). While there is a respectable niche for biology and biologists in most modern societies, their subject-matter, animals, particularly wild ones, have retained their atavistic cultural association with childhood.

In Africa and elsewhere there is also a strong and negative association between living in a landscape inhabited by wild animals and in what is perceived as an 'undeveloped' past or, worse still, present. Such primitive attitudes are still sufficiently widespread for biologists themselves to feel the need to dress their science in 'adult', 'developed' robes, voluntarily narrowing the real scope of their discipline. Biologists have also had to create recognized niches within the larger society. These pressures have been strongest in countries that are still dominated by non-scientific cultures. Among the consequences are an allocation of biological science and natural history to privileged enclaves: intellectual and physical ghettos outside the mainstream. As a result there are very few countries in Africa where animals, other than domesticated ones, merit much serious interest in the society at large.

For cultures that are intrinsically non-scientific (a majority), science has proved too useful for it to be proscribed, but a hostility that is specific to biology remains, strongly reinforced by predominantly non-secular education systems. Any serious enquiry into the existential meaning of the animal life that is all around us challenges pre-scientific orders of knowledge.It should not be forgotten that as recently as 1619, the Italian philosopher Lucilio Vanini, originally a Carmelite friar, was burned alive by the Inquisition for suggesting, among other heracies, that humans might have originated from apes: the institutions responsible for his murder are still influential (Namer 1980). The current guardians of non-scientific education/indoctrination know, and fear, biology's potential to subvert the dogmas of the past. They have a vested interest in keeping animals funny or grotesque and 'for children only' and their hostility to biology remains a huge obstacle to the scientific education and enlightenment of millions of young people in Africa and elsewhere.

Mammals of Africa is a handbook and inventory of the mammalian fauna of a continent. Other mammalian inventories of other continents have been published and are important for what they can tell us about the biological history and evolution of the modern world. Mammals, world wide, are of absorbing interest in many other ways but over and beyond the many insights that are essentially global, the mammal communities of Africa deserve special attention for several uniquely compelling reasons, some of which have been touched on above and in other introductory chapters.

For a start, the diversity of orders, families and genera is far higher than that of any other continent as our lists and tables confirm. Some likely explanations for such mammalian richness are discussed elsewhere (in several other introductory chapters), but much more remains to be discovered, so among the functions of these volumes are the advertisement of such lacunae and the provision of base-lines for just such studies.

The ecological and economic impact of mammals, especially of the many and numerous species of large mammals, impinges on rural African people in more significant ways than on peoples of other continents. Of all continents, Africa has the longest history of humans exploiting mammals as wild food. Hunting, both traditional and modern, legal and illegal, continues to play a significant part in human social relations, nutrition and recreation. When it comes to the husbandry of crops and livestock (activities that are much more recently developed than subsistence hunting), the impact of mammals is also pervasive. For centuries agriculturists have done their best to exclude wild mammals from their fields. As destroyers of their labour and of human subsistence most wild animals, quite naturally, were, and continue to be viewed collectively as, 'pests' or 'vermin'. Most field biologists will have encountered this total antithesis between their own fascination with the details of a wild animal's behaviour and the all-embracing disgust with which peasant farmers view the same subject. As agriculture expands, this ever-enlarging conflict of interest is still far from any sort of sensible resolution. On present trajectories it can only culminate in the extermination of many species of mammal, as has happened on all other continents.

Only a profound change in current cultural attitudes can deflect such a disaster. It has long been obvious that modern agriculture is rapidly destroying mammalian habitats right across the continent, yet, to date, no effective or compensatory demands have been required of large scale agricultural development. The ancient farmers' pioneering vision of 'reclaiming' fields and pastures from the 'wilderness' remains unchallenged in spite of two major innovations in a newly global civilization. One change is a vast amplification in the scale, scope and destructiveness of agricultural development due to science, technology and new agricultural machines, chemicals, genetic manipulation and the opening of international markets for African products. By the same token, science has emerged as humanity's most universal cultural achievement and our species' fastest-moving intellectual frontier. Yet, while science has also articulated many of the strongest arguments for conservation, it is still rare to find local people giving much value to any other 'use' of a landscape than for agriculture or commercial forestry. Provincial, nationalistic and sometimes precariously ephemeral, the political cultures of contemporary Africa have not been equipped to re-evaluate the role of modern agriculture. International institutions have been little better at re-thinking sensible limits for agriculture's and forestry's scope and role in the twenty-first century.

While it has always been known that all cultivated plants and animals have 'wild' ancestors, the complex ecosystems in which cultivars and domestic stock evolved still fail to elicit any serious interest from those who determine land-use policies. Likewise, recognition that our own 'wild' ancestors had evolutionary contexts that are profoundly relevant for humanity's future self-knowledge has yet to influence how we use land. Because uncontrolled hunting and the unimpeded progress of agricultural development is rapidly diminishing biodiversity and must eventually cut us off (together with all domesticated species and varieties) from our biological roots it is essential and urgent that agriculture's claims to monopolize land-use and land-use planning be challenged.

It is our generation that must take up that challenge and ensure that adequate samples of all natural ecosystems be preserved for future enjoyment, study and analyses by our descendants. We already know that they will have techniques infinitely superior to those available to us. It is the responsibility of our generation not only to put firm, scientifically defensible limits upon agricultural expansion. We must also initiate demands for revenue from the commercial use of land, on the argument that this is proper compensation for the valuable resources that have been and continue to be destroyed. Such revenue, demanded as of right, should fund the conservation of natural ecosystems and must eventually augment, perhaps even displace, the current dependence of conservation on 'charity'.

The title and content of this essay has contrasted the substantial reality of Africa's surviving mammals with the apparent insubstance and mutability of policies, books, records and thoughts. As all thoughts about animals must originate in actual encounters, there are a few easily listed situations in which people interact with animals directly, animal to animal. All the rest (and all that rest is mountainous) is second-hand, whether it be published science, hear-say, cultural baggage, camp-fire folk-lore or outright mythology. Reminding ourselves of the very limited ways in which humans can interact physically with animals helps keep the vast conceptual mountains we have erected around them in some sort of perspective. The practical utility of seeing just how few these physical interfaces are is that it helps highlight what challenges are faced when we try to learn or think about animals. (It also shows what difficulties conservation faces.) The interfaces may be physical, as often as not culminating in death, but each one also has great potential for distortion of ideas about what animals *are*:

1 Nuisances, pests and parasites: crop-raiding mammals and human measures to deter, punish or kill them. Distortion through anger or fear.
2 Hunter and prey: the animal pursued, killed, eaten and processed. Distortion through appetite, the thrills of the chase or social braggadocio.
3 The human as prey: predators finding meat where they may. Distortion through terror.
4 Aids to hunting morphing into companionship: artificial selection morphing wolves into dogs. Distortion through sentiment.
5 Domestication: tended livestock as stored meat, milk, transport, power or pest-control. Multiple sources of distortion.
6 Animals as avatars, vehicles for symbol or myth: animals or their trophies kept or hunted for attributed ideas. Multiple sources of distortion.
7 Animals as subjects of scientific study: observation, capture and/or experiment; excavation of fossils and extrapolation through comparative anatomy. Involuntary sources of distortion include abstraction and generalization but this is the one area where thinkers are most deliberately conscientious about minimizing distortion. It is also an area where the welfare of animals has sometimes been neglected, to great shame and loss of popular support for science.

In sum, by far the most significant difference between mammal communities in Africa and those of other continents is the fact that humans and most of their primate ancestors have been an integral part of these communities for many millions of years. We *are* African mammals. An ancient lack of awareness of Africa, certainly of evolution in Africa, once deprived people of any possibility of correctly answering central puzzles of human existence: 'where do we come from?' 'where is our ancestral home?' 'from what natural communities did we emerge?' and 'what is our place in the natural communities of the future?'

We cannot know how cultures will view or portray the mammals of Africa in the future but we can be certain there will always be people that seek truth in the concrete realities of living communities of animals and plants and in their decipherable histories. That quest will always be preferable to the culturally generated constructs and myths that have so greviously distorted our understanding in the past.

Science represents the single most radical and systematic departure from the myth-making of the past. It is a very specific manifestation of human thinking and one of its many triumphs has been analysis and formulation of the abstract principles that govern natural processes, including those that originated and sustain our own existence and survival. The educated public has had ample demonstrations that biodiversity has arisen by genetic drift and natural selection operating on populations that have somehow become separated and distinct. Behind such necessarily generalized insights the earth's many predictabilities have been punctuated by what can only be called accidents – asteroids arriving from outer space, the eruption of mountains and volcanoes, switches in ocean currents or the collision of continents. As Loren Eiseley remarked (in the opening to this essay) the catalogue of chaotic events that is prehistory has also created the very particular conditions, the specific opportunities, to which the ancestors of today's mammals adapted. It is in the very nature of life to be a triumphant response to chaos and happenstance.

One of the primary tasks science has set itself is to reconstruct the long drawn-out manoeuvres of a single mammal lineage, our own, through all this accident-strewn terrain. Science has brought us close to the starting point for an entirely new appreciation for the role chance has played in the evolution and differential survival of humans and hominins. The hominin radiation was no less a proliferation of experiments in adaptation than were the Pan-African radiations of antelopes, mice, mongooses and mangabeys. Science now has to address the question of why only one member of that fecund radiation now survives.

If 'strategic intelligence' played its part in the survival of *Homo sapiens* its beginnings are sure to lie in just a few of the innumerable permutations of climate and geography acting on just one of many populations. The many skills of other hominins were eventually trumped by those of just one species. Knowing that evolution takes place in real space and real time forces us to survey Africa's geography and dig her soils for clues as to exactly where the earliest ancestors of that singular lineage first emerged. Increasingly, the signs point to the territory south of the Limpopo but at this juncture it is, perhaps, the change in thinking that is as significant as the locality. We are discovering new significance, new depths of meaning, in the details of Africa's history and geography and in the minutae of behaviour and ecology. Sadly, opponents of science's questioning effectively deprive the majority of young Africans of contact and participation in an unprecedented and exhilarating expansion of knowledge in which Africa and its mammals have taken centre-stage.

Darwin, alone in seas of incomprehension even greater than in our own time, wrote

> No one ought to feel surprise at much remaining as yet unexplained in regard to the origin of species and varieties, if he makes due allowance for our profound ignorance in regard to the mutual relations of all the beings which live around us. … Yet these relations are of the highest importance, for they determine the present welfare, and, as I believe, the future success and modification of every inhabitant of this world.
>
> (Darwin 1859)

In trying to mitigate our ignorance and seeking answers to such questions in the light of contemporary science biologists are not alone. All thinking people now face a challenge that could only have surfaced in our time but it is a challenge that will not go away. Discovering our place in nature is a major frontier in contemporary thought and because of it we have no option but to be thinking mammals: our very survival depends upon it. New frameworks of thinking about mammals are developing and must go on developing in the minds and actions of the self-defined 'Thinking Mammal'.

In the quest to provide modern scientific answers to so many puzzles we hope that this work will make a contribution and promote a deeper respect for mammals, including illumination of our own status, as one among many other African mammals.

CHAPTER THREE

The Evolution of a Continent: Geography and Geology

Daniel Livingstone & Jonathan Kingdon

I look at the natural geological record as a history of the world imperfectly kept and written in a changing dialect; of this history we possess the last volume alone ... of this volume, only here and there a short chapter has been preserved; and of each page, only here and there a few lines.

Charles Darwin, *The Origin of Species*, 1859

Geographic background

Africa is a large continent. With Arabia, with which it formed one land mass through most of its history, it has an area of almost 33,000,000 km^2, greater than that of Europe, Australia and the United States combined. It extends over 73 degrees of latitude, almost equally split north and south of the Equator. It extends from one temperate zone to the other and includes the full array of tropical and sub-tropical climates in between.

As described in detail in Chapter 5, the Congo cuvette and West Africa contain, or recently contained, large blocks of evergreen or deciduous forest, and there are smaller areas of forest elsewhere, but woodland, wooded grassland, grassland and various kinds of wetland, thicket, scrub, alpine moorland and other types of vegetation cover most of the continent.

Africa is home to the largest and two of the driest deserts on earth, and, on the seaward slopes of Mt Cameroon, receives one of the earth's highest rainfalls. Rainfall over much of the continent is low, and not only in the sub-tropical trade wind belts: desert conditions occur within 2 degrees of the Equator in Kenya. High African mountains carry glaciers even close to the Equator (although they are melting fast in the face of global warming; Kaser & Osmaston 2002, Mote & Kaser 2007). During the Quaternary ice ages glaciers were more extensive and descended to much lower altitudes (Osmaston 1967, 1989 and see Chapter 4, p. 43). Furthermore, Africa has spanned the southern temperate, southern sub-tropical, tropical and northern sub-tropical belts throughout the last 100 million years.

The evolutionary history of Africa's fauna cannot be understood only in terms of present geography or environment but needs to be viewed within the perspective of the evolving geological and environmental history of the continent (King 1962, Fage 1963, Janis 1993, Adams *et al.* 1999, Denton 1999; and see Chapters 4, 5 and 6). The way in which the continent was formed shapes the physical surfaces that support life, and its tectonic history shaped the origins of its mammalian fauna (Maglio & Cooke 1978, Kumar & Hedges 1998). Here we first consider some of the gross features of Africa's physical geography and geology, interleaving discussion of some of the implications for mammalian evolution and biology.

Plate tectonics: separations and collisions of continents

No geologic development of the last hundred years has had so great an effect on our understanding of the earth as the discovery that continents have moved around on its surface. The continents are the surface expression of crustal plates, which sometimes move together, colliding to form super-continents and throwing up mountain ranges, or sometimes split apart to form an array of more or less separate land-masses such as we see in the world around us today. The remarkably exact match between Africa's western coastline and the eastern margin of South America has been noticed ever since they were first mapped. That correspondence was the first evidence for the longitudinal fracture of an earlier land mass and its break-up into drifting continents and islands. A compelling body of evidence now shows that not only Africa and South America, but also India, Australia, Antarctica and Madagascar were once parts of the single southern super-continent Gondwana.

Still earlier, in the Permian (223–190 mya) a single land-mass called Pangaea had formed, only to become bisected, around 180

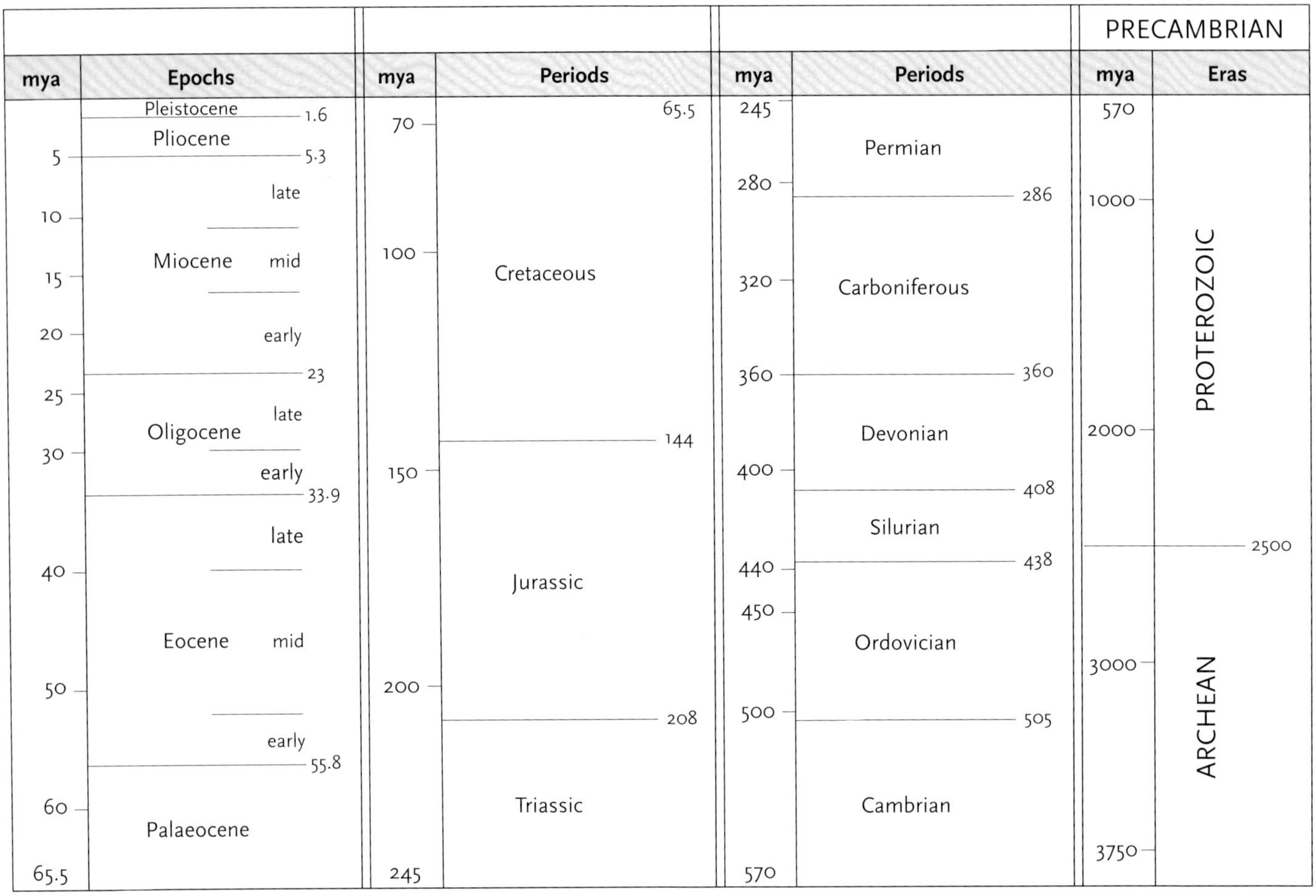

Timescale chart. Shows Eras, Periods and Epochs from the Archean (3750 mya) to the Present. Geological timescale currently estimated as: Palaeocene 65.5–55.8 mya; Eocene 55.8–33.9 mya; Oligocene 33.9–23.0 mya; Miocene 23.0–5.3 mya; Pliocene 5.3–1.8 mya; Pleistocene 1.8 mya to present day.

mya, by the Sea of Tethys to give rise to Gondwana in the south and Laurasia in the north. Where continents are moving away from each other, as the New World has moved away from the Old during formation of the Atlantic Ocean, their past positions are relatively clear. Lava welling up along the mid-oceanic ridge bears the signature of the earth's magnetic field at the time it solidified. That magnetic field has undergone a series of reversals, which have been mapped in great detail by oceanographers at sea and dated radiometrically by stratigraphers on land. We therefore have a good understanding of the times when Africa separated from North and South America and when it separated from the rest of Gondwana. We have also learned much about the break-up of that southern super-continent into Australia, Antarctica, India and Madagascar.

Things are much more complicated where plates collide, and in the case of Africa's relationship with Eurasia it is just such continental connections that are of central interest to mammalogists. Unfortunately there is no simple palaeomagnetic chronology to draw on, and the collisions disrupt or destroy geological evidence from the times of merger. The details of collision between Africa and Eurasia are, and are likely to remain, much fuzzier than the details of Gondwana's break-up (Dercourt *et al.* 2000). None the less, it is known that Africa's northward movement slowed or halted after about 37 mya and that some form of connection or corridor permitted faunal exchange by about 27 mya. Actual suturing between Arabia and Asia was as late as 16.5–15 mya.

Africa is unusual in having stayed very largely in the same attitude, and at similar latitudes, since the origin of mammals, with only some 20 degrees of northward movement and a small measure of counter-clockwise rotation. While North and South America drifted away from Europe and Africa, and while Australia, Madagascar and India split away from Antarctica and drifted to their present positions, Africa moved much less. Other continents and plates drifted away to collide with each other, throwing up such mountain ranges as the Andes, the Cordillera and the Himalayas, but Africa, by comparison, held its place. None the less, collision with Europe between 37–15 mya has thrown up one folded mountain range, the Atlas and Tell Mts, of the type so common elsewhere and, in Europe, the northward movement of Africa moulded the Mediterranean and played its part in pushing up and folding the Alps (Stampfli *et al.* 2001).

The splitting of continents is driven by currents deep within the mantle of the earth and its earliest manifestation on the surface is doming above 'plumes', which create 'hotspots'. There are several such domes in contemporary Africa, the highest and most extensive being Ethiopia (King & Ritsema 2000, Rogers *et al.* 2000, Pik *et al.* 2003, Goudie 2005). When domes eventually split open the cracks become rift valleys that gradually widen (Pritchard 1979). The South Atlantic Ocean is the best example of a long, north–south rift valley that opened up and let the sea in. The Red Sea, a rift valley that opened up to let in the Indian Ocean about 10 mya (Hempton

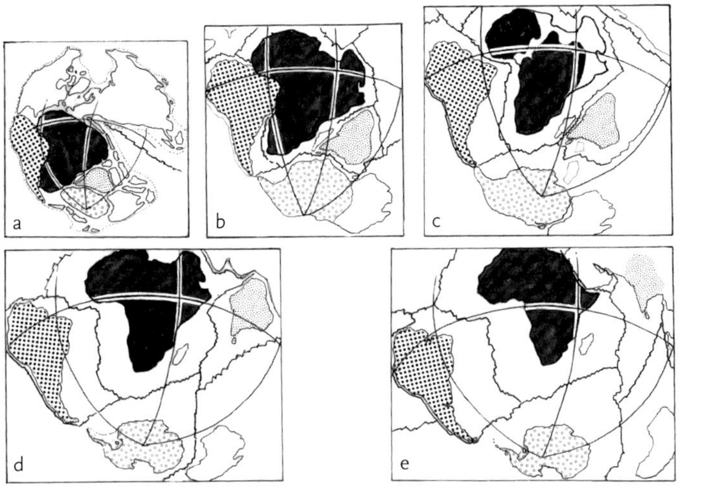

Reconstruction of Africa's tectonic history from circa 250 mya to the present. Africa's northward progression and Equator's southward shift and tilt indicated by three notional degrees of longitude in relation to Equator.
a. Pangea at circa 250 mya.
b. Gondwana breakup at 115 mya.
c. Africa disconnected on all sides and Saharan shallow sea at 80 mya.
d. Continental plates widen west, east and south of Africa at 39 mya.
e. Distribution of former Gondwanan continents and plates at Present. In part after Smith *et al.* (1994) and de Wit *et al.*(1999).

1987), exemplifies an early phase in the creation of oceans. It is also a reminder that the further fragmentation of Africa, the central core of ancient Gondwana, continues: today's rift valleys are future seas and oceans.

One result of continent formation is that the upwarps that once lifted the margins of rift valleys remain close to their parent land-mass's new shore-line. This is most vividly illustrated by both sides of the Red Sea, which are lined by long north–south mountain chains that end abruptly along typical rift valley escarpments. The margins of the Atlantic rift valley are also still visible in a long chain of raised ground along most of the length of Africa's western edge but its former escarpments have been worn away by more than 100 million years of erosion by rivers and modified by subsequent flexure of the land surface within Africa. During the Cretaceous these rift margins were still high mountains that yielded montane floras (R. Morley pers. com.). Fracture of the continent's eastern coast was still earlier, so former uplift is only perceptible in ancient hills and mountains close to the Tanzanian and Mozambique coast.

Sometimes the fusion of continents is permanent, as India's suture onto Asia; sometimes it is intermittent, as Afro-Arabia's connections with Eurasia. Collision of India with the Asian land mass began more than 50 mya but major uplift began about 23 mya (Aitchison *et al.* 2005). This had a profound indirect influence on Africa and its fauna because, by throwing up the Himalayas, this collision created a rain-shadow stretching from the Gobi Desert and Sinkiang through Persia and Arabia to the Sahara. This effect was augmented by plate movement further east, which disrupted the movement of heat from the Pacific towards India and Africa. It is thought that from the early Miocene onward moisture off the Pacific was intercepted by South-east Asia (Cane & Molnar 2001).

Whatever the causes, an extensive belt of dry land required all migrants into or out of Africa to cross an unforested, often arid or semi-arid corridor more than 30 degrees north of the Equator. The most significant tectonic events for our understanding of today's mammalian fauna were those that brought Africa into physical contact with Eurasia. These began no earlier than about 23–27 mya. The evolution of mammals was probably influenced strongly by closure of the Straits of Gibraltar between the Mediterranean Basin and the Atlantic Ocean between 5.6 and 5.32 mya (Clauzon *et al.* 1996, Duggen *et al.* 2003). Named the Messinian Salinity Crisis, the near-total evaporation of the Mediterranean created a vast, deep depression, floored with salt lakes and pans and bounded by the cliffs of continental shelves, that could have been even a more formidable barrier than the Mediterranean Sea itself. While the Messinian probably created a dry land corridor between continents it also saw the start or intensification of desert conditions in the Sahara, yet another obstruction for inter-continental migration.

While the position of continental plates is now well understood, the roles of mantle plumes, superplumes, downwellings and their influence on changing land and sea levels remain little known. None the less, the earliest formation of Antarctic ice in the early Oligocene (32–30 mya) was linked with falls of between 30 and 90 m in sea level. We still rely on records of the first appearance of fossils from newly immigrant fauna to suggest how Africa exchanged mammalian stocks with the rest of the world. Such new arrivals imply some form of physical bridge shortly before their fossils appear in Africa for the first time.

Continental origins of placental mammals

In recent decades it has become clear that primitive mammals and proto-mammals were a significant part of the vertebrate fauna of the world for a very long time during the Mesozoic (65.5–225 mya). Taking into account the sources of various ancestral mammals and the biotic communities in which placentals would later evolve, we should not forget the Mesozoic history of the continent, part of which was spent as an integral part of the megacontinent Pangea, and part as a component of the southern super-continent Gondwana.

Over all of this time mammals were less important than dinosaurs and other reptiles, and most of those ancient mammals were very different from modern placentals (Kemp 1982, Rose 2006). Mesozoic mammals known before the present decade were small, unspecialized, insectivorous and apparently nocturnal, but recent discoveries show that by the middle Jurassic some were highly specialized for aquatic life (Ji *et al.* 2006), by the late Jurassic some were fossorial (Luo & Wible 2005) and by the lower Cretaceous some were large enough to prey on small dinosaurs (Hu *et al.* 2005). Mesozoic mammals that have been collected in Africa so far do not show such flamboyant specialization (Jacobs *et al.* 1988, Krause *et al.* 2003, Kielan-Jawarowska *et al.* 2004).

Today there is a gathering consensus, fuelled mainly by new, robust molecular phylogenies, that extant superorders of placental mammals emerged at about 105 mya according to Springer *et al.* 2003 and Bininda-Emonds *et al.* 2007. The basic distinction between Eutherian (placental) and Metatherian (marsupial) stocks is even older; fossils of both have been found in lower Cretaceous beds with an age of 125 million years (Ji *et al.* 2002, Luo *et al.* 2003), essentially doubling the age of these two groups. Genetics and palaeontology now broadly agree that the fundamental split between marsupial and placental mammals was established by 125 mya (Bininda-Emonds *et al.* 2007, Murphy *et al.* 2007, Wible *et al.* 2007). Most early placental lineages have gone extinct but at least one that gave rise to the modern radiations survived. Where it survived and then diversified has been disputed but the bulk of current evidence suggests Asia, which is consistent with the sparse geological record for mammals in Africa. No marsupial survives in Africa, although marsupials are known from African beds of early Oligocene age (Crochet *et al.* 1992).

After South America's break-away about 100 mya, Africa (at that time including Arabia) began a long period of almost total separation from all other continents. The influence of this isolation on the history of African mammals has only begun to be appreciated in recent years (Kumar & Hedges 1998).

That protracted isolation underpinned the evolution of such striking indigenous elements as the Afrotheria (elephants, hyraxes, sirenians, aardvarks and others), all of which evolved in Africa during this period (Springer *et al.* 1997). This rather limited spectrum of mammals, all of which are very specialized today, looks much more like the progeny of a very early chance colonist than the vestiges of an indigenous placental parent population (Seiffert *et al.* 2004). Zack *et al.* (2005) have claimed that some Palaeocene hoofed mammals in North America are close relatives of one extant sub-group of Afrotheria, the macroscedelians (sengis or elephant shrews) but Tabuce *et al.* (2007) have shown that hyraxes and macroscelids were already well differentiated by the Eocene and that the Afrotheria must have evolved in Africa from a very much earlier common ancestor, possibly allied to a hypsodontid condylarth.

The established approach to the history of evolutionary divergence used to be totally dependent on the morphological fossil record. Unfortunately the absence of evidence for a particular lineage at a particular place and time is not evidence for its actual absence. The oldest known fossil of a mammal taxon shows that the mammal had evolved by its time, but leaves open the possibility of a long previous history that goes undetected, perhaps because conditions were not conducive to fossilization, or because relevant fossils have not yet been found and identified. Also, preservation of mammal skeletons is seldom close to complete (Maglio & Cooke 1978). Fossil teeth are singularly informative, but where only teeth are available, as is often the case (especially for the earlier part of mammalian evolutionary history), the identity and phylogenetic position of the mammal that chewed with them may be less than certain.

Until recently fossil mammals in Africa had mostly come from a very few sites, most of them Miocene or later and all of them after the K–T event at 65.5 mya. In effect the fossil record indicated that the orders of mammals were of Cenozoic age. Mesozoic mammals seemed to be small unspecialized creatures that came into their own after the demise of dinosaurs. Released from the competition and predation of those dominant reptiles, mammals were seen to have evolved and diversified very rapidly during the early Cenozoic.

The idea of such a late development of mammals lost some of its credibility when it was demonstrated that by the early Cenozoic, Africa, South America, Antarctica, Australia, India and Madagascar were separated by wide water gaps from each other and from the rest of the world. Molecular methods show that the fundamental divergences between the orders of mammals are much older (Bininda-Emonds *et al.* 2007, Murphy *et al.* 2007). Even such closely related orders as rodents and lagomorphs diverged close to the Cretaceous/Palaeocene boundary (Asher *et al.* 2005). Although the full-blown modern orders may be mostly Cenozoic in age, the underlying differences between those orders arose well back in the Cretaceous.

Estimates of the times of divergence can still be wildly discrepant. It is not uncommon for the molecular estimate for the separation of two evolutionary stocks to be very much older than that provided by the oldest known fossils. In mammals the discrepancy seems to get worse and worse as we investigate older and older divergences. This is particularly well illustrated by the primate tree shown in Volume II.

Where the details of fossil or extant anatomy have been well studied or reviewed in the light of new knowledge, the most recent molecular phylogenies have turned out to be broadly accordant with morphological and palaeontological ones (Archibald 2003). There have been several major surprises, notably recognition of the superorders Afrotheria, Laurasiatheria and Euarchontoglires (Hedges 2001, Madsen *et al.* 2001), and also some within lower-level taxonomic groups, such as the Carnivora. The cheetah lineage *Acinonyx*, which evolved in North America, probably as the main predator of American pronghorns and deer, seems to have gone through more than one intercontinental exchange (Johnson *et al.* 2006). Similarly, ancestral zebras and canids originated in North America. Such histories raise still-to-be-resolved questions of how and when such lengthy dispersals took place.

It is likely that two major taxa, notably the ancestors of today's caviomorph rodents and platyrhine anthropoids, crossed from Africa to South America. It has been suggested that these were the outcome of chance rafting events; a less likely explanation could be movement across oceanic rises when they were above sea level.

There were also movements between Africa and the separated parts of East Gondwana. Carnivores, tenrecids and primates moved between Africa and Madagascar during the Palaeogene. In the case of Madagascan strepsirrhine primates an Asian origin has been suggested (Marivaux *et al.* 2001) but is generally considered unlikely. Angiosperm plants suggest dispersal from Gondwanan Africa to India as it drifted north to Asia during the Cretaceous and Palaeocene (see Chapter 4), and it is remotely possible that some African mammals might have reached India as it drifted north to Asia; if so, no fossil or other evidence has been found. The only hint of an early placental mammal in Gondwana has been the recent discovery of some teeth in Australia that were attributed to a mesozoic erinaceid (over 100 million years old, Rich *et al.* 1997). Placental mammals appear to be of Laurasian origin.

Africa–Eurasia connections and disconnections

With such a history one might expect the original fauna of Africa to consist mostly of species inherited from Pangaea, and to resemble that of Gondwana more than that of Laurasia. There are, in fact, some vertebrates, such as lungfish, and some ancient plant lineages, that seem to be descended from Gondwanan ancestors. For the most part, however, the vertebrates of Africa are more like those of Eurasia than those of the rest of Gondwana (this similarity between African and Laurasian biotas is apparent even well back into the Mesozoic, when dinosaurs were a significant part of all the continental faunas), but a high proportion can be attributed to more recent invasions or exchanges. Gheerbrant and Rage (2006) have provided an enlightening treatment of the history and affinities of the African vertebrate fauna. They show that a continuous long-lasting land connection was not re-established between Africa and any part of the outside world until the Miocene, when intermittent attachments of Afro-Arabia to Asia were formed.

Mammalian exchanges in north-western Africa appear to have been small in scale and infrequent, which suggests that the water gap between north-west Africa and Iberia has functioned as a barrier; no direct exchanges between Europe and North Africa have been demonstrated before the Messinian (5.7–5.3 mya) (Kowalski & Rzebik-Kowalska 1991, Thomas *et al.* 1982). However, the palaeogeographic maps of Smith *et al.* (1994) do suggest the possibility of a route that did not neccesarily involve crossing a deep strait. On this restless earth there are few bits of land that have been such close and constant companions as Spain and North Africa. Sometimes bits of Africa have been attached to Spain, with the water barrier lying south of them. Other times bits of Spain have been closer to Africa than to Europe, with the water barrier lying north of them.

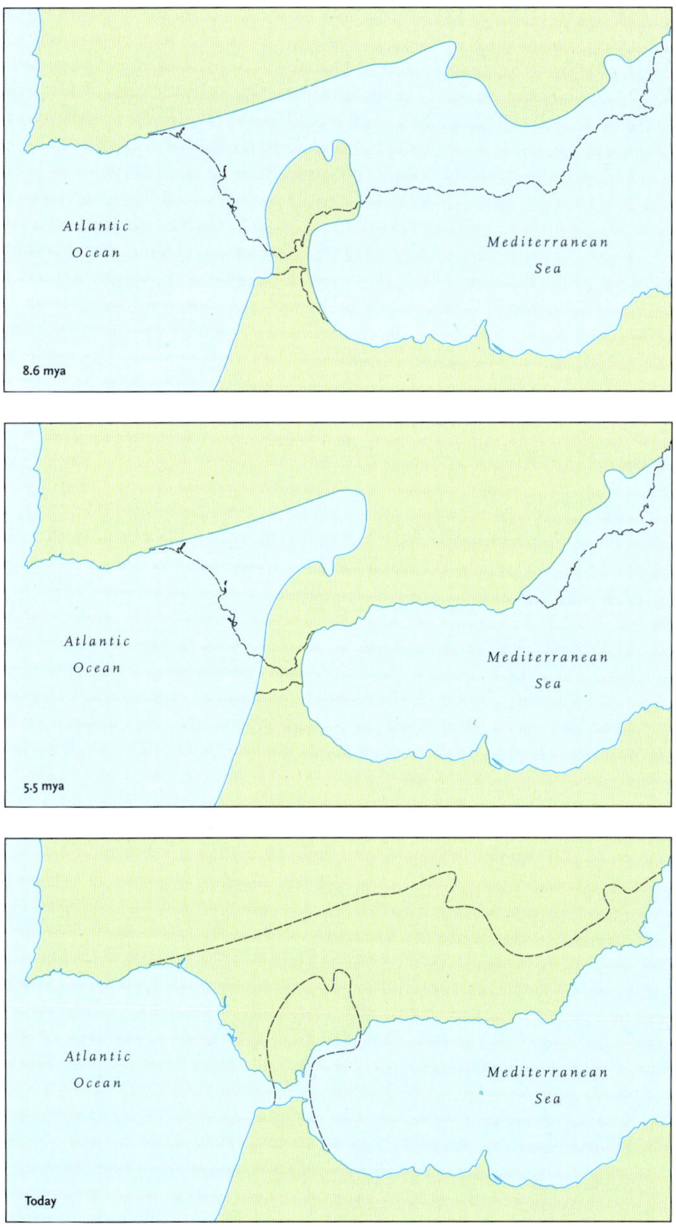

Sketch maps of supposed coastline of NW Africa (Morocco) with Eurasia (Spain) at 8–6 mya, 5.5 mya (Messinian) and Present. (After Smith *et al.* 1994 and others.) Black dotted lines indicate putative changes in coast lines.

Likewise, part of southern Italy and Sicily are the northward extension of a basin that in the late Miocene became detached from the North African shelf and sutured onto the rest of Italy (Stampfli *et al.* 2001, Rook *et al.* 2006). Like other Gondwanan fragments, notably India and parts of Indonesia and Australia, that have become embedded in or close to their northern continental neighbour, these chips off continents could have rafted placental mammals but, so far, there is no evidence that they did.

Apparently connections were never sufficiently sustained between Africa and Eurasia for a wholesale exchange of fauna, like the one that followed establishment of the Isthmus of Panama, but exchanges were frequent enough, and significant enough, to create many faunistic resemblances between Africa and Eurasia.

Biogeographers distinguish between three types of dispersal:

1 Along **corridors** that permit the passage of most or all of a biota.
2 Across **filter bridges** that selectively admit only a small part of the total biota.
3 **Sweepstake** dispersal over wide water gaps by accidents too rare to be observed but significant over geologic time and certainly very significant in terms of potential for genetic proliferation of the immigrant.

Sweepstake routes probably account for a very small number of colonizations, such as the ancestral founders of the afrotherians, the anthropoid and strepsirrhine primates, creodonts, anthracotheres and some rodent groups. All these appeared, or can be extrapolated to have appeared, in Africa well before solid land connections arose in the Miocene.

Since the start of the Miocene, Africa and Eurasia have sometimes been joined by filter bridges, via the Middle East (as they are now), at Gibraltar, and possibly via sills and islands to southern Europe. These later filter bridges selected for mammals that could live in grassland, desert, dry thickets or woodlands, such as antelopes and their predators. Although longer-lasting than bridges across the Mediterranean to Europe, the Middle Eastern route was always too dry for mesic vegetation and obligate forest dwellers. Such equatorial forest species as Okapi, Bongo and Gorilla had Eurasian ancestors that were all non-forest types and have adapted to a forest environment within the last 10 million years (Stewart & Disotell 1998, Kingdon 1990, 2003).

The three dispersal categories are not entirely discrete and can grade into each other. Intercontinental dispersal of flightless terrestrial mammals can sometimes be regarded as sweepstake events, especially the earlier ones, but most later biotic exchanges probably took place over filter bridges.

Rafting is the usual mechanism invoked to permit sweepstake colonization, and the most likely sweepstake migrants would be small and able to live in the branches of drifting trees or on floating islands. It should not be common among large mammals less well suited to long survival on a floating island and it has been rare even in the dispersal of squirrels (Mercer & Roth 2003). Sweepstake dispersal seems to have permitted chameleons to reach Africa from Madagascar and lemurs, carnivores and hippopotamus to go the other way. These dispersals were rare accidents – all the carnivores of Madagascar are descended from one immigrant species (Yoder *et al.* 2005). When several groups of mammals cross an intercontinental water barrier at about the same time there is reason to suspect that they may have used island way-stations to aid their passage. Islands too small and land bridges too transitory to appear on the maps of palaeogeographers could have permitted the movement of primates, creodonts and some rodents between Eurasia and Africa in the late Cretaceous or early Palaeogene.

Although there is no basic difference between crossing from one island to another and from one continent to another, distances between islands are commonly shorter, making sweepstake dispersal between islands more likely. It is worth noting, however, that Spain and North Africa have never been far apart since late-Cretaceous time. Water gaps between the various parts of Europe were commonly wider albeit less lasting (Smith *et al.* 1994). The earliest exchanges were possibly between western Europe and Africa. Then a second route opened up to Asia, and finally, with establishment of more frequent land connections, many more mammals passed between Asia and Africa.

Sea level changes further complicate the picture of inter-continental connections, especially where the sea is shallow and the submarine slope gentle. A small change in sea level can make an enormous difference in the ability of mammals to spread from one continent to another, or to island-hop between them. Smith *et al.* (1994) provide a sobering evaluation of the evidence on which such summaries as theirs are based. In addition to the problems of colliding plates and changing sea level, it is not always easy to recognize a past shoreline in the geologic record. Uncertainties too small to be of interest to plate dynamicists can be vital to mammalogists.

After about 23 mya faunal exchanges with other continents brought in additional ancestral mammalian stocks, such as a succession of carnivores, antelopes, giraffes, pigs and hippo-like mammals, all groups that are characteristic of Africa's fauna today.

When the fauna of a smaller system meets that of a larger one the smaller system tends to lose out. This is what has happened time and again to the fauna of oceanic islands (Cox & Moore 2000), but species that have evolved life strategies that are exceptionally well-adapted to local conditions tend to endure. When fauna from Asia entered Africa the outcome was not necessarily ruinous for the mammals already in Africa. Placental carnivores probably contributed to wiping out African creodont predators, despite their rich diversity, their specializations and the size advantage of the largest ones (which could be 3 m long! – Rasmussen *et al.* 1989). But among the Afrotheria, which had until that time succeeded only in sending tenrecids to Madagascar, the Proboscidea spread with great success to every continent but Australia and Antarctica.

Africa, of course, is a big and diverse system, so its fauna was adapted to many different habitats and to the competitive and predatory pressures that came with them (Grubb 1999). Among larger mammals, Africa's success in colonizing the outside world, though impressive, was limited mostly to three groups: elephants (Proboscidea), monkeys, apes and hominins (Anthropoidea) and, rather later, antilopine ancestors of the sheep and goats (Caprini).

Changing landscapes on a stable continent of ancient rocks

Africa appears to have moved rather little with respect to the earth's axis of rotation during the last 200 million years, and, while its northward movement was instrumental in creating the European Alps, since their formation Africa seems to have been relatively stable with respect to the convective circulation of the mantle. None the less it has had a highly dynamic geological history.

From at least 65 until 30 mya it was a relatively low-lying continent with widespread deep weathering (Schlueter 2006). Exposed Precambrian surfaces are of great extent and are known as the 'Precambrian Shield'. These surfaces are not always flat and have been subject to tectonic forces that have either perforated and overlaid them with lava, as in Ethiopia, down-warped them as in the Zambezi valley (Thomas & Shaw 1991, Goudie 2005) or up-warped them along numerous swells, notably those in East Africa and around the Congo basin.

In other continents large areas of low-lying land have previously been submerged beneath the sea by various combinations of rising sea level and tectonic warping. Africa, too, is thought to have had a well developed seaway during the late Cretaceous (the Trans-Saharan Seaway), which was caused by plate tectonic suture between the Niger Delta area and Libya, temporarily separating Africa into two terranes. The suture failed in the Maastrichtian/Palaeocene, when the Niger Delta began to form. Since 30 mya much of Africa has lain at too high an altitude to be flooded by either local or world-wide changes in sea level. Small areas of marine sediment occur here and there around the African coast, but marine sediment also occupies much of low-lying northern Africa, especially under the Sahara Desert, which is extensively underlain by marine sedimentary rocks. Most of the African continent, however, is composed of very old metamorphic, volcanic and plutonic rocks, covered in places by a thin and discontinuous mantle of much younger fluviatile, lacustrine and terrestrial deposits.

There is a sharp physiographic contrast between highland south-eastern Africa, most of which is a plateau 1000 m or more above sea level, and north-western Africa, much of which lies below 500 m (King 1962, Hamilton 1982). As summarized earlier, many of Africa's uplands, particularly those close to the continent's margins, derive from ancient uplift left over from the break-up of Gondwana, but some of these have been augmented by much later tectonics. In terms of understanding complex mammalian biogeography the distinction between ancient massifs and more recent volcanic mountains is crucial, even when the two are very close to one another. For example the base of volcanic Mt Kilimanjaro is less than 20 km from the relatively low but ancient Pare/Usambara mountain chain, yet the latter shelters very many endemic animals and plants while Kilimanjaro has very few, all recently evolved. Thus the 'older' Angola Pied Colobus *Colobus angolensis*, which occurs down the coast, also occurs in the Usambaras, while the 'newer' Guereza Colobus *Colobus guereza* has colonized Kilimanjaro from the north-west (Kingdon 1971). The Usambara and Pare mountains belong to a wide scatter of ancient continental margin hills and mountains, which, together with choice forested localities on the coastal littoral, including Zanzibar I., are rich in endemic organisms.

Map of Precambrian surfaces in Africa.

Present-day topography showing 1000 m contour.

Volcanism and the emergence of swells and basins

Before 30 mya continuous volcanism was largely restricted to Cameroon, although there was volcanic activity for about 10 million years close to the place where the southern Ethiopian Rift crosses the Kenya–Ethiopia border, and for shorter periods elsewhere. Around 30 mya volcanism began simultaneously in many parts of the continent wherever rising hot mantle plumes reached the crust (Burke 1996).

Plume-driven continental doming and rifting of Miocene and later age, especially in East Africa and Cameroon, punctuates the broader pattern of extensive surfaces of Precambrian Shield (Cox 1989, Goudie 2005). By thrusting up blocks of older rocks, such as Rwenzori, and by generating volcanoes such as Cameroon, Elgon, Kenya, Kilimanjaro and Rungwe, rifting has provided a sprinkling of high mountains across Africa, as well as creating its rift valleys and, through the formation of deep trenches, warpage and craters, many of its lakes. The high mountains rise from a plateau of much lower altitude, and are limited in areal extent (UNESCO 1971). Only in Ethiopia has Cenozoic volcanism produced an extensive highland area. All of the rift-associated highlands, extensive or not, date just from the mid-Tertiary, with domes first forming about 45 mya (Pik *et al.* 2003) but actual rifting starting about 30 mya (Burke 1996). The geologically young domes have been characterized as 'hotspots'; while the mantle plumes that lie beneath them have acquired geological names, Ethiopia has been thrust up by the Afar plume, the Drakensberg mountains rise above the Karoo plume and the Angolan Bie plateau lies above the Parana plume (Gilchrist & Summerfield 1990, Moore & Blenkinsop 2002). Other significant hotspots are the Hoggar, Tibesti and Darfur massifs, the Adamawa uplands of Cameroon and, of course, the string of uplands associated with the East African Rift Valleys (King & Ritsema 2000, Goudie 2005). The latter probably derive from more than one plume (Rogers *et al.* 2000) but the entire East African rift system has been attributed to the continent's northward movement over a single very extensive plume (Ebinger & Sleep 1998). The opening of the eastern rift has been dated to 15 mya (Grove 1983).

With the appearance of doming a pattern of basin and swell structures became established over much of Africa. Many of the swells are capped with volcanoes, and one might be tempted to attribute their height to accumulation of volcanic rocks. Some of them, however, such as those around the Congo and Kalahari basins, have no volcanoes on their crests; it seems likely that all the swells result from flexure due to low-density melted rocks beneath them.

The swells, volcanoes and rifting, all of which started at about the same time, might have been influenced by collision of the African plate with Eurasia (which may also have helped keep the African plate from moving farther north). However, King & Ritsema (2000) have shown that mantle plumes are directly responsible for most of the basin/swell topography of Africa. They are also likely influenced by the onset of new imminent plate boundaries along the East African Rift (Burke 1996) and the formation of constructive plate margins. Although there are similar structures on the surfaces of Mars and Venus, the swell and basin form of the African continent is unique on earth and may be a product of the formation of new plate/terrane boundaries.

Basins and lakes

The swell and basin topography of Africa has had a profound effect on the hydrology of the continent. During dry periods depressions tended to be basins of internal drainage, but during wet ones some of the basins overflowed and cut rivers to the sea (Beadle 1974).

Every major African basin has had one or more sumps where waters form freshwater or salty lakes (now or at some time in the past). Over geological time the location of such sumps can change with the wholesale tilting of land surfaces or they can be drained by deep-cutting gorges or down-warps; uplift tends to hide or dismember the evidence for previous drainage basins (Partridge & Maud 1987).

Lake 'Mega-Chad' has periodically occupied much of the Chad Basin. The Niger inland delta covers a sizeable area of the southern portion of the Eldjouf Basin while the smaller Taodeni depression lies much further north. Likewise, Tidikelt is the principal south-westerly sump for the Sahara (or Gabes) Basin while the lakes around Nefta drain the north-east (Griffin 2002). The Sudd swamps were, according to Salama (1987), a lake in the Tertiary and this region has remained a well-established sumpland for the southern part of the Sudan Basin. In south-western Africa the Okavango and neighbouring sumplands collect waters near the centre of the kidney-shaped Kalahari/Cubano Basin while the Etosha pan lies close to its north-western margin (Buch 1997). In East Africa the effect of swells and basins on river drainage has been greatly complicated by activity of the Great Rifts, all of which have a scatter of rift valley lakes along their lengths. Ponding of the water of rivers that formerly flowed eastward to the Indian Ocean began the process that created the Lake Victoria basin. Following uplift to the east, further uplift along the Western Rift reversed flow of the then-westerly flowing Nzoia-Katonga and Mara-Kagera rivers, eventually closing exits to east and west. Thus, intercepted by uplift associated with both Eastern and Western Rifts, these rivers and Lake Victoria eventually added their waters to the northward-flowing main Nile.

It was once thought that Lake Victoria was about 1 million years old, but Kendall (1969) showed that it was saline and fell below the level of the Nile outlet prior to 12,500 years. Stager *et al.* (1986) and Johnson *et al.* (2000) showed that the level of the lake had fallen so much that it could have held water only in the deepest part of the basin as mapped on the Admiralty charts. Johnson *et al.* (2000) believed that the Admiralty chart incorporated serious sounding errors, and this has been supported by further echo-sounding (R. Hecky, personal communication). The main lake was completely dry prior to 12,500 years ago, although there may have been some standing water in the estuaries of rivers running off Mt Elgon. Johnson *et al.* (2000) found a thickness of sediment consistent with an age of 400,000 years for the basin. A claim that the Nile outlet was not established until well into the Holocene was inconsistent with the data of Kendall (1969) and with the isotopic signature of downstream beds along the Nile (Talbot *et al.* 2000). Lake beds of Miocene age at Rusinga have no genetic connection to the modern lake.

Even shallow ephemeral lakes, such as the present Lake Victoria, the former Mega-Chad and the still earlier 'Lake Congo', can act as temporary boundaries of mammal ranges. The much deeper tectonic lakes of the Rift provide longer-lasting boundaries, probably for several million years, even though they have not been as large as they are today through all of that time (Scholz *et al.* 2007).

More significant has been the longer-term role of shallow basins in providing habitat for the evolution of swamp species such as the Marsh Cane-rat *Thryonomys swinderianus*, Marsh Mongoose *Atilax paludinosus*, Sitatunga *Tragelaphus spekii* and reduncine antelopes of the genera *Redunca* and *Kobus*. The habits, ecology and distribution of such animals provide clues as to their geographic origin. For example, the Sitatunga is a shade-loving browser and it ranges all through the Congo Basin swamps, where it probably evolved, but extended its range after becoming a swamp specialist. By contrast, take the distribution of the grass-eating Southern Lechwe *Kobus leche*, strictly limited by its swamp specializations to the sumplands of upper Zambezia while the Nile Lechwe *Kobus megaceros* is similarly restricted to the Sudd. The isolation of two Lechwe species, each equally specialized, in two well-separated regions only makes sense in the context of the Lake Victoria basin having been as extensive an area of grassy swamps as the upper Zambezi region is today. The open, grassy swamps of the Lake Victoria basin in its recent past may therefore have been the focal centre for the common ancestral Lechwe and just possibly represented their place of origin.

Thought to have held an extensive 'Lake Congo' in the Pliocene (Beadle 1974), the Congo Basin differs from all others in its immense extent (currently 3.7 million km^2 but once larger still) and in being the principal and perennial captor of Africa's equatorial rainfall. The significance of this can be gauged by the river's annual discharge, which is 1400 billion m^3. No other African basin can compare with the Congo for the amount of water falling on and flowing through it, even during relatively dry periods. The basin and river are central to African geography and to the dynamics of mammalian evolution because the river, wherever its course may have run in the past, is a major barrier to dispersal (currently from north southwards or from the south northwards) and different sections of the basin, with different climates, have been perennial foci for distinct communities of forest fauna and flora.

Today its immense water-capture spills over falls, narrows and rapids into the Atlantic Ocean, where it has created a spectacular submarine canyon cut into the narrow continental shelf and, beyond it, a wide deep-sea fan. Beadle (pers. com. to J.K.) thought that all these features were indicative that the present exit of the Congo R. was not very ancient and was due to backward erosion by a short Atlantic river that captured the Congo Basin's waters during a period when water or lake levels were particularly elevated within the basin. Burke (1996) suggested that this might have happened at some time during the last 30 million years but Beadle considered a Pliocene date more likely.

Most of the rain falling on equatorial Africa falls into the Congo Basin and this probably dates back to the continent's separation from South America more than 100 mya. The present course of the Congo R. may not have had enough time to build up a delta fan above sea level but past courses were, potentially, the major source of sediments along the Atlantic coast. It may therefore be significant that between 80 and 35 mya the present R. Niger's immense delta in the Gulf of Guinea was mainly fed by waters flowing westwards down the widely and deeply eroded Benue R. (Goudie 2005). Although there have been periods when the Chad Basin emptied into the Benue, it seems less likely that the waters that gouged out the Benue came down from the north than that the Benue then connected with the Congo, the major source of water and sediment out of Africa. Such a scenario is made less improbable when it is remembered that only a very slight uplift along the spine of today's Central African Republic would have been sufficient to interrupt and reverse northward flow by the Congo R.

Still earlier there might have been other exits for the bulk of Africa's equatorial rain. For example, the main Atlantic site for sediment during the upper Cretaceous was the mouth of the very deep-cut Ogooue R. (Uenzelmann-Nebel 1998). It is, therefore,

KEY

1. Domes (plumes) and Rift Valley swells

hp	Hoggar
tbp	Tibesti
dp	Darfur
rsp	Red Sea dome
afp	Afar plume
kyp	Kenya rift dome
tap	Tanganyika rift dome
map	Malawi rift dome
adp	Adamawa (Cameroon)
pap	Parana plume
nap	Namibian uplands
kop	Karoo plume

2. Pressure fold-belt ranges

ats	Atlas Mountain
mus	Muscat Ranges
cape	Cape Ranges

3. Basins

SaB	Sahara
L-EB	Libya-Egypt
Ed B	Eldjouf
CB	Chad
SuB	Sudd
IB	Iraq
RaKB	Rub-al-Khali
ZB	Congo-Zaire
Cu-KaB	Cubango-Kalahari

4. Depressions

taod	Taoudeni
titd	Tidikelt
gabd	Grabes
qd	Qattara
nigd	Niger
mcd	Mega-Chad
sud	Sudd
mesd	Mesopotamia
pgd	Persian Gulf
zd	Congo/Zaire
vd	Lake Victoria
etd	Etosha
okd	Okavango

5. Drainage lines*

RS	Senegal
RNG	Niger
RCR	Cross
ROG	Ogooe
RCZ	Congo/Zaire
ROR	Orange
RL	Limpopo
RZZ	Zambezi
RJ	Juba
RW	Webshebele
RNL	Nile

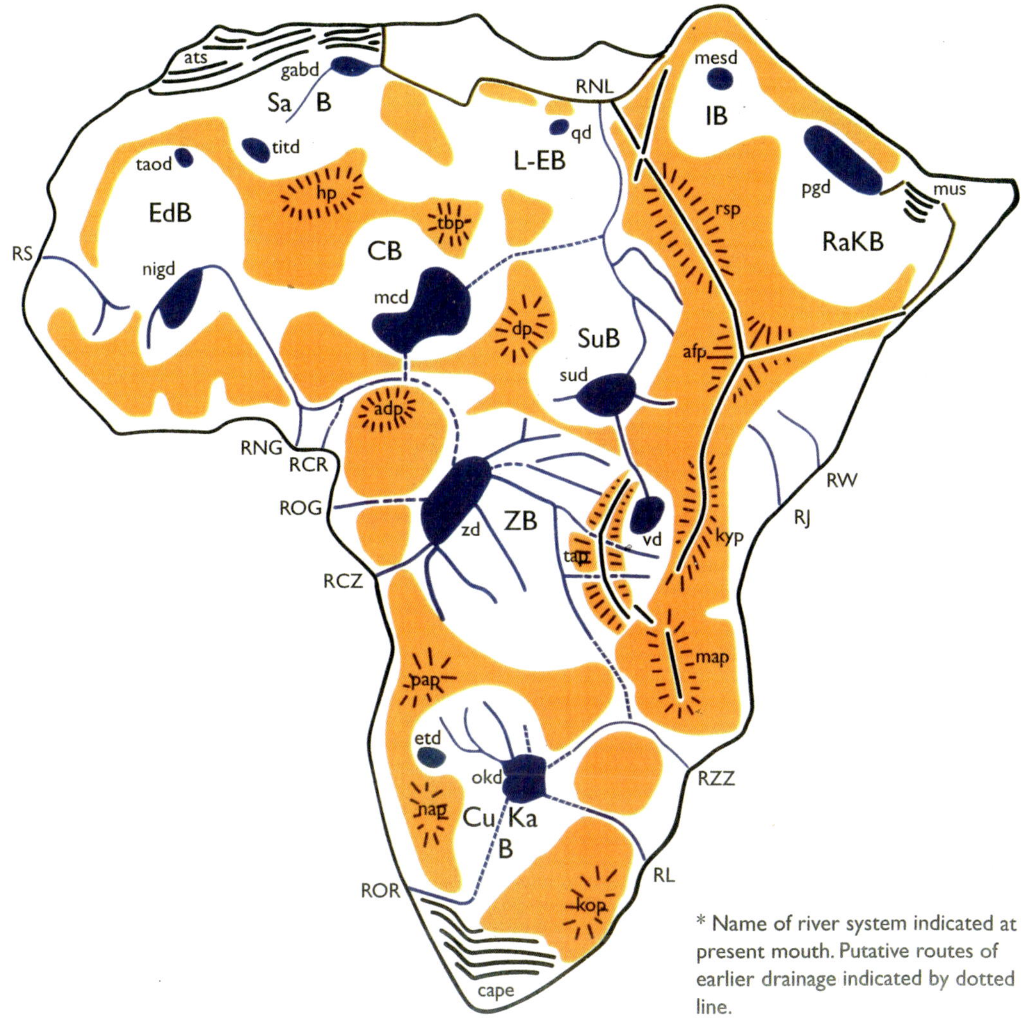

* Name of river system indicated at present mouth. Putative routes of earlier drainage indicated by dotted line.

Afro-Arabian landmass showing plumes, swells and three pressure fold-belt ranges. Also shown are depressions within nine major basins. Overspills are indicated by the courses of present and putative past rivers. Timespan: Palaeocene to Present. (In part after Holmes 1965, Burke 1996 and Goudie 2005.)

possible that the much less significant Ogooue Basin formerly connected with the main Congo Basin and captured its vast volumes of water. Goudie (2005) has sketched in a still earlier drainage pattern in which Congo Basin waters once flowed south-east, exiting in the Indian Ocean. This reversal of direction of flow would be consistent with the substantial uplift that has been associated with formation of the Rift Valley system, raising surfaces and blocking southward flow. Goudie (2005) has stressed that such scenarios for former drainage patterns are highly speculative but that they are informed by a new appreciation of how unstable Africa's crust has been and by new data from radar scanning by satellites.

Further south, the Cubango–Kalahari Basin, an internal basin during dry periods, probably spilt over, successively, in three different directions during wet ones. McCarthy (1983) has suggested a south-westerly pitch, at which time waters exited into the far south Atlantic at the mouth of today's Orange R. Drainage south-east into today's Limpopo R. has been suggested by De Wit *et al.* (2000). Down-warping in the 'Gwembe trough' (Caprivi–Kariba area) probably helps explain how the present Zambezi captured the Cubango drainage, as late as the Pliocene to mid-Pleistocene (Thomas & Shaw 1991, Goudie 2005).

African rivers

Africa's river basins have their origins in the development of broad depositional basins that are fringed by uplifted ridges or plateaux (Butzer & Cooke 1982). Beginning well back in the Palaeozoic, but continuing up to the present, uplift around the margins of these basins has been well illustrated by Holmes (1965).

Swells have changed the drainage patterns of Africa profoundly, with flows being directed to basins of internal drainage (Holmes 1965, Goudie 2005). Uplift anywhere leads to runoff, uplift along continental margins, inland domes and swells creates interior basins (Holmes 1965, Moore & Larkin 2001), and one of the most striking features of Africa's topography is its inland basins. The swells and basins have affected the drainage pattern of the continent, by diverting river flows into the basins, and by periodically shifting the course of rivers over the swells.

Falconer (1911) first observed that many contemporary river courses seem very youthful: this has been generally and widely confirmed. Burke (1996) believes that the drainage pattern of southern Africa has been changed by river capture due to rising swells, which has continued at least into the last million years.

In addition to the gross pattern of some ten or so major basins, some minor drainage basins have developed down the eastern side of Africa and their tree-lined rivers seem to be of significance for the dispersal and speciation of moisture-dependent mammals previously isolated in eastern African coastal or montane forests (Kingdon 2003).

Because rivers sometimes form the boundaries of mammal ranges, sometimes routes of dispersal, their age and history since initial establishment are of special interest to mammalogists. The Congo R. is by far the most important in this respect but, as the previous section showed, its history has been complex. None the less, most distributions can be related to its present course and topography and these are discussed shortly as well as in individual species profiles. So long as the Congo R. has flowed into the Atlantic (which is certainly throughout the rise of mammals) it has represented a barrier to movement between the areas north and south of it. There may have been periods that were sufficiently dry to bring arid habitats close to both left and right banks yet it is striking that non-volant arid-adapted taxa in southern Africa have closer ties with the Horn of Africa than they do with the Sahara, in spite of the latter being physically closer. Except for large powerful swimmers or chance 'rafters' the lower Congo R. has long represented a nearly impassable obstacle to any exchange between northern and southern arid-adapted mammals.

The Nile is illustrative of the difficulties of tracing the history of Africa's rivers. At 7000 km, the Nile is the longest river in the world, draining 3.2 million km^2 (Said 1981), but it was not always so and its great length is essentially a product of the very unstable, south–north East African rift system which has tended to block any flow to the east from central Africa.

The present lower Nile follows the western margins of the Red Sea up-warp and thus likely dates from the time of initiation of this up-warp. Prior to the initiation of the eastern Africa/Red Sea rift system the basin of the present Nile was probably covered by five or more separate hydrographic units (Butzer & Hansen 1968). Burke (1996) noted that there was no great river reaching the north coast of Africa in the area of the present Nile Valley during deposition of the late Eocene and early Oligocene sediments of the Fayum but that a great delta lay 200 km south of the present Nile Delta during the early Miocene.

Goudie (2005) lists three distinct stages in the Nile's formation, starting with a Saharan Gulf system between 40 and 24 mya. This was followed by the second Qena stage in which uplift of the Red Sea margin reversed the flow of the Wadi Qena. The third stage saw the development of the modern Nile system.

During the Messinian (5.6–5.32 mya, when the Mediterranean Sea evaporated, forming a colossal trench) the Nile cut a gorge, the Messinian Canyon, which was four times as deep as the Grand Canyon in Arizona. This downcut the Miocene Nile Delta as far south as 24 degrees and, when the Mediterranean refilled, deep marine sediments, from about 5.0 mya, formed in a long, narrow arm of the Mediterranean that reached Aswan (Krijgsman *et al.* 1999, Goudie 2005).

The present Nile, as a perennial river, is extremely young. Although most of the water in the lower Nile comes from the Blue Nile and the Atbara (where the north-westerly tilt of the Red Sea Rift and western Ethiopian highlands captures virtually all the rain), their flow is strongly seasonal. The river is sustained as a year-round stream by the much smaller but steadier discharge that leaves Lake Albert. The Nile became a seasonally flowing river at the height of the last glaciation between about 20,000 and 12,500 years ago when the level of Lake Albert fell below the level of the Nile outlet, so there was no discharge of water from the upper part of the Nile catchment (Harvey 1976, Richardson *et al.* 1978, Adamson *et al.* 1980). Discharge from the Blue Nile and the Atbara would have been reduced by the same climatic change that lowered the surface of L. Albert. There is reason to believe that similar circumstances prevailed many times during the Quaternary. There is also evidence for a much bigger river at other times in the Quaternary (Said 1981). From the point of view of a freshwater fish, and probably from the point of view of a mammal wishing to cross the Nile Valley without drowning or being eaten by crocodiles, the Nile is 12,000 years old rather than 30 million years.

It has been suggested from radar evidence that a westward-flowing river across the eastern Sahara formerly drained much of the Nile catchment to the Atlantic (McCauley *et al.* 1982), but Burke & Wells (1989) point out that this is inconsistent with the large delta of the Nile and the lack of one at the putative Atlantic outlet. Shared aquatic fauna implies physical connection between the Nile and Niger basins via Chad, probably most recently during the wet early Holocene, but we perceive no effect of such connection on mammalian biogeography.

Africa has two other major rivers, the Niger and Zambezi. The sedimentary record of the Niger Delta suggests that the delta has has been in place since the mid-Eocene (R. Morley pers. comm.). At present the Niger drains to the ocean, but in dryer times it probably lost some of its catchment. Urvoy (1942) thought that the upper

Niger had once run into the Saharan Lake Azaouad but then became deflected by the build-up of sand dunes. Talbot (1980) suggested that Lake Chad formerly emptied into the Benue R., the Niger's main tributary. That connection does not operate today, so the Benue has a relatively small catchment area, but the watershed between the Benue and Chad basins is very tenuous. A possible earlier connection between the Benue and the Congo has already been mentioned.

The Zambezi in its present form is another relatively new river. Its most westerly tributaries, arising on the margins of the Cubango basin, once drained south and then emptied to the west (McCarthy 1983) and De Wit *et al.* (2000) have suggested that this section was once part of the palaeo-Limpopo. The down-warping that channelled the present river down its present course and over the Victoria Falls took place, according to Thomas & Shaw (1991), between the late Pliocene and mid-Pleistocene (perhaps around 1.8 mya). It is just possible that the eastern limits of some species, such as the Cape Fox *Vulpes chama*, Meerkat *Suricata suricata* and the Yellow Mongoose *Cynictis penicillata*, became established or were influenced at a time when the Zambezi exited further south but it is more likely that ecological and competitive factors explain their present ranges.

Large rivers have acted as major barriers so that some non-swimming species that have dispersed from a place of origin north or south of, say, the Congo or east/west of the Nile, often have these rivers as a boundary. Some typical examples are the Giant Forest Hog *Hylochoerus meinertzhageni*, which is distributed from Guinea to Kenya and Ethiopia, yet is never found south of the Congo. Likewise there is no evidence that the Gorilla *Gorilla gorilla* ever existed south of the river and there are numerous other examples (White-bellied Duiker *Cephalophus leucogaster*, Giant Genet *Genetta victoriae*, some squirrels and others). By the same token, animals dispersing from a southern or south-eastern source fail to cross to the north bank. Typical examples are the Bonobo *Pan paniscus* and the Four-toed Sengi *Petrodromus tetradactylus*. There is the interesting possibility that some of the species restricted to north of the Congo might derive from ancestral stocks that had come into Africa and gradually adapted to the conditions found north of the river (which very approximately coincides with the Equator) and never succeeded in reaching or adapting to the south bank.

The Nile, in spite of its intermittent flows in the past, seems to have inhibited the most recent colonists from the east, such as the Bushy-tailed Jird *Sekeetamys calurus* and the Nubian Ibex *Capra nubiana*, neither of which have succeeded in invading land west of the river.

One very minor river that seems to mark the western or eastern boundary of some forest mammals is the Cross R. in south-eastern Nigeria: the list includes the Western Gorilla *Gorilla gorilla*, Olive Colobus *Procolobus verus*, Preuss's Red Colobus *Piliocolobus preussi*, the Drill *Mandrillus leucophaeus* and Grey-cheeked Mangabey *Lophocebus albigena*. While this probably has more to do with ecological factors the possibility should be borne in mind that this stream marks the course of a much more formidable barrier-river in the past.

Lesser river valleys often act as dispersal routes, not barriers, and this is particularly relevant for rivers connecting the highlands of eastern Africa with the Indian Ocean. Here an arid belt that periodically connected the Horn of Africa with the Kalahari has served to separate the coastal and far eastern mountain forests from those in central Africa. Galleries of forest growing on the banks or in the valleys of such rivers have been sufficient to create links, only subsequently to be broken. Such cycles of connection and disconnection via rivers have probably been major facilitators of speciation or subspeciation in tropical Africa and distribution patterns imply very long lasting results. Thus the Suni *Nesotragus moschatus*, a basal antelope that most resembles the ancestral Neotragini, has an eastern coastal distribution from the Equator to South Africa. This dwarf antelope lives well inland and up into montane areas, having dispersed along many riverine forests, including the Rufiji and Zambezi valleys. Ancestral populations gave rise to forest colonists that spread much further west, notably Bates's Pygmy Antelope *Neotragus batesi* and the Royal Antelope *N. pygmaeus*, that live within the equatorial forest belt. Some eastern African littoral mammals, squirrels (Sciuridae), monkeys (Cercopithecidae) and sengis (Macroscelidae) suggest comparable histories in which coastal or Eastern Arc forest forms have used riverine gallery forests to reach suitable habitats further west.

Such dispersals can flow both ways and the Two-spotted Palm Civet *Nandinia binotata*, which has a very extensive distribution in West and central African forests, still occupies riverine strips close to its effectively disjunct eastern African distribution. As both the Neotragini and Nandiniidae are ancient lineages it is clear that such peculiar ranging patterns could have origins going back to the Miocene.

The chemical composition of African lake and river waters is dominated by incongruent solution reactions between soil water and silicate rocks rather than the congruent solution of marine carbonates that is dominant in most of the world. Two consequences of this are likely to affect mammal distribution and abundance. First, African waters are rich in sodium and potassium, poor in calcium and magnesium, which would handicap cervids in their competition with bovids. Cervids have a foothold in Africa, but only in the Atlas Mts, where calcium-rich marine sediments are common. Secondly, sodium fluoride is three orders of magnitude more soluble than calcium fluoride, so African waters are unusually rich in that halogen. It is probably no coincidence that our own African species requires so much fluoride for dental health that we commonly suffer from caries on other continents (Livingstone 1963, Garrels & MacKenzie 1971, Kilham 1972, 1990).

Mountains and uplands as corridors and centres of endemism

Apart from the Atlas Mts, Africa lacks folded mountain belts but it does have a series of domes and raised uplands that run, in a slightly disjunct way, on both sides of the great north–south rift valleys and these provide opportunities for dispersal by upland mammals, or even specifically 'Rift' species.

As in the Americas, the distribution of African biota, including mammals, reveals that, during cool periods, many so-called 'temperate species' (some of recent Eurasian origin) have dispersed south along this raised continental spine while biotas that have evolved in temperate South Africa have dispersed to the north when conditions were right. Along the way there are pockets of upland habitats in north-eastern, eastern and southern Africa, which appear to conserve relictual pockets of once extensive populations.

Choice localities within these mountains and uplands have permitted species to accumulate because, with minimal movement, established and well-adapted species can survive under relatively consistent local climates in relatively stable habitats. The more long-established of these are generally described as 'archaeo-endemics'. Examples of such archaeo-endemics are readily found among the seven genera and many species of golden-moles (Chrysochloridae) of the South African Cape.

Centres of endemism can also attract 'pre-adapted' species from relatively widespread parental stocks. For example, high-altitude moorlands growing on ground recently released by the melting of glaciers in Ethiopia have been colonized by various temperate-adapted murine mice. Living under conditions that are unlike any elsewhere in Africa, these mice have speciated rapidly. Such recent arrivals, when identifiable, can be labelled as 'neo-endemics'. Examples of neo-endemic rodent genera from Ethiopia are *Stenocephalomys* and *Desmomys*. The Giant Root-rat *Tachyoryctes macrocephalus* is another rodent with ancestors of known recent Eurasian origin.

Like the Galapagos Is., the isolated East African Arc highlands exemplify a much wider scatter of 'Centres of Endemism'. The latter consist of very varied eco-geographic foci, all ecological 'islands' that enjoy the common feature of relative stability in a continent that has suffered many climatic and tectonic vicissitudes. An important feature of highlands such as the Udzungwa and Uluguru Mountains, which are sufficiently high and close enough to the Indian Ocean to enjoy relatively reliable rainfall, is the existence of sufficient physiographic and geological complexity for humid habitats to survive in a region where the overall climate has been very unstable. These habitats and their inhabitants tend to depend on the maintainance of a narrow spectrum of temperatures, rainfall and soil types. Whenever the global climate changed, the topographic complexity of these mountainous regions ensured that established ecological communities could endure simply by slipping up or down the catena or by retreating into particularly sheltered valleys. One of the most striking and exciting examples of this phenomenon is the very recent discovery, in the Udzungwa Mts, Tanzania, of numerous endemic species of animals and plants, all restricted to three isolated forests, Ndundulu, Luhomero and Mwanihana (Butynski & Ehardt 2003, Rovero & Rathbun 2005, Rovero *et al.* 2005, Burgess *et al.* 2007). Because of the detailed topography of these highest reaches of the Udzungwa mountain chain these forests have escaped the near total deforestation (mostly by fire) around them.

Rift valleys

Rift valleys are the product of uplift as well as downthrust. Oligocene marine beds were raised to 3000 m at the centre of the Ethiopian Dome (Butzer & Cooke 1982). This is the earliest and most northerly of a series of domes that have split open along their crests, mostly north–south. The most notable rift, L. Tanganyika, 650 km long with its surface close to 700 m above sea level, is 1434 m deep. Its waters extend almost as far below sea level as they do above and are underlain by deep deposits much of which was laid down after rifting began. This multiple half-graben basin (Rosendahl 1988), like L. Malawi, acts as a defining boundary for the distribution of many mammals.

These lakes, and the longitudinal rift valleys of which they are a part, act as physical boundaries. That role is modified by biological influence: most of the land to the west is warm, wet, forested lowland while most of the higher, drier, cooler land to the east is not forested. The forest/non-forest divide is a major defining separation for most organisms in tropical Africa and this ecological divide modifies the physical effect of the rift valleys as physical barriers. In Uganda, formerly extensively forested, many forest organisms skirted mountains, lakes and rift valleys to extend their ranges eastwards, some as far as the Nile, others as far as Mt Elgon or Mt Kenya. Within East Africa the Somali arid corridor seems to have been a more decisive barrier between western and eastern forest biota than the more permeable rift valleys and mountains, which also have approximately north–south alignments.

Regardless of whether the boundary of their range is primarily ecological or physical, many forest species get no further east than the very physical boundaries of Lakes Tanganyika, Kivu, Edward and Albert. To name but a few of the more obvious ones: Agile Mangabey *Cercocebus agilis*, Dent's Monkey *Cercopithecus denti*, Demidoff's Galago *Galagoides demidoff*, Beecroft's Anomalure *Anomalurus beecrofti*, Long-nosed Mongoose *Xenogale naso* and White-bellied Duiker *Cephalophus leucogaster*.

A very different physical and ecological boundary lies between the highlands of Ethiopia and those of East Africa, notably Mt Kenya and Mt Nyiru, which lie only about 250 km south of the Ethiopian piedmont. This region, the Marsabit district, is low and dry and Clayton (1976) has drawn attention to its significance as a major dividing line between African grasses. For mammals and many other organisms the primary importance of this divide is that during glacial periods Eurasian immigrants were able to move south into Africa along the raised ground that lines the western side of the Red Sea Rift (Kingdon 1990). While there were substantial glaciers on the heights of Ethiopia the surrounding uplands were so extensive that a variety of cool habitats were available for such immigrants and they remained and even became dominant long after the glaciations were over, becoming unique Ethiopian endemics in the course of time. The best known examples of Eurasian mammals that followed this route are the Ethiopian Wolf *Canis simensis*, the Walia Ibex *Capra walie* and root-rats of the genus *Tachyoryctes*.

The great height of the Ethiopian dome and of the string of lakes and grassy valleys that floor its 600 km-long rift has ensured that many tropical life forms are effectively eliminated during each glaciation. This 'sterilization' of the landscape probably favours cool-adapted biota that are relatively generalized but may be less resistant to intense competition. A great abundance of hominin fossils has been excavated along the Ethiopian Rift system; perhaps the bipedal apes were effective and rapid colonists of such habitats whenever glaciation ameliorated. The relative rarity of tropical primate species (and of their diseases) might have been among the keys opening the Ethiopian Rift Valley floor to hominins.

Another rather surprising traveller along Rift Valley walls, but in the opposite direction, is Smith's Red Rock Hare *Pronolagus rupestris*. Recent research (Robinson *et al.* 2002) has shown that hares belonging to this exclusively south-eastern genus are the most conservative of all hares and rabbits. Despite being early invaders from Eurasia, perhaps in the earlier Miocene, *Pronolagus* spp. found in the extreme south of Africa a refuge from the competition of more advanced hares and, perhaps, from hare-specific predators. During favourable climatic cycles, the marked preference of *P. rupestris* for steep rocky hillsides has made the recently slumped walls of the Rift Valleys an appropriate path for northward expansion.

Deserts

Morphologically, the desert areas of Africa have the same swell and basin substrate as the rest of Africa. The physical surfaces of this substrate, however, are profoundly altered by break-down into rocks, pebbles, sand and silt, creating very different types of desert and very diverse habitats for mammals.

The best known desert in Africa and, at over 10 million km^2, the largest in the world, is the Sahara. Recurrent desert conditions have prevailed there at least since 7 mya (Schuster *et al.* 2006). Evidence for xerophytic plants in the Sahara dates back to the Pliocene (Kowalski & Rzebik-Kowalska 1991). Daily fluctuations of temperature can exceed 35 °C. Nocturnal temperatures can hit –10 °C and diurnal ones can reach 56 °C. Typical precipitation in the centre of the Sahara is 3–12 mm but the 150 mm isohyet is commonly used to demarcate its southern border. All the major Saharan endemic mammals are true desert-adapted species but where there is evidence for their evolution this implies either post-Miocene Asiatic origin (some gerbilline and dipodid rodents and foxes) or *in situ* evolution since the onset of the Messinian (Addax and Oryx).

The various physical surfaces found in the Sahara provide habitats for mammals and all have Arabic names that have been widely adopted in English and other languages. Hills and mountains are called jebels; stony plateaux are hamadas; rocky basins that conserve springs or rock pools are guelta while other small depressions are daya; more extensive depressions, typically of gypsum or clay, are called sebkha while saltflats or ephemeral salt lakes are shotts. Alluvial plains are reg; seasonal water-courses are wadis and areas of sand, sometimes covering great areas and forming wind-blown dunes that can be hundreds of metres high are called ergs or great ergs. Although specific desert types can be extensive these names give expression to the diversity of habitats that exist within the Sahara.

In spite of its immense size, the Sahara may not, however, be the oldest arid area in Africa. Aridity in the extreme south-west of Africa may have fluctuated over time but the conditions that cause such aridity have been in place for at least the last 10 million years when the cold South Atlantic Benguela current first formed (Siesser 1980, Denton 1999). Indeed some aridity in south-western Africa was likely earlier still because the polar ice-sheet and circumpolar current developed about 23–20 mya, according to Barker & Burrell (1977). Even then, cool surface temperatures in the South Atlantic would have reduced precipitation in south-western Africa and encouraged general adaptive trends towards drought resistance. Namibian desert plants and animals tend to confirm the greater age of their adaptations compared with those in the Sahara. For example, the Sahara has no equivalent to the sand-burrowing Grant's Golden-mole *Eremitalpa granti* (which is an afrothere). The Namibian endemic Pygmy Rock Mice *Petromyscus* spp. derive from a particularly ancient rodent group, the Petromyscinae; their poorly approximate equivalents in the Sahara derive from much more recent murine immigrants.

True desert is of very limited extent in the Horn of Africa but many Somali animals and plants exhibit adaptations to generally arid conditions that are likely to originate well before the Messinian. For example, molecular clocks suggest that the arid-adapted Beira *Dorcatragus megalotis*, a local endemic, differentiated from its closest relatives about 9 mya and dik-diks *Madoqua* spp., small arid-adapted antelopes, have been a distinctive lineage for even longer.

Adaptations to desert-living are described in the profiles of many species. In addition to selection for physiologies that save water and resist heat or anatomies that facilitate escape from predators or heat by leaping, digging or aestivating, some desert animals have adapted to physical niches that allow them to combine shelter from the elements with escape from predation. For example, several species, from totally unrelated taxa, retreat into rock crevices in the desert. The most striking of these is the Noki (Dassie Rat) *Petromus typicus*, which has a broad, flat skull and malleable body selected for squeezing into very narrow spaces. Other rodents, such as some dormice species and the Pygmy Rock Mice *Petromyscus* spp., show similar flattening but a bat has taken this adaptation to its extreme. A Horn of Africa endemic, the Flat headed Bat *Platymops setiger* has a wafer-thin skull and body that almost approaches two-dimensionality!

Among larger mammals one of the least remarked-upon adaptations to desert living is physical mobility and modification of the feet for difficult substrates. Before their near extermination by humans, long-legged desert antelopes made extensive seasonal migrations between seasonal pastures and some species have hooves modified to cope with loose sand while the Beira has rubbery hooves that provide traction over pebbly hillsides.

In the past a belt of dry country has connected the arid north-east Horn of Africa with the Kalahari/Namib south-west (Kingdon 1971, Coe & Skinner 1993). At the present time this corridor is barely discernible but it would seem to owe its existence to a rain-shadow that periodically lay behind the chain of mountains that runs from the Usambaras, Ulugurus and Uzungwas on down both sides of L. Malawi to the Manica and other uplands of Zimbabwe. This long chain of higher ground intercepts moisture blowing in from the Indian Ocean and this desiccating influence is likely to have been amplified during arid climatic phases. Higher ground may have made this extremely narrow corridor less than absolute but it seems to have been an enduring and influential feature of African biogeography. This arid corridor constitutes a line of fracture between Central and Eastern forest communities and has been called 'Kingdon's line' (Grubb *et al.* 1999) and its influence is extensively discussed in other chapters as well as individual species profiles. Kingdon (2003) has argued that this barrier likely provided the essential separating mechanism between the ancestors of modern apes to its west and the earliest ancestors of modern people to its east. If that view proves correct this inconspicuous geographic barrier will deserve very much more attention than it has received to date. Some species common to both the north-eastern and south-western arid zones are the springhares *Pedetes* spp., Caracal *Caracal caracal*, Aardwolf *Proteles cristatus*, Bat-eared Fox *Otocyon megalotis*, dik-diks and the oryxes *Oryx* spp. None of these is adapted to true desert but their discontinuous distribution suggests that areas of very dry *Acacia* bush have, at various times, been sufficiently continuous between these two areas to connect them.

Fossil mammals and fossil sites

While physical and geological features of Africa's contemporary and past landscapes can be shown to have shaped many aspects of mammalian biology and distribution, the single most important interface between contemporary mammalogy and geology is the fossil record (Benton 2000). The palaeontological literature on African mammals is immense and good introductions to it are provided by Cooke (1968, 1972), Hopwood & Hollyfield (1954), Turner & Anton (2004), Werdelin & Sanders (2010) and Kielan-Jaworowska *et al.* (2004). Among the many contributions of that literature is the principal evidence for human evolution from primate ancestors in Africa. It has also revealed a very incomplete but very detailed record of first and last occurrences of extinct species and groups and, of course, provided concrete evidence for the first occurrences of incoming biota from Eurasia.

Of surviving placental lineages, Afrotheria dominate the scene, so our brief survey of fossils begins with Afrotheria and then follows the sequence of *Mammals of Africa*.

Afrotheria

The semi-aquatic, tapir-like *Moeritherium* from the Eocene/Oligocene beds at Fayum in Egypt offers many insights into the first beginnings of the elephants Proboscidea. The further evolution of proboscids into ever larger and taller forms is well illustrated by fossils of *Gomphotherium* and *Stegotetrabelodon* from several north-east African sites. Mastodon (*Mammutus*) fossils appear at about 22 mya but their lineage eventually went extinct. The living elephant genus, *Loxodonta* (in the form of *L. cooki*), first appears in the late Miocene. The dentally more advanced *Elephas recki*, closely allied to modern Indian Elephants, appears later, in the early Pliocene. Hyracoids first appear in North Africa but continue to be common and diverse in many later African deposits. A giant form, *Gigantohyrax*, lasted into the Pliocene in South Africa and a very robust form resembling living hyraxes, *Prohyrax*, is a common and widespread fossil.

Another fossil, named *Myohyrax* because it was initially mistaken for a hyrax, is actually an early, herbivorous macroscelid, hinting at the ultimate common roots of all Afrotheria. A fossil belonging to the modern Sengi genus *Rhynchocyon* (*rusingae*) has been found in 17 mya deposits on Rusinga I. in L. Victoria. Observers of modern *Rhynchocyon* should bear in mind that the very peculiar life-form that these sengis manifest has been around at least 17 mya! Aardvarks somewhat similar to the living species, *Orycteropus gaudreyi* and *Leptorycteropus guilielmi*, date from 7–5 mya.

Fossil sirenians that were already adapted to sea-going have been found outside Africa from about 50 mya. Such extreme specialization and divergence at this early date hints at a bench-mark for the likely pre-Cenozoic beginnings of afrothere evolution.

Primates

According to Sige *et al.* (1990), the earliest fossil primate appears in Africa at Adrar Mgom in Morocco, dated to about 60 mya. Other early primates, *Azibius trerki* and *Biretia piveteaui*, have also come from North Africa (Bonis *et al.* 1988), dated to the Eocene (55–34 mya). The earliest currently recognized anthropoid has been dated to about 49 mya and thereafter there is a rich primate fauna, including early anthropoids, throughout the Eocene/Oligocene beds at Fayum in Egypt. There have even been claims for a Tarsier, *Afrotarsier*, from about 35 mya. Apart from this single fragmentary fossil Tarsiers are an exclusively Asian group, so doubts have been raised as to its real identity.

In East and South Africa a rich record illustrates the differentiation between proto apes and proto monkeys and of the Colobine/Papionine split covering an almost continuous record from the early Miocene to the Holocene. *Victoriapithecus* spp., ancestors for most of today's monkeys, first appear about 15 mya in East Africa.

Rodentia

The great diversity of living rodent groups owe their variety to a succession of invasions out of Eurasia. The earliest fossil zegdoumyids and hystricognaths date from the Eocene (50–35 mya) in Fayum. Also from North Africa are Eocene records of phiomyids, notably *Protophiomys algeriensis*. It is therefore likely that more than one primitive rodent lineage was already present before the Miocene invasions. Among possible descendants of the Zegdoumyids are the anomalurids *Nementschamys lavocati*, *Megapedetes* and *Parapedetes*. Among the latter are numerous Ctenodactyla and Thryonomyidae.

Lagomorpha

It is probable that the ancestral stocks of *Pronolagus* and *Bunolagus* differentiated from a primitive stock entering Africa during the early Miocene, but currently the earliest hare is the mid-Miocene *Kenyalagomys mellalensis*.

Chiroptera

Although bats are known to have been numerous and diverse by the Eocene they are currently best represented by living genera in the middle Miocene (17–11 mya) of North Africa, with *Rhinolophus*, *Hipposideros*, *Tadarida* and various megadermids and vespertilionids all represented. Bat radiations are thought to be post-KT. No bat families are exclusive to Africa but Nycteridae almost certainly evolved in Africa, emerging from a common ancestor with the globally distributed Emballonuridae.

Erinaceomorpha and Soricomorpha

Hedgehogs and crocidurine shrews are known from the mid-Miocene onwards. One hedgehog, *Gymnurechinus comptolophus*, from Rusinga has been dated to about 17 mya. *Protechinus salis* from Morocco has a similar age.

Carnivora

The now extinct Creodonta were already present in the Palaeocene and some hung on after the arrival of modern lineages. Among the earliest of modern types are two that have gone extinct: 'half-dogs' (Hemicyonina) and 'bear-dogs' (Amphicyoninae).

Felids, in the form of various large or largish sabre-toothed forms, tend to dominate the early Neogene record. The ancestors of living forms have tended to arrive as separate invaders but it would appear that the Caracal *Caracal caracal*, the African Golden Cat *Profelis aurata* and the Serval *Leptailurus serval* may have differentiated within the continent from a single ancestor. The first fossil lion appears at 3 mya and the leopard at 3.4 mya.

Mustelids and viverrids were among the earlier invaders. Among the fossil mustelids a very large predator *Ekorus ekakeran* has been found at Lothagam (about 6 mya). A distinctive viverrid is *Kanuites*, which resembles *Genetta*. Bears made a brief appearance in the late Pliocene.

Canids, which originated in North America, were the last carnivore group to arrive in the late Miocene.

Pholidota

Pangolins are known from Eurasia well before their first appearance as fossils in Africa during the mid-Miocene in Kenya and Pliocene in Uganda.

Perissodactyla

The extinct chalicotheres were among the earlier perissodactyls and they survived in various forms from about 20 mya to 2 mya. Rhinoceros species have been successful and diverse members of the African fauna from the early Miocene onwards. Equids, all of North American origin, arrived as three-toed hipparions in the late Miocene while fossils of modern *Equus* species first appear at 2.5 mya.

Cetartiodactyla

Anthracotheres, a now extinct group but probably linked to hippos and whales, are among the earliest fossil mammals, being found in North Africa from the Eocene (Coiffait *et al.* 1984). The earliest pig fossils date from 17.5 mya and the dominant *Nyanzochoerus/Noyochoerus* lineages subsequently went extinct. The earliest fossil hippopotamids date to the early Miocene. Ancestral chevrotains and primitive giraffids arrived around 22 mya. Cervids only make an appearance about 1 mya. The first fossil of a primitive antelope, *Eotragus*, is recorded from about 18 mya (Thomas 1979), but it is likely that the ancestor of today's African antilopine radiation arrived still earlier. The first fossil of a tragelaphine bovine dates to 14.5 mya. All living members of the genus *Tragelaphus* probably derive from a common ancestor entering Africa at about this time.

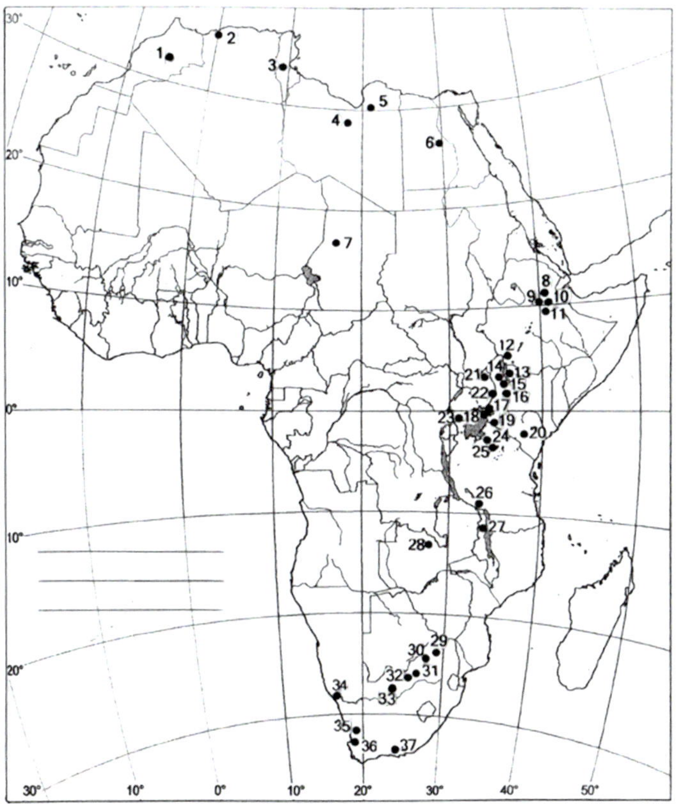

Fossil sites in Africa. 1. Beni Mellal 14 mya; 2. Bou Hanifa 11 mya (+ Ternefine 0.6 mya); 3. Bled ed Dourah 14–10 mya (+ Ichkeul 2 mya); 4. Jebel Zelten 22 mya; 5. Sahabi 5 mya; 6. Fayum 40–30 mya; 7. Totos-Menalla 7–5 mya (+ Koro Toro 3.2 mya); 8. Hadar 3 mya; 9. Middle Awash (Aramis) 4.4 mya; 10. Bodo 0.6 mya; 11. Bouri 2.5 mya; 12. Omo 4.5–1 mya; 13. Allia Bay 3.9 mya; 14. W. Turkana 1.5–3.5 mya; 15. Lothagam 7–5 mya; 16. Kanapoi 4 mya; 17. Maboko 15 mya; 18. Rusinga & Mwafangano 17 mya; 19. Fort Ternan 14 mya; 20. Olorgesaillie 0.8 mya; 21. Moroto 18–20? mya; 22. Bukwa 9? mya; 23. Nsongezi 0.9 mya; 24. Olduvai 2–0.2 mya; 25. Laetoli 3.8–3 mya; 26. Songwe 57–50? mya; 27. Chiwondo 4–1.6 mya; 28. Kabwe 0.2 mya; 29. Makapansgat 3 mya; 30. Gondolin 2–1.5 mya; 31. Drimolin 2–1.5 mya; 32. Sterkfontein Valley 3.5–1 mya; 33. Taung 2.5 mya; 34. Oranjemund 150,000–100,000 ya; 35. Oudrif Permian–Holocene; 36. Langebaanweg 5.2–2.5 mya; 37. Klasies 110,000–75,000 ya.

Fossil sites

Sites that contain important mammal fossils in Africa have a wide scatter but are mainly concentrated in a very few countries. The best-known sites for fossils are clustered in geologically favoured parts of East, north-east, southern and northern Africa. Many of these are Miocene to Holocene and have become well known because key hominin fossils have been found there. Apart from strictly commercial explorations, the search for human origins in Africa has driven much of what palaeontological activity there has been.

Sites yielding earlier mammal fossils are still rare, the best known being in Egypt, Algeria and Morocco. New sites, dating from the Palaeocene onwards, in the North Songwe valley, Tanzania, promise to yield new material representative of mammalian evolution closer to the centre of Africa, rather than from its peripheries. Some of the best known or most fossil-rich sites are listed below to demonstrate their uneven spread in space and time.

Adrar Ngom – Morocco	60 mya
Quarzatzate – Morocco	58 mya
Chambi – Tunisia	53 mya
Dra – Algeria	50 mya
Bir el Ater – Algeria	40 mya
Fayoum – Egypt	40–31 mya
Jebel Zelten – Libya	22 mya
Moroto – Uganda	20.6 mya
Beni Mellal – Morocco	17 mya
Testour – Tunisia	17 mya
Turkana Basin (Kenya and Ethiopia)	17–0.5 mya
Rusinga & Mfwangano – Kenya	17 mya
Maboko – Kenya	15 mya
Fort Ternan – Kenya	14 mya
Arris Drift – South Africa	14 mya
Bled ed Douarah – Tunisia	12–10 mya
Oued Zra – Morocco	10 mya
Lothagam – Kenya	7–5 mya
Chad Basin – Chad	7–5 mya
Sahabi – Libya	5 mya
Langebaanweg – South Africa	5 mya
Aramis – Ethiopia	4.4 mya
Kanapoi – Kenya	4.1 mya
Turkana Basin – Kenya	4.5–1 mya
Chiwondo – Malawi	4–1.6 mya
Laetoli – Tanzania	3.8–3 mya
Hadar – Afar Ethiopia	3–2.5 mya
Sterkfontein/Makapansgat – South Africa	3.5–1 mya
Olduvai – Tanzania	2–1 mya

STRATIGRAPHY AND DATING

The discipline of geology developed first in Western Europe and later in Africa. Despite enormous progress Darwin's 1859 assessment of the geologic record is still accurate all over the world, but especially in Africa.

Radiometric dating now makes it possible to correlate the eras, periods and epochs of Phanerozoic time (the last 542 million years) and also to date them. Note on p. 28 especially the great length of Cretaceous and of Miocene time. Some non-radioactive processes provide especially precise correlation between different bodies of rock. For example, periodic changes in the motion of the earth around the sun and shifts in orientation and intensity of the earth's magnetic field affect the whole world at the same time.

Fossiliferous marine sediments are not widespread in Africa so we are particularly dependent on the new methods to correlate sedimentary deposits in which our mammals are found. Had we a good sequence of widespread terrestrial deposits we would be able to erect a stratigraphy based on terrestrial fossils. Occasionally that can be done. Pig fossils helped to straighten out the confusing radiometric dating of hominid-bearing sediments around Lake Turkana (Cooke 1978, White & Harris 1977, Harris 1991). For the most part, however, we depend on radiometric dating of volcanic deposits below, above and between the rocks in which fossil mammals occur.

Several terms common in the literature are not included in our figure. **Holocene** is sometimes called **Recent**, **Holocene** combined with **Pleistocene** is frequently referred to as **Quaternary**, and the pre-Pleistocene part of the Cenozoic used to be called **Tertiary**. All of geologic time has been subdivided much further but in this chapter the only such subdivision we use is **Messinian** (5.75–5.3 mya), during which the Mediterrarean was cut off from the world ocean and evaporated largely or completely. Several Saharan desert-adapted mammal species seem to have evolved then.

Recent discoveries and future prospects

The current revision of the International Geological Map of Africa (Commission for the Geological Map of the World, 1985–1990) provides an overview of knowledge of the geology of the continent. Comparison of the enormous outcrop area of Mesozoic and even more of Cenozoic rocks with the very small number of sites from which most of the mammal record comes suggests immediately that the palaeontological exploration of Africa has hardly begun.

Not all of those rocks are fossiliferous, of course, and finding those that are is much easier in regions too dry for a continuous and heavy cover of vegetation. It is also easier to find significant fossils in a countryside where relief and erosion expose them for inspection. Less commonly in Africa than in some other parts of the world, discovery is also aided by the exposures produced by human engineering or mining activities.

Research driven by curiosity about human origins shows what can be achieved by serious prospecting. No doubt a comparable effort devoted to non-hominid mammals would bring similar rewards. Unfortunately the mineral resource in greatest demand, petroleum, requires direct searching of such small volumes of rock that it is unlikely to provide many fossil mammals, although it can provide useful background information about the mammalian environment. Some other minerals, such as alluvial gold and diamonds, and even humble limestone, have yielded a steady trickle of information significant to mammalogists. One might argue that the greatest contribution of economic geology to African mammalogy was recruiting L. B. S. Cooke to vertebrate palaeontology. In any case, there is no discernible current increase, nor any prospect of a spectacular future increase, in the flow of information about mammals and their past from economic geology.

The most spectacular progress of the past decade has come from the interaction of palaeontology and molecular genetics and the development of statistical methods for handling their data (see, for example the papers by Stanhope *et al.* 1998, Eizirik *et al.* 2001, Hedges 2001, Murphy *et al.* 2001a, b, Jaeger 2003, Springer *et al.* 2003, 2004, Jaeger & Marivaux 2005, Bininda-Emonds *et al.* 2007 and others already cited in this chapter). At the same time the classical methods of vertebrate palaeontology have continued to enrich our understanding of African mammals and their history. Note for example Suwa *et al.* (2007) describing gorilla-like teeth from a late Miocene ape in Ethiopia and McBrearty & Jablonski's description of a fossil chimpanzee (2005), Seiffert *et al.*'s (2005) account of Africa's higher primate radiation, Sige *et al.*'s discovery of a late Palaeocene omomyid primate from Morocco (1990), and Kappelman *et al.*'s use of Oligocene mammals from Ethiopia to constrain the time of faunal exchange between Afro-Arabia and Eurasia (2003).

Two features of the African landscape have special promise for understanding the history of African mammals and their environment. Deep tectonic, volcanic and meteoritic lakes contain a very detailed stratigraphic record covering up to about 5 million years of Cenozoic time. That is long enough to embrace all of hominid evolution and reach back to the separation of chimpanzee and human evolutionary lines. Cores have already been raised that cover the last million years at L. Malawi and L. Bosomtwe in Ghana (Scholz *et al.* 2007). Bones of mammals are not likely to be found in useful quantities in those slender cores, but the environment is well represented in them by fossil pollen grains, diatoms and grass cuticles. At its best, the sediment in these lakes provides a discernible record of each individual year. Using Darwin's analogy we are likely to read each individual letter of a record covering the last million years, although we are likely to find that even in meteoritic impact lakes, where the record is most complete, an occasional word is missing.

The extensive basins between plume-driven swells, such as the Congo and Kalahari Basins, are the other special African resource. At present there is no immediate prospect of tapping the rich mammal record they must surely contain. Most have not been uplifted and exposed as a consequence of collisions between continental plates, like the Cenozoic beds of Eurasia and America, and excavating deposits that are many thousands of metres in thickness is still prohibitively expensive.

Could we tap the possibilities of those basins, however, we would be able to answer questions about the ancestral stock of afrotherians and anthropoid primates, and possibly about filter bridge connections across the Mediterranean, that puzzle us today.

CHAPTER FOUR

Africa's Environmental and Climatic Past

Robert J. Morley & Jonathan Kingdon

When an icy mantle gradually crept over much of the northern hemisphere, the greater part of the animal life must have been driven southward, causing a struggle for existence which must have led to the extermination of many forms, and the migration of others into new areas. But these effects must have been greatly multiplied and intensified if, as there is good reason to believe, the glacial epoch itself... consisted of two or more alternations of warm and cold periods.

Alfred Russel Wallace, *Man's Place in the Universe*, 1903

In this review we have three primary concerns:

1 To reconstruct the broadest features of the environments that ancestral mammals had to adapt to, notably in relation to the major division of African fauna and flora into humid (or forest-adapted) and arid (or non-forest-adapted) species and groups.
2 Given the very high proportion of contemporary mammal groups that have ancestral roots in, or have a history of exchange with, Eurasia, to understand the nature of those past environments that facilitated or inhibited such exchanges between the two land masses. It is also important to correlate such environments with estimated times of connection and disconnection.
3 To identify those features of change that may help us understand the biology, distribution and immediate history and ancestry of contemporary species.

There are many very puzzling mammal distribution patterns as well as many unexplained peculiarities in the biology of living species that clearly have their roots in their ancestral pasts. To explain these alone requires deep perspectives in time: some can be explained on relatively short-term time-scales within the range of tens to hundreds of thousands of years, relating to the climatic perturbations of the Quaternary ice age, but others need to be considered on a much longer time-scale.

At generic or other higher taxonomic levels, time-frames on scales of millions or tens of millions of years need to be considered, taking us back to the mid-Cretaceous (105 mya) when placental mammals are thought to have emerged (Bininda-Emonds *et al.* 2007, Murphy *et al.* 2007), to later differentiation of the major lineages and further radiation of mammals at the beginning of the Tertiary period (65 mya) following the extinction of the dinosaurs. Newly found fossils in China and Mongolia reveal that eutherian mammals were already established during the early Cretaceous (96–130 mya), and thus diversified substantially earlier than was formerly thought (Lopatin 2006). We have therefore begun this review from the beginning of the Cretaceous period, from which time the flowering plants first appeared, and subsequently came to dominate the vegetation of the continent.

Today, the vegetation and climate of the African continent contrasts markedly with the archipelagic region of South-East Asia and the Neotropics in that it is characterized by widespread deserts, such as the Sahara, the Horn of Africa and the Kalahari (see map p. 53), and huge stretches of savanna and open woodland, but relatively smaller areas of forest (Moreau 1963a, b, 1969, Kingdon 1971, 1990, White 1983). Africa has only one-fifth of the global total of tropical rainforest, despite having by far the largest land area within the tropical zone. South America and South-East Asia, instead, have comparatively more extensive tropical rainforests and fewer dry regions.

The flora of the African tropics is also less diverse than its other tropical counterparts. For instance, Africa has fewer species of palms than the tiny island of Singapore, and is also poor in Annonales, orchids and rainforest undergrowth taxa, epiphytes and lianes; there is also only one species of bamboo. In contrast to the relative poverty of tropical African flora, the southern tip of the continent has a very rich and diverse temperate flora, characterized by the Fynbos of the western Cape, a diversity only approached elsewhere in temperate areas of Western Australia.

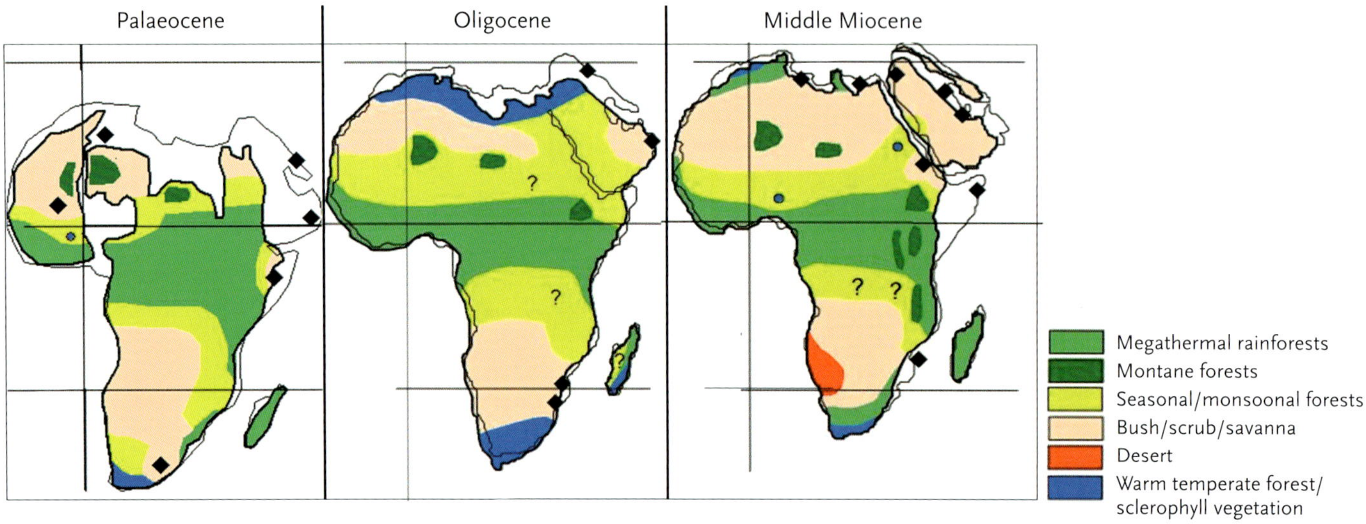

Tentative vegetation maps for the Palaeocene, Oligocene and middle Miocene (from Morley 2000 with modifications). (In all maps palaeogeography and palaeocoastlines after Smith *et al.* 1994.) Occurrences of evaporites and bauxites (diamond symbols) from Boucot (unpublished).
Left map: Late Palaeocene/early Eocene thermal maximum.
Centre map: Oligocene, following the terminal Eocene cooling event.
Right map: Middle Miocene, coinciding with the Miocene thermal maximum.

The vertebrate fauna adapted to African forests also tends to differ from its Asian and American equivalents in that numerous species and groups have non-forest sibling species (Moreau 1966, Kingdon 1971, 1990, Schiøtz 1975). Conversely, neotropical and South-East Asian forest faunas are largely derived from forest-dwelling ancestors. The scarcity of larger organisms that can be shown to have a long and unbroken record of forest-adapted ancestry implies a tenuous hold for this habitat in Africa's past. Another mammalian group, the golden-moles (Chrysochloridae), that has recently been shown to derive from the very earliest radiation of afrotherian mammals (Stanhope *et al* 1998), is best represented in temperate southern Africa, suggesting that long-term relative stability for temperate habitats in Africa may also be significant.

These contrasting differences between Africa and its tropical counterparts, and between tropical and temperate Africa, are the result of their different geological and climatic histories, involving a variety of time-scales (Morley 2000). The scenario presented here sets out to provide a background within which the evolution and diversification of African mammals can be investigated and visualized.

Palaeovegetation as the setting for mammalian evolution

This review of vegetational history and climate has been augmented by continent-wide vegetation reconstructions modified from Morley (2000), and constructed from palaeobotanical and palynological data, together with data from climatically sensitive lithologies, such as evaporites and coals, coupled with and utilizing climate modelling techniques of Parrish & Barron (in press). Data on the occurrence of bauxites from Boucot *et al.* (2008) have also been used in this reconstruction because bauxites form under hot, strongly seasonal climates and are therefore good climate indicators (albeit difficult to date). For generalized maps for the Palaeocene, Oligocene and Middle Miocene see above.

Cretaceous climate and vegetation

The African continent formed part of Pangaea during the Jurassic (180 mya) and became part of Gondwana in the earlier early Cretaceous (130 mya). During these periods it lay further south than today but still straddled the Equator (Smith *et al.* 1994). However, the climate of the region was relatively uniform (as shown by macrofossil and pollen floras summarized in global maps by Vakhrameev 1992). The vegetation of North and central Africa was characterized by woody members of the gymnosperm families Araucariaceae and Cheirolepidaceae (latter now extinct), perhaps with an understorey of ephedroid plants and some ferns. The absence of coal deposits (which reflect periods of ever-wet climate) for the entire African Jurassic and Cretaceous suggests that ever-wet climates over this period were absent, and Parrish *et al.* (1982) and Parrish (1987) have interpreted this as indicating the widespread occurrence of monsoonal climates (i.e. alternate wet and dry seasons) across the African land mass. Southernmost Africa, south of about 40° latitude, probably bore a more mesic vegetation with austral affinities, especially southern gymnosperms, such as *Dictyozamites*, *Nilssonia* and *Araucarites* (Vakhrameev 1992).

First appearance of flowering plants

A more distinctive climatic gradient became established across the continent in the later early Cretaceous (128–96 mya), until the early late Cretaceous (96–85 mya). This roughly coincided with the time of initial appearance and subsequent radiation of angiosperms across the region and their eventual rise to dominance over other groups. It is likely that the emergence and earliest differentiation of placental mammals was also taking place at about this time, probably in the Eurasian land mass (Beard 2004, Bininda-Emonds *et al.* 2007).

This period coincides with the global Cretaceous 'thermal maximum' (Barron & Washington 1985), when equatorial climates were considerably warmer than at present, with equatorial oceanic temperatures being 5° warmer than modern oceans (Zachos *et al.* 2003). There was a clear climatic gradient at about palaeolatitude 20° N and S (Crane & Lidgard 1990), in the vicinity of the subtropical high pressure zones, probably reflecting belts of drier climates (Morley 2000), and with more humid (but not ever-wet) climates becoming established in the region of the palaeoequator (Doyle *et al.* 1982). Comparison of climatic inferences drawn from mid-Cretaceous palynomorph data with computer-derived climatic models are often in conflict in that the climate models propose often moist, seasonal low latitude climates (Barron & Washington 1982), but fossil data infer uniformly drier climates (Herngreen *et al.* 1996). A possibility is that although moisture may have been freely available at this time, higher temperatures may have resulted in overall moisture deficiency, registered by plants as indicating aseasonal aridity (Morley 2000); equatorial climates at the time of the radiation of the flowering plants would therefore have no present day analogue.

The first flowering plants (magnolialian dicots and monocots) probably became established in ephemeral habitats such as river flood-plains, and eudicots, the main clade of non-magnolialian dicots, exhibited an important centre of radiation in Africa during the Aptian (113–108 mya) (Doyle *et al.* 1977). These diversified rapidly during the Albian (108–96 mya), to become dominant over other plants during the Cenomanian (96–88 mya). Possibly this diversification and expansion reflects adaptation of flowering plants from flood-plains to most other habitats. The earliest radiations of mammals (according to Bininda-Emonds *et al.* 2007, between about 100 mya and 90 mya) seem therefore to have coincided with the sudden appearance of many new habitats and food sources that could have provided the setting for mammalian emergence, adaptation and diversification. There is currently no fossil evidence for the presence of placental mammals in Africa at that time but molecular clocks (most notably Bininda-Emonds *et al.* 2007 and Murphy *et al.* 2007) imply that an early common ancestor of the Afrotheria and South American Xenarthra might have been present around 90 mya.

Development of the rainforest canopy

The development of canopied tropical rainforest is of central interest because one of the most fundamental distinctions among many African taxa, including mammals, is between forest and non-forest species. Resemblances between African and Asian forest biota tempted earlier scientists to posit direct rainforest connections or corridors between these distant regions, connections that are now thought to be false. While the need to explain the resemblances remains, answers based on the adaptability of migrants promise to be more interesting and complex than explanations based on hypothetical forest corridors.

During the late Cretaceous, following the separation of Africa from South America, and also parallel with a phase of global cooling, the pattern of monsoonal climates across Africa began to change, and a more zonal climate regime came into place with the stabilization of the intertropical convergence zone to the equatorial belt. This led to the development of ever-wet climates at equatorial latitudes, indicated, for instance, in Nigeria by the presence of coal deposits from the Campanian (84–74.5 mya) (and possibly Coniacian 88 mya) onward (Reyment 1965). These wet climates resulted in a remarkable change in vegetation, thought to reflect the first development, on a global scale, of the rainforest canopy, which is nowhere better illustrated than in West Africa. The appearance of tall trees is indicated by the presence of large-girth angiosperm wood (Duperon-Laudoueneix 1991), the occurrence of large-seeded angiosperms (large seeds being a necessary advantage for germination below the forest canopy) and also appearance of the first climbers, such as members of Passifloraceae and Icacinaceae (Chesters 1955), indicate that the main building blocks of the rainforest canopy were in place at this time, a development considered by Rubitski (2005) as one of the major stages in the development of all land plants. The gradual diversification of this vegetation is illustrated by the pollen record, which shows a steady increase in numbers of pollen types up to the end of the late Cretaceous (Boltenhagen 1976), and this has been proposed to reflect evolutionary adaptations that were associated with development of the forest canopy (Niklas *et al.* 2003).

Just a few years ago the idea that grasslands occurred in the late Cretaceous would not have been taken seriously. However, the recent discovery of diverse grass phytoliths in dinosaurian coprolites from India (Prasad *et al.* 2005) suggests that grasslands were probably extensive in the Indian subcontinent during the late Cretaceous, and it is therefore possible that grasslands were similarly present, and perhaps extensive, on the African Plate, but this is still a contentious conclusion.

During the late Cretaceous and early Tertiary (70–60 mya) Africa was an island continent, separated from South America by the widening Atlantic Ocean, and from Eurasia by the Sea of Tethys. However, the pollen record shows that numerous plant dispersals were taking place with South America (Morley 2003), presumably by island hopping or by physical drifting on 'floating islands'. Dispersals were also taking place with Eurasia, for the latest Cretaceous of North Africa yields common Normapolles pollen (Kedves 1971, Meon 1990), which forms the dominant pollen group of western Eurasia and eastern North America at this time (Herngreen *et al.* 1996). The absence of this group from equatorial Africa, and of more southern elements from the north, such as hexaporotricolpate pollen (probably derived from Didymelaceae, a family widespread across equatorial Africa in the late Cretaceous but now restricted to Madagascar), emphasizes that a strong climate zonation must have been in place across the African continent in the latest Cretaceous (Morley 2000).

While it is remotely possible that placental mammals had arisen earlier in Gondwana, it is now thought more likely that the first placental mammal got into Africa after the mid-Cretaceous. This would have been the ancestral afrothere (which might have been semi-aquatic) and presumably also arrived by island-hopping, sweepstake or rafting dispersal. The distant but exclusive affiliation of Afrotheria with South American Xenarthra suggests that the latter, in turn, rafted across the proto-Atlantic, then very narrow.

Widespread destruction following meteorite impact

The end of the Cretaceous period was heralded by a massive meteorite impact (Alvarez *et al.* 1980) in the Yucatan Peninsula in Mexico (Hildebrand *et al.* 1991). This collision, known as the K–T event, affected both fauna and flora globally, with the destruction of 38% of marine animal genera (Raup & Sepkowski 1984), including all ammonites, pliosaurs, mososaurs and ichthyosaurs, together with dinosaurs and pterodactyls on land. The destruction that followed this impact affected both vegetation and environment globally, with wildfires causing massive destruction and short-term climate change during which sunlight was, perhaps, cut out over several years (Alvarez *et al.* 1980). African tropical rainforests did not escape, indicated by a 40% loss of pollen types across the Cretaceous–Tertiary boundary (Morley 2000, Fig. 5.15c) and the total loss of equatorial lowland gymnosperms. The effect of this meteorite impact on the flora and fauna of Africa was clearly quite dramatic, and many new niches must have become available virtually overnight, especially with the removal of generalist dinosaurian herbivores and carnivores.

Clearly a wide range of placental mammals survived the event. According to the molecular clocks devised by Bininda-Emonds *et al.* (2007), Springer *et al.* (2003) and Murphy *et al.* (2007), at least 20 major placental lineages had already differentiated before K–T. This discovery contradicts previous assumptions, which were based on an apparent lack of fossils before the K–T event, which was generally supposed to have accelerated the diversification of mammals in the early Tertiary (but see Wible *et al.* 2007 for a contrary view). In Africa the only living placental lineage present before the K–T event is currently thought to have been restricted to the Afrotheria and, probably, bats (Seiffert 2006). If molecular clocks are any indication it is also just possible that early Anthropoidea had also arrived in Africa before K–T.

Earlier Tertiary (Palaeocene to Eocene) climate and vegetation

Data on the vegetation and climate for the Palaeocene and Eocene of Africa are sketchy, and reconstructing palaeoclimates and vegetation is problematic, and open to different interpretations of detail.

Based on pollen diversity data, following the K–T extinction the African flora went through a period of rapid diversification during the Palaeocene and Eocene (Morley 2000, Fig. 5.15c). Palynological and macrofossil data from northern, equatorial and southern Africa is best explained in terms of a climatically driven latitudinal zonation of vegetation across the continent through this period, and that the zonation was least marked during the late Eocene when global climates were cooling, and closed rainforests were extensive at equatorial latitudes, but best developed prior to this.

Palaeocene

During the Palaeocene (see left map p. 44), the equatorial climate is thought to have been less moist than in the latest Cretaceous. It was also probably seasonal, since coals, which were extensively present in the West African Maastrichtian (74–65 mya), are absent. Geomorphological evidence also points to seasonal climates, on the basis of the widespread distribution of bauxites and ferruginous crusts of Palaeocene–Eocene age across equatorial Africa (Guiraud & Maurin 1991). Around the coasts palynological data indicate that mangroves were well developed, and included the palm *Nypa*. Based on pollen data, other palms were also prominent at low latitudes.

Several lines of evidence suggest closed, mesic forests in widely scattered localities. Seward (1935) has described a mesophyll macroflora from the early Palaeocene of the Red Sea, an area which has also yielded wood of megathermal rainforest trees, such as *Myristicoxylon princeps* (Myristicaceae) (Boureau *et al.* 1983). Marine sediments of Palaeocene age from Yemen have yielded diverse *Ctenolophon* pollen (Krutzsch 1989), from the rainforest tree family Ctenolophonaceae (a useful ecological indicator type), which was also particularly prominent in Cameroun at this time (Salard-Cheboldaeff 1990). *Ctenolophon* pollen is, however, absent from the Palaeocene of the nearby Niger Delta, despite its presence there in the latest Cretaceous (Germeraad *et al.* 1968), and conditions there may have been less favourable for rainforests than during the latest Cretaceous. Other Palaeocene pollen records suggesting the presence of closed mesic forests are forthcoming from Senegal (Caratini *et al.* 1991, Morley 2000). It is also likely that open vegetation is indicated by the presence of pollen of Gramineae, which occurs in modest frequencies in Nigeria (Adegoke *et al.* 1978), although it is equally possible that Gramineae were primarily dwellers of swamps at that time. Data therefore provide limited evidence for closed forest and possibly open woodland within the equatorial zone, and that palm-dominated swamps were probably extensive in low-lying areas.

Limited evidence for Palaeocene vegetation to the north of the equatorial belt is available from fossil woods from Algeria and pollen from Tunisia. Louvet (1971) recorded wood comparable to that of the rainforest emergent *Entandrophragma angolense*, together with wood of several Leguminosae, from the Palaeocene of Algeria (in Boureau *et al.* 1983), but pollen analytical studies from North Africa suggest that Eurasian elements were also widespread components of the vegetation (e.g. Meon 1990). Floristic differences with the equatorial belt, shown by the absence of many characteristic low palaeolatitude pollen types in North Africa, suggest that the Saharan region formed a major biogeographic divide even at this time. This divide might have included upland areas, centred on the regions of Tibesti and Hoggar (Axelrod & Raven 1978) and might have been defined by arid climate.

Floras in southern Africa also differed from those of the equatorial zone, and included many southern hemisphere elements. In the Cape floral region, palynological studies at Arnot by Scholtz (1985) suggest the presence of open-canopied, dry, warm temperate forest with *Araucaria*, *Casuarina* and Proteaceae, and an understorey of Epacridaceae, Ericaceae, Restionaceae and Gunneraceae but without *Nothofagus*. Within the earliest Tertiary, the South African flora thus showed a close relationship with the Australian region, emphasizing the antiquity of the Cape Floristic Kingdom.

The only fossil placentals currently known to have been present in Africa during the Palaeocene are afrotheres, the primate *Altiatlasius*, a few supposed 'insectivores' (todralestids and, possibly, adapisoriculids) and some basal hyaenodontid creodont carnivores (E. Sieffert, pers. comm.). However, the only sites from this period are marine or shoreline so the samples may not be very representative.

Palaeocene/Eocene Tertiary thermal maximum

Evidence for African vegetation at the time of the late Palaeocene/early–middle Eocene thermal maximum is particularly meagre. However, there is good palynological data for floristic diversity change over this event from Colombia by Jaramillo *et al.* (2007) and Davis *et al.* (2005) and it is likely that, in the broadest terms, equatorial South America and Africa would have shown some parallels. The short-lived thermal maximum marking the Palaeocene/Eocene boundary

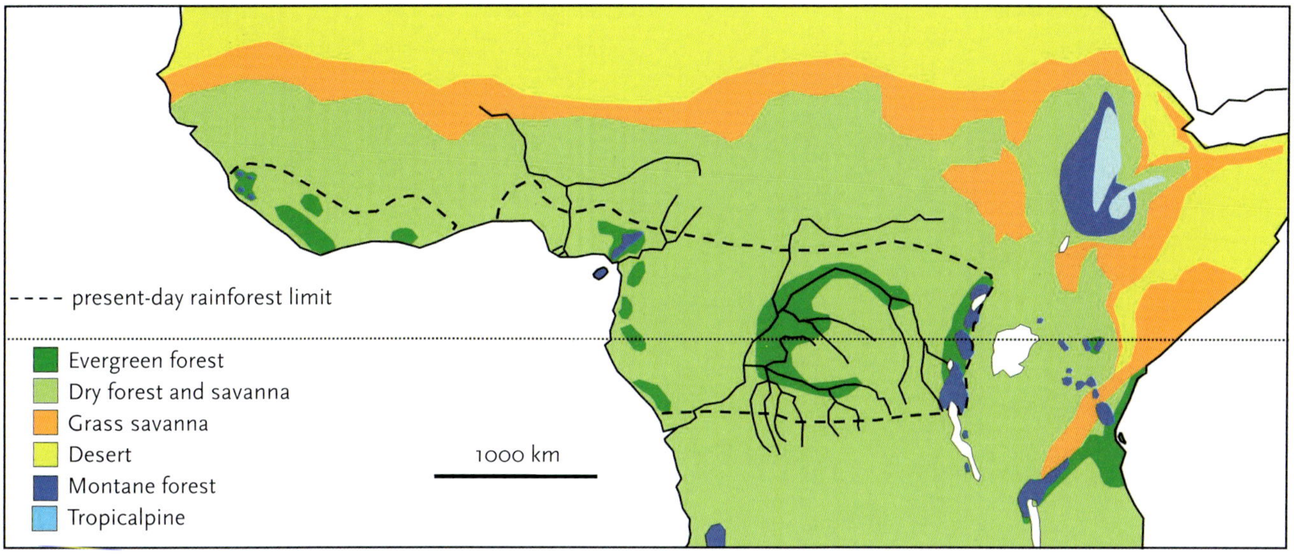

Vegetation of low latitude Africa for last glacial maximum. West and central African refugia from Maley (1996). Present text for eastern African refugia. Savanna and desert limits modified from Anhuf *et al.* (2006).

was caused by a sudden release of large volumes of the greenhouse gas methane from deep water hydrates, affecting temperatures globally (Pancost *et al.* 2007). The resulting rapid temperature rise at this time resulted in a further reduction of biotic diversity and the virtual elimination of the Colombian Palaeocene flora. Florisitic diversity then increased rapidly, with maxima during the middle and late Eocene (Jaramillo *et al.* 2007). A similar scenario is likely for equatorial Africa (see illustration p. 51).

Reconstruction of African low-latitude vegetation during the early and middle Eocene is open to different interpretations. Bauxites occur commonly at very low latitudes, and this suggests that monsoonal climates may have been in place (see left map p. 44). A leaf flora recently described from a mid-Eocene crater lake from Tanzania, about palaeolatitude 15°S, suggests the presence of wooded rather than forest vegetation with near-modern precipitation estimates for this area (Jacobs & Herendeen 2004). The plant community was dominated by caesalpinoid legumes and was physiognomically comparable to miombo woodland. Coetzee (1993) indicated that microfossil assemblages from the Palaeocene–Eocene boundary point to a significant change in the lowland vegetation, and suggested the co-existence of a mosaic of dry and humid forest, with savanna woodland extending over most of the Congo Basin. This suggestion is to some extent supported by the data of Salard-Cheboldaeff (1990) from Cameroun, which showed that Gramineae pollen was common there together with forest elements. A change to drier, probably strongly xerophytic, vegetation in Egypt is likely on the basis of palynological studies by Kedves (1971), who showed that the earliest Eocene vegetation was dominated by Gramineae and cycads.

Fossil wood and seed determinations from North Africa, discussed by Boureau *et al.* (1983), suggest that mangroves were well developed there in coastal districts, with records of the palm *Nypa* from various localities, and fossil woods from probable Rhizophoraceae and Sonneratiaceae from Libya. In this region a coastal forested zone bordered by mangroves gave way to a more open, seasonal forest inland, with gallery forests along rivers (Boureau *et al.* 1983). Palynological studies from North Africa, however, continue to emphasize floristic differences from the equatorial region, through the presence of a significant Laurasian warm temperate element (Kedves 1971, 1986). Although some pollen of tropical elements, such as *Crudia* or *Isoberlinia* and *Acacia* (Leguminosae), was represented, the majority of those taxa characteristic of equatorial Africa were missing.

Late Eocene

A major change occurred in the equatorial African flora during the late Eocene (37–34 mya) with the sudden appearance of pollen of many extant taxa, particularly of legumes, most of which occur, sometimes as dominants, in extant African rainforests. The new appearances included the bat-pollinated tree *Parkia*, suggesting that nectivorous bats were already well established at this time. At about 37 mya fossil anthracotheres first appear (E. Seiffert pers. comm.); as these Eurasian hippo-like animals may have been partially aquatic their appearance implies origin by some form of island-hopping or rafting. The rainforest tree *Ctenolophon* reappeared in Nigeria, after an apparent absence in the earlier Eocene and Palaeocene (Germeraad *et al.* 1968), suggesting a wetter climate, and Gramineae became less common in Cameroun (Salard-Cheboldaeff 1990). To the north, palynological analyses by Kedves (1971) show some significant changes in the vegetation of Egypt, although pollen of warm temperate taxa, such as caryoid and myricoid pollen, dominate, together with pollen of cycads. Pollen of tropical tree taxa are much more common than previously, and can be referred to Sapotaceae, Bombacaceae, Mimosaceae and Palmae.

Differences between North African and equatorial vegetation is also shown by mangroves, which were characterized by *Nypa*, Rhizophoraceae and ancestral Sonneratiaceae in North Africa. West African mangroves began to diversify; in addition to Nypoidae, they also included *Pelliciera rhizophorae* (Theaceae), now restricted to the Panamanian isthmus and Pacific Colombia, and probably the brackish water palm *Oncosperma*, now restricted to South-East Asia (reflected by the presence of the pollen type *Racemonocolpites hians*).

Oligocene to mid-Miocene climate and vegetation

In equatorial regions generally, there was a marked floristic change at the end of the Eocene, paralleling the disappearance of tropical-aspect forests from the mid-latitudes as a result of global cooling. In Africa the rainforest flora subsequently underwent a significant phase of modernization in the Oligocene, with the first appearances of many extant genera from this time (Morley 2000). However, the overall trend was of extinction (Morley 2000, Fig. 5.14b). Particularly notable was the disappearance of many palms, especially of the brackish water Nypoidae (*Nypa* and *Proxapertites*). However, the overall diversity decrease suggests that all vegetation types were affected by the terminal Eocene event. Detailed palynological datasets over the Eocene–Oligocene boundary that might explain these changes are not available for equatorial Africa, but from South-East Asia there is clear evidence for much cooler and drier climates during the Oligocene, driven by global temperature changes (Morley *et al.* 2003), which also resulted in a reduction in species richness there. For Colombia, palynomorph diversity data by Jaramillo *et al.* (2007) also shows a dramatic reduction in floristic diversity. A similar pattern is likely for the African equatorial flora.

Oligocene

Mapping Oligocene vegetation in Africa relies heavily on global climate models, which suggest that temperate climates and vegetation were extensively in place within northern and southern areas (see centre map p. 44).

Unpublished palynological data from the Niger Delta (summarized in Morley 2000, p. 140) suggest that drier climates characterized the late Oligocene (28.5–23 mya). Grass pollen maxima, suggesting a widespread expansion of grasses, coincide with periods of lower sea levels. The late Oligocene phases of increased representation of grasses were not so pronounced as those of the late Neogene, and it is most likely that grasses formed the understorey of open woodland, rather than forming open grasslands.

There is clear evidence for a significant drying at the end of the Eocene in Egypt. Kedves (1971) showed that whereas late Eocene palynomorph assemblages are characterized by Eurasian warm temperate and megathermal elements, these were replaced in the earliest Oligocene by probable Chenopodiaceae/Amaranthaceae and Gramineae, with all of the megathermal elements, except Palmae, being absent. It is therefore most likely that in Africa the terminal Eocene event resulted in the expansion of drier and cooler climates, and an associated widespread retraction of tropical forests toward the palaeoequator. The character of North African Oligocene vegetation is also revealed by wood and leaf fossils, particularly from Algeria, Libya and Egypt, reviewed by Boureau *et al.* (1983). They suggest that in the Oligocene, North Africa was characterized by a mosaic of savanna, open forests and gallery forests, with continuous forest developing locally in low lying coastal areas. Mangroves would have fringed coastlines.

Eocene–Oligocene mammals

Although there were many changes in plant communities from the late Eocene to early Oligocene, especially at low latitudes, faunal changes are poorly known. Seiffert (2006) found less change between faunas from Fayum in the late Eocene, and Oligocene (37–29.5 mya) and the late Oligocene (26.5–23 mya) from the Chilga region of Ethiopia (Kappelman *et al.* 2004). It has been suggested that African faunas therefore might have been less affected by the climate changes that elsewhere seem to have resulted in widespread extinctions (Prothero 1994). The dominant mammalian groups present at Fayum (and some at Chilga) were paenungulates (Embrithopoda, Proboscidea and Hyracoidea), arsinotheres, anthropoid primates and creodont carnivores. It is possible that other mammalian groups may have inhabited rain-forested areas at this time, of which there is no fossil record. This is particularly likely for hystricognath rodents, which first appear as fossils at the Eocene/Oligocene boundary (E. Sieffert pers. comm.) but may have made their first appearance in Africa earlier.

Early and mid-Miocene, Neogene thermal maximum

From the mid-Tertiary onward, the African land mass underwent a phase of uplift, with flexure of the pre-Miocene surface into a number of warps and basins (see p. 35) and the initiation of the East African Rift Valley (Baker *et al.* 1972). During the early Miocene, global temperatures once again began to increase, reaching a maximum within the earliest mid-Miocene.

In the earliest Miocene, prior to Neogene uplift and volcanism, when the continental divide was low, climates were moist over most of equatorial Africa, with rainforests extending more or less from coast to coast (Andrews & Van Couvering 1975). This is suggested by mammalian fossils preserved in early Miocene deposits east of the divide, that include forest dwellers. Palynological studies from southern Sudan suggest the presence of rainforest vegetation during the Oligo-Miocene (Awad & Brier 1993, Awad 1994), and a rich leaf flora of early Miocene age described from Rusinga I., L. Victoria by Chesters (1957), was thought to represent lowland rainforest by Andrews and Van Couvering (1975).

It was probable, however, that various biogeographic barriers existed between the east and west throughout the Tertiary, in order to account for the restriction of certain wood and leaf fossils to East and North Africa, especially to explain the presence of fossils of Asian Dipterocarpoidae in East Africa (Chiarugi 1933, Bancroft 1935) but to the west their absence, a divide that has a long history. Grubb *et al.* (1999) drew a line that represents the optimal separation of overlapping centre/western and eastern forest faunal elements. By analogue of Wallace's line in South-East Asia they termed this intra-continental biotic divide 'Kingdon's line'. This is discussed in more detail in Chapter 3 on evolution.

At the beginning of the early Miocene, West African mangrove swamps recruited two new taxa, *Rhizophora* (Rhizophoraceae) and the parent plant of the *Verrutricolporites rotundiporus* group of pollen (Lythraceae or Sonneratiaceae) that substantially modified coastal vegetation. Both of these taxa dispersed from South America (Morley 2000, 2003). *Rhizophora* subsequently became the most important element of West African mangroves, and pure swamps of this genus must have been extremely widespread, to account for the abundance of its pollen in marine sediments, especially during phases of Neogene rising sea levels. The parent plants of the *Verrutricolporites rotundiporus* pollen group are likely to have been derived from ancestral *Sonneratia* spp. (characteristic of South-East Asian mangroves), and thus early and mid-Miocene mangroves were much more diverse than those of the present day.

Early Miocene extinctions and the expansion of grasslands

The palynological record from both the Niger Delta and Cameroon suggests a gradual change in the character of low latitude vegetation during the course of the Miocene, with successive extinctions of rainforest taxa (Legoux 1978, Morley 2000). During this time, Gramineae pollen, suggesting open grassy woodland or savanna, increases in representation, especially at times of low sea levels, with the first evidence of burning (from the presence of charred Gramineae cuticle) at about 15 mya, in the Middle Miocene (Morley & Richards 1993), widespread natural burning becoming successively more common during the mid-Miocene. Drier conditions are suggested over a wide area of the Niger catchment, with the expansion of open woodland and savanna at the expense of rainforests.

Rainforests in East Africa also became greatly reduced during this period and were replaced by open woodland and grassland. The character of mid-Miocene grass-dominated vegetation and gallery forest in East Africa has been reconstructed by Retallack *et al.* (1990) and Retallack (1992a, b), through the study of plant fossils and palaeosols associated with the *Kenyapithecus* site, which is dated at about 14 mya. He interprets a mosaic of early successional woodland, grassy woodland and wooded grassland and observed that mid-Miocene grassland ecosystems differed substantially from modern grasslands in that although the grass subfamily Chloridoideae and supertribe Panicanae were common, there was no evidence for the supertribe Andropogonae, which is now dominant in seasonally arid, overgrazed and burned African grasslands. Tooth enamel studies of mammalian faunas from Fort Ternan also suggest that East African mid-Miocene vegetation differed from Serengeti-type wooded grassland (Cerling *et al.* 1997) in that most food sources consisted of a C_3 diet, whereas modern tropical grasslands are dominated by C_4 grasses. Janis (2003) indicated that mid-Miocene ungulates represent a diversity of browsers that is unlike that found in any present-day environment, which is thought to reflect vegetation growing under conditions of greater CO_2.

The change from dominance of forests to the widespread development of woodland and savanna during the Miocene has generally been attributed to global cooling. However, upwarp of the African Plate following the formation of the East African rift is also important, and has recently been discussed by Sepulchre *et al.* (2006).

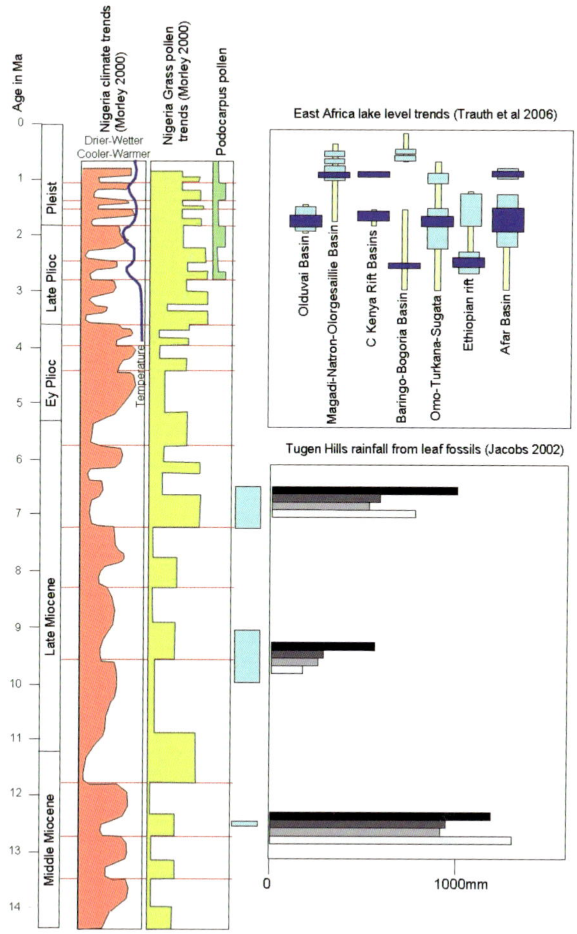

Some indicators for climatic change in Africa. The two columns on the left show temperature fluctuations since the mid Miocene as inferred from Gramineae and *Podocarpus* pollen abundance in Niger Delta (modified from Morley 2000).
Top right: East African lake level data from Trauth *et al.* (2005).
Lower right: Rainfall inferred from leaf fossils in Tugen Hills (Jacobs 2002).

Vegetation and climate in northern and southern Africa

There is little evidence for moist forests within North Africa during the Miocene, which from the pollen record of Kedves (1971) was characterized by woodland and savanna. However, in South Africa, mesothermal to megathermal, mesic forests extended as far south as the southern tip of Africa during periods of warmer climate and high sea levels (Coetzee 1978). She recorded two periods, which she interpreted as within the early Miocene, about 19 mya, and mid-Miocene, 14–16 mya, when palm-dominated vegetation was widespread at the Cape. These intervals are associated with the occurrence of pollen of the primitive angiosperm *Drimys* (Winteraceae) and of the now endemic Madagascan rainforest tree family Sarcolaenaceae (Coetzee & Muller 1984). This suggests that during warmer, moist periods, the Cape vegetation bore similarities to the present day humid forests of Madagascar. It is possible that conservative mammals, such as the Cape Grysbok *Raphicerus melanotis* (which displays characteristics of the very earliest types of antelope), became adapted to what were once more extensive habitats, but are now relicts inhabiting the drier end of the Cape vegetation spectrum (Kingdon 1982, 1990).

Malagasy-like forests were also probably widespread along the east coast of southern Africa, but not in the west, since the west coast experienced a very dry climate after the inception of the cold circum-polar current, which began to flow at about the Oligocene–Miocene boundary, about 23 mya (Barker & Burrell 1977) and development of polar ice in the Antarctic (Van Zinderen Bakker 1975). According to Denton (1999), the Benguela current in its present configuration formed about 10 mya.

During intervening times, temperate woodland, with southern hemisphere gymnosperms *Podocarpus*, *Microcachrys*, *Widdringtonia* and drier, sclerophyllous vegetation, with *Casuarina*, were widespread. Sub-tropical evergreen forest patches with Cunoniaceae and Proteaceae survive today as refugia in gulleys in the mountains of the Cape region, and may be relics of the tropical aspect rainforests which were widespread during the Miocene.

The extent of austral-affinity warm temperate vegetation in South Africa has been greatly influenced by the northward drift of the African

Plate into the southern hemisphere high pressure zone. In the Oligocene and early Miocene, most of South Africa would have occurred south of this zone (Morley 2000: Fig. 13.5), permitting the expansion of forests with common megathermal elements during the mid-Miocene thermal maximum. During the late Miocene and Pliocene, the combination of northward drift of the African Plate coupled with a strengthening of the southern hemisphere high pressure zone, opportunities for austral warm temperate communities of animals and plants became substantially reduced, with the result that today, they are restricted to just a few localities across the Cape region. Recent molecular studies on this flora suggest that the diversity of species can be partially attributed to rapid late Neogene evolution (Richardson *et al.* 2001, Linder 2003, Klak *et al.* 2004). It is probable that an enlarged source area for this flora, prior to the later phases of northward drift of the African Plate, was also influential in the development of this uniquely diverse flora. The diversity of Afrotherian golden-moles in southern Africa and the peculiarities of their distribution are also suggestive that the beginnings of that diversity lay in the once more extensive range of southern temperate environments.

Interchange of African and Eurasian mammals and its relation to plant dispersal

The first interchange of Eurasian and African mammals, long believed to have taken place in the mid-Miocene, following the collision of the African and Asian Plates, is now thought to have taken place in the earliest Miocene at about 23 mya (Jolivet & Facenna 2000) but immigration of single taxa, presumably by sweepstake or rafting, took place earlier, notably ancestral anomalurids and ctenohistricid/hystricognaths. It is possible that these taxa are uniquely African in origin, the former giving rise to extant anomalures and springhares, the latter to such ancient endemics as gundis, mole-rats and cane rats. Late Eocene fossil rodents, *Protophiomys algeriensis* (thought to belong to the cane rat lineage, Thryonomyidae) and an anomalurid, *Nementschamys jebeli*, have been found in Algeria (Jaeger *et al.* 1985) and are consistent with African origins (as estimated by molecular clocks), which imply an invasion during the early Eocene (Adkins *et al.* 2001, Montgelard *et al.* 2002, Steppan *et al.* 2004). These discoveries support suppositions for very early, sweepstake arrivals of single ancestors for major rodent radiations. Still earlier fossil ctenohystricoid rodents have been unearthed recently from 37 million-year-old deposits in the Fayum (E. Seiffert pers. comm.). These are currently undescribed and show little diversity, implying that their arrival in Africa might not have been much earlier (E. Seiffert, pers. comm.).

At about 23 mya the earliest documented major faunal exchange between Africa and Eurasia is likely to have brought in the common ancestor of all African antelopes (Antilopinae) as well as those of cricetomyine (and, perhaps, proto-gerbilline) rodents (Adkins *et al.* 2001, Steppan *et al.* 2004). Following this time, and particularly during the mid-Miocene (18–15 mya), many Eurasian mammals, such as tragelaphine bovines, shrews, carnivores, perissodactyls, Anthracotheres and giraffids, dispersed into Africa, and some Afro-Arabian mammals, such as some elephant, primate and porcupine emigrants, successfully migrated in the opposite direction.

Such interchanges are not apparent from the record of plant fossils, at least, with respect to tropical taxa. This is probably due to the presence of strong latitudinal vegetational zonation that restricted dispersal between higher and lower latitudes, a constraint that operated from the terminal Eocene onwards. Only the appearance of the conifer *Juniper* within the East African highlands (noted by Bonnefille 1984) and *Alnus* in North Africa (Kedves 1971) can be considered candidates for such dispersal. Latitudinal belts of arid climate and vegetation may have been breachable for the more versatile of mammals but it seems very likely that, for more sensitive species, dry habitats represented a barrier that was not easily crossed in either direction. For example, duikers, galagos and other forest mammals have never made it out of Africa and there are Eurasian forest mammals, such as tapirs, tupias and tarsiers, that were equally inhibited from colonizing Africa.

The situation for mammals may differ from plants in that most, perhaps all successful mammalian immigrations up to the Miocene probably derived from a very few, single ancestral invaders, not some major influx, and their ecological restriction may have been less constraining. For some mammal immigrants, particularly later ones, a broad latitudinal belt of arid land delayed or inhibited colonization of more southerly regions, thus shaping the pattern of evolutionary radiation in Africa, a phenomenon that is discussed in more detail in Chapter 6 on evolution.

As noted above, members of the group Afrotheria dominated Africa's fauna throughout the mid-Tertiary, but today South Africa has retained more afrotherians (notably the Chrysochloridae or golden-moles) than any other region. A possible dimension to this very marked biogeographic pattern is that the equatorial rainforest belt, acting as a barrier between afrotherians and their ecological competitors, inhibited Eurasian taxa from invading southerly niches (certainly up to the uplift of East Africa in the mid-Miocene). A similar dynamic may have operated with the bathyergid mole-rats (blesmols), which are less ancient than golden-moles, but none the less, represent an early endemic rodent radiation which is now predominantly southern and south-eastern in distribution.

Late Miocene and Pliocene climate and vegetation

Late Miocene

The late Miocene was a time of global cooling, with a widespread build-up of ice in Antarctica (Abreu & Anderson 1998, Zachos *et al.* 2001). A detailed record of the effects of global temperature fluctuations on the vegetation of equatorial Africa is forthcoming from palynological analyses of deep boreholes from the Niger Delta (Morley & Rosen 1996, Morley 2000). This delta, being located on the northern edge of the rainforest belt, between the Dahomey Gap and the Cameroon–Gabon rainforest block is in a very sensitive position for recording climatic change in West Africa (Poumot 1987, Morley & Richards 1993). From East Africa, sites display at best only an intermittent palynological record (Bonnefille 1984) but benefit from well-preserved leaf floras, which provide detailed palaeoclimate data for short time intervals (Jacobs 1999, 2002).

Palynological records from deep Niger Delta boreholes display alternating maxima of rainforest and Gramineae pollen (Morley 1996, 2000). These reflect successive periods of expansion of grasslands, possibly across the Niger Delta, and a parallel reduction in range of rainforests. The grasslands, probably open savannas, were maintained by natural fires, since the fossil grass pollen is generally

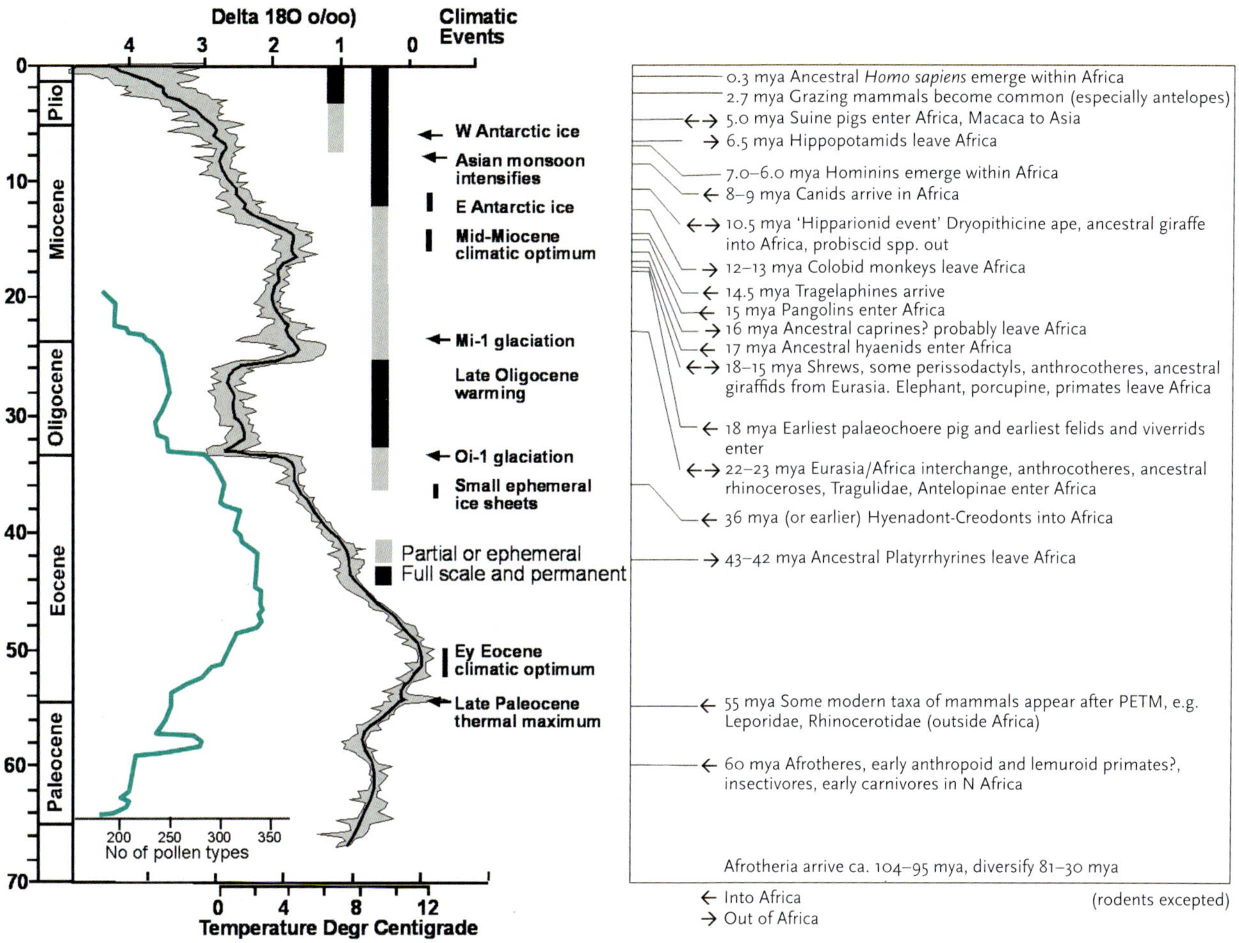

Global temperature curve based on C18 by Zachos *et al.* (2001), and South American (Colombia) pollen diversity by Jaramillo *et al.* (2007). The timing of main mammal appearances, migrations and emergences are tentatively indicated.

accompanied by charred Gramineae cuticle, which suggests burning (Morley & Richards 1993). The many fluctuations of pollen from savanna to rainforest closely parallel sea level fluctuations that have been identified by sequence stratigraphy (Morley & Rosen 1996), testifying to their significance operating on a regional scale.

The first major dry phase at the beginning of the late Miocene was characterized by the frequent presence of Gramineae pollen and by the common occurrence of pollen of Caesalpiniaceae, suggesting a combined expansion of savanna and open woodland, and this coincided with the most pronounced of the Neogene sea level falls at about 11.7 mya (Hardenbol *et al.* 1998). Characteristic rainforest and freshwater swamp taxa became common only after sea levels subsequently rose to their previous maximum levels, the climate became more moist and mangrove swamps expanded. During this time, mangrove communities actually underwent a remarkable *reduction* in diversity, losing sonneratioid taxa and also the possible brackish water palm *Oncosperma* possibly associated with phases of late Miocene sea level fall and increased aridity.

The late Miocene is an important period for the emergence of the common ancestor for humans and African apes. While this was long thought to represent an endogenous African development, it is now considered more likely that a species of Eurasian tree ape or woodland dryopithecine entered Africa at about 10.5 mya (Stewart & Disotell 1998, Heizemann & Begun 2001, Kingdon 2003). This was a major episode of intercontinental faunal exchange that is known as the Hipparionid event, when there appears to have been a good connection between Africa and Eurasia when sea levels were low. Among the incomers was the porcupine *Hystrix* and an ancestral murine rodent (Misonne 1969, Steppan *et al.* 2004), various carnivores, giraffes, caprines and tetraconodont pigs (Harris & White 1979, Pickford 1993): among the likely outgoing stocks were colobine monkeys (Stewart & Disotell 1999) and some afrotherian emigrants.

The alternating climates of this period are likely to have been of relevance for the speciation of many equatorial mammals, including the emergent hominins, because climate change fragmented both forest and non-forested areas into temporary islands that expanded and contracted. The moist equatorial islands, in particular, probably followed patterns that were repeated over and again with predictable major centres in the far west (Upper Guinea), Cameroon/Gabon and central Africa (i.e. eastern Congo Basin) (Moreau 1966, Kingdon 1971, 1990, 2003, Hamilton 1976 and see map p. 55).

The most obvious effects of such isolation of populations would have been to reinforce the distinctness of isolates but another dynamic also operated. Each time changing climates altered the vegetation of an ecological island from, say, savanna woodland to moist forest, its inhabitants became engulfed in forest and had to adapt to the new conditions or die out. Although former 'corridors' that might have isolated non-forest organisms within what are today forest areas and the adaptation of such organisms to true forest have to be extrapolated, there are numerous examples of 'engulfed' organisms, especially in the Congo Basin and Gabon/Cameroon region. Among primates, the example of the ancestral gorilla is discussed in more detail in Chapter 6. The Sun-tailed Guenon *Allochrocebus solatus*, a very localized, now forest-dwelling species from a small region of the Ogooue River Valley, also exemplifies just such an engulfed species inasmuch as this very terrestrial primate exhibits several hints of previous adaptation to drier environments (Dutrillaux *et al.* 1982, Kingdon 1997).

The broader continental pattern of interaction between past climates and habitats is also relevant to any attempt to reconstruct human origins and prehistory. At the core of any evolutionary radiation is the isolation of populations, and the bisection of Africa into an upland, cooler and drier south-eastern zone versus a lowland, warmer and wetter centre-western zone is germane to any consideration of continent-wide patterns of evolution. This is discussed in further detail in Chapter 6. Another corridor between separate arid regions occurs in East Africa; the Somali arid zone has probably been of biogeographic significance at least since the mid-Miocene uplift that created the East African highlands and their associated rift valleys. This corridor has long been one of the defining major features of African biogeography, with naturalists noting the close Horn/Namib affinities between such arid-adapted mammals as oryxes and dik-diks and, among birds, bustards, sandgrouse and larks (Lydekker 1908, Moreau 1952, 1963a, Kingdon 1971, 1990, Coe & Skinner 1993). The role of this corridor in separating forest communities has had less emphasis and the recognition that there was a temporal dimension, with older biota occupying the coastal region and ancient crystalline mountains of eastern Tanzania and younger forest biota on the volcanoes and moister areas of central and western Kenya and Uganda, is more recent (Kingdon 1971, 1990).

The East African dry corridor has periodically connected the Horn of Africa with the deserts of extreme south-west Africa and the entire eastern region can be predicted to have been strongly perturbed by almost every major episode of climatic change since the earliest Miocene (see profiles of *Madoqua* and *Oryx* spp. in Volume VI). Because equatorial lowlands are demonstrably the preferred habitat of all the larger diurnal primates many of them would probably span the continent had there not been climatic and tectonic interruptions, albeit periodic, that broke the continuity of these forests (see map p. 55).

Because many primates are dependent on a year-round supply of fruit, aridity can be predicted to have eliminated all but the hardiest of primates from much of the East African interior. The East African coastal region and montane piedmonts, by contrast, have been very favourable habitats for primates, even up to the very recent past, and it is here that the earliest woodland-adapted Miocene apes would have found a narrow but particularly favourable environment. The fossil record shows that between 6 mya and 1 mya most of the many populations of hominins had an eastern distribution. It is therefore plausible that the eastern coastal region and favourable habitats inland are the most likely settings in which mainly terrestrial, highly 'manual' (and eventually bipedal) habits first developed, in relative isolation, in a Miocene tree ape population (Kingdon 2003). Anthropologists have, for many years, identified climate change as a dominant force driving the evolution of hominins. The biogeographic, ecological and behavioural contexts in which hominins evolved have had less attention and, by the nature of the evidence, largely remain areas of conjecture and extrapolation (as with so much of hominin prehistory).

In West Africa, periods of expansion of open vegetation with grasslands became more pronounced after about 7.0 mya. The pollen record shows the presence of abundant Gramineae and Cyperaceae, and the presence of charred Gramineae cuticle fragments with the fossil pollen indicates that the grasslands were regularly subject to burning, and thus may have resembled modern savannas in many areas, with dry, or seasonal climate vegetation increasing in diversity.

Late Miocene leaf floras from the Tugen Hills in Kenya studied by Jacobs (2002) and Jacobs & Deino (1996) have been dated at 9–10 mya (from Waril) and 6.6 mya (from Kapturo), respectively. The older of the two localities suggests an open vegetation structure and a climate with a pronounced dry season, whereas the younger suggests a woodland or dry forest setting. Jacobs (2002) suggests from these data that there was not a unidirectional change from forest to open environments in the Kenya Rift Valley during the late Miocene (as is often proposed to explain the evolution of hominids in Africa). As with the Niger catchment, it is more likely that a succession of alternating wetter and drier phases occurred, with an overall trend toward cooler and drier climates, in line with global trends. The Waril locality would thus tie in well with the West African expansion of open vegetation from 9 to 11.7 mya, whereas the 6.6 mya Kapturo assemblage would tie in with the phase of wetter climates from 6.5 to 7 mya recorded from the Niger Delta. Intervening wet phases might also account for the occurrence of 'rainforest' elements far north of their current distribution, such as the occurrence of late Miocene fossils of a waterbird, *Podica*, at L. Chad, which is exclusive to forest-fringed waters (Louchart *et al.* 2005).

Isotopic, C_3/C_4 analyses from the Tugen Hills suggest that late Miocene vegetation in the East African rift was derived from a mix of C_3 and C_4 plants, and that open grasslands at no time dominated this portion of the Rift Valley (Kingston *et al.* 1994). They suggest that the vegetation through this period was ecologically diverse, and that there was no single shift from more closed C_3 to more open C_4 grassland habitats. Two factors need to be considered when interpreting such datasets: first, how complete is the stratigraphic succession (a sequence stratigraphic approach could bring to attention the likelihood of repeated stratigraphic gaps); and secondly, it must be borne in mind that the succession for this part of the East African Rift Valley may not be representative of African environments as a whole, especially since palynological data from the Niger Delta indicate that widespread grasses contributed to African vegetation long before the late Miocene (Morley 2000).

Pliocene and late Pliocene cooling and aridification

The early Pliocene (5–3.5 mya) was characterized by moist climates, the expansion and diversification of rainforests, and the contraction of savanna. The expansion of rainforests was greatest at the end of the early Pliocene, when they extended northward along the coast with palynological signals as far as northern Senegal (Leroy & Dupont 1994).

It should be borne in mind that rainforest elements reach Senegal even today, as extremely wet rain seasons counterbalance the effects of long intervening dry seasons. The late Pliocene (3.5–1.6 mya) however, saw much more pronounced climatic changes, with marked drying and cooling. A drying phase at 3.2 mya was followed by a much more pronounced period of desiccation at 3.0 mya, both being accompanied by the marked expansion of grasslands, reduced floristic diversities and numerous regional extinctions. At about 2.7 mya, coinciding with the onset of glaciation in the northern hemisphere, a sudden change to a cooler climate resulted in the expansion of *Podocarpus* into West Africa (Morley 2000, 2003), indicated by pollen records (first brought to attention by Knaap 1972). This was the first evidence for a gymnosperm to occur in West Africa since the end of the Cretaceous. These changes took place during successive wet–dry oscillations. Recent studies based on isotope data, by Trauth *et al.* (2005) for East Africa, clearly show the intervening periods of increased humidity associated with these oscillations, with periods of increased humidity from 2.7 to 2.5 mya, and 1.9 to 1.7 mya.

Studies of the last glacial/interglacial cycle (Dupont & Wienelt 1996, Maley 1996) suggest that *Podocarpus* was common in West Africa at times when climates are cooler than at present, but still moist, at which time cloud forest was probably widespread (Maley 1987, 1989, 1996), and it is thought that such conditions must have created a suitable pathway for the dispersal of *Podocarpus milanjianus* from upland areas in East Africa via Angola and Gabon. Opportunities for dispersal of this species may also have been assisted as a result of the late Miocene–early Pliocene uplift of the African Plateau (King 1962).

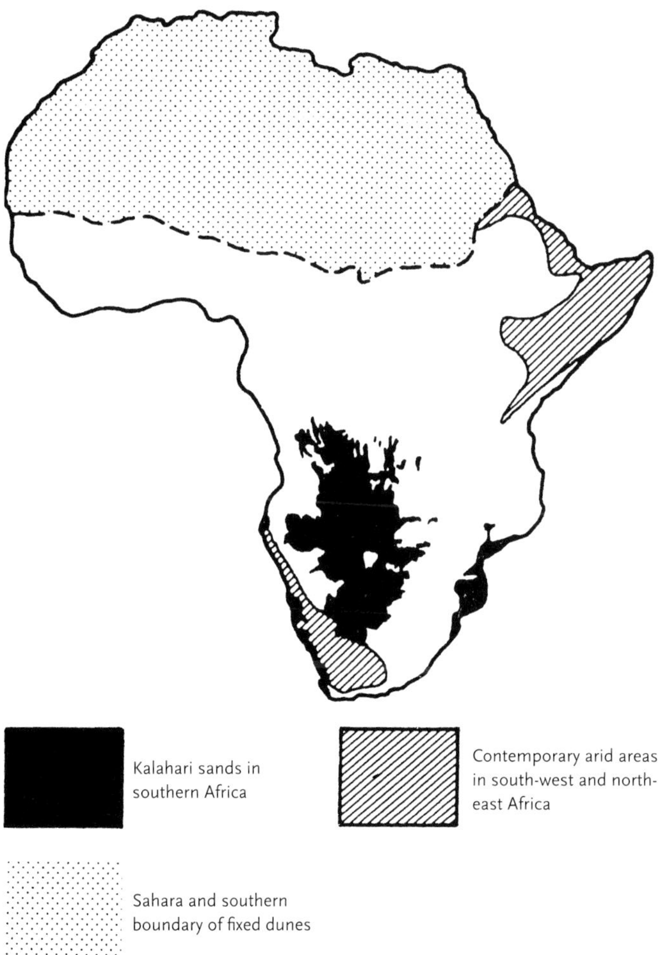

Evidence for extreme aridity in Africa. Wind-blown sands extend from the Sahara to form fixed dunes well south of their present limits. To the south, Kalahari sands underlie contemporary forest.

Aridification was a continent-wide feature of the late Pliocene. Studies of ODP site 658 offshore of Senegal by Leroy & Dupont (1994) indicate an increase in trade wind activity between 3.3 and 2.5 mya, with a marked phase of aridification after 3.2 mya. Further aridification occurred around 2.7 mya, with desert conditions developing after 2.6 mya. Similarly, Bonnefille (1984), using palynological data, records a change to much drier conditions after 2.5 mya for East Africa. The pattern of expansion of open, wooded and savanna habitats and of aridification, with major phases of expansion at the beginning of the late Miocene (after 11.7 mya) then again at about 7 mya, followed by the Messinian (5.75–5.3 mya) and yet again within the mid-Pliocene, after 2.7 mya, corresponds with fossil evidence that browsing mammals, notably antelopes, were widespread and numerous during the late Miocene, while grazers supposedly only became prominent in the fossil record from the mid-Pliocene (Turner 1995). However, the ability of particular ungulate lineages to digest grass evolved at different rates and times, with Perissodactyla, for example, becoming grass-eaters much earlier than other taxa. Grass-eating equids were well represented as fossils substantially earlier than the Pliocene.

Quaternary climate and vegetation

The role of past climates in determining distributions and limiting species to specific localities was recognized early (Lonnberg 1929, Wayland 1940). Discontinuities in the geographic distributions of African birds (Moreau 1963b, 1966) and rainforest plant species (Hamilton 1976, Brenan 1978, White 1983) were proposed to reflect the restriction of rainforests to refugia during the driest and coolest periods of the last glacial, from which they subsequently spread during the Holocene (Mayr & O'Hara 1986). For mammals, accumulations of species in favourable localities were attributed to multiple phases of connection and disconnection (largely driven by climatic changes), with representatives of successive faunal and climatic eras superimposed upon one another within such local centres (Kingdon 1971, 1990). Early palynological studies from the East African highlands provided little evidence either to support or refute such suggestions (e.g. Coetzee 1967, Livingstone 1967, Hamilton 1982); they mainly showed that during the last glacial maximum, montane vegetation zones were depressed altitudinally by 1000 m or more.

Quaternary distribution of forests and forest faunas

Subsequent palynological data from low altitude sites and from marine cores are, however, now providing clear evidence for the retraction of rainforests during the time of the last glacial maximum and of the occurrence of refugia (Anhuf *et al.* 2006). Lake Bosumtwi in Ghana (Maley & Livingstone 1983, Maley 1989, 1991, 1996) and L. Barombi Mbo in Cameroun (Brenac 1988, Maley 1996) are currently surrounded by closed forests, but pollen records show that the area surrounding L. Bosumtwi was unforested during the time of the last glacial maximum, whereas L. Barombi Mbo, close to the Cameroon–Gabon Refuge, was surrounded by open forest, with grassland at the same time. Similarly, palynological studies of a deep marine core offshore the Dahomey Gap, to the south-west of the Niger Delta, indicate that during the period of the last glacial, savanna was much more extensive in the Niger Delta catchment than today, and also that rainforests formed a continuous belt from Cameroun to Guinea during the early Holocene (Dupont & Wienelt 1996), tying in with wetter early Holocene climates from North and East Africa suggested from high lake levels (Street-Perrott & Grove 1976). Offshore cores from West Africa also show that during the last glacial maximum, the savanna belts substantially shifted to the south (Hoogheimstra & Agwu 1988), correlating with evidence for low lake levels across the Sahara, in Mauritania, Ethiopia, Sudan and the East African highlands (Street-Perrott & Grove 1976).

On the other hand, palynological studies from a core offshore Ghana recorded the presence of rainforests on adjacent coasts throughout the last glacial maximum (Lezine & Vergnaud-Grazzini 1993), and pollen analyses from Uganda (Sowunmi 1991) and Burundi (Bonnefille & Riollet 1988) provide evidence for forest continuity in areas proposed as forest refuges. In both these instances rainforest pollen assemblages contain a mixture of lowland and montane elements, suggesting that during glacial maxima, rainforests contained a mixture of lowland and lower montane taxa, as suggested also for the Amazon Basin by Colinvaux *et al.* (1999) and Bush & Silman (2004). A recent review of palaeoecological data from equatorial Africa by Bonnefille (2006) emphasizes that patterns of floristic diversity seen in rainforests cannot be explained by simple replacement of rainforest by savanna during glacial periods, but that more complex processes must have been in place that are currently poorly understood.

Studies from offshore cores are also beginning to show the nature of vegetation change in equatorial Africa during the last interglacial and beyond. The offshore Niger Delta deep-sea core studied by Dupont & Wienelt (1996) showed that rainforest vegetation during the last interglacial (Oxygen isotope stage 5e, 120,000–110,000 years ago) was very similar to that of the Holocene, and that at that time also, rainforests were probably continuous along the coastal region of the Gulf of Guinea. During stages 5d–a (110,000–74,000 years ago), however, although rainforests were still widespread, there was some expansion of woodland, but most notably, mountainous *Podocarpus* forest expanded in Cameroun and possibly Nigeria, as a result of some degree of lowering of temperatures, but without major changes in rainfall. It is unlikely, however, that rainforests extended far beyond their Holocene limits at any time during the Quaternary, for a core from L. Oku, which occurs in the Guinea Savanna belt, in a tectonic depression at the southern end of the Benue Trough (7° N) in Nigeria (Medus *et al.* 1988), has yielded a sedimentary succession of late Pliocene and Pleistocene age (Morley 2000), which shows rapid fluctuations of Gramineae and Cyperaceae pollen and provides evidence for fluctuations of savanna grasslands, marshes, gallery forests and montane vegetation over the last 2.7 million years, but without evidence for rainforests.

A legacy of 'glacial' forests with a mixture of upland and lowland taxa may find some reflection in the distribution of some mammal species, notably those that are equally well adapted to lowland rainforest and montane habitats, but such adaptability is far from universal. Examples of mammals that appear to be relatively insensitive to altitude are guenon monkeys belonging to the very widely distributed Gentle Monkey *Cercopithecus mitis* complex that commonly inhabit both habitats, sometimes, but not always, showing subspecific differences that correspond with altitude. Likewise, Lord Derby's Anomalure *Anomalurus derbianus* inhabits equatorial lowland and montane forests from the Atlantic to the Indian Ocean coasts.

The Bushbuck *Tragelaphus scriptus* and Two-spotted Palm Civet *Nandinia binotata* show similar versatility while several other species are more restricted in overall range but seem equally indifferent to altitude, notably the Chequered Sengi *Rhynchocyon cirnei*, the African Golden Cat *Profelis aurata*, Alexander's Cusimanse *Crossarchus alexandri*, the Giant Forest Hog *Hylochoerus meinertzhageni*, the Black-fronted Duiker *Cephalophus nigrifrons* and the Bongo *Tragelaphus eurycerus*. It could be significant that some of these species represent relatively conservative taxa that have survived a succession of climatic fluctuations and have presumably had ample time to adapt to cooler, drier uplands as well as humid hot lowland forests (but this is unlikely to be a universal explanation).

Two pied colobus species that are both found in lowland as well as montane forests, *Colobus angolensis* and *C. guereza*, exhibit a geographic separation that has strong temporal implications. The former species is, on morphological grounds, 'older' and has a somewhat spotty distribution from Angola to the East African coast, mainly south of the Equator. *Colobus guereza*, on the other hand, seems to derive more recently (probably from a common ancestor with *C. vellerosus* in Upper Guinea) and lives mainly north of the Equator (and north of the Congo R.), where its range is generally more continuous.

It is not always easy to discriminate between older and younger branches of a single radiation, but the biogeographic pattern of older forest fauna south of the Equator and more recent communities north of it seems to hold for a number of species (Kingdon 1971 and see pp. 85–6 this volume). None the less, 'older' species are not always in retreat and at least one species of the chequered Giant Sengis (*Rhynchocyon cirnei*), unlike other *Rhynchocyon* species, seems to have been able to recolonize ground lost to periodic aridity as well as make accommodations to less than total forest cover. (This situation is discussed in some detail in Chapter 6, p. 75.)

As described in Chapter 3, the equatorial Congo R. has served as a massively influential and long-established barrier to non-volant or non-aquatic animals. Coupled with former shifts in the position and extent of the Sahara Desert's southern boundaries, the distributions of forest mammals are likely to have been significantly affected. Thus, forest communities north of the Congo R. (and broadly north of the Equator) may have been impoverished or partially extirpated by climate changes in the past while forest areas south of the river were less affected by such changes. If forests sandwiched between the Sahara and the Congo R. were seriously degraded, they were probably open to relatively recent, eastward recolonizations from wetter 'refugia' closer to the coast and from the southernmost parts of western Africa. Expansions westwards from a central African refugium are also likely. Some such effects are discernible at the subspecific level among Bushbuck populations (Moodley & Bruford 2007).

Speciation and subspeciation may well proceed within larger rainforest blocks so long as the subdivisions enjoy quite specifically distinct environmental features such as peculiar seasonal patterns or regionally distinctive plant and animal associations (Kingdon 1990). In the case of guenons, every population is bounded by some identifiable form of permanent or temporary barrier. When such barriers are breached due to climatic shifts, some populations show signs of hybridizing when brought back into contact, while others appear to maintain their distinctness. Whether any given population maintains its genetic distinctness may have as much to do with species-specific behaviour, mediated through such local peculiarities as 'head-flagging' or signalling with mask-like facial patterns, as with external environmental factors (Kingdon 1981 and see pp. 83 & 116 this volume).

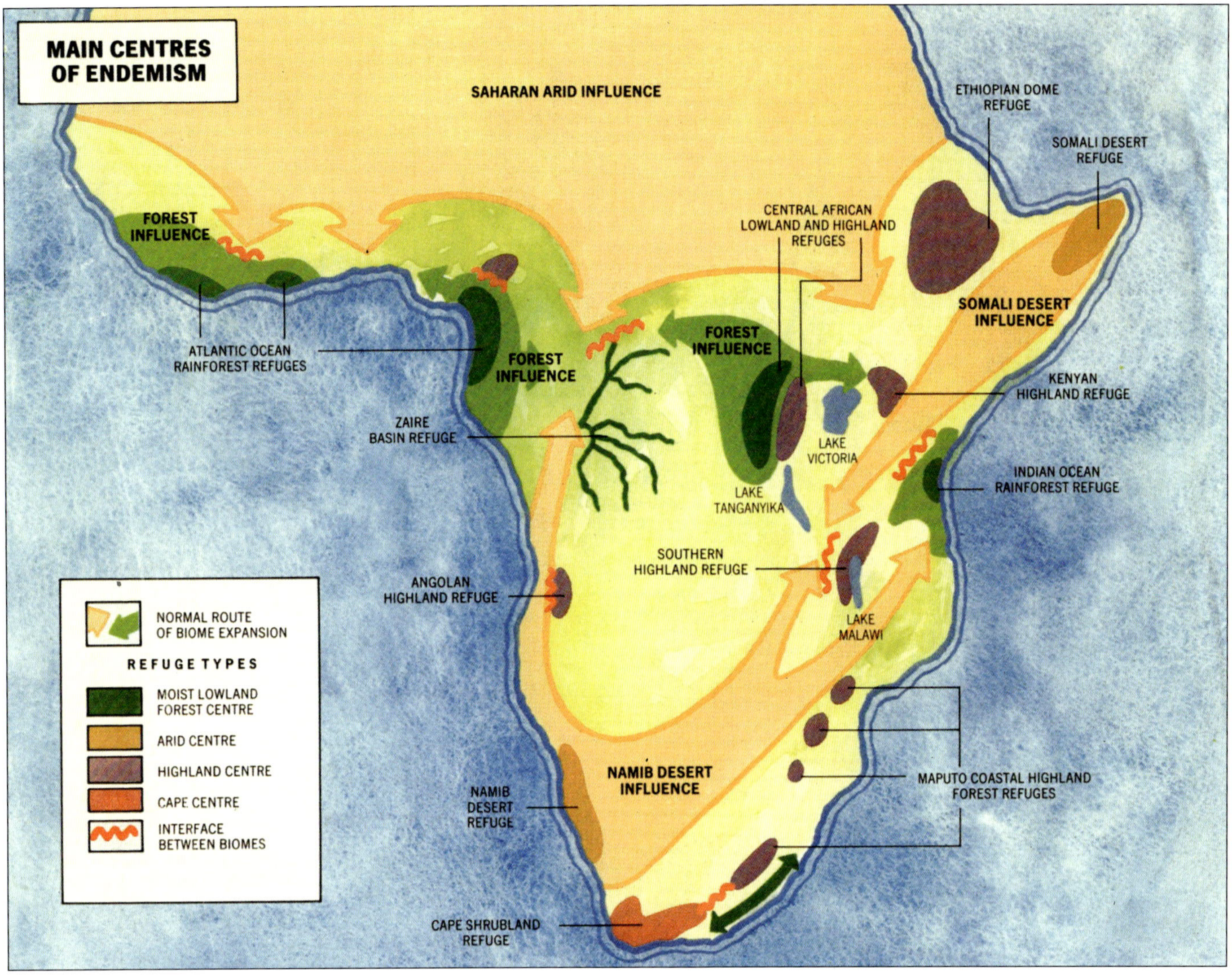

Main centres of endemism in sub-Saharan Africa. Showing N–S, S–N expansion and contraction of arid foci and E–W, W–E expansion and contraction of humid foci. Intermediate ecotypes (often narrow) are typically located along interfaces between these biomes and on piedmonts of upland massifs.

The influence of past climates in blurring or accentuating the lowland/montane forest divide is well illustrated by a complex of closely related squirrels from eastern Africa. Red Bush Squirrels, *Paraxerus palliates* are almost exclusively eastern African lowland forest squirrels ranging from Somalia to South Africa, but there is a single very distinctive form *P. p. swynnertoni* that lives in montane forests in the Manica highlands (Chirinda). Another exclusively montane squirrel, *P. vincenti*, is also closely related but lives only on Namuli Mountain, Mozambique. Further north and, for the most part, deeper into the interior, the speciating process has gone still further with a number of isolated montane *Paraxerus* populations that have, in the past, been very variously grouped as species or subspecies (*byatti*, *laetus*, *vexillarius*, *lucifer*). A plausible sequence of events was that a single ancestral squirrel population was widespread in eastern Africa during a period when forests were relatively extensive and the distinctions between montane and lowland forest were less marked than today. During one, or more likely, several fluctuations of climate, montane and lowland populations became physically separate and each subsequently adapted to the conditions within their respective enclaves.

Montane forests are generally cooler and less diverse in numbers of species, including plant foods, predators, pathogens and competitors. Of special current interest in this context is the 2004 discovery of a new primate genus, *Rungwecebus kipunji*, in the Southern Highlands of Tanzania (25° S) where it ranges up to 2500 m. A partial explanation for the survival of this very isolated, wholly montane and very puzzling population could lie in their former stranding, after changes in climate, in a region with fewer species of fruiting plants, much colder temperatures and fewer competitors. It would seem that this precariously rare species exists on the very edge of a tropical primate's environmental tolerance. If the consequences of diminished dietary resources but fewer other constraints in south-temperate or cooler montane areas are examined in the context of climatic change there are conclusions that are of central significance for an understanding of primates, for their conservation and for exploration of early human origins.

In respect to the latter, account needs to be taken of earlier periods than the Quaternary, but it is evident that early hominin fossils in South Africa are direct evidence for our own lineage escaping some of the constraints that govern many other African primates. It should be borne in mind that the role of climate might, in such instances, be over-ridden by the influence of other factors, such as changes in diets, the development of innovative food-getting behaviours, diminished competition and changes in susceptibility to diseases.

'Refugia' as centres of endemism

A recent review by Anhuf *et al.* (2006) emphasizes that African rainforests experienced much greater retraction during the last glacial compared to the same latitude in the Neotropics, and that rainforests retracted to a limited number of regions, which can be clearly mapped (see p. 55). There is unequivocal palynological evidence from equatorial Africa to support certain aspects of Pleistocene refuge theory, unlike the situation in South America. Five such forest refuges are now recognized (see p. 47), Upper Guinea in West Africa, Cameroon–Gabon (or Biafran Bight), the 'Central Refuge' in eastern DR Congo (Kingdon 1971) and the Congo Basin (Colyn 1991, Maley 1996) with an East African 'Zanj' (or 'Zanzibar–Inhambane') centre of forest endemism strung out along the East African coast (White 1983, Kingdon 1971, 1990a). The effects of climate change most likely help explain the paucity of 'wet climate' plant groups, such as palms, understorey shrubs, epiphytes and lianes. It is likely that this demonstrable loss of diversity has been caused by particular stresses that have hit communities adapted to ever-wet climates rather than those adapted to more seasonality. Several factors may apply in explaining this diversity loss. First, it is possible that during previous glacial maxima rainforest refugia may have become too small to maintain previous diversity levels. Second, the high relative elevation of the African continent (Morley 2000: Fig. 5.1), steep continental slopes and lack of continental shelves or low-lying littorals suggest that during cool/dry climate episodes lowland rainforests may have had limited opportunities to find refuge within appropriate low altitudinal zones (Morley 2000). These factors may also help to explain the scarcity of mammals in Africa that can be shown to have a long, unbroken, forest-adapted ancestry.

However, forest is not the only habitat to provide 'refuge'. At the opposite extreme desert-dwelling animals and plants are no more able to make rapid adaptation to an abundance of rain than forest organisms are to aridity. In between these extremes are more subtle, often seasonal or temperature-dependent regimes that also centre on refugial 'islands' (often mountain blocs) where there are less obviously specialized endemics. Many mammal distributions reflect the reality of diverse, climatically determined 'refuges' in which the successions of past changes have played crucial roles:

> It is the absence or moderation of exterminating climates that favours an accumulation of many species. Such accumulation has been most favoured in highly diversified or stratified landscapes. In such localities animals and plants could escape any minor change by shifting a few kilometres up or down some valley or further along some coastal plain. How specialized or different species become is influenced by how long that environment has been stable.
>
> (Kingdon 1990: p. 17)

That long-term stability (rather than simple allopatric speciation following forest fragmentation) has shaped speciation in 'refuge' areas has also been proposed for birds and plants (Fjeldsa and Lovett 1996, Morley 2000). The broad pattern of climatic change in the equatorial zone (with successive retractions and expansions of rainforests) has been in place back to at least the late Miocene, so the main 'refugia' are likely to be at least this old.

The major 'centres of endemism' (sometimes made up of clusters of 'refugia') are relatively few and their location can, in every case, be analysed in terms of their geo-climatic position on the continent and their relationship to climate changes in the past. The southernmost Cape of Good Hope centre of endemism has already been mentioned and numerous mammal species are endemic to this regional centre of ancient temperate climate, notably the Cape Mole-rat *Georychus capensis*, African White-tailed Rat *Mystromys albicaudatus*, Riverine Rabbit *Bunolagus monticularis* and Cape Grysbok *Raphicerus melanotis*. Adjacent, but distinctive in being an arid focus, are the Namibian deserts and sub-deserts, which also have a number of very distinctive mammalian endemics, exemplified by the Round-eared Sengi *Macroscelides proboscideus*, Damara Ground Squirrel *Xerus princeps* and Noki (Dassie Rat) *Petromus typicus*.

The closest biotic affinities of this centre are often in the Horn of Africa, where the Somali centre of endemism is notably rich in a variety of more or less arid-adapted species, often concentrated in narrow strata of specialized vegetation. Well-defined centres of endemism are sometimes focused on single mountain blocks such as the (dry) Ethiopian Dome and (wet) Rwenzori Mts, or more fractured mountain chains such as the volcanoes and scarps bordering the Western or Albertine Rift and the 'Eastern Arc' and 'Southern Highlands' mountains of Tanzania. There are also quite extensive uplands in Cameroon and the Angolan Scarp and smaller, but biogeographically significant isolated massifs at Nimba in upper Guinea, Namuli, Mulanje and Inyanga in Zambezia and the Saharan massifs of Hoggar, Tibesti, Aïr and Jebel Marra.

All these regions have distinctive bird or mammal endemics but the disparate mountains of eastern and north-eastern Africa are exceptionally rich in endemic mammals (and birds), many of which probably have histories of speciation and adaptation that go further back well before the Quaternary.

Correspondingly, for mammals of woodland, savanna and even drier habitats, the pattern of retracting rainforests would have opened up repeated north–south corridors across Gabon and East Africa during glacial periods. During particularly warm humid periods these routes could subsequently have been severed by the expansion of rainforests or gallery-laced habitats across much of the equatorial zone. Although some species, notably the larger, more mobile ones, such as bats, the larger carnivores and Roan Antelopes *Hippotragus equinus*, somehow managed to sustain single species, with (superficially) insignificant variation, across the entire extent of African savannas, other populations north and south of the Equator have been isolated by the forest belt, particularly in the west, or by lakes and rivers. This process would have encouraged speciation of woodland and savanna mammals within a similar time-frame. Such differences are common, as is well exemplified by members of the Hartebeest *Alcelaphus buselaphus* complex (see Volume VI), but, as might be expected for relatively recent changes, a high proportion of such differences between northern and southern savanna species is at the level of subspecies.

Correlating the distributions and biology of surviving mammal species with the climatic events that have shaped them is a science in its infancy. This review advertises many unresolved puzzles but further investigation of mammalian biology in the perspectives of new information about past environments and past evolutionary events represents a rich field of enquiry for future scholars.

CHAPTER FIVE

The Biotic Zones of Africa

David Happold & J. Michael Lock

Africa is a continent of diversity and contrast. The altitude varies from sea level to over 5000 m. The vegetation includes deserts, rainforests, woodlands, bushlands, shrublands, grasslands, alpine heathlands and grasslands, and swamps. There are numerous streams and rivers; most flow to the coast but some flow into huge swamps where the water evaporates. Virtually every part of the continent has a seasonal climate. This great diversity in altitude, vegetation and climate is one of the main reasons why there are so many species of mammals in Africa. Every one of them has evolved a set of characteristics that enables it to live in its environment, and the many environments in Africa have resulted in many ways of living. The environments of Africa can be broadly classified on their major characteristics. Such a classification provides a framework that allows an understanding of the distribution and characteristics of each species of mammal. Combinations of climate, vegetation and topography (which collectively determine the 'environment') enable Africa to be classified into biotic zones. Before describing these biotic zones, it is worth considering briefly the factors that determine their distribution.

The environment

Geology and soils

Much of the southern half of Africa is underlain by very ancient rocks (the 'Basement Complex') that have been exposed to weathering for many millions of years. Consequently the rocks themselves are only rarely visible at the surface. Because of this long period of weathering, the soils developed from these rocks are generally low in mineral nutrients for both plants and animals. In such areas, sites where minerals are concentrated ('salt-licks') are often sought after by animals. The long period of weathering also means that there can be large variations in soil type with topography – valley bottom, slope and ridge-top soils are often very different. The northern half of Africa mostly lies at lower altitudes and parts of it are covered with more recent sedimentary rocks. Even here, though, the effects of prolonged weathering often obscure the underlying rocks. Along the Rift Valleys, and in the Ethiopian Highlands, soils are often derived from recent volcanic materials and are much richer in mineral nutrients.

Climate

Because Africa straddles the Equator, the areas to the north and south are essentially mirror images of one another; corresponding areas have similar seasonal patterns, but displaced by six months. The extreme northern and southern parts of the continent have a winter-rainfall ('Mediterranean') climate with hot dry summers and cool moist winters. Moving towards the Equator, there is an arid zone in which rain is sparse and unpredictable. Because of the shape of the African continent and the proximity of Arabia, this zone is much more extensive in the north, and maritime influence is much greater in the south. In south-western and north-western Africa the general aridity is reinforced by cold water upwelling offshore. This reduces convective rainfall and can cause great local aridity, sometimes partially mitigated by regular fogs that condense on suitable surfaces.

The main part of Africa, here referred to as tropical Africa (i.e. between the Tropic of Cancer and the Tropic of Capricorn) has a climate that is influenced and determined by the Intertropical Convergence Zone. At and between the Tropics the sun is overhead at midday on at least one day each year. Briefly, the portion of the Earth's surface where the sun is overhead at midday is heated more than regions to the north and south. In this zone of greatest heating, the hot air rises and rain is generated as the air expands, cools and loses its water-holding capacity. The rising air is replaced by warm moist air that blows in from the north and south, but is deflected by the Earth's rotation to produce the north-east and south-east 'trade winds'. The air that rises at the Equator, having lost its moisture, flows polewards at high altitudes and finally descends as a dry air mass, producing the sub-tropical dry belts (e.g. the Sahara). Because of the tilt of the Earth's axis, the sun appears to move north and south during the course of the year. The sun is overhead at noon (12:00 h) on 21 December at 22° S (southern summer solstice), at the Equator on 21 March and 22 September (the equinoxes) and on 22 June at 22° N (northern summer solstice). This apparent movement of the sun produces a rain belt that moves with the sun (actually, a month or so behind it). This rain belt is a broad one, taking about three months to move across any point along its path. At the northern and southern ends of its passage, therefore, there is a single wet season each year. Thus there are two seasonal cycles, six months apart; the 'boreal cycle' in the northern hemisphere (with rainfall sometime

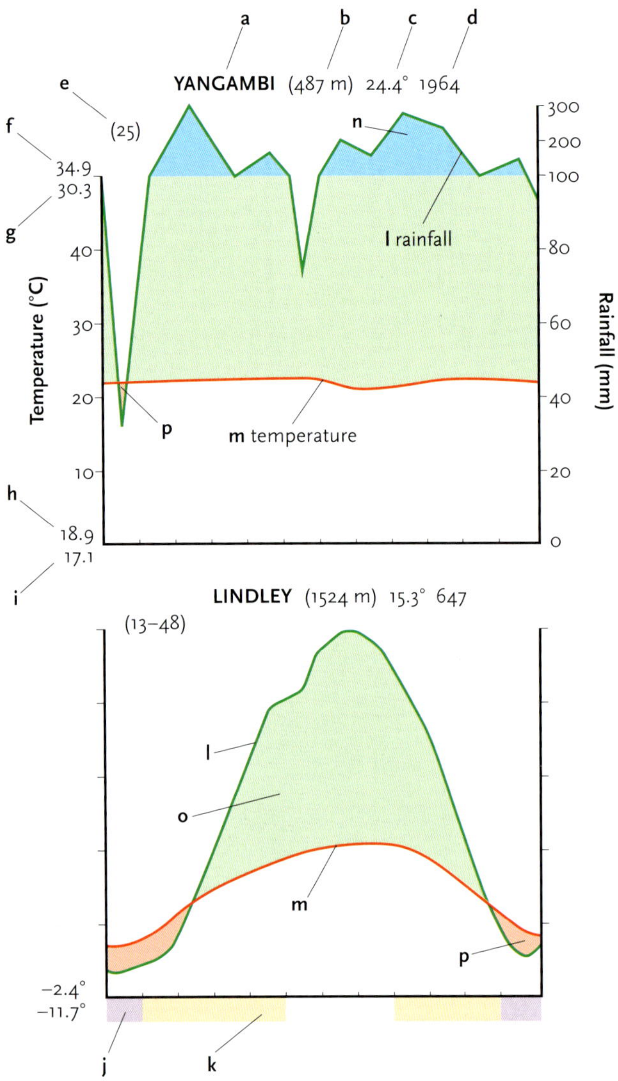

Figure 1. **Key to Climate diagrams.** Abscissa: Months of year (Northern Hemisphere, January–December; Southern Hemisphere, July–June); warmest seasons in the centre of diagram. Ordinate (left): Temperature in degrees Centigrade (each division = 10 °C). Ordinate (right): Rainfall in millimetres (each division = 20 mm from 0–100 mm, and 100 mm thereafter). Ordinate numerals are normally omitted from each graph. Data is not available for all locations.

(a) Name of locality. (b) Altitude of locality (metres above sea level). (c) Mean annual temperature (°C). (d) Mean annual rainfall (mm). (e) Number of years of observation. (f) Absolute maximum temperature. (g) Mean daily maximum of hottest month. (h) Mean daily minimum of coldest month. (i) Absolute minimum of coldest month. (j) Months with a mean daily minimum below 0 °C (purple bar). (k) Months with an absolute minimum below 0 °C (beige bar). (l) Monthly means of rainfall in millimetres (green line). (m) Monthly means of temperature (red line). (n) Months when monthly rainfall is more than 100 mm/month = perhumid season (blue shading). (o) Months when monthly rainfall is less than 100 mm/month and when the rainfall curve is above the temperature curve = humid season (green shading). (p) Months of relative drought when the rainfall curve is below the temperature curve (pale orange). Modified from Walther (1978) and White (1983).

between April and October, with a peak in about August) and an 'austral' cycle in the southern hemisphere (with rainfall sometime between December and April, with a peak about February). Along the Equator, there tend to be two wet seasons (Mar–May and Aug–Nov), corresponding approximately to the equinoxes. In areas of very

OPPOSITE: Figure 2. Climate diagrams for selected localities arranged according to Biotic Zones (see also overleaf). See Figure 1 and text for further details. All climate diagrams place the wettest season on the year in the centre of each diagram, even though localities north of the Equator are six months out of phase with localities south of the Equator.

high rainfall these may merge so that rain falls almost throughout the year, but in Africa there are virtually no areas that do not experience at least a brief dry season of four to six weeks during the year.

In most of tropical Africa, diurnal temperature variations are generally much greater than seasonal variations. Indeed, the climate of high mountains in tropical Africa has been described as 'winter every night and summer every day'. Temperatures tend to be highest at the beginning of the wet season, when cloud cover is low and early showers have washed the smoke haze out of the atmosphere. However, the diurnal temperature range is often greatest during the dry season, when low atmospheric humidity and low cloud cover allow rapid radiative cooling during the night. As is usual, temperature falls with increasing altitude; this 'lapse rate' is variable, but a figure of 0.5 °C per 100 m is a useful approximation. Within the tropics, frosts are virtually unknown at low altitudes, but occur regularly above about 2500 m (Hedberg 1964). Snow can fall above 4000 m, and glaciers descend to this level on the Rwenzori Mts.

The ways in which climatic variations interact can be shown graphically by the use of climate diagrams (Walter & Lieth 1967, Walter 1978). These are plotted so that 20 mm of precipitation is equivalent to each increase of 10 °C above zero (Figures 1 & 2). Using this convention, periods when the rainfall curve falls below that of temperature can be regarded as times of relative drought, and periods when the rainfall curve is above that of temperature, as relatively moist. Periods when the monthly rainfall exceeds 100 mm are regarded as perhumid, and here the scale is reduced so that one scale division equals 100 mm of rainfall. Each of these three conditions is indicated by different shadings in the diagrams so that the overall climate, and seasonal changes in climate, are easily recognizable. By and large, plant growth is likely to be relatively rapid during the moist and perhumid periods, and slow, spasmodic or absent during periods of drought. A glance at the climate diagram of a particular location (Figure 2) provides a good indication of the type of vegetation that may be expected, as well as showing the environment that is experienced by the mammals that live there.

Fire

In a highly seasonal climate, vegetation dries out and becomes flammable during the dry season. Fires are a regular feature of the more seasonal types of African vegetation. In most of seasonal Africa a high proportion of the land surface is burned each year. Some fires are ignited by lightning. However, there can be little doubt that fire frequency has increased since the advent of humans, first as humans learned to use naturally occurring fires, then again when they began to make fire themselves, and finally, even more with the introduction of the safety match (Lock 1998). Fires tend to be hotter and fiercer in regions of higher rainfall, as long as there is a marked dry season, because vegetation productivity is greater and hence there is a higher standing crop to burn.

Many plant species in regularly burned vegetation are highly adapted to fire. Many grasses have awned seeds; the hygroscopic movements of the awn buries the seed in the soil, often out of the

1 Mediterranean Coastal Biotic Zone

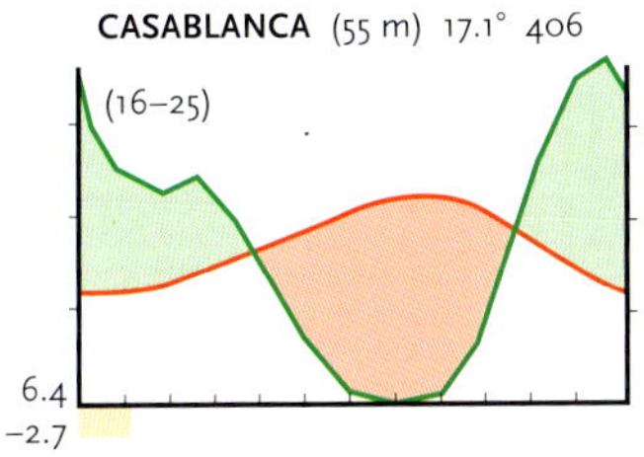

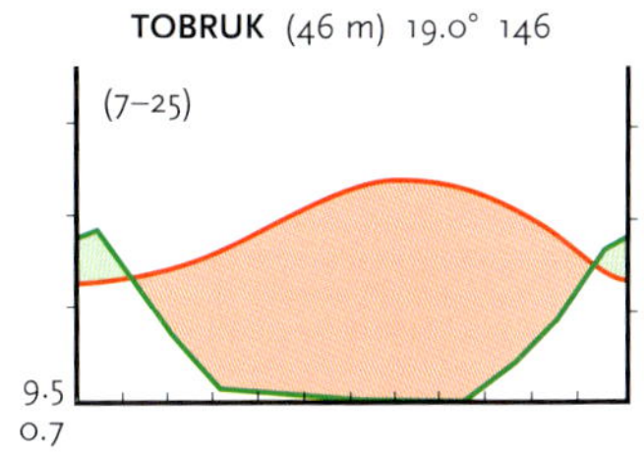

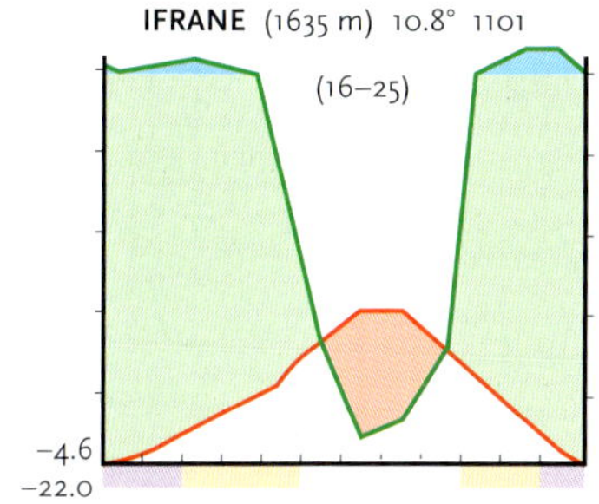

2 Sahara Arid Biotic Zone

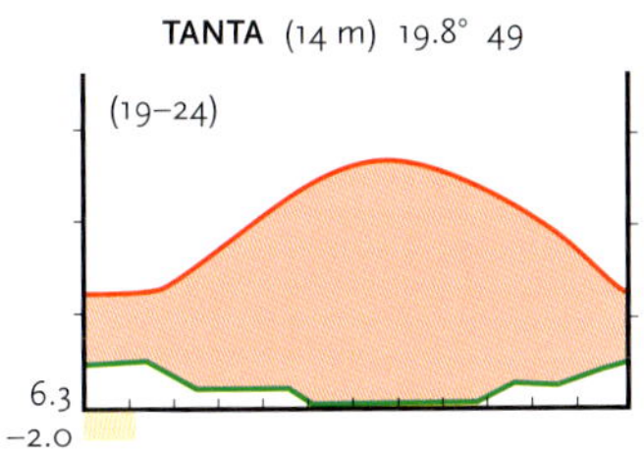

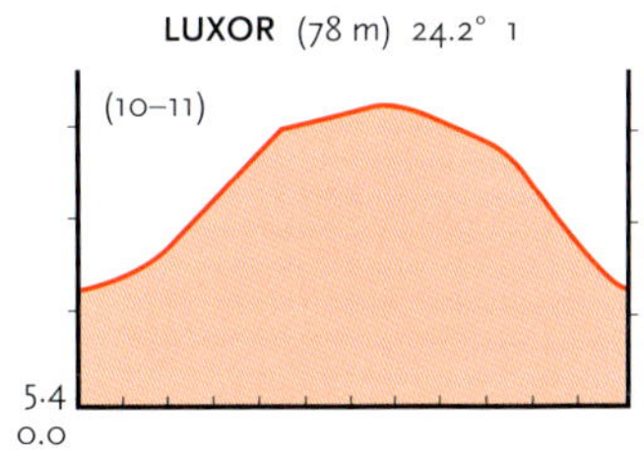

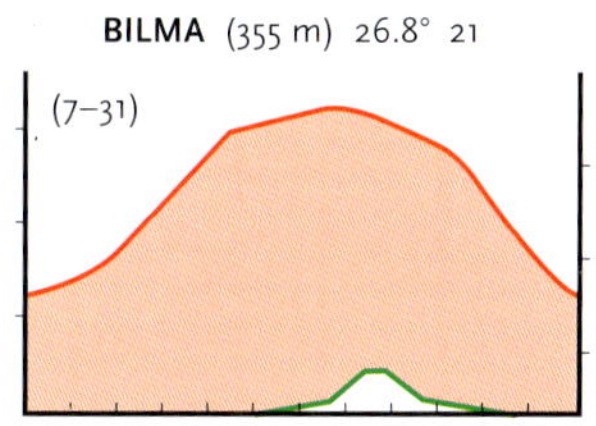

3 Sahel Savanna Biotic Zone

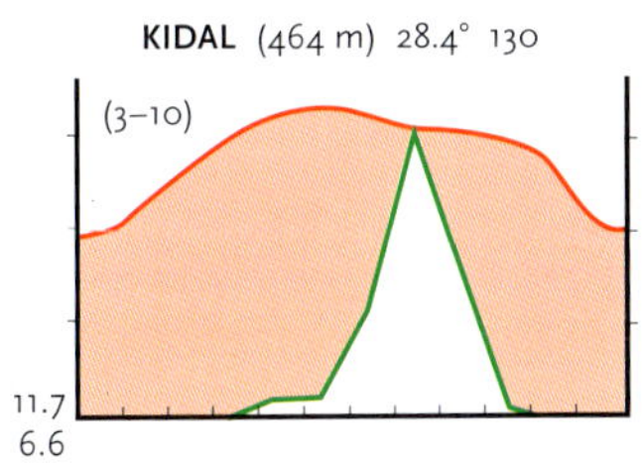

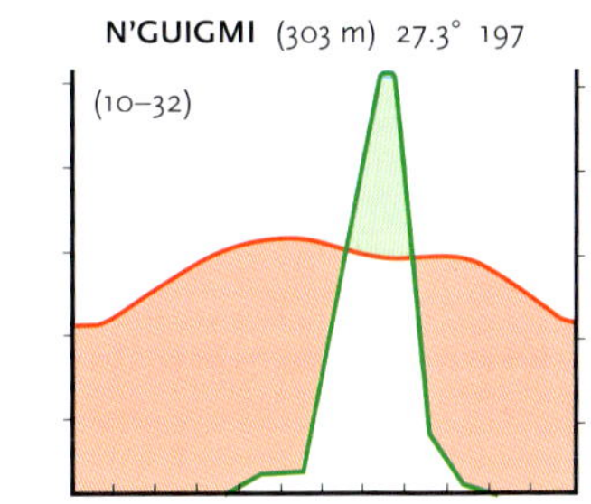

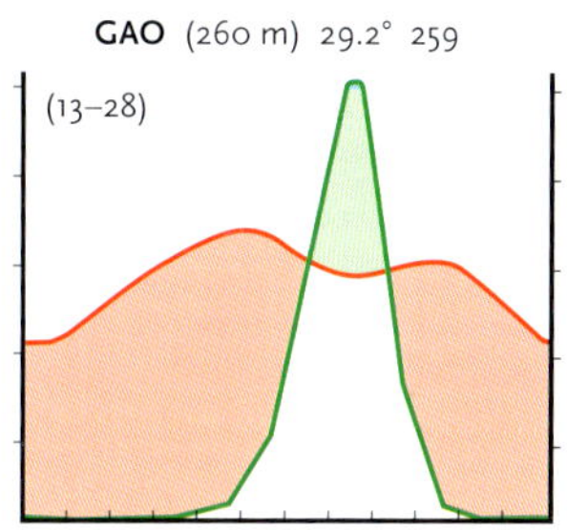

4 Sudan Savanna Biotic Zone

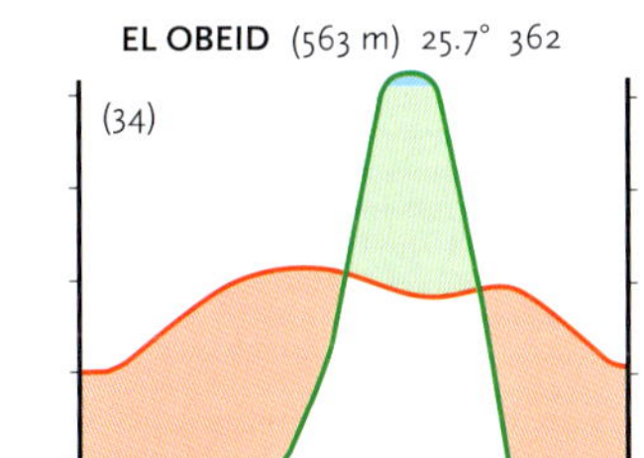

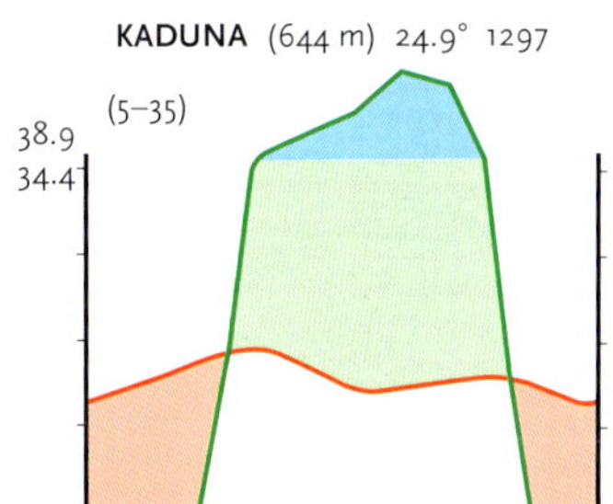

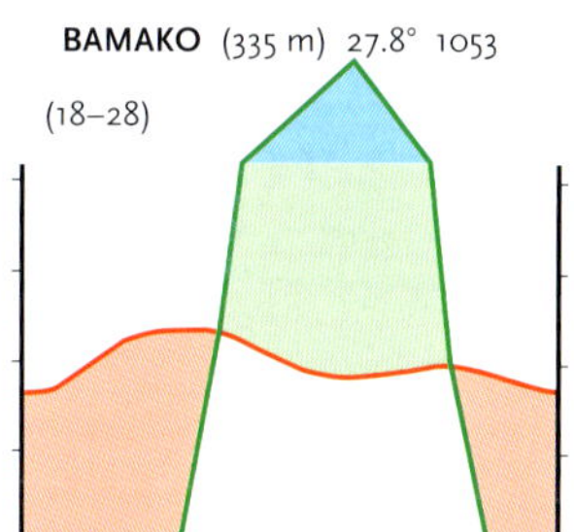

5 Guinea Savanna Biotic Zone

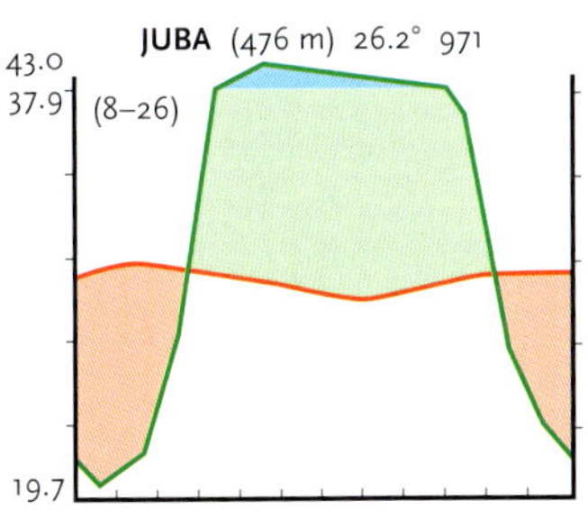

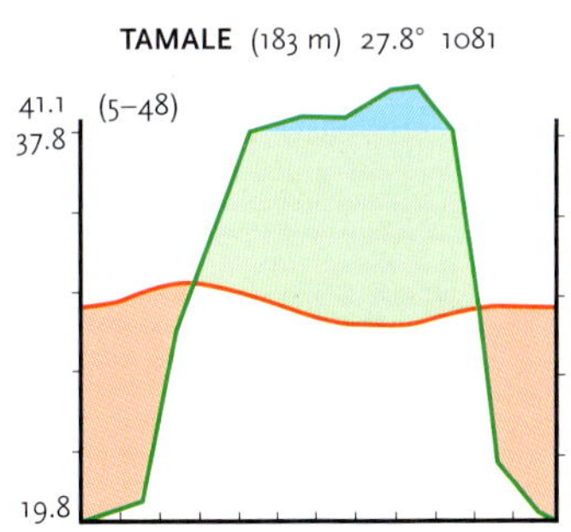

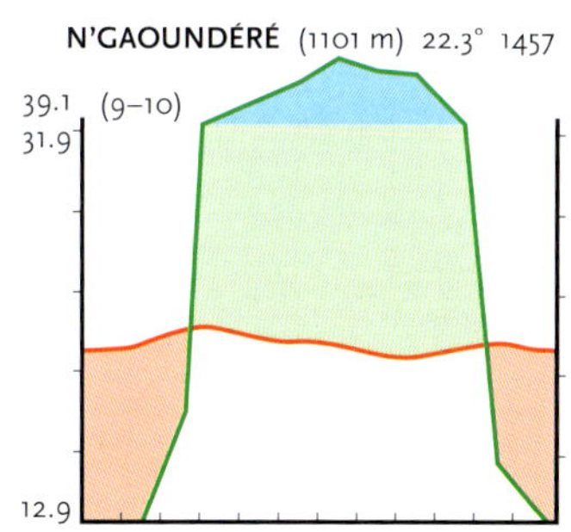

6 Rainforest Biotic Zone

STANLEYVILLE (415 m) 25.3° 1842

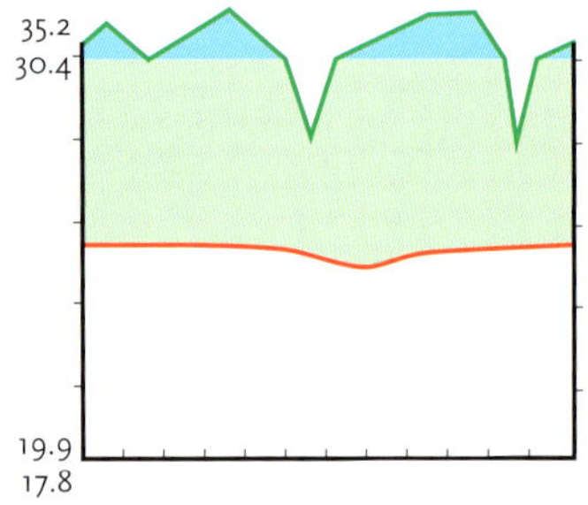

WARRI (6 m) 26.6° 2753

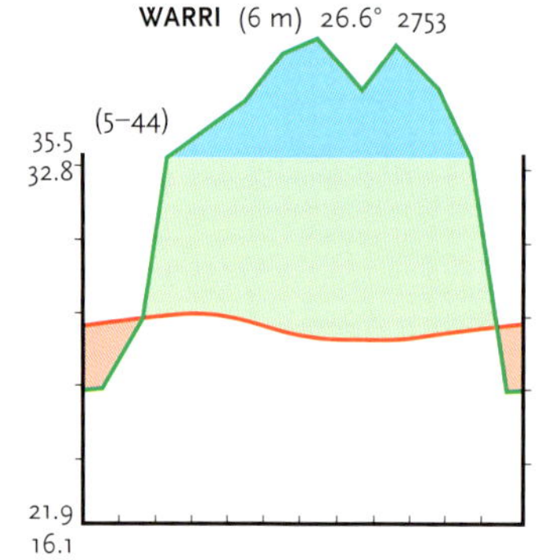

CONAKRY (7 m) 26.6° 4349

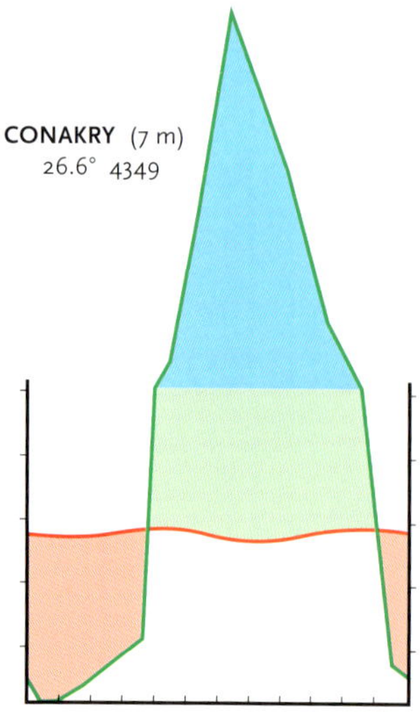

6a Rainforest–Savanna Mosaic

BANGASSOU (500 m) 26.1° 1764

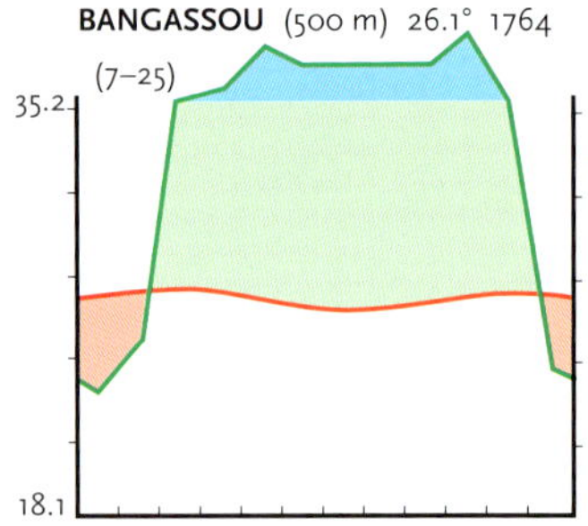

ARUA (1280 m) 23.1° 1361

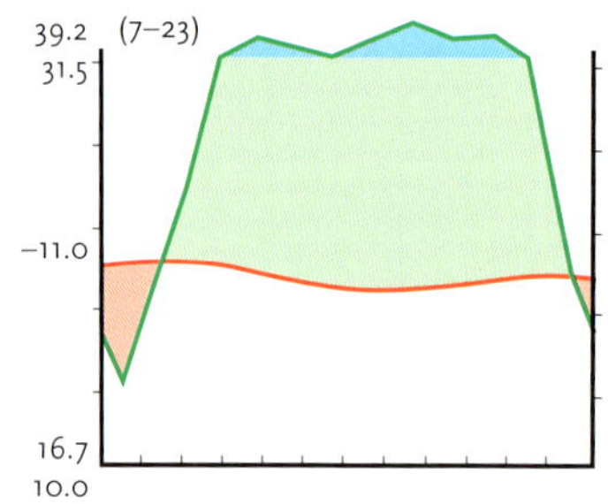

7 Afromontane–Afroalpine Biotic Zone

MBABANE (1163 m) 16.9° 1406

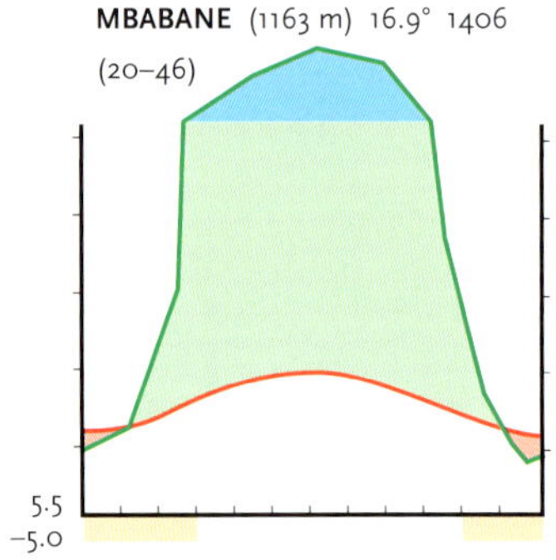

WOODBUSH (1528 m) 15.0° 1788

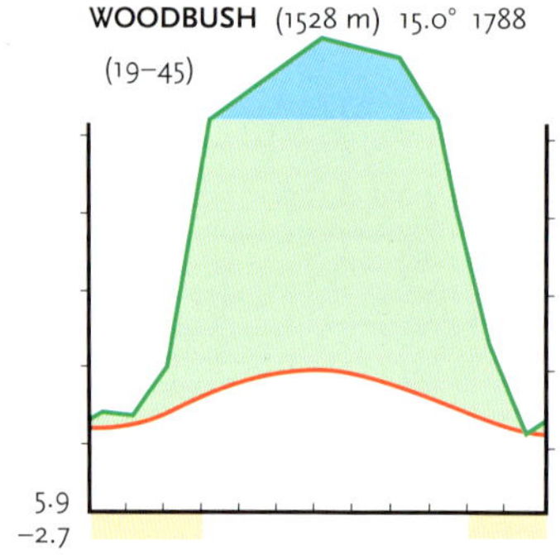

MOUNT NUZA (2032 m) 12.9° 1805

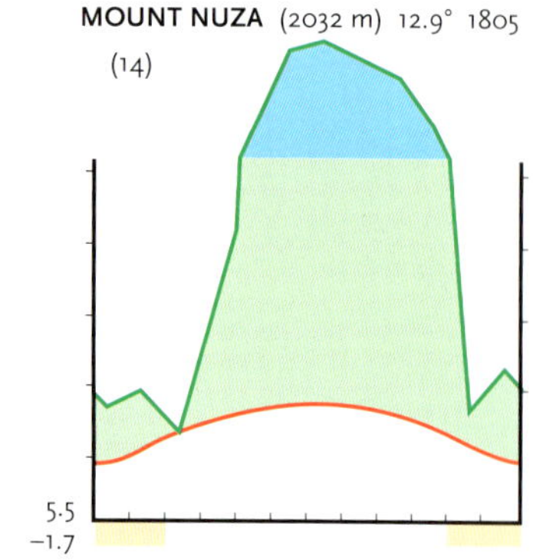

11 South-West Arid Biotic Zone

Namib

ALEXANDERBAAI (12 m) 15.4° 41

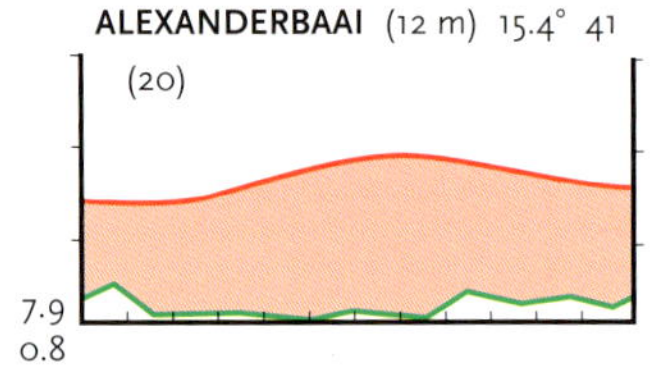

SWAKOPMUND (10 m) 15.3° 15

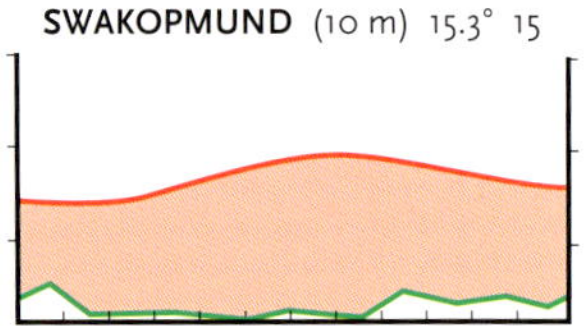

Karoo

BEAUFORT WEST (868 m) 17.9° 223

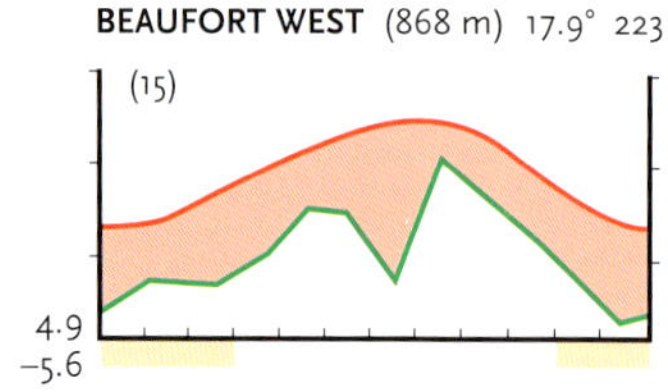

CALVINIA (981 m) 16.3° 209

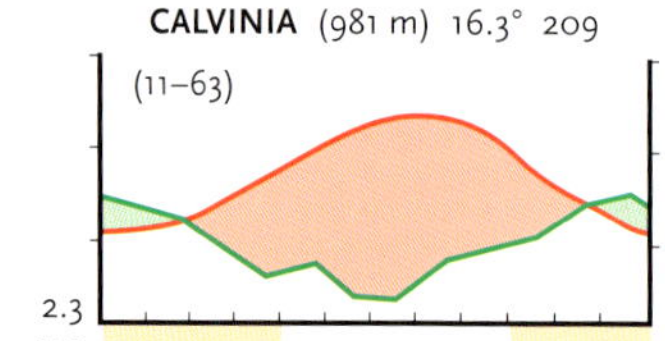

Kalahari

GHANZI (1131 m) 20.7° 470

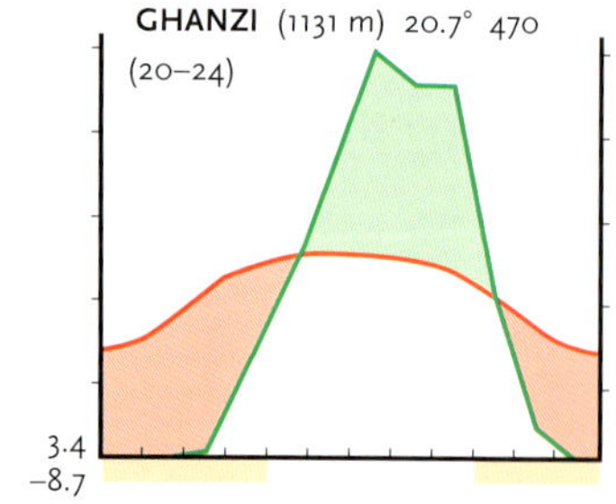

MAUN (942 m) 22.2° 457

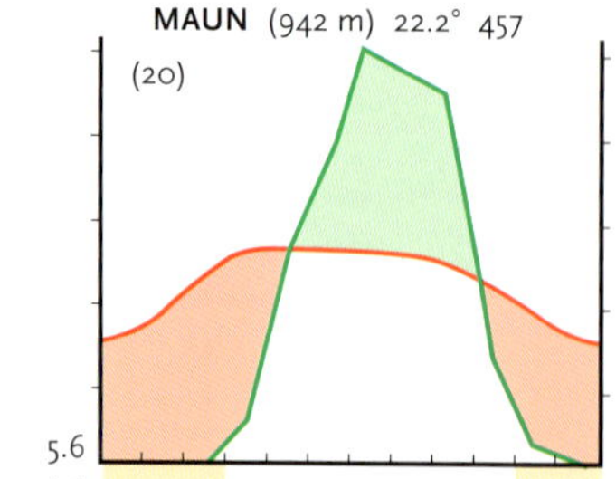

12 Highveld Biotic Zone

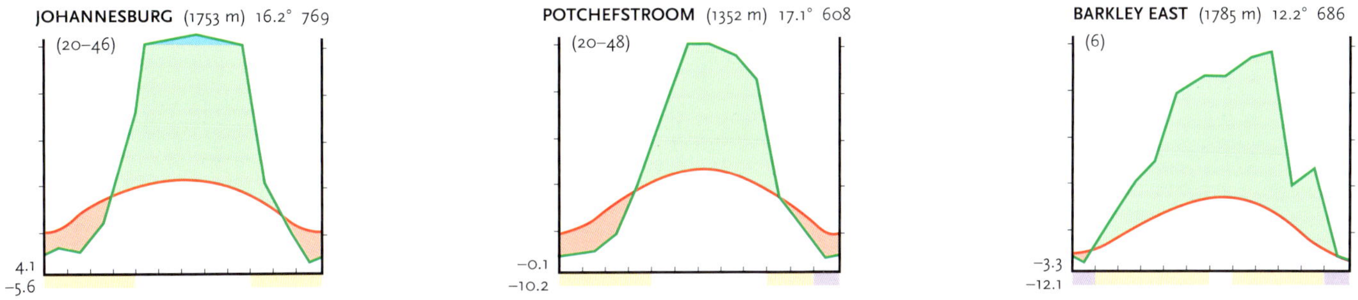

13 South-West Cape Biotic Zone

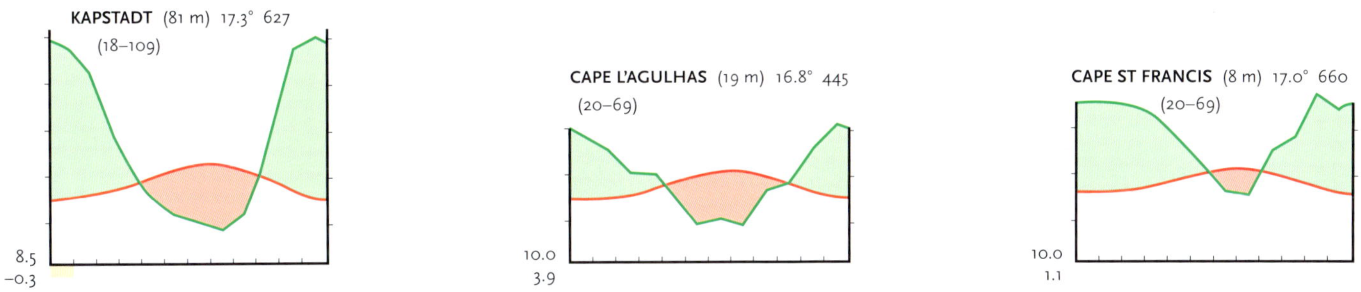

OPPOSITE AND ABOVE: Figure 2 (*continued*). Climate diagrams for selected localities arranged according to Biotic Zones.

heat of the fire (Lock & Milburn 1971). Clumped grasses have their growing points below ground level in the depths of the clump. Many trees have thick, furrowed corky bark that insulates the growing tissues from the heat of a fire. Some plants are stimulated by fire, and flower afterwards even if no rain falls. Others take advantage of the clearance of the vegetation to flower early before the taller species grow up. Adjacent areas of unburned forest and regularly burned savanna have very few species in common. Regularly burned areas, particularly those burned late in the dry season when the fires are hottest, have reduced densities of trees.

Fires have a major influence on mammals. A few may be killed by the fire and scavengers are often seen patrolling a burned area after the fire has passed. Large mammals can move much faster than a fire, while small ones can often take refuge in holes in the ground or inside termite mounds. Soil is an excellent insulator, and under more than 2 cm of soil there is almost no change in temperature during a fast-moving fire. However, once the vegetation cover has been removed and the sun shines directly onto the soil surface, soil temperatures are often much higher than prior to the fire. Fire destroys forage, but regrowth is usually fairly quick, and areas burned are generally not so large as to preclude local movements to nearby unburnt grasslands. It is often assumed that fire destroys valuable forage. In fact, by the early dry season, a very high proportion of the nutrients in the grass will have been transferred to the root system and stem bases, leaving the above-ground parts with a low nutrient content. Seasonal variations in grass growth and nutrient content are a major cause of some of the great mammal migrations of Africa, such as those in the Serengeti (Sinclair & Norton-Griffiths 1979) and southern Sudan (Howell *et al.* 1988). Small terrestrial mammals that cannot migrate exhibit seasonal changes in diet, seasonal changes in populations numbers (higher mortality when resources are scarce, high natality when resources are abundant) and limited local movements to preferred environments.

Biotic zones and mammalian biology

Africa may be divided into a number of biotic zones (Figure 3) and the distribution of mammals is often closely related to the distribution of these biotic zones. For example, the distributions of many species of desert rodents agree closely with the distribution of the arid biotic zones, and the distributions of rainforest arboreal species closely follow the distribution of rainforest. However, the relationship is not always exact. Many mammalian species occur in several biotic zones, indicating that they have a wide environmental tolerance. In contrast, others have a very limited tolerance and are found in only part of a biotic zone, suggesting that their distribution is determined by local conditions. For example, a species that requires cool temperatures and an arboreal habitat will occur only in montane forests. The limiting factors that prevent such a species expanding its distribution may be higher temperatures and an absence of trees in adjacent areas. Of course, such factors may not act directly on the species itself but rather on the plants upon which it is dependent. Precise analysis of such factors, as well as of competitive interactions with other species, is extremely difficult.

When a particular feature of the environment exceeds the level of tolerance of a species (e.g. the temperature on a long-term basis is too hot or too cold, the water supply becomes inadequate, the vegetation structure changes, or appropriate food is no longer available), populations may become locally extinct. The exact nature of 'limiting factors' is different for every species, and hence the geographic distribution of every species is different. The geographic distribution of a species (as shown in each distribution map) may be thought of as being composed of three parts:

1 'Core Area', where conditions for survival are optimal. Here, population numbers are maintained, and the birth rate approximates

Figure 3. The biotic zones of Africa. The numbers refer to the biotic zones as described in the text.

the death rate. There may be temporary peaks and slumps in population numbers, but, on average, population numbers are maintained at the carrying capacity of the area.

2 'Peripheral Area', on the edge of the geographic range, where conditions are not optimal all the time or in all locations. Sometimes, when conditions are temporarily favourable, the geographical range may increase; at other times it may contract.

3 'Relict Areas', where populations survive in isolated patches, sometimes far removed from the Core Area. Such patches were joined to the Core Area in the past, but are now isolated because of changes in climate and the activities of humans. Relict populations survive because their environment is the same as in the Core Area, even though now separated by unsuitable habitats. The best examples are several species of rainforest mammals that live in small relict patches of rainforest within savannas, sometimes several hundreds of kilometres from the present boundary of the rainforest zone. For all species, the outer limits of distribution are where environmental conditions exceed the limits of tolerance, or where geographical and historical barriers have prevented colonization.

The characteristics of any biotic zone have a profound effect on the biology and structure of mammalian communities. Thus for any biotic zone, there are communities and species that are 'typical' or 'representative'. Over the course of evolution, the mammals of a particular biotic zone have evolved characteristics (often shared by many species) that are of value for living in that biotic zone. In addition, different species have evolved to exploit different niches within a biotic zone, so there may be several sets of characteristics that are typical. In a rainforest, for example, there are many species that are arboreal; but other species are highly adapted for life on the ground. There are four major ways in which mammalian species and communities are affected by the characteristics of biotic zones:

1 The number of species, and the composition of the mammalian community, is strongly affected by the structure and resources of the biotic zone. If a biotic zone is very extensive, the species composition is likely to vary in different regions due to natural variations in vegetation and climate. In general, the richest biotic zones (i.e. those with many plant species, great variation in plant structure and high annual plant productivity) contain the highest number of mammalian species, and the least productive biotic zones have the smallest number of species. Because there are many mammalian species in most biotic zones, the interactions between the species and the environment, and between the species themselves, are numerous and complex. The expansion and contraction in the area of biotic zones (resulting from past climatic fluctuations), and the presence of rivers and mountains that have acted as barriers to movement, have also had significant effects on species composition.

2 The number of individuals and their biomass is also determined by the complexity of the biotic zone and the availability of resources. In order to exploit the varied food resources in biotic zones, mammals have developed many ways of ingesting and digesting food. Within a community, each species differs in its diet; the differences may be large (such as between a herbivore and a carnivore) or very slight (such as closely related species eating slightly different proportions of a particular food). The result is that many food resources are exploited, thus increasing the numbers of individuals and biomass that can be sustained of each species.

3 Many of the characteristics of species are determined by the biotic zone. Most desert mammals, for example, have evolved mechanisms to cope with shortages of water. These may be physiological, morphological or behavioural, but all of them enable survival in an arid environment. The relative importance of each of these mechanisms is usually associated with the size of the mammal – what is possible for a small mammal is not possible for a large mammal, and vice versa. In rainforests, many species are arboreal and have morphological adaptations for climbing, swinging and gliding – all of which are inappropriate for species living in a treeless region. The relationship between the characteristics of an animal and its habitat is so close that it is easy to infer where and how it lives from its physiological and morphological characteristics.

4 The period of the year when a species is able to reproduce is closely related to the seasonal changes in temperature and in the availability of water and food. Mammals time their reproduction so that the young are born when conditions are optimal for the survival of the mother and young. The timing of mating and length of time for gestation have evolved in relation to the size of the species, the length of time required for foetal development and the stage of development when the young are born. For some species, mating occurs at a time when environmental conditions are at their worst – but the young are born when conditions are at their best. Smaller species, with shorter gestations, are the most responsive to environmental change; larger species (because of their longer gestations) are the least responsive because gestation may last over several seasons. For most species, young are born at a well-defined period of the year that is more or less constant

from year to year. Some species give birth to their young over a period of several months, whereas other species give birth during a restricted period of only a few weeks. Populations increase rapidly when the young are born (particularly in those species that have large litters in quick succession) and then decline because of natural mortality, particularly when environmental conditions deteriorate. The reproductive strategies (and hence life-cycles) of species vary greatly depending on their phylogeny, size and diet, so that a community composed of many species also shows great variation even though all species live under the same environmental conditions.

5 For individual mammals, seasonal and annual changes in the characteristics of a biotic zone affect where the individual lives, whether it has to move from one place to another, whether it will reproduce, the composition of its diet, and whether it lives or dies. These changes affect populations in a similar way, controlling the increases and decreases in numbers and the very existence of the population itself. Each species profile in these volumes refers, directly or indirectly, to the biotic zone (or zones) where the species lives. All of the characteristics of a biotic zone impinge, to a greater or lesser extent, on the biology and life-style of every species of mammal.

The biotic zones

Concept and definition

Biotic zones, as used here, are defined by a combination of physiognomic vegetation type and phytogeographic area. Both of these defining terms are used here in the sense of White (1983). Biotic zones can be defined as follows (Rutherford & Westfall 1986):

1 They refer to broad continental patterns and large natural areas, and are normally mapped at scales of 1 : 1,000,000 or greater. They do not adequately describe small-scale local variations in the environment. They generally have a characteristic climate.

2 The climax vegetation of each biotic zone is characterized by a particular physiognomy. Thus, in forest, the vegetation is dominated by trees, arranged in several layers, usually associated with dependent growth forms such as epiphytes and lianas. The physiognomy of a plant community is independent of floristic composition, which may vary considerably within a small area because of slight variations in slope and drainage, and the consequent changes in soil type. Such catenary variation, for example, does not invalidate the general characteristics of a biotic zone. Successional stages of growth (as when forest is cleared and is in the process of regrowing) are also not considered as biotic zones, although they may be important to mammals. The major physiognomic vegetation types recognized by White (1983) are forest, woodland, bushland and thicket, shrubland, grassland and afroalpine. There are, of course, intermediates between most of these (such as wooded grassland) but for our purposes these need not be considered.

3 White (1983) divided Africa into a number of 'phytochoria' – areas with similar floras. The highest level of this classification is the Regional Centre of Endemism. To qualify as such, a region must not only have a high proportion of endemic species (White used 50%), but also a substantial number of these endemic plant species (White used a figure of 1000). Although these definitions have been applied somewhat elastically, they remain useful when used with caution. White's major phytochoria in Africa, which he termed 'Regional Centres of Endemism', are: Guineo-Congolian, Zambezian, Sudanian, Somalia–Masai, Cape, Karroo–Namib, Mediterranean and Afroalpine. The Afroalpine is exceptional in being discontinuous ('archipelago-like', to use White's expression). He recognized Transition Zones between many of these phytochoria, and also three Regional Mosaics (RM), where elements from several Regional Centres mingle – the Lake Victoria RM, the Zanzibar–Inhambane RM and the Tongoland–Pondoland RM. White only named and mapped Transition Zones where they were large, sometimes larger than some Regional Centres of Endemism. He defined them as having few endemic species and with the majority of their species occurring in adjacent phytochoria. The Regional Mosaics are more complex, with mosaics of different physiognomic vegetation types and floristic relationships, and more endemism than in the Transition Zones. Dowsett-Lemaire & Dowsett (1998) have shown that there is a close correlation between forest bird species distribution and the Guineo-Congolian and Afromontane chorological categories. They conclude that 'Frank White's phytochoria provide zoologists with an excellent environmental framework, a prerequisite for studies of biodiversity, endemism, and species conservation'. The names for biotic zones used in *Mammals of Africa* are those in common usage in the regions concerned; the equivalent in White's terminology is, however, also given (Table 5).

4 Within some biotic zones, human activity has altered the physiognomy of the vegetation, but for our purposes this is regarded as a temporary perturbation. Although the changes wrought by humans have had a detrimental effect on most species of mammals (and many other animals), they have favoured a few by allowing an increase in numbers and an extension of geographic range.

5 While the boundaries between biotic zones are drawn on any map as hard lines, one must always remember that a line on a map separating Zone A from Zone B will, on the ground, separate A with patches of B, from B with patches of A (Keay 1959a). The 'change-over' from one zone to another may take 50–100 km, or more; the changes may be subtle and only recognizable by a gradual change in the percentage occurrence of a few dominant plant species. Even where the plant structure between adjacent zones is very different, such as between rainforest and savanna, patches typical of one zone may be found deep inside the other.

The biotic zones of Africa

There have been many attempts to classify the vegetation of Africa. The most widely accepted classifications of vegetation for the whole of the continent are those of Keay (1959a) and White (1983). There have also been many regional classifications, such as those of Acocks (1975) and Rutherford & Westfall (1986) for South Africa, Greenway (1973) for East Africa, Rosevear (1953) for West Africa and Keay (1959b) for Nigeria. In some respects, the biotic zones of Africa north and south of the Equator are mirror images. The various biotic zones form bands that roughly follow the lines of latitude. The rainforests that span the Equator are followed by successive bands of savanna, then by deserts or arid conditions, and finally by temperate environments. This simple banding is modified by the effects of the oceans in coastal regions, and by the

Table 5. The biotic zones and mosaics of Africa as used in *Mammals of Africa*. The vegetation zones, vegetation number and vegetation types (following White 1983) are given for each of the Biotic Zones and Mosaics.

Biotic Zones and Mosaics	Vegetation Zone	Vegetation Type	
		No.	Name
1. Mediterranean Coastal BZ	A	10	Mediterranean sclerophyllous forest.
	B	55	Sub-Mediterranean semi-desert grassland and shrubland.
		49	Transition from Mediterranean *Argania* scrubland to succulent semi-desert shrubland.
		10	Mediterranean sclerophyllous forest (small patches only).
2. Sahara Arid BZ	C	71	Regs, hamadas, wadis.
		70	Desert dunes with perennial vegetation.
		67	Absolute desert.
		69	Desert dunes without perennial vegetation.
		72	Saharomontane vegetation (small patches).
		-	Oases
3. Sahel Savanna BZ	D	54a	Northern Sahel semi-desert grassland and shrubland
	E	43	Sahel *Acacia* wooded grassland and deciduous bushland.
		62	Edaphic grassland mosaic with *Acacia* wooded grassland.
		75	Herbaceous swamp and aquatic vegetation (around Lake Chad).
4. Sudan Savanna BZ	F	29a	Sudanian undifferentiated woodland.
		30	Sudanian undifferentiated woodland with islands of *Isoberlinia*.
		29b	Ethiopian undifferentiated woodland.
		62	Edaphic grassland mosaic with *Acacia* wooded grassland (in east).
		63	Edaphic grassland mosaic with communities of *Acacia* and broad-leaved trees (patches only).
5. Guinea Savanna BZ	G	27	Sudanian woodland with abundant *Isoberlinia*.
(The Sudd [included mostly into the eastern end of Guinea Savanna BZ])	I	61	Edaphic grassland in Upper Nile Basin.
		64	Edaphic grassland in Upper Nile Basin with semi-aquatic vegetation
6. Rainforest BZ	Ja	1a	Guinea-Congolian lowland rainforest (wetter types)
	Jb	2	Guinea-Congolian lowland rainforest (drier types).
	Jc	8	Swamp forest
		9	Mosaic of swamp forest and wetter lowland rainforest
6a. Northern Rainforest-Savanna Mosaic	H	11a	Guinea-Congolia/Sudania mosaic of lowland rainforest and secondary grassland.
6b. Eastern Rainforest-Savanna Mosaic	M	11a	Lake Victoria mosaic of lowland rainforest and secondary grassland.
6c. Southern Rainforest-Savanna Mosaic	N	11a	Guinea-Congolia/Zambezia mosaic of lowland rainforest and secondary grassland.
		14	Mosaic of lowland rainforest, Zambezian dry evergreen forest and secondary grassland.
		60	Edaphic and secondary grassland on Kalahari Sand.
		31	Mosaic of wetter Zambezian woodland and secondary grassland.
7. Afromontane–Afroalpine BZ (discontinuous; shaded black)	Y	19a	Afromontane undifferentiated montane vegetation
		38	Evergreen and semi-evergreen bushland and thicket (Ethiopian Highlands only)
	Z	23	Mediterranean montane forest and altimontane shrubland
8. Somalia–Masai Bushland BZ	K	54b	Somalia-Masai semi-desert grassland and shrubland.
	L	42	Somalia-Masai *Acacia-Commiphora* deciduous bushland and thicket.
9. Zambezian Woodland BZ	O	25	Wetter Zambezian miombo woodland dominated by *Brachystegia, Julbernardia* and *Isoberlinia*.
	Pa	28	*Colophospermum mopane* woodland and scrub woodland.
	Pb	26	Drier Zambezian miombo woodland dominated by *Brachystegia* and *Julbernardia*.
		29c	N Zambezian undifferentiated woodland.
		29d	S Zambezian undifferentiated woodland.
		22a	Mosaic of dry deciduous forest and secondary grassland.
		60	Edaphic and secondary grassland on Kalahari Sand.
10. Coastal Forest Mosaic BZ	Q	16a	East African coastal mosaic: Zanzibar-Inhambane.
		16b	East African coastal mosaic: forest patches.
		16c	East African coastal mosaic: Tongaland-Pondoland.

Biotic Zones and Mosaics	Vegetation Zone	Vegetation Type	
		No.	Name
11. South-West Arid BZ 11a	R	35a	Zambezian Transition from undifferentiated woodland to *Acacia* deciduous bushland and wooded grassland.
11a	S	44	Kalahari *Acacia* wooded grassland and deciduous bushland.
11a		56	Kalahari/Karoo-Namib transition.
11b	U	51	Bushy Karoo-Namib shrubland.
11b	V	74	Namib Desert.
11c	Wa	53	Dwarf Karoo shrubland.
11c	Wb	57b	Transition from Karoo shrubland to Highveld grassland.
11c		57a	Montane Karoo grassy shrubland.
11c		51	Bushy Karoo-Namib shrubland.
11c		52	Succulent Karoo shrubland.
12. Highveld BZ	T	58	Highveld grassland.
13. South-West Cape BZ	X	50	Cape shrubland (Fynbos) (with central island of 51 Bushy Karoo-Namib shrubland)

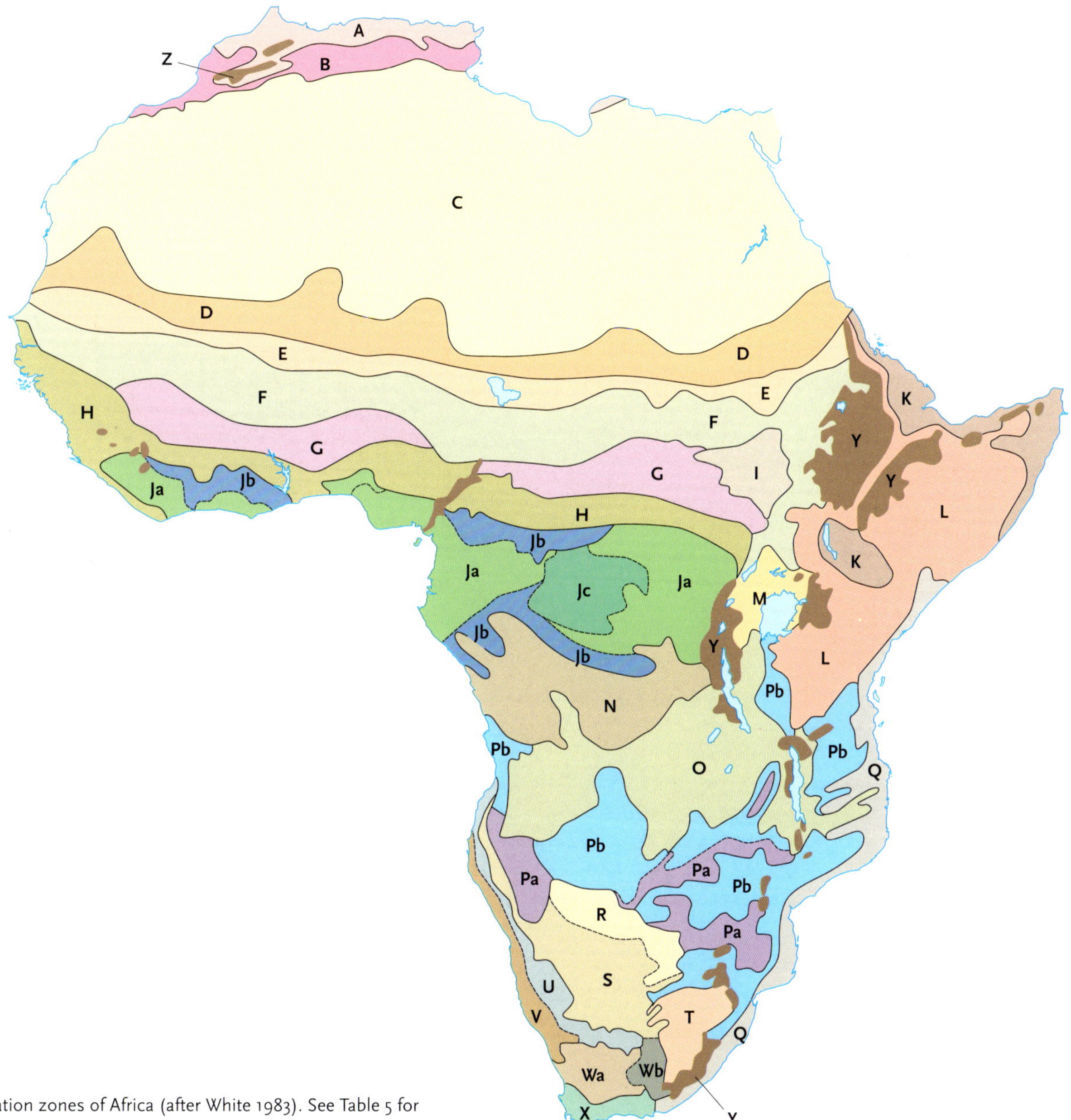

Figure 4 . The vegetation zones of Africa (after White 1983). See Table 5 for explanation of the lettering codes.

highlands and plateaux that are so characteristic of the eastern side of the continent. Overall, however, there is considerable similarity in environments at equivalent latitudes in both the northern and southern hemispheres of the continent. Hence it is not surprising that ecologically equivalent animals (though of different species) occur in both hemispheres in equivalent biotic zones; this is especially noticeable in the drier savannas and arid environments.

Thirteen biotic zones, and one azonal biome, are recognized; brief details of each are given below. In each zone, four principal characteristics are described: plant structure and composition, plant biomass and production, rainfall and water availability, and temperature. The biotic zones are mapped in Figure 3, and the climatic characteristics of each is represented in Figure 2. Further information on each biotic zone and its vegetation is provided in Table 1 and Figure 4. The biotic zone (or zones) where each mammalian species is known to occur is recorded in the 'Distribution' section of each species profile. The term 'biotic zone' is abbreviated to BZ in each profile.

(1) Mediterranean Coastal Biotic Zone

(White's Mediterranean Regional Centre of Endemism)

The Mediterranean Coastal zone occurs only north of about 31° N in Morocco and 33° N in Algeria and Tunisia. A small outlier occurs in Jebel el Aktar (formerly Cyrenaica) in Libya. The climate is typically Mediterranean, with hot, dry summers during which no rain may fall for 4–5 months (May to Aug–Sep). The winters, by contrast, are cool and wet. Most of the rain (700–1000 mm) falls in winter (Nov–Mar), when the temperature may drop to around freezing at night. The coastal plain may originally have been covered by forest in which *Pistachia atlantica* and *Celtis australis* were probably important, with evergreen oak *Quercus ilex* and cork oak *Q. suber* occupying the drier sites. Degradation of the forests by repeated clearance and an increase in fire frequency produces a shrubby thicket in which most of the species, such as mastic tree *Pistachia lentiscus* and wild olive *Olea europaea*, have small tough leaves.

Further inland is a chain of mountains and plateaux – the High Atlas, Middle Atlas and Saharan Atlas (collectively called the Maghreb by local people) – which stretches for about 2000 km, bordered on the north by the coastal plain and on the south by the Sahara Desert. Most of this land is over 1500 m, parts are above 3000 m and the highest peak (in Morocco) is 4165 m. The climate in the mountains follows the same seasonal pattern as on the plains, but in the winter there are frosts at night for 6–8 months of the year, and snow occurs on the higher peaks (and hence is more similar to that of the Afromontane–Afroalpine Biotic Zone). Above the lowland forests, from about 1400 m, is a belt of coniferous forest in which North African Cedar *Cedrus atlantica* and Aleppo Pine (*Pinus halepensis*) may dominate. Above the treeline (around 3000 m) there is often a zone of low spiny cushion-forming shrubs ('hedgehog heath').

The phytogeographic affinities of North Africa are generally with Europe and the rest of the Mediterranean basin, rather than with tropical Africa. Although parts of the central Sahara were considerably wetter several thousand years ago (so that at least some mammals were able to cross it), the great differences between the flora north and south of the desert suggest that it has long been a substantial barrier to movement of plant species.

(2) Sahara Arid Biotic Zone

(White's Sahara Regional Transition Zone)

The arid zone of the Sahara Desert stretches from the Atlantic Ocean to the Red Sea, and north to the Mediterranean in Libya and Egypt. Its physical characteristics are extremely varied and include ergs (seas of sand with little or no vegetation where the sand is mobile), hamadas (wide plateaux covered with small stones or boulders), regs (smooth sand or clay plains covered with fine gravel) and rocky massifs. Surface water is virtually absent, but scattered throughout the Sahara are natural oases where ground water reaches the surface. Originally these probably supported a shrubby plant community in which the doum palm *Hyphaene thebaica* and species of tamarisk *Tamarix* may have been prominent, but have now been replaced by planted date palms *Phoenix dactylifera*. Pools of water support typical marsh plants such as reed-mace or cat-tail *Typha latifolia* and reed *Phragmites australis*.

The climate throughout the whole of the zone is extremely harsh. Temperatures vary greatly annually and diurnally. In summer, air temperatures are typically 35–45 °C during the day and 15–20 °C at night; in winter, the night temperatures may be close to 0 °C. Rainfall is sparse, unpredictable and variable throughout, generally averaging 75 mm or less per year. On the northern margin of the zone, rain falls during the winter (Jan–Mar), and towards the southern edge, during the summer (Aug–Sep). In the central Sahara, between 18 and 30° N, the mean annual rainfall is less than 20 mm. Average figures are, however, almost meaningless because there may be no rain for years and then the equivalent of several years of the 'average annual rainfall' can fall in a few hours. When this happens, normally dry wadis flow for a few hours or days, and the water percolates deeply into the sand. The seeds of ephemeral desert plants such as *Neurada procumbens*, and grasses such as species of *Aristida*, germinate to give a temporary flush of green. This type of vegetation is important to animals because it dries quickly, retaining much of its nutritive value, and remains for years as 'standing hay'. Perennial plants are restricted to places where water accumulates. Shrubs and trees are rare and restricted to well-watered sites such as outwash fans. Here species of acacia (*Acacia tortilis*, *A. ehrenbergiana*) are often accompanied by the perennial grass *Panicum turgidum*. Phytogeographically, the Sahara is a transition zone. As one moves south, species of Mediterranean affinity, such as annual species of *Astragalus* (Leguminosae) are gradually replaced by species of African affinity, such as *Zornia glochidiata* (Leguminosae).

In the eastern Sahara, the Nile R. provides water throughout the year to a narrow strip of fringing vegetation and irrigated crops. The strip varies in width through the year from a few metres to a few kilometres with the annual rise and fall of the river dependent upon rainfall further south. Many species which live in mesic habitats to the north and south of the Sahara are found close to the Nile. The strip of vegetation along the river also acts as a passageway for birds that migrate between Europe and Africa.

(3) Sahel Savanna Biotic Zone

(White's Sahel Regional Transition Zone)

The Sahel Savanna, like the Sahara Arid Biotic Zone, stretches from the Atlantic to the Red Sea. It forms an intermediate zone between the arid environments of the Sahara and the mesic environments of sub-tropical Africa. The climate is semi-arid, with rain during the hot summer. The climate is very hot for most of the year (especially Mar–Jun), cooler during the wet season (see below) and coolest during 'winter' (Nov–Feb). There is considerable diurnal temperature fluctuation, especially during the winter. Annual rainfall varies from 100 mm falling in 1–2 months (Aug–Sep) at the northern edge, to 350–400 mm falling in 3–4 months (Jul–Oct) at the southern edge. Interannual variability in rainfall is high, but less so than in the Sahara. The variations are linked to the well-known El Niño – Southern Oscillation climatic cycles. There is considerable evidence for a long-term decline in the rainfall of the Sahel. The landscape is fairly featureless over much of the zone, with the exception of isolated mountains like Jebel Marra in Sudan.

The vegetation is mainly a sparse woodland or wooded grassland in which species of acacia (*Acacia tortilis*, *A. laeta*), *Commiphora africana* and members of the caper family (Capparidaceae) are the commonest woody plants. Most of the grasses are annuals; cram-cram *Cenchrus biflorus* and *Schoenfeldia gracilis* are common species. The perennial grass *Andropogon gayanus* has been much reduced by cultivation and overuse. Fires are relatively uncommon because there is usually insufficient fuel to support them.

(4) Sudan Savanna Biotic Zone

(the northern part of White's Sudanian Regional Centre of Endemism)

The Sudan Savanna stretches from the Atlantic Ocean to the Ethiopian Highlands and to northern Uganda. The climate is less arid than in the Sahel Savanna Biotic Zone. Seasonal temperature variations are not so extreme (the mean annual temperature lies between 24 and 28 °C), average annual rainfall is higher (500–800 mm), the wet season lasts for 4–6 months (May–Oct) and there is less annual variability in rainfall. Some months may have more than 100 mm of rain. Annual rainfall increases from north to south, as in the Sahel Savanna Biotic Zone. These slight differences in the climate from that of the Sahel Biotic Zone allow higher and more regular plant productivity.

Because of its higher and more reliable rainfall, the Sudan Savanna Biotic Zone is much more intensively farmed than in the Sahel to the north. Large areas are almost completely cleared of natural vegetation, leaving a 'farmed parkland' where useful trees such as *Parkia biglobosa*, baobab *Adansonia digitata* and apple-ring acacia *Faidherbia* (*Acacia*) *albida* are retained among the crops of maize, millet, sorghum and groundnuts.

Vegetation destruction by cultivation has been so widespread in this region that it is very difficult to be certain of the nature of the original vegetation. As far as can be ascertained from relict patches, it is essentially transitional between the Sahel to the north in which most of the trees have finely divided leaves, and the Guinea Savanna to the south in which most trees have broad undivided leaves (Clayton 1963). Areas in which the soils are too poor and stony for satisfactory cultivation often support sparse woodland where *Combretum* spp. and *Anogeissus leiocarpus* are usually the prominent trees. Grazing and cultivation may lead to the elimination of *Anogeissus* and its replacement by a low *Combretum*-dominated scrub. Other heavily used areas can develop into dense *Acacia* thicket during the fallow period. Seasonally wet valleys are characterized by the shrub *Mitragyna inermis* and a ground flora of moisture-loving grasses and sedges.

(5) Guinea Savanna Biotic Zone

(including the southern half of White's Sudanian Regional Centre of Endemism and the northern parts of his Guineo-Congolia/Sudania Regional Transition Zone)

The Guinea Savanna Biotic Zone is the band of savanna that lies immediately to the north of the rainforest. It extends from the edge of the rainforest in the west of West Africa to north-western Uganda. Some typical trees of the Guinea Savanna include *Daniellia oliveri*, *Isoberlinia doka* and *Cassia sieberiana*. East of the Nile in Uganda there are extensive woodlands, sometimes referred to as 'Tall grass–low tree savannas'. Although similar in physiognomy, these woodlands lack the species listed above; instead, the dominant trees are the shea nut tree *Vitellaria* (*Butyrospermum*) *paradoxum* and species of *Terminalia* and *Combretum* (which are also found in the true Guinea Savanna). This zone, together with the Sudan and Sahel Savanna Biotic Zones, may be collectively referred to as the 'northern savannas'.

The climate of the Guinea Savanna Biotic Zone is wetter than further north. Mean annual rainfall is 800–1500 mm and the wet season lasts for 7–8 months (Mar–Oct), sometimes with a 'short dry season' in July and August. The usual vegetation is open woodland, with a dense grass layer. The commonest grass species mainly belong to the genus *Hyparrhenia* and attain a height of 2–3 m by the end of the wet season. As the dry season progresses they wither and die. Intense annual fires are usual in much of this zone; fire reduces most of the grass to ash or charred stems, and stimulates the growth of new grass before the next wet season. This new grass provides nutritious food for herbivores at a time when the grass in unburnt areas is unpalatable and poor in nutrients.

Most trees are 10–20 m tall, with a few up to 30 m. Most of the trees have thick fire-resistant bark, and at least some have seed germination mechanisms that bury the growing point in the soil where it is protected from the heat of fires (Jackson 1974). Two forms of Guinea Savanna can be distinguished: northern and southern (Keay 1959b). The southern part is characterized by trees such as *Daniellia oliveri*, *Lophira alata* and *Afzelia africana*. All three species have similar 'sister species' in the rainforests. The northern part is characterized by trees such as *Isoberlinia doka*, *I. dalzielii* and *Monotes kerstingii*. These northern Guinea Savannas are physiognomically similar to those of the Zambezian Woodland Biotic Zone south of the Equator.

The balance between woodland and rainforest in the southern part of the Guinea Savanna Biotic Zone, close to the forest margin, is a delicate one. Long-term experiments (see Lock 1998 for a list and summary) show that burning early in the dry season, before the grass layer is completely dry, damages the trees rather little and allows the survival of tree seedlings, so that tree density slowly increases, even to the point where grasses are partially suppressed so that fires become much less intense. If there is a nearby source of seed from forest tree species, they may begin to invade and the vegetation will begin to change to forest. The complete exclusion of fire has a similar but faster effect, but is in practice difficult to achieve. On the other hand, fires that occur late in the dry season when the grass layer is fully dry are much hotter, often damage adult trees and also kill their seedlings, so that the balance of the vegetation shifts from trees to grassland. Further north, far from the forest margin, similar but much less marked changes occur; here, sources of forest tree seeds are usually absent, and the main change in the absence of fire is an increase in the density of savanna trees.

This region, intermediate between rainforest and savanna, with vegetation whose nature probably depends on human intervention (particularly fire frequency and timing) over a long period, is often called 'Derived Savanna' or 'Forest–Savanna Mosaic'. It is a very extensive area which surrounds the Rainforest Biotic Zone, but it is not a Biotic Zone in itself; here it is referred to as the Northern Rainforest–Savanna Mosaic, Eastern Rainforest–Savanna Mosaic and Southern Rainforest–Savanna Mosaic (labelled as 6a, 6b and 6c in Figures 2 & 3, and Table 1). It provides a mixture of forest and savanna habitats, often with sharp boundaries so that the plant species composition changes almost completely over a hundred metres (or less). Over the last few millennia the position of this boundary has certainly changed a good deal and its position continues to be labile. Swaine *et al.* (1976) show that, in Ghana, the nature of the mosaic may be linked to the underlying rocks. On sandstones, the forest extends into the savanna along streams in a dendritic pattern. On finer-grained rocks, there is often forest along streams but also on hilltops, with savanna restricted to the intervening slopes. The Mosaics contain a mammalian fauna comprising selected species from the Rainforest Biotic Zone (which live in the rainforest habitat) and from the Guinea Savanna Biotic Zone (which live in the savanna habitat); the mosaics are as important biogeographically as the zones themselves.

Forested habitats in the Rainforest–Savanna Mosaics contain a mixture of rainforest and savanna trees; they occur, in ribbon-like fashion, along the edges of rivers and streams (often far from the edge of the rainforest), and are variously referred to as 'riverine forest', ' riparian forest' or 'gallery forest'. Some occur as isolated patches (or islands) completely surrounded by savanna and are referred to as 'relict forest'.

(6) Rainforest Biotic Zone

(White's Guineo-Congolian Forest, with outliers in his Lake Victoria Regional Mosaic and Zanzibar–Inhambane Regional Mosaic, and elsewhere; montane forest is dealt with under the Afromontane Biotic Zone)

The Rainforest Biotic Zone extends for about 4500 km across West and central Africa, eastwards into western Uganda and western Tanzania. There are also areas of forest in the coastal regions of eastern and south-eastern Africa. Many wide rivers pass through the zone, and mammalian communities on one side of a river often differ from those on the other side, suggesting that some of these rivers may be important barriers to the movement of some species. On the basis of their mammalian fauna (as well as other faunas), the rainforest may be divided into regions and sub-region (Happold 1996; Figure 5).

The climate of the Rainforest Biotic Zone is warm and humid. The annual mean maximum temperature is about 30 °C and the minimum is about 20 °C; the daily range is only a few degrees, although this diurnal variation is greater than the annual between-month variation. The annual rainfall is 1600–2000 mm – rather low in comparison with areas of tropical rainforest in other parts of the world. Two areas on the edges of the zone receive a higher rainfall – a large area of Cameroon near Mt Cameroon receives more than 3000 mm per annum, and at least one small area receives 10,000 mm. Here there is a short and irregular dry season of 4–6 weeks in December and January. In Guinea, Sierra Leone and Liberia there is an area along the coast that receives 3000–4000 mm of rain annually; this falls during a wet season lasting 8–10 months, with a marked dry season between December and March. In the Congo Basin, and along the West African coast, there tend to be two rainfall peaks separated by two short periods of 1–2 months (Dec–Jan and Jun–Jul) of lower rainfall. At the wettest localities on the coast, and throughout much of the forests of the Congo Basin, rain falls in every month of the year with more than half the months recording more than 100 mm each month.

Although the general climate of the rainforest is uniform, there are important variations at different levels within the forest. Absolute temperatures, diurnal fluctuations in temperature and humidity, wind flow and light levels are much greater in the canopy than near the ground. In the understorey, temperature may not vary by more than a degree or two over 24 hours, humidity is uniformly high and air movement is minimal. Such differences effectively produce a variety of environments within a small area, and may go some way to explaining the great diversity of species in rainforest.

Forest vegetation is generally multilayered. The largest trees, 60–100 m tall and with broad crowns, project above the general canopy and are known as emergents. The main canopy trees, also with broad crowns, form a continuous upper layer, generally at 30–40 m above the ground. Below these are smaller trees with more vertically elongated crowns. Some of these are young individuals that may eventually reach the canopy, but some remain small even at maturity. Closer to the ground there are pygmy trees (with a single main stem) and shrubs (branched into several stems at the base), often 1–3 m tall. In undisturbed forests there are few herbs growing on the forest floor, and it is generally easy to walk. Disturbance in the canopy lets in more light and encourages the growth of herbs and young woody plants, so that walking can be difficult. Other plants that are important in the rainforest are the lianas (woody climbers). These often ascend to the crowns of the tallest trees, relying on them for support and linking the crowns of adjacent trees one to another. Epiphytes are frequent in rainforest, attaining their greatest abundance in the wetter forests, particularly those that have only a short dry season, or that grow in sites where mists are frequent. Many of the epiphytes are ferns or orchids, relying on their host tree only for support and drawing no nutrients from it.

Another tree growth habit is that of the strangler. In Africa, most stranglers are figs (*Ficus* spp.), whose fruits are an important food for many mammals. The seeds of figs germinate above the ground, often where a branch forks from the main trunk. The plant grows as an epiphyte at first, but also produces long roots that eventually reach the ground. Once rooted in the ground, the fig grows faster, and its roots enlarge to form a network around the trunk of the host. Eventually the host tree dies, leaving the fig as a free-standing tree with a hollow trunk.

The tree flora of the rainforest is extremely diverse. It is also different from that of Asian and South American rainforests. The family Dipterocarpaceae, which is well represented in South-East Asia, is virtually absent from Africa. The only group that can be regarded as characteristic of African forests is the subfamily

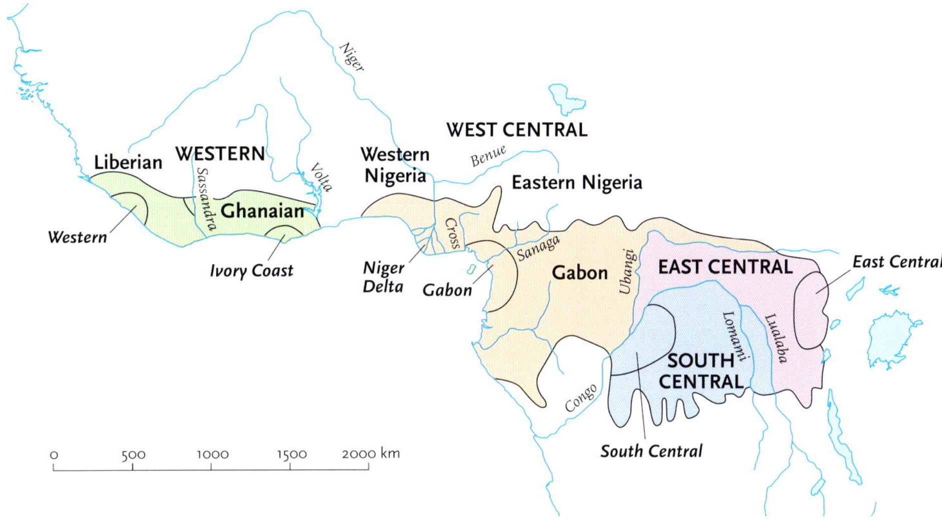

Figure 5. The Rainforest Biotic Zone showing the regions, sub-regions and refugia. Regions are indicated in capital letters; sub-regions are indicated in lowercase letters; refugia are indicated in lowercase italics (after Happold 1996 and references therein).

Caesalpinioideae of the Leguminosae. Members of this subfamily are common both in the canopy and in the understorey, and in some areas they form almost monospecific stands. The forests of Uganda ironwood (*Cynometra alexandri*) in western Uganda and the Ituri region of DR Congo, and *Gilbertiodendron dewevrei* in western DR Congo and southern Cameroon, are examples. Apart from these single-species dominated forests, it is not particularly meaningful to distinguish the normal and more diverse forest types by using the names of trees that are often rare. Classifications such as that of Hall & Swaine (1976, 1981) used moisture status, combined with the degree of deciduousness (e.g. Wet Evergreen; Dry Semi-Deciduous). Hall & Swaine (1976, 1981) found the greatest plant diversity in the wettest forests and the lowest in the driest; they also showed that their Ghanaian forests were less diverse than those of South-East Asia and this is true of African forests in general. They also found a relationship between tree height and rainfall, with the trees in somewhat drier and more seasonal forest type being tallest, and both drier and wetter forests having a lower maximum tree height. This may be due to nutrient availability; in areas of high rainfall many of the mineral nutrients in the soil are leached out.

Gaps caused by tree falls, as well as those caused by clearance for cultivation, occur throughout the rainforest. Old trees fall, often in the strong winds that accompany storms at the beginning of the wet season. Because tree crowns are often tied together by lianas, the fall of one tree may bring down others to produce a substantial gap. Initially the vegetation of gaps is dominated by quick-growing herbs, often of the ginger family (Zingiberaceae). This lush vegetation is attractive to elephants as well as to gorillas and other primates, and their continued foraging visits may prolong the life of the clearing. In general, however, colonization by fast-growing tree species such as *Musanga cecropioides* and *Trema orientalis* is rapid; both have fruits that are eaten by primates, birds and bats so that their seeds are distributed widely over the forest. These trees are, however, relatively short-lived (perhaps 20–30 years), and are rapidly replaced by slower-growing species that can regenerate in shade. Within a century, a former gap will usually be indistinguishable from its surroundings.

Within the Rainforest Biotic Zone, seasonally flooded or waterlogged valleys often support swamp forests in which species composition is completely different from that of dry-land and upland forests. Palms, including species of *Raphia*, and the climbing rattan, *Calamus deeratus*, are often common. Both have abundant fruits that are eaten by many animal species. Other specialized forest types include those found on and around granite inselbergs – isolated rounded rock outcrops – within the Rainforest Biotic Zone. Here, species characteristic of drier regions can often be found far outside their normal range, just as rainforest tree species occur around springs within the Guinea Savanna Biotic Zone, often far from the Rainforest Biotic Zone. Finally, there are mangrove forests, regularly inundated by seawater; these usually have only a few species, mostly belonging to the genera *Rhizophora* or *Avicenna*.

Seasonal changes in rainforest are not as great or as obvious as in the savanna, but most species of plants show seasonal cycles in relation to slight variations in climate. Deciduousness, in particular, is often used to describe forest types but in fact there is a complete gradation from fully evergreen to almost fully deciduous forests. Hall & Swaine (1981) found that even in the driest forest types the canopy trees were fully deciduous. Species of plant vary in their times of leafing, flowering and fruiting so that there is generally some form of plant food available for animals throughout the year. However, the few detailed studies of forest phenology (e.g. Tutin 1998) show not only that there are large variations in fruit availability in each month of the year, but also that this availability may vary greatly from one year to the next. Thus animals that rely on fruits for food will have to switch from the fruits of a particular suite of species in one month to those of other species in the next month. Swaine & Hall (1986) found in a Ghanaian forest that species with fleshy fruits showed less seasonal fluctuation in fruiting than those species with explosive or wind-dispersed fruits.

Recent destruction of rainforest by humans has resulted in many 'savanna-like' habitats within the Rainforest Biotic Zone, often linked to each other by roads and urban areas. These are often dominated by vigorous-growing elephant grass (*Pennisetum purpureum*) and spear grass (*Imperata cylindrica*). These, as well as the natural treeless patches sometimes found around rock outcrops, can provide habitats for some savanna species of animals and plants within the Rainforest Biotic Zone.

(7) Afromontane–Afroalpine Biotic Zone

(White's Afromontane Archipelago-like Regional Centre of Endemism, and Afroalpine Archipelago-like Region of Extreme Floristic Impoverishment)

This biotic zone is found on the mountains of Africa south of the Sahara. It differs from all other zones because it is broken up into a series of widely dispersed areas. In West Africa these are few and far between – the Fouta Djalon in Guinea, Mt Nimba on the Liberian border with Guinea and Côte d'Ivoire, Bioko I. (Equatorial Guinea) and Mt Cameroon and associated highlands to the north in Cameroon. The most extensive highland area in Africa lies in Ethiopia. The Ethiopian Plateau is volcanic in origin and lies mainly between 1500 and 3000 m, although some massifs reach 4000–4300 m. It is divided into eastern and western halves by the Eastern (or Gregorian) Rift Valley, and numerous deep and spectacular gorges cut into the margins of the plateau. The streams originating on the plateau are the sources of several large rivers – the Blue Nile, the Awash, the Juba and the Wabe Shebelle – that flow into the arid lowland environments that surround the plateau. Moving south, most of the highland areas are volcanic and associated with the Rift Valley. Mount Kenya, Mt Elgon, the Aberdare Range and Mt Kilimanjaro are close to the Eastern (Gregorian) Rift Valley. Mountains associated with the Western (or Albertine) Rift include the Virunga Mts and the Kungwe–Mahali Mts and the mountains bordering L. Kivu and L. Tanzania. The Rwenzori Mts on the Uganda–DR Congo border, although associated with the Western Rift Valley, are not volcanic. In eastern Tanzania there are also several massifs that are neither volcanic nor associated with the Rift Valley; these include the Usambara, Uluguru and Uzungwa Mountains, all rich in endemic animals and plants and collectively referred to as the Eastern Arc Mts. The Livingstone Mts in south-west Tanzania, and the Nyika Plateau (both associated with the Gregorian Rift Valley in Malawi) and Mt Mlanje (a large non-volcanic mountain, not associated with the Rift Valley, in south-east Malawi), may form 'stepping-stones' between the mountains of eastern Africa and the Drakensberg Range and the Knysna highlands of South Africa. Finally, although they barely attain 2500 m, the Angolan highlands near Huambo are rich in endemic plants and birds.

The environment of the Afromontane zone differs in a fairly predictable way from the lowlands. Mean temperature falls with increasing altitude at about 0.5 °C per 100 m. In the clear air of the high mountains there are very large diurnal fluctuations, such that the climate of the alpine zone (see below) has been described as 'summer every day and winter every night'. Above about 4000 m, snow is present throughout all, or most, of the year. Glaciers exist on the Rwenzori Mts, on Mt Kenya and on Mt Kilimanjaro. Rainfall increases with altitude up to a certain level (generally corresponding to the height normally attained by the daily cloud cover) and then declines. For example, the saddle between the two peaks of Mt Kilimanjaro has a very low annual rainfall of around 150 mm and is largely barren; it has been described as a high altitude desert (Greenway 1965). Within the montane forest zone (see below) precipitation occurs partly by rainfall, but to a greater extent by the interception of moisture droplets from mist.

Because of the change in climate with altitude, the vegetation of mountains is zoned. The position of each vegetation zone varies on each mountain depending on the height, aspect and latitude of the mountain. A particular vegetation zone may not occur on all mountains, and it may have different altitudinal limits on different mountains. Different aspects of the same mountain may show differences in the altitudinal limits of the vegetation zones.

Montane forest generally begins at about 1500 m but this is an arbitrary limit. As a rule, montane forest is shorter in stature than lowland rainforest. Lianas are often more abundant than in the lowlands, as are epiphytes because of the greater incidence of mist in montane forests. Characteristic tree genera of montane forest include *Podocarpus*, *Ilex*, *Ocotea*, *Nuxia* and *Olea*; many of these are widespread, occurring from the Cape (Knysna) to Ethiopia. On dry mountains such as those of northern Somalia and Kenya, there may be juniper *Juniperus* forests.

Above the montane forest there may be a zone dominated by bamboo *Arundinaria alpina*. Few other species grow with the bamboo, and the tangle of fallen stems makes them difficult to traverse. Bamboo flowers gregariously over large areas and then dies, replacing itself by seed.

Above the bamboo, or above the montane forest, there is a zone of thicket or small trees commonly called the giant heath zone. The commonest plants belong to the genus *Erica* (including the species formerly placed in *Philippia*), and some species may be 7 m tall. This zone is moist and misty, and the trunks and branches of the trees, and the ground, are usually covered with thick mats of mosses and liverworts.

The uppermost vegetation zone, the Afroalpine zone, is physiognomically remarkable, and is characterized by the giant lobelias *Lobelia* and giant groundsels *Dendrosenecio*, which are characterized by their huge leaf rosettes. Most *Lobelia* species do not form a tall trunk, although their inflorescences can be up to 5 m tall. Some *Dendrosenecio* species, on the other hand, form substantial trunks up to 6 m tall, covered either by a thick layer of dead leaves, or thick corky bark. The ground vegetation is varied, with grasses, broad-leaved herbs and small shrubs. Many of these belong to, or are closely related to, temperate genera.

The vegetation of the high plateaux of Ethiopia generally follows the patterns outlined above. However, the extensive areas of fairly level ground, the fertile soils developed from volcanic rocks, and the fairly reliable rainy seasons have encouraged human settlement and cultivation. In some parts, the natural vegetation has now been cleared and replaced with a patchwork of small fields of cereal and pulse crops. The remaining patches of natural vegetation can be found along streams and marshes, on rocky hills and on steep slopes and cliff ledges.

(8) Somalia–Masai Bushland Biotic Zone

(White's Somalia–Masai Regional Centre of Endemism)

This zone is centred on the Horn of Africa. To the north it is bounded by the foothills of the Ethiopian Highlands. To the south it merges into the Zambezian Woodland zone and to the west it bounds on to the moister woodlands of northern Uganda – an impoverished extension of the Guinea Savanna Biotic Zone (see above). A thin extension of the zone stretches northwards between the Red Sea and the Ethiopian Highlands and joins the northern part of the Sahel Biotic Zone. There is growing evidence that the Horn of Africa has a long history of both aridity and isolation, and both its flora and fauna are rich in endemics. It is sometimes distinguished as the Somali Arid Biotic Zone. The landscape varies between hot arid plains, rugged barren hills and dry savannas. The mountains along the northern edge of the Horn are an extension of the Ethiopian Highlands, but are lower and drier, with a tendency to winter rains. Southwards and westwards this region includes large areas of grassland and wooded grassland at altitudes of up to 1700 m; these areas include some of the major national parks of East Africa (e.g. Amboseli N. P., Tsavo N. P., Serengeti N. P.). These may appear very different to the much drier bushlands of the Horn, but are linked to them by a continuum of vegetation types and by a similar climate. They are, however, poorer in both endemic plants and endemic mammals.

The main climatic characteristic of this region is the unpredictability of the rainfall. Its equatorial position means that there are usually two wet seasons (Mar–May and Sep–Nov) but either or both of these can and often do fail. In general, the rainfall declines in both total quantity and reliability northwards and eastwards. The extreme tip of the Horn (Cape Guardafui) has a mean rainfall of 21 mm, and much of lowland northern Somalia has annual means of less than 200 mm. Further inland and at higher altitude, the rainfall may be much higher, with up to 1000 mm at Nairobi.

The dry bushlands of the Horn are dominated by two tree genera – *Acacia* and *Commiphora*. Both have many species in this region. *Commiphora* spp. are characteristic of the drier parts of the region. For much of the year they are grey and leafless, but they rapidly produce leaves as soon as there is rain. Acacia species are especially widespread and several form vegetation communities that are easily recognized and characteristic of the region, particularly the more mesic parts outside the Horn. The whistling thorn *Acacia drepanolobium*, with its swollen spine-bases ('ant-galls'), occurs in shallow seasonally waterlogged valleys on black soil. *Acacia tortilis* forms the flat-topped trees scattered in the open grasslands. *Acacia xanthophloea* is the yellow-barked 'fever-tree' that occurs beside rivers and lakes. The vegetation of the drier parts of the Horn tends to be heavily grazed. Perhaps in response to this, numerous plant genera that are herbs elsewhere in Africa have woody representatives in the Horn, and many species are spiny. Grasses in this part of the region tend to be annual, and appear after rain. In the highland and more mesic regions, however, the open grasslands are dominated by perennial species, particularly the red oat-grass *Themeda triandra*. This species has awned seeds that bury themselves in the ground (Lock & Milburn 1971), and it appears to be encouraged by regular fires.

The characteristics of the vegetation of this region that most affect animals arise from the unpredictability of the rainfall. A good wet season produces copious grass and a good flush of leaves and young stems on the trees and shrubs. There may, however, then be a period of many months with little or no rain during which animals either have to subsist on what is left or migrate. Trees and shrubs tend to carry nutritious leaves and young growth for longer than grasses and it is perhaps no coincidence that many of the larger herbivores in this area are browsers. In the wetter upland grasslands, the higher biomass of grass is able to support a high biomass of animals. In Serengeti N. P., the characteristic large herds of herbivorous mammals have regular migrations. The herds travel from area to area, exploiting each area when the grasses have attained their most nutritious stage of growth. As a result, the grasslands can support a higher biomass of animals than if the herds stayed at the same place throughout the year.

The Somalia–Masai Bushland Biotic Zone is assumed to have expanded and contracted during the dry periods of the Quaternary. The drier parts of the savanna south of the Ethiopian Highlands now merge into the easternmost parts of the Guinea Savanna Biotic Zone. When the rainforests were more extensive, they may have reached almost to the foothills of the Ethiopian Highlands, and would have blocked any movement of savanna and arid-adapted animals between northern and southern savannas. During dry periods, there was a wide band of savanna, perhaps rather dry (the so-called 'dry corridor' or 'drought corridor'), which linked the northern and southern savannas. In the past, the opening and closing of the 'dry corridor' helps to explain the present disjunct distributions of some species of savanna mammals.

(9) Zambezian Woodland Biotic Zone

(White's Zambezian Regional Centre of Endemism)

The savanna woodlands of Africa south of the Equator cover a vast area. They form the 'southern savannas' and are the southern equivalent to the Guinea and Sudan Savannas north of the Equator. To the north they interdigitate with the forests of the Congo Basin; to the north-east they merge gradually into the Somalia–Masai bushlands and grasslands and to the south they grade into the drier shrublands and grasslands of the South-West Arid Biotic Zone and the upland grasslands of the Highveld. There are a number of highland 'islands', as well as several large lakes, associated with the Rift Valley, and extensive areas of seasonal and permanent swamp in the area of the headwaters of the Congo and Zambezi rivers. Soils in the southern part of the region are particularly poor because they are underlain by Kalahari Sands. These are wind-blown deposits dating from the cold dry stages of the Quaternary period, when arid conditions extended much further north than they do now.

The climate is remarkably uniform throughout the Zambezian Woodland Biotic Zone. There is a single wet season that lasts for 5–7 months (Oct–Apr). Annual rainfall is generally 700–1200 mm. The dry season lasts for 5–7 months and coincides with the cooler months of the year. The seasonal pattern of rainfall is thus similar to that of the Guinea and Sudan Savannas, but displaced by six months. Temperatures in the southern savannas are strongly influenced by altitude and season. In the north, the daily minimum and maximum temperatures in *Brachystegia* woodland (Liwonde, Malawi, 500 m) are 21–33 °C in the hot dry season, 15–27 °C in the cool dry season and 23–30 °C in the wet season. These contrasting seasons each provide different environments for plants and animals; in the wet season, water is freely available and plant growth is rapid, but in the dry season water is scarce or unavailable and plant growth is limited. Some plants survive as seeds but the majority of the grasses and herbaceous plants survive as underground parts. Fires are common during the dry season, destroying much of the above-ground parts of the grass.

The vegetation that is most widespread in this region is woodland, often referred to by its local name of 'miombo'. The most abundant and widespread components of this woodland are species of *Brachystegia* and *Isoberlinia*, both trees belonging to the subfamily Caesalpinioideae of the Leguminosae. They generally form a light but fairly continuous canopy at 15–20 m. On the Kalahari Sands, however, there is a more mixed woodland from which *Brachystegia* and *Isoberlinia* are often absent and in which *Cryptosepalum pseudotaxus* and *Baikiaea plurijuga* are often prominent. In the areas of highest rainfall, a woodland physiognomically similar to Guinea Savanna and locally referred to as 'chipya' is widespread. Again, *Brachystegia* and *Isoberlinia* are absent, and the tree assemblage is more varied, including *Burkea africana*, *Parinari curatellifolia*, *Pericopsis angolensis*, *Pterocarpus angolensis* and *Combretum* spp. The ginger *Aframomum alboviolaceum* and bracken fern (*Pteridium aquilinum*) are widespread and characteristic, growing among the grasses.

The topography, developed for the most part on an ancient land surface, tends to be gently undulating, and the different parts of the landscape bear different vegetation types. Miombo woodland occupies the gentle slopes and plateaux of this region. The broad, shallow, seasonally waterlogged valleys, with black clay soils, have either grassland or sparse and open woodland of small *Acacia* trees standing in tussocky grasslands with numerous small ephemeral herbs between the tussocks. The hilltops are often crowned by inselbergs bearing specialized succulent-leaved plants and often surrounded by denser woodland or thicket that benefits from the run-off water from the rocks as well as the mineral nutrients derived from them. A further source of habitat diversity in the woodland is the presence of large termite mounds. Such mounds are common throughout most savanna zones of Africa, but here they are especially large and prominent. They are often covered by dense thickets, frequently less completely deciduous than the main woodland, and also much richer in species. As well as providing a less seasonal food source, the holes at the base of these mounds provide a valuable refuge for animals.

The trees of the main canopy are deciduous. The leaves fall early in the dry season, and as the humidity drops, so the pods on the trees burst open and release their seeds. Some weeks before the beginning of the wet season, the trees flush into new leaf, and many also flower. The ground vegetation begins its main period of growth at this time, although some species flower during the dry season, sometimes apparently stimulated by fire. During the wet season the grasses continue to grow, and most flower late in the wet season.

At the southern, drier edge of this zone, *Brachystegia* and *Isoberlinia* tend to be replaced by *Colophospermum mopane*, another tree in the same family (Fabaceae–Caesalpinioideae), which gives its name to the 'mopane' woodlands. Mopane woodland tends to grow in hotter and drier areas than miombo, and it can tolerate soils in which there are high concentrations of sodium. This is a tolerance rather than a preference – mopane grows better on ordinary soils but it is susceptible to competition from other species. There is usually little grass in mopane woodland and, indeed, if the woodland is opened up (by, for example, heavy browsing by elephants), an increase of grass cover can lead to fire damage to the trees.

(10) Coastal Forest Mosaic Biotic Zone

(White's Zanzibar–Inhambane Regional Mosaic and his Tongaland–Pondoland Regional Mosaic)

This biotic zone is a thin strip of low-lying land extending along the coast of eastern Africa between the ocean and the higher country of the interior. It begins in southern Somalia and ends in KwaZulu–Natal, South Africa. Much of this area is fairly flat but there are isolated hills. The lower reaches of large rivers such as the Tana, Rufiji, Rovuma, Zambezi, Sabi and Limpopo flow into extensive estuaries at the coast. Included in this biotic zone are the offshore islands, of which the largest are Pemba, Zanzibar and Mafia Is.

The climate is warm and humid for most of the year, and strongly influenced by the monsoons of the Indian Ocean. The annual rainfall is 800–1900 mm, according to locality. In the northern part of the area there tend to be two rainy seasons (Apr–Jun and Oct–Dec), but from southern Tanzania southwards there is a single wet season (Nov–May), which lasts for about six months. However, a little rain falls in all months of the year and the effect of the dry season is mitigated by the humidity from the sea. The proximity of the ocean also helps to maintain an even temperature throughout the year.

The vegetation of this biotic zone has been heavily cleared and cultivated for a long time, and little of the original vegetation remains. Many of the surviving fragments owe their preservation to being sacred groves ('kayas') that are preserved by local people for their mystical significance, as well as acting as a source of fuel, poles and medicinal plants. Forest was probably formerly widespread, and the fragments that remain suggest that it was originally a mixed forest. Many of the tree species are endemic to this biotic zone, but many also have close relatives in the Guineo-Congolian forests. Whether this indicates a former connection between these forests and the main Guineo-Congolian forests remains uncertain. Many of the species are endemic to the region; in the forests around Amani, in the East Usambara Mts, at 1250 m, about 40% of the plant species are endemic. The general stature of the vegetation is lower than that in Guineo-Congolian forest, with few emergents exceeding 35 m in height. Further south, the longer dry season leads to floristic impoverishment, so that the forests of Mozambique and KwaZulu–Natal are much less species-rich than those of northern Tanzania and southern Kenya. However, interesting endemic trees have recently been discovered in coastal central Mozambique. Swamp forests occur close to the river estuaries; those dominated by *Barringtonia racemosa* are somewhat salt-tolerant, and occur on the landward side of the true mangrove forests.

The drier areas bear thicket vegetation, spiny and often leafless during the drier parts of the year. In areas where there has been extensive cultivation, the vegetation is a mosaic of tall grasslands, crops and patches of fruit-bearing trees such as coconut *Cocos*, mango *Mangifera* and cashew *Anacardium*. There may also be isolated trees from the original forest, left either for shade or because they have some other use.

(11) South-west Arid Biotic Zone

(White's Karoo–Namib Regional Centre of Endemism, with part of his Kalahari–Highveld Regional Transition Zone)

The South-West Arid Biotic Zone is the southern equivalent of the Sahara Arid Biotic Zone, but because of the influence of the Indian and Atlantic Oceans, aridity is less marked here than in the Sahara where the land masses of Asia and the Arabian Peninsula intercept any rain that may be coming from the east on the trade winds. In southern Africa, the eastern side of the continent receives moist air from the Indian Ocean and is therefore far from arid. On the western side, however, the presence of the cold waters of the Benguela Current close inshore mean that any onshore winds bear little moisture, except as fog. Much of the region is arid throughout the year. As in the Sahara Arid Biotic Zone, the northern and southern edges have different patterns of rainfall. At the northern and eastern edges of the region, rain falls during the summer (Dec–Mar), while at the southern edge it does so during the cool winter season (May–Sep). At the southern margin there may be 200–300 mm of winter rain each year, and at the eastern margin, up to 400 mm of mainly summer rain. Throughout the region, and particularly in the areas with the lowest rainfall, there is great variation from year to year. Temperatures, in general, are not as high as in the arid zones north of the Equator, probably because of the moderating effects of the sea. Daily maximum temperatures range from 25–30 °C during the summer (Nov–Feb) and 10–15 °C during the winter (Jun–Jul). Away from the coast, there are large daily fluctuations in temperature and frosts occur in several months of the year. In the Kalahari Gemsbok N. P., for example, the daily mean maximum and minimum temperatures are 36 and 19 °C respectively in summer, and 22 and 0 °C in winter.

The South-West Arid Biotic Zone may be divided into three subzones: the Namib, the Karoo and the Kalahari. All are arid to varying degrees but in rather different ways. Briefly, the Namib has extremely low rainfall, the Karoo has low and rather unpredictable rainfall and the Kalahari has more rain but very permeable sandy soils that mean that surface water is very scarce. As in the arid zones north of the Equator, the landscape is varied, and includes extensive sand dunes, plains with very low sparse scrubby vegetation and seasonal river beds that contain water after heavy rain and have a more luxuriant vegetation.

The Kalahari Desert subzone (11a) occupies a great basin between the Highveld to the east and the Windhoek Mts to the west. Much of the basin lies between 850 and 1000 m and is largely devoid of hills, mountains or rock outcrops. The whole area is underlain by deep sands of varying age, and the drainage is largely internal. The Okavango R. flows into the northern part of the Kalahari and disappears into the sands. Surface water for drinking is very scarce. Annual rainfall is generally 250–500 mm and falls mainly in the summer. The winters are dry and cold, with frost at night. The natural vegetation of the sandy soils of much of the region is wooded grassland. Most of the grasses are perennial (although overgrazing may alter the balance towards annuals). The trees are mainly species of *Acacia* in the south, with more species with broad leaves (such as species of *Commiphora* and *Combretum*) in the north. Areas with more stony or gravelly soils are usually covered by bushland in which *Tarchonanthus camphoratus* is prominent.

The Namib Desert subzone (11b) forms a coastal strip extending from north-western South Africa through Namibia to the extreme south-west of Angola. For most of its length it is about 100 km wide. Over the whole area, annual rainfall averages less than 100 mm and over much of the area it is very much lower, with less than 10 mm per annum on the coast at Swakopmund. When rains occur, they tend to fall in the winter in the south and in the summer in the north. In the driest areas there may be virtually no measurable rain in some years. Particularly near the coast, this extreme aridity is mitigated by frequent fogs. These form offshore over the cold waters of the Benguela Current and are blown inland during the night. Condensation from these fogs can provide water in quantities sufficient to be useful to animals although it is only useful to plants where it is concentrated in some way, as near rock outcrops. The waters of the Benguela Current are highly productive, with some of this productivity being transferred to the land through the activities of birds and seals. The vegetation of the Namib Desert is extremely sparse. Gravel desert may support many lichens, but higher plants are mostly ephemeral succulents and grasses that appear only after years with exceptionally heavy rain. The extensive sand dune areas are usually virtually free of vegetation. However, drainage lines may carry perennial vegetation, even including trees in the most favoured sites.

The Karoo subzone (11c) occupies the northern parts of South Africa, mainly within the Northern Cape Province. It lies between the winter rainfall zone of the Cape and the summer rainfall of the Highveld and the Zambezian woodlands to the north. Even if the rain falls mainly in the summer or winter, there is often significant precipitation in the other parts of the year, and sometimes the normal rainfall distribution may be reversed. Much of the Karoo lies above 800 m, and frosts may occur during several months in winter. Much of the Karoo is fairly level and has clay-rich soils, which tend to accumulate salts.

The vegetation of the Karoo is mainly a low shrubland. Many of the shrubs and herbs have either succulent leaves or succulent stems. The main family in this vegetation is the Mesembryanthemaceae. These are usually small herbs with thick, succulent leaves, sometimes cryptic – they may be coloured and shaped so that they resemble stones. Flowering takes place mainly in response to rain. There are also scattered larger plants, some of them almost of tree stature; most of these grow among rocks or in other places where some extra water is available from run-off. Among these are species of *Aloe*, including *A. pillansii* that attains 10 m, and the more drought-resistant species of *Acacia* such as *A. redacta* and *A. erioloba*. There are also many annuals (more strictly ephemerals), many of them in the family Compositae, that germinate in response to heavy rain and produce the carpets of flowers for which the southern parts of the region are famous. Grasses are present, but not abundant; it is possible that they have declined with increasing grazing of these areas by domestic stock.

(12) Highveld Biotic Zone

(Most of White's Kalahari–Highveld Regional Transition Zone)
This biotic zone occupies the interior plateau of South Africa. Most of the plateau is gently undulating country of low relief, usually above 850 m, although at its eastern and western margins it reaches nearly 2000 m.

Its inland position means that it is generally a region of low rainfall, with an annual total of about 250 mm in the south-west and about 650 mm in the south-east. The rain falls in a single rather extended summer wet season between November and May. The seasons are well defined and the generally high elevation of the plateau means that average temperatures, even in summer, are relatively low. The mean maximum of the warmest month is 20–25 °C. Frosts occur throughout the zone between April and September, and may be severe.

The vegetation of the Highveld Biotic Zone, between 1200 and 2200 m, is grassland. There are many species, but the most widespread and conspicuous is *Themeda triandra* (red oat grass). It is possible that the dominance of this species has been increased by regular burning, which it withstands well. Heavy grazing can lead to the loss of *Themeda* and its replacement by grasses such as *Aristida* and *Chloris* spp. Species of *Hyparrhenia*, which are characteristic of the warmer seasonal grasslands of the Zambezian Woodland and Guinea Savanna Biotic Zones, are absent.

On the Kalahari Sand, which underlies much of the western and north-western parts of this biotic zone, the characteristic vegetation is wooded grassland. In the drier southern parts, the main trees are species of *Acacia*, with other genera including *Combretum* and *Commiphora* in the wetter northern parts. The main grasses are species of *Anthephora*, *Eragrostis*, *Panicum* and *Schmidtia*.

(13) South-west Cape Biotic Zone

(White's Cape Regional Centre of Endemism)

The South-West Cape Biotic Zone is a small but very distinctive biotic zone occupying the south-west corner of the African continent. In a climatic sense, and in the physiognomy of its main vegetation, it is the southern analogue of the Mediterranean Coastal Biotic Zone. The region is also diverse geologically and topographically, with several mountain ranges trending generally east–west, and a variety of underlying rocks including sandstones (such as that which makes up Table Mountain), granites, shales and limestones.

The climate of the southern and south-western parts of the zone is a winter rainfall season (Apr–Sep), with warm summers and cooler winters. Because of the proximity of the sea, there is some rain in every month and, probably for the same reason, the temperature extremes are not great. Annual rainfall is 400–600 mm in the centre of the zone, but considerably less to the north on the western side. Eastwards, the seasonal rainfall pattern changes, and at East London there is summer rainfall (Oct–Apr). Rainfall increases rapidly with altitude, so that the annual total for Table Mountain (761 m) is 1780 mm while that of Cape Town (12 m) is 630 mm. Frosts occur regularly inland and on the mountains, but not at the coast, and at the highest altitudes snow falls regularly but does not persist.

The Cape flora is extremely rich. Within White's (1983) Cape Centre of Endemism, there are over 7000 plant species (of which over half are endemic) in an area of 71,000 km^2. Some genera, such as the heathers (*Erica*), have undergone prolific speciation (now some 650 species). Many of the very species-rich genera are small shrubs, but geophytes (plants with bulbs and other underground organs) are also abundant.

The main vegetation type of this biotic zone is locally referred to as 'fynbos'. It is a dense shrubland 1–3 m tall, made up of plants with drought-resistant small thick leaves. The best-known components of fynbos are members of the Proteaceae, particularly *Protea*, as well as *Leucadendron* and *Leucospermum*. Members of the Ericaceae, Leguminosae-Papilionoideae and Compositae are also prominent. Grasses are not an important component of the vegetation now, but it has been suggested that they were commoner before the introduction of domestic stock. Many of the grass-like plants belong to a completely different family, the Restionaceae. Fynbos is very vulnerable to invasion by species from other parts of the world with similar climates: Australian species of *Acacia* and Mediterranean species of *Pinus* are particularly invasive.

Fire is probably an essential factor in the maintenance of fynbos. Although a burned area may appear devastated after a fire, many species sprout from the base of the stems, or germinate from seeds that are protected within fruits or inflorescences and are released when a fire occurs. Other species flower in response to a fire. Many species are now known to lose vigour or even die if they are not burned.

(14) Aquatic environments

Although not a biotic zone as such, aquatic habitats in Africa are an important ecological entity for mammals and other animals. Most of the rivers of Africa have a fringe of distinctive vegetation. In many savanna habitats, long strips of forest occupy the river banks and valleys, providing pathways from the main parts of the rainforest into the savanna. In drier regions, where rivers may not flow all the year, there is often a fringe of forest or thicket, surviving on water deep in the river bed. Fast-flowing rivers are usually devoid of submerged vegetation (except for the highly specialized moss-like Podostemaceae), although some plants may grow in the slower reaches if the water is not too deep or too turbid.

Lakes in Africa are essentially of two types. Most of the Western Rift Valley lakes, such as L. Albert, L. Edward, L. Kivu, L. Tanganyika, and L. Malawi, are deep and have steeply sloping shores with little shallow water. Such lakes often lack much of a distinctive vegetation fringe, and there is also usually little submerged vegetation. Other lakes, such as L. Victoria, L. Kyoga and L. George in Uganda, L. Chad on the Chad–Nigeria–Cameroon border, and L. Mweru wa Ntipa in Zambia are shallow and much more productive. The margins of shallow lakes have extensive beds of tall swamp plants such as papyrus *Cyperus papyrus*, reed *Phragmites australis* and reed-mace *Typha dominguensis*, which provide shelter, but are of little value as food. Animals can live in these reed-beds during the day, and move out to feed on swamp grasses during the night. Shallow water areas are also densely vegetated with species rooted in the bottom mud (e.g. water-lilies *Nymphaea*) or free-floating, such as the Nile cabbage *Pistia stratiotes* and the introduced water hyacinth *Eichhornia crassipes*. Again, these species are little used as food by animals, but they provide valuable shelter during the day, and also harbour large numbers of fish that are much preyed upon by some mammals.

Flood-plains are important for some African mammals. The major flood-plains are the Inland Delta of the Niger, the Sudd Region of the Nile in southern Sudan, L. Bangweulu and the Luangwa Valley in Zambia, and the Okavango Swamps in Botswana. In all of these except the Okavango, a central river flows throughout the year; but during the wet season, the river rises and bursts its banks flooding huge areas of the surrounding plains. In the Okavango, the river flows into the swamps, where it eventually disappears. River flooding may be supplemented by direct flooding from heavy rain, as many of these areas lie on impervious clay soils. The fringing swamps of the river are usually dominated by giant swamp herbs, such as papyrus. The seasonally flooded grasslands are gradually exposed during the dry season as the river levels fall. Grazing animals are able to follow the flood as it recedes, feeding as they go on the newly exposed grasses, most of which are specialized swamp species such as *Oryza* (wild rice) and *Echinochloa*. In southern Sudan, these seasonally flooded grasslands are a key dry season resource for several mammalian species, including the Nile Lechwe *Kobus megaceros*, White-eared Kob *Kobus kob leucotis*, Tiang *Damaliscus lunatus tiang* and several species of small rodents. Likewise, the seasonally flooded grasslands of southern tropical Africa are a crucial resource for the Red Lechwe *Kobus leche*.

A final type of aquatic habitat is the saline or alkaline lake. These are characteristic of the Eastern Rift Valley. Here salts are leached from the fresh alkaline volcanic rocks and are washed into basins of internal drainage where they accumulate. Most of these lakes are too saline to be used as drinking water (although there are often springs of fresh water at their margins). As such they do not provide a very useful habitat for many mammals although they may be used as a refuge and for wallowing and bathing. Scavengers may also find rich pickings along their shores.

CHAPTER SIX

Mammalian Evolution in Africa

Jonathan Kingdon

The theory of evolution by cumulative natural selection is the only theory we know of that is in principle capable of explaining the existence of organized complexity. Even if the evidence did not favour it, it would still be the best theory available! In fact the evidence does favour it ... Cumulative selection, by slow and gradual degrees, is the explanation, the only workable explanation that has ever been proposed, for the existence of life's complex design.

Richard Dawkins, *The Blind Watchmaker*, 1986

Beginnings

Australia suits kangaroos, Antarctica suits penguins. With 17 of the world's 20 orders of terrestrial mammals – more than any other continent – Africa certainly suits mammals. Does this imply that the first mammals were African? Is Africa the place to find evidence for the evolution of aardvarks, antelopes, apes, bats, giraffes and zebras from a common ancestor? On current knowledge the answer must be no. Placental mammals are unlikely to have originated in Africa and, if they did, the obliteration of any traces of early beginnings has been extraordinarily thorough. However, the fact that so many types have flourished and diversified, once their ancestors did get here, has to be an evolutionary theme of major significance.

While the mammals of Africa are celebrated for their distinctiveness and variety, the reasons for such diversity have been quite slow to emerge and one of the purposes of these volumes, and of our introductory chapters, is to synthesize a large body of new research and explanations drawn from a broad spectrum of disciplines. Our introductory chapters discuss the roles of geography and geology, vegetation, climatic changes, behaviour and morphology in mammalian biology. This chapter explores the continent as a very singular theatre of mammalian evolution, a perspective that complements its other purpose, which is to identify some fundamental features of the evolutionary process that seem particularly well illustrated by African mammals. In profiles of living species, with standardized topic headings, some aspects of evolution can be easily overlooked. Among them are considered discussions of selection, by predators, by disease, by sexual competition, by inter- or intra-specific competition or mate selection. The importance of co-evolution and the role of, say, elephants in shaping the habitats within which many *other* mammals have evolved also invite discussion; so size and co-evolution are among the topics that are discussed in this chapter.

What was the first placental like? Externally it probably differed scarcely at all from earlier non-placentals because its special physiological advantages were internal, not external. A generalized outline of the earliest mammal as a small, possibly very small, probably nocturnal and at least partially arboreal animal remains a plausible model in spite of that characterization having been around for well over a hundred years (Matthew 1904, Simpson 1937, Romer 1945, Cartmill 1974). Now there is a fossil to confirm this general outline. In 2002 Ji *et al.* described a 125-million-year-old proto-placental mammal, *Eomaia scansoria* (meaning literally 'climbing dawn-mother'), from China, which was clearly semi-arboreal (in some respects resembling a dormouse in size and appearance: see below). The disparate characteristics of supposed stem mammals are also consistent with some such ancestral placental emerging on the Eurasian land mass some time after150 mya (Bininda-Emonds *et al.* 2007). However, the discovery of *Eomaia* serves to remind us that all living placentals derive from just one of more than a dozen lineages, all but one of which have become extinct (Wible *et al.* 2007).

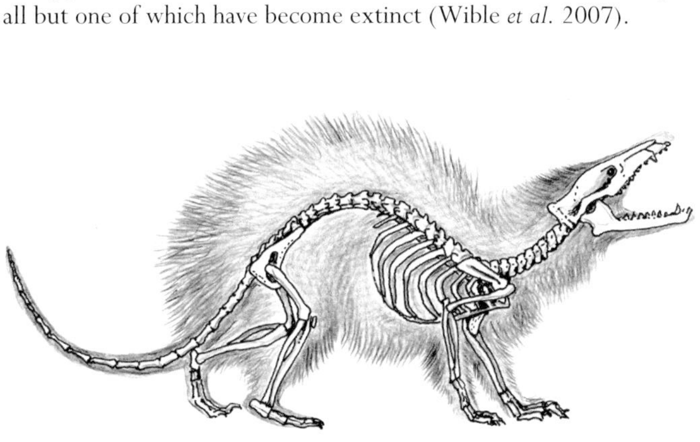

Reconstruction of *Eomaia scansoria*, a furry early Cretaceous placental mammal.

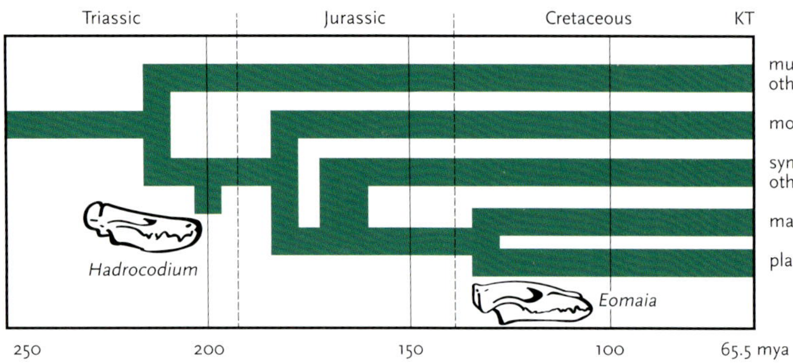

Tentative phylogenetic tree for placental mammal evolution before the KT event (65.5 mya).

What distinguishes the present state of mammal studies is that we are becoming better equipped now than at any time in the past to get past loose generalities and explore some of the specifics of mammalian evolution. Apart from fossils such as *Eomaia*, the main source of new information about mammalian origins has derived from the construction, by geneticists from around the world, of molecular phylogenetic trees for all the orders and families of mammals. For several major groupings such strong implications of continental origin have emerged that some now have continental names, notably the Afrotheria and Laurasiatheria. The more modern trees are accompanied by a time-scale deriving from molecular clocks. Fossils have provided crucial cross-checks or calibration and, when sufficiently abundant and diverse, have generally ended up supplementing and confirming the likely truth of some of the most recent molecular trees.

While the power of genetics and palaeontology to reveal evolution in action is incontestable it is important to remember that innovations in science, notably molecular techniques, take many years to develop and the conclusions or hypotheses that derive from a newly emergent branch of science can be subject to error and initially have to be treated as provisional. This is particularly so for dating, but none the less, the broad thrust of contemporary molecular science has been one of the most illuminating of all advances in the history of science. Some radically revised taxonomies have emerged from all this activity in contemporary genetics, and many of the taxonomic recommendations of these molecular scientists have been adopted or adapted in this work.

Changing time-tables for mammalian origins

As is detailed in the chapter on Africa's geological history, it has long been known that Pangaea, the supercontinent, had resolved into two major land masses by about 150 mya. Gondwana (of which Africa was the central and most stable part) lay to the south while Laurasia embraced the present three northern continents in what was initially a more or less contiguous continental mass. The sea of Tethys lay in between. Correlating continental tectonics with placental emergence is clearly central to understanding the timing of mammalian evolution in general but it is particularly significant for understanding mammalian evolution in Africa.

The Eutherian protoplacental fossil *Eomaia scansoria*, mentioned above, comes from the Yixian formation in southern China and has been dated at 125 mya (Ji *et al.* 2002). This extraordinary fossil, jacketed in fossil hair and without a single bone missing, post-dates the placental/marsupial divide and although its lineage is extinct it serves to reinforce the likelihood of a semi-arboreal and Asiatic origin for placental mammals. The age of this fossil accords with recent molecular clock datings, which suggests placental and marsupial mammals diverged 150–180 mya (Bininda-Emonds *et al.* 2007, Meredith *et al.* 2011). Both fossil and molecular dates are a lot earlier than was estimated in the twentieth century and make it more probable that primitive placentals emerged *after* Pangaea had begun to break up into northern and southern land masses. The Chinese fossils hint very strongly that Asia was the placentals' motherland, but even more persuasive has been the fact that a high proportion of phylogenetic trees and most of the 'stem' mammal groups root in Asia. Furthermore, all but one supercohort of extant African mammals can trace ultimate origins to non-African roots (and even those, the Afrotheria, might have still earlier origins in the northern continental masses).

Mammals of Africa has been compiled during a period of unprecedented scientific discovery. When the project was begun it was still possible to argue that Africa, its precursor mega-continent Gondwanaland, or even the still earlier supercontinent Pangaea, were equally plausible sites for the emergence of placental mammals. Yet within the few years that this book has been in preparation, such evidence as there is, and mainstream scientific opinion, has swung strongly against placental mammals originating in Africa or, indeed, anywhere in the southern continents, although that position is still robustly defended and the possibility must remain open (Montgelard *et al.* 2002, Murphy *et al.* 2007). Admittedly, the appearance of supposed placental fossil teeth and mandibles in Australia has given some support to the idea of southern origins. The teeth are said to have belonged to a hedgehog-like mammal from about 110 mya (Rich *et al.* 1997), and if correctly identified as a placental could imply a Gondwanan origin for all modern mammals. It would also imply their early total extinction and up to now there is no other evidence for placentals ever having lived in ancient Australia and further complicates our ability to interpret the significance of these few teeth. In Africa ancient placental fossils are even scarcer and the earliest reported so far only date from about 60 mya (Gheerbrandt *et al.* 2005). In spite of such tentative hints at southern beginnings, the weight of both fossil and molecular evidence is now in favour of Asia as the place of origin for placental mammals (Archibald 2003, Beard 2004, Robinson & Seiffert 2004).

The primary radiations

The taxonomic revisions that have emerged from contemporary genetics (Murphy *et al.* 2001a, b, 2007, Waddell *et al.* 2001, Huchon *et al.* 2002, Bronner *et al.* 2003, Bininda-Emonds *et al.* 2007) all recognize a radiation, within Africa, of the newly recognized and newly named 'Afrotheria' (Springer *et al.* 1997). This 'supercohort', a grouping of seven orders, derives from a single ancestor in Africa and is extensively discussed elsewhere in this volume

While the earliest molecular separations (dated to about 105 mya, by some authorities) were between ancestral Eurasian mammals, Afrotheria and the South American Xenarthra (another supercohort), this does not necessarily mean simultaneous continental separation. This is because molecular trees can only sample living species and putative Eurasian representatives of the early lineages that reached Africa and South America probably went extinct long ago. A current estimate of the time when a common afrotherian ancestor might have arrived in Africa suggests some time between 80 and 92 mya (Springer *et al.* 2003). The likelihood of a common afrotherian ancestor deriving from a gene pool that was already established in Africa is now thought to be very low. An ultimate Asian ancestry for the afrotheres and (much later) for the anthropoid primates is made less improbable by tens of millions of years being more than adequate time for the occurrence of chance raftings or island-hoppings between the Asian and African land masses, however far apart they might have been then.

Following these earliest radiations Eurasian mammals divided into two supercohorts: Laurasiatheria, named after the parental land mass, and the Supraprimates (or 'Euarchontaglires'), which allies two particularly ancient groups, primates and rodents. There has been considerable disagreement over extrapolated ages for these two Asian complexes but one of the most recent estimates for their divergence is about 97 mya (Bininda-Emonds *et al.* 2007). Until very recently placental mammals were assumed to have made their main radiations after the demise of the dinosaurs, at the K–T boundary, some 65 mya ago, a time when the broad configuration of today's continental masses was already recognizable (a view that is still held by some scientists, for which see Wible *et al.* 2007). It now seems more likely that the basal placental stock had begun to differentiate into distinct adaptive lineages well before any mammals had reached an Africa that was then surrounded by ocean on all sides. None the less, it is important to remember that the very earliest placental immigrant to Africa did not face an ecological vacuum. It is, indeed, more than likely that it arrived in a continent that was without placentals, but there were abundant dinosaurs and a variety of proto-mammals that would have narrowed the choice of niches into which the new arrivals could expand (Rose 2006).

The colonization of continents: Africa's afrotheres

When Springer *et al.* (1997) first revealed that the genes of what were once thought to be completely unrelated orders of living placental mammals had a common (and exclusively African) evolutionary history the association of elephants and their kin with sengis (elephant-shrews) and golden-moles was greeted with surprise, even disbelief. The Afrotheria embrace a very disparate assortment of seven major mammal orders: hyraxes, elephants, sirenian dugongs and manatees, the Aardvark, sengis (elephant-shrews), golden-moles and (disputedly) otter-shrews, and they are discussed in detail in the appropriate profiles. Although Afrotheria may well represent the oldest surviving group of all living placentals, the implication of a Gondwanan origin for all placentals does not necessarily follow (Montgelard *et al.* 2002, Springer *et al.* 2003, Murphy *et al.* 2007). If, as now seems most likely, modern placentals arose on the Laurasian land mass, the presence of the most primitive of all living types in the southern continents can only be explained by very early 'sweepstake' raftings. A continental bifurcation of the earliest placentals has been proposed by Murphy *et al.* (2007), the Eurasian branch labelled 'Boreoeutheria' and the Afrotheria/Xenarthra common lineage dubbed the 'Atlantogenata'. In this scenario a very small, and possibly semi-aquatic, proto-afrothere island-hopped or drifted from Laurasia to Africa across the Sea of Tethys. It has also been suggested that the Afrotheria might derive from some very early condylarth-like Euramerican ancestor (Tabuce *et al.* 2007).

On current evidence the afrotherian ancestor was the earliest placental to become truly endemic to Africa and its extraordinary radiation into such diverse forms is a clear manifestation of the African continent's prolonged isolation from other continents. Indeed, many afrotheres, both extant and fossil, show remarkable convergences with mammals that derive from totally different groups. Some early elephants in Africa appeared to resemble Eurasian and American tapirs in their anatomy (but probably more aquatic in their habits). Golden-moles resemble unrelated Eurasian moles *Talpa* spp. as well as marsupial moles *Notoryctes* spp. Otter-shrews *Potamogale* spp. resemble the Eurasian desmans *Desmana* spp., which are insectivorous, aquatic moles. In each instance such non-African equivalents of Afrotheria are unknown in Africa and were presumably unable to colonize that continent in the face of an established competitor. The hyraxes are also survivors of another ancient endemic lineage that was once a dominant herbivore type in Africa. In their case incoming Eurasian artiodactyls eventually replaced them in all but their rock and tree fastnesses, where most hard-hoofed antelopes could not compete.

As placental mammals with the longest history in Africa, it would seem that some Afrotheria acquired an adaptive advantage

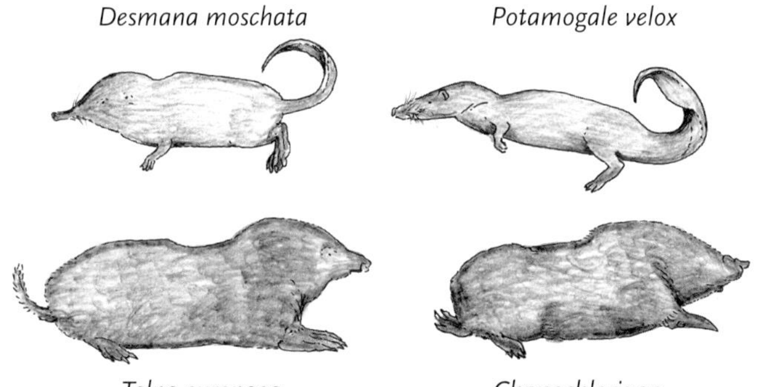

Comparisons between Afrotheres and unrelated species with convergent features: African Otter-shrew *Potamogale velox* and Eurasian Aquatic Mole *Desmana moschata*; African 'Golden Mole' *Chrysochloris* sp. and Eurasian Mole *Talpa europaea.*

because they occupied specialized niches that no later invaders could displace them from. This would seem to have been the case for sengis (elephant-shrews) Macroscelidea. Numbering some 16 extant species, all long-legged, long-nosed and insectivorous, the divergence between the smaller, more arid-adapted Macroscelidinae and the somewhat larger forest-adapted *Rhynchocyon* has been estimated at 42 mya (Douady *et al.* 2003). This has confirmed that all sengi species share an extremely conservative body form. (The *Rhynchocyon* radiation offers a fascinating expression of one of the basics of evolution, predator-selection, to be discussed shortly.)

Of other mammal groups that co-existed with the Afrotheria, the extinct *Arsinotherium*, creodonts, primitive rodents, two marsupials and primates, are the only ones known as fossils, although bats were almost certainly about. Anthropoid primates, long assumed to be archetypically African, now seem more likely to have arisen from a very tiny and very primitive Asian primate that somehow rafted to Africa between 45 and 60 mya (Sige *et al.* 1990, Beard 2004).

How do mammals speciate within Africa?

While it was rare geological and climatic events, and perhaps chance raftings, that established Eurasian mammals in Africa the founders of surviving lineages were selected somewhat less randomly. The exact environmental conditions within ephemeral, often narrow or discontinuous connecting corridors (or, perhaps, island chains) probably filtered out all but a few appropriately adapted species. The filtering effect was further biased in favour of populations that naturally lived close to the temporary corridor or bridge at the time the corridor opened.

Such restrictions on lineage founders would have contrasted strongly with the wide ranges of environments open to their now pan-African descendants but it would be wrong to envisage subsequent radiations spontaneously adapting to such a diverse choice of habitats. Various lines of evidence suggest that the most likely course of events is an initial wide-ranging prevalence eventually followed by fragmentation of that population, primarily in response to climatic changes. This situation was not only faced by an incoming immigrant mammal – any species that had temporarily become very widespread in the continent found itself in a comparable situation. Some predictable patterns of speciation emerge wherever we have some evidence for the break-up of any once widespread population. Near ubiquity can occur at any time and the accretion of species in Africa implies layer upon laminated layer of super-imposed radiations going back to the earliest mammalian invasions of Africa. The older the radiation, the fewer the surviving lineages become, thus it is among more recently successful groups that we find the clearest evidence for speciation.

Some initial conclusions about how radiations happen can be deduced from examining a series of widespread species (or closely related species groups) and relating their subdivisions to the continent's geography. When reliable phylogenetic trees can be dated and correlated with known fluctuations in climate, and when knowledge of adaptive traits within such entities are added to the equation we can conclude that adaptation to different environmental conditions must have been directly dependent on and a consequence of past physical and reproductive isolations under differing environmental conditions. These patterns are clearest when a radiation is relatively recent and when the founding ancestor's arrival can be dated.

Are all African mammals immigrants?

The comparison of dated fossils from Eurasia and Africa has revealed that virtually all the other species of living mammals in Africa descend from a succession of invasions (often accompanied by some two-way exchanges) that only assumed major proportions after Africa regained physical contact with Eurasia. Dates for the earliest connections are contested but the earliest clear evidence for a major, multiple influx of mammal taxa is as late as 27 mya, at which time fossils of several rodent, perissodactyl and carnivore groups and chevrotains, Tragulidae, appear for the first time. The dates for later tectonic connections and disconnections are becoming better known as are fluctuations in both global and local climates (see Chapter 4 on climate change). Wherever these events can be tied in to the first appearance of Eurasian taxa in African deposits it becomes possible to chart an increasingly diverse spectrum of mammals that has derived from a staggered succession of invasions.

We can now learn, for example, which groups have radiated within the continent and which derive more directly from immigrants. Perhaps most astonishing of all is to discover that because of Africa's extreme isolation during the most critical periods of mammalian diversification, a sizable proportion of today's mammals derive from single ancestral species that somehow found their way to this vast continental 'island' across substantial distances of ocean. The continent's mammalian diversity has therefore built up through a steady accretion, augmentation, replacement and extinction of species.

One type of accretion derives from a succession of invasions from Eurasia and these invaders, even after becoming distinct, tend to maintain most of the characteristics that ensured their success *before* they found their way to Africa. Many can be called neo-endemic (or, while still indistinguishable from their Eurasian stock, *actual* immigrant species). Another form of accretion has taken place as particular types of mammal have speciated *within* the continent. These can often be called true endemics (and the older ones archaeo-endemics). There are numerous intermediate grades between the oldest archaeo-endemic and the newest of immigrants but the polarities are real.

The existence of these two major sources of diversity becomes particularly significant in discriminating between (1) mammal types that show adaptations specific to some peculiarly African set of conditions and (2) mammals that live in ways that demonstrably evolved on other continents. Of the latter the most obvious are recent arrivals still restricted to northern Africa, carnivores such as foxes, jackals and cats and some generalized rodents. Of mammals that have evolved in Africa, humans are an obvious example, but while it is true that we are predominantly African there are undoubted Eurasian interludes to our animal past that involve both immigration and emigration and these, as well as our African-ness, will deserve serious attention if we are ever to come to grips with humanity's evolutionary history. Understanding speciation within Africa involves some sort of reconstruction of what happens to new immigrants once they have arrived.

The centre-west versus south-east divide

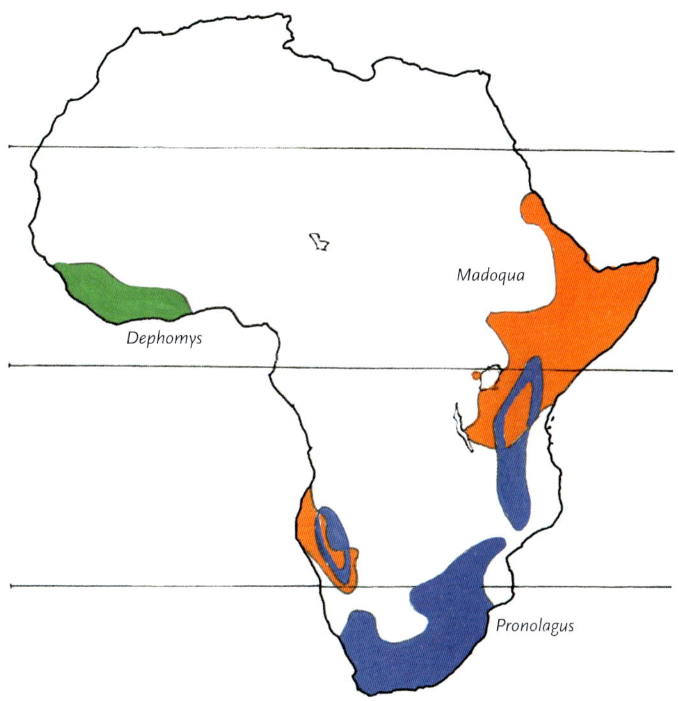

Simplified map showing regional foci of distributions of the genera *Madoqua*, *Pronolagus* and *Dephomys*.

One of the starting points in the process of speciation is the break-up of a previously single population into two or more separate parts (Mayr 1970). If each part maintains its genetic separation it will come to be different, if only through neutral genetic drift. Given that any mammal population in Africa exists in a landscape where no two places are ever exactly the same and given that habitats are altered by changes in climate, selection for locally adaptive traits will differ in each of two or more parts.

Time and again successful mammal types such as baboons, porcupines or giraffes have dispersed across Africa only to break up into recognizable regional subpopulations. This pattern repeats itself at numerous different levels from cryptic demes that are only now being revealed by molecular analysis, to distinct allopatric species-pairs. Within long-established taxa such as primates and antelopes, adaptation to local differences has eventually culminated in the evolution of genera (or even tribes), each now exclusive, or nearly so, to a region or ecotype. Examples are dik-diks *Madoqua* spp. in the north-east, rock hares *Pronolagus* spp. in the south and dephua mice *Dephomys defua* in the far west. Even among higher taxa, which are necessarily older, with histories that are more difficult to reconstruct, there can be hints that the differentiation between, say, one subfamily and another had its roots in gross regional separation followed by local adaptation.

Are there consistent patterns to this process? Has Africa's peculiar geography imposed its own rules upon the evolutionary history of mammals as a whole? There are, of course, innumerable patterns, but in sub-Saharan Africa one gross bifurcation repeats itself with many slightly different variations. It is a differentiation between populations occupying the latitudinally oriented equatorial centre-west and those in the longitudinally oriented south-east. Today this split roughly corresponds to the difference between lowland forest and upland non-forest, or, latitudinally, boreal and austral communities. Because of this, distribution patterns are commonly treated as direct expressions of adaptation to major contemporary habitat differences. While this may be quite correct there are other, less proximate dimensions to this gross subdivision of a continent. Outlines of differentiation, from family level downwards, suggest that this centre-west/south-east bifurcation (which broadly follows Africa's 'inverted L shape') has been profoundly influential in the evolution of mammals at every level and way back in time.

It might be expected that whatever split single populations into two would be a consistent and well-defined discontinuity between the two regions. To some extent the agencies for just such definition are there: the Nile and Congo Rivers, L. Tanganyika, the Rwenzori Mts, the Rift Valleys, the Somali arid zone or the main rainforest block itself. All can demonstrably act as physical or ecological barriers for various species. Which boundary or barrier is most decisive for a living population varies. Which dividing mechanism can be deduced as playing the major role in any one evolutionary history may differ from its contemporary expression as well as differing from one geological era to another.

The centre-west/south-east divide is exemplified in several antelope species, some rodent duos and two non-forest galago species that have different forms occupying broadly north-western and south-eastern savanna distributions. The Common Eland *Tragelaphus oryx* once inhabited most of southern and eastern Africa while its less derived cousin, the Giant Eland *T. derbianus* ranged through the moister wooded belts south of the Sahara. For these groups the forest belt, assisted by hurdles such as the Nile, the Rift Valley or the East African highlands, are obvious separating mechanisms.

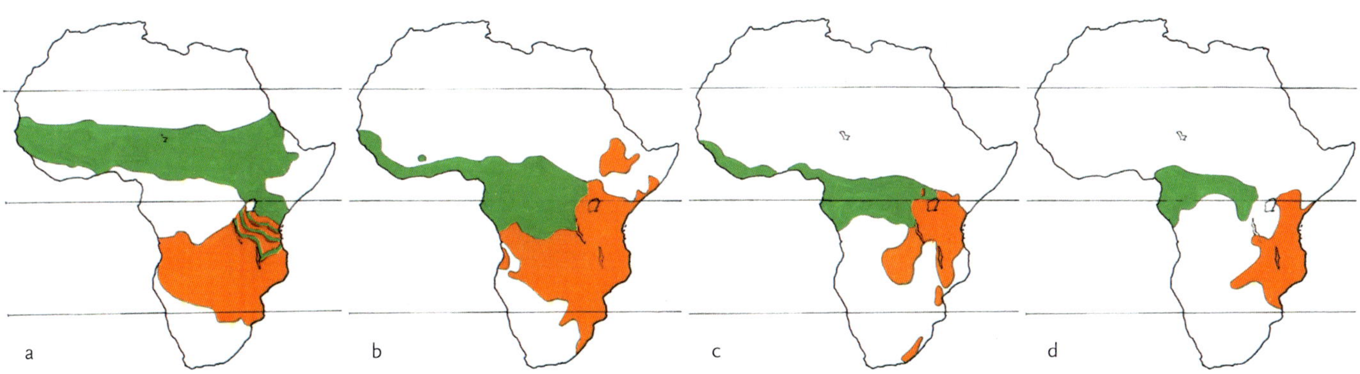

Simplified species distribution maps of (a) *Galago*, (b) *Potamochoerus*, (c) *Dendrohyrax* and (d) *Bdeogale* showing centre-west/south-east divides.

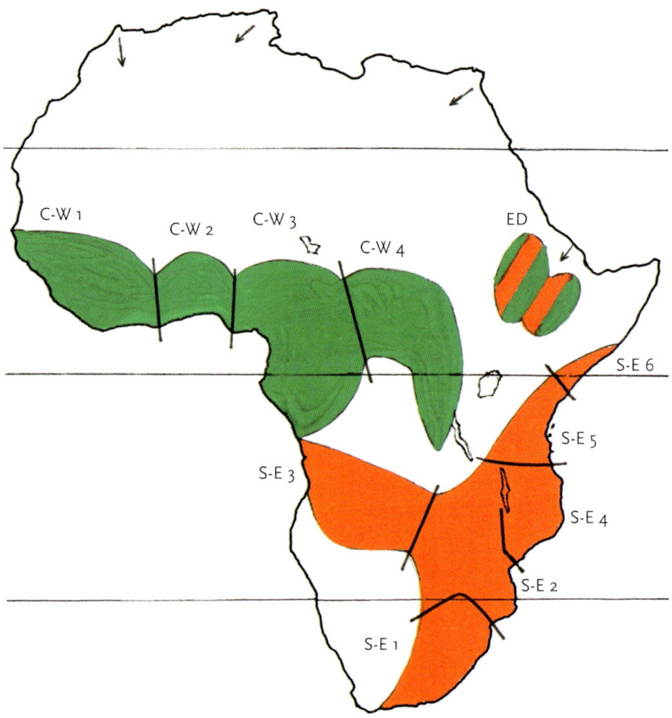

KEY
C-W 1. Upper Guinea
C-W 2. Ghana-Benin
C-W 3. Cameroon/Gaboon
C-W 4. Centre-Haut-Zaire
Arrows = supposed points of past connection with Eurasia
S-E 1. South Africa
S-E 2. Zambezia
S-E 3. Angola
S-E 4. Mozambique
S-E 5. Zanj
S-E 6. Somali littoral
ED Ethiopian Dome

Division of sub-Saharan Africa into two major evolutionary realms. 1. Humid, latitudinal 'Centre-West' (C-W). 2. Drier longitudinal 'South-East' (S-E).

It is postulated that populations originally confined within either realm may extend into intervening areas (most notably to the Ethiopian Dome) or may eventually spread more widely as new species.

Past changes in climate, augmented by geographic barriers, have subjected each realm to further (and changing) subdivisions or conjunctions.

Forest-adapted taxa also separate out along similar lines. The Red River Hog *Potamochoerus porcus*, which inhabits the equatorial forest, is essentially separated from its close relative the Bushpig *P. larvatus* by the Albertine Rift and by the southern edges of the rainforest (the latter a particularly blurry boundary). Likewise, the Western Tree Hyrax *Dendrohyrax dorsalis* ranges right across the main forest belt while the Southern Tree Hyrax *D. arboreus* is its south-eastern equivalent. Among duikers, the *Cephalophus callipygus/weynsi* group inhabit the main forest block, the *C. natalensis/harveyi* group the south-eastern woodlands and forests. There are centre-west and south-eastern populations of bushy-tailed mongooses *Bdeogale* spp., some sun squirrels *Heliosciurus* spp. and various smaller rodents. Some closely related genera also follow this pattern; for example the forest pygmy antelopes, the Royal Antelope *Neotragus pygmaeus* and Bates' Pygmy Antelope *N. batesi* have an eastern relative, the Suni *Nesotragus moschatus*. To merit generic differentiation this separation is clearly of very long standing, revealing a phenomenon with a history that can be measured in millions of years.

Entire groups of animals are wholly or nearly exclusive to one of these regions; notably sengis (Macroscelidae), blesmols or 'mole-rats' (Bathyergidae), rock-hares or the Bat-eared Fox *Otocyon megalotis* in the south-east; forest pangolins (Pholidota), rope squirrels *Funisciurus* and many primate groups in the centre-west. The list could go on and could easily invoke species, genera and families of reptiles, amphibians, birds and other biota. The exact boundaries of living species have less importance than the implication of a major bifurcation, which reasserts itself at almost every taxonomic level.

The persistence of this pattern suggests that there has been a perennial (but also permeable) divide for both forest and non-forest communities through most of the Cenozoic evolutionary history of African biota. I have argued elsewhere (Kingdon 2003), and summarize shortly, the idea that this divide, in one of its many permutations, was also central to the initial separation between ancestral apes and ancestral hominins. I have also argued elsewhere in this work (Volume II, p. 27) that the Oligocene split between hominoids and cercopithecoids and the Miocene parting between colobines and cercopithecines involved some such geographic polarization.

If we return to the starting point, a hypothetical, single primary population with sufficient adaptability to spread very widely (but not into waterless true desert), there are important differences between its centre-western and south-eastern wings that are due to Africa's intrinsic shape and to its tectonic history as well as to climatic and environmental differences. Whether forest-adapted or not, the centre-westerners live within a long but relatively narrow latitudinal belt of mainly humid lowlands with boreal seasonality to influence their life-cycles and breeding. The Atlantic Ocean (with water temperatures that are known to have fluctuated periodically) is the ultimate source of rainfall throughout. Habitats range from near-desert to rainforest with no clear or lasting boundaries between very long east–west strips.

By contrast, the south-easterners live along a north–south axis that can range over as much as 45 degrees of latitude of mainly hilly lands and plateaux with austral seasons and seasonal rainfall (coming off an Indian Ocean that has less labile temperatures). Mountain ranges tend to intercept much of the rain, creating rain shadows and a very varied ecological mosaic. Some of these highlands are Mesozoic, most are post-Oligocene but the region as a whole has been decidedly cooler and drier than in the moist lowlands of the centre-west.

Differences between these two regions go back well beyond the Oligocene and have been very marked for all the period in which mammals have been populating Africa. The climatic and ecological differences would have promoted differentiation within even the most ubiquitous population, but the discontinuities and physical barriers listed above have certainly served to accelerate local adaptation, separate gene pools and define boundaries and sub-boundaries.

Bifurcations upon bifurcations

Ideas about the driving forces behind speciation are subject to a large measure of scientific fashion and, in any case, modes of speciation are, in themselves, diverse (White 1978). Recent years have seen much emphasis on climate-driven 'turnover pulses' and sudden spasms of evolution, supposedly in direct response to changes in climate. Many of the latter have now been shown to have been so numerous as to defy our ability to tie any one climate change to particular evolutionary events. What has been neglected is awareness of sustained, incremental adaptation among

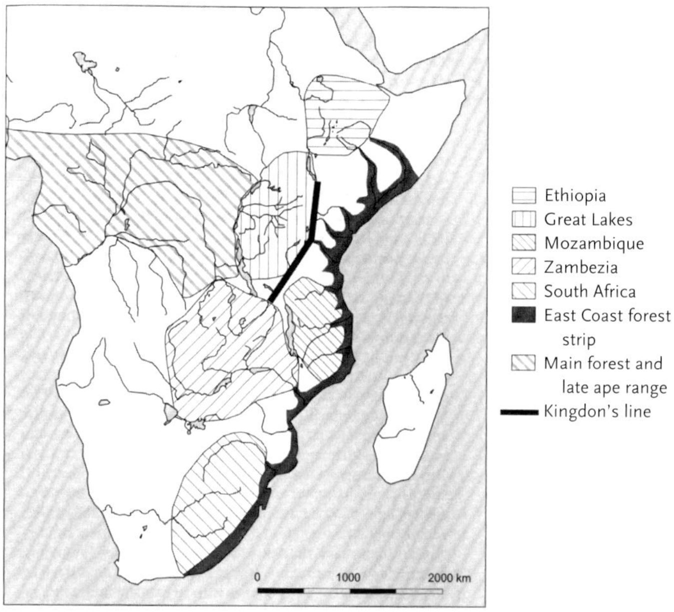

Map of east and central Africa indicating five ecologically discrete hinterlands or river basins inland from the east coast forest strip. Some hinterlands might have sheltered distinct types of hominins and other mammal populations (after Kingdon 2003).

pre-existent taxa to, say, wetter or drier conditions, or to more or less competition, disease or predation in some corner or block of their range. Because the evolutionary products of this bifurcation flow, ebb and change it is tempting to suppose that the sudden appearance of a novel form as a fossil represents a 'pulse'. Such sudden appearances are, instead, an artefact of the pin-prick sampling and semi-arbitrary dispersion of fossil sites in relation to two vast, interacting sub-regions of Africa and still further ecological fragmentation within them.

When new species suddenly appear in a layered bed of fossils it is therefore much more likely that changes in climate have permitted former regional isolates to expand their range than that these apparently 'new' species manifest a sudden evolutionary 'pulse'. It is in this context that the centre-west and south-east divide finds special significance. Each block can be further subdivided and within each such compartment there are evidently further opportunities for bifurcation, even multiplication. For example, the northern and southern ends of the south-eastern block are in separate hemispheres and 40 degrees of latitude apart. Likewise, the westernmost lowlands of the 5000 km-long centre-west block have suffered repeated ecological isolation due to fingers of Saharan aridity periodically probing southwards while its easternmost mountains remained perennially wet. In spite of these and other subdivisions assisting further speciation or subspeciation, the significance and tangible existence of a single major divide in Africa cannot be over-emphasized.

There are many inferences suggesting that 'pre-adapted' populations from one block, or some part of it, are able to spread out from their enclave when conditions favour them or shrink back when events turn against them. For instance, there are many examples of relictual extant fauna and flora that appear to have common and widespread fossil predecessors. Likewise, species that are montane isolates in eastern Africa are less constrained in temperate South Africa, with the clear implication that such species ranged more widely through the south-east during cool, glacial periods. While there are many more dimensions to evolutionary success than mere adaptation to more or less rain or higher/lower temperatures, adaptation to wet/warm and cold/dry polarities is a realistic way of throwing the main mechanisms of evolution into high relief. At its crudest, the centre-west/south-east divide represents, on the massive scale of an entire continent, a geographic and ecological polarization. It need surprise no one that no two species ever coincide exactly in their distribution and that patterns of spread, retreat and further speciation or subspeciation are as varied as there are taxa.

In the past, south-east and centre-west populations have been periodically separated by a barrier that is mainly climatic and geographic. In the context of the present discussion arid habitats and their inhabitants can be regarded as a labile, crude negative interposed between the two blocks. This, of course, debases the inhabitants of that dry corridor, which have also speciated with their own polar centres in the south-west and in the north-eastern Horn of Africa. The dynamics of speciation among these arid or semi-arid adapted species is discussed further in relation to geographic factors in Chapter 3 and climatic ones in Chapter 4. Here it is sufficient to point out that relatively small, but ancient arid foci in the Horn and south-west (and the much larger but more recent Sahara) have expanded and ebbed many times. At each expansion aridity has presented numerous physiological and environmental challenges to animals and plants adapted to more humid habitats and conditions. Periods of aridity have also served to reinforce the isolation of forest communities in more consistently moist enclaves or refugia.

One of the most instructive explorations of incipient speciation, subspeciation and genetic diversification has been provided by a recent study of the Bushbuck *Tragelaphus scriptus* by Moodley & Bruford (2007). The choice of this pan-sub-Saharan species for such a study was apt because it is very unusual in the number of its regional subpopulations, in the extent of its range, in its relatively sedentary habits and in its tolerance of a wide spectrum of temperatures, rainfalls, altitudes and habitats, including forest. The Bushbuck's 'generalist' tragelaphine niche and relatively small body size (about 50 kg) corresponds to an optimum found in many of the most successful species of antelopes and deer, world-wide.

The gross partitioning between what is commonly called the 'harnessed' bushbuck (*scriptus* type) in the centre-west and the 'sylvan' bushbuck (*sylvaticus* type) in the south-east has been known and discussed for a century or more, but Moodley & Bruford's revelation of the very complicated patterns of affinity within each of these two polar mega-populations is unprecedented. Gene profiles reveal that some of these discrete subpopulations must have retreated into small enclaves while others seem to have expanded their ranges. See-sawing interactions along the boundaries between *scriptus* and *sylvaticus* types imply that boundaries are mobile, with environmental changes favouring, say, some types of Sahelian *scriptus* at one time, or particular East African *sylvaticus* groups at another.

This study is complicated by uncertain genetic affinities between bushbuck and the south-eastern Nyala *T. angasi* and centre-western Bongo *T. eurycerus*, but it has demonstrated that terminal level haplogroups have diversified most at the centre of each mega-

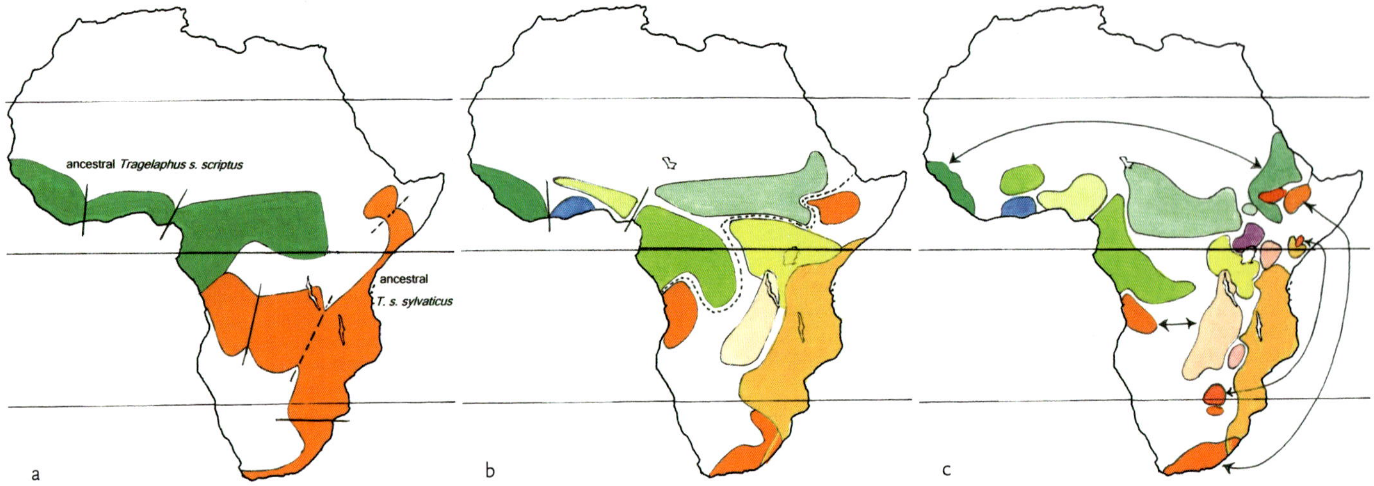

a. Hypothetical map of ancestral *Tragelaphus s. scriptus* and ancestral *T. s. sylvaticus*.
b. Hypothetical map of later subdivisions of both ancestral *Tragelaphus scriptus* populations (based on data extrapolated from Moodley & Bruford 2007).
c. Distribution of 23 *T. scriptus* haplogroups identified by Moodley & Bruford (2007).

population's range while, very significantly, the outermost extremities of the Bushbuck's overall range shelter basal haplogroups that are indicative of a more generalized ancestral state. This is because the haplotypes most isolated from later genetic contact have the least complex history and so tend to retain more of the oldest genetic signature. Thus the most westerly of all the *scriptus* haplotypes is effectively closest to the ancestral type. On the south-eastern half of the continent conservative *sylvaticus* haplotypes occur in three outlying distributional 'polyps', namely in the extreme south, the extreme north-east and the extreme north-west of *sylvaticus* range (see diagrammatic maps of this process above).

Regional diversification of Bushbuck populations is of intrinsic interest because it exemplifies and holds within it the potential for even further speciation. A mega-bifurcation can lead on to still further bifurcations and then still more. This is the process that has generated the extraordinary diversity of species in Africa.

The Bushbuck pattern closely resembles what seems to have happened in other widespread groups. Most notable in this regard is a pattern of morphological resemblances and differences among Gentle Monkeys *Cercopithecus* (*nictitans*) group, which awaits an ongoing comparative genetic study that will, hopefully, be comparable with that of the Bushbuck. In this complex of widely scattered regional populations distinctions between full species, incipient species, subspecies and regional demes are particularly difficult to draw: a radiation that graphically illustrates evolution as a very finely tuned, on-going process. It is already clear that it was an earlier, smaller-bodied outlier of this group that gave rise to the *C.* (*cephus*) radiation. Other populations now subsumed within the *C.* (*nictitans*) group may eventually be shown to be scarcely less distinct.

Recent molecular studies imply that the species that most closely links the *C.* (*nictitans*) group to the rest of the *Cercopithecus* guenon radiation is the little-known and secretive Owl-faced Monkey *C. hamlyni*, living mainly in the eastern Congo basin (Dutrillaux *et al.* 1988, Tosi *et al.* 2005). This has several interesting implications. One is confirmation that *C.* (*nictitans*) was likely to have shared a common origin in the Congo basin forests. Another implication is that *C.*

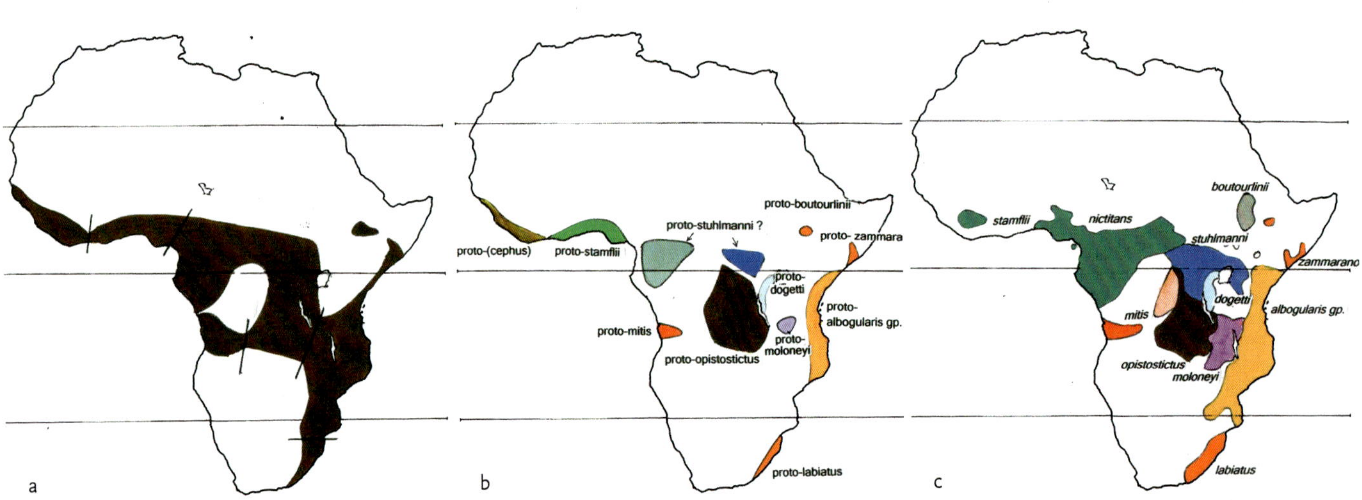

a. Hypothetical map of common ancestral population to the *Cercopithecus* (*nictitans*) and *C.* (*cephus*) radiations.
b. Hypothetical map of later subdivisions of above population.
c. Distribution of currently recognized populations in the *C.* (*nictitans*) group.

hamlyni, having continued to refine local species-specific adaptations within its area of origin, might represent some sort of modified survivor of a previous common ancestor linking *C.* (*nictitans*) with the rest of the guenon radiation.

To complicate matters, there may be yet further layers of relictual populations left over from the spread of *C.* (*nictitans*) over a range that now stretches very far from its place of origin. For example, there are reasons to suppose that, after *C. hamlyni*, another layer of relict populations might be represented by the swamp-forest-loving *C.* (*n.*) *m. opistostictus* and *C.* (*n.*) *m. heymansi*. Both are currently classed as subspecies of *C.* (*nictitans*) *mitis* but a difference in chromosome count in the former (Dutrillaux *et al.* 1982), if confirmed, might merit recognition as a separate species.

The evolutionary layering of populations does not stop here. Like the Bushbuck, Gentle Monkeys have centre-western populations (*C.* (*n.*) *nictitans* and *C.* (*n.*) *mitis*) and a south-eastern complex (commonly grouped under *C.* (*n.*) *albogularis*). Also like the Bushbuck, Gentle Monkeys seem to have reached out, at some very early stage of their expansion, to the furthermost extremities of forest. Thus *C.* (*n.*) *n. stamflii*, from West Africa, *C.* (*n.*) *a. labiatus*, from South Africa, *C.* (*n.*) *boutourlini* from Ethiopia and the dwarfed *C.* (*n.*) *a. zammaranoi* from the Juba R. in Somalia, closely resemble the Bushbuck pattern in that all these four extremities probably shelter more conservative populations than those from more central regions. These relictual populations out on the margins of *C.* (*nictitans*) overall range imply a genetic conservatism that must qualify any attempt to cluster taxa on purely geographic criteria.

New species emerge from changes in size: dwarfing

Among functional traits that are selected for in marginal areas during hard times is diminished body size. Large bodies can be a severe constraint under such conditions, and in the case of Gentle Monkeys it seems likely that a trend towards dwarfing in far western Africa resulted in emergence of the most recently evolved of all forest monkeys, members of the *C.* (*cephus*) group, which are also among the smallest. Regional diversification of Gentle Monkeys is of intrinsic interest but hypothetical dwarfing in the extreme west and demonstrable dwarfing in the extreme north-east serve to illustrate that even within a single species, or species-group, a potential exists for smaller bodied animals to make their own radiations. For the *C.* (*cephus*) group, small size opened new opportunities and the radiation of some ten regional populations is discussed in the *C.* (*cephus*) species-group profile in Volume II. Suffice here to point out that the extraordinary diversity of facial patterns in this group would seem to be the product of population fragmentation taking place, due to an accident of forest contraction at the very moment when selection for visual distinctiveness was at its most critical (Kingdon 1980). Such elaborations of facial patterns not only seem to have a large element of arbitrariness in their designs, their very diversity seems to be largely an accident of timing.

Selection for dwarfing is best known on islands; rapid and drastic reductions in the size of elephants have been well documented on Mediterranean islands such as Sicily, Cyprus and Malta. In such circumstances it has been plausibly argued that small individuals survive best because resources are less consistently available in bulk and/or because smaller animals can sustain themselves best on small home-ranges. Dwarfing is one manifestation of a more general characteristic, which is commonly summarized as character displacement (Hutchinson 1959). All members of the forest guenon genus *Cercopithecus* exemplify displaced variations on a common guenon template; the effect has been to pack many more monkey species into the African forest canopy. Variables other than size concern growth patterns, dietary preferences, physiological tolerance levels, relative speed of movement or alertness and other less obvious differences. The range of sizes among arboreal forest monkeys tends to be strongly constrained and narrowed by the weight-bearing tolerances of canopy and other branches.

Size can alter functional behaviour and social signals. For example, among bats that emit ultrasonic bursts of sound through their nostrils it has been shown that the frequency of pulses correlates with body size (actually nose-size as expressed in the distance between the two nostril emitters). Thus Horseshoe Bats, Rhinolophidae, produce CF wavelengths that measure twice the distance between their nostrils (Mohres 1953). It seems improbable that living bats would preserve evidence for a phylogenetic size differentiation along centre-western or south-eastern axes but there are differently sized species of *Rhinolophus* that do conform with such geography and this perspective might merit further examination.

Reduced body size and the centre-west/south-eastern divide are both involved in the evolution of duikers (see Volume VI). The Suni has already been mentioned as a typical south-easterner while its even more diminutive relatives, dwarf antelopes of the genus *Neotragus* are centre-westerners, inhabiting parts of the rainforest belt. In this instance it was probably forest shade that indirectly first induced extreme dwarfing through the scarcity of young foliage on the forest floor. A longer-term evolutionary effect was to entrench tiny bovids in a habitat where a very significant dietary change could take place. Such surviving dwarfs illustrate a dynamic that also played out in the evolution of duikers. It is understandable that leaf-eating antelopes in a forest should become restricted to the rare spots

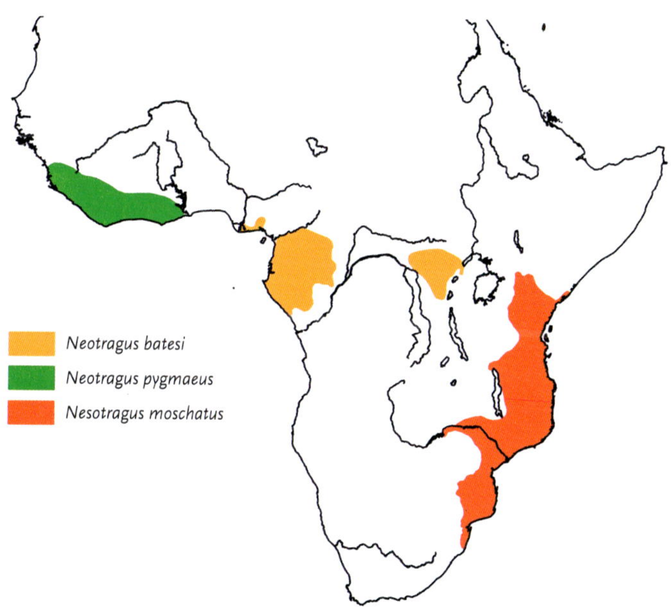

Distribution map of Suni *Nesotragus moschatus* with the ranges of its nearest relatives in the forest (after Kingdon 1997).

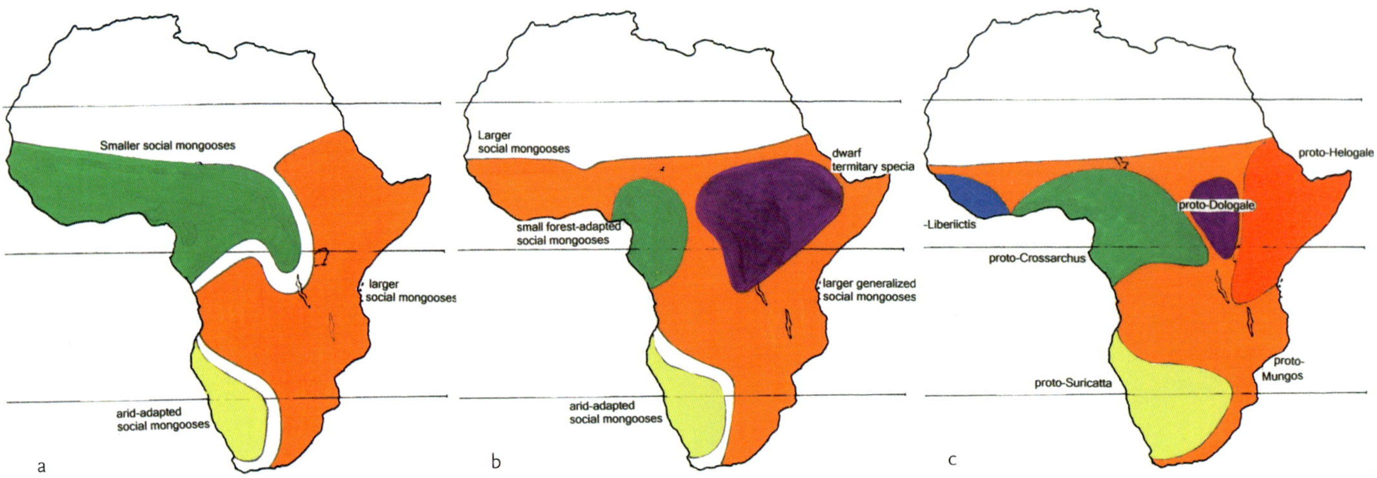

a. Hypothetical map of a common ancestral population of social mongooses separating into three major divisions.
b. Hypothetical map of further regional and ecological specialization among social mongoose populations.
c. Schematic map of postulated regional origins for six genera of social mongooses.

where ground-level herbage is fed by sunlight penetrating the canopy and that very small bodies are more easily sustained by relatively scarce and patchily distributed food. However, the most abundant fresh green growth on the forest floor consists of fragments dropped by arboreal animals in the canopy. Scavenging for these rather than browsing directly off plants would have introduced the very earliest proto-duikers to a mode of feeding that was mainly dependent on guilds of canopy animals. Because fallen fruits and other products of the canopy are both more abundant and more nutritious than sun-starved leaves, selection would have favoured those miniature antelopes that became more frugivorous or reliant on dropped plant-parts. The extant spectrum of duikers is a living demonstration that appropriate behavioural, physiological and anatomical adjustment developed by small increments. Once such adjustments to frugivory and omnivory had been made these very small antelopes could then become larger and the course of their radiation is followed in some detail in the appropriate profiles.

Another example of dwarfing involves a phylogenetic decline in size among several genera of social mongooses. Other than living in places with both predators and abundant bolt-holes, sociality in mongooses has no other obvious ecological correlates and various authors (Rasa 1977, Rood 1983, 1990, Macdonald & Nel 1986, Clutton-Brock *et al.* 1999) have demonstrated that social living has, indeed, evolved in response to selection by predators, especially raptors. Attrition in social mongooses tends to be severe (Creel 1996) and social mongooses produce large numbers of young and expend much energy in maximizing reproduction, sharing care for the young and sharing socially beneficial vigilance. Apart from alarm calls and flight, anti-predator behaviour includes 'rescues' and coordinated responses to predators (Rood 1975, 1986, Kingdon 1977).

Recent molecular studies have shown that *Mungos*, the genus to which the common and widespread Banded Mongoose *M. mungo* belongs, has its closest affinities with the rainforest-adapted and apparently relictual Liberian Mongoose *Liberiictis kuhni* (Veron *et al.* 2004). Both these long-clawed excavator-foragers average about 400 mm in head-and-body length and weigh about 2 kg and they live in complex social groups (Booth 1960, Ewer 1973, Rood 1975, 1986, Kingdon 1977).

According to Veron *et al.* (2004), the common ancestor of the *Mungos–Liberiictis* lineage also gave rise to another species of social mongoose, which was probably smaller and had a very wide range. Its descendants split along the classic centre-west/south-east divide: cusimanses *Crossarchus* became forest animals while the Dwarf Mongoose *Helogale parvula* occupied the south-east. The centre-westerners have several resemblances with the Liberian Mongoose but have head-and-body lengths of about 350 mm and weigh about 1 kg; adult *Helogale* measure about 250 mm and weigh 300 g – real dwarfs!

The point has been made that social mongooses, as a class, are not tied to specific habitats, so the common ancestor of *Crossarchus* and *Helogale*, like the Bushbuck, can be predicted to have been a single very wide-ranging species. It is likely that the relict offshoot of an early ancestor to *Helogale* (and closer to its shared ancestry with *Crossarchus*) survives in the form of yet another genus, the rare central African Pousargues' Mongoose *Dologale dybowskii*. Intermediate between *Crossarchus* and *Helogale* in size and general morphology, this species also bridges both regions geographically and ecologically. Very significantly, the few observations of this species suggest a close association with termitaries (Hayman 1936, Kingdon 1977). This observation is relevant to the apparent paradox that *Helogale parvula*, the species with the greatest exposure to predators, is the smallest, but Rasa (1977), Creel (1996) and others have pointed out that there are special ecological and behavioural reasons for this. Even more than other social mongooses, *Helogale parvula* is assailed by very many predators: the profile in this work names 13 as recorded instances, but there are many more, including raptors, reptiles and larger carnivores, and escaping them calls for vigilance, a quick response to warning cries and speedy decamp to a nearby refuge. Their dispersal through the landscape correlates closely with a high density of termitaries. Of several constraints on size, the major one is probably the need to fit or flit down narrow termitary ventilation chimneys (serving as ubiquitous bolt-holes as well as social dens). In other habitats, notably in the forests of the centre-west, refuges come in many forms and sizes whereas ventilation shafts have quite narrow ranges of variance. If termitary chimneys are the evolutionary determinant of *Helogale* body size their pre-existence was essential to the evolution of dwarfing. In effect evolution of the two *Helogale* species, and their dwarfing, were tied to the behaviour and activity of other social animals, termites. Termites provided essential but highly structured refuges from the raptors and carnivores that would otherwise have exterminated the mongooses. The pre-existence of other animals and the ecological impact of their behaviours is a much neglected topic in evolutionary biology. Wherever they occur in Africa termite-mounds are an integral part of the landscape, while the termites themselves possibly represent the largest biomass of any single class of animal, but, in the past, elephants must have run them a close second.

Pioneers in gigantism: afrotherian elephants – elephant activity defines habitats

It is well known that the first motor roads in Africa, as well as countless traditional tracks, followed the courses of elephant paths. Even today, in the few areas where elephants survive in landscapes that are still friendly to them, there are networks of paths that clearly express the logic of elephant needs. Over tens of thousands of years paths to water and to separate dry and wet season pastures have followed subtleties of relief as expertly as ever did any Roman engineer. Traditional elephant paths are familiar landmarks in such landscapes, as are 'stations', to which many generations of elephants have returned, again and again, for social or operative reasons, or as broadcast/reception posts on strategic ridges or hilltops.

In the evolution of hominids, especially flesh-footed, soft-toed hominins, elephant paths were probably as important to their survival as termitary chimneys have been for the evolution of dwarf mongooses. The association between elephants and specific localities is still expressed in numerous place-names, Njoro, Njoko, Ndoto, Embu, Mbungu, Nyamandhlovu, to name but a few, (even in Europe titles such as Kaiser and Tsar are said to ultimately derive from an ancient North African name for the elephant, Casius). Another respect in which elephant species, living and extinct, are likely to have shaped the lives of other mammals is their 'gardening' of the landscape. Many, probably most large plant species are unable to survive sustained and heavy browsing by elephants. Some have evolved chemical or mechanical deterrents in their leaves, bark or trunks, others, like several *Acacia* species, have developed a tolerance for very severe wounding and can re-sprout from stumps or mangled coppicing. The long-term effects of such 'gardening' were to limit the range of available plant species in elephant-dominated landscapes, creating communities of plants and animals that were much less diverse than they might otherwise have been. This might have been especially important for hominins because plant impoverishment greatly reduces the scope for both frugivorous and folivorous primates, reducing both competition and susceptibility to primate diseases in hominins. Species such as browse rhinos *Diceros bicornis* have been observed to decline under the influence of elephants steadily degrading the woody vegetation on which the rhinos were dependent. Among the effects, very numerous elephants would have opened up the vegetation and fast-growing grasses and herbs would have been favoured over vulnerable trees and bushes. There can be little doubt that many grazing antelopes and rodents were among the beneficiaries and it is possible that the southward expansion, perhaps even speciation, of arid-adapted antelopes, such as Thomson's Gazelle *Eudorcas thomsonii* and those of the Grant's Gazelle *Nanger (granti)* complex might have been assisted by the modification of such habitats by past populations of proboscids.

It has been widely observed that when the impact of numerous elephants is/was augmented by fire the trend towards open grassland accelerates. Elephants alone, or fire alone may be insufficient to destroy some very resistant woodland types, but when both combine grassland tends to become progressively more open. Given that humans and their ancestors are likely to have been the major setters of fires certain combinations of past behaviour in *Homo* and *Loxodonta* probably shaped the ecology of substantial areas of Africa and may even have influenced the very recent evolution of the Hartebeest *Alcelaphus* as the dominant grassland antelope species. The presence of elephants can also deter predators so that much smaller animals take advantage of elephant herds at, say, waterholes, to gain access to water. In all such examples it is not always possible to separate opportunism from the potential for co-evolution, but the important point is that the gigantic size of elephants has undeniably helped shape the conditions under which other plants and animals have evolved, including human ancestors.

Mammalian radiations within Africa

When Eurasian immigrants entered Africa they commonly brought with them a heritage of adaptation to temperate climates and ecotypes. That legacy would sometimes have put constraints upon expansion, but where the immigrant had decisive advantages over established populations it would eventually widen its range and diversify. Most immigrants probably began by favouring the habitats that most resembled those of their region of origin or of the bridging corridor that allowed them into Africa. As a consequence, drier, cooler zones, especially in uplands or on mountains could be expected to (and demonstrably do) shelter some of the more conservative members of immigrant lineages. As remarked earlier, highlands also harbour species that flourished and probably had wider ranges during earlier periods of cooler climate.

For mammal immigrants that had successfully entered northern Africa and moved south, a significant challenge was presented by the continent's equatorial girdle, the single largest area of equatorial habitat on earth. The difficulties involved were far from trivial: there were established competitors and predators of which the immigrant had no evolutionary experience and the environment was dominated and shaped by large numbers of giant animals, notably endemic species of elephants. There were also physiological and behavioural challenges for animals entering moist, hot habitats from relatively dry cool ones.

Greater than any other obstacle was the existence of numerous tropical diseases. Disease is a primary agent in evolution so the relative susceptibility of mammals to Africa-specific (or extra-African) diseases, will, undoubtedly, become a very significant topic for research in the future. When immigrant populations encountered the habitats of equatorial Africa they met diseases against which their recent evolutionary history offered no defence. Both evolutionary and biogeographic patterns in Africa are likely to have cryptic subtexts in which disease has played a critical, if not central role (W.D. Hamilton, pers. comm.).

The tragelaphine example

To illustrate the dynamics of adapting to novel habitats, take the relatives of the Bushbuck, the Tragelaphini or spiral-horned antelopes. Now widespread and diverse in sub-Saharan Africa, they are known to have speciated from a common ancestor that left Eurasia some 14 mya. A recent molecular study (Willows-Munro *et al.* 2005) has provided a genealogical tree that offers clues to the sequences and eco-climatic determinants behind this very successful radiation and some suggestive correspondences between the conservativeness or innovative adaptedness of species and their eco-geographic distribution. Of this distinctive group of mainly browsing bovines, two very similar species, the Nyala *Tragelaphus angasii* and the Lesser Kudu *T. imberbis*, are close to the stem of the entire radiation. Extrapolating from tragelaphine morphology and fossils we can infer that these two are probably more like their Asiatic ancestor than any other extant species. Their separation, in dry enclaves at opposite ends of eastern Africa, is probably of long standing but thicket-dwelling and leaf-browsing plus, perhaps, their strongly residential habits, are likely to be common, conservative retentions. Taken together, this closely related pair can be regarded as contemporary relicts still occupying favoured pockets of a once more extensive south-eastern range.

Other tragelaphines have moved further from their ancestral niche and, of these, the species that derive most directly from their common ancestor are the Greater Kudu *T. strepsiceros* and the Mountain Nyala or Gedemsa *T. buxtoni*. An appreciable increase in size is apparent in these two species and fossil relatives of the Greater Kudu were bigger still. This trend towards larger size can be correlated with the development of more mobility, larger ranges and a wholesale shift out of closed woody environments. It is a trend that culminates in the two species of eland, which can weigh close to one ton, lead nomadic lives in savannas or relatively open woodlands and have already provided an illustration of the centre-west/south-east divide (for further discussion see the species and genus profiles and Kingdon 1982, Matthee & Davis 2001 and Willows-Monro *et al.* 2005).

The more conservative tragelaphines, Lesser Kudu and Nyala, conform to the expectation that early descendants of incoming immigrants prefer the drier and sometimes cooler habitats of eastern and southern Africa. Overcoming the adaptive challenges of lower, warmer and wetter habitats would have taken more time. The spatial dimension of such an expansion had to be westwards along the centre-western axis and this is what two apparently more advanced species did. Willows-Munro *et al.* (2005) identified the Bongo *T. eurycerus* and Sitatunga *T. spekii* as members of a tragelaphine 'closed forest group' and confirmed that the two are particularly closely related (for example they produce fertile hybrid offspring in captivity). Although their distribution ranges overlap extensively, the two species have differentiated both morphologically and ecologically in substantial and very significant ways (see the species profiles).

Today the shaggy Sitatunga inhabits deep swamps, most notably those in the forests of the Congo Basin, and it has evolved long thin legs and splayed hooves to aid it in walking over mud and compacted vegetation in search of herbaceous and fresh leafy growth. The ancestors of the Sitatunga were unlikely to have moved into such habitats 'voluntarily'. Instead, like the Dwarf Mongoose's reliance on termitaries, swamp or swamp forest would have been adopted by relatively slow, defenceless animals as a refuge from predators:

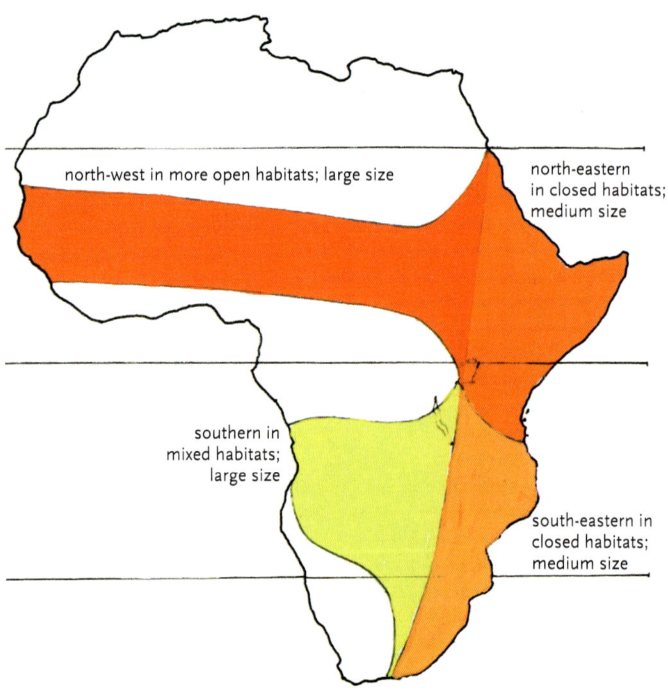

Map 1. Schematic map of adaptive trends in early tragelaphines to Boreal/Austral, drier/wetter, more open/closed environments. Note larger body sizes in drier west (retained by *Tragelaphus derbianus* in N-W, *T. strepsiceros* in S-W.); medium size in eastern thickets (retained by *T. imberbis* in N-E, *T. angasii* in S-E).

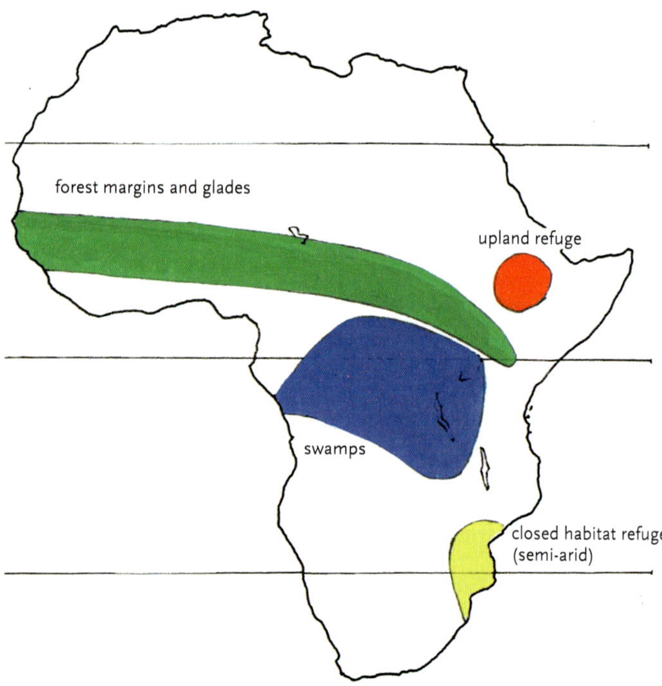

Map 2. Schematic map of adaptive trends in later tragelaphines to: 1. forest margins and glades along northern margins of forest belt (retained by *Tragelaphus eurycerus*); 2. swamps in or near forest (retained by *T. spekii*, especially south and east of forest belt).

selection through predator-evasion. Only slowly, step by step, have swamps become a permanent, preferred and exclusive habitat and the Sitatunga's long-term survival must have been facilitated by an important property of swamps. They are a relatively predictable environment and persist because they are dependent more on impeded drainage than on local climate. In a vast, relatively shallow river basin where many tributaries flow, sluggishly, from east to west, swamps stretch, as an enduring and predictable habitat, from central Africa to the Atlantic. While extreme climatic fluctuations had the capacity to eliminate entire communities over very extensive areas, and probably did, swamp communities still survived wherever the drainage profile was right.

Another type of ecological continuity exists in the same central African forests. Ground-level growth needs sunlight and the agencies that fell forest trees to produce fresh new plant growth are predictable. Trees fall for many reasons and they create 'chablis' or temporary clearings that attract browsing animals. These open areas are rarest and most temporary in the areas of highest rainfall; they are commonest on or close to forest margins, where elephants and fires knock the forest back, or beside impervious substrates where trees find it harder to grow. Bongos *T. eurycerus* have learned to exploit ephemeral but very nutritious sources of new growth around forest clearings and along forest margins. However rich they are at any one time, such resources are unpredictable and they require exceptional mobility of their consumers. In the process of acquiring just such mobility, the Bongo has modified both its behaviour and its morphology, some of which parallel strategies adopted by elands (for which consult the species profiles).

The adaptive shifts by ancestral sitatungas and bongos into specialized forest subtypes therefore seem to have taken place along identifiable ecological (and latitudinally aligned) gradients. One has taken place at the wettest end of the catena while the other follows 'forest edges' or, more generally, 'the forests' edge'. The common ancestor of the Bongo and Sitatunga probably resembled the extinct, but formerly widespread *T. nakuae*, and molecular trees suggest that the Bushbuck, which currently occupies the 'generalist' tragelaphine niche, has derived from the same stock.

It is interesting that two duikers have followed a similar trajectory to the Sitatunga/Bongo bifurcation in exactly the same region. The splay-hooved Black-fronted Duiker *Cephalophus nigrifrons* favours Congo Basin swamps while its closest relative, the 'edge-loving' Red-flanked Duiker *C. rufilatus* prefers secondary growth along the northern margins of the same river catchment. Both species occupy the rich, complex forests of the Congo Basin where the resources available to duikers are partitioned out between multiple species and both are classic centre-westerners. As with the tragelaphine Bongo/Sitatunga pair, the genes of *nigrifrons/rufilatus* suggest they are among the most recently evolved species in their group and in both instances, only a less specialized common ancestor could have given rise to such contrasting life-styles. Their earlier origins conform with the centre-west/south-east model in that their known closest relatives are the generalized, non-specialist (and therefore more conservative) Natal/Harvey's duiker complex, *C. natalensis* and *C. harveyi*, which live in the patchy, impoverished forests of south-eastern Africa.

Tracing the origins as well as the fate of more generalized ancestors is central to understanding how adaptive niches proliferate in African habitats (particularly when the adaptor has immigrant ancestry).

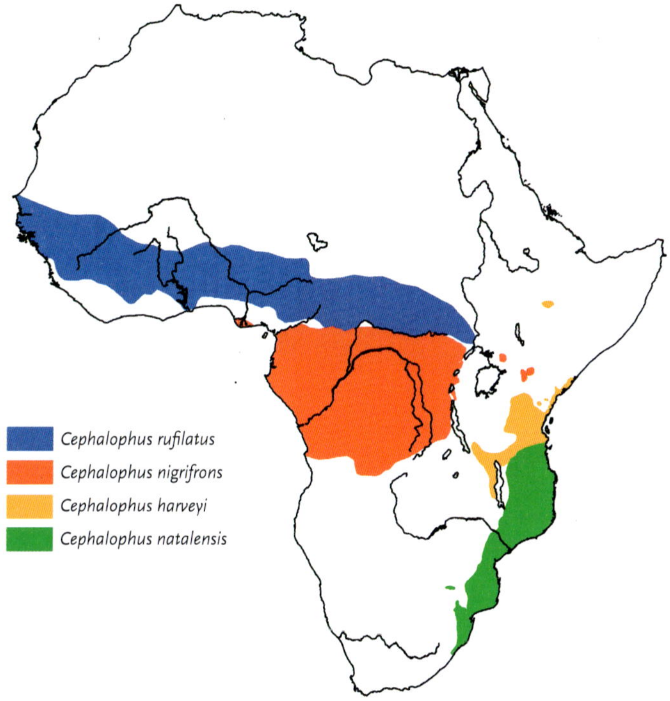

Eco-geographic speciation in duikers: *Cephalophus nigrifrons/rufilatus* in west; *Cephalophus harveyi/natalensis* in south-east.

Recent immigrants and how they fare in Africa

The most recent arrivals can be identified by their close affinity with relatives in Eurasia. For example, the (Barbary) Red Deer *Cervus elaphus* has been present in north-west Africa for less than a million years, likewise, various rodents such as the Long-tailed Field Mouse *Apodemus sylvaticus,* Garden Dormice *Eliomys* spp., the Four-toed Jerboa *Allactaga tetradactyla* and some Asian gerbils (Gerbillinae), the Palestine mole-rat *Nannospalax ehrenbergi* and the Red Fox *Vulpes vulpes* are all Eurasian species with marginal extensions of range into northern Africa. The polecat and two species of weasels (all *Mustela* spp.) are also recent arrivals that have failed to penetrate beyond the northern fringes of the Sahara, essentially outliers of species that are widespread and successful in Eurasia.

Eurasian immigrants that have yet to push their frontiers very far south are the Rhyzomyidae (root-rats), mainly a Himalayan and South-East Asian group. Their African representative, *Tachyoryctes*, has speciated into a variety of forms and is locally very successful and abundant yet only known at present from the upper reaches of some north-eastern African uplands. One constraint may be their own recent Eurasian origins; another inhibition to expansion might be the presence of competitors, in this case from an Africa-evolved equivalent, the blesmols, Bathyergidae, which are known to be a very ancient African 'mole-rat' lineage.

A corollary to the fortunes of immigrants was their impact on established species and communities. Each new incursion introduced new competitive challenges and, in many instances, new predators. A long sucession of immigrant carnivores presented challenges to all but the largest mammals as Eurasian herpestids, mustelids, ursids, viverrids, felids and finally canids found their way into Africa. Here were new fields for their particular, often highly specialized, predatory talents, so, with more excavation of dated fossil mammal assemblages, we can expect to see the rise and fall of both predators and their prey documented in increasing detail. While canids, in particular, were likely to have hit all those species that were vulnerable to such fast and intelligent predators, not all canids were a universal success.

All canids in Africa have relatively shallow roots in the continent but the Ethiopian Wolf *Canis simensis* may have the youngest pedigree of all because its closest affinities have been ascribed to Eurasian wolves and its arrival can be linked with one or other of the more recent glacial periods. It remains restricted to tundra-like habitats on the highest reaches of the Ethiopian Dome. Other immigrants with unequivocally northern temperate roots have managed to establish enclaves in the extreme south. These include the southern Cape Fox *Vulpes chama*, with close relatives in Eurasia and North Africa. Both these last examples emphasize how forbidding the equatorial region can be for northern-adapted mammals, yet other mammal lineages have eventually succeeded in adapting to the diverse habitats of tropical Africa, many of them emphatically different from anything in neighbouring parts of Eurasia. The restriction of particular species to identifiable vegetation types encourages the assumption that the primary influence on adaptation is environmental. In the proximate sense this may be true but it can be demonstrated that the technique that an established lineage brings to the task of making a living is decisive in its success. For example, Leopards *Panthera pardus* are successful over a wide spectrum of habitats and their ability to over-ride local climatic differences is due to the efficacy of their generalized predatory habits. Regardless of where they originated, leopards are no less successful in Africa than in Eurasia.

There are some apparent anomalies; bears have been widely successful in both temperate and tropical Eurasia as well as the Americas, but not in Africa. Why? An extinct fossil bear *Indarctos* occurred in Libya and a rather athletic brown bear *Agriotherium africanus* is known from the very early Pliocene in Ethiopia and South Africa. Both north-east and southern Africa have consistently temperate climates but the brevity of bear presence in Africa correlates with, and could have much to do with, the emergent radiation of bipedal apes or hominins in the same areas. Fossil hominins seem to have particularly favoured these cooler regions and, as relatively large omnivores, they were already well established in Africa at the time of these bear invasions. It is therefore plausible that hominins and bears shared similar dietary and climatic tastes, implying that ecological overlap between them was substantial and led to the bears being eventually out-competed by hominins.

Hominins and bears – competitors? Extinct African bear *Agriotherium africanus* and Nutcracker Hominin *Paranthropus boisei*.

The Ethiopian and East African uplands resemble South Africa in being an essentially temperate region, but differ in straddling and extending north of the Equator. In the case of Ethiopia's higher reaches these represented an extreme southern extension of northern glaciated habitats at the time when colder temperatures probed deep into Africa. During northern ice ages any incoming Eurasian invaders would have found upland north-east Africa especially inviting because environments existed there that were broadly comparable with those in the invader's region of origin and also had the virtue of less competition from established mammals. Actual glaciers are inimical to all life, but in the aftermath of extensive glaciation, especially over high upland plateaux, their retreat exposes extensive areas of sterilized, new and 'empty' land. In particular, the higher regions of the Ethiopian Dome became available for animals that could tolerate low temperatures. The example of invading root-rats of the genus *Tachyoryctes* has been mentioned.

Because habitats on the higher reaches of Ethiopia are subject to very distinct climates and are isolated from equivalent habitats both outside and within Africa, many animal and plant populations have speciated there with exceptional rapidity. During glacial maxima the glaciated peaks would have been effectively sterile, but tundra-like belts would have radiated out over all the surrounding mountain belts and piedmonts, creating corridors for cool-adapted species. The more mobile of these colonized other uplands and a few (notably birds such as Wattled Cranes *Grus carunculatus* and bald ibises *Geronticus* spp.) got as far as temperate South Africa. During warmer interglacial periods inter-connecting corridors disappeared, leaving smaller subpopulations of temperate-adapted species stranded on higher massifs. The massifs included Cameroon in the west, Ethiopia in the north-west and a wide scatter of mountains southward down the eastern side of Africa, mainly following the Rift Valley. The flow of species was in both directions, with incomers drifting south and a few southerners drifting north. A well-known example (of a predominantly southern montane species) is the Mountain Reedbuck *Redunca fulvorufula*, which favours dry uplands between about 1500 and 5000 m on mountains as far apart as South Africa, northern Tanzania, Ethiopia and Cameroon.

A long succession of climatic swings, connecting and disconnecting populations, has provided the primary mechanism for local populations to become distinct. As a result most members of these montane communities are likely to have deeply laminated genealogical histories that current research techniques are still unable to untangle. None the less, geographic foci for distinct speciations can be identified. For example, in north-eastern Africa niches for small herbivorous rats are dominated by four closely related 'grass rat' genera, *Arvicanthis*, *Pelomys*, *Mylomys* and *Desmomys*. As entire biota ebbed and flowed up and down the flanks of the vast Ethiopian Dome, 'grass rat' populations must have made opportunistic excursions back and forth from this major region and these pulses from an identifiable focal point must have influenced the evolution of some 15 species of 'grass rats'. A similar north-eastern bias is evident in some of the omnivorous murid genera. Other aspects of these murid radiations are discussed further below.

Rodent serial invaders

The special opportunities that Africa offers to incoming mammals and their subsequent fortunes over time can be illustrated with a brief summary of the likely history of rodents on this continent, as reconstructed from fossils and recent molecular studies. There is now persuasive evidence that the rodent fauna of Africa breaks down into about a dozen invasions, of which at least five or six are major groupings, each of which plausibly derives from a single ancestral species and each of which invaded Africa at a different period and then radiated. The groups can be listed (in the sequence of their supposed time of arrival in Africa with dates approximated from Adkins *et al.* 2001, Huchon *et al.* 2002, Montgelard *et al.* 2002, Jansa & Weksler 2004, Steppan *et al.* 2004, and E. Seiffert pers. comm.).

Zegdoumyids, possible anomalurid ancestors, are first found fossilized at 50 mya, so a Palaeocene arrival is possible. After a major gap in the fossil record, definitive anomalurids are found in Africa at 37 mya. Protophiomys-grade hystricognathous rodents are also present by 37 mya (at Fayum and Bir el-Ater) and a major radiation of phiomyid and thryonomyid rodents seems to have taken place between 34 and 29 mya in the Jebel Qatrani Formation (E. Seiffert pers. comm.). Then, after another major gap in the African record, an ancestral Nesomyid arrives, probably between 25 and 27 mya. Soon after this immigration, Proto-Gerbillinae arrive about 21–23 mya and, finally, at about 10.5 mya African murines arrive, to produce the largest of all rodent radiations. In each case, likely arrival would have taken place well before occurrence as fossils.

Given the quite numerous total of immigration events, it is tempting to conclude that opportunities for successful immigration were frequent, easy and involved the incursion of multiple ancestral species each time. Such a conclusion does not square with the fossil evidence and the sheer diversity of descendants deriving from single ancestors implies that such events were rare and generally quite widely spaced. The earliest radiations of rodents took place outside Africa, possibly as early as the late Cretaceous, 66–88 mya (Adkins *et al.* 2001, 2003, Bininda-Emonds *et al.* 2007, Meredith *et al.* 2011). By the Eocene rodents had already split into proto-anomalure, proto-squirrel and proto-porcupine (hystricognath) lineages. The last vestiges of the anomalurid radiation survive in the gliding anomalures (*Anomalurus* and allies) and hopping springhares (*Pedetes*),

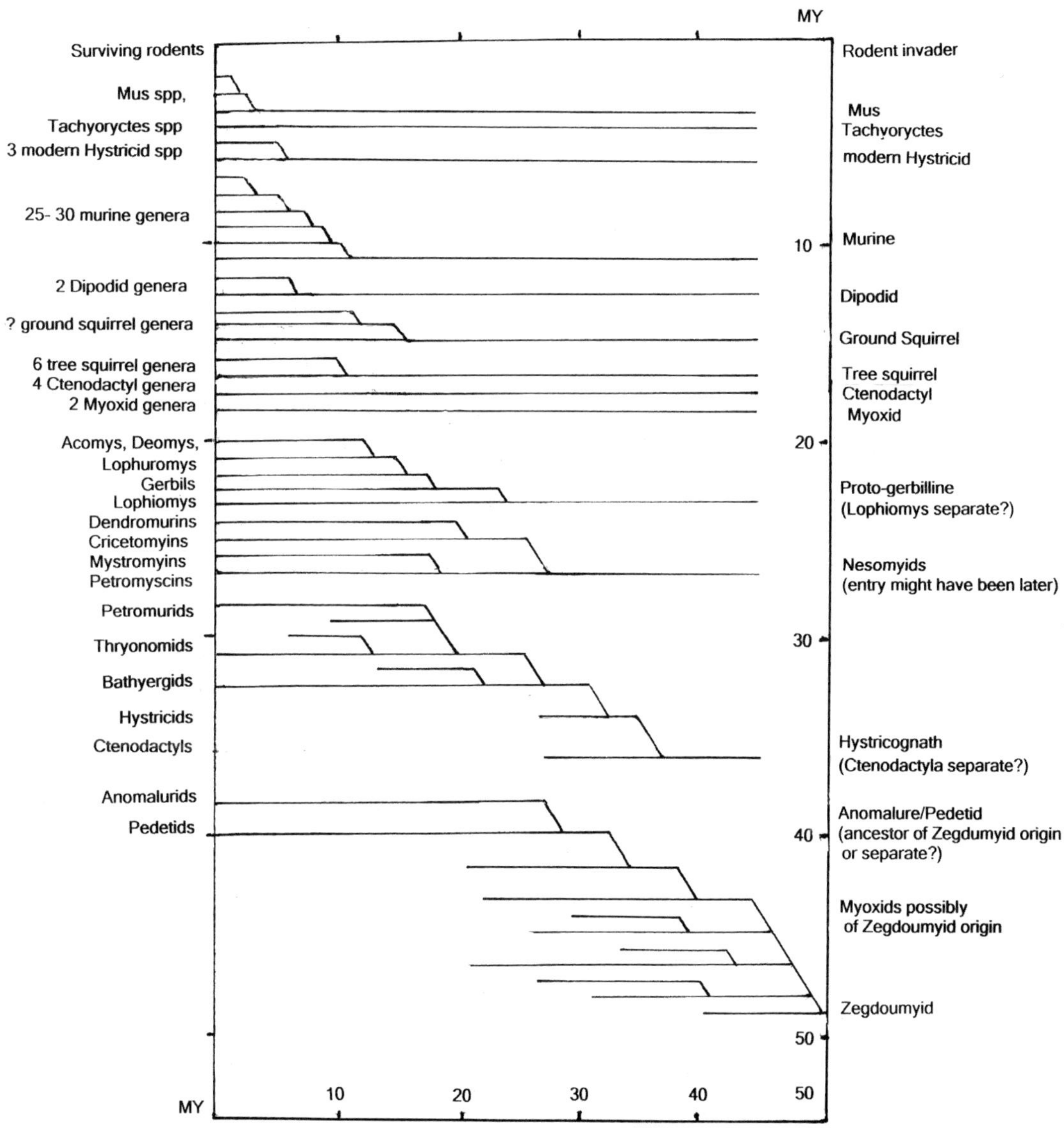

Schematic table of procession of rodent radiations from supposed single invaders. Note the late entry of modern hystricids, ctenodactyls and myoxids, possibly preceded by much earlier African radiations (in part after Montgelard *et al.* 2002, Jansa & Weksler 2004, and courtesy of Erik Seiffert, Louise Roth, John Mercer and David Happold.)

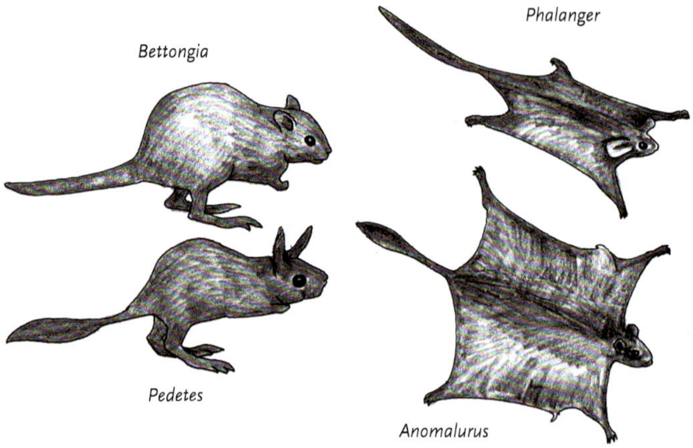

Silhouettes of Australian marsupial *Bettongia* and African springhare *Pedetes*, and Australian marsupial *Phalanger* and African anomalure *Anomalurus*, illustrating convergence between the arboreal and terrestrial members of ancient Australian and African taxa.

both of which are herbivorous. (Since the Eocene, Australian arboreal marsupials have evolved gliding forms several times and, at an equally early date, the same phalanger lineage has also given rise to the hopping bettongs (*Bettongia*), so the anomalure–springhare duo are less than unique in the breadth of their ecological span! It is, perhaps, not surprising that subsequent rodent invaders have been unable to supplant such super-specialists, particularly since several anomalurids are of relatively large size for a rodent.

Another rodent radiation, of which five families survive, mostly largish herbivorous specialists, are grouped together in the Ctenohystrica or Nesomyids (Montgelard *et al.* 2002). This disparate group may or may not include desert rock-dwellers, the gundis (Ctenodactylidae) (genetically thought to be the most primitive type); the porcupines (Hystricidae); the subterranean and sometimes eusocial blesmols or African mole-rats (Bathyergidae) (Jarvis *et al.* 1994); the flat-skulled Noki (Dassie Rat) *Petromus typicus* (single survivor of a once diverse group, the Petromuridae); and two species of large, valley-grass-eating specialists, cane rats of the family Thryonomyidae.

Fossil precursors of the Noki were unremarkable rodents, yet the only survivor of this once abundant lineage has transformed its anatomy by flattening and widening both body and skull, betraying thereby how strong selection has been for survival in a relatively open, predator-rich terrain, where the only shelter is in narrow horizontal rock crevices. The Noki also illustrates why older lineages are more likely to be found in marginal or demanding, difficult habitats. In such places selection is severe and operates on many fronts so that, over time, a large number of incremental changes are needed to survive predation. Appropriate adaptations then become partly a function of time lived in such habitats. Both the Noki and some Saharan gundis are extremely localized, a restriction that is consistent with long tenancy and the likelihood that their ancestors were long ago out-competed in more generalized rodent niches (with those competitors generally being the descendants of later rodent immigrants).

Another rodent species arrived in Africa, possibly at the end of the Oligocene, this time a small mouse; it was probably somewhat of a climber and adapted, initially, to the relatively benign climatic conditions of the period. If its descendants are anything to judge by, they were a lot more sluggish in their movements and activity than advanced rodents (such as common rats). The descendants of this mouse penetrated most habitats and, having reached Madagascar, became dominant rodents there, a status that their descendants, the Nesomyinae, still retain. Like other rodent groups that arrived relatively early, their African survivors are now highly specialized, or, in two cases, *Leimacomys* and *Delanymys*, relicts on the verge of natural extinction.

The more successful specialists have taken advantage of small size to become very adept climbers on grass or fine herbage stems (the climbing mice *Dendromus* spp. and allies) or exploiters of tiny crevices in desert rocks (pygmy rock mice *Petromyscus* spp.). The one lineage that has increased in size, the pouched rats or Cricetomyinae, have evolved a unique foraging strategy that seems to compensate for their relatively slow reactions and movements. Very large cheek pouches permit these rats to gather volumes of food at the safest time (generally at night) and return to a secure refuge, where the gatherers may spend many days consuming their store. The largest of these are the celebrated Giant Pouched Rats *Cricetomys* spp. All these rats and mice are mainly vegetarian.

Head of Gambian Giant Pouched Rat *Cricetomys gambianus* showing partially filled cheek-pouches.

Once established in Africa, emigrant Ctenohystrica went on to make a sweepstake colonization of South America, possibly in the late Eocene, where they have been particularly successful, while an early member of the Old World porcupines 'returned' to Eurasia and speciated still further. Later in this discussion, I draw attention to the 'return' of previously African primate lineages to Asia and then a still later 're-return' to Africa. The porcupines have followed a similar peripatetic history, with the most recent returnees giving rise to the largest, most prickly species, *Hystrix cristata* and *H. africaeaustralis*.

In spite of their spiny armour being so familiar an example of animal defence, the selective transformation of rodent dorsal fur into a barrage of long sharply pointed spikes on porcupines is still astonishing. As a vivid demonstration of selection being driven by predators the entire transformation can still be seen in a comparison of living porcupine species. The least prickly of species remains a relatively nondescript and smallish rodent while the largest is the most fiercely spinous, implying, incidentally, that predators may have been the main constraint on rodents getting very large. Porcupines are not the only rodents to shield themselves with prickles and another example is discussed shortly.

Some time after the Giant Pouched Rat's ancestor got into Africa, another immigrant arrived, probably about 20–22 mya. This was a

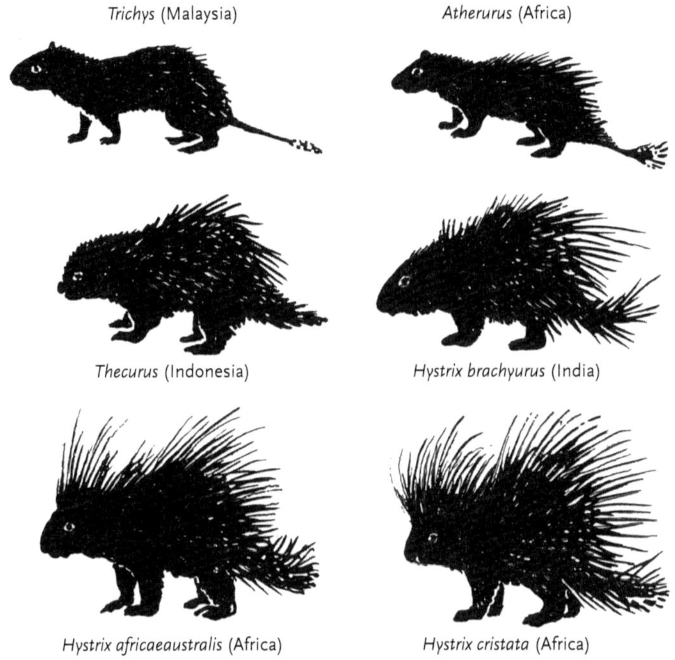

Silhouettes of African and Asian porcupines illustrating a gradient in relative spinyness.

mouse that could be described as an unspecialized proto-gerbil. This time the ancestral form was likely to have had more omnivorous habits, was probably already adapted to dry conditions and, again, judging by many of its descendants, was fast and active. Initially, a continental divide might have kept proto-gerbils (Gerbillinae) in Eurasia from the proto-spiny mice (*Acomys* and allies) that are more nearly wholly African. The gerbils have continued to cross back and forth between Eurasia and Africa, a mobility that must be helped by resistance to the heat and drought typical of the north-east African corridors with Asia. Today gerbils are common and very successful inhabitants of all the drier habitats of Africa and western Asia. In their progressive adaptation to the difficulties presented by drought, heat and multiple predators their evolutionary history probably parallels some features of the Noki and pygmy rock mice, but these are more fecund, faster and more generally adaptable rodents.

The revelation of a genetic relationship between the short-legged, short-muzzled and mainly arid-adapted spiny mice *Acomys* spp. and the long-legged, long-muzzled and forest-dwelling Link Rat *Deomys ferrugineus* came as a surprise to field biologists (Jansa & Weksler 2004). None the less there are resemblances in the springy, flattened, finely grooved (and probably predator-deterrent) hairs, large, mobile ears, pointed muzzle and adept use of the hands to handle invertebrate foods. Furthermore, the molecular clock suggests more than 12 million years have elapsed since divergence of the spiny mice and Link Rat lineages (Steppan *et al.* 2004). This discovery is of considerable interest because it shows how animals can be totally transformed by the adoption of new niches. In this case a rodent with dry-country gerbil-like ancestors has (in common with some of the brush-furred mice *Lophuromys* spp.) adapted to equatorial forest. Within the forest, *Deomys* ranges through relatively dry areas during the rains but descends into valley bottoms during the dry season, where its long legs are clearly an advantage in shallows and it can hunt invertebrates along swamp-forest margins.

The food resources in this rather linear habitat were clearly the primary incentive for this rodent's shift into forest. At the overall landscape level, many swamps dry out seasonally and tend to be dependent on the vagaries of climate but, within the forest zone, where evaporation is reduced by a canopy, or along the larger forest galleries, even during the driest climatic cycle, there are always obstructions to the free flow of water. Sometimes these dams, which hold back water and create swamp forest, are of substantial extent, sometimes they are extremely local, but they serve as refuges and 'concentration zones' for a very wide range of water-dependent organisms. It is the long-term durability and predictability of this habitat that ensures a rich invertebrate fauna and attracts such predators as can cope with a wet and densely obstructed substrate and find prey that is seldom on the surface. *Deomys* shares a preference for invertebrate foods (including indigestible ants) with its closest relatives, *Acomys* and *Lophuromys*, but it has enlarged their common sensory apparatus; the whiskers are almost shrew-like in their length and fineness, the ears are enormous and highly mobile and the long barrel of a nose is extremely sensitive to scent clues; this barrage of detectors clearly enhances its success as a hunter of invertebrates. While the divergence of *Deomys* from its more arid-adapted cousins involves an ecological shift, the biogeographic dimension of its earliest beginning brings us back to the south-east/centre-west divide, with *Deomys* ancestors likely inhabiting the centre-west and proto-*Acomys* the south-east and north.

The evolution of this seasonal swamp-forest specialist from more generalized *Acomys*-like cousins is doubly interesting because there are extremely close parallels among some of the descendants of yet another rodent colonist, the last to enter and radiate widely in Africa. Furthermore, there is good evidence that the most *Deomys*-like of the later arrivals already puts substantial competitive pressure on *Deomys* wherever the two species overlap in range.

An even more extraordinary adaptive history surrounds the Crested or Maned Rat, *Lophiomys imhausi*. According to Jansa & Weksler (2004), this sluggish long-haired rodent, which can weigh nearly a kilo, probably shares ancient origins with the proto-gerbillines and, in common with the spiny mice and relatives, has modified hairs. In the case of *Lophiomys* its flank hairs are probably the most complex of all mammalian hairs. Tips and bases are solid, normal hairs but the central shaft is inflated with a strong outer cylinder perforated by abundant vacuoles. This outer cylinder encloses numerous long, fine fibres that together act as a 'wick'. These wicks ensure rapid absorption of secretions and once hairs are saturated the open lattice-work ensures that the hairs cannot be touched without direct contact with the secretion. These hairs grow in tapered, leaf-shaped tracts running from behind the ears across each flank to the groin. The animal can expose these tracts by means of specially modified dermal muscles, which erect the animal's long, externally dull grey fur upwards above it and deflect it downwards below it. This flaring of the fur, which is triggered by any external interference or attack, reveals bold black and white bands on the longer hairs that grow above and below the shorter spongy flank hair. The effect of this striking pattern is somewhat zorilla-like but effectively serves to outline the flank tracts. The rat pulls its head back into its shoulders, hisses loudly and turns its flared tract towards its adversary, as if actively inviting a bite!

This visual display is not the only hint of anticipating an attack – the cranium of *Lophiomys* has evolved a 'double hull' or cranial 'helmet'

a

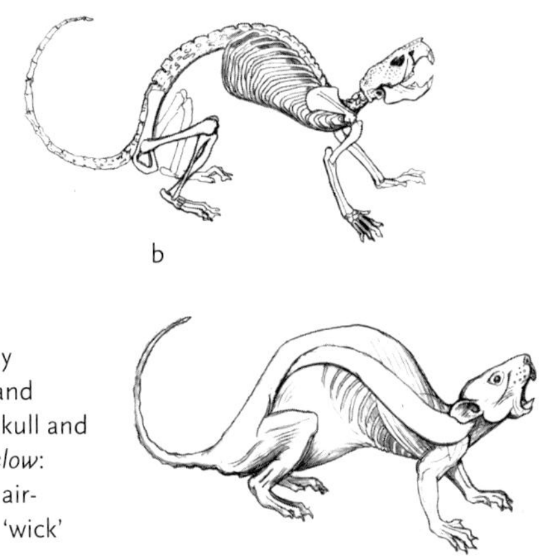
b

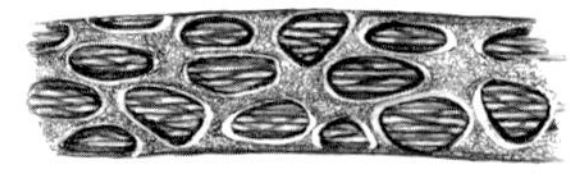
c

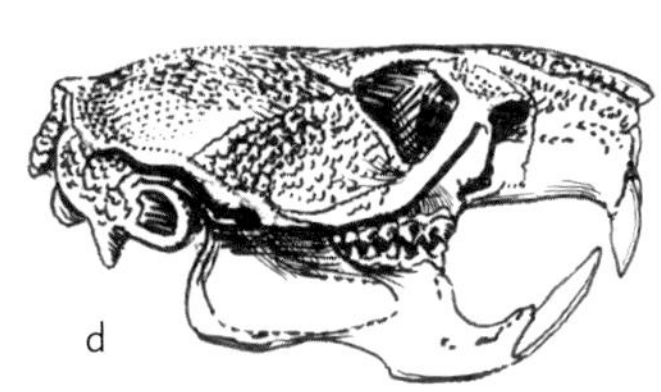
d

Crested Rat *Lophiomys imhausi*.
a. External warning (aposomatic) pattern in defensive posture. Display serves to expose a specialized hair tract and focus attention on neck and shoulder. b. *Above*: Skeleton showing damage-mitigating features of skull and skeleton: cranial 'helmet', 'floating' scapulae and robust vertebrae. *Below*: Schematic myology showing position of area underlying specialized hair-tract. c. Vacuolated hair-shaft enclosing numerous fine filaments that 'wick' secretion. d. Reinforced, 'helmeted' skull.

The molecular and physiological details of this mammal's toxin-management systems have significant future implications for human health, making it potentially one of the most important and interesting rodents in Africa.

of reinforced bone. The temporal muscles (which attach to the top or sides of the cranial capsule in all other rodents) are, in this case, totally encased under a canopy of dense, carrunculated bone. Rugose, 'pimpled' bone extends over the bridge of the nose, the back of the bullae and, significantly, over most of the external surfaces of the occiput. This skull reinforcement is unique to this species. *Lophiomys* has an extraordinarily dense, tough and close-textured dermis, which is resistant to all but the sharpest of teeth, beaks or claws. It has an exceptionally robust, flexible and elongated vertebral column (with three extra thoracic and one extra lumbar vertebrae), all with enlarged bodies and short spines apparently designed to enhance the amplitude and flexibility of lateral movements. Further peculiarities are near-atrophy of the clavicle, effectively freeing the broad scapulae, apparently to act as shield-plates over the neck and thorax when the shoulders are hunched. These 'floating shoulder-blades' are made all the more mobile by particularly well-developed attachment muscles, notably the trapezius, latissimus dorsi, levator anguli scapulae and brachiocephalicus (see Kingdon 1974, 1983). The lips and tongue are also prominent and well developed. Beneath all these specific external features are yet further adaptations unique to *Lophiomys*. Its stomach has weakly sacculated into five compartments (Vorontsov 1967), implying a complex digestive physiology. Consistent with the role of some mammal salivas in producing proteins that can bind to plant polyphenols, *Lophiomys* has large salivary glands. Circumstantial evidence suggests that both the salivary glands and stomach might be adapted to metabolize plant toxins.

This apparently random assortment of adaptations is connected by the single evolutionary thread of predator-selection. Dogs that attempt to bite *Lophiomys* exhibit every symptom of being poisoned by contact with the hairs and, in some cases, die with startling rapidity, apparently from heart failure. How could a rat, which seldom shows much propensity to bite back itself (in spite of much hissing and snapping), inflict such a deadly response? And how could such a fast effect on the predator have evolved? The answer lies in an astonishing case of association between this mammal and very toxic plants belonging to the Apocynaceae, notably *Acokanthera* spp. In East Africa *Lophiomys* is found in rough or hilly areas, mostly between 1100 and 2300 m where these locally very common evergreen trees grow in thickets, low-canopy forests or in stony grassland mosaics and gullies. In common with primates and birds, *Lophiomys* feed on the non-poisonous berries of *Acokanthera* but unlike them *Lophiomys* tolerates chewing the bark and roots, which contain higher concentrations of Ouabain, a cardiac glycoside or cardenolide that precipitates pain and cardiac dysrhythmia in many vertebrates. Natural selection has led *Lophiomys* to exploit the susceptibility to Ouabain in its principal predators, mammalian carnivores, while sharing a general insensitivity to Ouabain among many fellow rodents. *Lophiomys* has evidently enhanced its own tolerance of the same toxins to the point where it can chew up and slaver Ouabain/saliva colloid for absorption into its flank hairs (Kingdon *et al.* 2011).

There are abundant instances of insects and some birds utilizing plant toxins to deter or poison their predators, but this is the first instance of something comparable evolved by a mammal. So, while further details of the exact relationship between rat and plant await further research, the genetic basis for Ouabain-control metabolism has obvious implications for therapy in humans and other mammals. Even the barest outlines of this toxic defence hint at the power of predation to select for very unusual defences in prey species.

Another peculiar example of a predator-driven defence strategy has evolved in a founding population of the last, and most extensive radiation from a Eurasian rodent invader. Its ancestor was essentially a generalist and it arrived, together with a number of other Eurasian mammal immigrants, at about 10.5 mya. On entering Africa it found a continent that was populated by the descendants of up to nine previous rodent radiations, but this Eurasian murine clearly possessed many of the attributes that have ensured the almost world-wide success that *Rattus*-like rats enjoy today.

According to Misonne (1969), this ancestral stock has a direct and morphologically little-changed descendant in the acacia rats *Thallomys* spp. Consistent with their semi-arid Eurasian origins these rats range through the drier parts of eastern and southern Africa, from Ethiopia to the Kalahari, preferring thorny acacia woodlands where they are both terrestrial and arboreal. In spite of their generalized and conservative morphology these *Acacia* folivores have evolved an energy-consuming behaviour that must have been predator-selected no less than the prickles of porcupines or the poisonous-ness of *Lophiomys*. Like miniature beavers, Acacia Rats expend much time and energy gnawing, cutting and carrying thorny twigs of *Acacia*,

Thorn 'corridor' built by *Thallomys paedulcus* along branches of Acacia tree. *Enlarged*: a single typical thorn twig (10 cm long) with nine 8 cm thorns, cut, carried and woven into 'corridor' by *T. paedulcus.*

typically *A. tortilis* and *A. xanthophloa*, each twig being up to 15 cm long (about the rat's body length). The rat then transports these twigs and interlaces them in dense self-made thorn tangles that snake over the branches of their home trees and connect arboreal nests with terrestrial burrows. The rats travel along these thorn corridors and it would seem that this corridor-making behaviour has served to artificially extend the one pre-existent property of their food-trees that protects them from predators, principally genets and owls – a barricade of thorns. If Misonne was correct about their conservative genotype then it would seem that the main evolutionary innovations in this species have been behavioural and digestive, leaving the morphology of an already successful invader largely unchanged.

The ancestor of all African murines soon diverged into two lineages separated by dietary preference. For one, the *Praomys* group, adaptable omnivory was probably the established norm and was retained or refined. For the other, the 'grass rats', *Arvicanthis* + allies, the vegetable resources of a continent were sufficient to induce a return towards the primary adaptive niche of rodents: the more-or-less exclusive gnawing and eating of plants.

In the centre-west region, mainly in rainforest, another, related branch of omnivorous rats/mice, *Praomys* and *Hylomyscus* spp., emerged (the latter becoming mainly arboreal) together with several much more specialized terrestrial relatives. I have mentioned how the long-legged, long-headed *Deomys ferrugineus*, aberrant ally of the *Acomys* lineage, became transformed by forest dwelling. Among the *Praomys* group an equivalent transformation took place. The swamp-forest-dwelling Long-footed Rat *Malacomys longipes* has very similar habits and morphology to *Deomys* and the convergence is so close that it clearly involves competition. In a comparison of the status of rats and mice in two Uganda swamp-forests, I found *Deomys* captures accounted for 28.5% of omnivorous rodents where *Malacomys longipes* was absent and only 5% in a swamp-forest where the latter was present (Kingdon 1974).

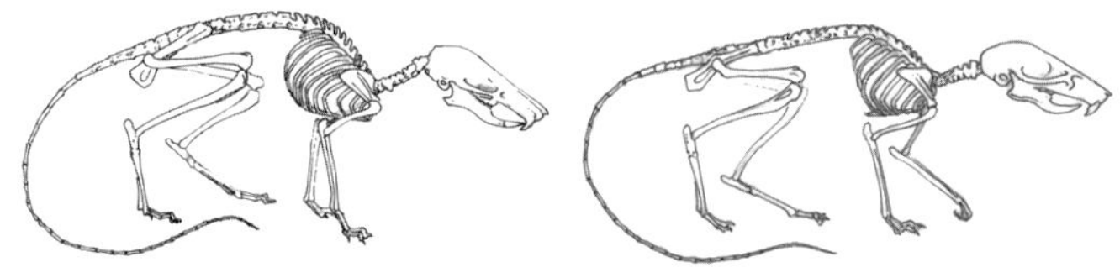

Skeletons of *Deomys ferrugineus* (*left*) and *Malacomys longipes* (*right*) illustrating close convergence in proportions of limbs and skulls.

The general conclusion to be drawn from these examples (and many less obvious ones among other mammals) is that most immigrant colonists begin as successful generalists because the narrowness of bridges into Africa seldom favour specialists. Initially the main species to suffer from the new invader are generalists that established themselves after earlier colonizations. In the case of advanced murines, they are more fecund, faster, tougher, more active and aggressive than their nesomyid and proto-gerbilline precursors and possibly have more efficient digestive and immune systems. As the recent invaders penetrated more and more habitats and speciated they became adapted to a steadily widening range of specialized environments, often in well-defined biogeographic zones. As specialization proceeds, competition with earlier forms can have many interesting outcomes, some of them culminating in extinction for the more archaic competitor. Others appear to lead on to still greater specialization and sometimes, especially in older species, further contraction into narrower or more difficult geographic and/or ecological zones (as has happened with *Thallomys* in genet and owl-infested thorn bush).

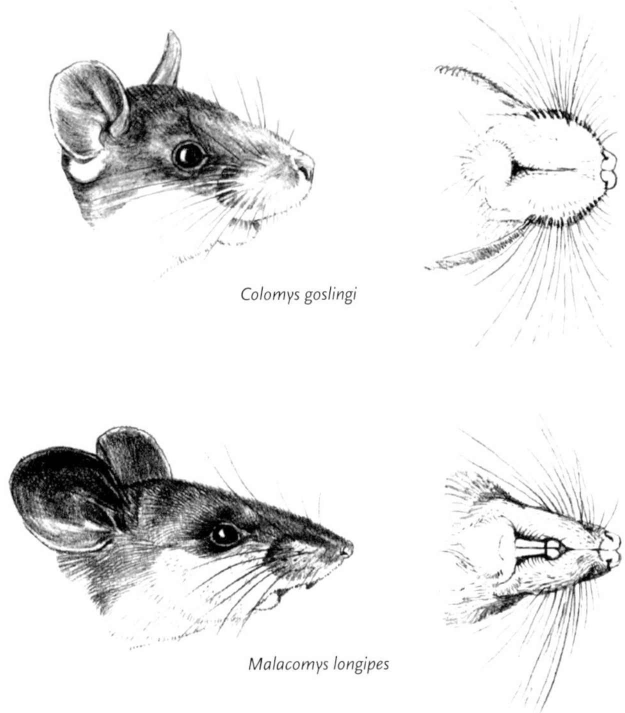

Comparison between *Colomys goslingi* and *Malacomys longipes* mouths and whisker patterns.

In a very recent adaptation to aquatic habits, yet another murine, the velvet-furred African Water Rat *Colomys goslingi* exemplifies a trend that has gone much further in the Australian Water Rat *Hydromys chrysogaster*. Even so, its muzzle has become hugely swollen and its very long whiskers arrayed into an otter-like splay that contrast strongly with those of its closest relative, the less aquatic *Malacomys*. Given more time, *Colomys* would probably evolve into a still more otter-like form and further complicate the ecology of water-side and swamp-adapted small mammal communities. Similar patterns of adaptation can be discerned in the biology of other mammals but patterns of competitive replacement are seldom as clear-cut as they seem to be among rodents.

Stranded in the south, stranded in the east

A factor that is discussed more fully in Chapter 3 is the northward migration and tilting of the Afro-Arabian land mass. This continental drift has diminished the area of Africa that lies south of the Tropic of Capricorn and this history may help explain why numerous endemic animals and plants are today restricted to very small ranges in the Cape of Good Hope. The ranges of ancient temperate-adapted species must have been greater in the past simply because a higher proportion of the continent lay within the southern-temperate zone. This can be taken to imply that formerly there were also greater numbers of species in these southern temperate groups (and fossils provide some limited support for this). The timing of evolutionary radiations within these southern endemics is currently unknown, but their speciation is likely to have been quite protracted.

Of contemporary mammal groups with overall ranges that are likely to have contracted the best examples are the ancient afrotherian Chrysochloridae or golden-moles and the archaic 'mole-rats' or blesmols Bathyergidae. If external catastrophies, such as major volcanic eruptions, have played their part in the survival of ancient groups then it could be no accident that both these groups are subterranean. Of seven genera of Chrysochloridae (embracing 21 species) only two genera occur outside southern Africa and both are montane or relictual. Of the remainder, all occupy precise, often very localized South African habitats with rather little overlap of ranges. In parcelling out a very limited area into numerous specialized species-ranges these southern golden-moles therefore resemble 'island' taxa (Birds of Paradise on Papua, say). The implications and effects of progressive contraction in a region that once covered a significant proportion of an entire continent deserves much more intensive examination than it has received so far. As their range gradually contracted did the number of golden-mole taxa (or lineages) remain much the same? Did they become fewer, or have they speciated more recently and become more diverse? Further knowledge of the biology of contemporary species will hopefully help answer such questions.

Golden-moles are by no means the only group of African mammals that have a well-defined 'regional centre of endemism'. If latitude and cooler, more seasonal climates have made southern Africa into an ecological enclave for some ancient afrotheres, long established peculiarities of climate and topography have created a similar set of small enclaves in tropical East Africa. The periodically semi-arid Somali–Kalahari corridor has served to cut off forest and montane-adapted species on or close to the eastern seaboard from forest communities in central and western Africa. The eastern coastal forests are intimately connected with the montane forests that grow mainly on eastern faces of mostly ancient massifs scattered from the L. Malawi region north-eastwards to the Usambara Mts and Taita hills in Kenya.

It is at the more tropical end of this montane/coastal complex of forests that another relictual group of afrotherians, like some golden-mole species in the Cape, maintain a precarious hold on existence. There are five species of giant sengi: Black-and-Rufous Giant Sengi *Rhynchocyon petersi*, on Zanzibar I. and parts of the adjacent coast; Golden-rumped Giant Sengi *R. chrysophygus* and an as yet undescribed species *Rhynchocyon sp. nov.* on the Kenya coast; *R. udzungwensis*, a newly discovered species known only from three small forest blocks in the Udzungwa Mts; and the widespread Chequered Giant Sengi *R. cirnei* (*R. udzungwensis* and *R. sp. nov.* are too recently described to be profiled in this work).

These five species illustrate how predators can drive evolutionary change through differential selection on their prey in different environments. This finds external expression in very interesting differences in coat colouring: relatively uniform in some populations, highly polymorphic in others.

Colour and pattern selection by predators

Before exploring this complex situation, the role of predator-selection can be exemplified in its simplest terms by another Macroscelid, the so-called Rufous Sengi *Elephantulus rufescens*. Over much of its range this species is mildly polymorphic in colour, its pelage exhibiting various shades of sandy, tinted browns.

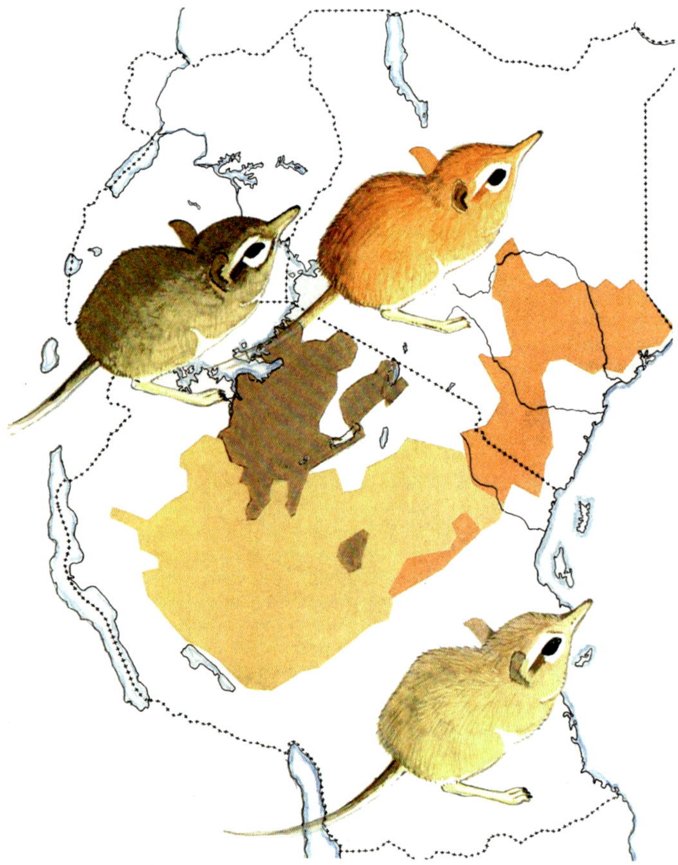

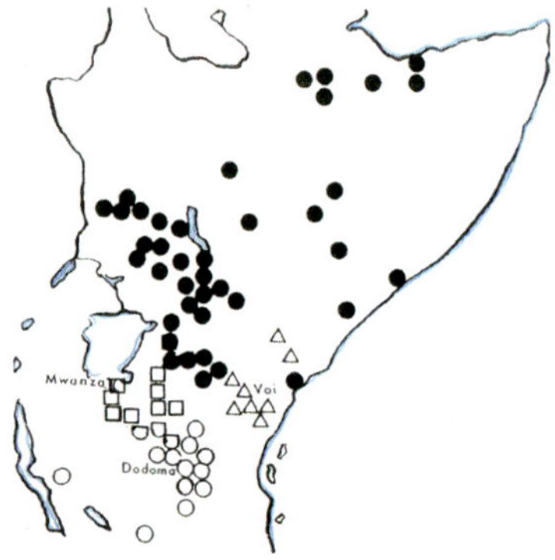

Distribution of dominant soil colours (*above*) and localized types of *Elephantulus rufescens* from Voi (red), Mwanza (grey) and Dodorna (yellow).
Local distribution map (*left*):
black circles = mildly polymorphic populations;
open triangles = rufescent population;
open squares = grey population;
open circles = yellowish population.

However, in three localized populations in the southern part of this species' (north-eastern) African range, dorsal coat colours exactly match local soil colour. This mainly diurnal species inhabits relatively open habitats with a scatter of acacia bushes and thickets and the ground becomes bare during the dry season, the period when predation from raptors is most intense. This indicates that predators selectively remove the least cryptic individuals. Over time they strongly influence and progressively refine the colour traits of the survivors.

All sengis are difficult to see and their behaviour is singularly well suited to being cryptic. At the slightest disturbance they 'freeze', only darting away at great speed when they have located the source of a sound or movement. Unlike *Elephantulus*, *Rhynchocyon* species are predominantly forest floor ant-eaters and are the largest and longest-nosed of all sengis. Their colouring and patterning is varied but there are ecological and biogeographic contexts that suggest their diverse patterns of black, red, yellow and grey have also been evolved by selective predation, but in a more nuanced way than in *Elephantulus rufescens*.

The ranges of *R. petersi*, *R. udzungwensis*, *R. sp. nov.* and *R. chrysopygus* are small, discrete but quite tightly clustered. Like many other organisms endemic to the forests east of the Somali arid corridor, they only survive in very small enclaves on the strip of equatorial coast opposite and north of Zanzibar, or in moist pockets on the mountains commonly described as 'the East African Arc'. Researchers on a wide range of biota agree that the fragmentation of these relictual forests is of very long standing (Moreau 1952, 1963, Hamilton 1982, Lovett & Waser 1993, Burgess & Clarke 2001, Burgess *et al.* 2007). None the less, the four species of forest *Rhynchocyon* are sufficiently alike for them to have presumably differentiated from a common ancestor that had flourished when forests were more widespread and continuous. That the period may have been very remote is made more plausible by the existence of this genus as a fossil in the early Miocene some 20 mya. Of the four species, *R. udzungwensis*, which is a substantially larger animal, is likely not only to be the most conservative species (as deduced from some suggestive features of its coat pattern) but its provenance appears to put it close to the geographic centre of the entire *Rhyncocyon* radiation. It is therefore plausible to suggest that *R. udzungwensis* retains more features of a common ancestor than any other species.

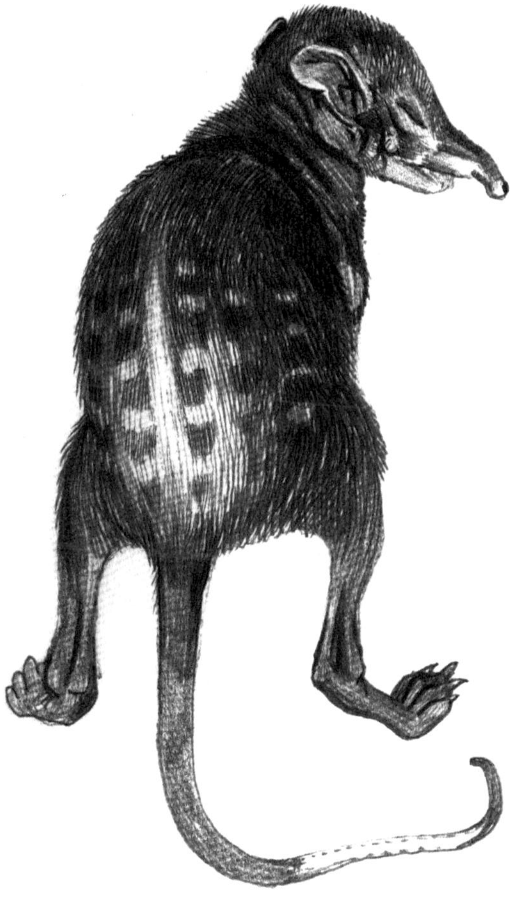

Pattern elements common to most *Rhynchocyon* spp. illustrated by dorsum of new-born Golden-rumped Giant Sengi *R. chrysopygus*.

Both *R. udzungwensis* and *R. petersi* live under high, thick, evergreen canopies and the absence of both species outside small enclaves of permanently well-shaded forest implies that they are restricted, probably to a large degree, by predators. *Rhynchocyon chrysopygus* and *R. sp. nov.*, isolated in a narrow northern tongue of low canopy coastal forest, close to an arid hinterland, live under broken, partly deciduous canopies through which some light commonly penetrates. Adults of *R. chrysopygus*, of the Zanzibar population of *R. petersi* and of some polymorphic *R. cirnei*, have pronounced crests on their necks and crowns. The crest is longest (and most mobile) in *R. sp. nov.* (B. Agwanda pers. comm.). Rathbun (1979a) has suggested that the yellow patch may also serve as an intra-specific signal for *R. chrysopygus*.

The fifth species of giant sengi, *R. cirnei*, is completely unlike the four species locked into their enclaves. Instead it has a very extensive range that breaks down into three very large geographic blocks, *each* about 500,000 km^2 in extent. In two of these blocks the local morph is, in each case, quite consistent in colouring but in the third, which covers Tanzania south of the Rufiji–Kilombero Rivers and all of Mozambique north of the Zambezi, there is an astonishing variety of patterns, which I once mistakenly attributed to hybridization (Kingdon 1974). It now seems more likely that these patterns are influenced by predators *failing* to select for a single colour pattern (just as Rufous Sengis *E. rufescens* living on mosaics of differently coloured soils do not match any one soil-type). The polymorphism of Chequered Sengis *R. cirnei* in this region is likely to be strongly influenced by their living under very varied light levels in an extremely diverse mosaic of habitats. These habitats include grassland, if only seasonally. Mozambique and southern Tanzania differ from many other parts of Africa in having six months of rains (that derive directly from the adjacent Indian Ocean) falling on both coast and uplands in the interior. The entire region is therefore webbed by mostly perennial rivers that are substantial enough to sustain narrow gallery forests within a patchwork of thickets, woodlands and savannas.

Unlike other regions of Africa where large areas have types of plant cover that respond quite uniformly to climatic changes, the basic diversity of the Mozambique mix is likely to have been maintained over many fluctuations in climate because of its proximity to the Indian Ocean. Under such consistently varied conditions raptor-selection was less likely to favour any single type of pattern. Thus some individuals living in dense forest in coastal Tanzania are almost entirely shades of red or have black or mahogany rumps and red flanks. These Sengis often exhibit patterns of dark longitudinal streaks between paler red dorsal blotches. In drier, more exposed areas there has been selection for lighter spotting and paler agouti streaking.

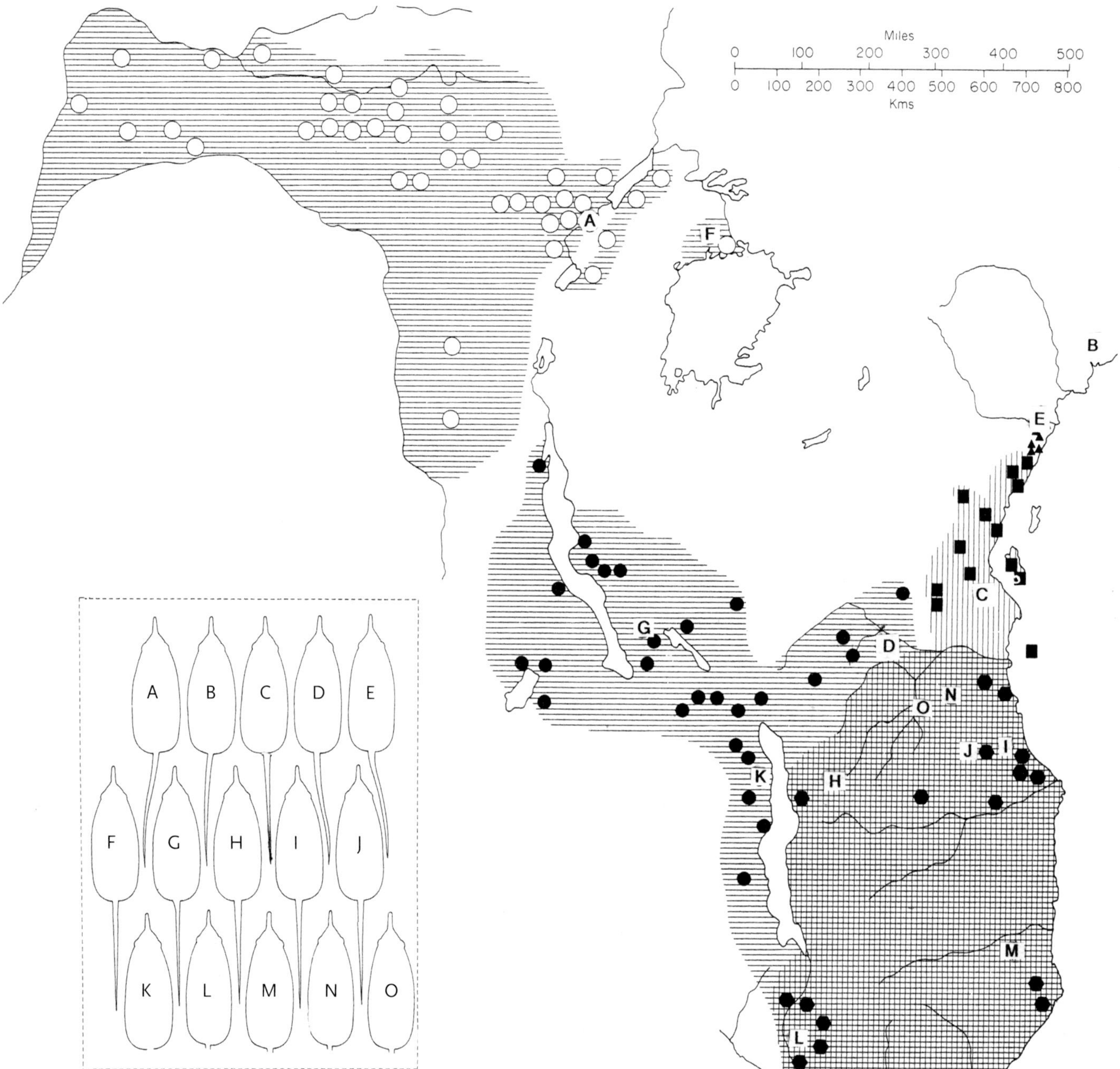

A. *R. cirnei stuhlmanni* humid forest form of *R. cirnei*
B. *R. sp. nov.* peripheral isolate (*nomen nudum*)
C. *R. petersi* coastal eastern isolate
D. *R. udzungwensis* large mountain isolate, relictual?
E. *R. chrysopygus* Kenya coastal isolate
F. *R. cirnei stuhlmanni* Uganda isolate of A.
G. *R. c. reichardi* woodland and gallery form of *R. cirnei*
H. *R. c. macrurus* = 'Mozambique polymorph' (Songea)
I. Mozambique polymorph (Lindi)
J. Mozambique polymorph (Rondo)
K. *R. c. hendersoni* dark montane form of G.
L. Mozambique polymorph (Shire)
M. Mozambique polymorph (Nampula)
N. Mozambique polymorph (Utenge)
O. Mozambique polymorph (Shuguri)

B *R. sp. nov.*
D *R. udzungwensis*
E *R. chrysopygus*
A. F. *R. c. stuhlmanni*
G. *R. c. reichardi*
C. *R. petersi*
Mozambique polymorphs

Pelage patterns and corresponding map and key of *Rhynchocyon* spp. distribution.

A
B
C
D
E
F
G
H
I
J
K
L
M
N
O

The occurrence of redder colouring between the Rufiji and Rovuma Rivers and more agouti towards the Zambezi was once interpreted as a north–south cline (Corbet & Hanks 1968), and these authors' observation that northern morphs are most like their northern neighbours has acquired new significance in light of the very recent discovery of *R. udzungwensis* and *R. sp. nov.* Francesco Rovero (pers. comm.) has recorded traces of spotting and striping on the backs and flanks of some individuals of *R. udzungwensis*, demonstrating the existence of minor pattern polymorphisms even within this very small and confined population. Some individuals have a particularly close resemblance with some *R. cirnei* morphs living in its immediate vicinity, implying that both species originally sprang from a common parental population restricted to the eastern seaboard and 'Eastern Arc' mountains. Indeed, *cirnei* may well have begun as the southernmost extension of an exclusively tropical forest-adapted animal with a quite limited eastern distribution.

I contend that the explanation for ecological success in *R. cirnei*, as well as the polymorphism of its easternmost population, lies in the likelihood that riverine forest galleries in Mozambique and southern Tanzania were maintained over many climatic cycles, probably over many millions of years and over the greater part of this very substantial area. This provided long-term continuity of a predictable, reliably stable habitat for a large population of giant sengis. The crucial distinction, for this discussion, is that forest galleries are unlike solid blocks of forest in being linear and are nearly always emarginated by drier, more open country. During the long wet season, habitats beyond the forest margin become a rich source of invertebrate prey, with dense grass cover making the margins only slightly less safe than the forest itself. Contemporary *R. cirnei* do, indeed, move out into grassland during the wet season (J. Kingdon pers. obs.) but it has to be significant that they encounter more than three times as many raptor species; all fifteen or so of them potential predators.

Convergence in predator-selected pattern

So what of the two other *R. cirnei* populations? Inland from ecologically diverse Mozambique/southern Tanzania there is an upland region where the temperatures are cooler, rivers are narrower and the savannas and miombo woodlands that surround them are more uniform and less diverse. It is here that a much more consistently patterned morph, *R. cirnei reichardi*, has evolved. The ground colour in this subspecies is a freckled light olive agouti that is similar in tone and micro-structure to that of numerous other cryptic mammals. The dark longitudinal bars that occur in various permutations on the polymorphs from further east are, in *reichardi*, ordered into six 'strips' that run from shoulder to rump. Each dark strip is punctuated by five to seven white or off-white 'spots' (actually irregularly shaped blotches) that observably derive from very variable lighter-toned blotches in the eastern polymorphs. This disposition of light and dark markings closely resembles the pattern evolved by various American ground squirrels and by some African maculated grass-mice of the genus *Lemniscomys*. There are even some lizard and bird species that have feather or scale patterns organized along very similar optical principles. The patterns of all, sengis, squirrels, mice, lizards and birds, conform to highly abstracted principles of camouflage in which the disposition of three tones, light, dark and intermediate, are ordered within systematic grids that reflect a structured equivalent to the probabilities of light-distribution in real environments. Pattern elements that are mutable in the polymorphs have, in *reichardi*, become consistent and 'geometricized'. During their evolution by natural (aka predator) selection, all the taxa listed above have shared a common vulnerability and all have been weeded out according to how closely their patterns match the statistical breakdown of three tones in their environment.

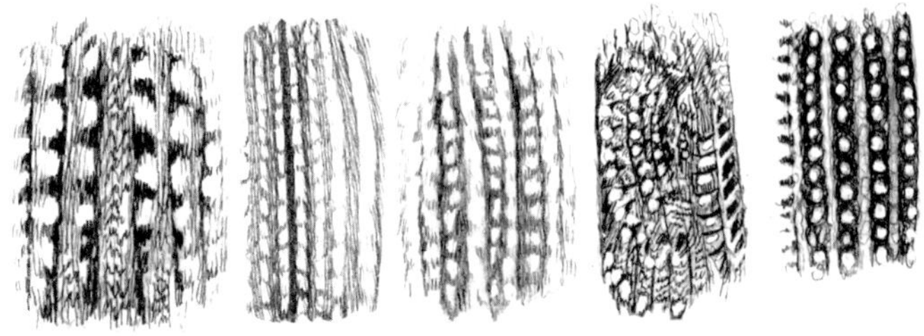
a b c d e

Convergences in cryptic, predator-selected patterns. a. *Rhynchocyon cirnei reichardi* (Afrotherian sengi). b. *Lemniscomys macculus* (African murid rodent). c. *Spermophilus tridecemlineatus* (American ground squirrel). d. *Caprimulgus pectoralis* (African nightjar bird). e. *Cnemidophorus* sp. (whiptailed lizard).

Evolution of the grid structure is probably favoured by more than one factor, genetic and physiological ones included. I contend that this development correlates with a narrower range of environmental settings within which the sengis are vulnerable to hawks. In this respect the complex grid pattern of *reichardi* has an outcome that does not differ in principle from selection for soil-matching in monotonous *E. rufescens*. It is just that the larger animal must break up its colouring and body outlines to achieve similar 'invisibility' in a more cluttered setting.

It would seem that once the genetic coding for a systematic grid-like pattern has been established, predation very quickly fine-tunes it. Predators select against any mismatch with the average light-levels of the environment so that the overall tonality of the pattern goes down or up according to the probabilities of lighter or darker backgrounds in different localities. This helps explain why, within the overall range of *reichardi*, much darker forms occur in montane forests (typified by *R. cirnei hendersoni*, for which see figure K on p. 97).

The tonal versatility that is implicit in the *reichardi* pattern-type has helped it to spread over very extensive areas west of L. Malawi and around both shores of L. Tanganyika. Outwitting hawks may not be the only advantage that *R. cirnei* has over its strictly forest-adapted cousins; presumably it has physiological adaptations to drier, perhaps colder conditions as well. Whatever its hidden attributes, I contend that its ecological advantages have permitted this type of sengi to reach, and then colonize, an enormous block of lowland rainforest north and east of the Congo R. Here I see giant sengis 'returning' to an ecological setting and light levels that are comparable with those in which its *udzungwensis*-like ancestors evolved. In effect, the tonalities of all three components of the grid pattern have darkened to the point where the whole animal is as dark, sometimes even darker than *R. udzungwensis* and *R. sp. nov*. The type for this Congolese population is called *R. cirnei stuhlmanni*, but even within this major population there are lighter morphs along the drier north-eastern margins of its range. Paler brown, with more visible spots, the tonality of *stuhlmanni* from Uganda presumably correlates with higher light levels in evergreen forests that are less consistent in the density of their shade.

Although the north-eastern Congo Basin and Uganda are demonstrably good habitat for *Rhynchocyon* today, the radiation pattern outlined here implies that this extensive equatorial region might have been devoid of giant sengis before the arrival of *stuhlmanni*'s *reichardi*-like ancestors. At the grossest level of explanation past changes in climate could be implicated, but this also implies phases of extreme vulnerability to past climatic changes in *Rhynchocyon*. If so, this only serves to emphasize what unique combinations of circumstance must have allowed *R. udzungwensis* to survive in its tiny enclave.

Finally, the ecological success of *reichardi* has allowed it to spread to the point where its range now actually surrounds the three forests where *udzungwensis* survives. Indeed, *reichardi* now occurs in forests that are said to be indistinguishable from those in which *udzungwensis* lives and only 25 km distant.

The presence of such a successful, and presumably newer, species so close to such a conservative and rare type could imply several things. The most likely is that the forest blocks in which *udzungwensis* survives have an unbroken history that goes back for immense stretches of time while those in which *reichardi* occurs may be 'regrown' secondary forests. A more unsettling possibility is that *reichardi* has been steadily advancing at the expense of *udzungwensis* as the extent of forests has steadily shrunk around them. In any event the differing status of these two sengi species makes them into very valuable indicators for both past and current change in African environments.

This summary analysis of a hypothetical but ecologically and morphologically plausible evolutionary radiation suggests that the details of a living species' biology and distribution can help shed light on events that have taken millions of years to unfold. The example of an African radiation following the incursion of a Eurasian primate will be discussed shortly but, first, consider some more general features of the African continent as a theatre of primate evolution.

Primate histories

If Africa suits mammals, Africa certainly suits primates, and things about this continent have favoured the radiation and multiple speciation of bush-babies, colobus monkeys and Cercopithecinae, or 'cheek-pouch monkeys', and, in the past, apes or proto-apes. Today, living higher primate species are at their most diverse in Africa and it is also clear that all the anthropoid primates that currently live in Asia are derived from African ancestors that reached Asia during the Miocene by means of at least three, possibly more, emigrations. The likely timing of those events and the classes of emigrants can be summarized as follows:

Approx. 20 mya	At least one, possibly two, species of proconsul-like 'proto-ape' left Africa and gave rise to the gibbon, orang-utan and related lineages in Eurasia (this almost certainly included the ancestors of extant African hominoids).
Approx. 13 mya	An ancestral colobine ancestor left Africa and gave rise to Asiatic Leaf-monkeys.
Approx 10.5 mya	Eurasian apes very abundant in Europe: at least one form likely to have entered Africa and given rise to modern African apes and hominins.
After 10 mya	Macaque ancestor left Africa and radiated across Asia.

As was emphasized earlier, continental exchanges are generally rare and commonly involve no more than a single species, which then radiates within its new continental home. Of even greater importance for primate evolution, especially for our understanding of the evolution of modern African apes and humans, has been the return, back into Africa, of a Eurasian primate. A return passage is still hotly contested but the evidence for a Eurasian lineage of ape giving rise to both modern African apes and humans has been cogently and persuasively argued by Stewart & Disotell (1998). These authors have shown that alternative explanations for the indisputable common genetic ancestry of African hominids and the orang-utan would all involve more exchanges between Asia and Africa than are plausible (or necessary) to explain the facts.

Of the known invasions out of Eurasia that have been listed above, one, at about 10.5 mya, brought in several examples of large-bodied Eurasian fauna, which implies a particularly solid, broad land connection at that time (Hempton 1987). This is the most likely time of arrival for a Eurasian dryopithecine tree ape. When it came to penetrating the more tropical regions of Africa with their well-established and diverse primate fauna, a Eurasian ape coming from a temperate or semi-temperate background would have suffered similar delays or constraints to other mammals. Expansion south was easiest at times when the pattern of seasonal changes most resembled those of the Mediterranean (where dryopithecines were particularly abundant between about 12 and 9 mya). Such expansion was therefore favoured most during cool, dry periods, when arid or semi-arid corridors cut into the forest belt, sometimes linking northern arid-adapted communities with southern ones. This scenario is central to attempts to understand the evolutionary origins of gorillas.

Interpreting their distribution, biology and history poses many questions, but my explanation for their present pre-eminence in central Africa begins with an early proto-gorilla population (almost certainly of dryopithecine ancestry) entering just such a temporary 'corridor' that passed through today's Cameroon and Gabon (Kingdon 2003). When the climate became warmer and more humid these apes would have become engulfed in forest and this engulfed population became the putative ancestors of today's gorillas. Gorillas are still most abundant in this region, where it is proposed that their ancestors first adapted to true forest-living. The situation is not without precedent; there are other primates that seem to have non-forest ancestors, notably drills *Mandrillus* spp., Sun-tailed Guenons *Allochrocebus solatus* and the Squirrel-Galagos *Sciurocheirus* complex, and all of them now inhabit this region, most of them exclusively. This evolutionary pathway from non-forest into forest is but a detail in a much larger and much repeated feature of Africa's biogeographic history.

Geographical axes and the biogeography of speciation under changing climates

I began this chapter by pointing out that the success of mammals in Africa, like that of many other animals, was partly due to such a large portion of the continent being equatorial and tropical but not necessarily permanently forested. I went on to discuss the south-east and centre-west continental divide. Beyond pointing out that climatic changes subdivide these blocks and alter the boundaries between them, I have deferred until now discussion of the mechanism that has given rise to extraordinary levels of diversity in groups such as primates, rodents, mongooses and antelopes. It is a mechanism that is driven by climate change. Dry habitats expand or contract on a north–south axis under the influence of dry, cold climates. These are associated with northern ice-ages and a substantial lowering of sea levels. It is at these times that arid-adapted communities in northern and north-eastern Africa make contact and exchange fauna and flora with arid south-western Africa: Oryx and dik-dik antelopes and bat-eared foxes are much cited examples (Lonnberg 1929, Moreau 1963, Kingdon 1971, Coe & Skinner 1993, and see map p. 55).

At the opposite extreme, forests in Africa expand and contract along an east–west, equatorial axis. Warm, wet periods cause forests and forest communities to expand and join up only to retreat into enclaves, large and small, when the droughts return. The (*cephus*) monkeys provide a particularly vivid example of the diversification of forest-adapted mammals across the main forest zone (see Vol. II) and red colobus *Piliocolobus* spp. are another example of speciation within today's apparently continuous forest belt. There can be little doubt that the isolation of forest blocks and of the animal populations that live in them has been influential in causing evolutionary change. However, it is not only forest and desert animals that expand and contract. Populations belonging to every adaptive permutation between true desert forms and exclusive forest forms shift the boundaries of their ranges during periods of climate change. This frequently involves isolation in discrete pockets of territory, especially on the slopes of mountains and uplands. These periodic disjunctions and reconnections between populations have clearly provided a major mechanism for speciation in Africa and help explain the extraordinary diversification of mammals that was emphasized at the beginning of this chapter.

The diversity of African mammals and primates includes *Homo sapiens*, a species that has evolved during the same period, over the same continental mass and by the same processes as other mammals (Cartmill 1974). Even more significant, it is a species that is the product of the same back-and-forth exchanges between continents that have brought several 'returnees', such as porcupines, some caprine and cat lineages and rousette bats, 'back' into Africa from Eurasia.

The dryopithecine tree ape that is the most likely ancestor of African great apes and humans (Stewart & Disotell 1998, Kingdon 2003) would have entered Africa during a well-known period of connection: the 'Hipparionid Event' at about 10.5 mya. This was precisely the time that the *Rattus*-like ancestor of today's murine rodents is thought to have entered Africa (Steppan *et al.* 2004). Like that immigrant rodent generalist, it was probably able to spread widely at first, especially in the northern and eastern parts of Africa that most resembled their western Eurasian homeland. Even as its range expanded, it is to be expected that differences between the south-east and centre-west, overlaid by fluctuations in climate, soon led to some differentiation in that founding stock. The earliest speciating event that can be plausibly reconstructed was that of proto-gorillas in the Gabon–Cameroon 'corridor' (as summarized earlier). At much the same time, or perhaps a little later, increasing aridity was likely to have isolated another population of dryopithecine-like apes east of the Somali–Kalahari arid corridor that was discussed earlier.

Reconstructing the subsequent fortunes of African ape and hominin descendants has to be, by the sparse nature of the evidence, a highly speculative enterprise, but whatever the true course of events, it must have followed similar patterns to those that underlie the history of other speciating organisms in Africa. In any event, the very eastern distribution of fossils implies a marked preference for this region by early hominins after their differentiation from the ancestors of African apes. An eastern bias is still reflected in the distribution of many fauna and flora and, among mammals, the Suni and Red Bush Squirrel *Paraxerus palliatus* are species that are limited to the long, narrow strip of rather dry forests, thickets and mosaics that are sustained by rain coming off the Indian Ocean. Even some birds with immigrant origins, for all their mobility, seem capable of retaining a preference for those regions or habitats closest to those of their source. Thus the Trumpeter Hornbill *Bycanistes bucinator*, a bird with close affinities to some Oriental hornbills, has a mainly eastern and south-eastern distribution. Likewise the monarch flycatchers *Erythrocercus holochlorus* and *E. livingstonei* have a close Oriental relative, *Culicicapa*, and an eastern coastal forest distribution, and there are other similarly suggestive avian distribution patterns.

For forest-adapted eastern littoral species the main pathways across the semi-arid belt and on into the interior are up the forested banks of major rivers. Because major rivers tend to be well spaced out, what could be called 'basin evolution' is not easily separable from latitudinal stratification. In effect the inhabitants of major river basins, such as those of the Limpopo, Zambezi, Rufiji and Web Shebelle, can become distinct populations. 'Basin evolution' has been examined for its explanatory powers in relation to the proliferation of hominins and a sequence of speciation events (that are broadly consistent with the fossil record) has been proposed in a previous work (Kingdon 2003).

If that model of evolution is correct, the emergence of hominins could be said to have hinged on nothing more than sudden aridification of a long tract of land that became interposed between our ancestors in the east from those of apes in the west. So far as we can tell, whatever has happened to those two lineages over the last 7 million years has been predicated upon a 'mere' accident of geology and climate.

> Every creature alive is the product of a unique history. The statistical probability of its precise reduplication on another planet is so small as to be meaningless. Life, even cellular life, may exist out yonder in the dark. But high or low in nature, it will not wear the shape of man. That shape is the evolutionary product of a strange, long wandering through the attics of the forest roof and so great are the chances of failure that nothing precisely and identically human is likely ever to come that way again.
>
> (Eiseley 1946)

However much we live for the present, we need history and prehistory to retrace that 'strange, long wandering', if only in our minds. For the first time ever, we have the technical tools and intellectual frameworks to begin that journey. It was a wandering that wove in and out of the mammalian communities of Africa. We need all the descendants of those communities in order to understand how mammals, including humans, have speciated in Africa and evolved their particular characteristics. This is a science in its infancy and only a lot more excavation and analysis of genes, fossils and behaviour will reveal the true depth and breadth of the process that has given us, and all other African mammals, our life and existence.

CHAPTER SEVEN

Classification: A Mammalian Perspective

Colin Groves & David Happold

Classification in its widest sense is the ordering of organisms into groups on the basis of their relationships (Simpson 1961). Two other terms are frequently used in relation to the classification of organisms: 'systematics' is the scientific study of the kinds and diversity of organisms, and 'taxonomy' is the theoretical basis for classification. Actually, these three concepts grade into one another, in practice if not in principle, and they are sometimes regarded as loose synonyms. In particular, taxonomy is no longer regarded as merely a theoretical study, and 'a taxonomy' is used in very much the same way as 'a classification'. It is hard to insist on different meanings for the three terms any more, though it remains correct to say that they imply different degrees of inclusiveness. A fourth word commonly used in the context of biological classification is 'nomenclature'. In this case, the concept has a different sphere of concern, and its distinctness must be carefully maintained; nomenclature is a system of rules to determine what name shall be used for a taxon.

Species Concepts

The basis of all biological studies is the species. According to the Biological Species Concept, a species is a population (or series of populations) that does not interbreed under natural conditions with other such populations. The qualification 'under natural conditions' is important because there are well acknowledged species that can be persuaded to breed in captivity but that do not interbreed in the wild. In fact, some species that are considered to be genuinely distinct may commonly cross to form fertile hybrids (for example, between the two species of wildebeest, *Connochaetes taurinus* and *C. gnou*; Fabricius *et al.* 1988); so it is not true that hybrids between distinct species are necessarily sterile.

In some situations, closely related species that do not usually interbreed and whose geographical ranges join or overlap may occasionally form hybrids. When this happens, there are narrow zones where hybrids occur infrequently, but since there is little backcrossing, there seems to be little or no gene-flow between the parent forms. Examples of hybridization in the wild between African mammals of distinct species, or what many would consider distinct species, include a hartebeest *Alcelaphus*–tsessebe *Damaliscus* hybrid (a single instance but noteworthy because it is intergeneric; Selous 1893) and several records of baboon *Papio*–gelada *Theropithecus* hybrids, also intergeneric (Jolly *et al.* 1997). A further example is the presence of a number of hybrid zones between different taxa of baboons (Jolly 1993). The difficulty with taxa that hybridize is whether they should be regarded as two species or as a single species that exhibits considerable geographic variability. In the case of the baboons, Jolly (1993) considered that the taxa were conspecific (i.e. they belong to a single species, namely *Papio hamadryas*) whereas Groves (2001) maintained that they were separate species. These examples illustrate that an exact definition of what constitutes a species under the BSC is not as simple as it seems.

Two other concepts – the Recognition Species Concept and the Phylogenetic Species Concept – have been developed to accommodate those situations that seem to be contrary to the Biological Species Concept and on which the concept offers no guidance. The Recognition Species Concept assumes that members of a species share a common fertilization system. The meaning of 'a common fertilization system' can be interpreted in a variety of ways. In a study of galagos (Galagidae), it has been shown that vocalizations are the main determinators in bringing potential mates together, and so may be regarded as the principal mechanism for this 'common fertilization system'. Although the characters that may comprise a 'common fertilisation system' are debatable, this method can only be employed in the field and can not be used for museum specimens.

Both the Biological Species Concept and the Recognition Species Concept are *process* concepts, which recognize a species according to the process by which it is maintained. But it is not always – in fact, not even usually – easy or possible to detect that process. But it is possible to recognize a *pattern* that may identify a population as a species which differs diagnostically from other populations. This pattern concept is known as the Phylogenetic Species Concept. A diagnosable taxon is one that can infallibly be recognized from all others (as far as the available evidence goes), implying that there are fixed genetic differences that separate it from any other taxon. Any character, as long as it can be reasonably inferred to have a genetic basis, is suitable, e.g. morphology, vocalization, karyotype, nuclear DNA sequences, etc. Interbreeding between species, or its absence, is not a criterion under the Phylogenetic Species Concept; the examples mentioned above, even the baboon hybrid zones, would not affect the status of the hybridizing taxa as full species, as they are none the less diagnosably distinct.

Under the Phylogenetic Species Concept, many more taxa merit species status than under the Biological Species Concept – perhaps twice as many in some groups of land vertebrates. Under the Biological Species Concept there are no criteria for deciding on the species status of allopatric populations; no conceivable evidence will permit an objective evaluation. Under the Recognition Species Concept, the criteria are more often available and potentially more nearly objective, but they change from group to group, and it may need an unattainable degree of knowledge to decide what counts as a suitable series of criteria. Under the Phylogenetic Species Concept, the categorization is as clear-cut as the available evidence allows.

A species, consequently, is a population (or series of populations) diagnosable by one or more unique, fixed, heritable character states.

Subspecies

Two populations within a species may be separable because some or many of the individuals that comprise each of these populations differ in body size, body proportions or colouration, and it may be worthwhile distinguishing them taxonomically. Such populations may be considered to be subspecies within a species. Just how small a proportion of individuals can differ for the populations to be deemed subspecifically distinct? Most taxonomists would probably follow the rule-of-thumb proposed by Mayr (1963): that is, 75% of individuals in a population should be distinguishable from all other members of the same species. Statistically, this is equivalent to a 90% non-overlap between two populations, and is especially useful in morphometric comparisons.

Subspecies replace one another across the geographic range of the species; they never overlap and hence are never sympatric. They are not variants within a single population in a single location. A species does not have just one subspecies: it has two or more, or it has none. If a species has no subspecies, it is termed monotypic; if it is divided into subspecies, it is termed polytypic.

If, for example, two taxa differ only on average (according to the 75% rule), they may be termed subspecies within a single species; but if they differ absolutely so they do not overlap in one or more characters they should be named as separate species. It is quite normal in classification for a taxon originally described as a subspecies to be reclassified as a species in the light of new information, and for two species to be replaced by a single species with two subspecies using one of the names originally designated. For example, the Forest and Savanna Elephants were long considered to be two subspecies of a single widespread species, the African Elephant (*Loxodonta africana*), and were called *L. a. cyclotis* and *L. a. africana* respectively. A recent revision (Grubb *et al.* 2000), subsequently supported by genetic evidence (Roca *et al.* 2001, Rohland *et al.* 2010), showed that they differ diagnostically, and hence they are now classified as two species (*L. africana* and *L. cyclotis*, respectively).

Subspecies, of course, are merely points along a scale of differentiation at which it becomes convenient to provide separate names, whether subspecies or species. The objective demarcation is between subspecies and species, not between subspecies and populations with less divergent gene frequencies.

The Hierarchy of Classification

Organisms are classified in a descending series of less and less inclusive categories called 'ranks'. These categories are 'nested', that is to say each one belongs only to one higher category. There is general agreement that in some way taxonomy must reflect their phylogenetic affinities.

The tenth edition (1758) of Linnaeus's *Systema Naturae* is taken as the starting point for all biological nomenclature. The hierarchy of taxonomic ranks, from highest to lowest, is: Kingdom, Phylum, Class, Order, Family, Genus and Species. These are referred to as the obligatory ranks, and it is mandatory to classify each species with respect to them. An example is as follows. The class Mammalia are in the phylum Chordata (which also includes the fish, amphibians, reptiles and birds, as well as the so-called 'protochordates', the sea-squirts and lancelets). The class Mammalia is, in turn, divided into many orders – 29 according to the most recent reckoning (Wilson & Reeder 2005), including the Lagomorpha (hares), Rodentia (rodents), Primates (humans, apes, monkeys, lemurs), Carnivora (carnivores), Proboscidea (elephants), Chiroptera (bats) and others. Each one of these orders is further divided into one or more families, each family into one or more genera, and each genus into one or more species. Thus the order Lagomorpha has two living families, the Leporidae (hares and rabbits) and Ochotonidae (pikas). The Leporidae contain ten genera including *Lepus* (hares), *Oryctolagus* (rabbits) and *Pronolagus* (rock hares). In Africa, the genus *Lepus* contains five or six species, the genus *Oryctolagus* contains one species and the genus *Pronolagus* contains four species. When a classification is viewed from the species upwards, the classification of the Cape Hare *Lepus capensis* is species *capensis*, genus *Lepus*, family Leporidae and order Lagomorpha. The name of any species is a binomial name, which includes the genus name and the species name (see Nomenclature). All species within a genus have a set of shared derived character states that places them in a single higher taxon (a genus in this case), indicating that they are the exclusive descendants of a common ancestor. A family contains one or more genera, which, likewise, have a set of shared derived character states. Thus all species in a genus are more similar, and more closely related, to each other than they are to species in another genus in the family (or indeed any other genus in any other family); the implication of 'closely related' is that they are a monophyletic group (the exclusive descendants of a common ancestor).

Sometimes finer levels of discrimination are required and so it is necessary to insert subordinate ranks below obligatory ones; these are typically 'sub-' ranks, with 'infra-' ranks below that. This results in a more complicated (but more precise) classification. The 29 orders of mammals are not evenly inter-related, and the Mammalia are divided into two subclasses: the Prototheria (for the order Monotremata alone) and Theria for the other 28 orders. The Theria are in turn divided into two infraclasses: Metatheria (for the marsupials, of which there are seven orders) and the Eutheria (for the placentals, of which there are 21 orders). A classification can be very unbalanced and there is no necessity to have even approximately equal numbers of subordinate categories in a major taxonomic group.

Other subordinate ranks can be used in some circumstances. Superorders can be used to group related orders together (and come below infraclasses); in the past, other ranks (such as megorders, parvorders, grandorders, mirorders, cohorts) have been used when needed. By far the most common of these splitting practices is to insert the ranks of 'tribe' and 'subtribe' below the subfamily in order to group related clusters of genera, and subgenera to group related clusters of species.

Any taxonomic group, at whatever level, is referred to as a taxon (*pl.* taxa). The order Lagomorpha is a taxon, so is the genus *Lepus*, so is the species *Lepus capensis.*

Splitting and lumping

Disagreements about how to classify species are not uncommon. A 'splitter', who divides up a genus into several different genera, can be accused of failing to see the wood for the trees; while a 'lumper', who combines several genera into one all-inclusive genus, risks missing worthwhile information. At the species level, splitting and lumping have rather different connotations from the generic level or above, and because at the species level we have a hope of introducing much more objectivity, it is better not to use the terms 'splitting' and 'lumping' for species taxonomy, but to reserve them for more restrictive and more inclusive arrangements of genera and families.

Nomenclature

The naming of animals (by their scientific names) is controlled by the *International Code of Zoological Nomenclature*, now on its fourth edition. The *Code* 'has one fundamental aim, which is to provide the maximum universality and continuity in the scientific names of animals compatible with the freedom of scientists to classify animals according to taxonomic judgements' (International Commission on Zoological Nomenclature, 1999). After taxonomists have completed a taxonomic investigation, any changes to names or status of a taxon are subject to the rules laid down in the *Code*. 'Changing the names', which so many non-taxonomists complain about, is generally a result of an advance in taxonomic understanding, which thereby *automatically mandates* a nomenclatural decision, which may involve changing some names.

The basic principle of zoological nomenclature is priority: the earliest name given to a taxon is the one that should be used for it, so long as the name is *available*. To be available, a name must pass certain criteria:

- It must have been published.
- It must not predate 1758, as the tenth edition of Linnaeus's *Systema Naturae* is the official starting point of the system.
- The work in which the name was introduced must have used binomial nomenclature.
- It must have been accompanied by a description or a bibliographic reference to one (otherwise the name is called a *nomen nudum* and it cannot be used).
- It must be unique: two species in the same genus cannot bear the same specific name; two genera in the animal kingdom cannot bear the same generic name. Two nominal taxa with the same name are called *homonyms*.
- From the year 2000 onwards, it must be accompanied by a set of formalities: it must be specifically stated to be new, and must have a *type specimen* (see below).

Ironically a name does not have to be in any way appropriate. The name of the Hartebeest *Alcelaphus buselaphus* means 'elk-stag ox-stag', but the animal does not in any meaningful way resemble an elk, an ox, or a stag!

The *Code* provides rules and guidelines for the names of a taxon within a species-group (species and subspecies), genus-group (genera and subgenera) and family-group (superfamilies, families, subfamilies, tribes and subtribes). Some of these are described in the following paragraphs. There are no rules for other ranks such as orders.

In nomenclature, a species has two names (a binomial): the name of the genus followed by the name of the species. A subspecies has three names (a trinomial): the generic, the specific and the subspecific. The name of a genus, or of a species or subspecies (the binomial or trinomial), is either written in italics or is underlined. The genus name always begins with a capital letter, but species and subspecies names are always in lower case even if the name commemorates a person (*wilkinsoni*) or a geographic region (*africanus*). If a large genus is divided into subgenera, the subgenus name may be included in parentheses (e.g. *Tadarida* (*Mops*) *condylura*). One of the subspecies – the one found at the type locality and which includes the type specimen, is nominotypical, and repeats the species name (e.g. *Pan troglodytes troglodytes*; *Cephalophus nigrifrons nigrifrons*; *Otomys sloggetti sloggetti*); other subspecies that are found elsewhere are given a separate trinomial name (e.g. *Otomys sloggetti turneri*, *O. s. robertsi*). Under the principle of coordination, the names given to species and subspecies are interchangeable, so if a subspecies is found actually to be a full species it retains the same name, but as a binomial not a trinomial. Some of the bushbucks that have been described as subspecies of *T. scriptus* may, in fact, be distinct species; if so, the subspecies *Tragelaphus scriptus sylvaticus* would change its name to *Tragelaphus sylvaticus*. Likewise, if a genus is divided into subgenera, one of them is the nominotypical subgenus and bears the same name as the genus. If a family is divided into subfamilies, one of them is the nominotypical subfamily and bears the same name as the family, except that it ends in –inae instead of –idae; and there may also be nominotypical tribes and subtribes (which end in –ini and –ina, respectively).

When a species or subspecies is first named, it must be represented by a type specimen (the original specimen that was used for the original description of the taxon) and a type locality (the place where the type specimen was collected). Every genus, species and subspecies has a type locality and a type specimen. For each of the genera and species described in *Mammals of Africa*, the full citation of the taxon is given. The citation consists of the binomial scientific name of the species, the name of the describer and the date (year) when the description was published, the name of the publication where the description was published, and finally the details of the type locality. Thus for the Golden-rumped Sengi, the full citation is: '*Rhynchocyon chrysopygus* Günther, 1881. *Proceedings of the Zoological Society of London* 1881:164. Mombasa, Kenya'.

From this information, the interested reader can find the original description, which give the details of the taxon (size, pelage colour, skull characteristics, etc.) and the distinctive characters, which, in the describer's view, separate the taxon from other similar and related taxa (though subsequent research will often have modified the describer's original ideas).

The specific name is unique to any organism, which is why biologists always use the scientific name of a species rather than its common or vernacular name (see later). For example, the squirrel named *Heliosciurus gambianus* is a unique name. There is only one *gambianus* in the genus *Heliosciurus* and there is only one *Heliosciurus* in the family Sciuridae (or any other family or order). A specific name may however be used in conjunction with another genus; besides *H. gambianus*, there is *Cricetomys gambianus* (a rat), *Mungos gambianus* (a mongoose) and *Nycteris gambianus* (a bat); and *gambianus* is used as a specific name in many other non-mammalian animals.

Names in a species-group, a genus-group and a family-group are separate: a specific name (the second name in a binomial) cannot be 'upped' to generic rank. On the other hand, by the *principle of coordination*, a name given to a subspecies is deemed to have been given to a potential species, and vice versa; a name given to a genus is deemed to have been given to a potential subgenus and vice versa; and a name given to a family is deemed to have been given to a potential subfamily, tribe, subtribe or superfamily. So subspecies can be 'upped' to species rank or species downgraded to subspecific rank – and so on.

In some instances, the name of the describer and the date are placed within parentheses, which indicates that the species was originally described in a different genus to that in which it is now placed. For example, *Bunolagus monticularis* (Thomas, 1903) means that Thomas in 1903 ascribed *monticularis* to a genus other than *Bunolagus* (in fact, he described it as *Lepus monticularis*), whereas *Lepus capensis* Linnaeus, 1758 has no parentheses because Linnaeus ascribed it right from the start to the genus *Lepus*. Two authors with the same surname are usually distinguished by a single initial. Thus in *Pronolagus crassicaudatus* (I. Geoffroy, 1832) the 'I.' distinguishes Isidore Geoffroy St.Hilaire (usually shortened just to Geoffroy) from his father Étienne, who would be referred to as E. Geoffroy.

Scientific names are Latin or latinized; so a specific name, if the species name is an adjective, must agree in gender with the generic name. Thus, when the Savanna Elephant was transferred to a new genus, the former *Elephas africanus* became *Loxodonta africana*. Specific names may also be nouns in apposition and so not subject to gender agreement (so *pardus* in *Panthera pardus* does not become '*parda*'); or they may be genitives, as in *Galago gallarum* ('galago of the Galla people'). When a species is named after its discoverer or someone it is intended to honour, the name is a reflection of this: if the person is a male, *-i* is added to the person's name (*smithi*); if a female, *-ae* ; and if two or more people, *-orum*. In the past, other terminations were occasionally used (e.g. *Tragelaphus oryx pattersonianus*, which, by the rules of nomenclature, cannot now be corrected to *pattersoni*).

The suffix at the end of the higher ranks indicate the status of the rank. A family name ends in –idae, a subfamily in –inae, a tribe in –ini and a subtribe in –ina. A superfamily ends in –oidea. A family-group name is formed from the (latinised stem of) the name of one of the included genera: so, Sciuridae from *Sciurus* and Bovidae from *Bos* (stem *Bov-*). The genus concerned does not have to be a valid genus, though usually it is; nor does it have to be the earliest named genus.

The *Code* is not entirely problem-free, and the International Commission on Zoological Nomenclature is constantly being asked to adjudicate on certain tough problems such as when a well-known name is threatened by the discovery of a long-forgotten name (*nomen oblitum*) which is thought to have priority. Applications to the Commission, and the Commission's Opinions, are published in the *Bulletin of Zoological Nomenclature*.

Types

A type specimen is the one to which the name of a taxon in a species-group is irrevocably attached. Usually, the type was designated or otherwise fixed as the type when the taxon was described (and is known as a holotype); but there may have been two or more specimens used by the original describer, and just one of these may have been designated to be 'the type' subsequent to the original description of the taxon (in which case it is known as a lectotype). Occasionally it is necessary, to sort out some vexing problem of nomenclature, to provide a type specimen retrospectively, using a specimen unknown to the original describer: this is known as a neotype. Every species and subspecies has a type locality, the locality at which the type specimen was collected. If the type locality is unknown, a subsequent reviser may fix it by fiat. A topotype (which may be collected later than the holotype) is any specimen from the type locality of the taxon.

There are types for a genus-group and family-group, but these are taxa rather than specimens. The type of a genus is a species; the type of a family is a genus.

Type specimens are extremely valuable because they are the basis for the naming and description of a species. Some type specimens are very old (dating from two centuries ago, or even more), others (for newly described species and subspecies) are barely a few years old. In museums, type specimens are kept under lock and key and are identified as types by red labels. Ironically, some types are not typical of the species they represent – because, by chance, the collector collected an individual which was atypical of the population as a whole. But once a type has been selected, it always remains as the type specimen.

Synonymy

Whereas homonyms are the same name given to two different taxa, synonyms are different names given to the same taxon. The earliest synonym is the senior synonym; the others are junior synonyms.

Up to about 1850, international communications were so poor that it was not uncommon for different authors to name the same taxon independently, even (in the case of taxa of the species-group) using the same type specimen. These names, objectively referring to the same taxon, are called objective synonyms.

Names given to putative taxa, which at the time were considered to be distinct but which subsequently were considered not to be distinct, are subjective synonyms. The senior subjective synonym is the name to be used for the taxon, but it is of course possible that one or more of the junior subjective synonyms may be removed from synonymy if subsequently it is considered that it represents a valid taxon. This illustrates the difference between nomenclature, an objective, artificial but nevertheless very useful system, and taxonomy, a subjective study reflecting how we view the natural world. Because a currently recognized species may have one or more junior synonyms (by which the species was referred to in previous works), the names of junior synonyms are given for each species described in *Mammals of Africa*. The full citation for a synonym includes the name, describer's name and date of publication; in *Mammals of Africa* only the synonym name is given; additional details (describer's name, date, etc.) are provided by Allen (1939), Ansell (1989), Wilson & Reeder (1993, 2005) and Hoffmann *et al.* (2009).

A word is necessary here about the nomenclature of domestic animals. For fairly obvious reasons, the domestic form of a particular animal was often described and given a scientific name before any wild representatives were known, so that the name given to the domestic form has priority. This, however, creates enormous problems for subspecific taxonomy, biogeography and so on, and a ruling of the International Commission on Zoological Nomenclature published in 2003 stipulates that a name given to a domesticate should not be used for a wild taxon; the domestic species should either be included within the wild species or, preferably, treated for convenience as a separate species (which it is in one way, and is not in another). The history of the problem, the meaning of the ruling and the philosophy behind it, are recounted by Gentry *et al.* (2004). It explains, for example, why the African Wild Ass is called *Equus africanus* (von Heuglin & Fitzinger, 1866), rather than *Equus asinus* Linnaeus, 1758 – the latter name applies to the domestic ass or donkey.

The higher categories

The higher categories are, in essence, phylogenetic branches. To say that *Civettictis* and *Genetta* belong to the Viverridae, and *Crocuta* to the Hyaenidae, is to say that *Civettictis* and *Genetta* share a more recent common ancestor than either does with *Crocuta*. It is astonishing that, until very recently, there were no criteria for deciding on precise *ranks* above the species level. In general, these were set by some authority's fiat in the nineteenth or even the eighteenth century, and changed by incremental creep thereafter. The question has very rarely been asked, for example, why the Bovidae is ranked as a family, rather than as an order or a genus?

Recently, the idea of a relationship between a taxon's rank and its time depth (the time since it became separate from its sister-group) has been revived. For example, one can ask 'When did the family Lorisidae originate?' This question can mean either 'When did divergence between the family Lorisidae and its sister-group (probably family Galagidae) occur?', or 'When did the last common ancestor of the modern genera of Lorisidae live?' Goodman *et al.* (1998) suggested that, in mammals, a genus should have separated from its sister genera at least 4–6 mya (back to the Miocene–Pliocene boundary), and a family from its sister families about 22–23 mya (i.e. close to the Oligocene–Miocene boundary). Their elaborate scheme gives expected time depths for orders, semiorders, suborders, infraorders, superfamilies, families, subfamilies, tribes, subtribes, genera and subgenera. A simpler arrangement is to restrict the idea to the obligatory ranks (order, family, genus, as total groups) only (Groves 2001); the subordinate categories are inserted where needed merely to divide up a large, unwieldy obligatory taxon. Avise & Johns (1999) broadened such a scheme to comparisons with other animal groups, specifically the cichlid fish of L. Victoria, which have a far shallower time depth than primates at corresponding ranks, and fruit-flies, which unexpectedly have a far deeper one. These authors envisaged either a gradual convergence of ranking schemes, or else the development of a new temporal-banding system alongside the Linnaean system.

It is probably too soon to expect even a mammal-wide consistency in fitting rank to time depth, but if any progress is to be made beyond the present state of subjectivity, then some such system must come. A problem, of course, is that it could not possibly apply to species, and a break between the taxonomy of species and that above the species level will become accentuated.

Methods of taxonomy

A – Collecting

Taxonomy has traditionally been the preserve of morphologists. Most of the world's natural history museums are devoted to storage cabinets full of specimens of animals and plants (preserved dry or in alcohol and not on exhibit to the public). Here generations of biological taxonomists have spent their careers, measuring and examining specimens with the naked eye, under the microscope, or by X-ray. In the case of mammals, the specimens are overwhelmingly skins and skulls; postcranial bones are scarcer and soft parts scarcer still. In the eighteenth and early nineteeenth centuries, mammalian skins were usually stuffed and mounted in life-like postures. Later, when storage space came to be increasingly scarce, they came to be stored either as flat, opened-out skins or as slightly padded-out skins (called 'puppet skins') filled with cotton-wool or tow.

The morality of collecting biological specimens has rarely been discussed in the literature. Until the early twentieth century, much of the mammalian material in museums was supplied by people who were, in effect if not by designation, big-game hunters; it was opportunistic and unsystematic, and of course large mammals predominated. Most specimens, in fact, were not made available to museums but were kept as trophies, or even left where they fell. Major P.H.G. Powell-Cotton was one of a very few collectors who kept his entire bag (mainly African), which is today still available for study. Even Powell-Cotton, however, collected mainly what interested him (Primates, Carnivores and Ungulates) and in places that he enjoyed visiting (Cameroon, Sudan, Ethiopia, and DR Congo). Collectors in the twentieth century tried to address the imbalance between large and small mammals by collecting many of the smaller species such as rodents, insectivores and bats. This was perhaps appropriate in another way, too, as the tide of public opinion was turning – people who would loudly deplore the ruin of an elephant would have little objection to the collection of a dozen or so mice. At the same time, wildlife conservation was becoming a necessity, and collecting large species was becoming less and less ethical.

Given that we cannot do without collections, what should our guidelines be? It goes without saying that we should avoid collecting anything, whether large of small, that is of conservation concern, even if it is legal in the country concerned. It also goes without saying that collection methods should always be humane, but this is a much harder area to define. Should we avoid collecting intelligent species, whose cognition is closer to our own? It may seem anthropomorphic, but it seems reasonable to conclude that species whose mental condition is more like ours – such as chimpanzees and gorillas, perhaps some other primates, and maybe elephants – have more claim to our moral concern than do other species. Other questions that should be considered concern an individual mammal's position in its social group, and its reproductive status. The human social setting must also be taken into account; it does not set a good example if foreigners collect and hunt when the local people are enjoined not to hunt (and if they do so, suffer legal penalties).

B – Sources of evidence

There is no substitute for minute analysis of morphological features when making a careful taxonomic analysis. Do, for example, individuals from different geographic regions have different patterns of basicranial foramina, shapes of the naso-frontal suture, or molar cusp arrangement? If they do, and the differences are consistent, then we classify them as different species; according to the Phylogenetic Species Concept, a diagnostic (100%) difference in just a single character, as long as there is a reasonable inference that the difference has a genetic basis, is sufficient to confirm their status as separate species. If they differ on average, but not absolutely, in one or more characters, then we might conclude that they are separate subspecies. Disentangling sympatric species (i.e. species with overlapping geographic ranges) can be much more problematic; here, we look for congruent patterns of variation in two or more characters.

At the level of species and below, morphometrics is often an extremely useful technique. Univariate or bivariate metrical comparisons may help in the construction of useful identification keys, but great care should be taken to exclude purely phenotypic effects, which may be caused by variations in diet, rainfall and temperature in different habitats. Similarly, differences in shape may be simply consequences of size differences, because of unsuspected allometric relationships. Multivariate analysis may not altogether get over this problem but, if phenotypic plasticity can be safely excluded, different techniques of multivariate morphometrics can be extraordinarily revealing.

At and above the species level where by definition character states are reticulate, cladistic analysis is essential for organizing basic phenetic information into a phylogenetic framework. Such an analysis results in a matrix of many different characters (often up to quite a large number), each of which is coded by its presence or absence or 'state'. A number of computer programs are available for analysing large sets of data, resulting in a cladogram, which is not itself an evolutionary tree but can be readily converted into one. The cladogram can in turn be converted into a taxonomic classification using the principle that taxa ought to be ranked according to the recency of common ancestry.

Behavioural characters have always been used in taxonomy, but have come into their own mainly since the 1930s, after the genesis and spread of ethology. A book such as Estes (1991) is an invaluable starting point for the use of ethological characters of African mammals in taxonomy. There are many examples where behavioural characteristics (e.g. displays, vocalizations, social organization) are species-specific and hence can be shown to be important in maintaining species boundaries and preventing wholesale hybridization between closely related species. Two taxa that appear morphologically similar (at least to humans), and which may have been considered as populations within a single species, may exhibit quite different behavioural characteristics, which would indicate that they are indeed separate species. Behavioural characteristics should not be used solely as a criterion for species determination, but in conjunction with other evidence.

Chromosome morphology, amino-acid sequencing and DNA sequencing have been successively added to the methodology of the taxonomist since the 1960s. It is necessary to emphasize that these new methods have not replaced traditional methods, but have been added to them. There has sometimes been a distressing tendency to assume that if no difference can be found between two well-established taxa in a 100+ base-pair sequence, then the craniodental or pelage character differences somehow do not exist, or do not count. Presumably most gross morphological differences do have a genetic basis; failure to find differences in a studied sequence means simply that this basis lies elsewhere within the genome. Claims that a certain level of genetic distance between two taxa indicates a specific difference (while less than this indicates only subspecific, and more than this indicates generic difference) must be taken as a guideline only, because, as discussed above, whether a taxon ranks as a species or not depends not on 'degree of difference' but on diagnosability.

It must be acknowledged that phylogeography (the study of the principles and processes governing the geographic distributions of genealogical lineages [Avise 1994]) has proved extremely fruitful in enlarging our understanding of the processes of geographic differentiation within species, although so far it is mainly mitochondrial DNA that has been used as evidence. When karyotypes or DNA sequences of a particular taxon are published, it is essential there be some guarantee that the taxon has been correctly identified. A photograph, or a note of a museum voucher number, would serve to do this.

Case studies of species and subspecies

African elephant

Grubb *et al.* (2000) examined the case of the African elephant. The Savanna Elephant and the Forest Elephant have, at least since the mid-twentieth century, always been classified as two subspecies of a single species – *Loxodonta africana africana* and *L. a. cyclotis*, respectively. They are strongly different, and many of the differences are fixed and diagnosable. Each subspecies occupies a wide geographic range, over which it is homogeneous: they make no morphological approach to one another towards the boundaries of their ranges. By the criteria of the Phylogenetic Species Concept, they are distinct species, and Grubb *et al.* (2000) designate them as such: *Loxodonta africana* and *L. cyclotis*. Examining specimens from the very border of their ranges, Groves & Grubb (2000) found that in some regions they almost certainly interbreed, while in other regions there is no evidence that they do. Because their geographic ranges do meet, the Biological Species Concept interbreeding criterion can in principle be applied (suggesting only one species), but the results are inconsistent. There is no equivocation, however, according to the Phylogenetic Species Concept.

Bushpig and African buffalo

The taxonomy of two other species, Bushpig and African Buffalo, is analogous to that of the African elephant (Groves 2000). The Bushpig (*Potamochoerus larvatus*) and the forest-living Red River Hog (*P. porcus*) are not known to interbreed where their ranges meet; and hence they rate as distinct species under both Phylogenetic Species Concept and Biological Species Concept. Among the African buffaloes (*Syncerus*), the Cape Buffalo and Forest Buffalo, like the elephants, seem to interbreed in some regions and not in others and hence it seems reasonable, once again, to refer them to two distinct species under the PSC: *Syncerus caffer* and *S. nanus*, respectively. The case has been confused because the normally forest-living *S. nanus* extends well into the West African savannas and has there developed a subspecies, *S. c. brachyceros*, which converges, but only superficially and only to a slight degree, towards *S. caffer*.

Bushbuck

The most difficult case is that of Bushbuck. Traditionally, only one species has been recognized, *Tragelaphus scriptus*, but Groves (2000) suggested, on the basis of an earlier analysis by Grubb (1985), that there are actually several species of Bushbuck (and see Moodley & Bruford 2007). There is a small species (*T. scriptus*) with well-marked white stripes and spots throughout adult life, found in both rainforest and in the bush savanna in West Africa, and as far east as the Nile; and there is a large species (*T. sylvaticus*), in which the white marks nearly disappear in adults and the males become very dark, found throughout most of southern and eastern Africa. Their ranges not only abut but even interdigitate in Sudan, Uganda and north-eastern DR Congo, and it is remarkable that they have never, since the early twentieth century at least, been recognized to be distinct species. There are two complicating factors. The first is the existence of three other distinctive taxa (*decula*, *fasciatus* and *meneliki*) in the Horn of Africa and along the East African coast. While perfectly distinct from the two widespread species and from each other, two of them show superficial convergence with one or other of the widespread species (*fasciatus* has white marks like *T. scriptus*, *meneliki* is very dark like males of *T. sylvaticus*). Under the Phylogenetic Species Concept, there would in fact be as many as five species. The second complicating factor is the presence of a population (*ornatus*) in southern DR Congo and western Zambia that, on the face of it, looks like an intergrade or widespread hybrid between *T. scriptus* and *T. sylvaticus*. Yet the East African evidence suggests five species. Occasionally, the useful system formulated by Linnaeus, which has served so well for 250 years, comes up against the reality that evolution is a dynamic process.

Vernacular or common names

Most non-scientists prefer to use the vernacular (or common) name of a mammal rather than the scientific name. The problem with vernacular names is that a species may have several vernacular names, even within a single language, and several species may have the same vernacular name. When the vernacular names in different languages are taken into account as well, a species that is known unambiguously by its two-word scientific name, may be known by 10, 20, or more vernacular names. It is clearly impractical to include all vernacular names in *Mammals of Africa*. Because of the widespread use of the English language in Africa, the preferred English vernacular name is given for each species. For some species, alternative English vernacular names are given in parentheses. French and German vernacular names are also provided if available; but many of the smaller species do not have vernacular names in these languages. In general, it is much better and more precise to refer to a species by its scientific name.

The English vernacular names used in these volumes are not necessarily the most well-known names. Using as many published works as possible, and especially the names proposed by Wilson & Cole (2000) and Wilson & Reeder (2005), *Mammals of Africa* has chosen what is considered to be the best and most accurate vernacular name for each species. Whenever possible, the following criteria have been used: (a) part of the vernacular name of a species should include the vernacular name of the genus, a name that is shared by all members of the genus; (b) no two genera can share a generic vernacular name, but additional words can be added to provide uniqueness (e.g. African, Gambian, etc.); (c) a vernacular name should be the translation of scientific name (e.g. *Myotis* = Mouse-eared); and (d) a vernacular name, likewise, should be a translation of the scientific name of the species and hence reflect the wishes of the describer. Well-established vernacular names may, however, be retained if they are firmly engrained in the literature. It has not always been possible to follow these criteria, but in general, the vernacular names used in *Mammals of Africa* bring a measure of consistency and logic to a rather chaotic naming system.

In the past, vernacular names have normally been written in a variety of ways. For example, the vernacular name for *Epomophorus gambianus* has been written as Gambian epauletted fruit bat, Gambian epauletted fruit-bat, or Gambian Epauleted Fruit Bat (note the variation in the use of capital letters and hyphens). But there are ambiguities; is this a fruit-bat with epaulets, or a bat that feeds on epaulletted fruits? The modern convention, used here, is that all words in a vernacular name begin with a capital letter, except where two words are joined by a hyphen (as in Fruit-bat), and that hyphens are used where the first word is adjectival to the second. Wherever possible, the vernacular name of the genus is included as part of the name of the species, so that all species in a genus are linked together by part of their vernacular names. The vernacular name of a particular species is adjectival to the vernacular name of the genus. Thus the genus *Dendromus* has the vernacular name of 'Climbing Mouse' and the vernacular names of some species within this genus are Kivu Climbing Mouse (for *Dendromus kivuensis*), Grey Climbing Mouse (for *D. melanotis*) and Nyika Climbing Mouse (for *D. nyikae*). In some instances, when these criteria were applied, a vernacular name became too long and complicated, so a shortened easier-to-remember name has been used. Likewise, well established and well-known vernacular names were maintained to ensure continuity and simplicity, e.g. Lion, Serval, Gerenuk, etc. However, in the case of some species, it has been necessary to add a qualifier to avoid any possibility of ambiguity, hence Greater Kudu (for *Tragelaphus strepsiceros*) and Lesser Kudu (*T. imberbis*), and Common Warthog (*Phacochoerus africanus*) and Desert Warthog (*P. aethiopicus*).

Classification and *Mammals of Africa*

The classification used in *Mammals of Africa* mostly follows that given by Wilson & Reeder (2005), but authors were given the liberty to follow a different classification if there were good, and published, reasons for doing so. Hence the genera, species and subspecies may, in some instances, differ from those given in earlier publications on African mammals and in Wilson & Reeder (2005). Of course, taxonomy is often controversial (as explained above), so parts of the classification used in *Mammals of Africa* may in time prove to be incorrect. This is inevitable as new information comes to light and our knowledge of African mammals increases.

The sequence of orders presented in the six volumes mostly follows the 'classical' tradition that has evolved during the last two centuries. There have however been some significant changes. Until about 100 years ago, for example, the Lagomorpha (hares and rabbits) were included within the Rodentia (squirrels, rats, mice and their allies) but are now placed in their own order. The Insectivora is no longer considered as a natural group, and is now divided into four separate orders: the Macroscelidea (elephant-shrews or sengis), Soricomorpha (shrews, moles and their allies), Erinaceomorpha (hedgehogs and their allies) and Afrosoricida (golden-moles, tenrecs and otter-shrews). The sequence of presentation of families within orders, and genera within families, mainly follows Wilson and Reeder (2005) and reflects in a very general way the phylogenetic relationships between them. But for some taxa, the sequence of presentation is pragmatic rather than phylogenetic, e.g. the genera within the family Muridae (Rodentia) are arranged by subfamily, and then alphabetically by genus; within each genus, species are also listed alphabetically.

It does not necessarily follow that a new hypothesis in respect of relationships (and hence classification) is correct; in the 1980s it was suggested, on the basis mainly of brain structure and neurology, that the fruit-bats (Megachiroptera) were more closely related to primitive primates than to the microbats (Microchiroptera). If proved to be correct, it would mean that the Chiroptera were diphyletic in origin rather than monophyletic. A great deal of fascinating research was generated as a result of the hypothesis (see e.g. Pettigrew 1986, 1991); but in the end, the similarities of fruit-bats to primitive primates was shown to be due to convergence and not to a common ancestry. Hence the hypothesis was discarded and the Chiroptera are still regarded as monophyletic.

The Proboscidea (elephants), Hyracoidea (hyraxes) and Sirenia (manatee and dugongs) have, for a long time, been placed next to each other because of their unexpectedly close phylogenetic relationship. Recent research has shown that these three orders, together with the Afrosoricida (golden-moles, otter-shrews and tenrecs), Macroscelidea (elephant-shrews or sengis) and Tubulidentata (Aardvark) belong to a superordinal grouping called Afrotheria. The Afrotheria, as the name suggests, is thought by its proposers to have originated and diversified within Africa, and hence to be a very 'African' group of mammals. Although the earliest fossils so far assigned to the Afrotheria are actually North American (Asher *et al.* 2003), the superorder today is more diverse in Africa and Madagascar than elsewhere; in fact, two of the orders (Tubulidentata and Macroscelidea) are entirely confined to the African continent, one occurs only in Africa and Madagascar (Afrosoricida), one extends beyond Africa only as far as the Levant and the Middle East (Hyracoidea), and two (Proboscidea and Sirenia) are more widespread as well as occurring in Africa. The Afrotheria hypothesis was proposed by Stanhope *et al.* (1998) on molecular grounds and subsequent molecular studies have substantiated it; on the other hand, Asher *et al.* (2003) could find no morphological characters that would unequivocally characterize such a grouping although there did not seem to be any glaring incongruities, while most recently Wible *et al.* (2007) succeeded in validating Afrotheria on the morphological level.

CHAPTER EIGHT

Behaviour and Morphology

Jonathan Kingdon & Fritz Vollrath

Morphology is the most interesting department of natural history and may be said to be its very soul. What can be more curious than that the hand of a man, formed for grasping, that of a mole for digging, the leg of the horse, the paddle of the porpoise and the wing of the bat, should all be constructed on the same pattern, and should include the same bones in the same relative positions? ... The explanation is manifest in the theory of the natural selection of successive slight modifications. ... The innumerable species, genera, and families of organic beings, with which this world is peopled, have all descended, each within its own class or group, from common parents, and have all been modified in the course of descent.

Charles Darwin, *The Origin of Species*, 1859

How does morphology relate to behaviour?

Humans, golden-moles, zebras, sea cows (sirenians), bats as well as many more mammals are profiled in considerable detail in the volumes that follow. Grasping, digging, galloping, paddling and flying are patterns of behaviour that have given rise to all these modification of limbs following 'the course of their descent from a common parentage'.

Contemporary evolutionary science brings many new perspectives to Darwin's words and insights, and one concerns the relationship between function and form, or, more precisely, the relative primacy between behaviour and morphology. It would seem counter-intuitive to allocate a primary formative role and priority to ephemeral, insubstantial behaviour (often belittled as 'mere' behaviour) over the physical substantiality of living animals and their bodies, bones and fossils. And yet it is so. Many authors within *Mammals of Africa*, under the heading 'Adaptations', have described specific morphological peculiarities and linked them with the species-specific behaviours of which they seem to be the agents. This is appropriate because, over evolutionary time-spans, behaviour in its most comprehensive sense has driven morphology. In addition, as the fossil record so vividly illustrates, morphology, too, is not the static state of being that it seems to be from an immediate, contemporary perspective. Rather it is the manifestation of an organism's adaptive condition at a given moment in time and as such it is, to a greater or lesser degree, fluid and subject to continuous change, but at a much slower rate than is behaviour, which is inherently flexible and able to respond immediately to challenges.

This chapter aims to demonstrate the link between what animals *do* and the structures that have evolved to increase the efficiency and effectiveness of that behaviour. Given the great diversity of expressions of form, we also seek to reconcile that extraordinary plasticity with the common body plan that underlies all mammal morphology. To this end we present a selection of behavioural and mechanical dimensions to 'form and function' as well as providing a guide to the terminologies employed in the description of anatomical and morphological features.

Instead of using its conventional near-synonym, the more ancient and more anthropocentric 'Anatomy', Darwin used the word 'Morphology' to create an important new association. From bald description and operative comparison of shape Darwin went on to add functional explanations. These, of course, required both implicit and explicit integration of the only rational explanation for the great diversity of morphologies, which was the mechanism that brought them into being – natural selection.

'Morphology' was a relatively new word in Darwin's time. Indeed, the 'science of shape' had been founded by the great German naturalist, poet and thinker Goethe, in 1790, not long before Darwin's birth (1809). At the time it was a particularly inviting field of study in Europe and America because museums, zoos and wealthy naturalists were making ever more diverse collections of animals (or their cadaverous relics) available for scientific study. The specimens, in turn, were interesting and exciting because many were newly discovered, exotic and in need of examination, classification and explanation. While the specimens were relatively easy to collect and export, their behaviour and ecology were left behind where they came from, typically unrecorded or with only marginal comments. One privilege of our own generation has been to be among the first to study animal behaviour, in depth, and in the knowledge that this apparent 'insubstance' has been a driving force in the evolution of animals and their wonderful diversity of form.

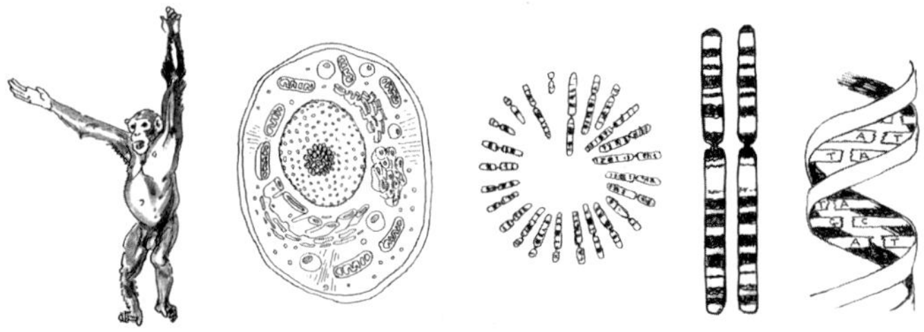

Today, expressions of the 'science of shape' embrace internal organs, tissues, cells, enzymes and molecules. A living mammal contains multiple 'shapes', each of them expressions of self-regulating systems that range from gene to whole body. A chimp body contains about 100 trillion cells; each cell has a nucleus; each nucleus has 48 paired chromosomes; of every pair, one chromosome is from each parent; chromosomes consist of packed strands of DNA. Genes are DNA segments coded to make proteins.

Knowledge of exotic animal behaviour was rudimentary in the nineteenth century but major differences in, say, locomotion, were so obvious that Darwin could confidently choose grasping, digging, galloping, paddling and flying as examples of activities that could become more efficient through the 'selection of successive slight modifications' in the limbs that served these activities or behaviour patterns. As our opening quotation emphasizes, linking form to function was integral to Darwin's thinking. A more contemporary expression of that preoccupation – working out how functional behaviour gives rise to morphological form – remains central to the interest and significance of an inventory of mammalian diversity such as *Mammals of Africa*.

In addition to relating anatomical form to activities and behaviour, Darwin also closely observed how morphologies had specific regional contexts (islands, of course, being of special interest) and he demonstrated how the shapes of animals and plants were adapted to what we would now call ecological niches. He also observed and compared the morphology of living and fossil forms. As we begin to refine our understanding of African environments and explore their many changes over time we can begin to appreciate how mammals have had to modify their behaviour and, ultimately, their anatomy to survive change, competition and other vicissitudes. For Darwin, the soul of natural history was comparative morphology. For contemporary scientists and naturalists his metaphysical metaphor would have to include behaviour and genes, the 'insubstances' that so closely approximate to traditional concepts of soul or spirit.

It is worth remembering that in the nineteenth century exploring and comparing morphologies and making deductions from the clues they provided was as novel and revealing as DNA fingerprints are today. The comparison of DNA patterns has become a new material base for contemporary taxonomy. We can already see that the 'science of shape' is gaining many new and important insights from the 'science of the shapers of shape'. Some of these essentially new connections will be explored in this chapter and in the volumes that follow.

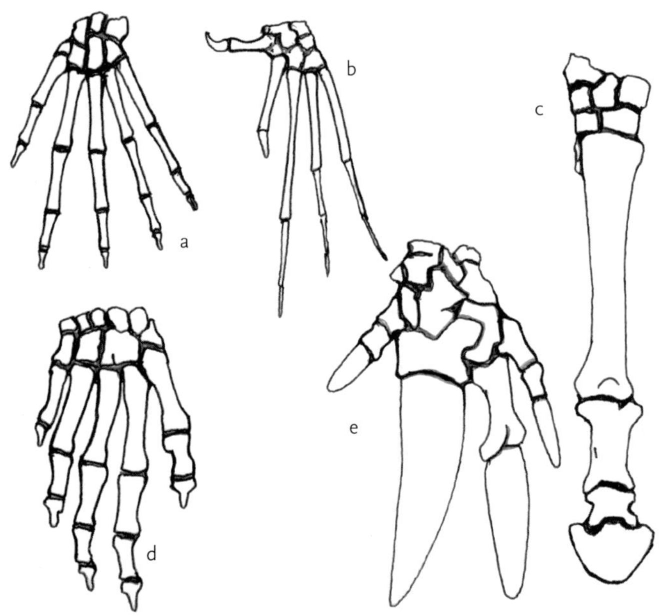

Manus: a. human *Homo sapiens*, b. fruit bat *Stenonycteris lanosus* (foetal), c. zebra *Equus quagga*, d. Dugong *Dugong dugon* and e. golden-mole *Chrysochloris* sp.

Behavioural shifts give rise to morphological change

The ancestral five-digit mammalian forelimb has been morphed through infinitely small incremental changes into all the very different behavioural functions that Darwin listed and many, many more. This shift by one utilitarian morphology into another (probing finger to pounding hoof) demonstrates that within specific physiological constraints most gross features of mammalian anatomy are capable of surprising modification and, ultimately, total functional transformation. In trying to understand and recreate the many steps in such transformation, comparisons cannot be superficial. Taking Darwin's example of primate hands; lizards, frogs and even salamanders have fingered hands and feet, the latter usually having five digits. These resemblances *do* denote evolutionary relationship, but at levels so deep as to confirm that morphology is only comprehensible if grounded in the broadest ethological, physiological, anatomical and genetic contexts and in an explicitly evolutionary framework. Non-scientific systems that make superficial associations between like and somewhat alike, especially on the basis of a single feature such as 'hands', would link humans with newts and geckos while limbless lizards would join eels and worms. 'All true classification is genealogical; that community of descent is the hidden bond that naturalists have been unconsciously seeking, and not some unknown plan of creation, or enunciation of general propositions, and the mere putting together and separating of objects more or less alike' (Darwin 1859: 420). Only rigorous scientific methods can emancipate us from primitive, traditional systems that based their

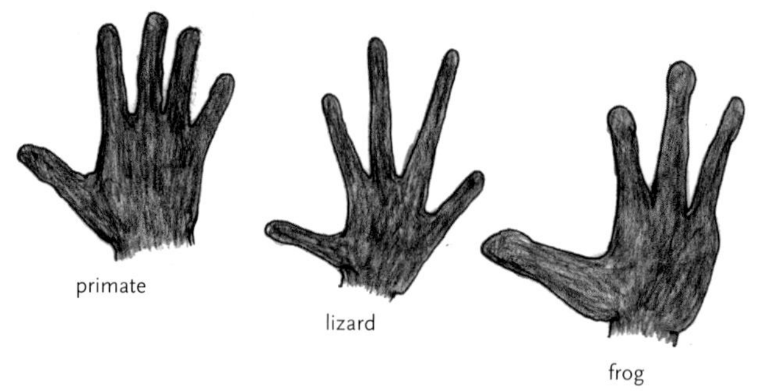

Outlines of primate, lizard and frog hands.

taxonomies on single or arbitrary criteria, such as 'hands', which are not that different from basing classifications (and value systems) on ownership, economic worth, castes and 'clean' or 'unclean' animals.

Once the complex of characteristics that unites mammals and distinguishes them from other vertebrates have been identified (the diagnostic features of mammals are dealt with elsewhere), it is the mutability of virtually every feature that becomes of central concern for the 'science of shape'.

Of course the most obvious diversity of form (in mammals as in other organisms) concerns differences between different orders and families of mammals. Throughout the class Mammalia, diversity of form is based upon the modification of already existent structures. This can involve elaboration, such as the extension of digits into the struts of a bat's wing, or reduction, as in the paring away of all but one digit in a horse or zebra's limb.

How does this happen? What are the forces at work that can turn a little shrew-like hand into the bat's spreading wing over a few million years? Begin with one, apparently rather obvious observation: morphology tends to be conservative but behaviour is not; our children bear our likeness but their behaviour differs from ours in innumerable and trend-setting ways. Furthermore, as conditions are constantly changing, the existence of behavioural plasticity becomes essential, particularly when changes are sustained generation after generation. Individual adaptability, together with natural selection for appropriate behaviour are fundamental to the evolutionary process and to the evolution of diversity. Since behaviour is the primary strand of our argument we can categorize it under headings similar to those used in these volumes to describe the biology of species: locomotion, foraging, communication, social facilitation, predator avoidance etc. Consider then, some examples of how changes in structures have derived from shifts in specific behaviour patterns.

Innovations in locomotion or limb-use

It was, perhaps, because humans make their own bodies the vehicle and prototype for most of their ideas about nature that naturally thumb-less monkeys so horrified early naturalists that they called them *colobus*, Greek for 'cripple'! The history of science's slow progress towards some sort of objective detachment is littered with this sort of visceral subjectivity. The real reasons for suppressing the thumb have to do with the adoption, by one lineage of ancestral monkeys, of a leaf diet and an elegant energy-saving form of arboreal locomotion. No digital manipulation is required of a leaf-eating monkey because, for the most part, it is easier to eat floppy material directly off the tree rather than handle it. A further, and related reason for doing without a thumb is that the remaining fingers have been aligned into a robust, weight-bearing *hook* that can support a body that has become exceptionally heavy because the stomach has been transformed into a copious, leaf-fermenting vat. A monkey that must catapult its barrel of a body across breaks in the canopy puts very different stresses on its hands than a smaller, slimmer primate that combines rapid traverses across the tops of fine branches with fast capture and manipulation of small insects.

Even within a single closely related group of the latter type of monkey, relatively modest differences in proportion can be instructive. African guenons, of the genus *Cercopithecus*, include a large-bodied,

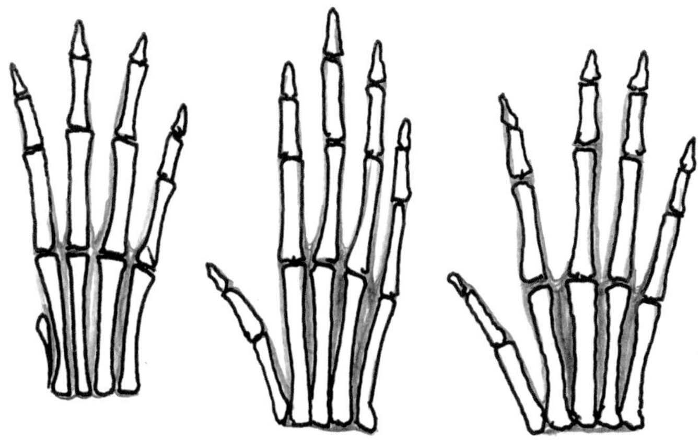

Right manus of, from left to right, *Procolobus rufomitratus*, *Cercopithecus nictitans* and *C. pogonias*.

relatively slow species, Putty-nosed Monkey *C. nictitans*, which has long, slender fingers and a short, weak thumb (probably for similar reasons to the colobus) and Crowned Monkey *C. pogonias*, a small-bodied, fast omnivore with a compact, small hand and a relatively strong thumb. The latter is an adept snatcher of agile insects on thinner branches and twigs in the canopy, the former is a heavier animal that eats both fruit and leaves and prefers thicker branches in the shade where less active arthropod (animal) prey are discovered under bark or on trunks (Gautier-Hion 1978, Struhsaker & Leland 1979, Kingdon 1980, 1988).

Detailed studies of the behavioural ecology of primates can reveal proximate, species-specific adaptations of this sort but they can also help resolve larger evolutionary questions. Fully arboreal African monkeys were long assumed to be archaic while more terrestrial species were regarded as later, more derived forms that had descended from the prototypical arboreal niche (Schwartz 1928). A combination of detailed comparative anatomy and molecular studies has shown this to be an inversion. Instead, it was semi-terrestrial guenon ancestors that effectively re-colonized the forest canopy (Kingdon 1971, 1988, Dutrillaux *et al.* 1982). In the process they modified (or even *re*modified) the proportions of their hands.

The ancestors of one semi-terrestrial monkey took to exploring potentialities for life along the 5000 km southern borders of the Sahara desert. This exposed successive generations to an ever-fluctuating frontier where trees were frequently scarce. By seeking to make a living in steppes where trees were rare, alongside a variety of fleet predators, the ancestors of the Patas Monkey *Erythrocebus patas* had to become extremely fast *in spite of* possessing typically primate hands and feet. This ecological and behavioural shift led to elongation of the entire length of the limbs, including the bones of the hands and feet. The Patas Monkey has become a sort of 'primate greyhound' and, in adopting an almost wholly terrestrial existence, has paralleled some of the modifications typical of other fast, light quadrupeds, such as jackals *Canis* spp. and the Cheetah *Acinoynx jubatus*. Elongation has not been the only change. Because it takes no weight during running, the thumb has greatly reduced but has been saved from disappearing altogether because of its residual usefulness for manipulating food.

Elongation of limbs and simplification of extremities are evident in all fast cursorial mammals but the trend has been most extreme in horses and zebras. Fortunately the equids have an exceptionally rich and continuous fossil record; and the progressive reduction and

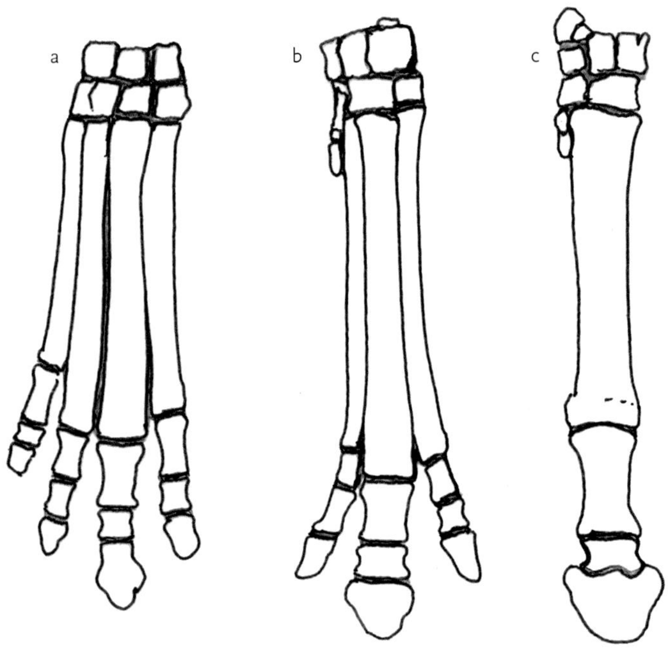

Evolution of zebra foot. a. *Hyracotherium* (early 4-toed equid); b. *Miohippus* (Miocene 3-toed equid); c. *Equus* (modern 1-toed equid).

eventual loss of all but one digit is beautifully illustrated by fossils that range from the Eocene to the Holocene.

The most extreme differences in the proportions of 'hands' are to be found among bat wings. The shortest wing, relative to the size of the bat, belongs to Moloney's Mimic Bat *Mimetillus moloneyi*, the longest to the Large-eared Giant Mastiff Bat *Otomops martiensseni*. Moloney's Mimic Bats are small and agile, living under loose bark, with body and skull appropriately flattened to this end. Nowhere is the formative role of behaviour in shaping morphology clearer than in *M. moloneyi*. In scurrying in and out of their loose bark refuges or in clambering up tree trunks to get to an elevated launching pad the short forearm and stubby little thumb are almost as active as the forelimb of a true quadruped. This agility is achieved by a unique folding pattern of the digits, the more distal of which have been secondarily shortened. The proximal digits of the wing are of the same length as the forearm and fold neatly under it in such a way as to minimize any risk of snagging. Behind these digits the very short distal ones fold away within the tightly furled umbrella of this very versatile hand-wing. These bats can open up to become highly mobile on their one-fingered forearm or, like a deck-chair, fold up into a flat, inconspicuous and easily stored package. (There are some similarities here to the Central American vampire bat, *Desmodus*, which has to sneak up on, and run away from, bigger mammals to suck blood.)

Sloughing tree bark represents a significant microhabitat for bats but it is intrinsically temporary and subject to predation from primates, birds and snakes. Under such constraints Moloney's Mimic Bat has developed a forelimb that combines flight with some of the mechanical advantages of a climber's limb. Among the costs are a need to take frequent rests (because such a short wing requires very fast and tiring wing-beats) and the need to take off from a height, because of the initial drop before lift can be achieved. The Large-eared Giant Mastiff Bat, instead, has long, pointed, sickle-shaped wings that allow it to fly very fast, very high, over great distances. It is the migratory swift, *Apus*, of the bat world, with wings that have the highest aspect ratio of any African bat and the greatest elongation of digits in any mammal.

Mimetillus moloneyi sketches.

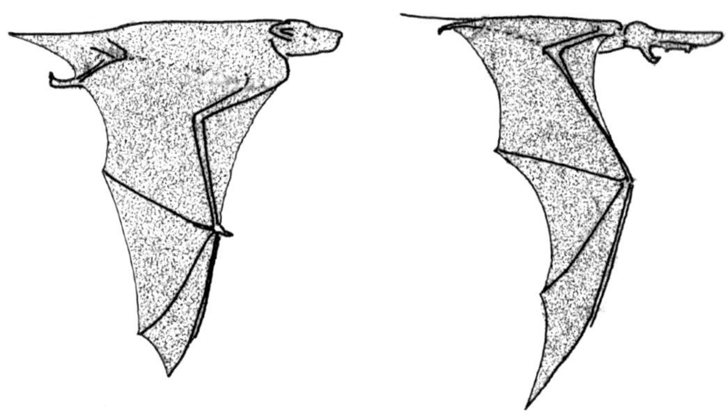

Wing outlines of *Mimetillus moloneyi* (*left*) and *Otomops martiensseni* (*right*).

Changes in foraging or food processing

Limbs and bodies can change proportions for many reasons and elongation of any feature is always a significant change. Lamark had the idea that giraffes grew long necks by stretching them. Lamark was right: but not in the way he imagined. Lengthening necks by stretching them is not the inherited adaptation that Lamark envisioned. However, inhabiting an environment and adopting a habit where there are rewards for neck elongation could very well have become a trait that spread through some ancient ancestral population of proto-giraffes. This behaviour could be consistently rewarded if it offered unique access to plentiful food resources that were untapped by other herbivores.

Giraffes stretch their necks because they seek to browse from the tips of high branches or from the tops of lower canopies. This is their foraging niche. It is a niche where they have few competitors and no competitor at all within their current morphological guild, which is that of heavy-weight, selective browsing ruminant.

Diagram of head of Giraffe *Giraffa camelopardalis* showing ossicones, tongue, vertebrae, etc.

Incidentally, necks also power uniquely giraffid 'necking' contests that help settle inter-male competition for females (Coe 1967, Simmons & Scheepers 1996). The weight of male heads increases by osteoblastic processes very similar to the evolutionary ones that gave rise to ossicones in the first place (Kingdon 1979).

Even today giraffe necks vary individually in relative length. Length is governed by the time neck-growing genes are in operation. Thus necks are actually lengthened by extension in the activity of genes that control growth and length.

Browsing, like all food-getting, is behaviour, and the earliest, possibly okapi-like, giraffes would have had to stretch their necks to reach the choicest sprouts and buds. Perhaps this behaviour was learned, perhaps it became inherited, or, more likely, the two became concerted, as with the giraffe-antelope or Gerenuk *Litocranius walleri*, which reaches for nutritious herbage on higher branches by standing on its hindlegs, thereby augmenting its already exceptional reach. Proto-giraffes were also likely to have inherited a proclivity for out-reaching other browsers but the nutritional rewards allowed an already large animal to get still larger. One advantage of eating young, nutritious growth that other browsers could not reach would have been less of the toxins that many plants generate in response to browsing and more of the concentrated nutrients of sprouts that were always exposed to the sun. Such specific preferences allowed higher-reachers to crop better forage and hence breed more and stronger offspring with just those traits that had enhanced the survival of their parents and their lineage.

Continued natural selection over very many generations would have fine-tuned foraging behaviour together with a longer and more muscular tongue and, eventually, and in concert with a longer neck, a stronger heart and valved arteries (to cope with sudden surges of blood pressure). All these morphological changes would have been the consequence of minor initiatives in foraging behaviour, not the other way round, so that it is logical to conclude that behaviour was the innovator while pre-existent morphology was the constraint: behaviour was the evolutionary motor while morphology was a sort of anchor, serving to ensure that the many other advantages of being a two-toed ruminant were retained. The constraint of pre-existent structures is nowhere more apparent than in the giraffes' laryngeal nerve, which originally linked the brain and larynx, a distance of a few centimetres. Because this nerve always loops around an aorta that has remained close to the heart, elongation of the neck in evolutionary time has required an ever longer detour of the recurrent laryngeal nerve until in modern giraffes it runs several metres down the entire length of the neck before ascending all the way back to reach the larynx.

While giraffes exemplify the role of foraging behaviour in shaping an ecological niche and the role of ecology in shaping morphology, it is only the most conspicuous example for an impressive radiation of African herbivores. Notably the ruminants, which, in addition to the giraffes, embrace those most typically African animals, the bovids, which encompass about a dozen distinct lineages, including the many forms of antelope.

Bovids are interesting because they demonstrate a great diversity of foraging niches that are the result of very early behavioural specializations coupled with morphological adaptations that then became the hallmark of their descendant lineage. They range from the rabbit-sized Royal Antelope *Neotragus pygmaeus* to the cow-sized Common Eland *Tragelaphus oryx* and they occupy quite distinct foraging niches. These centre on a combination of dietary types, ecological strategies and physiological refinements that have allowed them to radiate into a variety of special habitats. For example, large-bodied African Buffalo *Syncerus caffer* are mobile large-scale grazers in moist but climatically unstable habitats whereas small-bodied dik-diks *Madoqua* spp. are high-quality browsers able to survive on small, permanent territories in waterless scrublands because of their special physiology. By contrast, oryx and kin (Hippotragini) are relatively large-bodied mobile grazers able to live in arid or impoverished habitats but at low densities, while the Impala *Aepyceros melampus* is a medium-sized specialist in living along woodland/grassland ecotones, able to switch from grass to browse and sustain locally high densities.

Ever more refined, behaviour-led adaptations are evident in the duikers (Cephalophini), where individual species specialize in following particular patterns of fruit-fall. Thus a broad-mouthed species, the Bay Duiker *Cephalophus dorsalis*, ranges far and wide at night, seeking larger, rarer, fruits, while the diurnal, very localized and territorial Weyn's Duiker *C. weynsi* concentrates on the richest, most diverse forest zones where, by being attentive to the movements of primates and hornbills, they can rely on a year-round rain of nutritious plant-parts, particularly fruits and flowers.

The entire radiation of African bovids can be understood in terms of behavioural adaptation to different ecological situations and the adoption of modes of making a living, often within the constraints of a very particular and demanding habitat.

Quite minor shifts in foraging technique can give rise to interesting changes in anatomical details. For example, needle-clawed galagos *Euoticus* spp. have become different from their closest relatives through a greater dependence on tree exudates as their major source of sustenance. As a preponderance of gums ooze from tree trunks and very large branches the galagos need to spread-eagle their limbs and get a tight grip on the bark while levering chunks of gum away with their incisor teeth. To be able to accommodate this specialist foraging behaviour, originally flat, blunt finger-nails have developed sharp points allowing a firm, embedded grasp.

It is common enough to see shape vary among individuals, but Liers *et al.* (1990) recorded skull growth patterns varying seasonally within a local population of the common Multimammate Mouse

Mastomys natalensis. In the same population, Fadda & Liers (2009) measured significant differences in skull proportions between entire generations born during wet and dry years. They found that a prolonged drought had lasting effects on the shape of adult skulls. While most of the skull of the much smaller drought generation was simply scaled-down this did not extend to the brain, which was larger, relative to the rest of the skull, making for different proportions between the two generations.

Suppose that a discrete population of such 'large-brained' mice lived under drought over a very long period while a geographically separate population enjoyed a consistently benign climate. On morphological criteria alone the two might merit being treated as separate species. Indeed, with enough time and enough geographic separation speciation could be the outcome.

On the other hand, without the two elements of time and genetic separation such drought-induced differences represent little more than an entire generation becoming runts. Nevertheless, this example implies that, given enough time and isolation, species can emerge from common and natural physiological responses to environmental difference.

Carnivore skulls and teeth provide particularly good examples of foraging (in this case scavenging or killing) techniques that shape morphological form. For example, compare the skull shape, chewing muscles and teeth of, say, the insect-swallowing hyaenid Aardwolf *Proteles cristatus* with those of a bone-crunching hyaena or put the throttle-clamp of a Cheetah next to the killing bite of a Lion or the slicing stab of the recently extinct African Sabre-tooth *Megantereon*.

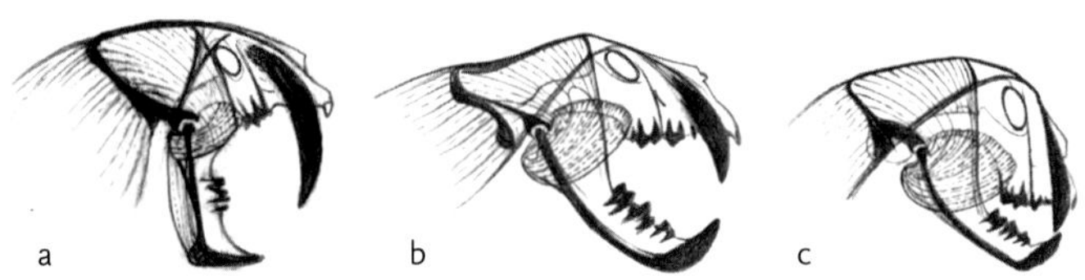

Skulls and tooth orientation in three felids, indicating different killing techniques.
a. *Megantereon* sp. Sabre-tooth. Stabbing strike by enlarged upper canines. Principal force exerted by neck muscles.
b. *Panthera leo*, Lion. Canine bite. Deep pincer action by long canines in both jaws.
c. *Acinonyx jubatus*, Cheetah. Sustained clench by short, bunched jaw muscles. Clamp on throat of prey enhanced by steep angle of face and down-turned maxilla (after Kingdon 1977).

Such comparisons of the species-specific buttressing of skulls display the mechanics of leverage around the front or back ends of the toothrow (or the atrophy of both in the Aardwolf). Remembering that every structure is 'the sum of many contrivances', each with a long evolutionary history, all predicated on highly specific behaviours, every head shape can be seen to manifest the physiology and mechanics of a species-specific killing or food-gathering technique. Lion, hyaena and other heads are telling illustrations of Darwin's insight into the dynamics of carnivore evolution. 'Some carnivores, ...being enabled to feed on new kinds of prey, either dead or alive; some inhabiting new stations, climbing trees, frequenting water, and some, perhaps, becoming less carnivorous. The more diversified in habits and structure the descendants of our carnivorous animal became the more places they would be enabled to occupy.' How Darwin would have relished watching all these species going about their predatory businesses, wild yet unafraid of their human observers, in Serengeti or Kruger National Parks!

Shifts in environment/habitat

All habitats have borders and some, like rivers, are essentially linear. It is along these margins or ecotones that mammals living in neighbouring zones most readily (and literally) shift from one habitat to another, but it is probably climatic change that has had the greatest influence in forcing animals to adapt to different habitats. Our earlier example of the Patas 'greyhound' is but one of many. At the other extreme, in a continent that has been dry for much of its history, swamps have tended to be ephemeral or, if of longer duration, have been quite localized. The one antelope group that has been best placed to invade swamps has been the waterbuck tribe, or Reduncini, which made an early accommodation to living along grassy drainage lines. By shifting down the catena into the permanently flooded sumplands (perhaps mainly to mitigate predation), this type of antelope has given rise to the lechwes, *Kobus leche* and *K. megaceros*. Aside from cryptic physiological adaptations, their most obvious adaptation is the development of long, splayed hooves. These limit the animals' ability to live anywhere other than in marsh and it is significant that lechwes of any sort have not survived or spread outside Africa's two most extensive regions of swamp.

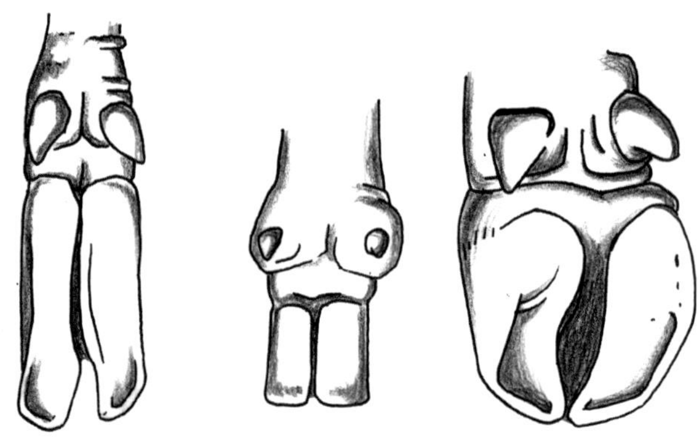

Hooves of, from left to right, *Kobus megaceros, Oreotragus oreotragus* and *Syncerus caffer* compared.

Another antelope that is quite severely constrained by the development of specialized toes is the Klipspringer *Oreotragus oreotragus*. These animals are also likely to have made their initial accommodation to an ostensibly impossible habitat; steep, rocky country (probably in an extensive region of deeply dissected mountains, such as Ethiopia). Their invasion of such a difficult habitat for a hoofed animal was probably driven as much by the need to find refuge from predators as the rewards of forage up among the crags, kloofs and kopjes. Indeed, in some parts of its range the Klipspringer tends to descend and leave these refuges to feed, yet it always returns to ruminate, rest or sleep among the rocks and cliffs. Its behaviour of leaping about cliffs has led to the development of rock-gripping rubbery hooves, a specialist adaptation that can be seen to a much lesser extent in other antelopes that also live in stony habitats.

Shifts or innovations in communication and in the balance of senses

Fruit bats are dependent on a dispersed food supply that is often narrowly seasonal in availability as well as being highly unpredictable in abundance between seasons and years. For some fruit bats efficient exploitation of such a shifting mosaic of fruit demands quite substantial seasonal movement and dispersal over an extensive terrain. The resulting semi-nomadic 'camping' life-style poses problems for bats that require to keep in touch with one another.

Their primary solution has been enhanced vocal communication, and three species of epomophorine bats illustrate three types of morphological adaptations to facilitate the amplification of advertising calls. Wahlberg's Fruit Bat *Epomophorus wahlbergi* has a loud clinking call, somewhat like a particularly resonant frog's croak. Franquet's Fruit Bat *Epomops franqueti* has a substantially louder and rather 'musical' call. The Hammer-headed Fruit Bat *Hypsignathus monstrosus* has such a loud blaring honk that on a still night it can be heard over several kilometres. In each instance the vocal apparatus devoted to the generation and amplification of sound is enlarged. *Epomophorus wahlbergi* has a larger voice-box than other bats but it comfortably fits in the throat in the space between chin and chest. *Epomops franqueti* has a voice-box sufficiently large to invade the chest and has clearly reorganized its breathing to increase the air pressure passing through the enlarged structure. In the case of *H. monstrosus*, the largest bat in Africa, the vocal apparatus has so enlarged that it has displaced heart, lungs and diaphragm, with the vast larynx doubled up like a tuba. To further amplify and project the sound, there are sacs on each side of the neck and a two-chambered pouch that is anchored on the enlarged nasal rostrum. *Hypsignathus* males have effectively become flying loud-hailers and the morphological modifications to this end have so reshaped their heads as to earn their 'monstrous' epithet.

While the last species seems to rely solely on a vocal signal, both *Epomophorus* and *Epomops* have augmented the call with scent and visual signals. Because vocalizing is the major signal mode, scent and vision are subordinate. In this instance this subordination is particularly obvious because these bats have developed scent-pouches containing evertible white 'plumes' or soft hair tufts on their shoulders. These 'epaulettes' effectively advertise the call. The logic behind the 'choice' of such a peculiar site to develop a visually advertised glandular area becomes obvious when the bat makes its call. As a diaphragm-pumping muscular spasm forces air through the voice-box it simultaneously flexes the shoulders and this spasm opens the epaulette, thereby dispensing a puff of scent from the brilliantly white hairs that are rooted in the glandular pocket that underlies it. In this way the bat sends out vocal, visual and olfactory signals more or less simultaneously, with the last two advertisements totally predicated upon the first, primary vocal signal. The interaction of behavioural selection (for a louder call) with a physiological constraint (having to 'heave' the chest and flex the shoulders) have here led to the evolution of a novel way to signal with scent and then on to a visual device associated with both the vocal and olfactory signals' source.

Many scented signals are distributed by way of faeces, and/or related methods. Tails are obvious dispensers for anal scents and many mammals with anal glands frisk or wave their tails in a variety of ways. Indeed, the tails of most genet species are boldly banded in black and white (in similar fashion to a barber's pole), strongly suggesting a coupling of visual and olfactory advertisement. Often these gestures, too, are advertised visually by conspicuous colouring or movement and nowhere is this more obvious than in the squirrels. In the case of giant squirrels *Protoxerus*, the upper side of the tail has narrow white and broad black bars that run from side to side whereas the underside is marked with equidistant, longitudinal black and white stripes. This difference in pattern indicates that a depressed tail sends an opposite visual signal to an elevated one.

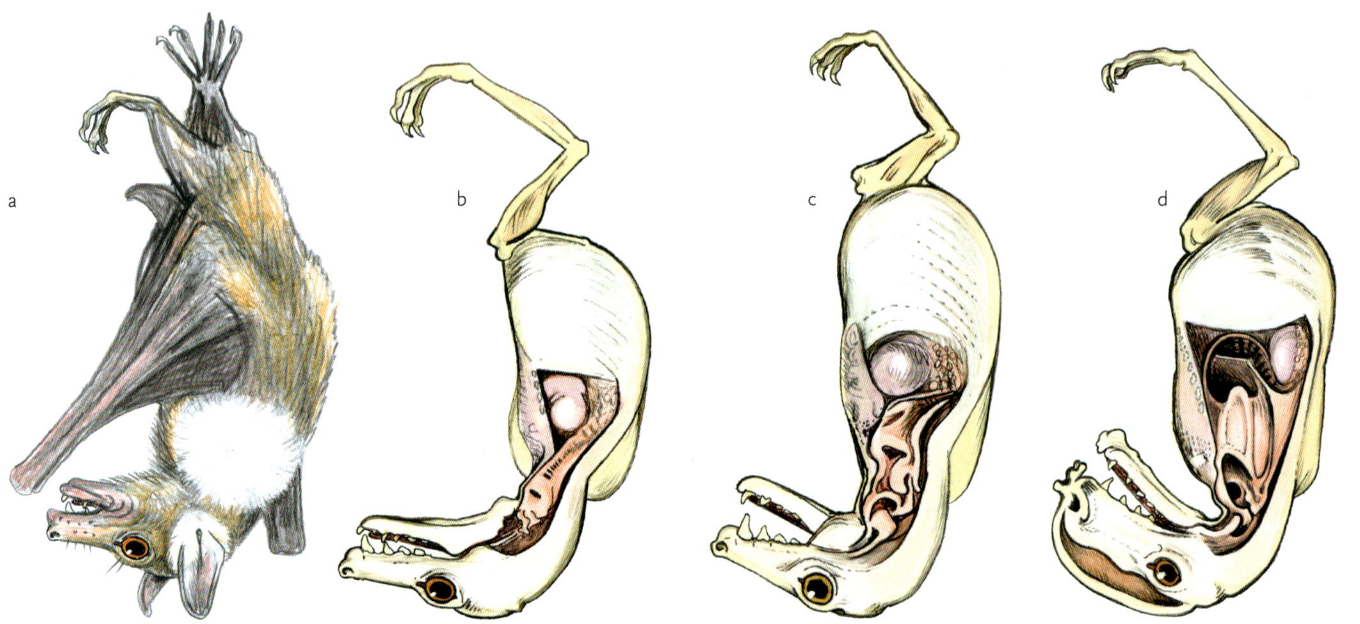

The evolution of a flying loud-speaker. Shape determined by signal-generating apparatus. Bisections of heads and thoraxes of:
a. & b. *Epomophorus*, showing some elaboration of the vocal tract and eversion of the white 'epaulette' signal.
c. *Epomops*, showing significant enlargement of larynx.
d. *Hypsignathus*, showing larynx enlargement has displaced the entire chest cavity.

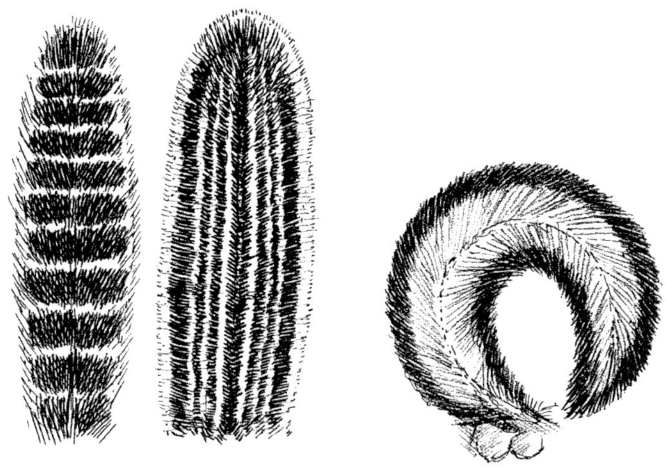

Tail patterns of *Protoxerus stangeri*, upper and lower sides, and *Funisciurus pyrrhopus* symmetrical tail curl.

Shifts in behaviour, i.e. the functional balance of senses (or changes in the relative importance of any one sensory activity), are given away when one member of an identifiable lineage exhibits a diminution or enlargement of its ears, eyes or nose. Thus the disproportionately big ears of the Serval *Leptailurus serval*, which lives much of its life in long grass, betray a dependence on hearing scurrying rodents, while the large eyes of Black-footed Cats *Felis nigripes* signify exclusively nocturnal habits. Likewise, the long snout of the earthworm-finding Liberian Mongoose *Liberiictis kuhni* contrasts dramatically with the short muzzle of the partially arboreal and more 'visual' Slender Mongoose *Herpestes sanguineus*.

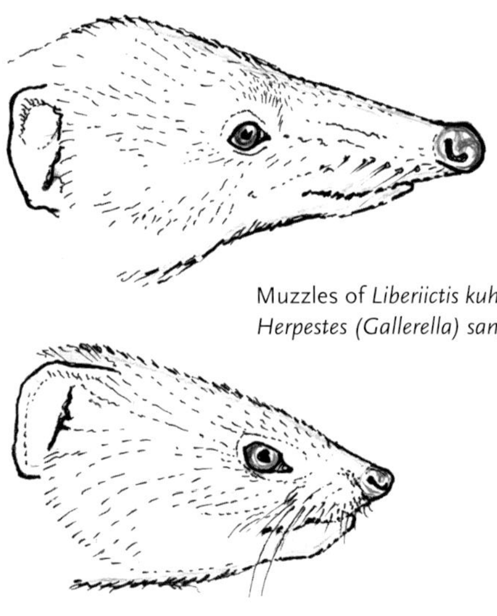

Muzzles of *Liberiictis kuhni* (*above*) and *Herpestes (Gallerella) sanguineus* (*below*) compared.

A more subtle manifestation of behaviour-driven sensory shift is exemplified by a group of small-bodied, very fast, guenon monkeys that garner much of their food on the smaller, outermost branches of the forest (mostly small fruits, flowers, budding leaf growth and active arthropods). By shifting out into this more exposed habitat, members of the moustached or red-tailed monkey group, the *Cercopithecus* (*cephus*) species complex, have gained access to a rich (but generally more dispersed) food supply but have also acquired substantial problems. An obvious hazard is greater exposure to eagles, which demands greater vigilance by all members of a social group and heightened visual awareness. Abundant but scattered resources require a dispersed foraging pattern and this, in turn, elicits problems for social behaviour because group members are spatially separated and this physical separation compromises a primary social greeting among guenons. This greeting consists of turning round and 'presenting' (an often distinctively coloured) genital region for both visual and tactile inspection. Enlarging the distance between individuals not only diminishes the value of tactile communication but (because grooming and genital signals have less utility) dispersal also facilitates female escape from males. Male frontal approaches are all too easily seen as threatening to females, which makes male inspection and mating with females more difficult, thus complicating efficient reproduction. Given these anti-social traits in *C.* (*cephus*) groups, any behaviour that can offset centrifugal tendencies is likely to be favoured by natural selection. 'Presenting' becomes all the more difficult and impractical while group members are scattered through a lattice of thin branches and it is futile to turn round, on precarious twigs, to present the genital pole to an out-of-reach partner!

Unlike many other guenons, *C.* (*cephus*) monkeys lack coloured or patterned signals on their rear ends, and they seem to pay less attention to that pole of the body than other monkeys. Instead, these monkeys have elaborated highly conspicuous face patterns, which act as 'flags' when they flick their faces in ritualized 'eye-avoidance' gestures. Significantly, flicking the head to avoid staring is an appeasing gesture that is roughly equivalent to 'presenting' as a reassurance. When the behaviour and morphology of these monkeys is compared with that of other guenons it would seem that *C.* (*cephus*) monkeys have switched a major primate social signal flag from 'back to front' (see profile of this super-species). This interpretation of *C.* (*cephus*) biology implies that specific behaviours, namely, foraging for small, nutritious items in the outermost foliage of trees, has favoured small bodies and a dispersed foraging pattern; this, in turn, has devalued tactile genital signals but favoured visual frontal ones. Thus the previously minor or inconspicuous gesture of eye-avoidance (which helped dampen aggression and ease inter-sexual communication) has been up-graded and visual signals have developed on the face to advertise the gesture. Morphological characters, such as coloured skin or deflected hair tracts, have evolved in the *C.* (*cephus*) group as a direct consequence of their behavioural shift from predominantly tactile to visual communication.

There are many other examples of a species making an apparent ecological shift that has demanded changes in what the animal does to survive in a new setting. These changes, in turn, have led to the evolution of interesting anatomical or physiological peculiarities. For example, any enlargement of a species' body size or of its home-range can severely challenge pre-existent adaptations. Thus the Oribi *Ourebia orebi* shares facial and other skin glands with various, much smaller, dwarf antelopes (Neotragini) that inhabit forested or thicket mosaics. Among the dwarf antelopes these glands are sufficient to mark out relatively small, exclusive territories in relatively stable, closed habitats. Oribi, instead, have come to inhabit more open grasslands that are subject to annual fires and are decidedly unstable. The Oribi's evolutionary response seems to have been to enlarge and intensify the effectiveness of skin glands, to multiply their number and to diversify their positions on the body. Thus Oribis have enormous face glands, very active

Oribi *Ourebia ourebi* indicating multiple sources of scent signals.

Male Lion *Panthera leo* strut displays body in tip-toed side-view, but head and mane head-on.

inguinal and pedal glands, scented knee-tufts and a patch of 'hot-plate-type' glandular skin below the ear. Each of these gland types dispenses and distributes scent in a slightly different way. Extended territories demand more frequent and dispersed scent marks but each territorial male Oribi also accompanies his female to ensure that she is exposed to a continuous barrage of his various scents (Monfort & Monfort 1974). This intensification of tending behaviour probably helps to deter defection by the female and may offset increased demands on the male's capacity to mark his territory. The sheer diversity of scent-production in a male Oribi combined with persistent monitoring of the female may also habituate the latter to prefer the scent of their shared living space and thus remain rather than defect and choose another male.

Changes in social structure

When previously solitary mammals become more social many residual details of ancestral behaviour can become problematic. Thus territorial species must modify or suppress territorial behaviour, as a result offsetting the centrifugal force of intra-specific aggression. When such aggression is gender-specific then interesting differences can appear in the appearance or morphology of the sexes, especially when aggression is, paradoxically, enlisted in the task of becoming social!

Take Lions *Panthera leo*, which are very frequently social and probably derive from ancestors with habits that would have resembled those of the solitary Leopard *Panthera pardus*. Lion sociality is built upon the retention and intensification of female-offspring bonds coupled with a fierce antagonism, amounting to xenophobia, from all pride members towards unrelated Lions, especially female ones (Schaller 1972, Bertram 1975). Thus young females remain with their mothers and aunts, having every reason to avoid wandering while being under no pressure to leave. Males, instead, leave their natal pride as they mature and then wander, often in sibling pairs or groups, until they can oust the males of another pride and impose themselves upon that pride's resident females (and typically start by killing all young, thus bringing the newly acquired females into oestrous as fast as possible; Bertram 1973). Imposition of male supremacy is demonstrably helped by the males' larger size but it would seem that the sheer ferocity of female xenophobia makes any ambiguity about gender something to be avoided.

Stressing that male Lion manes serve a primarily visual, not a physiological function, Kingdon (1977) correlated manes with a lateral, head-up, tip-toe display that is exclusive to males and mainly directed at females. In this behaviour strutting makes the actor look as tall and as large as possible while manes become primarily a head-enlarging device. Noting that male Lions are the principal source of fresh genes and the manes of prime-age males come into particular prominence during the 'lion strut', Kingdon (1990, 1997) suggested that manes and strutting combine to serve as an intimidating, intra-specific, anti-incest mechanism. Morphology and behaviour combine to assist mobile males to identify themselves as non-females and to impose themselves on the more sedentary and matrilineal females.

Other species that have become social have developed very different, almost opposite, trends. For example, the Bongo *Tragelaphus eurycerus* is, almost certainly, a social descendant of relatively solitary ancestors in which the horned males were larger than the hornless females. The development of horns in Bongo females probably reinforces female hierarchy within the herd while also ensuring some ability to protect the young. Similar trends are apparent in elands, *T. oryx* and *T. derbianus*, where females are also well horned and of imposing size in spite of remaining smaller than males.

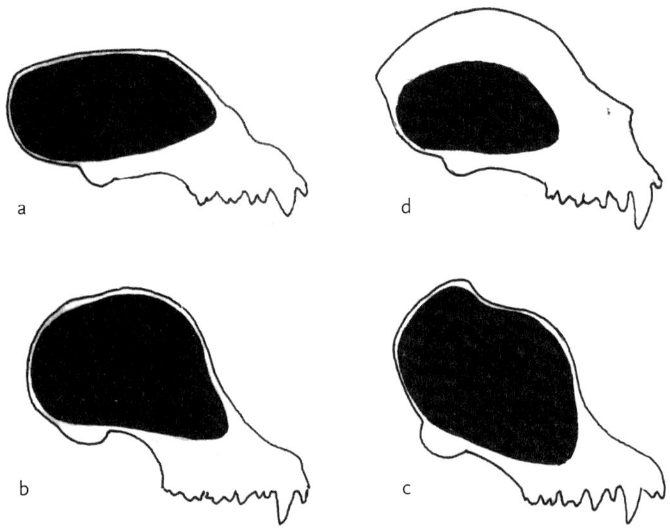

Relative brain-size in four bats. Exceptionally large brains are typified by *Cardioderma* (a), *Kerivoula* (b) and *Miniopterus* (c). A relatively smaller brain is typified by *Hipposideros commersoni* (d).

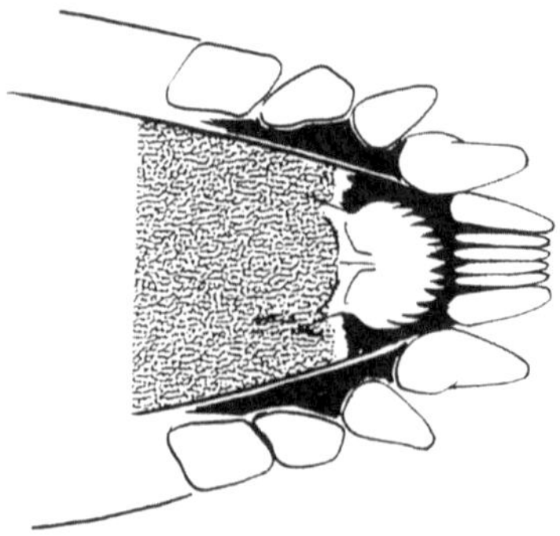

Incisor comb and sublingual in galagos.

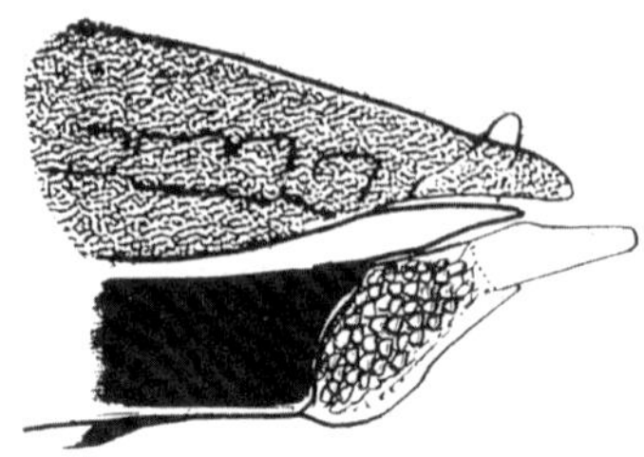

Increasing social complexity: brains and bonding

If changes in all morphological structures are initiated by changes in behaviour, alterations in the size or shape of brains can be no exception. Some echo-locating bats have, compared to the size of their bodies, very large brains. Among those with larger brains are the Heart-nosed Bat *Cardioderma cor*, species of woolly bats *Kerivoula* and long-fingered bats *Miniopterus*. What these very different bats have in common is that all are social and all have complex but emphatically different ways of finding and catching food. In *Miniopterus schreibersii*, a migratory species that ranges from Africa and Europe to China and Australia, young are born with large brains, develop very fast and are deserted by their mothers at about six months (Dwyer 1968) and disperse very widely, whereas adults are more sessile. Able learners (in orientation and the capture of food as well as in social interaction) are likely to have been selected for. Thus improving the ability to acquire, process and store information, from an early age, must have been one of the primary reasons that brains enlarged in any lineage. Primates, like bats, are already large-brained animals but still further enlargement in the ape and most especially in the human lineage signifies a fundamental shift in which foetal rates of brain growth continue for about a year after birth and brains still keep growing until puberty. Among humans there has been especially pronounced elaboration of the visual apparatus of the brain (Le Gros Clark 1959) as well as those parts of the brain that service the hands (Penfield & Rasmussen 1950).

One of the primary behavioural mechanisms that promotes sociality in mammals is one-on-one grooming of the fur with teeth or fingers. In primates, as with most mammals, the primary purpose of grooming, especially self-grooming, is the maintenance of a healthy skin and pelage. In the case of bush-babies (Galagidae) and pottos (Lorisidae), the development of 'incisor-combs' has increased the efficiency of their grooming and fur-cleansing behaviour. In the more social higher primates grooming is probably as important socially as it is in pelage-maintenance. Indeed, some monkeys seem to produce musk-scented sebum flakes at the roots of the fur, expressly to serve as 'grooming-bait' (Kingdon 1971). A much more elaborate development, putatively of 'visual bonding' out of physical nibble-grooming, is described in the profile of subgenus *Hippotigris* spp.

Reproductive behaviour shaping sexual attributes

The most common outcome of male competition for females is through direct trials of strength and endurance. When there are few constraints on increasing size and a big pay-off in terms of numbers of females inseminated there can be very rapid selection for gross sexual dimorphism. This is most marked in the Cape Fur Seal *Arctocephalus pusillus*, where very large numbers of females come ashore to give birth and copulate in choice 'rookeries' off the coast of south-west Africa. Here very many females are mated by a small number of powerful males. Each generation of seals reinforces selection for larger size and greater vigour but seal numbers are known to fluctuate widely, so it is likely that relative measures of sexual dimorphism also fluctuate. In *A. pusillus* the current level of male dimorphism by weight is approximately five to one.

The frequency with which males fight is another measure of male competition, and this is often an expression of the relative density of males during the rut. Thus substantial differences in the weight and shape of antelope horns can be correlated with relative density at this time. For example, within the Bohor Reedbuck *Redunca redunca* (generally well spaced out in territories) one locality on the banks of the Nile supports a dense concentration for the duration of the rut. Here horns are appreciably longer and heavier than anywhere else in the overall range of Bohor Reedbuck. Likewise, among Hartebeest populations, the Lelwel *Alcelaphus buselaphus lelwel* and the Kanki *A. b. major* often live at very high densities and have ultra-stout, tightly twisted, hook-like horns on extended pedicels while populations living at lower densities tend to have lighter, less complex horns that have shorter pedicels. At a grosser level antelopes that are small, well

Female Gelada Baboon *Theropithecus gelada* showing sexual swellings and vesicle-bordered chest and neck.

spaced-out and territorial thus experience encounters with a very few, and typically well-known, neighbours. These species tend to have small spiked horns and their contests tend to take the form of a regular testing of condition; their social structure tends toward well-established pairs, as typified by the dik-diks. By contrast, the males of harem species, such as Impala *Aepyceros melampus* and Kob *Kobus kob* have evolved heavy, lyre-shaped horns, which are used in frequent, bruising, and sometimes fatal, fights.

Among many extraordinary examples of behaviour shaping unique, species-specific structures is the transfer in female Gelada Baboons *Theropithecus gelada* of an anatomical signal indicating oestrus from the back to the front of the animal. In many higher primates, including chimpanzees and baboons, females advertise receptivity with conspicuous genital swellings. The most likely reason for female Geladas developing a partial 'necklace' of pink vesicles on their chests is the fact that they forage squatting and shuffling along on their haunches. Shuffling conserves warmth and saves energy in a cold, nutrient-poor environment but removes from view a prime indicator of reproductive condition. The conspicuous vesicles wax and wane with oestrus, thus providing an alternative signal of receptivity.

Many other African monkeys elaborate sexual differences in sex-specific structures, in size and in visual appearance, but few exceed drills and mandrills, *Mandrillus*, in the exaggeration of masculine features. In both species, the much larger males posture and strut with dominant individuals staring down potential rivals. Elaborate symmetry in the frontal aspect of a healthy, fully mature male Drill, *M. leucophaeus*, suggests that intra-specific selection (and more exactly, selection through the pattern-extracting susceptibilities of visual neurones in viewers) has favoured geometry and the evolution of an impressive and uniquely specific black, white and red facial mask.

Predator pressures and avoidance

Both predators and their prey are shaped by evolutionary 'arms races': among the morphological features are improvements or changes in camouflage. The pelage of many stalking predators has been selected to be as inconspicuous as possible so that they can sneak up closer to their prey. Likewise brownish agouti patterns, flecked and spotted pelage have evolved in rats, sengis and hares to aid blending in with the surroundings. Likewise, another strategy to escape predation is to minimize exposure, for example, by collecting and storing food in caches. Many rodents employ this strategy but the pouched rats and pouched mice (Cricetomyinae) have taken this to extremes, with cheek pouches that can hold as much as, or more, than their stomach. The need, by a slow, arboreal/terrestrial rodent (the Gambian Giant Pouched Rat *Cricetomys gambianus*) to minimize the time it spends foraging has led to particularly capacious cheek pouches. The 'cheek-pouch monkeys' (Cercopithecinae) have evolved similar structures and techniques. The practice of transporting food in bulk for later, more leisurely consumption is also found among ruminants who quickly accumulate forage, which is later processed by 'chewing the cud' at leisure, often hidden away from the feeding grounds where they may be more vulnerable.

Both the bodies and the skulls of the petromurid rodent *Petromus* and several other crevice-dwelling species, notably the bats *Mormopterus* spp., *Platymops setiger* and *Mimetillus moloneyi*, have flattened to accommodate to narrow refuges, as was outlined earlier. As for more direct protection against predator attack, enlargement of dorsal hairs to the point where they become spines reaches its extreme in crested porcupines, but hedgehogs (Erinaceidae) and some spiny mice *Acomys* spp. are less extreme examples. Scales have evolved to provide pangolins (Manidae) with a protective armour and the Zorilla *Ictonyx striatus* has evolved impressive anal stink-glands together with conspicuous black and white colouring, which advertises its noxious presence.

Male Drill *Mandrillus leucophaeus* illustrating the symmetrical geometry surrounding the muzzle and mouth.

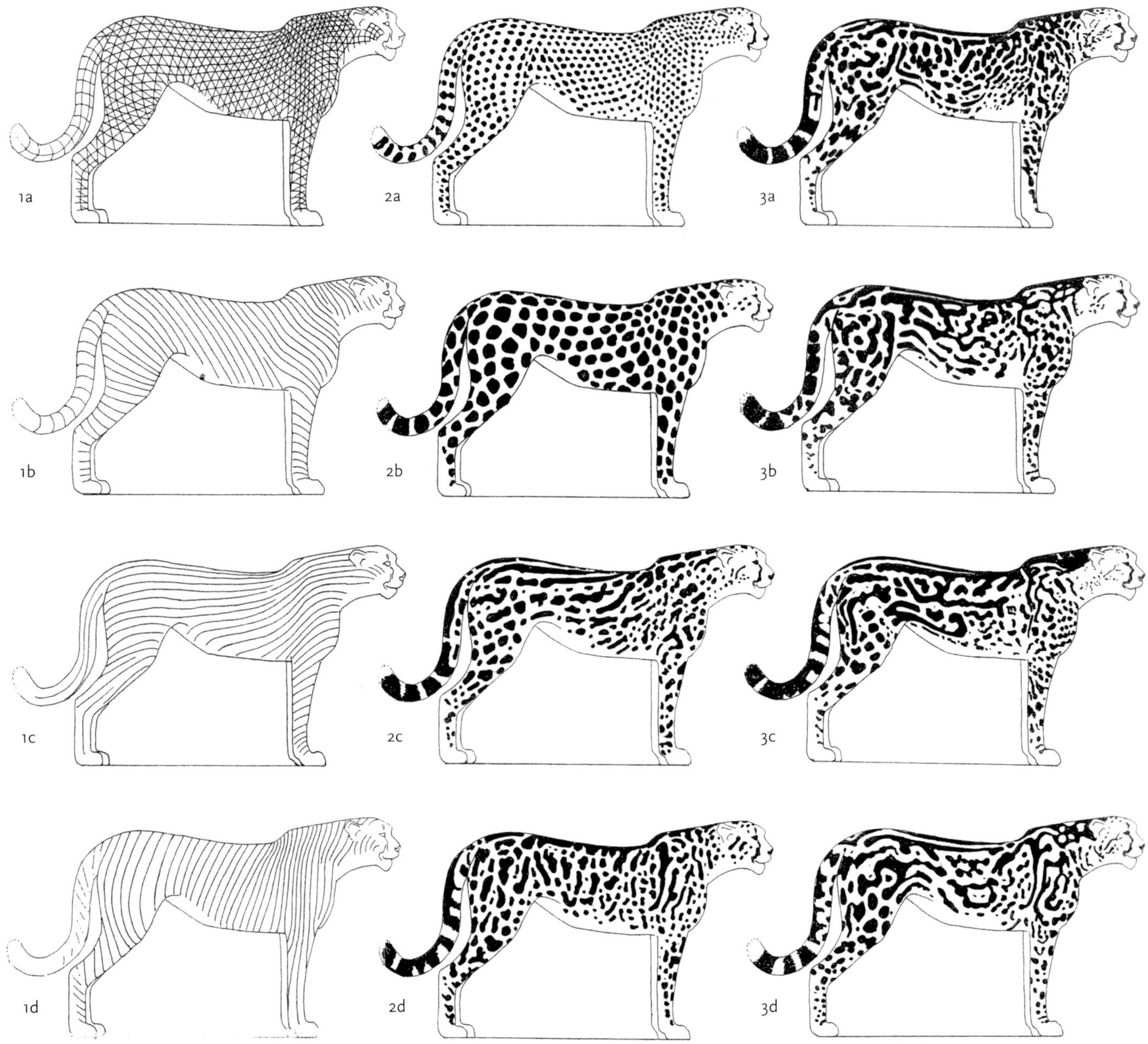

Spot patterns of the Cheetah *Acinonyx jubatus* serve to disrupt the visual impact of body mass and outline. Stalking prey mainly during daylight in open, well-lit habitats, the optimum size and frequency of spots is roughly equivalent to the average distribution of pixel spots on digitized images of Cheetah habitat. The Cheetah's pelage is effectively a visual abstraction of its environment. Approximately 60% of the Cheetah's total spotted areas are light coloured. Dark tones comprise about 40%.

Cheetah spots are distributed over all surfaces exposed to the view of potential prey. Spot sizes vary and counts range from 8 to 24 spots per 100 cm^2. Analyses of normal Cheetah patterns reveal that spots are aligned along a multi-stranded grid.

1a. shows a grid superimposed upon the outline of a Cheetah profile and shows the tight, relatively even, multi-directional spacing of alignments. 1b. isolates alignments that are mainly diagonal, 1c. horizontal alignments and 1d. mainly vertical alignments.

Analysis of aberrant patterns among so-called 'King Cheetahs' reveals that the genetic control of such patterns effectively consolidates single small spots into larger amalgams without changing the broad ratio of 60% light to 40% dark.

Were all smaller spots to be amalgamated within the same equal-distance grid that governs normal Cheetah patterns, this would result in the hypothetical polka-dot pattern shown in 2b., which would be highly conspicuous and is unknown in nature. Instead, break-down in the genetic control of spot-spacing results in spots amalgamating along particular linearities of the overall grid as well as at random.

2c. and 2d. are collages in which horizontal (2c.) and vertical (2d.) pattern elements have been extracted from photos of 15 different King Cheetah morphs.

3a. to 3d. show actual patterns on four King Cheetah individuals. Note retention of very small areas of normal spotting (3a.), some horizontal consolidation (3b.), some 'polka-dotting' (notably on haunches of 3c.) and typical 'marbling' (3d.).

Note that normal Cheetah patterns show less regional variance in spot counts and tonal ratios compared to taxa with broader habitat tolerances (such as Leopards and some genet species, Kingdon 1977).

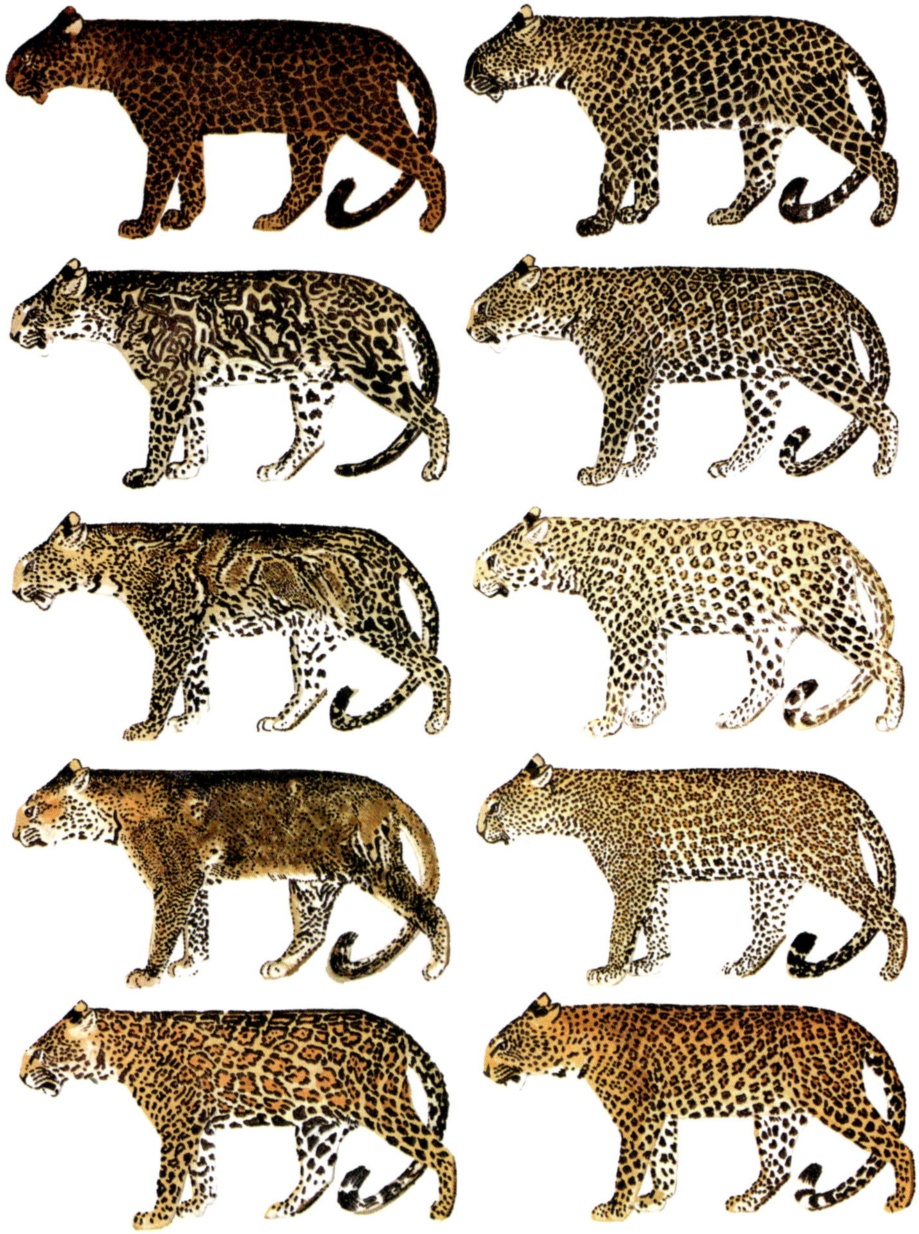

Pattern 'maps' of Leopards *Panthera pardus*.
Left-hand column = polymorphism (from top):
Melanistic morph (Black Panther);
'Marbled' morph;
'Panelled' morph;
'Freckled' morph;
'Jaguar-like' morph.

Right-hand column = regional types (from top):
Amalgamated rosettes, Ethiopian Highlands;
'Mosaic' pattern, Rwenzori Mts;
Pale, open network, Somalia;
Multiple, small rosettes, Zanzibar Island (exterminated);
Most widespread type (Mount Elgon).

Diversity and transformations of camouflage

In direct apposition to aposomatic signals are the many manifestations of 'camouflage'. Predators that are effective stalkers of their prey must evolve patterns that complement their stalking behaviour. In the case of Leopards, pattern variations show some correspondence with the very varied regions that they inhabit. Thus Leopards from the arid Horn of Africa have widely spaced, open-centred rosettes on a pale yellowish background. By contrast, the black spotting on rosettes of Leopards from dark, moist montane forests in Ethiopia have amalgamated into densely packed blotches (see figure above).

In each case the basic grid of rosettes would seem to have been modified in correspondence with very different vegetation types and light levels. The particular mimesis of the Leopard's coat appears to be the dapple of leaves and their shadows, the setting within which Leopards spend most of their time. Dark and light areas of the coat are dispersed in an abstract conformation that breaks tones up into broad averages of area and contrast. The groupings of spots into rosettes resembles the ink-spot clusters found in colour photographs after passing through a mechanical screen or scanner (Kingdon 1977).

The typical Cheetah pattern of isolated spots follows similar principles but has no relationship to leaf mimesis. It also represents an abstract mathematical averaging-out of relative tones in the Cheetah's more grassy habitat (see figure opposite). Among both Leopards and Cheetahs there are individual variants commonly described as 'aberrant'. Yet it can be demonstrated that natural selection for just such marbled variations can take place. For example, some so-called 'King Cheetahs' as well as some 'aberrant' Leopards have patterns that approximate to the Asiatic Marbled Cat *Pardofelis marmorata*. Leopard variants, in particular, can individually resemble the patterns of Clouded Leopards *Neofelis nebulosus*, Ocelots *Leopardus pardalis* and Jaguars *Panthera onca*. These non-African felids also vary, both regionally and individually, in their coat patterns.

Pelage pattern formation in African Wild Dog *Lycaon pictus*. a. Generalized, semi-cryptic formation in genus *Canis* as expressed in various jackals and wolves. b. Similar pattern becoming more conspicuous through enhanced tonal contrast. c. *Lycaon* pelage in which *Canis*-like format has been dislocated but elements are still perceptible. d. *Lycaon* pelage in which dislocation has generated typical 'marbling' (individual from Longido, N Tanzania).

In these instances, selection would still seem to function as 'camouflage', yet there is one more carnivore example in which marbled or blotched patterns vary with every individual as well as showing regional trends. It would seem likely that the swirling, blotchy black, white and yellow patterns of African Wild Dogs ('Painted Dogs') *Lycaon pictus* have evolved by a process analogous to that of the 'aberrant' felids. If this analogy holds, the *Lycaon pictus* pattern may represent the semi-random elaboration of ancestrally more structured and quieter patterns, such as the flank and dorsal stripes of jackals and more subtle colouring on the faces, legs and tails of Wolves *Canis lupus* and Dholes *Cuon alpinus*. Unlike the felid examples, African Wild Dogs are coursers, pursuing their prey in the open and cannot, by any ordinary use of the term, be described as 'camouflaged'. Furthermore, elaborate social behaviours in *Lycaon pictus* provide a plausible explanation for the evolution of their extraordinary tri-coloured mottle pattern, which resembles the sort of polymorphism of some domestic animals. Schaller (1972) interpreted this pattern as helping scattered pack members to keep in visual contact and maintain the group cohesion, but did not volunteer how this might have evolved or operate. More specifically, one of us has linked the adaptive value of a pattern that breaks up the body's contours directly to the social disciplines that maintain pack cohesion (Kingdon 1977).

Significantly, for a species where the vulnerable young are left in a den while the adults go hunting, wandering pups elicit strong responses from adults: observers have noted that provisioning adults returning to the den not only tend to disgorge meat in favour of those pups that stick together, they may even bite or harass stragglers. The behaviour of adults, especially provisioning ones, would therefore seem to influence survival in *Lycaon pictus* and it has been suggested that this interaction is sufficient to select for a unique form of 'camouflage': one that favours those pups that 'blend in' best with the rest of their noisy, hyperactive litter (Kingdon 1977). At the moment of maximum reward and maximum risk, a pup's immediate and actual environment becomes a cluster of prostrate begging siblings. The mud walls of a den's mouth are not the pups' immediate visual setting; instead it is the tactile, mobile, alive mass of fellow pups. In this parent-selector-driven context Kingdon (1977) argued that the mottled pattern *is* an aberrant form of camouflage, but one that instead of being predator-selected has its origin in social behaviour.

By begging, small pups, from the age of one month, can coerce adults to disgorge meat and this behaviour continues into adult life as a social greeting and a system whereby the weaker can influence the stronger without either party resorting to unrestrained biting (Kuhme 1964). Thus, prostration is a tactic that combines access to food with insurance against attack; both the beggar (paradoxically the 'aggressor') and the provider (inhibited from aggression by the beggar's behaviour) co-operate in a ritualized expression of interdependence. Schaller (1972) suggested such exhibitions of 'friendly aggressiveness' induced cohesion, especially during 'meets' or 'social rallies'. Kingdon (1977) contended that an 'aberrant' and highly conspicuous, contour-breaking pattern serves as the main visual cue for such cohesion. At such times adult dogs readily switch their behavioural role from typical 'dependent beggar' to typical 'coerced provider' and then back again. In the process, temporary alliances are formed between individuals and the alert, excited conditions that precede embarking on a hunt seem to be facilitated (Kuhme 1965). The pups' collective dependence on adult providers seems to have carried over into a mechanism that helps maintain interdependence and pack cohesion. Coat patterns, by 'dissolving' the individual contours of these most social of dogs, are integral to the signalling environment in which they interact and survive as a visually undifferentiated group.

The complex genetic recipe that encodes for the coat patterns of African Wild Dogs, Marbled Cats, Clouded Leopards or domestic 'sports' must include, within its complexity, residues of symmetrical regularities in distant ancestors. The larger message is that every genetic history embraces increment upon increment of selection for 'aberrant' individuals. In evolutionary biology, aberrant is a word to be used with caution.

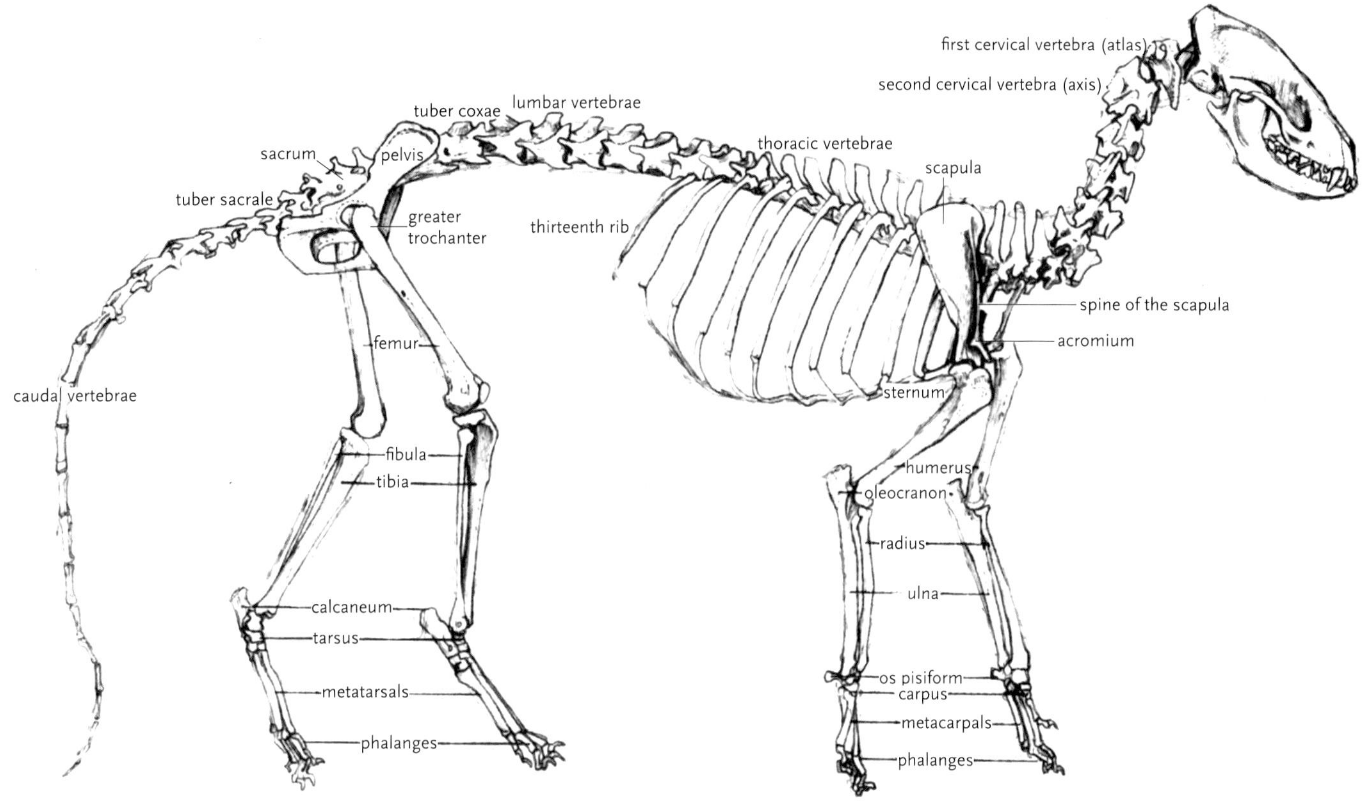

African Civet *Civettictis civetta* skeleton (*above*) and skull (*right*).

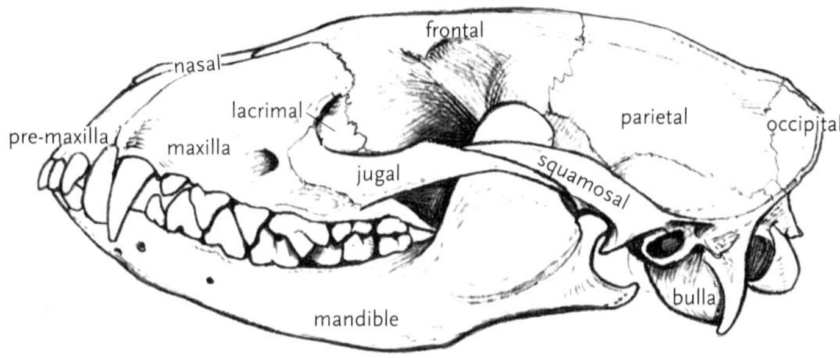

Tool-language and morphological metaphors

All examples given so far demonstrate that it is not possible to describe mammals, let alone discuss their structure, without invoking the relationship between form and function (thus we have used similes such as umbrellas and deck-chairs). In relating functional behaviour to the complex structures that the behaviour has helped engender, it is often difficult to avoid technical language and we have provided pictorial guides to body parts, to the names of all major bones and to the larger and more obvious muscles of mammals. Sometimes jargon can be side-stepped, but it is significant that attempts to modernize or Anglicize terminology are often little more than simple restatements of the Greek or Latin words we have inherited to describe the many and various parts of an animal. The reasons for this convergence lie in the difficulty of finding words that translate our understanding of form into functions, mechanisms and geometries that we can readily envision and which we already understand. Latin names actually help comprehension by preventing too literal a mental match while at the same time expressing the link between *nomen*, form and function. Thus *pelvis* is Latin for basin, *bulla* is bubble, *fibula* means clasp or attachment and *masseter* simply means chewer; all are technical terms that describe and (more importantly) suggest familiar artefacts as well as functions.

At a higher level of metaphor, biologists have struggled to conceptualize how the total architecture of organisms, from the gene to the protein and tissue, relates to the structured living behaviour of individuals that belong to those entities that we call 'species'. Theoretically minded biologists have invented a telling simile for all living organisms in designating them as 'self-regulating machines' or 'homoeostatic systems'. Although easily dismissed as overly mechanistic, these are examples of what the great biologist J. Z. Young called 'tool language' (Young 1957). Tool language is a large part of the terminology of modern biology and is almost inescapable whenever we try to formulate ideas about natural phenomena in general and functional morphology in particular.

When it comes to naming whole classes of mammals, especially extinct ones, palaeontologists sometimes have rather little to go by, but the names they invent must still allow comparison with more familiar animals. For some little-known, very early extinct species, teeth are the *only* part of the skeleton to have survived so that some important groups of early mammals are actually named after the shape of their teeth: creodonts ('flesh-teeth'), tribosphenids ('three cusps') and multituberculates ('many cusps'). Among living orders several have also been named in reference to their teeth, notably the rodents ('gnaw-teeth'), edentates ('toothless') and tubulidents ('mini-tube-teeth'). Even when limited to diagnosis or nomenclature, dental and mandibular structures are intensely useful but, as Darwin emphasized, dental, cranial or general anatomy only becomes central to biology when the evolutionary and functional meaning of structures

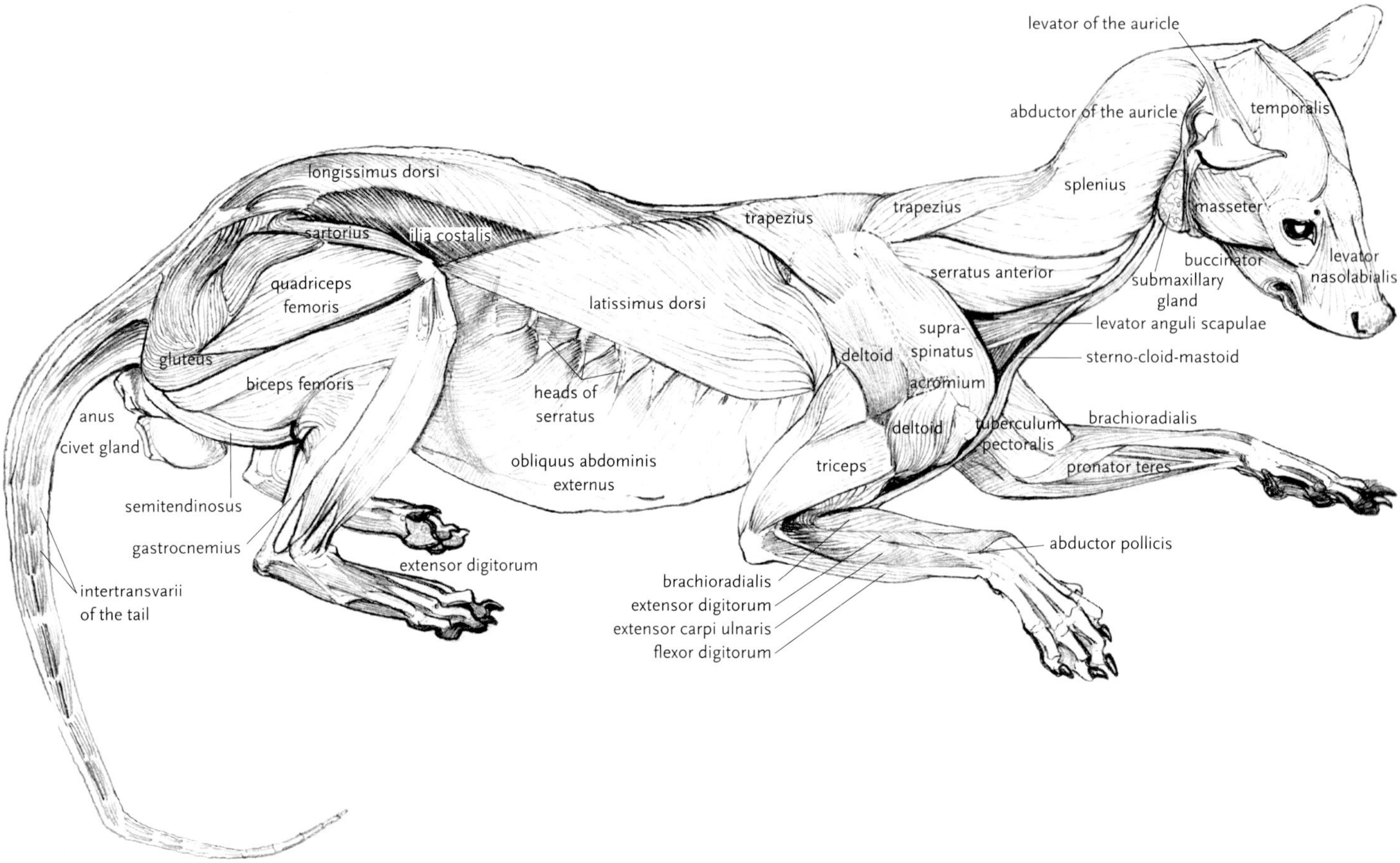

African Civet *Civettictis civetta* myology.

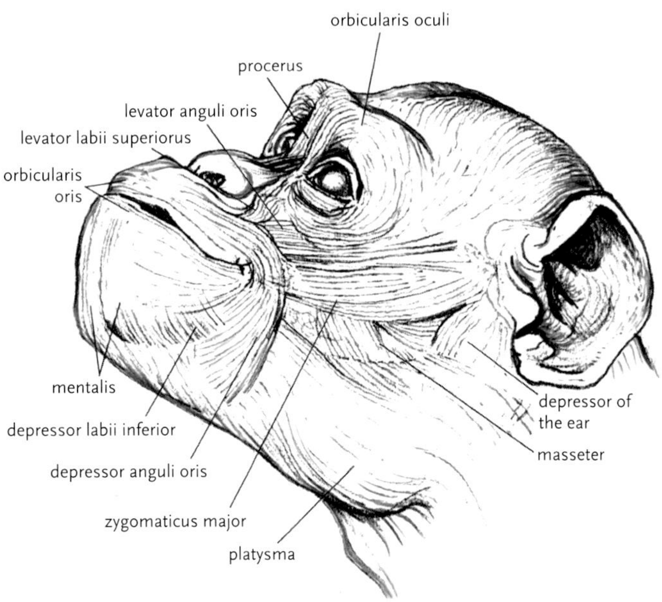

Chimpanzee *Pan troglodytes* head myology.

are revealed. And for teeth, again this means also the behaviour of choosing food as well as grabbing and chewing it.

Tool language is often best applied to component parts where there are extraordinary elaborations of limbs, heads, faces and tails. Less easily changed structures do exist, for example, the vertebral column, but even this most fundamental of vertebrate structures varies in interesting ways. An examination of the backbone and the varied relationship between body mass and limbs can serve as a useful introduction to the architecture of mammals, as we shall consider next.

The mechanics of morphology: struts, levers and bridges

In a typical quadrupedal mammal the vertebral column is arched between supporting legs, which, when static, have some analogy with bridge piers while the form of the vertebral column itself can be compared with a suspension bridge. The bodies of the vertebrae are the bridge's beam, or compression member, while the muscles and ligaments of the back serve as its cables or tension members (Thompson 1911). The vertebrae are generally heaviest and the muscles greatest in the small of the back, especially if the load is not spread out over ribs, limbs or pelvis. The vertebral spines and interspinous ligaments represent a web of struts and ties that give both strength and lightness to the back. The spines vary in height, the tallest being where the bending moments are concentrated. In many ungulates the head is cantilevered out on a long neck and, because of this extra load, as much as three-fifths of the animal's total weight may be taken on the forelegs. The spines are generally longest over the shoulders, where they serve as struts, absorbing the weight of the head through the tension members of the *ligamentum nuchae* and *sacrospinalis*. Variations on this pattern can be illustrated by wildebeest, giraffes, rhinos and Cheetahs.

The cantilevered head and tall thoracic spines of the Common Wildebeest *Connochaetes taurinus* provide the closest analogy with

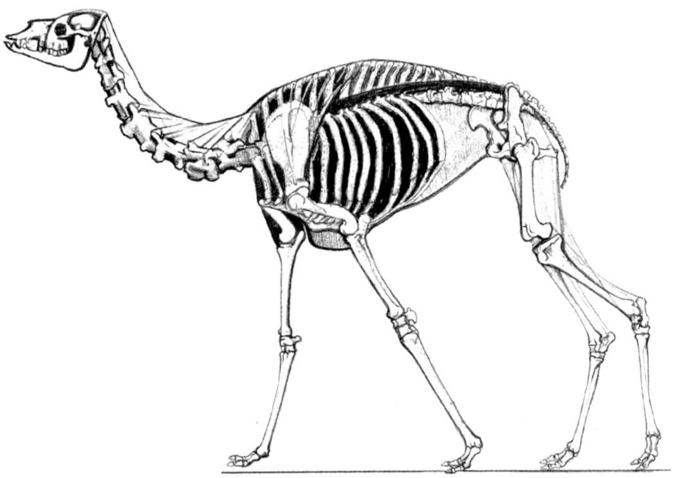

Camel skeleton and neck tendons to show 'suspension-bridge' structure. 65% of camel's weight is carried on its forelegs (after Kingdon 1990).

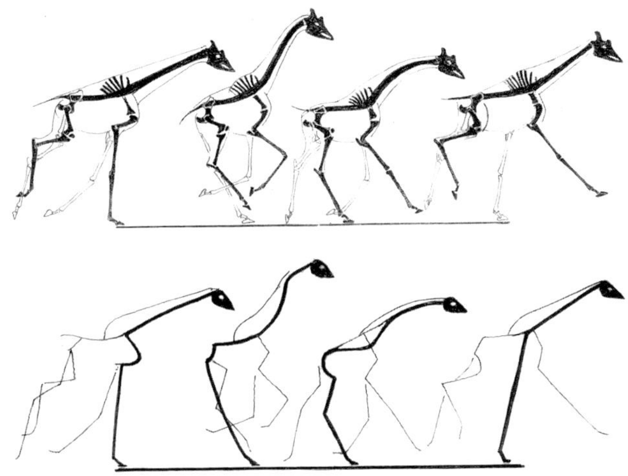

Silhouettes of cantering Giraffe *Giraffa camelopardalis* illustrating mechanical leverage of mammalian limbs.

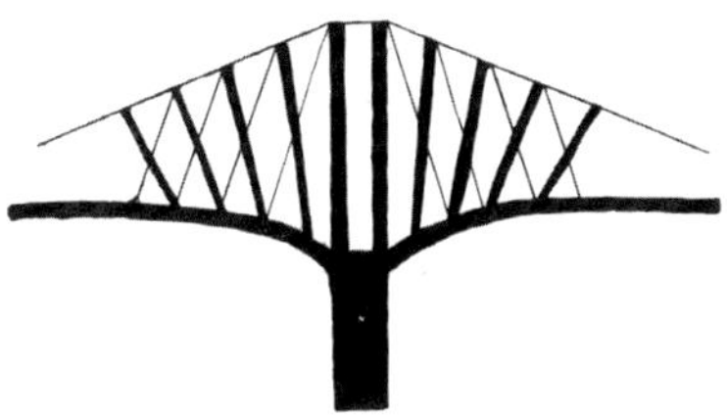

Silhouette of skeleton of Gnu *Connochaetes taurinus* to compare with diagram of cantilevered girder as used in bridge building.

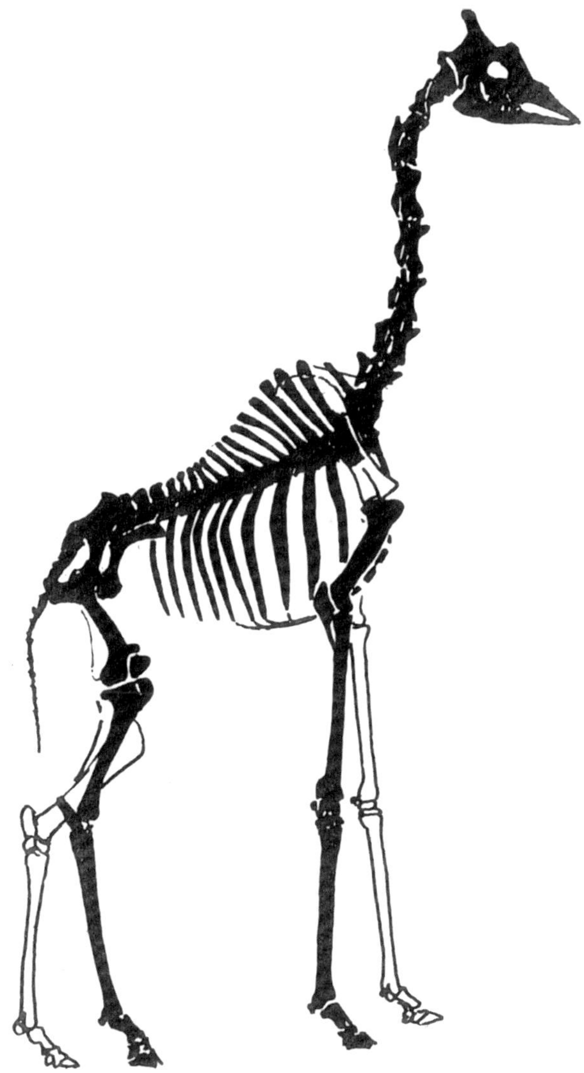

Silhouette of skeleton of Giraffe *Giraffa camelopardalis*.

a bridge. In engineering terms, if not in superficially aesthetic ones, the wildebeest is one of the most elegant and economically built of mammals while the giraffe is one of the most eccentric. Demonstrating that vertebral columns do not have to be horizontal, the weight of a bull giraffe's heavy, bony head is borne, for the most part, vertically, down the robust, elongated neuchal vertebrae and on down the forelegs but also out along the sloping back. However, the neck is not always under vertical compression, indeed, walking and browsing, certainly running, can cantilever the head and neck out quite horizontally. This exerts immense forces along the very highly developed and powerful *ligamentum nuchae*. The thoracic spines help spread this weight backwards and thus keep the animal in equilibrium. None the less, balance in such a long-necked and long-legged animal can be problematic and I have seen a running Giraffe that was jostled by a neighbour go crashing to the ground. Likewise, one contestant among neck-fighting males can occasionally get knocked down.

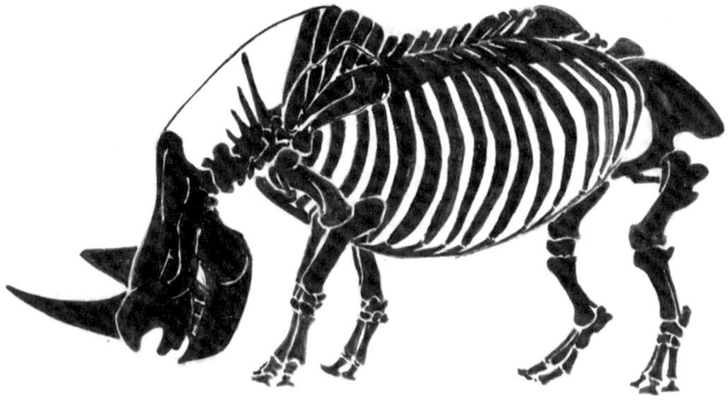

Silhouette of skeleton of White Rhino *Ceratotherium simum*.

The White or Grass Rhinoceros *Ceratotherium simum* possesses one of the heaviest of quadrupedal arches and the proportion of the animal's overall weight that is taken on the hindlegs is greater than in Giraffes, camels and other front-heavy animals. This is partly due to large, heavy muscles and bones in the back legs having to provide much of the propulsive power when the animal runs or gallops. The double demand for strength and flexibility in the White Rhino's shoulder is manifest in the broad, blade-like thoracic spines, which have slots to accommodate the spine behind when the vertebrae are compressed.

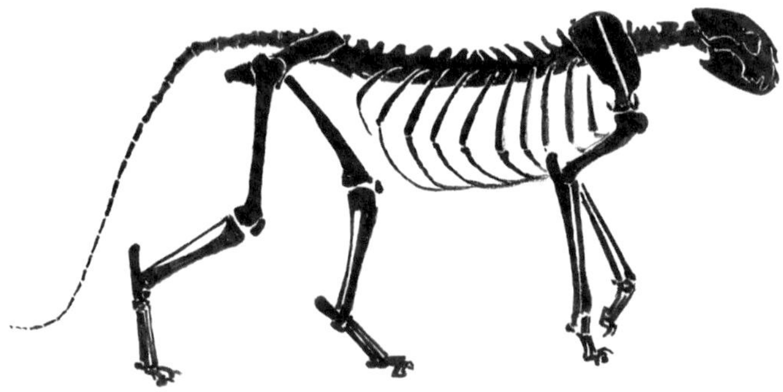

Silhouette of skeleton of Cheetah *Acinonyx jubatus*.

The Cheetah *Acinonyx jubatus* represents a lightly built extreme of the quadrupedal bridge, with flexibility of the back at a premium. When the animal is standing and relaxed the back tends to hang slightly in a long, loose arc.

Silhouette of skeleton of Ground Pangolin *Smutsia temminckii* in bipedal action (*left*) and silhouette of skeleton of quadrupedal Giant Pangolin *Smutsia gigantea* (*right*).

Silhouette of skeleton of Southern African Springhare *Pedetes capensis*.

The front part of an animal is not the only end capable of being a cantilever. The vertebrae of the tail, notably in the pangolins (Manidae), can be the most numerous and the most heavily built in the entire vertebral column. This is because the tail acts as a protective sheath when the animal curls up, a supportive strut while digging, a principal limb in the climbing forms and a posterior cantilever in the bipedal Ground Pangolin *Smutsia temminckii*. Thus tails can serve as clubs or protective shields while other mammals use them as flags, limbs or fly-whisks. In common with many other animals, the tail of the Southern African Springhare *Pedetes capensis* is a balancer. This essentially bipedal animal can, in mechanistic terms, be described as a single balanced cantilever. Unlike most mammals, it is the lumbar and sacral spines, not the thoracic, that are the longest and strongest. This is because it is the hindlimbs, not the tiny folded-up forelimbs, that absorb all the pressures of hopping and jumping.

The most drastic reorientation of the vertebral column has taken place in the evolution of true bipedalism in humans and in many species of now extinct hominins. The driving force behind the adoption of such a rare and finely balanced posture and gait was almost certainly due to the hands and manipulative behaviour having become the indispensable interface between hominin ancestors and their environment. These ancestors were likely to have been terrestrial 'squat-foragers', feeding in a somewhat similar fashion (but with a wider range of foods) to modern Geladas (Jolly 1970). Some confirmation of this squat-foraging phase in hominin evolution comes from 'platform-like' feet (with angled big-toes serving as stabilizing struts) in an early hominin, *Ardipithecus ramidus* (Lovejoy *et al.* 2009). Like all the other morphological specializations reviewed above, our ancestors' too, were behaviour-driven; further details (and alternative explanations) concerning this extraordinary transformation of the primate skeleton are explored in Kingdon (2003).

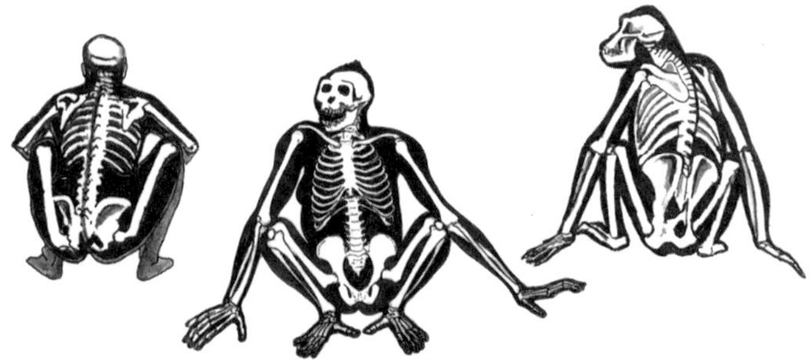

Left to right: Skeletons of Human *Homo sapiens*, 'Ground Ape' *Ardipithecus* and Gorilla *Gorilla Gorilla* in squatting positions showing differing proportions and morphology of pelvis, lumbar column, limbs and feet (in part after Kingdon 2003).

Attaching limbs: balls and sockets, bandages and other ties – different origins for fore- and hindlimbs are significant

Mammalian limbs, while acting as levers and supports for the body, have assumed a great variety of secondary functions, such as seizing and holding, feeling, digging, hanging and, in some mammals, flying and swimming. The body is supported and levered along by a variety of means, on toes or toe, on the soles or even on the elbows and wrists in some bats. The limbs are balanced by forces exerted by the weight of the body (and that of the upper part of the limb itself) bearing down through the limb and against the ground. During movement, strains within the body and limbs shift continuously and these strains are met by muscles that act as instantly adjustable ties and braces. Each muscle has an antagonist, another muscle or a ligament. The shape of these is determined by function, those for fast action being long and parallel, those exerting force being bundles of short-grouped fibres. Because limb bones are under compression, they are effectively struts. Their attachment to muscles can be direct or through tendons that can be attached on either side of the hinges or fulcrums that allow individual bones to join others in manoeuvrable, crane-like limbs.

The weight-bearing, supportive limbs of mammals are a distinct improvement on the lateral, only temporarily supportive ones of, say, reptiles and amphibians. The morphology of mammalian limbs has incorporated the evolutionary changes that were necessary to accomplish weight-bearing. In the forelimb, when the elbow swung back under, or in line with, the body, the hand unit had to rotate forwards. This phylogenetic shift forced the radius to twist around the ulna, a modification that the muscles also had to accommodate to. In the hindlimb no substantive change to a reptile-like ball and socket joint was needed but the swing of the knee forward and beneath the body necessitated a sharp bend in the orientation of the femoral head. The hindlimbs carry the weight of the body directly, bone to bone, through the femoral–pelvic socket joint.

The forelimbs have no direct bony connection with the main body mass, except for those mammals (primates, bats and digging animals) that have retained a collar bone or clavicle. This bone serves as a tie, stopping compression of the two limbs across the thorax. Both the clavicle and the scapula are modified remnants of the early synapsid shoulder girdle. In mammals that dig, fly, swim or run the stress of those behaviours are absorbed onto the barrel of the body via the scapulae. In typical, heavily quadrupedal mammals the scapulae transfer the stress of the body's weight to the supporting limb via muscles. The trunk of the body is attached to the scapulae by a hammock-like muscle-complex called the *serratus anterior*. The shoulder is pulled back by a fan-shaped sheet muscle, the *latissimus dorsi*, that links the humerus to the dorsal fascia. The entire shoulder is pulled forward by the *pectoralis*, which connects the humerus, scapula and clavicle to the median line of the chest; this is the main muscle powering flight in bats, digging in digging species and thumping an opponent in human boxers. Among bats the size of pectoral muscles is a direct measure of flight power and a strong, long-distance sustained flyer such as the Midas Free-tailed Bat *Tadarida midas* has much thicker, longer chest muscles than a short distance, short-flight bat, such as Moloney's Mimic Bat. The scapula is held to the body on its upper edge by the *rhomboid*, bound over by the thinner, more superficial *trapezius*. Other strap or 'bandage' muscles are the *levator scapulae*, which connect the scapula to the upper nape and skull. The *sternocleidomastoid* is a sheet muscle running from the back of the skull to the shoulder joint and the *basihumeral* is a strap running from the skull to the humerus and clavicle. Scapula and humerus are bound together by ligaments and also by the *supraspinatus* and *infraspinatus*. The *deltoid* both moves the foreleg and helps bind the shoulder into a compact functional mass. The shoulder joint is given extra stability and control by the *biceps*, *triceps teri* and *teres major*. The *biceps* and *triceps* muscles run from shoulder to elbow and help to flex or extend the entire forelimb. The humerus is generally broad and flat at its lower end, providing a powerful hinge joint with the two bones of the lower arm, ulna and radius. The olecranon, or 'funny bone', of the ulna protects and locks this hinge while the slender, rounded top end of the radius simply rotates at its junction with the rounded capitulum of the humerus. This combination of complex, large-headed ulna and simple, small-headed radius is reversed at the bones' lower ends. Here the radius broadens and provides the main attachment with the hand while the ulna slims down to a relatively feeble connection on the outer margin of the wrist. Because of this arrangement mammal forelimbs are unlike those of any other animal. The upper arm is effectively attached to the ulna, the hand or paw is attached to the radius and the two long bones swivel around each other so that the hand can be prone, supine or any position in between.

In primates and other animals with relatively conservative, highly manoeuvrable forelimbs the radius and ulna are two separate bones. In most ungulates the radius and ulna are fused and the muscles below the elbow have become semi-tendinous and then extend as tendons to the toe or toes, thus reducing the weight that has to be moved in running. The bones of the hand or forefoot have an original arrangement of three proximal carpals, a central carpal, five metacarpals followed by five digits of two or three phalanges. As Darwin's 'curiosities' exemplify, mammal forefeet range from single-toed horses to bat wings and porpoise flippers.

The hindlimbs usually mimic the forelimbs but in spite of their similarity they have quite different origins. Parts of the forelimbs derive from structures that were integral to the early vertebrate head, thus the pectoral fins were actually tethered to the head in early fishes. When tetrapods developed forelimbs from these fins, which became detached from the skull, one pair of gill arches was dragged away to become the scapulae. Forelimbs are mainly served by cervical nerves, close to the brain. From early vertebrate history, forelimbs helped adjust sense-driven decisions about the pace and direction of forward movement. By contrast it was vertebrae and associated bones from towards the tail that first differentiated into hind fins and eventually hindlimbs and pelvis. Like the rest of the rear end, their primary function was propulsion and that function is still evident in the hindlimbs of a majority of modern mammals.

The pelvis is a light, strong, balanced structure, the ilia sometimes fused onto the sacrum (which is part of the vertebral column) but sometimes capable of independent flexion. Together with the sacrum, the paired iliac bones form a ring through which sexual and excretory ducts pass. Ball and socket joints on either side allow the hindlimbs plenty of movement without loss of strength. The most powerful movement is that of 'kicking-off' and a complex of flexors and tensors concentrated in the upper leg exert power and control of movement. These are the *gluteus*, *biceps*, *gracilis*, *semimembranosus* and *semitendinosus*. The lower leg muscles are greatly reduced in most species and are primarily tendinous in all running species.

The morphology of heads and their evolutionary origins

The concentration of disparate activities that are welded together in vertebrate skulls would seem extraordinary were it not so familiar. The further elaboration of head shapes, especially by horns, antlers, tusks and ossicones, has provided humans with symbols and trophies from the earliest times, yet remain a poorly explored aspect of morphology. It is partly our human fascination with heads and partly the superior survivorship of teeth that has ensured that fossil heads dominate palaeontological collections and provide the data for much of our knowledge of the evolution of heads.

Because the fore-ends of primitive chordates and vertebrates were the first to encounter both food and obstacles, basic sense organs and a 'mouth' developed there. The polarity of organisms and their symmetrical organization crosses all the higher taxa (Wainwright 1988). The need to respond to light, chemical, electric or vibrational signals led to the differentiation of cells until they developed into eyes, nose and ears, each encapsulated in compartments that shared the upper part of what became the skull. The lower section of this structure became a hinged mandible and eventually both jaws developed teeth. In many predatory vertebrates, from the most primitive to the most advanced, the size of jaws closely reflects the size and toughness of the prey. Furthermore, a gross expression of animal proportions is the size of heads relative to the size of bodies.

Apart from whales, the most extreme examples of disproportionately large or small heads are to be found in bats. This is partly because the head and jaws get little or no heavy-duty help from the limbs in the behaviour patterns of capturing and processing prey. Thus Moloney's Mimic Bat, a species that only takes small, soft-bodied prey, has no need for a large head while the Heart-nosed Bat *Cardioderma cor* seizes robust, often vertebrate, prey with its large jaws and must quickly subdue it with deep, damaging bites powered by massive jaw muscles. The head of the former is about one-seventh of the combined head–body volume, whereas *Cardioderma* heads are closer to a quarter!

One of the most fundamental expressions of cranial morphology concerns the sizes of sensory activities in the head. It is in the relative sizes of compartments and in permutations of connecting bridges, struts and welds that the species-specific morphology of mammal heads becomes obvious. Wherever sufficient data exist on the ecology and behaviour of species, we find superb and detailed examples of forms evolving morphological modifications that serve very precise behavioural functions. This is particularly obvious in the relative sizes of, say, orbits, olfactory equipment and auditory bullae in mammal skulls. Even more explicit are the functional shapes of eyes, ears, noses and sensory whiskers in mammals as different as a galago, a Bat-eared Fox *Otocyon megalotis*, an Aardvark *Orycteropus afer* or an Aquatic Genet *Genetta piscivora*.

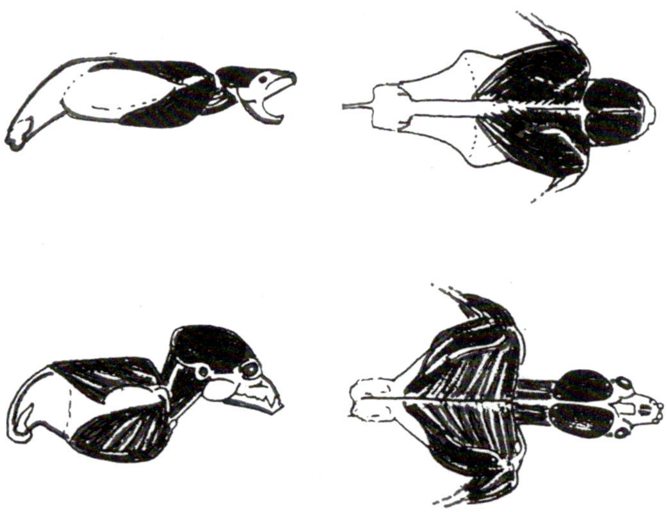

Head/body proportions in Mimic Bat *Mimetillus moloneyi* and Heart-nosed Bat *Cardioderma cor* (after Kingdon 1974).

Heads to illustrate super-development of different sensory faculties.

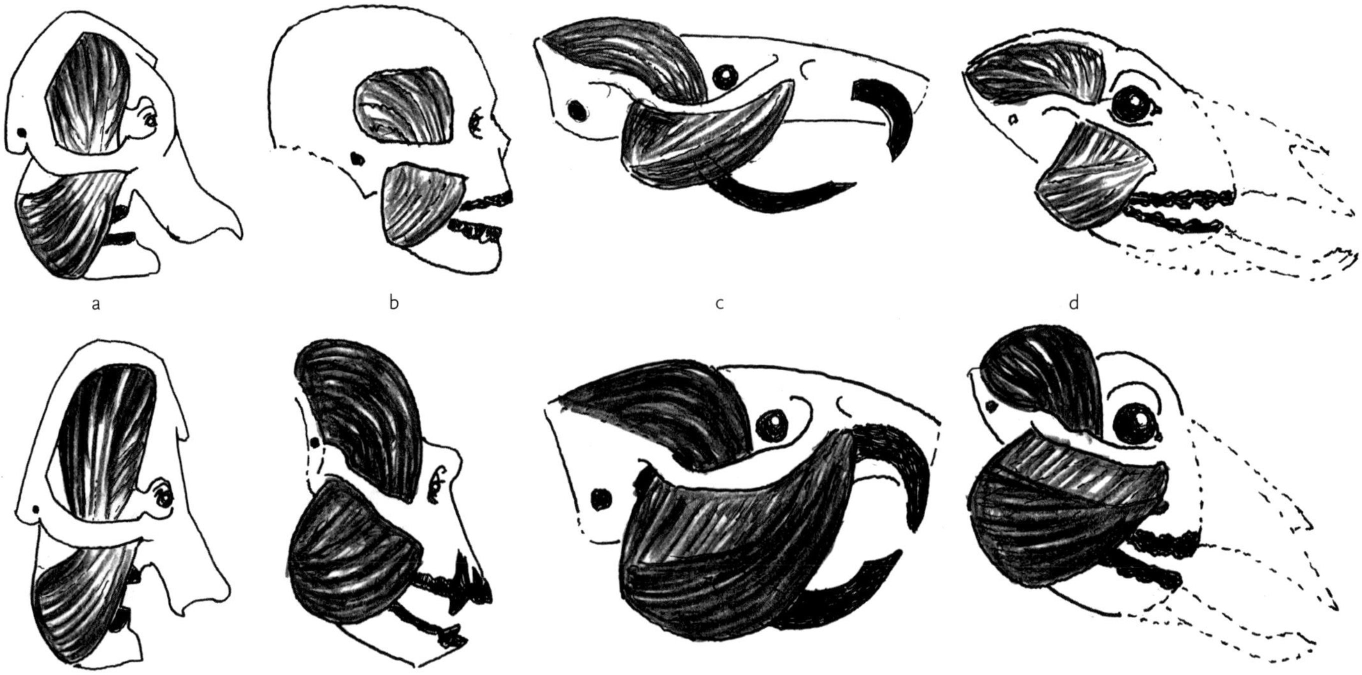

Skull outlines of four pairs of related mammal species, showing the role of chewing muscles (masseter and temporalis) influencing skull shape. Weaker chewers (*top row*) have shallow muscle attachments on relatively gracile skulls; powerful chewers (*lower row*) have robust skulls with deep muscle attachments.

a. African Elephant *Loxodonta africana* and Pleistocene Elephant *Elephas recki*. b. Human *Homo sapiens* and Western Gorilla *Gorilla gorilla*. c. Link Rat *Deomys ferrugineus* and Giant Squirrel *Protoxerus stangeri*. d. Bushbuck *Tragelaphus scriptus* and Mountain Reedbuck *Redunca fulvorufula*.

Head-shape and chewing

The mechanics of jaw function are nowhere more simple and explicit than in the Savanna Elephant *Loxodonta africana*, which has a back-and-forth mode of chewing with very little sideways action. The short, heavy mandible hangs from the *temporalis* muscle as if it were the chair of a swing. Here the 'ropes' of the *temporalis* are at right-angles to the lower tooth rows that resemble the chair's seat. The swing seat is pulled back towards the chest, in a loose, easy arc, by the *digastric* muscle. Coming forward again it closes over the mouthful of food and, clamping hard against the upper toothrow, grinds forward under the combined force of *masseter* and *temporalis* muscles.

In one of those confounding inversions of expectation, modern African elephants are, in terms of dentition and diet, relatively primitive. Until quite recently (about 19,000 years ago) the commonest and most widespread elephant species in Africa was *Elephas recki*, a close relative of the Indian elephant. Its fossils are abundant and widespread, from South Africa to the Sahara. *Elephas recki* (like its Indian congener *E. indicus* and its Nordic relative, the mammoth, *Mammutus*) was a dentally advanced species with very deep, multi-plated molars that could masticate coarse, abrasive foods more efficiently than *Loxodonta africana*. However, it is not just teeth and their bony buttressing that take the strain of chewing. More force must be exerted on tougher foods and that force must be exerted by larger, longer muscles. And increased muscle forces reshape the bones onto which they are attached, such is the adaptability of bone form and function. In the case of elephant skulls, this resulted in the phylogenetic elevation of paired, honeycombed arches that absorb all the forces of chewing and cowl the braincase like a bonnet (tool language again). Although *E. recki* had a rather small skull relative to its body, its forehead was immense and a large part of this enlargement can be confidently correlated with the behavioural need to enlarge the temporal muscles in order to exert more force on a formidable battery of chewing teeth. Comparing the skulls and teeth of *L. africana* and *E. recki* provides a vivid illustration of how changed diets (often triggered by climate change) drive modifications to tooth and skull structure. New stresses and strains on teeth call for changed mechanics in the skull, the external expressions of which are distinctive head shapes and silhouettes. Understanding the mutability of bones is, in turn, impossible without taking into account changes in the muscle forces exerted upon the bones. The interplay between behaviour, diet, teeth, the mechanics of mastication and the gross morphology of bones and muscles are evident in comparison between a dissection of an extant Savanna Elephant's head with a reconstruction of the same muscles on an *E. recki* skull.

Another simple demonstration of 'function and form' emerges from comparing the head proportions of two similar-sized antelopes: the Bushbuck *Tragelaphus scriptus* and Mountain Reedbuck *Redunca fulvorufula*. The former has a slender head with long-lipped mouth; the latter's head is short-mouthed and wedge-shaped. These differences begin with tooth proportions. The soft-foliage-eating Bushbuck has modest, shallowly rooted masticatory teeth set in equally shallow jaws, but the broad-leaved shrubs and herbs that it eats need a wide maw. The tough-grass-eating Mountain Reedbuck instead has heavily crenellated cheekteeth that are firmly rooted in a well-reinforced, deep-jawed skull and mandible. It is the expansion of teeth and their rooting, together with massive *masseter* and *temporalis* muscles, that open up that mandibular wedge. It is the nibbling of tough grasses that calls for that tight, short-lipped, muscular mouth. The external expressions of such dietary differences are an elongated sleek face in the Bushbuck, a relatively short, chunky face for the Mountain Reedbuck (see drawings).

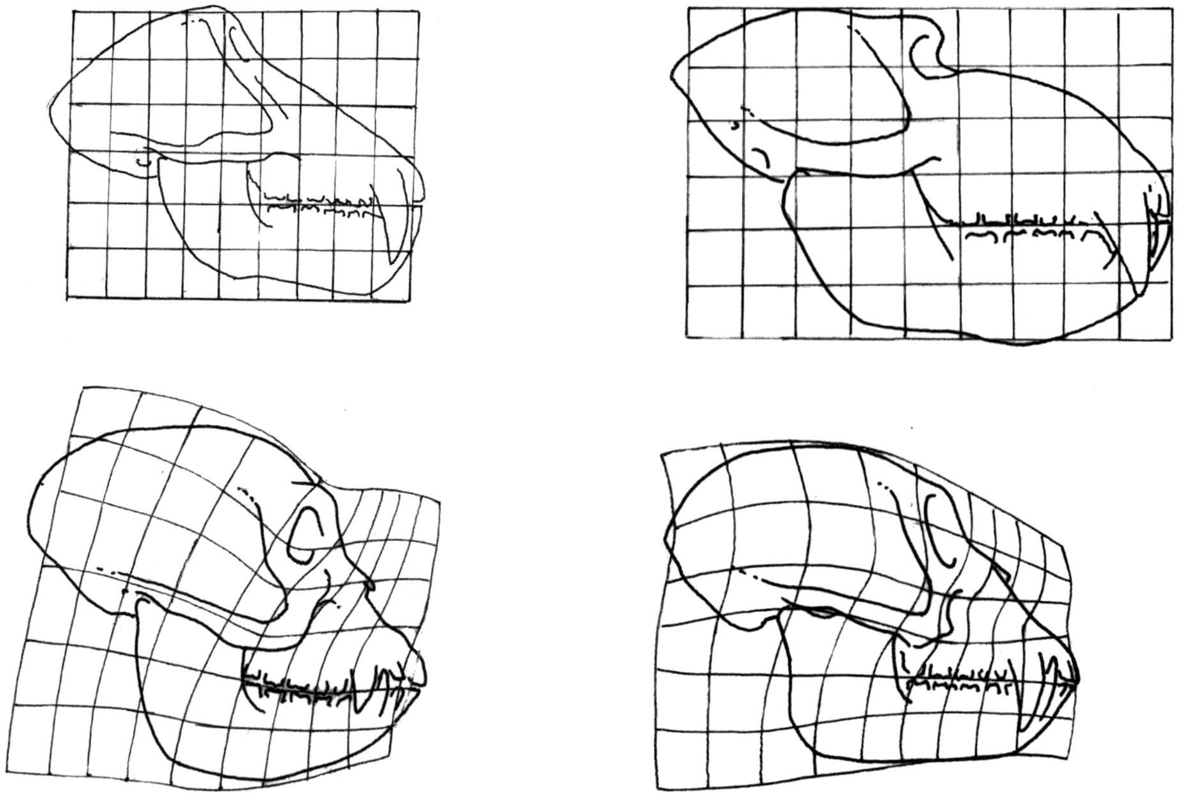

Cartesian co-ordinates on skull profiles illustrate phylogenetic retraction of the muzzle of the Baboon Mangabey *Lophocebus* compared with its closest relative, the baboon *Papio* (*top left*), and Drill Mangabey *Cercocebus* compared with its closest relative, the Drill *Mandrillus* (*top right*).

Linking morphological form and function with genetics

At a time when genetics is the primary conceptual preoccupation of biology, and in a culture that is visually dominated by surface photography, general interest in the functional morphology of animals and the elucidation of their 'hidden' anatomy has gone into a decline that surely would have disappointed Darwin. Yet many new discoveries in molecular science only pique a broader interest in their implications. Eventually genetic peculiarities need to be understood in terms of the structures for which they code and the behaviours that employ them. Eventually we will want to relate gene codes to whole animals living in real landscapes and in the context of deep evolutionary time.

For example, among the genes that govern the production of the muscle protein myosin there is one that serves to build powerful jaw muscles in many mammals. This gene, *MYH16*, is fully intact in most higher primates but has recently been shown to have mutated in humans in such a way as to disable some of its muscle-building capacity (Stedman *et al.* 2004). Furthermore, the application of molecular clock techniques has dated the first appearance of this disabling mutation to 2.7–2.1 mya, broadly the same period when some fossil hominins show a substantial reduction in relative jaw size. Since the action and development of muscles are major determinants of bone shape and size this sudden decline in the architecture of hominin mandibles some 2.5 mya can now be ascribed to mutation in an identifiable, muscle-building gene. The *MYH16* mutation could only spread in those prehistoric populations because selection favouring powerful *masseter* muscles must have been relaxed. Less need for powerful chewing muscles must have been influenced by significant shifts in foraging patterns, possibly the use of fire, tools and dietary changes. Although this is the first time that genetics, palaeontology and anatomy have been linked up in this way (Currie 2004), we can be confident that many more such instances will be discovered in the future.

To illustrate how contemporary science might integrate genetic information with morphology, diet, ecology and social behaviour, consider monkey jaws. Humans are not the only primates to have suffered a phylogenetic decline in the size of their muzzle. Two monkey lineages with even more prominent muzzles than apes, i.e. baboons *Papio* spp. and drills *Mandrillus* spp., have incisors and canines that project out from the male cranium about as far as is structurally possible without risking frequent fracture. Independently, ancestral populations belonging to each of these large, predominantly terrestrial monkey types have given rise to smaller-bodied, arboreal lineages, namely the baboon-mangabeys *Lophocebus* spp., the drill-mangabeys *Cercocebus* spp. and, probably, the newly discovered Kipunji *Rungwecebus kipunji* (Cronin & Sarich 1976, Davenport *et al.* 2006). The arguments in support of this phylogenetic shrinkage of jaws are presented in Volume II. Morphologically, these changes have involved remodelling of the skull as phylogenetic, backward migration of the entire toothrow has forced the jugal plate (on which the molars were previously dependent for support) to buckle backwards and inwards, creating peculiar suborbital fossae. Once thought to be the principal diagnostic marker for a single mangabey lineage, its functional purpose was earnestly puzzled over. Now that separate origins have been demonstrated for the three genera by genetic analysis, it is apparent that suborbital fossae are simply convergent by-products of evolutionary change. Buckling and dimpling of the jugal plate is an artefact of shifting skull parts and the 'fossae' are apparently without any direct functional significance in themselves. Some of the best examples of comparable remodelling of the skull can be found among whales, but these need not concern us here.

How muscles shape bone morphology

As demonstrated by the examples of elephants and antelopes, the size of a muscle can have a profound influence on the shape of the bones to which it is attached. Nowhere is this more pronounced than in the relationship between chewing muscles and the architecture of muzzles and skulls. The comparison between jowly early hominins and weakly jawed *Homo* is paralleled by clearly defined differences between two closely related porcupines. In the African Brush-tailed Porcupine *Atherurus africanus* the evolution of a weakly developed temporal muscle in this fruit-eater was probably as much due to softer foods as it was in *Homo*. It can be predicted that this decline was also mediated by the rodent equivalent of the *MYH16* gene. When the skull morphology of this porcupine is compared with that of an Asian equivalent, the Long-tailed Porcupine, *Trichys fasciculata*, there are precise correlations between the size of temporal muscles and the development of angular and temporal processes: prominent development in the strong gnawer *Trichys* and absence in the weak masticator *Atherurus*.

On a short-term (rather than the evolutionary) scale, behaviour-driven morphological change can be demonstrated through experiment. The role of temporal muscles in drawing out angular processes on mandibles and erecting temporal ridges on top of the cranial vault has been investigated in rats. Rats fed an exclusively soft diet have chewing muscles up to 13% lighter than those of rats on a hard diet and the angular process is also measurably reduced in size (Moore 1965). When the temporalis muscle is removed in very young rats both temporal ridges and angular processes entirely fail to develop (see right from Horowitz & Schapiro 1951).

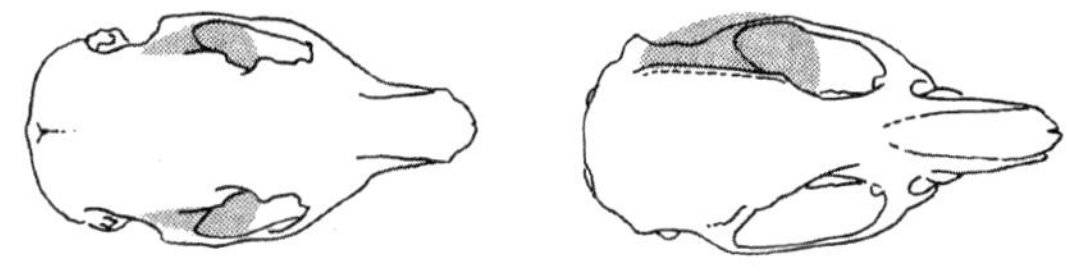

Dorsal views of skulls of Brush-tailed Porcupine *Atherurus africanus* (*left*) and lab rat *Rattus rattus* (*right*) in which the right temporal muscle was removed in infancy.

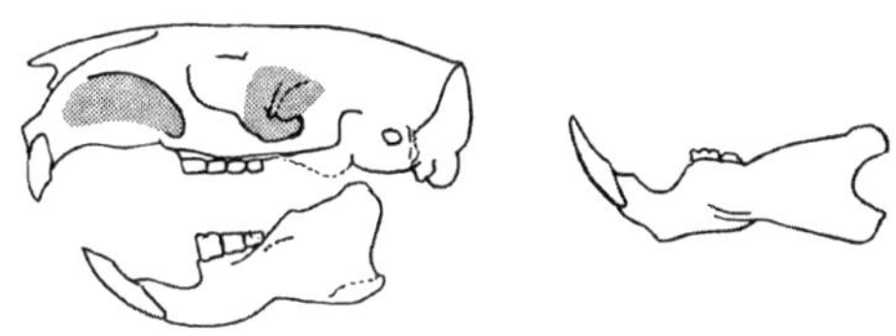

Lateral views of skull and mandible of Brush-tailed Porcupine *Atherurus africanus* (*left*) and right mandible of lab rat *Rattus rattus* (*right*) in which the temporal muscle was removed in infancy.

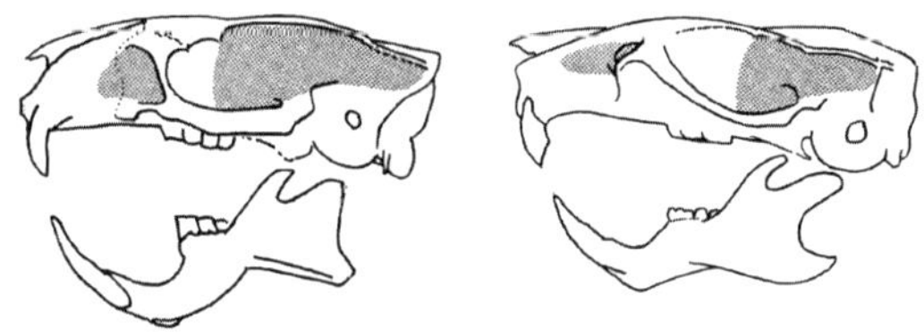

Lateral views of skulls of Asiatic Porcupine *Trichys fasciculata* (*left*) and lab rat *Rattus rattus* (*right*).

Describing morphology and profiling adaptation in *Mammals of Africa*

Practical concern for morphology and anatomy is expressed by the authors and editors of these volumes in various ways. To begin with, describing appearance requires verbal and conceptual skill, even to convey a minimal appreciation of an animal's shape and substance. The task is eased when the proportions, colours, shapes and sizes of one species are compared with those of another or others; indeed, the description of morphology leans heavily on the comparison of forms. However, any comparison of 'unfamiliar' animals with 'familiar' ones involves assumptions, not only about what species are familiar but also about what attributes constitute familiarity. Furthermore, verbal descriptions suffer from being linear processions of itemized attributes that all too easily become purely prescriptive: a cobbled amalgam of parts passing for a whole. Because any single individual of a single species possesses an immense, almost uncountable number of attributes, all descriptions have to be drastically condensed to a conventional summary of supposedly diagnostic attributes. Offsetting the atomistic limitations of language, and the apparent sterility of listing itemized components, are prime reasons to invoke the 'familiar'. Authors forced to itemize often seek to retrieve some sense of the whole animal by referring to other animals that already exist as whole entities in the psychological gestalt of their readers. The problems faced by authors trying to translate their data and perceptions of a unique type of animal form into a verbal summary are therefore very real, and they are no easier to solve within the prescriptions of a handbook than those of field guides (to appreciate the perils see Primates and Barbary Macaque texts in Van den Brink 1967). Of course, field guide texts have to be drastically pruned but the condensation is merely a matter of degree; even the most comprehensive descriptions fall far short of reality.

Most biologists and naturalists share Darwin's desire to explore the biological 'meaning' of animal design: they seek to interpret form in terms of function and selective advantage. 'When we regard every production of nature as one which has had a history: when we contemplate every complex structure and instinct as the summing up of many contrivances, each useful to the possessor; … when we thus view each organic being, how far more interesting, I speak from experience, will the study of natural history become' (Darwin 1859). It is partly in response to this Darwinian exhortation that all species profiles in *Mammals of Africa* include the heading 'Adaptations'. Herein authors are free to describe and interpret the behavioural, morphological and physiological adaptations that are unique to the species that is being profiled. Many of these adaptations involve secondary, tertiary, even umpteenth modifications of already intricate structures.

That the mammals of Africa should take so many diverse forms, yet be constructed on the same pattern and by a comprehensible process of natural selection, beginning with small changes in behaviour: what could be more curious? What could be more intellectually thrilling?

Illustrating morphology in *Mammals of Africa* (a postscript by Jonathan Kingdon)

Comprehension of mammalian biology can be deepened and supplemented by a variety of graphic media from maps, diagrams, histograms, gene charts, videos, photographs to all sorts of pictorial illustration. These modes of representation are sometimes self-sufficient for the sort of information they seek to propagate. However, Mitchell (1994) has pointed out that representation (in both modern and postmodern senses) becomes a conceptual issue when there are perceptible ruptures between image and text, and that illustrated texts can be better conceived not as a juxtaposition of two separate media but as composite synthetic works.

In these volumes the text is augmented by maps and drawn illustrations. A large number of the latter derive from a very long-term enterprise that has generated very many thousands of drawings (Kingdon 1971–1982, 1983, 1997). One of the reasons for preparing my *East African Mammals: An Atlas of Evolution in Africa* and my field and pocket guides was to enlist graphic imagery as a mode of exploration and discovery in itself. With my pencil I set out to record, both in field and laboratory, the morphology, functional anatomy and behaviour of many little-known mammals in the confidence that my audience would share with me a Darwinian excitement in making discoveries. Many of the smaller mammals that I observed or trapped had never been drawn, let alone photographed before and some of my work actually had the character of 'first contact' documents. Some 40 years after publication of the first volume of my *Atlas* these exploratory documents have been greatly augmented, often in close collaboration with the authors of this work, who have helped extend this enterprise into another millennium. We have been assisted by new forms of graphic aids that were inconceivable when I was drawing animals in the 1950s and 1960s (among the most exciting are camera-trap and space-tracking systems).

None the less, the greater part of an animal's structure is hidden under its coat, and photographs, however beautiful, seldom offer more than a hint of what lies beneath the surface. Of course, single-view drawings are similarly constrained but there is much greater scope for the isolation and presentation of those features that are most diagnostic of a species' appearance. Where the production of drawings has included detailed anatomical records of dissection and progressive stripping down to the skull and skeleton, such representations gain in authority. In the corpus of work from which these illustrations have been selected, there were more than 300 such published dissections, and several hundred skulls and other anatomical details were also drawn. The skull drawings have been augmented here by many hundreds more but these have been rendered in a more conventional format to assist comparability between taxa.

Verbal language is not our only artefact in the effort to study and conceptualize 'meaning' in the physical existence of mammals. Anatomically correct drawing, particularly when backed up by dissection and field sketches of ephemeral behaviour and postures, can augment description with a useful type of non-verbal functional analysis. Unfortunately, most expressions of contemporary visual media are designed for instant absorption and the habit of taking time to carefully examine both real objects and man-made pictorial images has become exceptional. Most of the drawings in this work are the outcome of sustained and time-consuming contemplation and analysis of representative individuals of particular species, with particular attention to the relative proportions of functional components. Many of the drawings were made under quite trying conditions in the field, during floods, heat-waves, downpours, out in the open, under canvas, in contention with innumerable insects or in the relative comfort of a vehicle. In the final production of coloured illustrations I have sought to recognize and display as many subtle intimations of uniqueness as are possible in a single image. In *East African Mammals* these were augmented by line drawings of relevant details of anatomical form and sheets of sketches that illustrated behaviour and posture. Similar augmentation has been applied to some of the species profiled in this work.

Sketches of *Tadarida thersites* showing the mutability of ear shape.

We now know that small changes in diet or in the techniques whereby food is processed are tied in with evolutionary changes in ecology and behaviour, and, ultimately, with the morphology of teeth, heads, limbs and backbones. At a more general level the comparison of species, one with another, helps suggest how physical aspects of the environment (such as the composition of food or an ecological shift) elicit changes in behaviour, which drive the evolution of appropriate adaptive structures as expressed in the details of animal form.

The problems of representation are nowhere better exemplified than with the Afrotherian golden-moles. In spite of the pretty metallic fur that has given them their name, their appearance within museum drawers or even in photographs, is of featureless oval blobs. Such apparent 'shapelessness' to superficial photographic or human lenses

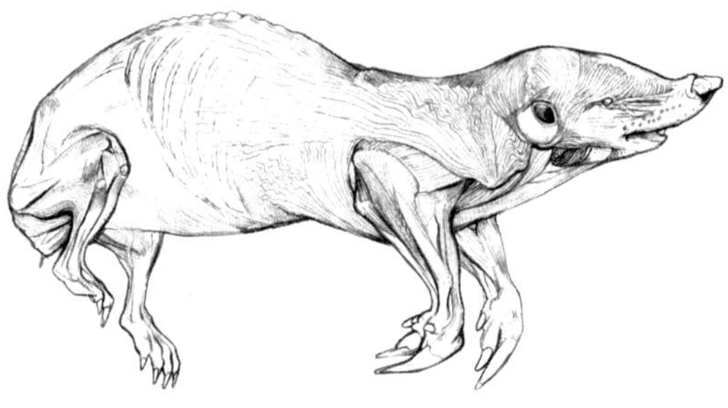

Myology of Stuhlmann's Golden-mole *Chrysochloris stuhlmanni*.

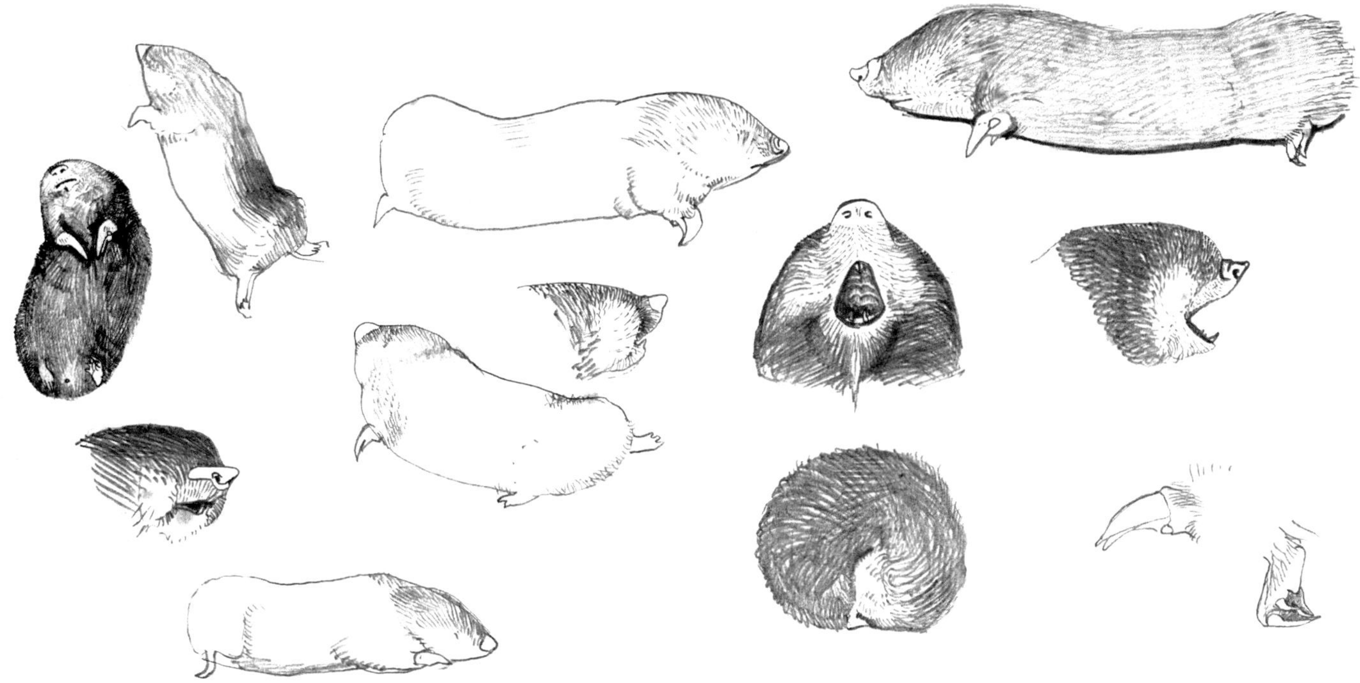

Action drawings of Stuhlmann's Golden-mole *Chrysochloris stuhlmanni.*

is profoundly misleading. Golden-moles are no less highly evolved than Aardvarks or elephants. Closer inspection of living animals, let alone careful dissection of their myology, reveals an energetic digger perfectly adapted to a subterranean existence.

Exposed outside their burrows, tiny clawed feet row their bodies over the surface without ever lifting their bellies off the ground. The spade-shaped snout can be forced deep into the soil or into interstices, whereupon one or two fingers, armed with sharp claws, are brought forward into the nose-made crevice and a very powerful opening-up action follows as the robust cranium and nose push up and the hard claws tear down and back.

Something of the golden-mole's vitality can be conveyed by sketches of the living animal *in natura* but simplified anatomical diagrams can enhance appreciation of the fitness of highly modified bones and muscles for probing and digging functions. A quick sketch of a Giant Golden-mole *Chrysospalax trevelyani* skeleton (see overleaf) positioned to display its digging action illustrates the streamlined anatomy of a superbly designed digging machine beneath that metallic, blobby oval. Golden-mole genera vary in sizes, in the proportions of heads in relation to body length and in the details of skulls, claws and limbs. Appreciation of such differences can be enhanced by matching colour images of their relatively bland external appearance with schematic 'x-ray' equivalents (see overleaf) in which diverse body sizes and body proportions become obvious and bleached skull outlines hint at interesting differences that are specific to each genus.

For a while, conservative biologists rejected the recent discovery that golden-mole genetics placed them unambiguously within the (almost as recently discovered) 'Afrothere' radiation. However, such contention has dropped away as more and more analyses have demonstrated the intrinsic power of the burgeoning new science of molecular phylogeny.

As the physiological and anatomical functions of genes are explored perhaps we should trust that one day it will be molecular science

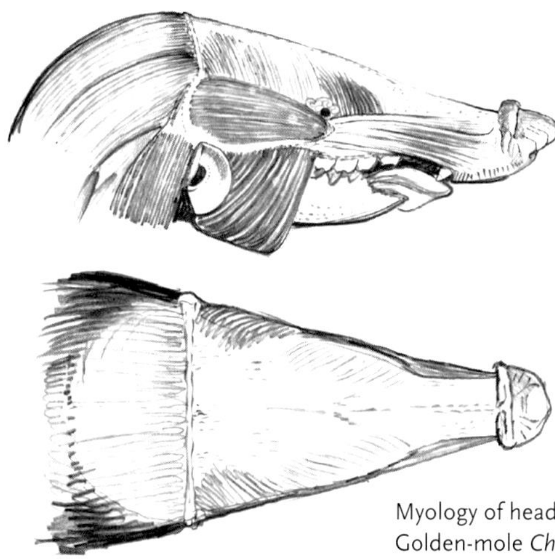

Myology of head of Stuhlmann's Golden-mole *Chrysochloris stuhlmanni.*

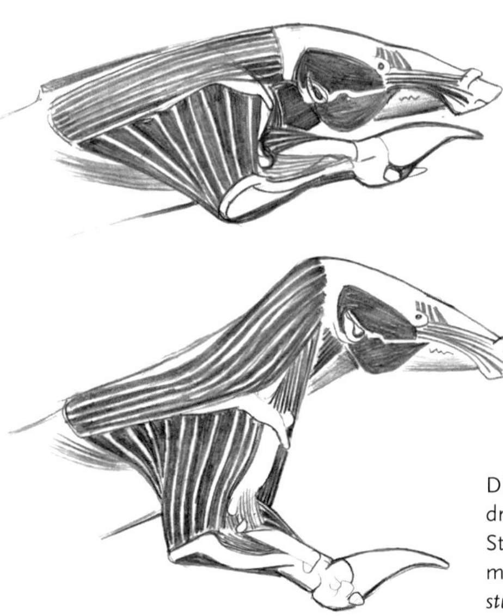

Diagrammatic myology drawings of a digging Stuhlmann's Golden-mole *Chrysochloris stuhlmanni.*

LEFT AND BELOW LEFT: a. Rough-haired Golden-mole *Chrysospalax villosus.* b. Cape Golden-mole *Chrysochloris asiatica.* c. Hottentot Golden-mole *Amblysomus hottentotus.* d. De Winton's Golden-mole *Cryptochloris wintoni.* e. Yellow Golden-mole *Calcochloris obtusirostris.* f. Grant's Golden-mole *Eremitalpa granti.* g. Duthie's Golden-mole *Chlorotalpa duthieae.*

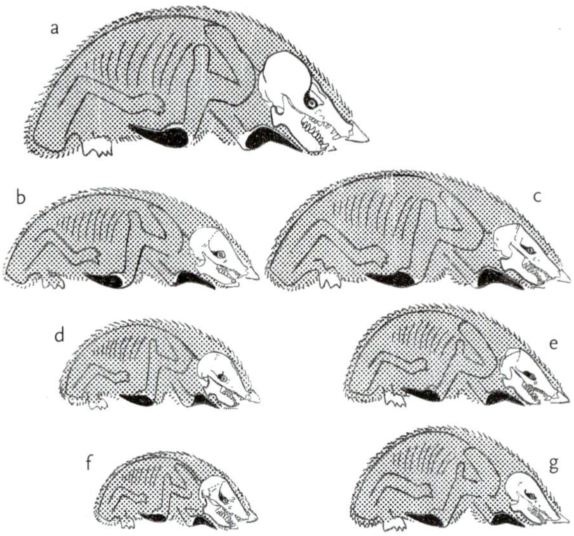

BELOW: Giant Golden-mole *Chrysospalax trevelyani* digging.

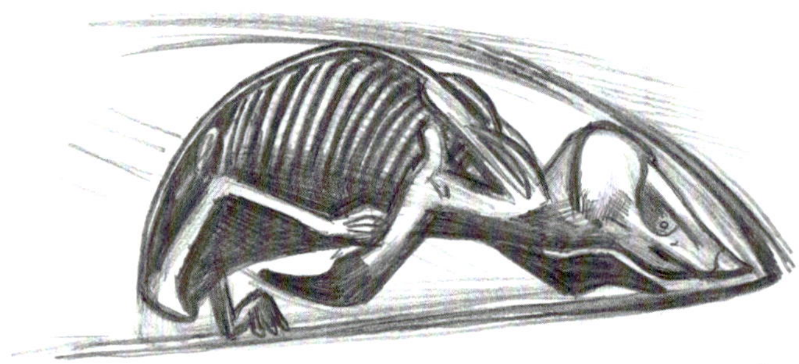

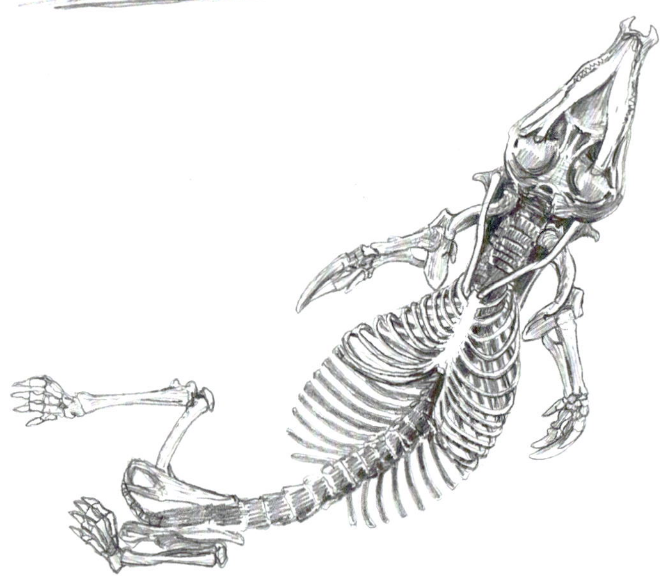

Ventral view of the skeleton of Stuhlmann's Golden-mole *Chrysochloris stuhlmanni* showing strut-like clavicles.

that leads a renaissance in morphological studies and on to a deeper appreciation of how form is shaped by function and functional behaviour!

As the richest and most diverse region for mammals on Earth and the context for our own emergence and existence as human beings, Africa's living and fossil fauna can be expected to offer many new insights into evolution and a great diversity of absorbing aesthetic experiences. We can expect numerous new ways of representing biological insights to be developed in the years to come but I hope the pioneering intentions of the drawings/documents on offer here will continue to help viewers formulate a deeper and more contemplative appreciation for their subjects.

Class MAMMALIA

Few people today can deny that they, too, are mammals. Yet ancient and pre-scientific traditions, that drew un-crossable mental lines between humans and other mammals, still persist, thereby barring the most direct and simple way of defining and appreciating what it is to be a mammal. In all essentials for life and reproduction they are like us and we are like them.

Suckling female Gorilla *Gorilla gorilla*.

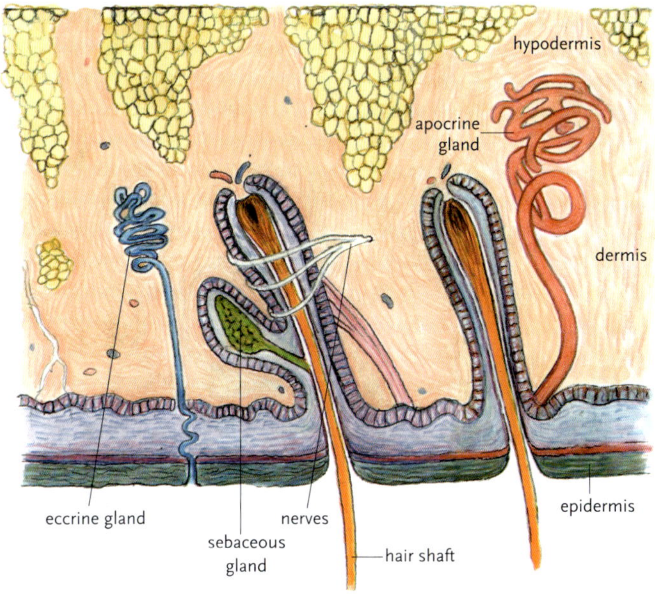

Microstructure of a typical mammalian hair and associated nerves, sebaceous, eccrine and apocrine glands in dermal tissues.

Such a starting-point can bring both objective scientific analysis and some measure of empathy to bear on the task of defining what it is to be a mammal. The most fundamental diagnosis includes the meaning of the name itself: mammals are animals whose mothers have mammary glands – mammae that have the evolutionary origin of being modified cutaneous glands (Darwin 1859). Being suckled is actually but a detail of the two adaptations that have ensured the world-wide success of mammals: one is drawing energy and nutrients from the mother's body both before and after birth. The other is having an energetic warm-blooded body, a property called homeothermy. Mammals have a higher metabolic rate than other animals (excepting birds) because they have evolved internally stable temperatures, mediated by warm blood, which circulates very efficiently thanks to several unique adaptations. This makes them rather independent from the disciplines of the environment; much more so than most other animals, and hence mammals are also found at climatic and geographic extremes. Homeothermy allows consistently faster and sustainable movement, so that mammals are among the fastest of all running, leaping, flying and swimming animals.

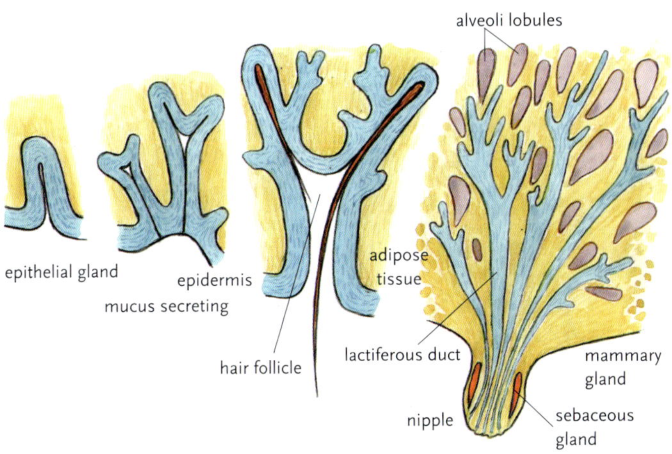

Diagram of possible evolutionary origins of mammary glands. Following the suggestion that mammae were derived from cutaneous glands (Darwin 1859), Oftedal (2002) suggested that milk-secreting glands derived from modified hair-follicles associated with apocrine-like glands. Vorbach *et al.* (2006) have shown that milk-production probably derived from an inflammatory response that involved immuno-protective proteins in skin glands.

The most obvious external manifestation of temperature-regulation consists of fur, hair or bristles that take many different forms, including scale-like and horn-like structures and may even have degenerated, leaving some mammals partly or wholly naked (see illustration p. 206). A less labile and more universal and fundamental regulator for the high metabolic turnover associated with high, endotherm body temperature is the mammalian four-chambered heart. This complex muscular organ keeps out-going, lung-oxygenated blood separate from blood that returns, spent of its oxygen, after energizing working muscles and actions. Immediately below the lungs is a thick, membranous diaphragm that is tightly attached to the inner walls of the thoracic cavity. Assisted by rhythmic contractions of the rib-cage and its muscles, the diaphragm helps pump the lungs, drawing fresh air in to oxygenate the blood and then expels 'used' air that includes carbon dioxide.

All African mammals are placental and the evolution of this mode of reproduction (which probably originated in Eurasia) is unique to placental or 'Eutherian' mammals. It overcomes several disadvantages

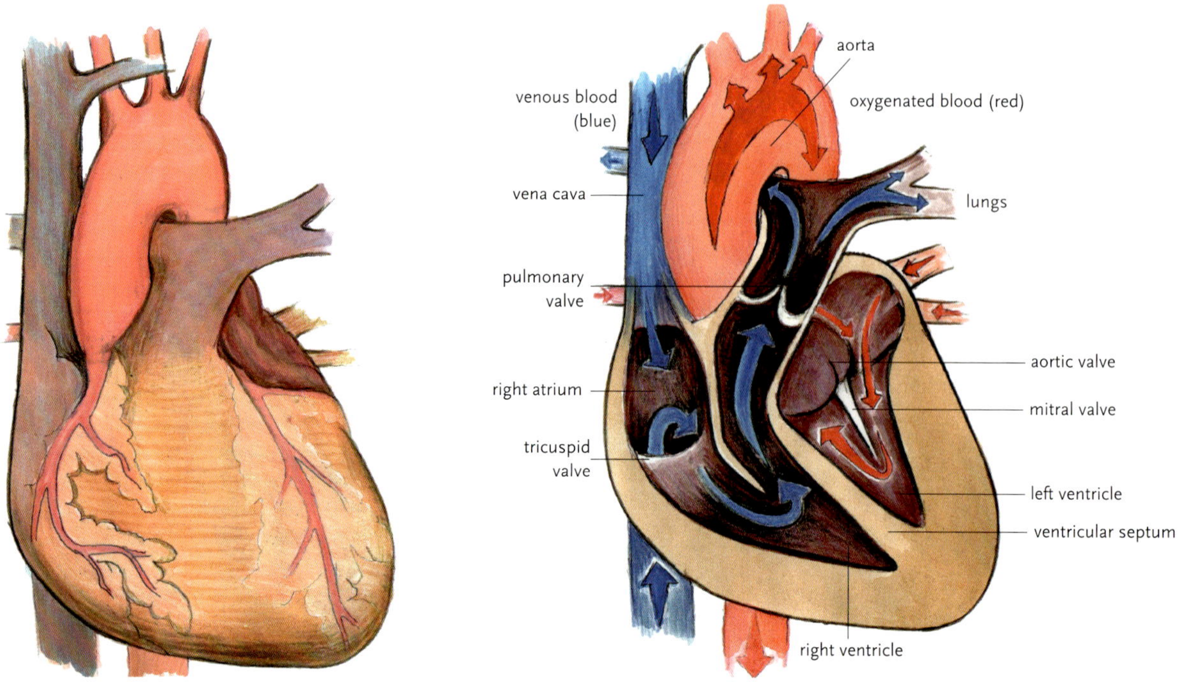

Typical mammalian heart showing venous blood (blue) flowing into right atrium through tricuspid valve into right ventricle, then via pulmonary valve and pulmonary arteries into lungs. Oxygenated blood (red) returns into left ventricle via mitral and aortic valves into aorta and on to arterial blood supply.

that are inherent in the eggs that are laid by other vertebrates: notably the egg's store of nutrients is finite, the period of development within an egg is relatively brief, the embryo is totally separated from the mother's metabolism at a very early stage of development and the developing embryo cannot void metabolic wastes through the egg's shell or membrane (Fox 1999). This combination inhibits or delays the building of energy-costly structures, such as brains. Indeed, Martin (1990) has pointed out that the resources supplied through the placenta are crucial in determining the eventual size of the brain. The existence of foreign (paternal) genes in the embryo may have been one reason for separating the embryo from the mother in a self-contained egg thus insulating the foetus from being rejected by the mother's metabolism as a foreign body. Exactly how the placenta overcame this last problem during its evolution remains a controversial topic (Cohen & Larsson 1988), but it is central to what the placenta represents as an evolutionary advance. Placentas are, of course, the defining adaptation for placental mammals.

The placenta begins as a disc-shaped tissue that issues from the earliest stages of embryo development and acts as the interface between foetus and mother. It secretes hormones that prepare the uterus wall to accept the foetus and it produces immunosuppressive proteins that are thought to inhibit maternal rejection of the foetus (Larsson *et al.* 1994). Placental cells fuse to develop a complex, multinucleated tissue called placental syncytium, which has multiple projections that

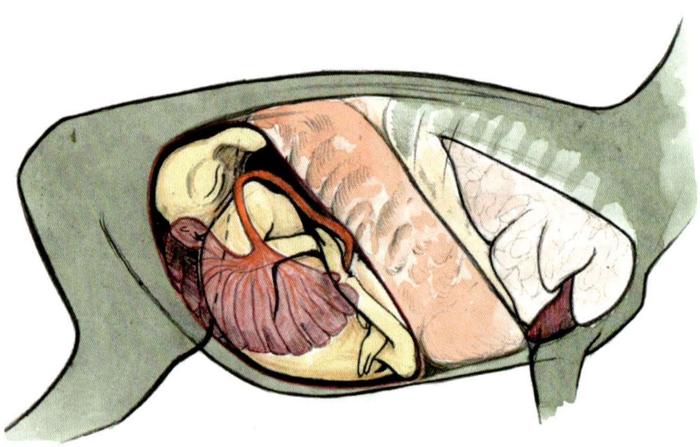

Placenta and foetus of an antelope (*Madoqua* sp.) showing position, size and form of placenta at near full-term.

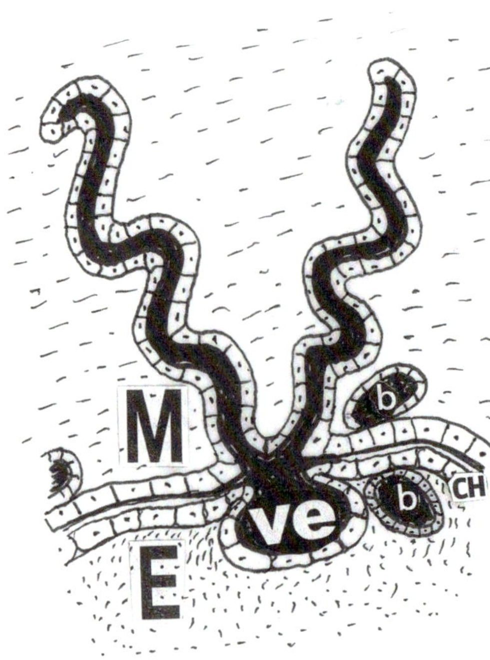

Details of interlacing between the maternal (M) and embryonic (E) tissues in the placenta. b = blood vessels on both portions of placenta; CH = chorion outer membrane of foetal portion; ve = vesicle/invagination below villi accessing maternal blood supply.

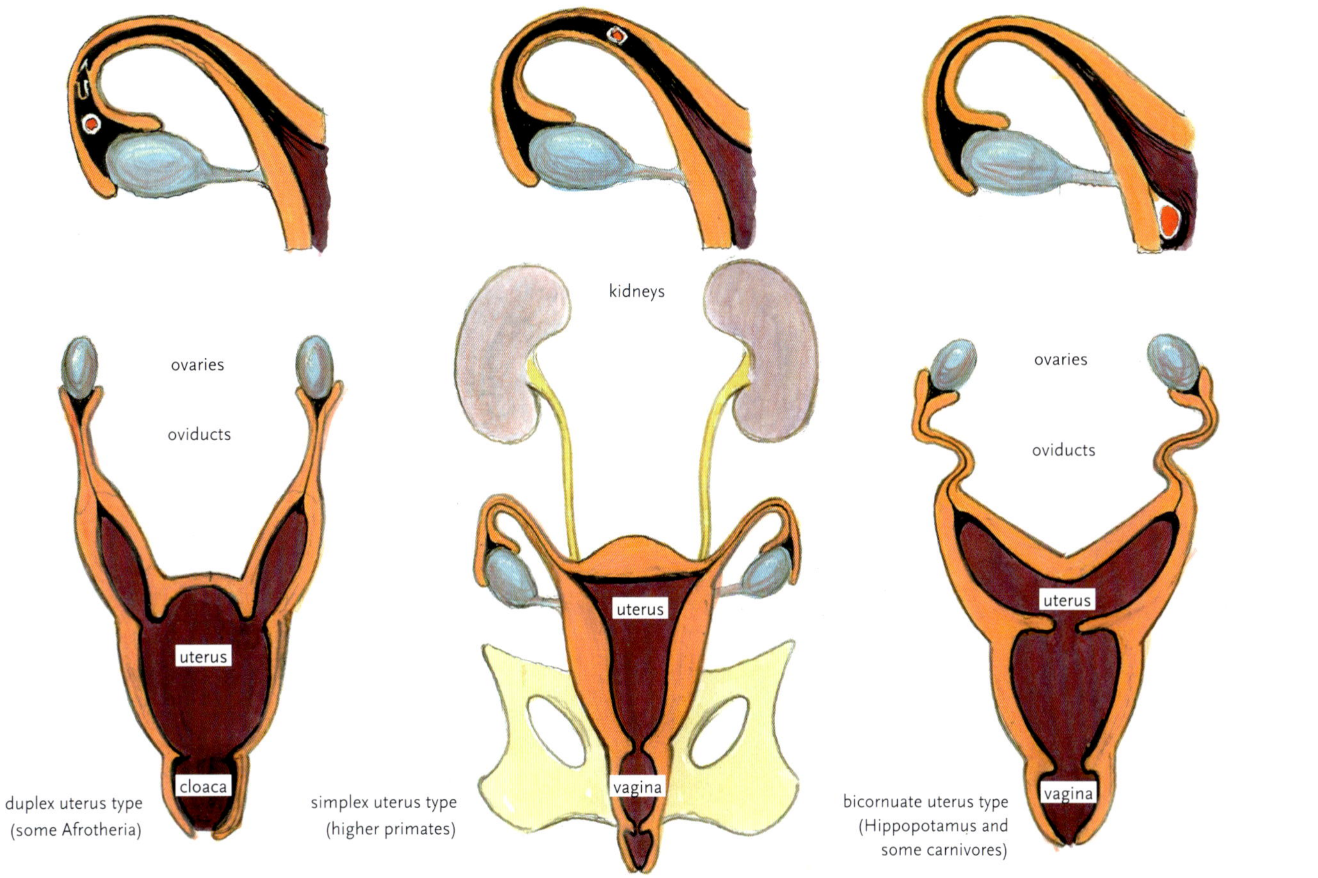

ABOVE: Course of the ovum (red) from the ovary (blue) to implantation in the uterus. BELOW: Types of female reproductive tracts.

probe deeply into the uterine lining and actually invade the mother's blood vessels. Once it is firmly embedded in the wall of the uterus, the placenta develops a rich network of blood vessels that taps into the female's vascular circulation. Connected to the placenta via its umbilical cord, the foetus not only absorbs oxygen and nutrients from its mother, it also discharges its own metabolic wastes into her system via the placenta (see figure, opposite below right). As well as allowing the foetus to share its mother's nutritional, respiratory and excretory circulation, the placenta produces multiple hormonal controls, including oestrogen, progesterone and gonadotrophin. It has been suggested that an embryonic association with an endogenous retrovirus (ERV) might have mediated the earliest beginnings of the placenta by becoming encoded into mammalian DNA (Larsson *et al.* 1994). This could explain why the mother is inhibited from rejecting the foetus, an idea that has been hotly contested, but research on the properties and origins of the placenta continue, with the gene coding for syncytin now identified in the human genome (Mallet *et al.* 2004). Immediately after birth, the placenta is shed.

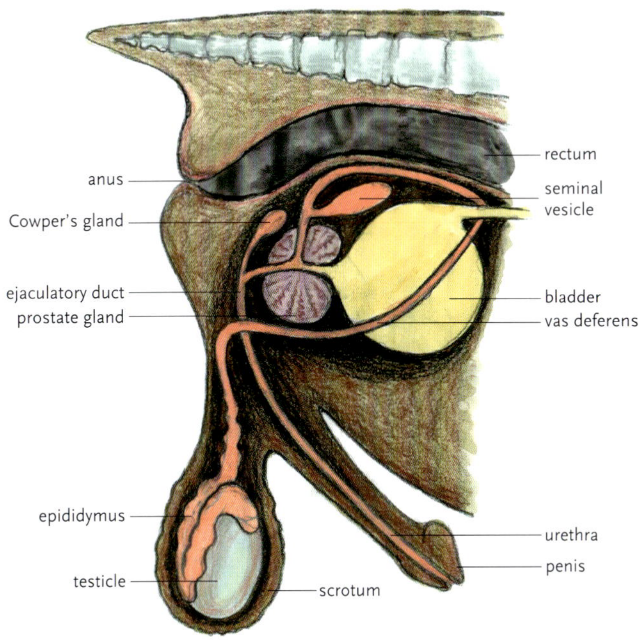

Male mammalian reproductive tract.

Mammals have specialized in maternal care: first there is the womb and all its placental properties. This is followed by another key feature (but shared with the marsupials that are so conspicuously absent in contemporary Africa) that is secreting milk from the mammae. This necessitates days, months (or even years in the case of elephants) of direct maternal contact and protection. In many species, protection (sometimes bi-parental) extends well beyond weaning and effectively augments the physiological adaptations that make mammals relatively independent of the environment. Social structures in some mammal species mimic or extend parent-like security to most members of the group. In humans, the mammalian traits of (i) detachment from ecological systems and from many rigours of the environment, and (ii) social dependency have been extended by technology, psychological management and social institutions well past adulthood, even into old age. Thus it can be argued that humans have evolved to become the most mammalian

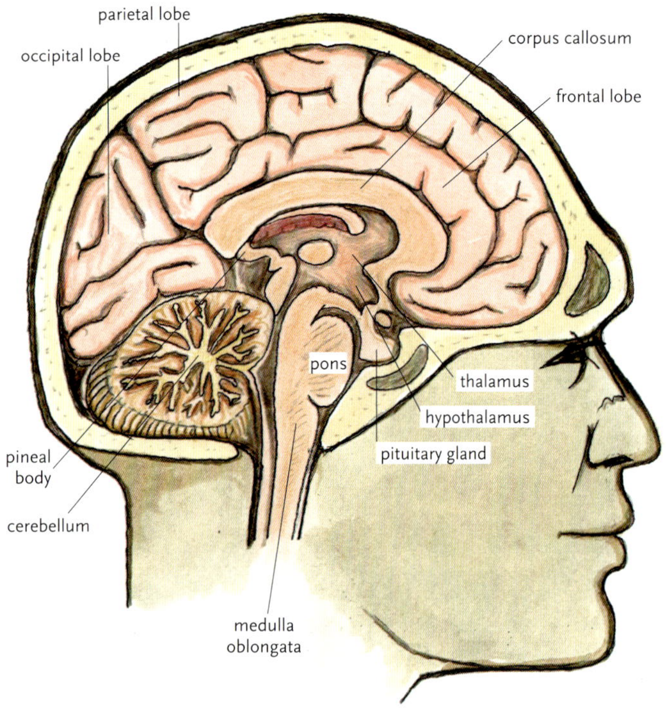

Principal components of the mammalian brain (typified by *Homo sapiens*). Signals are exchanged between the brain and the peripheral nervous system via the spinal chord.

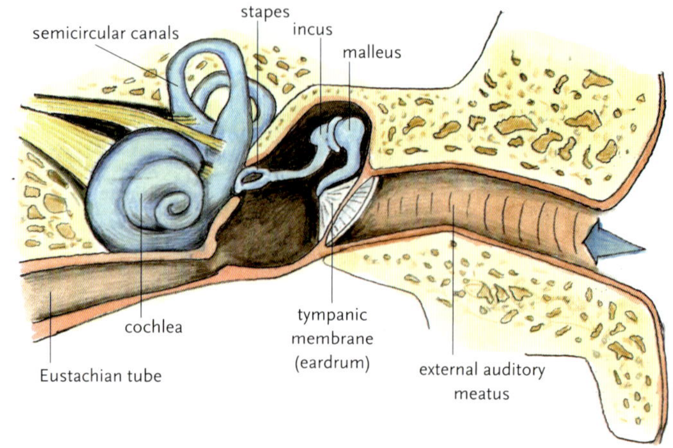

Mammalian auditory system. Incoming sound waves registered on the eardrum are transmitted via the malleus, incus and stapes to the cochlea, where auditory nerve endings translate sound waves into electrical signals for transmission to the brain. Of all biota, mammalian auditory capacity is the most refined. There is exceptional sensitivity to very high and very low pitch.

of all mammals by prolonging parental/social dependency and by extending relative independence from the environment – both of which features are, after all, typically mammalian strategic traits!

While a mammal shares a highly adapted brain with many other types of animal, the mammalian brain is relatively larger and more complex (especially the cerebellum). Perhaps more than any other morphological trait, the brain and its multiple functions is most developed in primates. And, within the primates, the brain has evolved its largest relative size in the humans. Other mammals, especially some cetaceans, elephants and bats, also have highly developed brains with capacities and ranges of functions that still remain largely unknown. However, none of these brains, well adapted as they have to be, matches the cognitive capacities of the human brain.

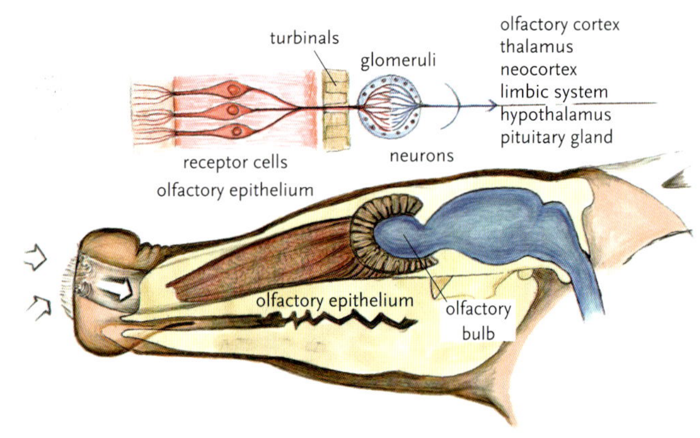

Mammalian olfaction pathways typified by Aardvark *Orycteropus afer*. Odour molecules entering nasal cavity bind to hair-like terminals (cilia) on trigeminal receptor cells embedded in olfactory epithelium covering thin turbinal bones.

Neuron extensions in receptors transmit olfactory information through the cribriform plate to the brain's olfactory bulb. Glomeruli within the olfactory bulb transform information coded in receptor cells and permit exchange with mitral and granule cells. Mitral and other neurons send information to the brain and activate appropriate responses in limbic, endocrine and autonomous nervous systems.

Mammals share their basic sensory systems with birds and reptiles; and are actually inferior to many bird and lizard species in visual acuity and in sensitivity to optical wave-range. However, most mammals (excepting primates and some bats) have very superior olfactory apparatus to birds and reptiles and consequently are able to navigate in a world of scent. In concert with this specific sensitivity, mammals have developed numerous glands that are rich in olfactory information.

Mammals also have evolved great sensitivity to sound, and their hearing capacities are far superior to reptilian equivalents. The evolutionary origins of the hearing apparatus have been reconstructed from a rich fossil record of the ancestry of mammals. This documents the transformation of bony components (originally part of the mandible) and the migration of these modified bones and their attachment to the base of the cranium. Indeed, the development of refined hearing in mammals is one of the most telling of all examples of the modification of one set of structures, by infinitely slow, incremental steps, to serve an entirely new purpose. In mammals, hearing not only serves in the finding of food, the identification of conspecifics and the detection of predators, it also has been greatly (and independently) developed and modified to serve as super-sensitive sonar in the bats and cetaceans. The range of sensitivity to sound signals has been greatly extended at both ends of the sound spectrum in rodents, insectivores and elephants (to mention a few of the best-known examples) as well as bats and cetaceans. Thus sound, in mammals, has been co-opted to serve in orientation, spatial awareness and social cohesion as well as numerous other communicative purposes.

The exact configuration of senses differs from one mammal taxon to another. Scent, hearing, sight and touch are balanced in permutations that are unique to each species, and that balance finds a gross expression in the shape of mammal skulls, heads, skeletons and myology (see Chapter 8, p. 109). Each of these faculties has developed by small increments from the less specialized sensory

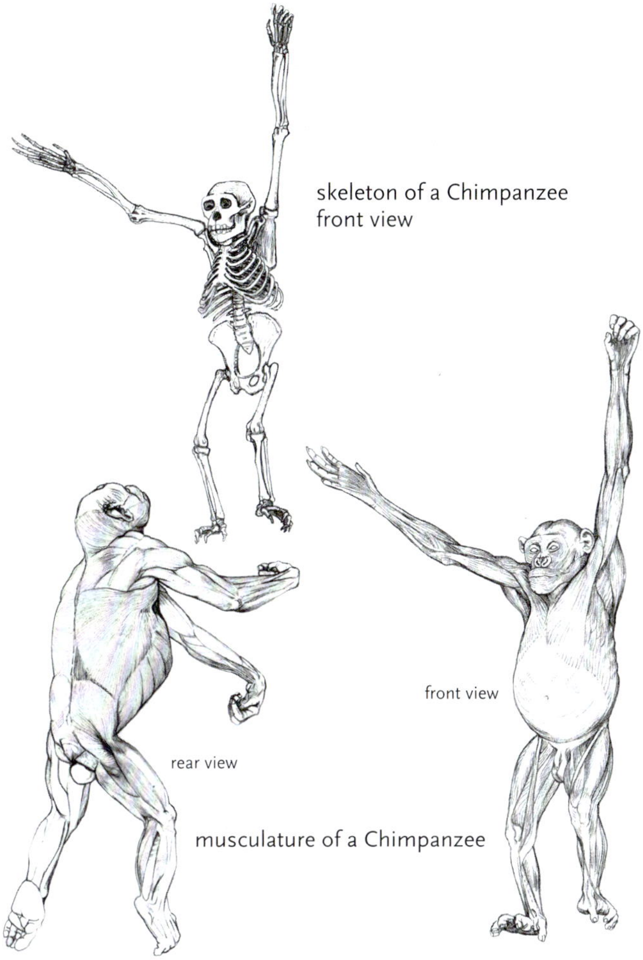

A mammal's musculature, skeleton, sensory, neural and vascular systems, reproductive, urinary and digestive tracts are all discrete 'homeostatic systems' contained within its skin and all obfuscated by its pelage.

apparatus that preceded them. Even at quite a refined level the proportions of subtle differences in the skull proportions of primates, rodents, foxes, antelopes and other speciose groups can all be related to similar shifts in the balance of senses. Likewise, the relative sizes and functional activities of mammal bodies and musculature are expressed in their proportions and gross anatomy.

Contours of bodies conceal ever more complex metabolic systems, respiratory, vascular, hormonal, digestive, excretory, reproductive and nervous systems, each subjects for distinct scientific disciplines. During the course of evolution the interweaving of these self-regulating systems has generated odd but essential structures such as a secondary palate and a valvular pharynx (both of which serve to keep nasal and oral passages separate). A series of schematic illustrations (below, mostly drawn from a single individual, a young chimpanzee) can only hint at the maze of interwoven pipes, cables, tubes and organic machines that lie beneath *any* mammal's bland exterior.

Teeth and the details of their structure have long been a staple in the diagnosis of mammals (McKenna 1975, Hillson 1986). Much effort has been expended on reconstructing the structure of ancestral mammalian teeth. A good fossil record of cynodont and early mammaliaforms shows that the differentiation of peg-like teeth into distinct classes of specialized form, namely incisors, canines, premolars and molars, long predates the emergence of mammals. Such differentiation is known as heterodonty (Novacek 1986).

A practical reason for emphasis on teeth is that they survive well after death, both in extant and extinct taxa. Their usefulness for phylogeny also derives from the fact that dentition is usually species-specific and is always highly diagnostic at higher taxonomic levels. Even the rise of molecular phylogenies cannot displace the

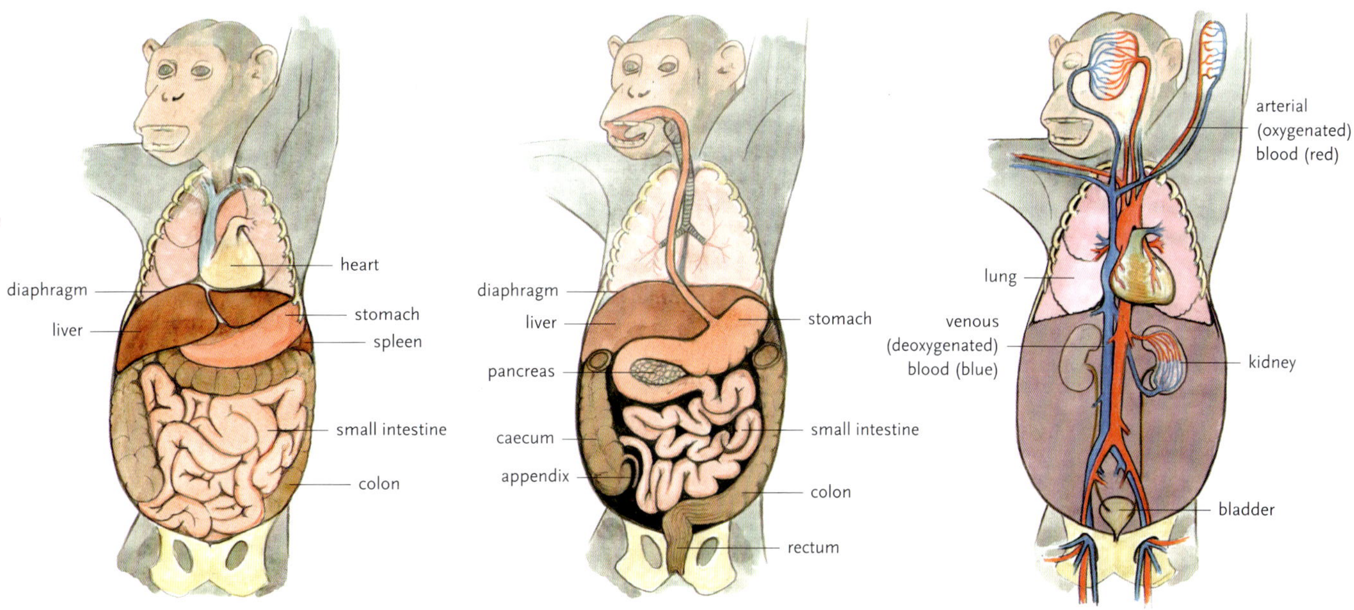

LEFT: Schematic representation of viscera: shows relative positions and sizes of heart, lungs, liver, stomach and gut in a Chimpanzee.

CENTRE: The respiratory and digestive tracts in a Chimpanzee, showing bronchial trees at lower end of trachea reach all parts of lungs. Pathway of oesophagus (connecting mouth and stomach), stomach and pancreas, small and large intestines (mid-section of latter cut away), appendix and colon/anus. Note glottis serves to separate oral and nasal passages in pharynx (all simplified).

RIGHT: Schematic representation of circulatory and renal systems in a Chimpanzee. Heart, lungs, diaphragm and blood vessels combine to send oxygenated blood (red) to body tissues, including kidneys. Veins return de-oxygenated blood (blue) to lungs where carbon dioxide is removed and the cycle repeats.

importance of teeth as indicators of affinity. Differences in the shape of teeth, no less than any other aspect of morphology or behaviour are outcomes from the struggle for subsistence itself. Getting access to, or processing food demands teeth, and their description has even led to the naming of entire groups, such as rodents or tubulidents.

The primitive dental condition in early placental mammals can be expressed in the dental formula of $I\,^{3}/_{3}\ C\,^{1}/_{1}\ P\,^{4}/_{4}\ M\,^{3}/_{3} = 48$. This represents a reduction from earlier conditions (Novacek 1986) but the details need not concern us here. The molar teeth of nearly all extant mammals derive from an ancestral tribosphenic structure (Bown & Kraus 1979). None the less, this origin can be wholly lost in the complex transformations that have evolved since the Mesozoic. Teeth, both deciduous 'milk' teeth and adult 'permanent' teeth, are therefore essential to mammalian diagnostics and dental descriptions and illustrations are an important part of the group and species profiles that follow (see figure below).

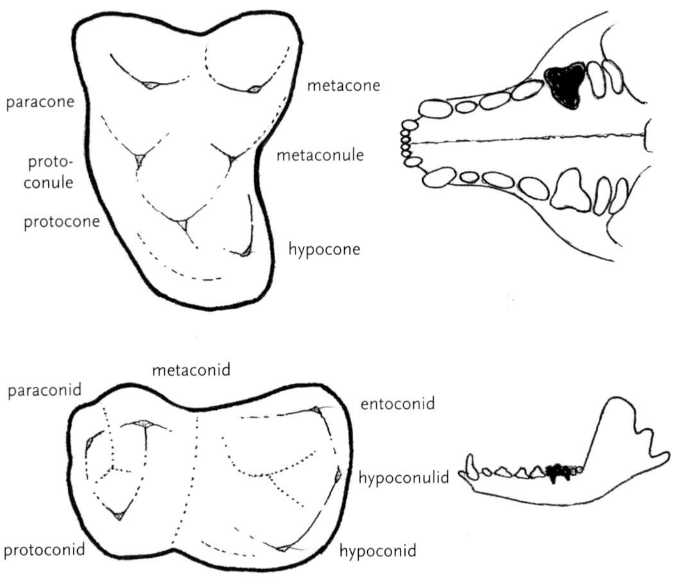

Terminology for dental cones on a generalized mammalian molar: maxilla above, mandible below.

Comparing anatomical, physiological and functional systems with those of other animal forms is part of defining the Mammalia but, given that reptiles, amphibians and birds all share a common ancestry with us, how long have mammals been distinct and when did they diverge?

What distinguishes all mammals from, say, reptiles and birds, is a common heritage that stretches back over 300 million years (Cifelli 2001, Rose 2006). Placental mammals parted from the marsupial mammals some time between 145 and 135 mya (Luo *et al.* 2002, Wible *et al.* 2007). To appreciate the magnitude of these time frames, and of the shared distinctiveness of both mammalian clades, we may remember that it is only about 9–6.5 million years since humans shared a common ancestry with our closest living primate relatives (Sarich & Wilson 1967) and that the dinosaurs were supreme until only 65.5 mya.

Today the diagnostic characteristics that distinguish modern mammals from reptiles and birds are numerous and clear-cut. The differences between any pair of living animals (an ostrich versus a lizard, a lion versus a crocodile or a human compared with a galago) encompass an enormous array of genetic, anatomical, physiological and psychological details. These are strong differences between end-members of long-separate lineages but fossils suggest that the differences were not always so clear-cut. The levels of comparison that are possible between living survivors diminish once lineages that have become extinct are taken into consideration. And immense numbers of extraordinary and interesting animals, especially types of mammals and proto-mammals, have gone extinct (see Hopson & Crompton 1969, Kemp 2005, Rose 2006). Luckily the accidents that have given us the rich 'History Book' of fossils do at least allow mineralized extinct teeth to be compared with teeth in living mammals, fossil bones with living bones and many other extrapolations can nowadays be made about the behaviours that gave rise to so many peculiarly shaped anatomies.

The more ancient of our mammalian (or reptile-like) ancestors, were they not all extinct, would have seemed somewhat similar to one another, with their resemblances deriving from their common ancestry among the amphibian-like Protothyrids (Carroll 1988). Until quite recently it was common to find mammals described as descending from reptiles. This is incorrect, in spite of many close resemblances in the earliest ancestors of both groups. Mammals and reptiles have had separate lineages since the Carboniferous, about 300 mya (Rose 2006).

The synapsids, the ancestors from which all mammals descended, used to be called 'mammal-like reptiles' (Romer 1945). However, synapsids are now known to have been a separate lineage from that of the reptiles. Synapsids, during more than 200 million years of separate evolution, radiated into a great variety of distinct forms (Hopson & Crompton 1969, Kemp 1983, Hopson 1994, Benton 1999). Within this phylogenetic 'bush', cladistic analysis of fossil morphology has traced the ancestors of modern mammals through primitive synapsids to pelycosaurs, then therapsids and, by the Triassic, to the cynodonts, which, while being more mammal-like and ancestral to modern mammals, were still not classed as mammals (Rose 2006). By the late Triassic, multituberculates had evolved and were to become the dominant mammaliaforms throughout the Jurassic. It was at this time and later, during the Jurassic, that a great variety of mammal types, including egg-laying monotremes (which survive in Australasia) and symmetrodonts (all now extinct), evolved.

Among the earliest and best preserved fossils of an early mammal is *Hadracodium*. One of the smallest mammals ever known, it was estimated (Luo *et al.* 2001) to weigh little more than 2 g, close to the weight of a tadpole! Possibly a crevice-dweller, its skull was peculiarly flat and wide. The outline of its skull and its 195 mya position on the mammalian tree are shown on p. 76. By the early Cretaceous there was a great diversity of mammalian types, including the ancestors of placental and marsupial mammals, which diverged at about this time (Bininda-Emonds *et al.* 2007, Wible *et al.* 2007).

The best fossil of an early placental mammal is a Chinese find, the 125-million-year-old *Eomaia scansoria*, in which not one bone is missing and even the dense fur has been preserved (Ji *et al.* 2002, illustrated on p. 75). This animal was partly arboreal and, although not directly related to any living lineage, its presence in the Xixian formation strongly supports suggestions that the earliest eutherian mammals were of Asian origin. The outline of its skull and position on the mammalian tree is also shown in our early mammalian tree.

No fossils of anywhere near this age have been found so far in Africa. In Morocco there is a single good Palaeocene site at Ouarzazate (about 60 mya). The evolution of mammals in Africa and the sparse fossil record is discussed in our chapters on African

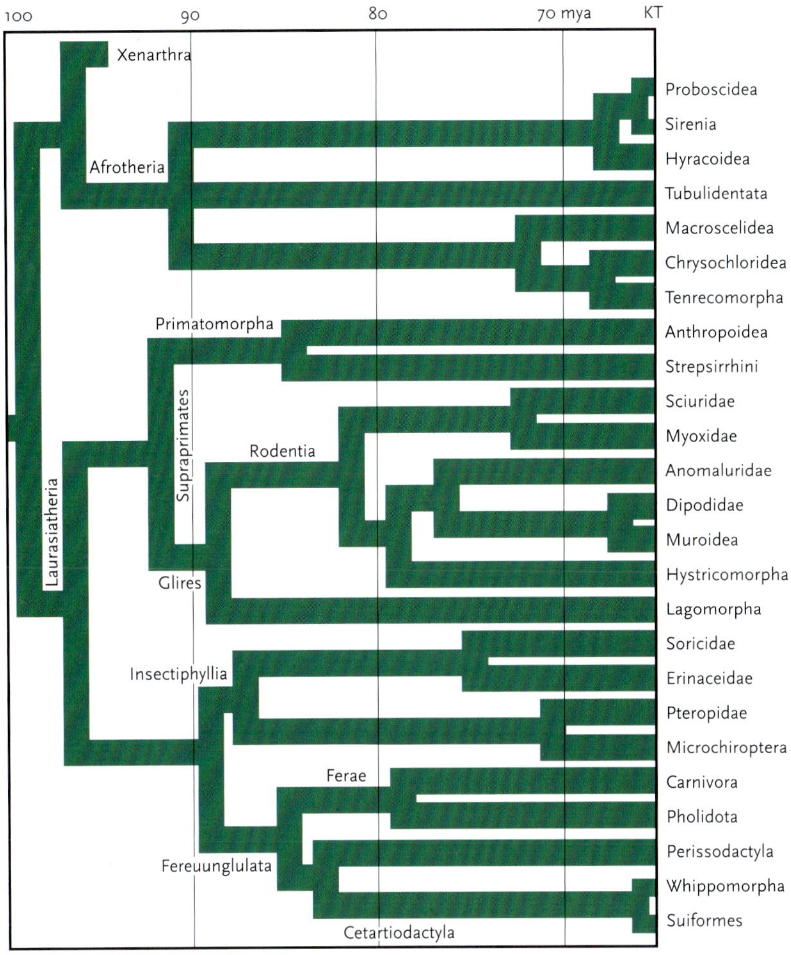

The 'short-fuse model', one of three alternative phylogenetic trees for placental mammals before the KT event (65.5 mya) (after Bininda-Emonds *et al.* 2007). Some dates are modified by O. R. P. Bininda-Emonds (pers. comm. 2009).

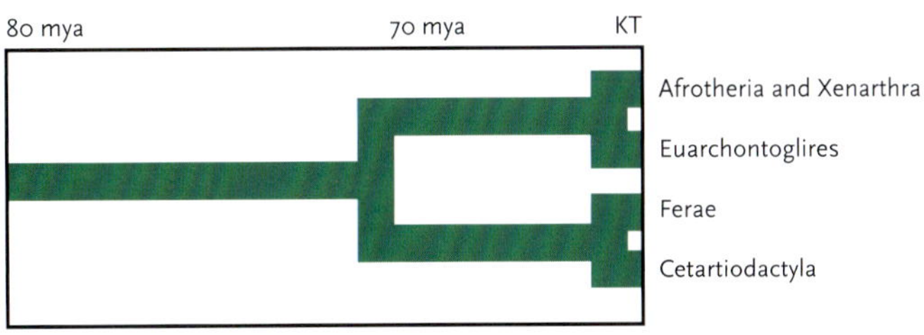

The 'explosive' model, one of three alternative versions of mammalian radiation before the KT event (65.5 mya) (after Wible *et al.* 2007).

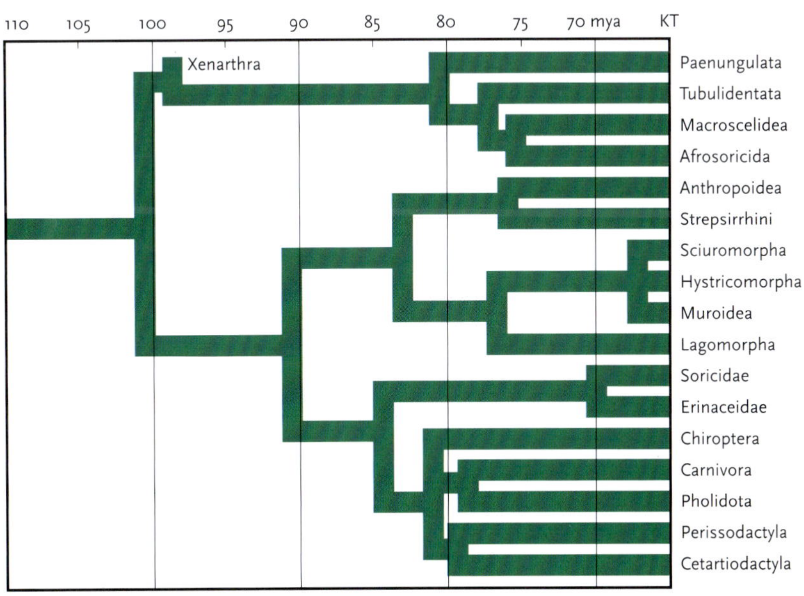

The 'long-fuse' model, one of three alternative versions of mammalian radiation before the KT event (65.5 mya) (after Meredith *et al.* 2011).

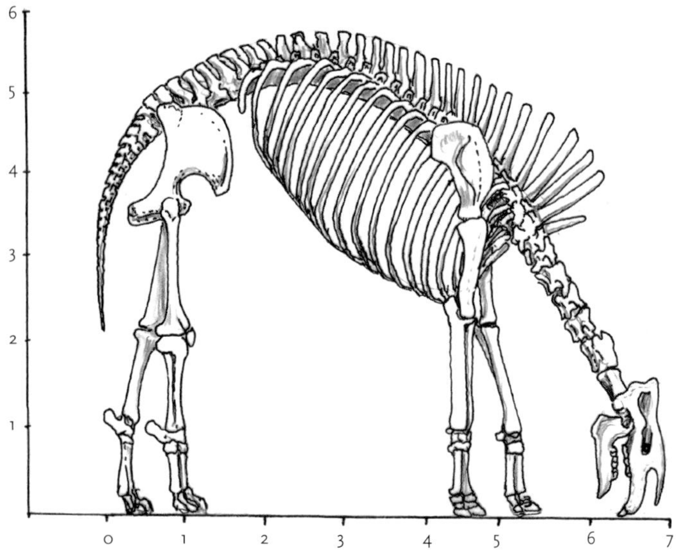

The largest land mammal known: the rhinocerotid *Paraceratherium* (height and length indicated in metres).

geology and on the evolution of African mammals (pp. 27 and 75). However, it must be appreciated that much remains to be discovered about the evolution of mammals, especially the timing and speed of their radiations. Very different concepts have been elaborated, and in keeping with the palaeontologists' geological background, the three principal models for mammalian evolution have acquired rock-breaking names: the Explosive, the Short-fuse and the Long-fuse models!

The Explosive model has become identified with the geological evidence for a late and sudden explosion of mammal types after the K–T asteroid impact of 65.5 mya (best known for being the event in which dinosaurs became extinct) (Alroy 1999, Benton 1999, Foote *et al.* 1999). The post-K-T 'explosive' model is best exemplified by Wible *et al.* (2007). The Short-fuse model depends upon molecular clock interpretations of the primary diversification of placental mammals taking place during the Cretaceous (Kumar & Hedges 1998, Bininda-Emonds *et al.* 2007). The Long-fuse model is an amalgam of the previous two in suggesting that the stem taxa for most of today's orders probably diverged during the Cretaceous but most of the elaborations we see in modern mammals are after the K–T event (Springer *et al.* 2003, Douady & Douzery 2003, Meredith *et al.* 2011).

Inasmuch as we have tried to illustrate the most likely radiation of mammals, we have often referred to Bininda-Emonds *et al.* (2007). Detailed comparisons of the genotypes of all known orders of mammals, as pioneered by Bininda-Emonds and by other molecular biologists, have permitted the erection of entirely new supraordinal taxa: superorders, cohorts and supercohorts. Their adoption has been far from universal and both the new taxonomic names and their existence are contested. However, their identification has been backed up by robust and authoritative arguments accompanied by molecular clocks that vary greatly in their calibration of divergence times. These phylogenetic trees have served to emphasize that the evolutionary branching of mammal orders, families and species has taken place in real time and in real continental or local settings. Many of these assessments will need revision, but we have included three tentative trees (see p. 141) as pioneering efforts to understand the mammalian radiations.

Under this phylogeny, Mammalia breaks down into four supercohorts, namely Afrotheria, Xenarthra, Supraprimates (also known as Euarchontaglires) and Laurasiatheria. These, in turn have been subdivided into cohorts and superorders, descending ultimately to the species, portrait profiles of which are our ultimate product and purpose. In the text that follows each supercohort merits a profile (excepting Xenarthra, which is exclusively American) and our opening texts and volume concern the Afrotheria: the one group that has best claim to be endemic to the African continent.

The primacy of maternal care and prolonged dependency has had particular emphasis in the definition of Mammalia: this emphasis is also evident in much of the behaviour of mammals, especially where it concerns their social lives. Unlike most reptile and bird mothers, a mammalian mother must care for her offspring; that is, after all, why she has a womb and mammae! To ensure that the young (and their lactating mothers) have access to the best resources that are available and at the optimal times of year, there has been selection for a wide range of very different types of social systems. In spaced-out, often residential patterns of land-use, several or single females enter or share the territories of single males. Competition among the latter tends to ensure that the best resources have been won by vigorous territory-holders. Where resources are more dispersed or seasonal, females can enjoy enhanced protection and access to food within groups that have enlarged male hierarchies. Yet another strategy has been either a sustained or a semi-permanent association with a male or males that actually help raise offspring (typically their own only). Long-term pair-bonding has developed in some species (notably wild dogs *Lycaon*, where the sexes are of similar size and appearance).

Male competition can have conspicuous consequences for male external appearances. Weaponry, in the form of horns, tusks or antlers, accompanied by loud calls and/or pungent scents and emphatic displays of patterns or structures have been developed to defend territories or rank. Age-graded gigantism has evolved in the males of hierarchical species, such as gorillas, fur-seals, giraffes and elands.

African mammals have provided much material for evolutionary theory, from Darwin, noting the abundance of spotted and striped felids to invoke just such an evolutionary ancestry to explain patterned lion cubs (Darwin 1859) to Richard Dawkins using gazelles and Cheetahs to illustrate an evolutionary 'arms-race' (Dawkins 1986). Jenny Jarvis has revealed fundamental connections between the ecology and behaviour of blesmols (mole-rats) and the evolution of complex social structures (Jarvis *et al.* 1992, Jarvis 1993). John Crook and Robin Dunbar have explored many socio-biological aspects of evolution from their studies of African primates (Crook & Gartlan 1966, Dunbar 1988) while Dorothy Cheney and Robert Seyfarth have begun to plumb the depths of primate communication with baboons and vervet monkeys (Seyfarth & Cheney 1984, Cheney & Seyfarth 1990, 2007). The list could run on and on.

The diversity of ways in which mammals have evolved ways of living, become social or communicated subtle information with one another are all areas of active ongoing research. We report on this explosion of interest in African mammals and on the multifaceted attempts to understand how the extraordinary variety of mammals has evolved. Here are important strands of thought among the many that are discussed in the volumes that follow.

Jonathan Kingdon

Supercohort AFROTHERIA

Afrotheria Stanhope, Waddell, Madsen, de Jong, Hedges, Cleven, Kao & Springer, 1998. Proc. Nat. Acad. Sci. (USA) 95: 9971

Modern placental mammals have not always inhabited Africa, and very recent fossil discoveries have raised, once again, the central question of just where the first placentals arose. At the centre of this controversy is the very new revelation, initially from molecular evidence, of a single and unexpectedly diverse evolutionary radiation of mammals that is so unequivocally African it has been named the Afrotheria ('African mammals'). This radiation includes three orders that have long been recognized as being related and of African origin, namely the elephants (order Proboscidea), manatees and dugongs (order Sirenia) and hyraxes (order Hyracoidea). They have now been joined by four other groups: the Aardvark (order Tubulidentata); sengis or elephant-shrews (order Macroscelidea); golden-moles (family Chrysochloridae) and tenrecs and otter-shrews (family Tenrecidae), now grouped in the new order, Afrosoricida (Stanhope *et al.* 1998). Some authors have argued that this new order should be called Tenrecoidea or Tenrecomorpha, but both of these names present taxonomic difficulties (Bronner & Jenkins 2005).

Are these animals modern derivatives of the very earliest placental mammals or are they later products of the Afro-Arabian landmass's known physical isolation? We know now that Afro-Arabia became separated, sequentially, from Eurasia, Indo-Madagascar and then South America by plate tectonics that opened up the Tethys Sea and proto-Atlantic in the later Mesozoic. Understanding the origins of placentals and of the afrotherian radiation is inseparable from the geological history of continents. That mammals confined to a single continent should radiate into highly diverse forms is entirely consistent with a very lengthy isolation, but was the afrotherian common ancestor native or immigrant? This question can only be answered with absolute certainty by more fossils from appropriate periods, but in the interim the biogeographic significance of both placental and afrotherian origins continues to be a matter of debate.

Some scholars have argued that the presence of two Gondwanan branches (Afrotheria and the endemic South American order Xenarthra) favours a Gondwanan, and possibly even African, origin for placental mammals (Murphy, Eizirik, O'Brien *et al.* 2001). Others cite the presence of much more primitive Late Cretaceous placentals and marsupials in Asia as evidence for a northern origin (Archibald 2003, Robinson & Seiffert 2004, Wible *et al.* 2007). If placental mammals did originate in Asia, as now seems most likely, then basal afrotherians must have dispersed to the Afro-Arabian landmass some time between the predicted origin of placentals (~108 mya) and the putative origin of afrotherians (~80 mya) (Springer *et al.* 2003).

A broad acceptance of Afrotheria has rendered traditional taxonomies obsolete, but it is interesting to retrace some earlier insights and intuitions. The present work is one of the first to attempt to come to grips with some of the implications and introduce readers to the many new questions that these discoveries raise. As for traditional ideas about relationships, they had to be founded on whatever evidence was available at the time, which often did not amount to much. The evidence is still incomplete but the molecular revolution that has unearthed the reality of Afrotheria is but part of a global effort to construct genealogical trees for all biota.

People familiar with wild animals, especially hunters and herders, in Africa and elsewhere, have often recognized that similar species or similar attributes imply ancestry. Thus fishermen, finding ivory tusks in elephant-like Dugong skulls sometimes called them 'elephants of the sea'. Scientific comparisons of the anatomy of elephants and hyraxes with dugongs and manatees, living and fossil, led to the recognition that these superficially very different animals shared a common ancestry (Simpson 1945). This conclusion had been greatly facilitated by the recovery of numerous fossils of these large mammals, primarily from the late Eocene beds of the Fayum Depression in Egypt (Andrews 1906). Following Simpson (1945), this taxonomic clustering is now referred to as the Paenungulata, and has subsequently come to be strongly supported by a vast array of genetic data (Springer *et al.* 1999, Murphy *et al.* 2001a, Amrine-Madsen *et al.* 2003, Murata *et al.* 2003, Nikaido *et al.* 2003, Nishihara *et al.* 2005, Meredith *et al.* 2011).

Determining the affinities of the Aardvark, sengis, otter-shrews and golden-moles has long been much more difficult because their early fossil record is either scarce or non-existent. The Aardvark shares a number of morphological features with similarly myrmecophagous pangolins (order Pholidota) and xenarthrans, and as such was historically aligned with these taxa, although more recently there have been suggestions that it might be more closely related to 'ungulates' – paenungulates, perissodactyls and/or artiodactyls (Novacek 1986, 1992, Shoshani 1986).

While the oldest possible fossil of Macroscelidea is from the early Eocene, about 50 mya (Hartenberger 1986, Tabuce *et al.* 2007) and that of primitive tenrec and/or golden-mole relatives may be Palaeocene or Eocene (Seiffert 2010, Goswami *et al.* 2011), aardvarks do not appear in the fossil record until the early Miocene, about 20–18 mya (MacInnes 1956, Patterson 1975). From the time of their initial discovery, these groups have always puzzled biologists, and have been very unsatisfactorily allocated to various other mammalian higher taxa, never with any confidence. Apart from the rarity of their fossils, a major reason for much of this confusion has been the combination of apparently primitive features and extreme specializations (or 'autapomorphies'), and their resulting transformation into swimming, digging and leaping 'insectivores'.

Insectivora has long served as a taxonomic waste-paper basket into which all small, apparently primitive, invertebrate-eating mammals were thrown. Thus, sengis, golden-moles and tenrecs once joined solenodons, shrews, moles, hedgehogs (and, early in the history of taxonomy, even tree shrews and flying lemurs) in the Insectivora (Wagner 1855). The menotyphlan insectivores (i.e. those with caeca – the flying lemur, tree shrews and sengis) were subsequently elevated to their own orders (Gill 1872, Butler 1956), leaving behind the remaining insectivorans, which were also placed in their own order, Lipotyphla (Butler 1972). However, the lipotyphlan assemblage has long been an unstable one. Following their eviction from the Insectivora, sengis came to be aligned either with the rabbits, hares and pikas (order Lagomorpha) (McKenna 1975, Szalay 1977) or with lagomorphs and rodents together (Glires) (Novacek 1986). Only with the recovery of early fossil macroscelideans

(Hartenberger 1986, Simons *et al.* 1991) was it recognized that their origins might lie with some 'ungulate' group.

Fortunately, technological advances have allowed scientists to rapidly sequence and compare large amounts of DNA, and a much more accurate, objective and less obfuscated analysis of the affinities of all biota has become possible. As noted, these data have confirmed the paenungulate clustering of elephants, hyraxes, and dugongs and manatees, but, much more importantly, have also revealed that paenungulates are the larger representatives of an extremely important and very ancient endemic African radiation that includes many very small mammals as well. A close relationship between aardvarks and paenungulates was first suggested by an analysis of eye lens crystallins (de Jong *et al.* 1981) that would also later provide the first biochemical evidence for the sengis' close relationship with paenungulates (de Jong *et al.* 1993). Analyses of mitochondrial and nuclear gene sequences subsequently expanded this group to include the golden-moles (Lavergne *et al.* 1996, Springer *et al.* 1997) and tenrecs (Stanhope *et al.* 1998). An ancient, single origin for this assemblage is now supported by numerous protein and nucleotide sequences (Amrine-Madsen *et al.* 2003, Meredith *et al.* 2011) as well as rare genomic changes such as protein sequence signatures (Van Dijk *et al.* 2001), unique deletions (Madsen *et al.* 2001, Scally *et al.* 2001), short interspersed nuclear elements or 'jumping genes' (SINEs, Nikaido *et al.* 2003, Nishihara *et al.* 2005) and chromosomal rearrangements (Robinson *et al.* 2004).

Afrotheria has attracted a considerable amount of controversy or scepticism because its members share so little superficial anatomical similarity (Springer *et al.* 2004). Currently, about the only morphological characters uniting Afrotheria involve the reproductive tract, including undescended testicles (the testicond condition) in males (Werdelin & Nilsonne 1999) and a four-lobed allantoic sac in females (Mess & Carter 2006), although some homoplasy exists in both traits. Sanchez-Villagra *et al.* (2007) have argued that afrotherians are specialized in having more thoracolumbar vertebrae than other placental mammals. There has also been intense speculation about the sequence and adaptive significance of phylogenetic branching within the supercohort (Seiffert 2002, 2007, Robinson & Seiffert 2004, Asher & Seiffert 2010). This interest stems primarily from the fact that living and extinct macroscelideans share a number of apparently specialized craniodental and postcranial features with paenungunlates that are not seen in tenrecs and golden-moles. The position of macroscelideans relative to these taxa depends on whether molecular or anatomical data are considered. Phylogenetic analyses of afrotherians that use morphology suggest that sengis are more closely related to paenungulates than to the other afrotherian insectivores (Seiffert 2003, 2007), and imply that the lipotyphlan features of golden-moles and tenrecs were likely to have been present in the last common afrotherian ancestor. However, recent analyses of various types of genetic data (Amrine-Madsen *et al.* 2003, Waddell & Shelley 2003, Robinson *et al.* 2004, Nishihara *et al.* 2005, Meredith *et al.* 2011) support a fundamentally different arrangement that aligns aardvarks and sengis with golden-moles and tenrecs in an assemblage that has been named 'Afroinsectiphillia' (Waddell *et al.* 2001). If this latter result is correct, then it could be the case that the resemblances between sengis and paenungulates represent primitive features within Afrotheria and that the afrotherian common ancestor would have been more like a small paenungulate than a lipotyphlan.

The above possibility is attractive in that it implies morphological support for Afrotheria, but the ultimate answer is unlikely to be so clear-cut, because at least some of the shared features of paenungulates and sengis are sure to be due to convergent evolution within the ancient afrotherian radiation. The best way to test these competing hypotheses will be to search for Late Cretaceous mammals in Africa from beds that span the time period since the living afrotherians are thought to have first appeared – about 80 mya (Springer *et al.* 2003, Meredith *et al.* 2011).

If Afro-Arabia had been sufficiently isolated from Eurasia throughout the Cretaceous, then the afrotherian common ancestor could have arrived on an African continent that was otherwise devoid of placental mammals and, presumably, of marsupials as well. Unfortunately, at present, we really do not know just how isolated Afro-Arabia was during this time period because very few vertebrate fossils (and no mammals – aside from a single tailbone!; Nessov *et al.* 1998) have been found in Late Cretaceous sediments in Africa. Were the Afro-Arabian continent as decisively isolated as is currently thought, its placental afrotherian colonist must have had some tolerance for exposure at sea. Whether this implies possession of semi-aquatic habits must remain conjecture.

However, the extraordinary morphological variation observable among past and present members of Afrotheria would appear to be consistent with an original, ancestral dispersal into a Late Cretaceous Afro-Arabia that had few, if any, eutherian mammal competitors. Estimates of divergence dates within Afrotheria indicate that the living orders radiated quickly (Springer *et al.* 2003), presumably invading vacant niches that would also come to be occupied by distantly related placental mammals on other continents. Several mammal lineages from outside Africa and members of the Afrotheria share many similar or convergent adaptations. These are listed and discussed further below, but they include myrmecophagy (various ant-eating mammals versus afrotherian Aardvarks), fossoriality (various mole-like animals versus afrotherian golden-moles), the zalambdodont pattern of molar cusps (solenodons versus afrotherian tenrecs and golden-moles), large-scale aquatic habits (whale and hippo-like mammals versus afrotherian sirenians), small-scale semi-aquatic faunivory (several placentals and one marsupial versus afrotherian potamogales), cursoriality (at small body size) (a diversity of mammals versus afrotherian sengis), and spines (true hedgehogs versus afrotherian tenrecid hedgehogs in Madagascar).

Likewise, hyracoids share enough detailed morphological similarities with perissodactyls that a close relationship between the two orders has been championed by morphologists until very recently (Prothero *et al.* 1988, Fischer 1989). Another convergence has only been revealed recently; this shows that the diverse radiation of extinct hyraxes contained cursorial bovid-like forms (Rasmussen & Simons 2000). Early proboscideans such as *Moeritherium* were likely to have had life-styles directly comparable with those of somewhat aquatic tapirs (Kingdon 1979). The most parsimonious explanation for the evolution of detailed morphological convergences is that these adaptations evolved in response to similar selection pressures on disjunct landmasses. Otherwise, direct competition between taxa with such similar evolutionary trajectories would have led to character displacement or early extinction in their evolutionary histories. It is for these and other reasons, we are unconvinced by the recent arguments of Asher *et al.* (2003) and Zack *et al.* (2005) that posit afrotheria orders originating on northern continents, alongside demonstrably convergent clades.

A major argument against Holarctic origins for Afrotheria is the implication that most afrotherian lineages dispersed to Afro-Arabia independently, via crossings of the Tethys Sea, at times when such dispersals would have been highly unlikely. Although it is clear that primates had arrived in Afro-Arabia by the late Palaeocene, ~56 mya (Sigé *et al.* 1990) and rodents had invaded Africa by the early or middle Eocene, perhaps as early as 50 mya (Vianey-Liaud *et al.* 1994, Marivaux *et al.* in press), these colonizations do not offer compelling evidence for a 'sweepstake' route that could also explain multiple afrotherian arrivals. Primates and rodents are, for whatever reason, remarkably adept at colonizing distant landmasses; for instance, members of both groups managed to subsequently colonize the distant South American landmass from Africa whereas no other mammals have. Primate and rodent groups have also crossed the Mozambique Channel to colonize Madagascar.

In evaluating the hypothesis of extra-African origins for Afrotheria, the likelihood that all the different members of a diverse, closely related group of mammals such as Afrotheria would have coincidentally (and successfully) crossed the Tethys Sea independently, is almost infinitesimal. In spite of morphological resemblances, hypotheses that taxa such as hyopsodontids, phenacodontids and leptictids should be aligned with afrotherians now have to be rejected in spite of the fact that such extinct taxa have left behind no DNA. Affinities suggested on the basis of a few widely distributed morphological features are no different from the taxonomic arrangements that have been erroneously erected over the course of the last two centuries. Had perissodactyls, lagomorphs, pholidotans or solenodontids gone extinct and left no DNA perhaps their few resemblances could have bolstered claims that these taxa too were afrotherians, aligned with hyracoids, sengis, aardvarks and tenrecs, respectively? It may also be the case that previous phylogenetic studies simply did not adequately sample the morphological information that has been provided by the radiation of living and extinct mammals. This possibility would appear to be supported by the recent outgroup- and character-rich analysis of Wible *et al.* (2007), which, unlike the studies of Asher *et al.* (2003) and Zack *et al.* (2005), placed North American taxa such as *Hyopsodus*, *Meniscotherium* and *Phenacodus* far outside of the afrotherian radiation rather than placing them as stem paenungulates.

The oldest, undisputed members of afrotherian orders appear in the Palaeocene of north-west Africa, where the earliest proboscidean *Eritherium* is now known (Gheerbrant 2009). However, Palaeocene African mammals are still scarce, deriving largely from a few micromammal sites in Morocco. These have produced numerous insectivores (Gheerbrant 1992, 1994, 1995, Gheerbrant *et al.* 1998), and a few isolated specimens of creodonts (Gheerbrant 1995, Solé *et al.* 2009) and primates (Sigé *et al.* 1990), as well as some indeterminate fragments of condylarth teeth. It is possible that some of the late Palaeocene insectivores are actually aligned with the tenrecid-chrysochlorid clade (Seiffert 2010, Goswami *et al.* 2011) but it is not yet possible to identify possible late Palaeocene stem members of the Tubulidentata on the basis of isolated teeth, because the only undoubted fossil aardvarks appear in the early Miocene with essentially modern, specialized, enamel-less, peg-like teeth (MacInnes 1956). One possibility is that the order traces back to the peculiar ptolemaiids, which are now known from the late Eocene, early Oligocene, and early Miocene of Africa (Cote *et al.* 2007, Seiffert 2007).

The remarkable diversity that is now evident among Palaeocene and early Eocene proboscideans (Gheerbrant *et al.* 2002, 2005, Gheerbrant 2009) is consistent with molecular estimates of divergence dates within Paenungulata, which would place proboscidean origins well back into the Palaeocene (Springer *et al.* 2003). For the next 35 million years or so, paenungulates would dominate Africa's large mammal fauna with morphologically diverse hyracoids, proboscideans and the extinct embrithopods occupying a variety of browsing and grazing niches. Afrotherian dominance was first challenged (apparently, not too seriously) by a trans-Tethyan dispersal of the semi-aquatic anthracotheriid artiodactyls, which first appear in Africa as fossils during the earliest part of the late Eocene, about 36 mya. None the less, paenungulates continued to be the dominant large-bodied herbivores in Africa until at least the latest Oligocene, about 25 mya, shortly after which the first major exchange of fauna took place between Africa and Eurasia. There is still no evidence for any other artiodactyls or perissodactyls having dispersed from northern continents before this time (Kappelman *et al.* 2003, Rasmussen & Gutierrez 2009). There is still no evidence for lagomorphs or eulipotyphlans in Africa until the Miocene. No other group of placental mammals is known to have existed in Africa before the Afrotheria. Still important elements of the continent's fauna, they represent a continuous presence for at least 50 million years and perhaps as much as 80 million years.

The molecular and palaeontological data provide a broad view of the phylogenies and biogeographical relationships of the different clades making up the Afrotheria, but these data fail to tell us much about the ecological relationships of current afrotherians. Two features of the extant afrothere fauna are striking. First, compared with other placental radiations, species diversity is low (with the exception of the tenrecs on Madagascar and, to a lesser degree, the golden-moles in southern Africa). Secondly, the extant species are

Comparisons between an Afrothere and an unrelated species with convergent features: Marsupial Banded Ant-eater *Myrmecobius fasciatus* (*above*) and Giant Sengi *Rhynchocyon* sp. (*below*).

very highly specialized in terms of their morphology or ecology or both, and this specialization is probably related to low species diversity and to the advantages, for such lineages, of a long head-start in adapting to difficult niches.

Currently, there are about 78 species of extant afrotheres, including three elephants (though there is some controversy regarding current recognition of two African species), four sirenians (dugongs and manatees), five hyraxes, one aardvark, 15 sengis, 30 tenrecs and 21 golden-moles. As was pointed out earlier, many, if not most of these, have ecologically convergent forms on other continents, but some do not and there are generally quite plausible reasons for both the presence and the lack of convergences. For example, the afrotherian golden-moles are remarkably similar to the marsupial moles (genus *Notoryctes*) and various Holarctic talpid moles, adaptive niches in which an early occupation of the continent has to be particularly advantageous, given how much body modification is necessary. The afrotherian Aardvark is broadly comparable to the Giant Anteater *Mymecophaga tridactyla* of South America in size and habitat, while afrotherian hyraxes, particularly the higher altitude forms, have many ecological similarities with the pikas (family Ochotonidae) in Eurasia and North America. The three afrotherian species of Potamogalinae are convergent with various aquatic soricid shrews and desmans in the Holartic region and with the marsupial Duck-billed Platypus *Ornithorhynchus anatinus* – in Australia. Interestingly, the only tenrecs on the African continent are the three highly specialized aquatic potamogales while the 27 others are confined to the island of Madagascar, where they occupy ecological positions similar to a diversity of Australasian marsupials, Caribbean solenodons and a variety of shrews and hedgehogs from other areas and continents. Although the larger sengis (*Rhynchocyon*) superficially resemble bandicoots and the solenodons of Cuba and Espanola, the smaller species (e.g. *Elephantulus*) do not have many resemblances with other small mammals. In addition, sengi life history sets them apart from any other group of mammal, with features that are uniquely African and might best be thought of as a cross between a miniature antelope and small anteater! Despite such specializations, sengis occupy the extremes of terrestrial habitats – from gravel plains of the Namib Desert and boulder fields in north-west Africa to tropical forests of central and eastern Africa.

All the smaller afrotheres continue to be restricted to the Afro-Arabian region, plus Madagascar, but the two clades with representatives with the largest body mass dispersed to other continents. The elephants were well represented in North America and Asia and the sirenians radiated into the world's tropical oceans at a very early date, in addition to the North Pacific, and the Amazon River system. Although elephants and sirenians are wide-ranging, they are both morphological specialists. In addition, sirenians are obligate aquatic herbivores and as such have no ecological equivalents outside the afrotherian radiation.

The palaeontological record suggests that the highly specialized nature of extant afrotheres may not have been so marked in the past. For example, the diversity of sengis in the Miocene included six subfamilies, which included several herbivores, based on the morphology of their dentition. Only two subfamilies are extant, and all taxa in these two subfamilies have retained somewhat hypsodont dentition as well as a caecum, which strongly implies herbivorous ancestors. The herbivorous macroscelideans became extinct in the Mio-Pliocene; perhaps they could not effectively compete with newly arriving rodents, hares and ungulates coming in from the north. The extant sengis seem to have escaped competition with these invaders by secondarily becoming invertebrate specialists, especially anteaters, where a head-start in adapting to ant chemistry must be an advantage. Likewise, the palaeontological history of hyraxes suggests that only those species that were rupicolous or arboreal were able to escape competitive extinction with hares, rodents and ungulates.

A similar argument does not explain the high species diversity, in the very recent past, of elephants and dugongs and manatees compared with their current sorry status, and most available evidence suggests that prehistoric humans might have played a decisive role in some of their extinctions. The diversity of tenrecs on Madagascar apparently suffered less from the impact of incoming rodents, undoubtedly because there was no competition for food between the herbivorous rodents and insectivorous tenrecs. Golden-moles similarly escaped competition from other insectivores by being fossorial. It thus appears that an important factor in the relatively small number of extant afrotheres is their ecological specialization, which allowed them to avoid extinction in the face of invading faunas from outside Africa. Although this specialization has served the extant afrotheres well in the past, their low species diversity, often accompanied by highly restricted distributions and ecological specialization, makes many forms especially vulnerable to another wave of extinctions – at the hands of humans. This is especially the case with the potamogales, several golden-moles and the forest-dwelling sengis. These taxa all occupy highly restricted habitats that are being increasingly degraded by human activities. Because of the low species diversity of the extant afrotheres, extirpations will have an especially severe impact on an already depauperate group.

Jonathan Kingdon, Erik R. Seiffert, Blair Hedges & Galen Rathbun

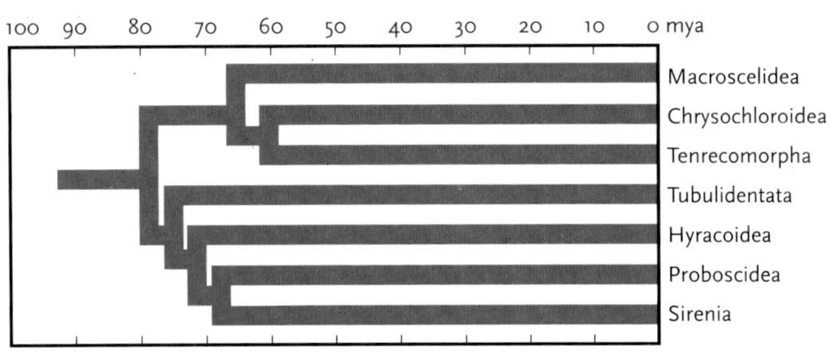

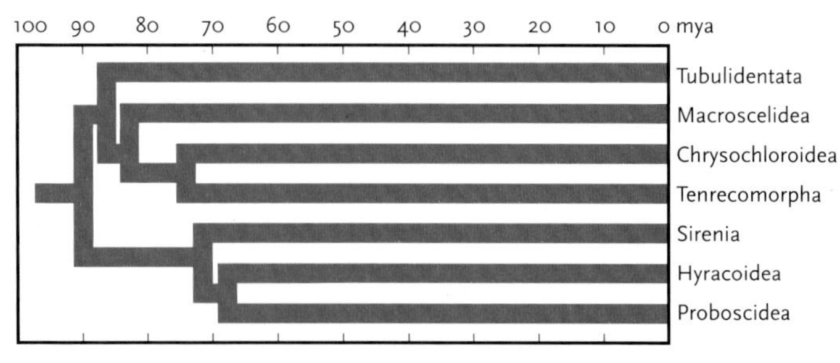

Tentative phylogenetic relationships of afrotherian mammals (*left*) based on a combined analysis of DNA from living species and morphology of living and extinct species (after Seiffert 2007), and (*right*) based on analysis of genomic data alone (after Meredith *et al.* 2011).

Cohort PAENUNGULATA

Cohort Paenungulata Simpson 1945. Bulletin of the American Museum of Natural History 85: 131.

A close relationship between hyraxes, manatees and dugongs, elephants and the extinct Palaeogene order Embrithopoda was first explicitly recognized by Gregory (1910), thanks in large part to fossil discoveries made in the late nineteenth and early twentieth centuries in the Fayum Depression of northern Egypt (Andrews 1906). Simpson eventually coined the name Paenungulata for this clade, but also included a variety of other extinct taxa whose affinities are now believed to lie elsewhere. The taxon Paenungulata has since been used by different authors to include very different assemblages of living and extinct placentals (Lucas 1993), but is now generally recognized as the afrotherian radiation that produced the orders Hyracoidea, Sirenia, Proboscidea, and the extinct Embrithopoda and Desmostylia (e.g. Gheerbrant *et al.* 2005a). In the past, many morphologists aligned paenungulates with perissodactyls in a clade called Altungulata or Pantomesaxonia, and some authorities have even favoured a closer relationship of hyracoids to perissodactyls than to tethytherian paenungulates (sirenians and proboscideans), but molecular data have now firmly established that the derived morphological features that perissodactyls share with paenungulates are evolutionary convergences (Springer *et al.* 2004).

The tiny Palaeocene proboscidean *Eritherium*, from Morocco, is the oldest undoubted paenungulate (Gheerbrant 2009). More derived proboscideans (*Daouitherium* and *Phosphatherium*) and the oldest known hyracoid (*Seggeurius*) have been found in earliest Eocene sediments of the same basin (Ouled Abdoun) (Gheerbrant *et al.* 2003, Gheerbrant *et al.* 2005b). By the earliest Eocene, hyracoids are morphologically distinct and there is already considerable morphological diversity within Proboscidea, suggesting that the evolutionary histories of both clades will eventually be traced much farther back into the early Palaeocene. As yet basal sirenians are missing from the early African record, and the oldest forms are from the early middle Eocene of Jamaica (Savage *et al.* 1994). However, by that time, sirenians already exhibit clear morphological adaptations for an aquatic existence.

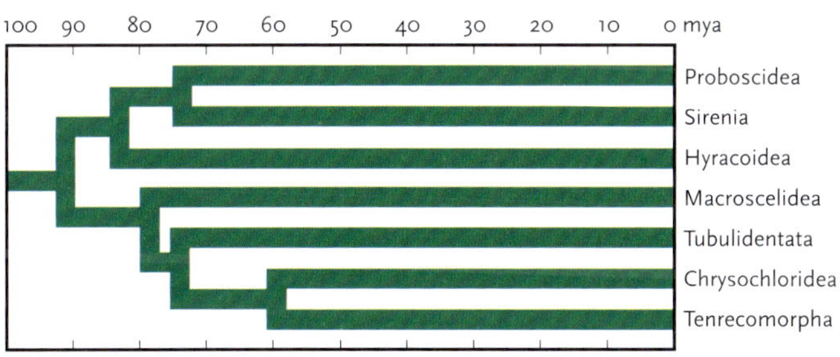

Alternative tree of phylogenetic relationships of afrotherian mammals based on a combined analysis of DNA from living species and morphology of living and extinct species, showing recovery of Tethytheria within Paenungulata (after Asher *et al.* 2003).

With the recognition of Afrotheria, a number of the morphological features that have been proposed as evidence for paenungulate monophyly now must be re-evaluated. For instance, one of the key features that was thought to align paenungulates to the exclusion of other 'ungulates' is a serial arrangement of the carpal and tarsal bones (i.e. no contact between the astragalus and the cuboid in the foot, and no contact between the lunar and unciform in the wrist) (Rasmussen *et al.* 1990). However, there is also no astragalar–cuboid contact in most other afrotherians (aardvarks, tenrecs, golden-moles), and there is no lunar–unciform contact in tenrecs, golden-moles and primitive sengis or elephant-shrews. As such, the serial carpus and tarsus could either be primitive within Placentalia or could represent afrotherian synapomorphies, but these characters can no longer be confidently interpreted as paenungulate synapomorphies. Other features, such as testicondy, a zonary placenta, the cup-like astragalar cotylar fossa (which articulates with an enlarged medial malleolus of the tibia), enlargement of the central upper incisors and caudal extension of the jugal to the glenoid fossa, are also seen in various other non-paenungulate afrotherians, suggesting a more ancient origin for these characters. Placement of the orbit over the premolars, which is seen in living and extinct tethytherians and extant procaviid hyracoids, is absent in primitive fossil hyracoids and so likely evolved convergently within Paenungulata.

Paenungulates can be distinguished from other afrotherians by a number of derived dental features and muscular arrangements (Shoshani 1993), in addition to postcranial characters such as loss of the clavicle and metacromion, an increase in the number of rib-bearing thoracic vertebrae, loss of the humeral entepicondylar foramen (though present, perhaps secondarily, in the early proboscidean *Numidotherium*), and, possibly, wide and flaring iliac alae. Amastoidy remains as a probable cranial synapomorphy of Paenungulata, while other possible synapomorphies include loss of lacrimal–palatine contact, presence of an alisphenoid canal, presence of a piriform fenestra rostral to the petrosal, a weakly excavated subarcuate fossa and a vertical to anterior orientation of the mandibular ascending ramus. However, these, and many other, candidate cranial synapomorphies of Paenungulata must be tested with fossil evidence from early stem and crown paenungulates.

Erik R. Seiffert

Order HYRACOIDEA – Hyraxes

Hyracoidea Huxley, 1869. An Introduction to the Classification of Animals. Churchill, London. viii + 147 pp.

Myology of *Dendrohyrax dorsalis.*

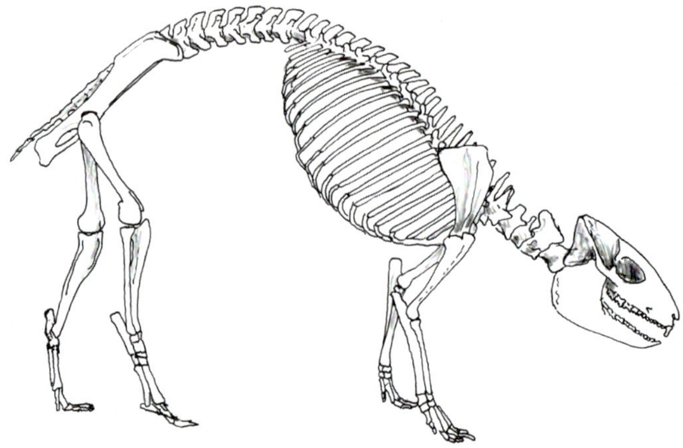

Oligocene hyracoid, *Saghatherium.*

The order Hyracoidea includes five families: the wholly extinct families Geniohyidae, Saghatheriidae, Pliohyracidae and Titanohyracidae (Pickford 2004), and the family Procaviidae, which contains both living and extinct genera (Hahn 1934, Meyer 1978, Jones 1984, Carroll 1988). The extant taxa include three genera: *Dendrohyrax* (three species), *Heterohyrax* (one species) and *Procavia* (one species), although the number of species in each genus is a matter of contention (Bothma 1971, Schlitter 1993, Shoshani 2005).

Hyraxes are small mammals (1.8–5.5 kg), the size of a guinea-pig, but without a visible external tail. The ears are small and rounded. Colour varies from light grey to dark brown. In the centre of the back there is a dorsal spot, with long (probably scent-dispersing) hairs that are of different colour than the rest of the body (white, cream, pale orange, russet-brown, or black), and that surround a glandular structure. The degree of piloerection of the dorsal hairs functions as a signal of alarm and threat while the dorsal gland secretions have general social, and possible sexual, functions (Sale 1970a). Length of the naked dorsal gland has been used as a diagnostic, albeit not always consistent, feature for distinguishing species (Bothma 1971). Tactile hairs or vibrissae up to 80 mm long are widely distributed over the body (Sale 1970a). The limbs are plantigrade; the forefoot has four digits and the hindfoot three. All the digits end in broad, flat hoof-like nails. The inner toe of the hindfoot has a long curved claw-like nail utilized for grooming. The soles are naked and covered by thick epithelium kept moist by glands (Dobson 1876, Sokolov & Sale 1981). Special arrangements of muscles enable shaping of the soles into air-tight cups for improved gripping. There are numerous thoracic and lumbar vertebrae (27–30) and 20–21 bear ribs. There are no clavicles (Sclater 1900). Nipples usually include one pectoral pair and two inguinal pairs, but variations are known (see profile Procaviidae).

In profile, the procaviid skull is relatively high, with the mandible contributing more height than the cranium; there is no prominent sagittal crest, and the rostrum is blunt. Crania of adults measure 65–118 mm long, 30–35 mm tall and 50–63 mm wide. Unique to Hyracoidea, the parietal contributes to the dorsal postorbital process (Meyer 1978); the lingual process of the hyoid apparatus is derived from the basihyoid (Flower & Lydekker 1891, Gasc 1967). The cranium has the postorbital bar complete (*Dendrohyrax*) or incomplete (*Heterohyrax* and *Procavia*), and temporal ridges that converge or form a sagittal crest (*Procavia* and *Heterohyrax*), or are far apart (*Dendrohyrax*). The dental formula is $I\ ^{1}/_{2}$, $C\ ^{0}/_{0}$, $P\ ^{4}/_{4}$, $M\ ^{3}/_{3} = 34$, although *Procavia* often has the lower first premolar absent. The upper incisors are long, growing from persistent pulps, and are uniquely triangular in cross-section; the lower incisors are chisel-shaped. The cheekteeth are lophodont and separated from the incisors by a large diastema (Thomas 1892). Molars show some general similarities with perissodactyls (Osborn 1907, McKenna 1975, Fischer 1989). Fossil forms retained primitive double-rooted canines and have dental formulae similar to primitive eutherian mammals, $I\ ^{3}/_{3}$, $C\ ^{1}/_{1}$, $P\ ^{4}/_{4}$, $M\ ^{3}/_{3} = 44$ (Osborn 1907).

Members of the order are characterized by several unusual features, including: a stomach divided into two chambers (a cardiac, non-glandular section and a pyloric, glandular section); two large conical supplementary caeca (a structure encountered frequently in birds); a highly subdivided liver; absence of a gall bladder; duplex uterus; and abdominal testes (Flower & Lydekker 1891, Sclater 1900, Glover & Sale 1968, Rahm & Frewein 1980). The unbranched caecum acts as a fermentation chamber that produces large amounts of volatile fatty acids that serve as an energy source, and about which there has been much discussion in the literature (Clemens 1977, Clemens & Maloiy 1978, Engelhardt *et al.* 1978, Leon 1980, Eloff & VanHoven 1985). Hyraxes have very efficient renal function, and have a high capacity for concentrating urea and electrolytes and excreting large amounts of undissolved calcium carbonate (Meltzer 1973, Rübsamen *et al.* 1982). All species defaecate in specific spots

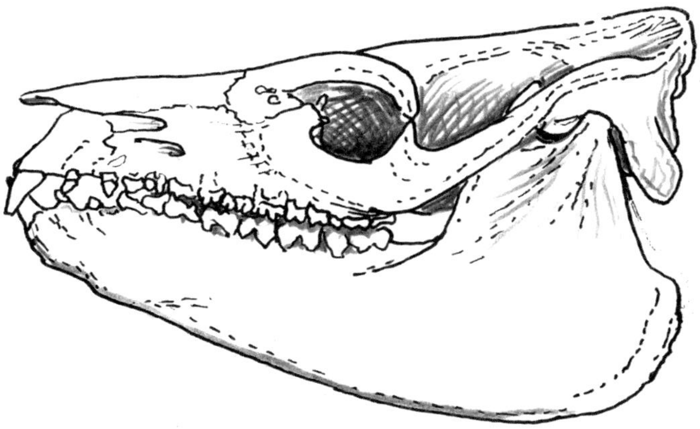

Skull of *Megalohyrax*.

and many species use latrines (Kingdon 1997). The accumulated deposits of crystallized calcium carbonate whiten the cliff faces below latrines; these crystals were used as medicine (hyraceum) by several South African tribes and by Europeans (Hahn 1934). Hyraxes have a poor ability to regulate their body temperature and a low metabolic rate for their body size. Body temperature is maintained mainly by behavioural thermoregulation, including gregarious huddling, long periods of inactivity and basking (Taylor & Sale 1969, Sale 1970b, Bartholomew & Rainy 1971). The pupil of the eye in several species (*Procavia*, *Heterohyrax*) has an umbraculum, a shield that allows a basking individual to stare into the sun (Millar 1973) and thereby detect aerial predators. All three genera are highly vocal (Fourie 1977, Hoeck 1978a). A long gestation period (6–8 months) is another peculiar characteristic (Sale 1965a).

The early fossil record of Hyracoidea is restricted to Afro-Arabia and begins in the earliest Eocene (~55 mya) of Morocco, where a single lower molar of the genus *Seggeurius* has been found (Gheerbrant *et al.* 2003). Fragmentary jaws and a few upper teeth of *Seggeurius* and *Microhyrax* are known from younger (~50 million-year-old) sites in Algeria (Court & Mahboubi 1993, Tabuce *et al.* 2001). *Seggeurius* and *Microhyrax* were small species and had very simple, unmolarized premolars; they lack the coronoid canal found in later hyracoids. Much larger contemporaries of these early forms are only known from a few upper molars (Sudre 1979), and suggest considerable early diversity. The latest Eocene was witness to the acme of the hyracoid radiation; a single locality of this age in Egypt has revealed remains of eight hyracoid species ranging in size from small procaviids to small horses. This ~34 million-year-old community shows great morphological disparity, and included cursorial bovid-like forms, tapir-like species and a few taxa with greatly inflated, hollowed out mandibles (Meyer 1978, DeBlieux *et al.* 2006). One of the latter genera, *Thyrohyrax*, was otherwise very similar to later fossil procaviids in its craniodental and postcranial morphology and might have been broadly ancestral to that group. Hyracoids remained the dominant small- to medium-sized mammalian herbivores in Afro-Arabia until the early Miocene, when many species were replaced by immigrant artiodactyls and perissodactyls (Kappelman *et al.* 2003). Early and middle Miocene species included archaic large genera such as *Afrohyrax* and *Brachyhyrax* (Pickford 2004), and more specialized genera such as *Parapliohyrax* (Pickford 1996, 2003), and the smaller, more procaviid-like *Prohyrax*. Some hyracoid lineages migrated out of Africa in the later Neogene (e.g. Chen 2003), but ultimately went extinct. Several specimens of *Heterohyrax* have been described from the late Miocene in Namibia, including *Heterohyrax auricampensis*, which has an estimated age of 10–9 mya, and whose molar teeth can barely be distinguished from living *Heterohyrax* species (Rasmussen *et al.* 1996).

Jeheskel Shoshani, Paulette Bloomer & Erik R. Seiffert

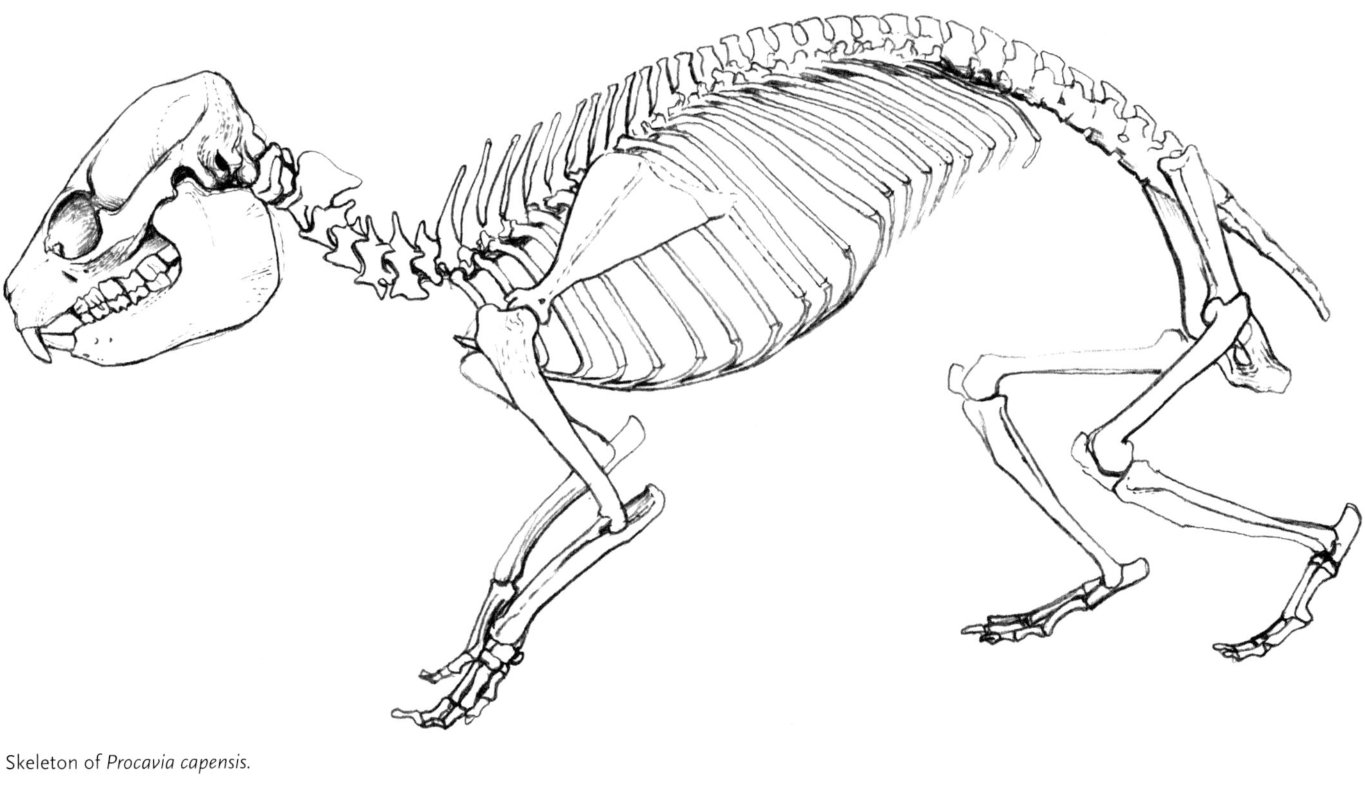

Skeleton of *Procavia capensis*.

Family PROCAVIIDAE

HYRAXES

Procaviidae Thomas, 1892. Proc. Zool. Soc. Lond. 1892: 51.

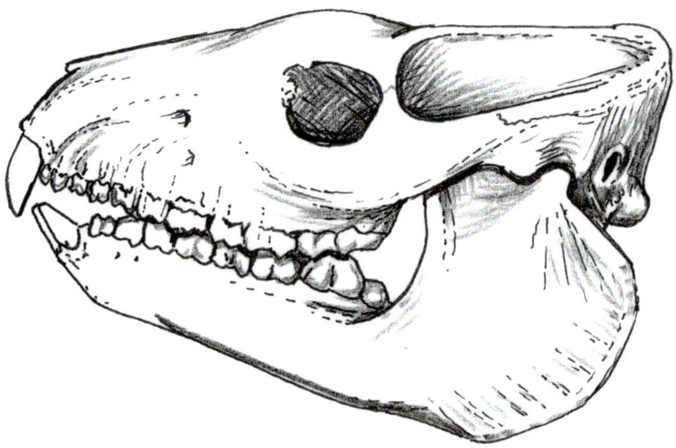

Skull of *Myohyrax hendeyi*.

The odd appearance of the hyrax has caused considerable taxonomic confusion. The superficial similarity of hyraxes to rodents led Storr (1780) mistakenly to link them with guinea pigs of the genus *Cavia*, and he thus placed Rock Hyraxes in the genus *Procavia* (meaning 'before the guinea pigs') and in the family Procaviidae. Later, the mistake was discovered and the group was given the equally misleading name of hyrax, which means 'shrew mouse'. The name 'dassie', colloquially used in English in parts of southern Africa, is Afrikaans and is derived from the Dutch *das* or badger, the name applied by the Dutch settlers in the Cape to this species.

In Phoenician and Hebrew, hyraxes were/are known as *shaphan*, meaning 'the hidden one'. Some 3000 years ago, Phoenician seamen explored the Mediterranean, sailing westward from their homeland on the coast of Syria. They found land where they saw small mammals, which they thought were hyraxes, and so they called the place 'I-shaphan-im' – Island of the Hyrax. The Romans later modified the name to Hispania. But the animals were really rabbits, not hyraxes, and so the name 'Spain' derives from faulty observation (Hahn 1934)!

Most authors recognize three distinct extant genera: *Procavia*, *Heterohyrax* and *Dendrohyrax* (Allen 1939, Swynnerton & Hayman 1950, Roberts 1951, Bothma 1971, Hoeck 1978a, Meester *et al.* 1986, Schlitter 1993, Shoshani 2005), although *Heterohyrax* has sometimes been treated as a subgenus of *Dendrohyrax* (Ellerman *et al.* 1953, Roche 1972, Ansell 1978, Corbet 1978). Generic delineation has traditionally been based on differences in dentition and skull characteristics, although colouration, mammary formulae, penis structure, behaviour and ecological differences also distinguish the three genera (see Table 6, p. 151). Numerous species and subspecies of hyrax have been described, and there remains little consensus regarding the number of species in each genus (Bothma 1971, Schlitter 1993, Shoshani 2005). Preliminary molecular phylogenetic data indicate high levels of intraspecific variation within *Procavia* and *Heterohyrax*. The three genera are genetically distinct and *Dendrohyrax* appears to be basal (Prinsloo & Robinson 1992, Prinsloo 1993).

Sketches of *Procavia capensis habessinicus*.

Table 6. Major differences among the three genera of hyrax (modified after Shoshani 1992, based on sources given below).

	Procavia Rock hyrax	*Heterohyrax* Bush hyrax	*Dendrohyrax* Tree hyrax
Distribution	Africa and the Middle East	Africa only	Africa only
Habitat	Rocky outcrops	Rocky outcrops and trees	Trees
Food	More grazer than browser	More browser than grazer	Mostly browser
Activity period	Diurnal	Diurnal	Nocturnal
Sociality	Gregarious, associates with *Heterohyrax*	Gregarious, associates with *Procavia*	Less gregarious, little association with other genera
Head + body length (cm)	39–58	32–56	40–55
Body weight (kg)	1.8–5.5	1.3–3.6	1.6–4.0
Body colour	Yellowish to greyish, brown and dark; buff underparts	Brown mixed with white and black; whitish underparts	Dark brown to blackish; greyish underparts
Facial marks	Dark patch around eye	White patch above eye	No marks around eye
Cranium, post-orbital bar	Not complete	Not complete	Complete
Cranium, profile on top	Flat	Flat	Depression above orbit
Cranium, temporal lines	Converge	Do not converge, with narrow gap between them	Do not converge, with wide gap between them
Cheekteeth	Hypsodont (high crown)	Brachyodont (low crown, intermediate)	Brachyodont (lower crown than *Heterohyrax*)
Cheekteeth	Length of upper molar toothrow (M^{1-3}) greater than length of premolar toothrow P^{1-4}	Length of upper molar toothrow (M^{1-3}) about equal to length of premolar toothrow P^{1-4}	Length of upper molar toothrow (M^{1-3}) shorter than or about equal to length of premolar toothrow P^{1-4}
Lower first premolar	Sometimes absent	Present	Present
Mandible, ascending ramus	Broad	Narrow	Narrow
Mandible, coronoid process	Same level as condyle or slightly above it	Higher than in *Procavia*	Higher than in *Procavia*
Uterus type	Duplex (similar to that found in rodents, lagomorphs and Aardvark)	Bicornuate (similar to that found in ungulates)	Bicornuate (similar to that found in ungulates)
Penis: distance from anus	34–36 mm	65–82 mm	17–25 mm
Penis: form	Simple, widens towards tip, flattened in cross-section	Complex, with an appendage at its tip, round in cross-section	Simple, curved, narrow towards tip, flattened in cross-section
No. of nipple pairs	1 pectoral, 2 inguinal	1 pectoral, 2 inguinal (or 2 inguinal pairs only)	1 inguinal (but sometimes 1 pectoral only, or 1 inguinal and 1 pectoral, or 2 inguinal)
Litter size	2–4	2–4	1–2

Sources: Hahn 1934, Mendelssohn 1965, Coetzee 1966, Glover & Sale 1968, Dorst & Dandelot 1970, Bothma 1971, Roche 1972, Meltzer 1973, Hoeck 1978a, 1982a, 1989, Jones 1978, Corbet 1979, Smithers & Wilson 1979, Hoeck *et al.* 1982, Olds & Shoshani 1982, Meester *et al.* 1986, Dor 1987, Yom-Tov & Tchernov 1988, Kingdon 1997, Barry & Shoshani 2000, and unpublished observations of J. Shoshani and J. M. Milner

Procavia is the most widely distributed of the genera and the only representative outside the African continent, being present in parts of the Arabian Peninsula and the Middle East. *Procavia* and *Heterohyrax* are rock dwelling, gregarious and diurnal, while *Dendrohyrax* is arboreal, less gregarious and nocturnal (Bothma 1964, 1971, Hoeck 1978a). *Procavia* is predominantly a grazer whereas *Heterohyrax* and *Dendrohyrax* mostly browse (Turner & Watson 1965, Hoeck 1975, Lensing 1983, Kingdon 1971, 1997). These dietary preferences are reflected in the dentition, with *Procavia* having hypsodont molars, and *Heterohyrax* and *Dendrohyrax* having brachyodont molars. *Procavia* also has an upper molar toothrow (M^{1-3}) that is longer than that of the premolar toothrow (P^{1-4}, with P^1 sometimes absent), whereas in the other genera the length of P^{1-4} is just less than or equal to length of M^{1-3} (*Heterohyrax*) or exceeds the length of M^{1-3} (*Dendrohyrax*) (Bothma 1971, Meester *et al.* 1986). The penis of *Heterohyrax* is more complex than those of *Procavia* and *Dendrohyrax*, and the distance between the anus and the preputial opening in an adult ♂ is 65–82 mm, which is two to three times longer than in the other genera (Coetzee 1966, Hoeck 1978a); mean distance between anus and preputial opening is 35 mm in *Procavia* and 20 mm in *Dendrohyrax* (Coetzee 1966, Hoeck 1978a).

The origin of Procaviidae remains something of a mystery. Eocene–Oligocene *Thyrohyrax* is probably more closely related to procaviids than to other Palaeogene genera (Rasmussen & Simons 1988, Seiffert 2003), but this genus also exhibits some bizarre specializations, such as an internal mandibular chamber, that appear to exclude it from procaviid ancestry. In some ways Miocene *Prohyrax* helps to bridge the morphological gap between *Thyrohyrax* and extant procaviids, but it too appears to be only a distant relative of the procaviids. Various extinct members of the procaviid crown group have been identified, including species of *Heterohyrax* (Rasmussen *et al.* 1996), *Procavia* (Churcher 1956) and the extinct genus *Gigantohyrax* (Kitching 1965), which was about three times larger than the extant procaviids.

Jeheskel Shoshani, Paulette Bloomer & Erik R. Seiffert

GENUS *Dendrohyrax*

Tree Hyraxes

Dendrohyrax Gray, 1868. Ann. Mag. Nat. Hist., ser. 4: 1–48.

This genus includes three species of hyraxes: the Eastern Tree Hyrax *D. validus*, from montane parts of East Africa; the Western Tree Hyrax *D. dorsalis*, from West Africa and central Africa; and the Southern Tree Hyrax *D. arboreus*, ranging from C Kenya to the Eastern Cape Province in South Africa. The validity of *D. validus* as a distinct species from *D. arboreus* has been questioned by some authors (e.g. Bothma 1971), and Shoshani (2005) considered the latter species to include *D. validus*. Compared with other hyraxes, all members of the genus are arboreal, less gregarious and nocturnal (Bothma 1964, 1971, Hoeck 1978a). Calls of all three species show a degree of geographical variation (Roberts 1999). Length of upper molar toothrow (M^{1-3}) is approximately the same length as that of upper premolar toothrow (P^{1-4}).

Paulette Bloomer

Dendrohyrax arboreus SOUTHERN TREE HYRAX (SOUTHERN TREE DASSIE)

Fr. Daman d'arbre; Ger. Baumschliefer

Dendrohyrax arboreus (A. Smith, 1827). Trans. Linn. Soc. Lond. 15: 468. Forests of Cape of Good Hope, Western Cape Province, South Africa.

Southern Tree Hyrax *Dendrohyrax arboreus ruwenzorii*.

Taxonomy Eight subspecies were recognized by Bothma (1971). All appear similar, but are distinguished by altitudinal range, habitat or geographic area, and possibly by vocalizations. Montane populations tend to be more distinctive in pelage characteristics. Taxonomic confusion has led to several subspecies having synonyms within the genera *Procavia* and *Heterohyrax*. Recognition of subspecies here is provisional, following Bothma (1971), pending a more detailed investigation employing vocalizations and molecular data. Shoshani (2005), who did not recognize subspecies, included Eastern Tree Hyrax *D. validus* as a synonym of this species. Synonyms: *adolfo-friederici*, *bettoni*, *braueri*, *crawshayi*, *helgei*, *mimus*, *ruwenzorii*, *scheelei*, *scheffleri*, *schubotzi*, *stuhlmanni*, *vilhelmi*. Chromosome number: 2n = 54 (Prinsloo & Robinson 1991).

Description Superficially guinea-pig-sized and shaped, with short, sturdy legs, no external tail, and soft, dense pelage. General colouration grey or brown, but quite variable, and may appear grizzled due to buff band below black tips of guard hairs. Individuals from high rainfall areas have darker pelage (Bothma 1967). Head often darker than rest of the body and ventral pelage very pale, creamy or white. Short white hairs fringe the ears. Black whiskers up to 80 mm in length, and a patch of long hairs on the eyebrows. Distinct creamy white dorsal patch of longer erectile hairs (45–50 mm) in the middle of the back marks the large bare dorsal apocrine gland. The length of the bare dorsal gland has been suggested as a diagnostic feature for this species (e.g. Bothma 1971, Jones 1978, both of whom cite a length of 23–30 mm). However, animals from Rwanda have a dorsal spot ranging in length from 17 to 48 mm (n = 19)(J. M. Milner pers. obs.), suggesting large overlap in length between Southern Tree Hyrax and Eastern Tree Hyrax and some overlap between large Southern Tree Hyrax individuals and Western Tree Hyrax *D. dorsalis*. Soles of feet are padded, generally with black skin although pink feet have been reported from Ngorongoro (H. N. Hoeck pers. comm.). Females have varying number of nipples: usually there is one inguinal pair, but sometimes there is only one pectoral pair, or one inguinal pair with the addition of one pectoral pair, or otherwise two pairs inguinal (Rudnai 1984a).

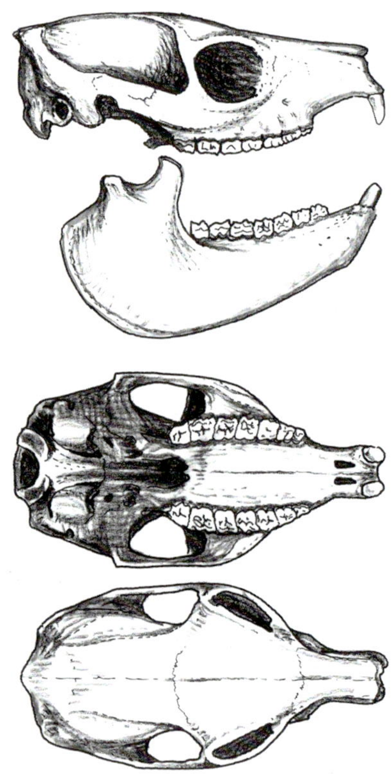

Lateral, palatal and dorsal views of skull *Dendrohyrax arboreus*.

Upper incisors are widely separated, long, curved and triangular in section. Outer pair of lower incisors retains tricuspid condition in adults. Upper diastema 15–17 mm, much wider than in *Procavia* but similar to *Heterohyrax*.

Geographic Variation

D. a. adolfi-friederici (including *schubotzi* and *helgei*): montane E DR Congo, Rwanda, Burundi, SW Uganda.

D. a. arboreus: Eastern Cape Province and KwaZulu–Natal (South Africa), C Mozambique.

D. a. bettoni (including *scheffleri* and *vilhelmi*): S Kenya.

D. a. braueri: NW Zambia and S DR Congo; also presumably the form in NE Angola.

D. a. crawshayi: C Kenya.

D. a. mimus: NE Zambia, W Malawi, WC Tanzania.

D. a. ruwenzorii: Rwenzori Mts. In open rocky habitat; one pair inguinal and one pair pectoral nipples.

D. a. stuhlmanni (including *scheelei*): SE DR Congo, Kenya, Tanzania. One pair inguinal and one pair pectoral nipples.

Kingdon (1971) drew attention to a possible hybridization zone between Southern Tree Hyrax and Western Tree Hyrax in S Uganda. He interpreted this in terms of a recent expansion of range, by the Western Tree Hyrax, eastwards as far as the Victoria Nile. Both species, as well as apparent hybrids, are known from this area.

Similar Species

D. dorsalis. Generally has shorter, coarser pelage; yellowish-white dorsal spot; longer dorsal patch (42–72 mm); rostrum naked; conspicuous white spot beneath chin; one pair inguinal mammae. Sympatric or parapatric in S Uganda and along the western border of DR Congo, where hydridization may occur.

D. validus. Yellow or russet-brown dorsal spot; naked rump patch 20–40 mm (though see notes above); one pair inguinal nipples. Isolated montane and coastal forests of Tanzania/Kenya, as well as Pemba and Zanzibar Is.

Heterohyrax brucei. Sympatric in NE DR Congo and NW Uganda. For further distinguishing characteristics see Table 6, p. 151.

Procavia capensis. Distributed widely in sub-Saharan Africa, with isolated population in North Africa, in a wide range of habitats. For further distinguishing characteristics see Table 6, p. 151.

Distribution Endemic to Africa. Patchily distributed in forested areas of central and eastern mainland Africa, from C Kenya as far south as Eastern Cape and KwaZulu–Natal provinces of South Africa. Occurs relatively widely in Uganda, C and SW Kenya, Tanzania, E and SE DR Congo, Rwanda and throughout Zambia (excluding the areas west of the Zambezi R.)(Kingdon 1971, Ansell 1978). Although occurs widely in Malawi (Ansell & Dowsett 1988), distribution east of L. Malawi is uncertain. Several authors (e.g. Skinner & Chimimba 2005) comment on the lack of records from Angola, but the species certainly does occur, having been noted earlier by Hayman (1963) and De Barros Machado (1969); distribution in the north-east of this country is mapped by Crawford-Cabral & Veríssimo (2005). Shortridge (1934) makes reference to this species from the Caprivi, but this appears unsubstantiated, and Smithers (1971) did not record this species from Botswana. May occur in NW Mozambique, otherwise confined to the central area south of the Zambezi R. (Smithers &

Dendrohyrax arboreus

Lobão Tello 1976). In South Africa, recorded from the southern and southern-central forests of KwaZulu–Natal and along the coast of the Eastern Cape, as far west as the Sundays R. (about 27° E). Also never recorded in Zimbabwe, although they almost certainly do occur (Smithers & Wilson 1979), or the Limpopo Province of South Africa.

Habitat Associated with well-developed woodland or forest. In South Africa occurs in Afromontane forests and better-developed forests and thickets of the Eastern Cape and KwaZulu–Natal provinces (Lawes *et al.* 2000). At the western coastal limit, occurs in milkwood-dominated coastal forests (Gaylard 1994), while further north in C Mozambique occurs in lowland evergreen forests as well as in the evergreen riverine forests of the Save R. (Smithers & Lobão Tello 1976, Skinner & Chimimba 2005). In East Africa also occurs in drier *Acacia* woodland and in rocky alpine and sub-alpine habitats. Across its range, found from sea level to sub-alpine areas of the Bufumbira and Rwenzori Mts (Kingdon 1971).

In both southern and eastern Africa this species is dependent on tree cavities, epiphytes or dense matted forest vegetation for shelter (Milner & Harris 1999b, Lawes *et al.* 2000, Gaylard & Kerley 2001, Skinner & Chimimba 2005). A decrease in numbers in southern Africa has been attributed to loss of structure within habitat, rather than forest size (Castley & Kerley 1993), although forest patches should be over 5 ha and less than 0.25 km from other larger forest patches to ensure a high probability of occupancy (Lawes *et al.* 2000). Preferred trees for denning are usually canopy trees in an intermediate stage of decay (i.e. >50% of the tree's material alive), and that have multiple cavity entrances. In the Eastern Cape den trees were 4–8 m in height (and often the tallest trees at a site), with diameters typically 40–80 cm and trunks angled between 45 and 67°; they were also preferred food trees. Preferred species include *Podocarpus falcatus*, *Schotia latifolia*, *Sideroxylon inerme*, *Rhus chirindensis*, *Andrachne ovalis* and *Apodytes dimidiata* (Gaylard & Kerley 2001). In East Africa selected trees are significantly larger (height and diameter) than other trees in the area. Preferred species in SW Kenya include *Podocarpus latifolius*, *Schefflera* spp., *Juniperus procera* and *Olea africana* (Milner 1994), and *Hagenia abyssinica* in the Virunga Mts (Milner & Harris 1999b), due to their tendency to form cavities and support epiphytes. Southern Tree Hyraxes have been recorded sheltering in termitaria in East Africa (Kingdon 197, Rudnai 1992).

Abundance Heard calling more frequently than they are seen. Southern Tree Hyraxes spend long periods inactive in the high canopy or tree holes and so often escape notice. Locally abundant in the Virunga Mts (13.4/ha; Milner & Harris 1999a) and Rwenzori Mts, where they live at a high density and many apparent family groups live close together (Kingdon 1971). Such high densities contrast strongly with other parts of East Africa. A rough estimate in SW Kenya suggests a density of at least 1.2/ha (Milner 1994). In southern Africa, where *D. arboreus* is considered rare (Lawes *et al.* 2000), relative density has been estimated using playbacks of recorded vocalizations, as well as by means of counts of latrines in cavity trees (i.e. Catch Per Unit Effort; Gaylard 1994): 0.07–0.29 latrines/man hour searching were found in the Eastern Cape Province of South Africa.

Adaptations Predominantly nocturnal, perhaps as an adaptation to human disturbance or diurnal avian predators (Rudnai 1984b), although in some parts of the range they are active by day. Generally quite inactive, likely a consequence of a low metabolic rate and a low-quality folivorous diet. Energy requirements are also minimized by a thick pelage and by thermoregulatory behaviour such as sun basking and the use of tree cavities, especially during inclement weather. In contradiction to this, ♂♂ in particular may be active in the canopy at night, despite low temperatures. *Dendrohyrax arboreus* is thought to be thermolabile, like other hyracoids.

Like other members of the genus, Southern Tree Hyraxes are arboreal and well adapted for climbing, having excellent agility and feet that have both a strong grip and can be easily supinated. They spend most of their time above ground level and feed principally in the high canopy (Gaylard & Kerley 1997). However, they have been observed on the ground foraging during daylight hours, giving birth and when sick (J. Rudnai, pers. comm.).

Most other adaptations are as for other hyracoids. Long white hairs around the large dorsal gland are erectile and raised under conditions of excitement (Sale 1970a). Hyraxes may assume a characteristic defensive position when threatened, turning the back and rump towards the enemy and spreading the hair around the dorsal gland to expose the naked glandular area. The animal also protects itself from smaller predators by furiously biting with its very large incisors. During aggressive interactions the upper lip is curled and the teeth are bared. Rubbing of the dorsal gland on branches has been observed in dry season, presumably as a means of territorial marking.

Foraging and Food Exhibits a dietary preference for mature foliage, hairy leaf petioles and woody tips of branches; eating poor quality, but abundant, items may minimize search effort whilst maximizing intake. In the Virunga Mts, leaves of *Hagenia abyssinica*, *Hypericum revolutum*, *Senecio maranguensis*, *Galium ruwenzoriense* and *Pleoppeltis excavata* formed the bulk of the diet (Milner & Harris 1999a). Important dietary species include *Podocarpus latifolius*, *Schefflera volkensii*, *Ilex mitis*, *Ficus natalensis*, *Acacia albida* and *Juniperus procera* in East Africa (Hoeck 1978a, Milner 1994) and leaves of *P. falcatus*, *Schotia latifolia*, *Cassine aethiopica*, *Euclea natalensis*, *Eugenia capensis zeyheri* in the Eastern Cape Province of South Africa (Gaylard & Kerley 1997). A low diversity of dietary plant species is consumed. In the Ngorongoro Crater, Southern Tree Hyraxes have also been recorded feeding on fig fruits (H. N. Hoeck pers. comm.). There is no evidence that they feed on insects. A primary feeding peak takes place in the evening, after dark; in the Virunga Mts, a second feeding peek occurs later in the night and appears to be more common among ♂♂, whilst ♀♀ are more likely to feed during the daytime.

Social and Reproductive Behaviour Southern Tree Hyraxes are solitary animals, although ♀♀ with offspring (often sitting on the mother's back) may be encountered; occasionally may be seen in family groups or pairs. The same ♂ has been observed with several ♀♀. In the Virunga Mts, home-range size varies between 150 and 2550 m^2, with young animals occupying the smallest areas (Milner & Harris 1999b). Tree density may influence range size. Female territories overlap, often with individuals of a different generation, possibly parents or offspring. Male territories overlap those of ♀♀, but the extent of male/male overlap is not clear. Seasonal variation in range size has not been investigated. Evidence from ranging patterns suggests that the mating system may be one of facultative polygyny; an exception to this general pattern is the semi-social rock-dwelling *D. a. ruwenzorii* that lives at an altitude of 3900 m, above the treeline.

Southern Tree Hyraxes defaecate in regular sites forming latrines, which are often found inside hollow trees. Each animal will use more than one latrine and several individuals may share each latrine. An average of 2.2 latrines per hyrax were found in the Virunga Mts (J. M. Milner pers. obs.). The social or territorial significance of latrines is unknown.

Southern Tree Hyraxes have a very loud and distinctive call, thought to be of territorial significance. It consists of a series of deep croaks (winding up) followed by a loud penetrating scream. Anecdotal reports suggest loud calls vary regionally, presumably by subspecies. Calls seem to be highly seasonal, and are heard regularly in the dry season in the early evening and middle of night. Calls can be heard as a wave passing through the forest as individuals respond to their neighbours. Both ♂♂ and ♀♀ call, ♀♀ only in the absence of a mate (Rudnai & Frere 2000). Whilst occasional calls may be heard outside the dry season, they tend not to elicit a wide response. Calling activity is not influenced by lunar phase. A large number of other less conspicuous sounds are also made (Rudnai & Frere 2000).

Courtship lasts 2–10 days in captivity (Rudnai 1998). Copulation has been observed in the dry season, preceded by the ♀ snarling at the advancing ♂, with the dorsal spot erect.

Reproduction and Population Structure Timing of the birth season is uncertain. Data from the Rwenzori Mts suggest a birth peak in Apr/May (Sale 1969), which agrees with observations of mating in Oct in the Virunga Mts (Milner & Harris 1999a) and the seasonal nature of calling and territorial marking. However, O'Donoghue (1963) found no apparent birth season in the Rwenzori Mts. Taylor (1998) collected two gravid ♀♀ in KwaZulu–Natal during Mar and Apr with two and three foetuses, respectively.

The gestation period is very long for such a small mammal, in the order of 7.5–8 months (Kingdon 1971). At birth, young are highly precocious, and weigh approximately 200 g. Young are weaned at 3–7 months and reach maturity at 20–30 months (Rudnai 1998). The reproductive strategy appears to be one of few young, with small litter-sizes (1–3; Rudnai 1984a, Taylor 1998) and low mortality rates, leading to a low ratio of juveniles to adults (0.2 : 1 in Virunga Mts; Milner & Harris 1999a). Miscarriage rate in captive animals is high (Rudnai 1998). A female-biased sex ratio (0.3 ♂♂ to 1 ♀ [n = 17] in the Virunga Mts) is presumably caused by differential dispersal and mortality rates. Longevity is at least 14 years in captivity (J. Rudnai pers. comm.).

Predators, Parasites and Diseases Known predators include Leopards *Panthera pardus*, African Golden Cats *Profelis aurata*, Servals *Leptailurus serval*, Caracals *Caracal caracal*, genets (*Genetta* spp.), African Civets *Civettictis civetta*, pythons (*Python* spp.), African Hawk-eagles *Hieraaetus spilogaster*, Crowned Hawk-eagles *Stephanoaetus coronatus* (e.g. see Boshoff *et al.* 1994) and large owls.

Southern Tree Hyraxes are subject to infection by a number of ectoparasites from the family Mallophaga (Roberts 1951). These parasites are host-specific and the presence of particular species of the parasitic family Mallophaga, in particular, may be used to identify hyracoid species (Roberts 1951). One captive animal died of *Toxoplasmosis* (Olubayo & Karstad 1983).

Conservation IUCN Category: Least Concern. CITES: Not listed.

Threats to this species include snaring for meat and skins (see, for example, Milner 1994), although the extent of this has not been evaluated, and loss of forest habitat. Populations of Southern Tree Hyraxes are now very fragmented, with limited gene flow among populations. Overall, the species has a relatively wide distribution, is present in several protected areas, and is generally not believed to be at immediate risk of extinction.

Measurements

Dendrohyrax arboreus
HB: 502 (441–566) mm, n = 14
T: 0 mm
HF: 62 (45–69) mm, n = 14
WT: 2.00 (1.17–2.65) kg, n = 14
Rwanda, excluding juveniles (Milner & Harris 1999a)

HB: 475 (428–520) mm, n = 9
T: 0 mm
HF: 63 (59–67) mm, n = 10
E: 32 (29–36) mm, n = 8
Southern Africa (Bothma 1967)

Key References Gaylard 1994; Gaylard & Kerley 1997, 2001; Hoeck 1978a; Kingdon 1971; Milner & Harris 1999a, b; Rudnai 1984a, b, 1998; Skinner & Chimimba 2005.

Jos M. Milner & Angela Gaylard

Dendrohyrax dorsalis WESTERN TREE HYRAX

Fr. Daman d'arbre; Ger. Westlicher Baumschliefer

Dendrohyrax dorsalis (Fraser, 1855). Proc. Zool. Soc. Lond. 1854: 99 [1855]. Bioko, Equatorial Guinea.

Taxonomy Six subspecies have been recognized (Rahm 1969, Bothma 1971, Jones 1978), though these are in need of revision. Synonyms: *adametzi*, *aschantiensis*, *beniensis*, *brevimaculatus*, *congoensis*, *dorsalis*, *emini*, *latrator*, *marmota*, *nigricans*, *rubriventer*, *stampflii*, *sylvestris*, *tessmanni*, *zenkeri*. Chromosome number: not known.

Description Small stocky animal, not unlike a guinea-pig, but larger. Coat thick, coarse, dark-brown and black with diffused lighter yellowish hairs; long sensory hairs (vibrissae) are scattered throughout the pelage. Pale cream morphs are also known (Kingdon 1971). Rostrum naked. Ears small and rounded and may be tipped with white. There is a conspicuous white spot beneath the chin, and an obvious large, yellowish-white dorsal spot concealing a naked dorsal scent gland (42–72 mm long). Tail short and does not extend past the end of the body. Foot-pads are black, ridged and flexible. One pair of inguinal nipples.

Western Tree Hyrax *Dendrohyrax dorsalis*.

Geographic Variation

D. d. dorsalis: Bioko I.
D. d. sylvestris: coastal forests from Sierra Leone to the lower Niger R. in Nigeria.

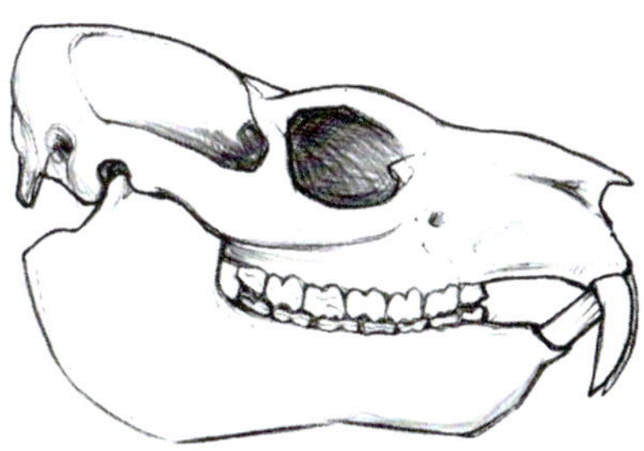

Skull of *Dendrohyrax dorsalis*.

D. d. nigricans: lower Niger R. in Nigeria to N Angola.
D. d. latrator: C DR Congo.
D. d. emini: N and E DR Congo.
D. d. marmota: N Uganda. Some hybridization may occur with the Southern Tree Hyrax *D. arboreus* in S Uganda. Kingdon (1971) suggested that the softer fur of *D. d. marmota* might be a product of long-term hybridization with *D. arboreus*.

Subspecific boundaries are not well established, but geographic variation in call structure is discernible, even between fairly close populations, though not as marked as between more isolated populations of *D. validus* (see *D. validus* profile). Call structure of different populations differs enough to possibly serve as a useful character for distinguishing subspecies. Inter-individual differences in calls may also be marked enough to allow discrimination of individuals based on calls (A. Gautier-Hion pers. comm.). Representative sonograms of calls from three *D. dorsalis* populations – forest of Minwo, Ebom, S Cameroon; Abidjan, S Côte d'Ivoire; and Makokou Forest, NE Gabon – illustrate population differences (Roberts 1999).

Similar Species

D. arboreus. Longer and softer pelage; creamy-white dorsal spot; naked rump patch given as 23–30 mm, though some Rwandan specimens to 48 mm (J. M. Milner pers. comm.); one pair inguinal nipples, but sometimes one pair pectoral and one pair inguinal or one pair pectoral or two pair inguinal). Sympatric or parapatric in S Uganda and along the western border of DR Congo, where some hydridization may occur.

D. validus. Yellow or russet-brown dorsal spot; naked rump patch 20–40 mm; one pair inguinal nipples. Not sympatric, being confined to isolated montane and coastal forests of Tanzania/Kenya, as well as Pemba and Zanzibar Is.

Heterohyrax brucei. Sympatric in NE DR Congo and NW Uganda. For distinguishing characteristics of genus see Table 6, p. 151.

Procavia capensis. Distributed widely in sub-Saharan Africa, with isolated population in North Africa, in a wide range of habitats; F. Dowsett-Lemaire (pers. comm.) recorded Rock Hyrax in sympatry with *D. validus* in Kalakpa Reserve, E Ghana. For distinguishing characteristics of genus see Table 6, p. 151.

Distribution Endemic to Africa. West and central Africa, from Sierra Leone through to S Sudan and N Uganda and southwards to Cabinda (Angola) and C DR Congo. Although Jones (1978) gave the range as 'from Gambia to the Niger River', Grubb *et al.* (1998) note there is no confirmed record west of Sierra Leone, but the species is recorded from near Somoria in NE Guinea (Ziegler *et al.* 2002). Absent from the Dahomey Gap, there being no records from east of the Volta R. in Ghana or from Togo (Grubb *et al.* 1998). Present on Bioko Is.

Habitat Usually found in moist lowland forests and moist savannas to an altitude of around 1500 m, but recorded in montane habitats up to 3500 m (Malbrant & Maclatchy 1949). Also from small and degraded forest fragments.

Abundance Individuals maintain territories, but population densities and structure are poorly known. Observations from Taï N. P., Côte d'Ivoire, provide a rough estimate of 1–2 individuals per square kilometre based on nocturnal calling records (S. Shultz pers. obs.). However, it is likely that actual densities are similar to other species in the genus (see *D. arboreus* and *D. validus* profile).

Adaptations Western Tree Hyraxes are extremely competent climbers and are able to ascend smooth tree trunks up to 50 cm in diameter (Richard 1964). Their limbs and body are flexible and can be contorted to navigate through complex branches (Rio & Galat 1982). The grip is strong, and the feet can be easily supinated. Captive animals have been observed climbing on vines and wires, and even up door frames.

Grooming, using the lower incisors and the inner digit of the hindfoot, is common, and captive animals groom each other regularly, particularly on the face and neck (Jones 1978). Kingdon (1971) has suggested that the dorsal gland is important for marking territories and home-ranges, as well as for interspecific identification. Captive animals have been seen to rub the gland against objects such as doors or trees, and ♀♀ excrete cinnamon-smelling oil from their dorsal gland prior to mating. When disturbed or exhibiting aggression, these animals exhibit piloerection of the hair around the gland, at the same time emitting odoriferous secretions.

Foraging and Food Herbivorous, consuming mostly fruit, twigs, shoots, bark and leaves (Rahm 1966, Kingdon 1971). Most activity occurs in the canopy, but they descend to the ground to forage and move between trees. They are predominantly nocturnal, emerging regularly at dusk with another period of activity shortly before daylight (Richard 1964).

Social and Reproductive Behaviour Primarily solitary but groups of two and three can be found (typically mother and subadult young). Spends the day resting in holes at the top of large trees (Malbrant & Maclachy 1949). Tree hyraxes have small home-

Dendrohyrax dorsalis

Sketches of *Dendrohyrax dorsalis*.

ranges, with each defended male territory overlapping those of several smaller female ranges. Individuals use middens, defaecating repeatedly at the bases of trees.

As with other tree hyraxes, Western Tree Hyraxes produce very loud, distinct calls. Call structure is characterized by long cries, repeated between 22 and 42 times at gradually increasing amplitude and intervals, reaching a loud climatic crescendo at end. Among captive animals, Rahm (1969) observed that at the beginning of each call there was a sequence of very faint, almost inaudible units that only showed up faintly on oscillograms. Both ♂♂ and ♀♀ call, the latter more often when solitary. Hyraxes call throughout the night, but with marked peaks in late evening (20:00–22:00h) and early morning (04:00–05:00h), corresponding to activity patterns (Rahm 1969). There is some seasonal variation in both calling frequency and schedules. They are also heard to call during the day, typically after being disturbed.

Tree hyraxes have been observed raising their dorsal crest, 'tutting' and licking their lips when alarmed (F. Maisels pers. comm.). Captive animals are often quite aggressive and will charge and snap at other individuals (Jones 1978).

Reproduction and Population Structure Both mating and birth peaks tend to coincide with the dry season (Kingdon 1971), but offspring may be born throughout the year. In Gabon and Cameroon, births are primarily during Mar and Apr, and from May to Aug in S and W DR Congo, while in the eastern part of the range (Uganda) young are born throughout the year (Kingdon 1971). Gestation period is eight months. One or two young are born; they are precocious, fully furred and fairly large, with birth-weights of 180–380 g (Mollaret 1962, Roche 1962). Jones (1978) recorded one captive animal attaining adult body size by 120 days and another at more than 200 days. Longevity in captive animals is likely similar to the Southern Tree Hyrax *D. arboreus*.

Predators, Parasites and Diseases Known predators include Crowned Hawk-eagles *Stephanoaetus coronatus* and possibly also larger eagle-owls *Bubo* sp. or hawk-eagles *Hieraaetus* sp. The species has been documented in very low proportion in the diet of both Leopards *Panthera pardus* (1.4%; Hoppe-Dominik 1984) and Crowned Hawk-eagles (approx. 2%; Shultz 2002) in Taï N. P. Chimpanzees *Pan troglodytes* have also been documented capturing and killing adults (Hirata *et al.* 2001), but have not been seen to eat them.

West African specimens have been found to have various nematode parasites (*Crossophorus collaris*, *Libyostrongylus alberti*, *Hoplodontophorus flagellum*, *Theileriana brachylaima*) (Dekeyser 1955). Fain & Lukoschus (1981) recorded a new species of psoroptid tick from DR Congo.

Conservation IUCN Category: Least Concern. CITES: not listed.

Apparently widespread and common in large forest tracts, and present in a number of protected areas, such as Taï N. P. and the National Park of Upper Niger (Guinea); however, they are probably sensitive to intensive habitat degradation and fragmentation. They are also killed for their fur and for food, and have been recorded in a number of bushmeat markets; Fa *et al.* (2000) actually recorded a significant increase in the number of carcasses of this species found in bushmeat markets in Bioko I. between 1991 and 1996. The status of this species should be closely monitored in the future.

Measurements
Dendrohyrax dorsalis
HB: 440–570 mm, n = 14
HF: 70–90 mm, n = 14
E: 21–30 mm, n = 14
WT: 1850–4500 g, n = 14
Origin unknown (Jones 1978); mean not given

Key References Jones 1978; Kingdon 1971; Malbrant & Maclatchy 1949; Rahm 1969.

Susanne Shultz & Diana Roberts

Dendrohyrax validus Eastern Tree Hyrax

Fr. Daman d'arbre; Ger. Östlicher Baumschliefer

Dendrohyrax validus True, 1890. Proc. U. S. Nat. Mus. 1890: XIII: 228. Mt Kilimanjaro, Tanzania.

Eastern Tree Hyrax *Dendrohyrax validus*.

Taxonomy Four subspecies have been recognized in Tanzania (Swynnerton & Hayman 1950), namely *D. v. validus*, *D. v. terricola* (including the form *vosseleri*), *D. v. neumanni* and *D. v. schusteri*. Bothma (1971) recognized three subspecies (including the form *schusteri* in *D. v. terricola*), but also listed *D. v. vosseleri* as distinct, although he indicated it was probably synonymous with *D. v. terricola*. Based on variation in loud-calls, there are distinct vocal profiles that correspond with at least three 'call zones' (see below), which have been identified by Roberts (2001). These may represent subspecific differences or even species-level divisions. Tree hyraxes, previously unknown there, have now been recorded in Ethiopia. Distinguished by apricot brow-spots and mottled white and apricot undersides, a new species or new subspecies of *D. validus* is likely (A. Mekonnen, pers. comm.). This brings into question the current taxonomy of the species, suggesting that a major revision of this group is overdue, all the more so since Shoshani (2005) considered this species a synonym of the Southern Tree Hyrax *D. arboreus*. We provisionally follow Swynnerton & Hayman (1950) in listing four subspecies, recognizing that more than one full species may emerge with further study of the genes and vocalizations of local populations. Synonyms: *adersi*, *neumanni*, *schusteri*, *terricola*, *validus*, *vosseleri*. Chromosome number: not known.

Description Small, robust mammal, similar in shape to a guinea-pig, but typically slightly larger. No discernible tail; dusky-brown feet with blunt, nailed toes, and a distinctive dorsal scent gland 20–40 mm long marked by a yellow to russet-brown, or cinnamon-coloured patch of erectile hairs (True 1890, Bothma 1971, Kingdon 1971). Hairs around the nostrils, eyes, feet and outsides of the ears are dusky-brown. Ears rounded, small to moderately sized and internally have a tuft of yellowish-white hairs. Hair covers the muzzle as far as the nostrils and there is a narrow, hairless, border around the nostrils extending to the margin of the lip (True 1890). Pelage is dense, soft and long-furred (Kundaeli 1976a, b). Colour varies greatly within the species, though generally dorsal pelage is cinnamon-brown darkening to dusky-brown or black, especially around the head. Most dorsal hairs are muted chocolate-brown at base, with a bright cinnamon subterminal ring and brown or black tip. Among these are numerous longer, straight, shining hairs coloured entirely dusky-brown or black. Around the head, subterminal rings on hairs are paler, giving the forehead and cheeks a greyer tinge. Undersides are paler, with hairs coloured chocolate-brown at the base and tinged at the ends cinnamon-brown to paler yellowish-white, especially between the hind legs (True 1890). One pair inguinal nipples only.

Skull depressed, slightly expanded at the posterior with an elongated muzzle and rectangular nasal bones. Orbit is completed by the fusion of the processes of the frontal and zygomatic bones. The coronoid process of the mandible is rectangular and angled forward forming a 45° angle with the molars, with the upper margin almost in line with the margin of the ramus posterior to the condyle (True 1890).

Geographic Variation

D. v. validus: Mt Meru and Mt Kilimanjaro.

D. v. terricola (including *vosseleri*): Taita Hills, Pare and Usambara Mountains.

D. v. schusteri: Uluguru Mts and (possibly) Udzungwa and Rubeho Mountains.

D. v. neumanni (including *adersi*): Zanzibar, Pemba and Tumbatu Is.

Compared with Southern and Western Tree Hyraxes, the calls of the Eastern Tree Hyrax show radically more apparent and dramatic geographical variance. Roberts (2001) identified three distinct vocal profiles, falling into three 'call zones': offshore islands, including Zanzibar and Pemba; southern mountains, including the Uluguru, Udzungwa and Rubeho Mountains; and northern mountains, including the Taita Hills, Pare and Usambara Mountains. No data were available from Mt Meru and Mt Kilimanjaro, although J. Kingdon (pers. comm.) suggests call patterns from these populations are akin to those recognized from the Taita Hills and Pare Mts.

The clear differences in vocal profiles between call zones can best be illustrated by comparing spectograms of vocalizations. On Zanzibar and Pemba Is, one call type has been recorded: a discrete (i.e. having an essentially constant pattern each time it is heard) 'knock' sequence with a distinctive crescendo and decrescendo pattern and characterized by an initial unit of relatively high amplitude followed by 2–4 less powerful (sometimes almost inaudible) units, prior to the main element of the call.

In the Uluguru, Udzungwa and Rubeho Mountains there are two frequently heard call types: (1) 'hac-phrases', which occur as

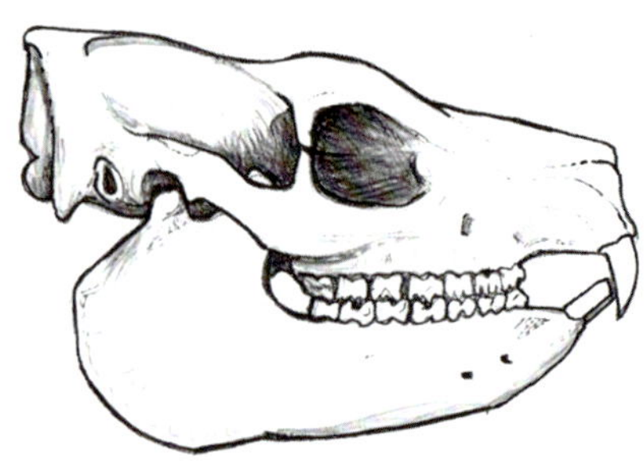

Skull of *Dendrohyrax validus neumanni*.

volleys of relatively monotonous short phrases of five or six units, constituting a short, sharp resonant 'hac' sound, which animals appear to exchange on an ad hoc basis; and (2) diagnostic 'hac-pingpong' sequences, in which animals emit a series of 'hac-phrases', building to a longer climactic 'pingpong' sequence which, as the name suggests, sound very similar to a bouncing pingpong ball and which seem to occur above a certain level of arousal. J. E. Topp-Jørgensen (pers. obs.) reports hearing high-pitched screams in the Udzungwa Mts uttered during fighting and another call type that starts with two rapid series of between 6 and 12 single 'hacs' followed by single or double 'hacs' uttered for up to three minutes. The latter call correlated with increasing density and could be related to increased competition.

In the Taita Hills, Pare and Usambara Mountains, call patterns vary more compared with the other two, the most diagnostic being the 'strangled-thwack' call-type, with a unique phrased-unit structure, i.e. composed of audible and observable sub-units, rather than individual units as in the other zones.

Similar Species

D. dorsalis. Generally has shorter, coarser pelage; yellowish-white dorsal spot; longer dorsal patch (42–72 mm); rostrum naked; conspicuous white spot beneath chin. Rainforests of West and central Africa.

D. arboreus. Generally grey or brown colouration, may appear grizzled, and head frequently darker than body; creamy-white dorsal spot; naked rump patch given as 23–30 mm (J. M. Milner pers. comm.), so not a reliable diagnostic character. One pair inguinal nipples, but sometimes one pair pectoral and one pair inguinal, or one pair pectoral or two pairs inguinal. Forested areas of central and eastern mainland Africa as far south as the Eastern Cape Province, South Africa.

Heterohyrax brucei. Recorded in montane forest habitat within the range of *D. validus* (Topp-Jørgensen *et al.* 2008; A. Perkin pers. comm.). For further distinguishing characteristics see Table 6, p. 151.

Procavia capensis. Distributed widely in sub-Saharan Africa, with isolated population in North Africa, in a wide range of habitats. For further distinguishing characteristics see Table 6, p. 151.

Distribution Endemic to Africa. Of all tree hyrax species, the Eastern Tree Hyrax has the most restricted and patchy geographic range, being limited to montane forests on the slopes of Mt Kilimanjaro and Mt Meru, and the Eastern Arc Mountains – a chain of isolated forest remnants on crystalline mountain blocks running from the Taita Hills in S Kenya to the Udzungwa Mts in S Tanzania; also present in coastal forests of Tanzania, S Kenya and offshore islands (Seibt *et al.* 1977, Pakenham 1984, Kingdon & Howell 1993, Burgess *et al.* 2000, Cordeiro *et al.* 2005). Swynnerton & Hayman (1950) documented *D. v. validus* on Mt Kilimanjaro and Mt Meru; *D. v. terricola* in the East Usambara (Amani, Monga), West Usambara (Lushoto, Magamba), and both North and South Pare Mountains; *D. v. neumanni* in coastal forests on the islands of Pemba, Unguja (Zanzibar) and Tumbatu; and *D. v. schusteri* in the Uluguru Mts (Bagilo, Mkarazi, Nyange, Nyingwa, Vituri). Burgess *et al.* (2000) indicated the species occurs on Mafia I., presumably a lapsus for Pemba (Kock & Stanley 2009). Seibt *et al.* (1977) reported the first record of the Eastern Tree Hyrax in Kenya, recording a population inhabiting a fossil reef area close to the shore at Vipingo, a small village north of Mombasa. Recently, between 1993 and 2005, the species has also been recorded from Nguu, and Rubeho Mts (Tanzania) (Eltringham *et al.* 1998, Cordeiro *et al.* 2005, J. Kingdon pers. comm., A. Perkin pers. comm.). It possibly occurs in Ethiopia.

Dendrohyrax validus

Habitat Confined to moist lowland and montane forest, and occupies a wide altitudinal range from sea level to 3070 m on Mt Kilimanjaro, where it occupies a continuous belt of forest between 1700 m and 3070 m (Kundaeli 1976b). Favoured tree species on Kilimanjaro include *Ocotea usambarensis*, *Schefflera volkensii*, *Nuxia congesta*, *Podocarpus* spp. and *Ficus thonningii* (Kundaeli 1976a, b, Grimshaw *et al.* 1995). In the Eastern Arc Mts, Cordeiro *et al.* (2005) recorded this species from several forest reserves at elevations between 900 m and 2000 m. In the Nilo F. R. it seemed most common in forest at lower elevations (900–1300 m), but was rarer in stunted forest on ridges and at 1500 m (where the absence of large, mature trees for nest holes may have been a limiting factor). Similarly, at Kindoroko F. R. it was commonly heard at 1500–1800 m, but less so above these elevations. *D. validus* is also reported to den where there are large boulders on Mt Kilimanjaro (Umbwe route), and in the Eastern Arcs and adjacent lowland sites (e.g. Kihansi) (N. Cordeiro pers. comm.). It possibly occurs in Ethiopia.

Abundance Locally abundant. As with other tree hyrax species, Eastern Tree Hyraxes are rarely seen, but can frequently be heard calling. Based on circular plot counts of calls, densities of up to 17 ind/ha have been estimated in undisturbed, closed-canopy forest in the Udzungwa Mts, and up to 12 ind/ha in moderately disturbed, closed-canopy forest (Topp-Jørgensen *et al.* 2008). Abundance varies greatly and seems to correlate negatively with open-canopy structure and hunting (Topp-Jørgensen *et al.* 2008). In forests at higher altitudes, where predation levels appear to be low, Eastern Tree Hyraxes can become the dominant herbivore in especially favoured patches, with biomass densities in the order of 75 kg/ha (J. Kingdon pers. comm.).

Evidence suggests the species is highly susceptible to disturbance. In the Udzungwa Mts, logging has a significant impact on population numbers, especially where hunting also occurs (Topp-Jørgensen & Pedersen 2001). In disturbed forests, individuals call less frequently, a feature associated with either severely reduced population densities or stress-induced behavioural change. Overall, high population densities (i.e. calling frequency) appear to be linked to isolated, undisturbed forest patches (Topp-Jørgensen *et al.* 2008). These findings support those of Kundaeli (1976b) on Mt Kilimanjaro: highest densities were estimated at an elevation of 2310 m, where disturbance from logging was low. Food within the forest was found to be generally abundant and not a factor regulating population density. However, access to, and availability of, specific cavity-bearing tree species for shelter was found to limit densities, and at this elevation preferred tree species, such as *Ocotea usambarensis* on the southern slopes, and *Nuxia congesta* on the western slopes, were abundant, though at lower elevations were exploited intensively. Using the density of defaecation sites as a measure, population numbers were higher on the southern and western slopes – average of 23 individuals/ha and 13 individuals/ha, respectively – compared with the northern (seven individuals/ha) and eastern (where only one defaecation site was recorded) slopes. This again was linked to variance in the abundance and presence of preferred shelter trees. Population density on Mt Kilimanjaro is also regulated by other factors, such as the solitary nature of these hyraxes and high intraspecific intolerance, and the fairly constant thermal and microclimatic conditions in the forest (Kundaeli 1976b).

Adaptations Like other tree hyrax species, the Eastern Tree Hyrax appears to be secondarily nocturnal, as indicated by its reliance on vocal and olfactory signals for communication, intensely solitary behaviour and observed activity patterns. However, these hyraxes have no specific physical adaptations to a nocturnal existence (e.g. there is no *tapetum lucidum* for enhanced vision or enlarged ears). They are predominantly arboreal, and very adept climbers, spending most of their time feeding and resting above the ground, but come down when looking for a mate or sometimes to forage. Thick pelage, long periods of inactivity and thermoregulatory behaviour (e.g. use of tree cavities and sun basking) minimize energy requirements. Kundaeli (1976b) suggests they are reliant on a constant environment and are unable to cope with rigorous or changeable climatic conditions.

Foraging and Food Eastern Tree Hyraxes are browsers, feeding almost entirely on leaves. They feed predominantly above the ground, but occasionally come down to forage on soft-tissued climbers and herbs (Kundaeli 1976a, b). Some feeding may occur during the day, but intense feeding occurs at dusk and again just before dawn.

At the top of its altitudinal range in the Udzungwa Mts, this species can live at very high densities (estimated from calls to be in excess of 20 per ha) in low canopy (15–20 m) mist forest where a very rare and localized species of African Violet *Saintpaulia grotei* grows abundantly, often rooted in decayed hyrax faeces. The plants appear to escape the very heavy browsing of hyraxes in this habitat, suggesting that the herb is distasteful to the hyraxes. The plants seem to derive competitive, as well as nutritional, advantages from their association with the hyraxes, suggesting that the herb and herbivore might have co-evolved (J. Kingdon pers. comm.).

Social and Reproductive Behaviour Essentially solitary, with animals nesting, foraging and travelling independently, but social relations are maintained by vocal calls and scent marking signals. On Mt Kilimanjaro most shelter trees are occupied by only one animal, except for ♀♀ with young or where a tree has more than one independent cavity (Kundaeli 1976b). No data are available regarding male and female territories, male–male behaviour, dispersal patterns, and behaviour of young.

Calling appears to be the dominant means of communication. Eastern Tree Hyraxes call throughout the night, with peak periods just after dusk and before dawn (Cordeiro *et al.* 2005). It is unclear whether there are seasonal differences or whether climate or moon phase affects calling activity. No large differences were noted in calling frequency over an eight-month period between Jul and Mar, although on misty days calling may be delayed by up to 15 mins (J. E. Topp-Jørgensen pers. obs.). Sometimes vocal during the day, but mainly inactive, resting very still on a platform near an escape hole (Kundaeli 1976a, b). Like other tree hyrax species, defaecation and urination takes place from the same vantage point, forming middens, usually at the base of trees (Kundaeli 1976a). In the Udzungwa Mts middens can cover several square metres and cover trunks and the surrounding ground with a very strong-smelling, thick, tarmac-like coat. As a result of hunting, daytime call frequency, sun basking and the use of middens is reduced even in lightly disturbed areas (Topp-Jørgensen & Pedersen 2001).

Reproduction and Population Structure Peak in births is probably around Aug, and both mating and birthing seasons appear to coincide with dry seasons at least for the Kilimanjaro area. There is a very long gestation period of around 7.5 months (Kundaeli 1973, cited in Kundaeli 1976a, b). A single young is the norm (but occasionally two are born), and the young are highly precocious. No data are available on suckling behaviour, age of sexual maturity, lifespan and population structure.

Predators, Parasites and Diseases Known predators include Leopards *Panthera pardus*, genets *Genetta* spp., Two-spotted Palm Civets *Nandinia binotata*, Servals *Leptailurus serval*, African Civets *Civettictis civetta*, Crowned Hawk-eagles *Stephanoaetus coronatus* and African Rock Pythons *Python sebae*. There is no known information on parasites or diseases.

Conservation IUCN Category: Least Concern. CITES – Not listed.

The major threats to this species are severe forest loss and fragmentation, and hunting. Individuals can persist in closed-canopy forests of less than 1 square kilometre; however, logging, including selective logging of large trees, removes potential shelter trees, destroys arboreal pathways and makes animals more vulnerable and prone to ground trapping (Topp-Jørgensen *et al.* 2008). Clear pathways through the undergrowth (runnels) also appear over time where tree hyraxes are active, which make them vulnerable to local hunters. Eastern Tree Hyraxes are hunted for their meat and skins, and easily trapped using snares set at the head of a runnel near the base of a tree. They may also be clubbed, speared, or run down by dogs having been 'smoked out' or following the felling of a den tree (a common method employed in the Udzungwa Scarp F. R.) or extracted from holes between boulders using a stick or spear (Topp-Jørgensen & Pedersen 2001).

Kundaeli (1976a) recorded 4708 legally caught hyraxes between May 1972 and July 1973 on Mt Kilimanjaro. Hunting was banned in Tanzania in 1973; however, it still continues and most, if not all, hunting of tree hyraxes in Tanzania is illegal. They are trapped for their meat and pelts to make karosses ('hyrax blankets') (Kundaeli 1976a), which, while much reduced in availability since the 1970s, can still be bought in Arusha, Tanzania. In the Udzungwa Mts both over-harvesting by local people and felling of den trees suggest significant effects on population numbers (Topp-Jørgensen & Pedersen 2001).

Protected areas from which this species has been recorded include the Udzungwa Mts and Kilimanjaro National Parks, Arusha National Park as well as several forest reserves in the Eastern Arc Mts of Tanzania, such as Nilo, Kindoroko and Nguru North Forest Reserves (see Cordeiro *et al.* 2005).

Measurements

Dendrohyrax validus

HB: 470–588 mm (n = 5)

HF: 58–64 mm (n = 5)

E: 12.5–15.5 mm (n = 5)

WT: 2.5–3.0 kg

Measurements: Kilimanjaro and Taveta, Tanzania (True 1890)

Weight: Kilimanjaro, Tanzania (Kundaeli 1976a)

Key References Cordeiro *et al.* 2005; Kingdon 1971; Kundaeli 1976a, b; Seibt *et al.* 1977.

Diana Roberts, Elmer Topp-Jørgensen & David Moyer

GENUS *Heterohyrax*

Bush Hyrax

Heterohyrax Gray, 1868. Ann. Mag. Nat. Hist., ser. 4, 1–50.

Heterohyrax is a monotypic genus, represented only by the Bush Hyrax, or Yellow-spotted Rock Hyrax, *H. brucei*. Two additional forms, *antineae* and *chapini*, have been considered distinct species (e.g. Bothma 1971; and see Schlitter 1993), but the former is a *Procavia* (see Hoffmann *et al.* 2008). Like *Procavia*, *Heterohyrax* is rock dwelling, gregarious and diurnal. Relative to other hyraxes, the penis of *Heterohyrax* is more complex than those of either *Procavia* or *Dendrohyrax*, characterized by an appendage at its tip and being round in cross-section, and the distance between the anus and the preputial opening in an adult ♂ is 65–80 mm, some two to three times longer than in the other genera (Coetzee 1966, Hoeck 1978a). Length of upper molar toothrow (M^{1-3}) is less than the length of the upper premolar toothrow (P^{1-4}) (Bothma 1971).

Paulette Bloomer

Heterohyrax brucei Bush Hyrax (Yellow-spotted Hyrax)

Fr. Daman d'arbuste; Ger. Buschschliefer

Heterohyrax brucei (Gray, 1868). Ann. Mag. Nat. Hist., ser. 4, 1: 35–51. Ethiopia.

Taxonomy Although the type species was originally designated *Dendrohyrax blainvillii*, *Heterohyrax blainvillii* was proposed as a provisional name (Gray 1868; see Barry & Shoshani 2000). Allen (1939), Roberts (1951) and others considered *Hyrax syriacus* the prior name for this species, but Ellerman *et al.* (1953), Bothma (1971) and Meester *et al.* (1986) referred this form to *Procavia*, such that *brucei* becomes the earliest available name.

More than 20 subspecies have been described from Africa, and two additional forms, *antineae* and *chapini*, were treated as distinct species by Bothma (1971). *Heterohyrax antineae*, from the Ahaggar Mts, was maintained as a distinct species by Bothma (1971) and Schlitter (1993), and included in the synonymy of *H. brucei* by Barry & Shoshani (2000). However, this form is likely conspecific with an earlier described form of *Procavia* (*bounhioli*) (Hoffmann *et al.* 2008). *Heterohyrax chapini* (described by Hatt 1933) from Matadi in SW DR Congo, not far from the mouth of the Congo R., supposedly characterized by the presence of only two pairs of inguinal nipples, is also now more commonly treated as conspecific with *H. brucei* (Roche 1972, Schlitter 1993, Barry & Shoshani 2000, Shoshani 2005). Barry & Shoshani (2000) and Shoshani (2005) recognized 25 subspecies for the African continent, including both *antineae* and *chapini*. Although no consensus exists on whether all of these subspecies are conspecific, sequences of the mitochondrial DNA

Bush Hyrax *Heterohyrax brucei*.

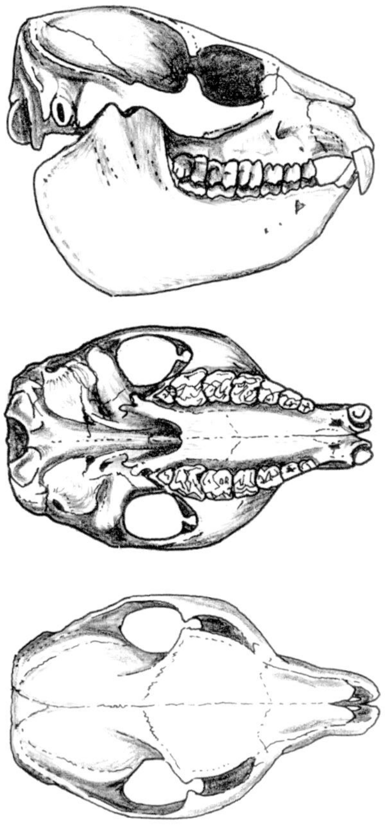

Lateral, palatal and dorsal views of skull of *Heterohyrax brucei*.

cytochrome *b* gene indicate that at least *H. b. hindei* in the central portion of the range and *H. b. ruddi* and *H. b. granti* from Zimbabwe and South Africa are highly distinct and may represent cryptic species (P. Bloomer pers. comm.).

Synonyms: no fewer than 34 synonyms are listed by Shoshani (2005), many of which have been considered subspecies (see Bothma 1971, Barry & Shoshani 2000). Chromosome number: 2n = 54 (Prinsloo & Robinson 1991). Karyotype marked by 20 acrocentric, two subtelocentric, two submetacentric and two metacentric autosomal pairs. The X chromosome is the largest submetacentric and the Y is a very small acrocentric.

Description Small to medium-sized, short-legged mammal with rudimentary tail, round ears, and rabbit-like appearance and movement. Lateral and dorsal colour ranges from grey in arid regions to dark reddish-brown in more mesic climates (Skinner & Chimimba 2005). Some individuals in isolated populations can be white-spotted (Hoeck 1982a). A small, linear creamy to yellow spot in the midline of the back surrounds the dorsal gland, but in some subspecies it may be lacking. Ventral colour white or creamy in distinct contrast to the sides, back, head and rump. Eyebrows strikingly white to creamy (paler than in *Procavia*) and conspicuous at a distance. Vibrissae (up to 90 mm long on the snout) evident on the snout, above the eyes, under the chin, along the back and sides, on the abdomen and on fore- and hindlimbs. Ears more prominent than in other hyraxes, the eyes bulge and the head is flat dorsally. Guard hairs are softer than in *Procavia*, black-tipped, and up to 30 mm long. Underhairs are basally brown or grey and terminally buffy. The anal-preputial distance in ♂♂ is 65–82 mm, two or three times longer than in other genera. Anatomy of the penis is complex, with an appendage at the tip and round in cross-section, and measures greater than 60 mm when fully erected (Coetzee 1966, Glover & Sale 1968, Hoeck 1978a, c). Females typically have one pair of pectoral and two pairs of inguinal nipples (although in some animals, such as the isolated form *chapini*, the pectoral pair is absent) (Hatt 1933, Roche 1962, Bothma 1971, Hoeck 1977a). Males and ♀♀ are, on average, of similar size, although ♀♀ are sometime larger than ♂♂ (Smithers & Wilson 1979).

The single pair of tusk-like upper incisors is ridged or triangular in cross-section in ♂♂ (Thomas 1892), but rounded in ♀♀. A diastema (10–16 mm long in adults; Thomas 1892, Bothma 1971) precedes the short-crowned molariform cheekteeth (Meyer 1978) that bear transverse cusps (lophodont condition, resembling those of the family Rhinocerotidae; Skinner & Chimimba 2005) adapted for a herbivorous diet. The length of the upper premolar toothrow, P^{1-4} is just less than or equal to that of the upper molar toothrow, M^{1-3} (Bothma 1971).

Geographic Variation Substantial geographic variation in coat colour has been recorded from across the range of this species, and has led to a proliferation of described forms. Bothma (1966) recorded the high degree of individual variation in coat colour in southern Africa (and see Skinner & Chimimba 2005). We provisionally list the 24 subspecies and their country type localities, following Barry & Shoshani (2000) and Shoshani (2005), but excluding the form *antineae*, which is here considered a *Procavia*. Many will probably not hold up to taxonomic scrutiny:

H. b. albipes: Kenya.
H. b. bakeri: NE DR Congo, S Sudan and NW Uganda.
H. b. bocagei: Angola.
H. b. brucei: Ethiopia.
H. b. chapini: DR Congo.
H. b. dieseneri: Tanzania.
H. b. frommi: Tanzania.
H. b. granti: South Africa.
H. b. hindei: Kenya.
H. b. hoogstraali: Sudan.
H. b. kempi: Kenya.
H. b. lademanni: Tanzania.
H. b. manningi: N Malawi, N and E Zambia.
H. b. mossambicus: Mozambique.
H. b. munzneri: Tanzania.
H. b. princeps: Ethiopia.
H. b. prittwitzi: Tanzania.
H. b. pumilus: Somaliland.
H. b. ruddi: Mozambique.
H. b. rudolfi: Ethiopia.
H. b. somalicus: Somaliland.
H. b. ssongeae: Tanzania.
H. b. thomasi: Sudan.
H. b. victoria-njansae: Tanzania.

Hatt (1933), in describing *H. chapini* from near Matadi, just north of the Angolan border in SW DR Congo, considered this form to be not closely related to the Angolan hyraxes, later (1936) commenting that *bocagei* 'is quite different in its longer, thicker pelage, smaller size, broader teeth, proportionately broader skull, elevated supraorbital ridges, shorter muzzle and flatter basicranium.'

Similar Species

Procavia capensis. Sympatric across much of its range in sub-Saharan Africa, except in West Africa where *Heterohyrax* is absent. For further distinguishing characteristics of genus see Table 6, p. 151.

Dendrohyrax arboreus. Sympatric in S Kenya and N Tanzania. For distinguishing characteristics of genus see Table 6, p. 151.

Dendrohyrax dorsalis. Sympatric in NE DR Congo and NW Uganda. For distinguishing characteristics of genus see Table 6, p. 151.

Dendrohyrax validus. Restricted to isolated montane and coastal forests of Tanzania/Kenya, as well as Pemba and Zanzibar Is. For distinguishing characteristics of genus see Table 6, p. 151.

Distribution Endemic to Africa. Widely distributed in Africa from Sudan and Eritrea through the Horn of Africa south to the Limpopo Province in South Africa (Barry & Shoshani 2000, Skinner & Chimimba 2005). Isolated populations in SW DR Congo (*H. b. chapini*; Hatt 1933) and Angola (*H. b. bocagei*; Crawford-Cabral & Veríssimo 2005). Records of this species in North Africa, specifically Egypt, along the Red Sea coast, and from the Hoggar Mts of Algeria, are erroneous and relate to *Procavia* (Hoffmann *et al.* 2008).

Habitat Restricted to rocky kopjes (inselbergs), krantzes and piles of large boulders with openings of at least 11 cm in height and 1 m^2 of floor space (Sale 1966). Kopjes provide a constant, moderate temperature (17–25 °C) and humidity (32–40%) and protection from fire (Turner & Watson 1965). Sometimes found also in forests along rivers in East Africa (H. N. Hoeck, pers. obs.). In East Africa, individuals range to 3800 m altitude (Kingdon 1971).

Bush Hyraxes are frequently found in the company of Rock Hyraxes *Procavia capensis*, sometimes even inhabiting the same rock crevices (Hoeck 1975, Barry 1994, Skinner & Chimimba 2005). This heterospecific association varies seasonally in Zimbabwe, but is especially evident during synchronous parturition in Mar (Barry & Mundy 2002).

Abundance Often conspicuous and common in appropriate habitat. Densities in the Matobo N. P. in Zimbabwe ranged from 0.5 to 1.1 individuals/ha (1.2–2.6/ha of kopje) from 1992 to 1996 (Barry & Mundy 1998), and rose to 1.9/ha (4.5/ha of kopje) in 1998 after several years of good rains (R. E. Barry pers. obs.). In Serengeti N. P., densities reached 75/ha of kopje in allopatry and 28/ha in local sympatry with the Rock Hyrax (Hoeck 1982a, 1989, H.N. Hoeck pers. obs.). Rainfall, through its effect on fecundity, appears to be the factor primarily responsible for annual fluctuations in abundance (Barry & Mundy 1998).

Adaptations A dorsal gland lies beneath a raised patch of skin approximately 1.5 cm long in adults that is surrounded by the dorsal spot of erectile hairs. The gland in sexually active, mature adults consists of lobules of glandular tissue, each lobule constituted by 25–40 alveoli of secretory epithelium surrounding a lumen (Sale 1970a). The gland is odoriferous and may function in mating and recognition of the mother by young (Sale 1965a, 1970a). During courtship the ♂ erects the hairs of the dorsal spot, exposing the dorsal gland. Pilo-erection of the dorsal spot functions as an alarm or threat signal to hyraxes and other animals nearby; dorsal hairs are erected during mating behaviour and when animals are aroused.

As with all hyraxes, the digits have flat, hoof-like nails except for the second digit of the pes, which has a long, curved claw for grooming; in addition, the four lower incisors are comb-like for grooming the fur (Hoeck 1982b). The soles have thick, rubbery pads with numerous skin glands that increase the grip for climbing (Dobson 1876, Meyer 1978, Sokolov & Sale 1981); Bush Hyraxes are agile climbers and good jumpers (Hoeck 1977b). The pupil of the eye, as in Rock Hyrax, houses an umbraculum, a shield that allows a basking individual to stare into the sun (Millar 1973) to detect aerial predators.

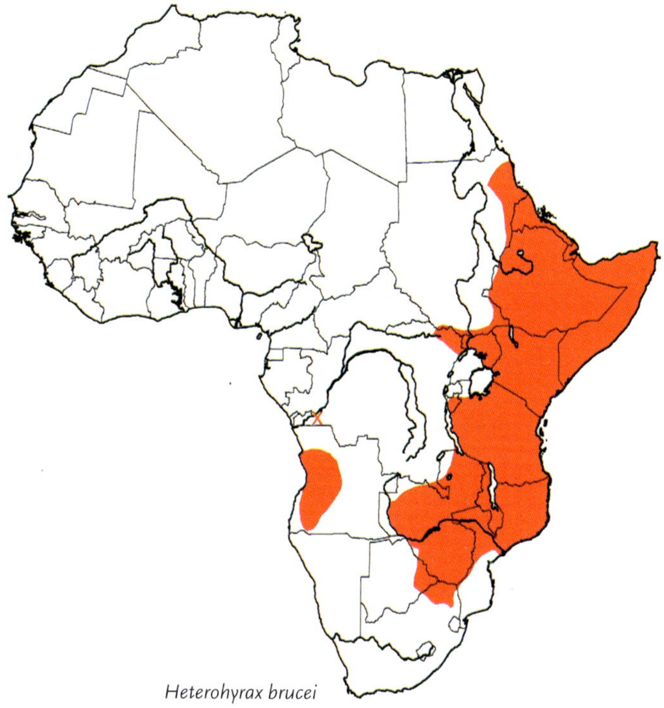

Heterohyrax brucei

The weight-specific metabolic rate is low, with a thermoneutral zone of 24–35 °C (Bartholomew & Rainy 1971). Body temperature typically ranges from 35 to 37 °C, but fluctuates up to 7 °C in response to changes in air temperature. At air temperatures above 25 °C body temperature is maintained by evaporative water loss from the nostrils, soles of the feet, panting, salivating and grooming (Taylor & Sale 1969, Bartholomew & Rainy 1971). Water is conserved by the production of small volumes of urine and faeces (Maloiy & Eley 1992). The highly concentrated urine, together with faeces at communal latrines, forms a dark, crystalline residue called 'klipstreet' or 'hyraceum' (Turner & Watson 1965, Eley 1994). Little free water is consumed because of the low metabolic rate, low urine volume, and thermal lability (Maloiy & Eley 1992); most water is obtained from their food.

Huddling and stacking of individuals conserves heat (Sale 1970b). Behavioural thermoregulation is achieved by early morning and late afternoon basking on the surface, and by retreating to the rocks and shaded areas in midday to avoid heat and desiccation (Sale 1966, Taylor & Sale 1969).

Foraging and Food Bush Hyraxes are obligate browsers (DeNiro & Epstein 1978, Walker *et al.* 1978), spending more than 80% of foraging time browsing on twigs and bark of woody species and leaves, buds, flowers and fruits of trees, bushes and forbs (Hoeck 1975, 1977b, 1982c); they only very rarely consume grass. Plants most commonly foraged in Serengeti N. P. include *Acacia tortilis*, *Allophylus rubifolius*, *Cordia ovalis*, *Grewia fallax*, *Hibiscus lunarifolius*, *Ficus glumosa*, *F. ingens*, *Iboza* sp. and *Maerua triphylla* (Turner & Watson 1965, Hoeck 1975). In Zimbabwe the plants most frequently eaten were *Combretum molle*, *Elephantorrhiza goetzei*, *Flueggia virosa*, *Strynchos usambarensis*, *Kirkia acuminata*, *Croton gratissimus* and *Mundulea sericea*; *Rhus leptodictya* and *Commiphora marlothii* also are taken by juveniles (Barry & Shoshani 2000).

Activity pattern is diurnal, with most feeding occurring between 07:30 and 11:00h and between 15:30 and 18:00h (Kingdon 1971, Hoeck 1975), but occasionally to 21:00h (Turner & Watson 1965). Individuals may feed alone or in a group (Sale 1965b). Group feeding can occur up to 50 m from the centre of the colony, although casual feeding rarely occurs at distances more than 20 m from the den site. Feeding bouts average 20 min and last no longer than 35 min. Individuals can climb vertical trunks of trees and balance on thin branches (Hoeck 1977b) in order to feed on the outermost leaves (Hoeck 1982c).

Social and Reproductive Behaviour Bush Hyraxes are gregarious (Sale 1970b), and group size can reach 34 individuals. The social unit is a polygynous harem, with a territorial adult ♂, up to 17 adult ♀♀, and juveniles (Hoeck 1977a, 1982b, Hoeck *et al.* 1982). Territorial ♂♂ threaten other ♂♂ by movement or changes in posture (raising head and shoulders), showing large incisors, grinding molars, growling, snapping, chasing or biting, and erecting hairs around the dorsal gland (Glover & Sale 1968, Hoeck *et al.* 1982). Appeasement is communicated with the hair and body flat and rump presented (Kingdon 1997). Latrines, located near sleeping quarters, are revealed as visible white stains resulting from deposits of urine rich in calcium carbonate (Estes 1991).

Bush Hyraxes have a highly structured repertoire of calls, indicating predator detection, contact, threat and distress; some of these calls are recognized by Rock Hyraxes (Hoeck 1982b) and Klipspringers *Oreotragus oreotragus* (Kingdon 1997). Loud territorial calls are frequent during the mating season (Hoeck 1982b); the territorial call of adult ♂♂ is shrill and long, lasting about 1.5 sec, and given repeatedly for up to 5 min (Hoeck 1978a).

Territorial ♂♂ copulate more often than non-territorial ♂♂ and mate preferentially with ♀♀ older than 28 months of age. Peripheral ♂♂ exhibit a dominance hierarchy and mate more often with young ♀♀ (Hoeck *et al.* 1982). The ♂ emits a shrill cry as he approaches to mate, and the ♀ erects her dorsal spot hairs. The ♂ sniffs the female's vulva, rests his chin on her rump, then slides onto her back as he makes thrusting movements followed by intromission in 3–5 minutes. A second copulation may occur after 1–3 hours (Hoeck 1978a, c).

Mothers suckle only their own young (Hoeck 1977a, 1982b). Young play with conspecifics, and sometimes with young Rock Hyraxes at heterospecific nurseries, by nipping, biting, climbing, pushing, fighting, chasing and mounting (Hoeck 1978b, Caro & Alawi 1985, Barry 1994). Young in nurseries are attended by their own mothers, and by mothers of other young, non-maternal conspecifics, or even Rock Hyrax individuals (Barry 1994). Females join the adult female group at sexual maturity (approximately 16 months of age), but male offspring disperse between 12 and 30 months of age (Hoeck 1982a). However, dispersal of ♀♀ also has been recorded in the Serengeti N. P.; dispersing ♀♀ prevent inbreeding and are mainly responsible for long-distance gene transfer between kopjes (Gerlach & Hoeck 2001).

Reproduction and Population Structure Females come into oestrus once or twice per year for up to three days, perhaps repeatedly during four weeks. Oestrus in ♀♀ in a family group is synchronized. Gestation lasts 26–30 weeks. In Serengeti N. P., two distinct birth seasons, after the long and short rains, were observed (Hoeck 1982a). Birth pulses occur from Feb to Mar, just before the rains, in Kenya (Sale 1969), and in Mar, two months after peak rainfall, in Zimbabwe (Barry 1994).

Sketches of *Heterohyrax brucei*. Note erectile fur tract above dorsal gland.

Litter-size averages 1.6 in Tanzania (Hoeck 1982a), 1.7 in Kenya (Sale 1969) and 2.1 in Zimbabwe (Barry 1994; and see Smithers & Wilson 1979). Females may bear young every other year (Barry 1994). Young are precocious at birth and weigh 220–230 g (Skinner & Chimimba 2005). Young nurse for 1–5 months and sexual maturity is achieved at 16–17 months of age (Hoeck 1977a). Maximum recorded life-span in the wild is a little more than 11 years in ♀♀ (Hoeck 1989), similar to records in captivity (Weigl 2005).

Sex ratios ranged from 1 male : 1.6–3.2 females (Hoeck 1982a, Hoeck *et al.* 1982). However, in Zimbabwe sex ratios of captured individuals and those that were prey to Verreaux's Eagles *Aquila verreauxii* did not differ from 1 : 1 (Barry & Mundy 1998). Density of this population was estimated at 0.5–1.1 individuals/ha (1.2–2.6/ha of kopje) over a five-year period. This population comprised 19.4–27.5% juveniles, 7.2–13.1% subadults and 62.9–73.7% adults. Juvenile mortality was estimated at 52.4–61.3%.

Predators, Parasites and Diseases The principal predator is Verreaux's Eagle, with more than 90% of the diet in this species made up of Bush Hyraxes and Rock Hyraxes; in Zimbabwe adults were selected disproportionately by Verreaux's Eagles (Barry & Barry 1996). Other predators include large snakes, Leopards *Panthera pardus*, Martial Eagles *Polemaetus bellicosus* and other raptors (Turner & Watson 1965, Wilson 1969, Grobler & Wilson 1972, Smith 1977, Hoeck 1982a, Gargett 1990, Aumann & Chiweshe 1995).

Bush Hyraxes are susceptible to viral pneumonia and tuberculosis (Sale 1969). The sarcoptic mite that causes mange can heavily reduce colonies (Hoeck 1982a). Individuals can harbour the flagellate *Leishmania* (Ashford 1970, Ashford *et al.* 1973) and nematode *Crossophorus collarus* (Barry & Shoshani 2000). Their role as hosts of *Leishmania* is thought to be of epidemiological significance in the transmission of leishmaniasis to humans, especially in Ethiopian villages close to rocky inselbergs (Ashford 1970). Ectoparasites collected from live-captured individuals in Zimbabwe included ixodid ticks (*Rhipicephalus distinctus* and *Haemaphysalis leachii*), fleas (*Procaviopsylla*), lice (*Prolignognathus*) and ear mites (*Acomatacarus*) (Barry & Shoshani 2000). Individuals often dust-bathe to remove ectoparasites.

Conservation IUCN Category: Least Concern. CITES: Not listed.

Bush Hyraxes are readily snared and, in parts of southern Africa, their pelts are used to make karosses (blankets), with resultant dramatic declines in the density of local populations. However, overall the species is widespread and present in several well-managed protected areas across its range.

Measurements

Heterohyrax brucei
TL (♂♂): 434 (382–469) mm, n = 28
TL (♀♀): 430 (325–472) mm, n = 65
WT (♂♂): 1.75 (1.40–2.10) kg, n = 57
WT (♀♀): 1.84 (1.30–2.41) kg, n = 133
Serengeti N. P. (Hoeck 1982a)

TL (♂♂): 497 (465–530) mm, n = 5
TL (♀♀): 516 (485–560) mm, n = 12
HF s.u. (♂♂): 68 (65–70) mm, n = 5
HF s.u. (♀♀): 69 (65–73), n = 12
E (♂♂): 32 (30–33) mm, n = 5
E (♀♀): 32 (29–34), n = 12
WT (♂♂): 3.01 (2.72–3.18) kg, n = 5
WT (♀♀): 3.01 (2.32–3.63) kg, n = 12
Zimbabwe (Smithers & Wilson 1979)

Key References Barry & Shoshani 2000; Hoeck 1982a, 1989; Hoeck *et al.* 1982; Skinner & Chimimba 2005.

Ronald E. Barry & Hendrik N. Hoeck

GENUS *Procavia*

Rock Hyrax

Procavia Storr, 1780. Prodr. Meth. Mamm., p. 40.

Several authors (Olds & Shoshani 1982, Schlitter 1993, Shoshani 2005) considered the genus *Procavia* to be monospecific, and the precedent set by these authors has been provisionally followed in this work. However, other authors have recognized additional species in the genus, including: Cape Hyrax *P. capensis*, Abyssinian Hyrax *P. habessinica*, Johnston's Hyrax *P. johnstoni* and Western Hyrax *P. ruficeps* (Hahn 1934, Kingdon 1971); Bothma (1971) added a fifth species, the Kaokoveld Hyrax *P. welwitschii*, which had been treated as a form of *Heterohyrax* (Shortridge 1934, Roberts 1951). More recent studies on geographic variation in mitochondrial DNA in South Africa indicate that there are at least two species within what conventionally has been regarded as *P. capensis* (Prinsloo & Robinson 1992) and mtDNA gene sequences also suggest the distinctiveness of the more northern taxa, *P. syriaca* and *P. johnstoni* (Prinsloo 1993, P. Bloomer & T. J. Robinson unpubl.). The monospecificity of the genus *Procavia* is, therefore, debatable and further research is required to resolve the taxonomy.

Like *Heterohyrax*, *Procavia* is rock dwelling, gregarious and diurnal, but unlike other hyraxes, it is predominantly a grazer, and has hypsodont molars. The length of the upper molar toothrow M^{1-3} is much greater than that of the upper premolar toothrow P^{1-4} (Bothma 1971).

Paulette Bloomer & Hendrik N. Hoeck

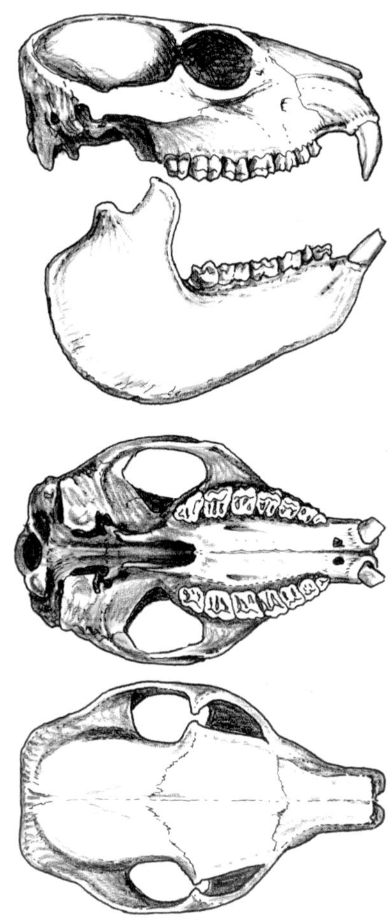

Lateral, palatal and dorsal views of skull of *Procavia capensis*.

Procavia capensis Rock Hyrax (Klipdassie)

Fr. Daman de roches; Ger. Klippschliefer

Procavia capensis (Pallas, 1766). Misc. Zool., p. 30. South Africa, Western Cape Prov., Cape of Good Hope.

Rock Hyrax *Procavia capensis*.

Taxonomy Most taxonomic treatments recognize 17 subspecies (Roche 1972, Olds & Shoshani 1982, Shoshani 2005), although the status of many of these subspecies is dubious. Several of these subspecies have been elevated to the level of distinct species. Synonyms: more than 65 synonyms are listed by Shoshani (2005). The form *antineae* from Algeria, often considered as a form of *Heterohyrax*, is a *Procavia* (Hoffmann *et al.* 2008; and see Olds & Shoshani 1982). Chromosome number: 2n = 54 (Hungerford & Snyder 1969). The X chromosome is the largest, with a submedian centromere; the Y chromosome is a very small acrocentric.

Description Small- to medium-sized, solidly built, small mammal. There is extensive variation in coat colour, which varies widely throughout geographic range from a yellowish-buff to a dark brown; this variation has been associated with mean annual rainfall patterns (Bothma 1966; and see Geographic Variation). Pelage dense, up to 25 mm long, with grizzled appearance due to banding of the hairs (dark at the base, with a lighter band of varying width and a black tip). Underfur short, soft and thick. Underparts are lighter in colour than the upper, and the hair is slightly longer and lacks the banding. Long black vibrissae (tactile hairs) 60–70 mm in length (though longer on the face) are widely distributed over the body, probably for orientation in dark fissures and holes (Sale 1970a). Rock Hyraxes have a dorsal gland (see Social and Reproductive Behaviour), surrounded by a creamy, yellow-coloured (typical of *P. c. ruficeps*) or brown to black (typical of *P. c. capensis*) margin of hairs that can be fanned when the animal is excited; this dorsal margin is not conspicuous in the Rock Hyrax of southern Africa. Females have three pairs of nipples: one pair pectoral and two pairs inguinal (Hahn 1934). Mean distance between anus and penis is 35 mm (n = 41) (Coetzee 1966, Hoeck 1978a, 1982b). Males and ♀♀ are approximately the same size.

Characteristic features of the skull of this species include: widely spaced and anteriorly situated eye sockets; well-developed interparietal; small tympanic bullae; premaxillae form a tubercle between the incisive foramina; coronoid process is small and recurved; hyoid bone is unusually scoop-shaped in structure. Zygomatic arches are broad and heavily built, indicative of a powerful set of masseter muscles that operate the lower jaw (Olds & Shoshani 1982, Skinner & Chimimba 2005). In contrast to the typically hyracoid dentition, dental formula is typically $I\,^1/_2$, $C\,^0/_0$, $P^4/_3$, $M^3/_3 = 32$, although specimens from northern parts of Africa often have the first lower premolar present (as in Bush Hyrax *Heterohyrax brucei*). The length of the upper molar toothrow M^{1-3} is much greater than that of the premolar toothrow P^{1-4} (Bothma 1971). The two upper, ever-growing incisor teeth, one on each side, are separated by about the width of one tooth. The upper incisors are tusk-like, ridged or triangular in cross-section in ♂♂, but rounded in ♀♀ (Hahn 1934).

Geographic Variation As noted earlier, no fewer than 17 subspecies have been recognized (Olds & Shoshani 1982, Shoshani 2005). The validity of many of these is doubtful, while some may actually represent distinct species. Recognized subspecies on the African continent and the origin of their type localities include:

P. c. antineae: Algeria.
P. c. bamendae: SW Cameroon.
P. c. capensis: Cape of Good Hope, South Africa.
P. c. capillosa: Western Bale Province, Ethiopia.
P. c. erlangeri: Somalia.
P. c. habessinicus: NW Ethiopia.
P. c. jacksoni: C Kenya.
P. c. johnstoni: C and S Malawi.
P. c. kerstingi: Togo–Benin border.
P. c. mackinderi: C Kenya.
P. c. matschiei: N Tanzania.
P. c. pallida: N Somalia.
P. c. ruficeps (Hemprich & Ehrenberg, 1832, not Thomas, 1892): NE Sudan.
P. c. scioanus: CE Ethiopia.
P. c. sharica: C Chad.
P. c. syriaca: 'Syria' (but possibly Lebanon).
P. c. welwitschii: SW Angola.

The average size of adult Rock Hyraxes varies greatly across Africa, and seems to be closely linked to average annual precipitation, which in turn affects the availability of food (Klein & Cruz-Uribe 1996); size increases up to a mean annual rainfall of 700 mm and decreases thereafter. On the other hand, Yom-Tov (1993) found size variation in the skulls of Rock Hyraxes from different regions to be positively correlated with temperature, indicating that this species conforms to Bergmann's Rule.

Similar Species

Dendrohyrax arboreus. Forests in East and central Africa south to the Eastern Cape Province of South Africa. For distinguishing characteristics of genus see Table 6, p. 151.

D. validus. Isolated montane and coastal forests of Tanzania/Kenya, as well as Pemba and Zanzibar Is. For distinguishing characteristics of genus see Table 6, p. 151.

D. dorsalis. Forests of West and central Africa. For distinguishing characteristics of genus see Table 6, p. 151.

Heterohyrax brucei. Sympatric across much of the range, though absent from West Africa. For distinguishing characteristics of genus see Table 6, p. 151.

Distribution Rock Hyraxes have a wide distribution in Africa. Their range is frequently given as from Senegal and S Mauritania through S Algeria and Libya to Egypt and then southwards to southern Africa, but their distribution is very discontinuous (and see Habitat). The precise northerly limit is not clearly defined, though the species is recorded in S Algeria (see De Smet 1989, Kowalski & Rzebik-Kowalska 1991) and parts of Libya, such as the Akakus Mts and Libyan Tibesti (Hufnagl 1972). In Egypt, the species only occurs east of the Nile R. Previously unrecorded from Zambia (Ansell 1978), until Osborne's (1987) record from the south-east on the border with Mozambique. In southern Africa, Rock Hyraxes are absent from much of Botswana except the eastern parts, while in Zimbabwe they occur only in the southern parts (Smithers 1971, Smithers & Wilson 1979). There is what appears to be a very distinct break in distribution between S Malawi and N Tanzania (see Kingdon 1971, 1997; not shown in Olds & Shoshani 1982 or Skinner & Chimimba 2005), and this potential discontinuity requires further investigation.

Extralimital to the African continent, Rock Hyraxes occur in S Sinai, Oman, Yemen, Saudi Arabia (probably occurring throughout the mountainous western regions), Jordan, Israel and Lebanon (Olds & Shoshani 1982, Harrison & Bates 1991). Reports of their presence in Turkey are in error (Kryštufek & Vohralík 2001). Although the form *syriaca* was described from 'Mount Lebanon, Syria', the occurrence of this species in Syria remains equivocal (D. Kock pers. comm.).

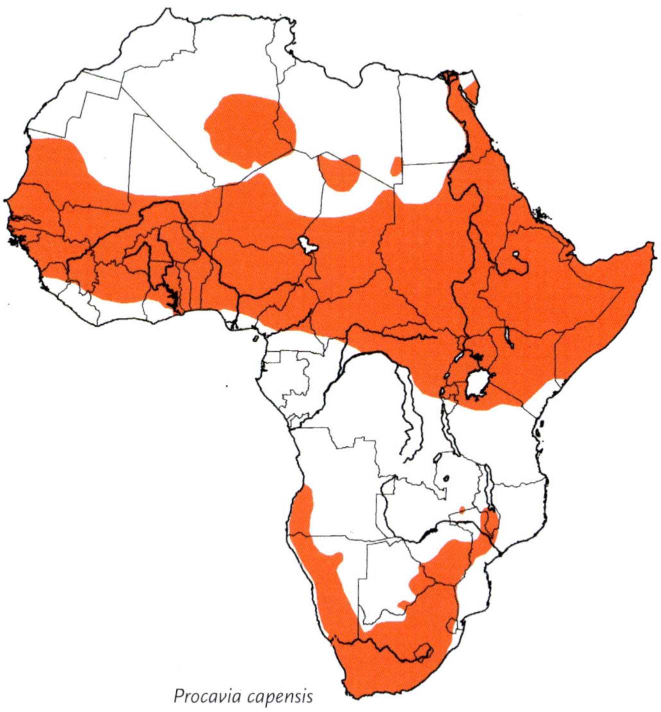

Procavia capensis

Habitat Rock Hyraxes occupy a wide range of habitats, from arid deserts to rainforest, and from sea level to the alpine zone of Mt Kenya (3200–4300 m; Coe 1962, Young & Evans 1993). However, as their name implies, Rock Hyraxes are dependent on the presence of rocky outcrops (kopjes), mountain cliffs or loose boulders that provide suitable refuge in the form of crevices and crannies in which to shelter ('The high mountains are for the wild goats; the rocks are a refuge for the conies'; Psalms 104: 18). The nature of the refuge environment differs substantially across Africa, but rock outcrops appear to be favoured due to the extensive networks of crevices and fissures, access to safe foraging areas and good vantage points (Turner & Watson 1965, Davies 1994). Overall the refuge environment provides stability compared with the surrounding habitats where conditions are more extreme (see Adaptations). However, Rock Hyraxes have also been found in erosion gullies (e.g. in the Karoo, a habitat they have colonized recently), in culverts under roads and holes in stone-walls, and even in the holes of other species such as Aardvark *Orycteropus afer* and Suricate *Suricata suricatta* (Roberts 1951). These refuges seem to be common in areas where overpopulation of rocky habitats occurs, and at these times individuals they may traverse considerable distances between areas of suitable rocky habitat (Kolbe 1967, Lensing 1983).

In several parts of Africa (e.g. the Serengeti N. P. in Tanzania, Matobos N. P. in Zimbabwe and the northern parts of South Africa), Rock and Bush Hyraxes occur together and live in close associations on rock outcrops (Hoeck 1975, 1982a, b, Barry & Mundy 1998, 2002, P. Bloomer pers. obs.).

Abundance Rock Hyrax densities depend upon a combination of several abiotic (rainfall and availability of holes and hiding places) and biotic factors, such as interspecific and intraspecific competition for food, predation and parasites (Hoeck 1989). In Serengeti N. P., Rock and Bush Hyraxes are the most important resident herbivores of the kopjes. The population density of Rock Hyraxes ranged from 5 to 56 animals/ha and group size varied from 2 to 26 individuals (Hoeck 1982a), while in the Matobo N. P. densities were 0.73–0.94/ha (Barry & Mundy 1998). In South Africa, major differences have been reported between regions, with density estimates ranging from 35/km^2 in the semi-arid Karoo N. P. (Davies 1994) to 376/km^2 in the Mountain Zebra N. P. (Fourie 1983). On Mt Kenya, where Rock Hyraxes are considered the most conspicuous mammals, average density has been estimated at 20–100 animals/km^2 (Young & Evans 1993).

As observed by Meltzer (1967) and Hoeck *et al.* (1982), larger and often more mobile family groups with associated peripheral ♂♂ (up to 80 individuals) are possible in areas where resources are favourable. Long-term observations in the Serengeti and Matobos show that Rock Hyrax populations fluctuate and while some kopjes show stable occupation over more than 20 years, small colonies are prone to extirpation (Hoeck *et al.* 1982, Hoeck 1989, Barry & Mundy 1998). These cycles of extinction and recolonization have also influenced the levels of genetic variability within and among Rock Hyrax colonies in the Serengeti, with Rock Hyraxes showing very low allelic diversity and heterozygosity at microsatellite loci (Gerlach & Hoeck 2001; and see Reproduction and Population Structure).

Adaptations Rock Hyraxes are well adapted to life in a rocky environment. Their feet are not adapted for digging, but rather the padded soles ensure increased traction and allow Rock Hyraxes to negotiate steep and often smooth rock surfaces with great ease; while the animal is running, the feet sweat, which greatly enhances its climbing ability (Sale 1970a, Fischer 1992). Hyraxes are able to move through very narrow crevices and crannies: George & Crowther (1981) found that the mean slot height (151 mm [n = 69]) through which they move is just over twice the mean skull height (66.1 mm [n = 45]). Hole/crevice size may bear some relation to the size of predators in the local area (Sale 1966). Rock Hyraxes are also good tree climbers.

The eye of the hyrax has a peculiar shield called an 'umbraculum' that allows them to stare at the sun; Millar (1973) has suggested this has evolved in response to predation by Verreaux's Eagles *Aquila verreauxii*. Hyraxes are the predominant prey of these eagles (Gargett 1990, Davies 1994, Barry & Barry 1996; and see Predators, Parasites and Diseases) and, being somewhat thermolabile, they often bask in the sun, thereby exposing themselves to attack (Millar 1973).

The extent to which the Rock Hyraxes meets energetic requirements depends on the interaction of ambient temperature, food quality and availability, and foraging efficiency in the presence of predators. The extent to which each of these factors influences behaviour differs on a daily and seasonal basis (Sale 1970b, Louw *et al.* 1972, Brown & Downs 2005). Basal metabolic rate, lower critical temperature and thermal conductance all vary inversely with body size, and are intimately related. The metabolic rate climbs precipitously with decreasing body weight. Hyraxes conserve energy by having a low metabolic rate and a labile body temperature. The metabolic rate is 30% lower than that predicted on a weight basis while the labile body temperature (which can drop by several degrees at night; Sale 1970a, Louw *et al.* 1972) suggests a strategy adopted by larger animals such as the Common Eland *Tragelaphus oryx*. The labile body temperature is activated by acclimatization and not by a rhythmic daily drop in body temperature (McNairn & Fairall 1984). Body temperatures are regulated mainly by gregarious huddling, long periods of inactivity, and basking. Although their physiology allows Rock Hyraxes to exist in very dry areas and use food of relatively poor quality, they are dependent on shelters (boulders and tree cavities) that provide relatively constant temperature and humidity (Turner & Watson 1965, Bartholomew & Rainy 1971, Rübsamen *et al.* 1982). Turner & Watson (1965) succinctly described kopjes as 'islands of constant environmental conditions surrounded by a sea of environmental extremes'. The Rock Hyrax form on Mt Kenya has much longer and darker fur than its lowland relatives, apparently an adaptation to extreme cold (Coe 1962).

Diet includes a significant amount of grass, and as grass is a relatively coarse material (because of phytoliths or plant opal), the molars and premolars are hypsodont, i.e. they have high crowns with relatively shorter roots. Hyraxes do not ruminate, but because of the complex morphology of their gut they are able to digest fibres as efficiently as ruminants. The morphology of the digestive tract of Rock Hyraxes differs from most other mammals. Its design is well suited for digestion of fibrous diets by means of microbial fermentation. The stomach is divided into a non-glandular part with very slow movements and a glandular part that rapidly mixes the digesta. The large intestine has two fermentation chambers: the caecum, which rapidly mixes the digesta, and the colonic sac, which efficiently, but slowly, mixes digesta. Between these chambers runs the connecting colon. No retrograde transport is observed in any part of the large intestine (Rübsamen *et al.* 1982, Bjornhag *et al.* 1994). There is no gall bladder (Olds & Shoshani 1982).

Rock Hyraxes are typically independent of free drinking water, obtaining their moisture requirements from their food. However, when water is available, they drink regularly. Their efficient kidneys allow them to exist on a minimal moisture intake (Louw *et al.* 1972). In addition, they have a high capacity for concentrating urea and electrolytes and excreting large amounts of undissolved calcium carbonate (Rübsamen *et al.* 1982).

Food and Foraging Diet comprises a variety of grasses, forbs and shrubs, favouring new shoots, buds, fruits and berries (Sale 1965b, Hoeck 1975). In Serengeti N. P., Rock Hyraxes were observed feeding on 79 plant species. The animals have a high seasonal adaptability: in the wet season they showed a high preference for grasses (78%), but in the dry season when grasses became parched and poor in quality they browsed (57%) extensively, and more or less in proportion to the foliage density of each vegetation class. They were observed to feed on 24 grass species such as *Panicum maximum*, *Pennisetum mezianum* and *Themeda triandra* and dicotyledonous plants such as *Cordia ovalis*, *Maerua triphylla*, *Hoslundia opposita*, *Iboza* sp., *Hibiscus lunarifolius*, *Ficus ingens*, *Solanum incanum*, *Grewia fallax* and *Acacia tortilis* (Hoeck 1975).

In studies in southern Africa, Rock Hyraxes exhibited similar seasonal preferences. In Namibia, a study of diet composition from two areas found that grasses were eaten only in significant quantities at the end of the hot, dry season and at the beginning of the wet season (Lensing 1983). Some 35 grass species were consumed, including: *Anthephora pubescens*, *Aristida congesta*, *Cynodon dactylon*, *Enneapogon scaber*, *E. brachystachyus*, *Eragrostis trichophora*, *Stipagrostis hirtigluma* and *S. uniplumis*. The most important dicotyledonous plants in the diet were *Acacia mellifera* and *Ziziphus mucronata*. Fourie & Perrin (1989) found that although they used a wide variety of plant species, ten species constituted more than 80% of the dietary biomass: *Acacia karroo*, *Olea europaea*, *Felicia filifolia*, *Grewia occidentalis*, *Cussonia paniculata*, *Maytenus heterophylla*, *Pentzia* spp., *Clematis brachiata*, *Lycium oxycarpum* and *Diospyros lycoides*. Leaves of trees and shrubs formed the major portion of the diet (being browsed throughout the year), whereas grazing was largely dependent on the seasonal availability of grasses. Examination of $C^{13} : C^{12}$ ratios of carbonate and collagen fractions of bone (DeNiro & Epstein 1978) and microwear patterns of the molariform teeth (Walker *et al.* 1978) confirm that Rock Hyrax switch between grazing and browsing at different times of the year. Rock Hyraxes are thus highly adaptable in contrast to the more specialist browsing Bush Hyraxes. In areas of sympatry, differential feeding behaviour of the two species, especially in the wet season, may limit interspecific competition (Turner & Watson 1965). However, Hoeck (1975) did not find clear niche separation between the two species and reported no instances of interspecific aggression even during the dry season when both species browse.

Sale (1965c) reported an instance of Rock Hyraxes feeding on a poisonous plant *Phytolacca dodecandra*, and they are also known to eat *Lobelia* spp. that are also toxic. However, there are a number of plant species that they avoid, such as *Anthoxanthum nivale*, *Sedum ruwenzoriense* and *Carduus keniensis* (Coe 1962, Young & Evans 1993).

Rock Hyraxes are predominantly diurnal, though they are sometimes active at night. It is not unusual to find them feeding at any time of the day, although there are peaks in the mid-morning and mid-afternoon during warm periods. In winter, peak periods are later, or may become a single extended period (Sale 1965b,

Hoeck 1975, Davies 1994, Brown & Downs 2005). Feeding activity mostly comprises group feeding during which the group often assumes a fan-like orientation that may serve to avoid conflict or to spot predators (Sale 1965b). Often a few individuals act as sentinels during group feeding because feeding away from their refuges makes them vulnerable to attack by predators (Davies 1994). Group feeding is intensive, usually only lasting on average 20 min. Rock Hyraxes usually do not spend more than two hours feeding each day (Sale 1965b). Kingdon (1997) suggested that the disproportionately large jaws are an adaptation to enable intensive feeding. Davies (1994) observed that the daily average distance travelled during group feeding was 169–572 m in the Karoo N. P. (South Africa), but Rock Hyraxes seldom feed more than 15–20 m away from shelter. Kotler *et al.* (1999) showed that the sentinel behaviour during group feeding can allow hyraxes to feed further away from their crevices. Casual feeding by single individuals occurs more sporadically and at any time of the day, usually only a short distance away from the protection of crevices (Sale 1965b, Davies 1994).

Social and Reproductive Behaviour Rock Hyraxes are gregarious (Sale 1970a), living in cohesive and stable family groups or colonies numbering as many as 80 individuals and consisting of 3–7 related adult ♀♀, one adult territorial ♂, dispersing ♂♂, subadult ♀♀ and the juveniles of both sexes. Their numbers vary depending on the size of the kopje: smaller kopjes or rocky outcrops support only a single colony, but larger kopjes may support several family groups, each occupying a traditional range.

There are four classes of mature ♂: territorial, peripheral, and early and late dispersers. Territorial ♂♂ are the most dominant, and repel all intruding ♂♂ from an area largely encompassing the females' core area. Their aggressive behaviour towards other adult ♂♂ escalates particularly in the mating season (when the weight of their testes increases 20-fold; Millar 1972) and when they monopolize all receptive ♀♀. Peripheral ♂♂ are those unable to settle on small kopjes, but on large kopjes they can occupy areas on the periphery of the territorial male's territories. Males are solitary, and the highest ranking among them takes over a female group whenever a territorial ♂ disappears; some studies have reported regular replacement of the dominant ♂. The females' home-ranges are not defended and may overlap. Rarely, an adult ♀ from outside a group will be incorporated into the family group (Hoeck 1982a, Fourie & Perrin 1987a, Davies 1994).

The early dispersers – the majority of juvenile ♂♂ – leave their birth sites at 16–24 months old, soon after reaching sexual maturity. The late dispersers leave a year later, but before they are 30 months old. Individual Rock Hyraxes have been observed to disperse over a distance of at least 2 km, although it has been suggested that gene flow over distances greater than 10 km is unlikely (Hoeck 1982a, b, 1989). However, the further a dispersing animal has to travel across open country, where there is little cover and few hiding places, the greater its chances of dying, either through predation or as a result of its inability to cope with temperature stress (Hoeck 1982a, b, 1989, Gerlach & Hoeck 2001; and see Predators, Parasites and Diseases). Fourie & Perrin (1987a) reported shorter dispersal distances, indicating that the nature of the rocky habitat and the intervening vegetation between habitat islands would strongly influence dispersal distance.

More than 90% of the day is spent resting. Heaping, where several individuals are stacked on top of each other, is observed inside crevices and also outside during very cold conditions. Young can often be observed showing this kind of behaviour with their mother. Even though it can also be observed inside crevices, huddling behaviour is the most common interaction during group resting. This is especially observed early in the morning when the hyraxes first emerge from their crevices to sun themselves. During warmer times of the day, inactive periods consist of solitary resting (Sale 1970b). Hyrax behaviour is directly linked to daily temperature fluctuations and their physiology (Sale 1970b, and see Adaptations). Some of the resting time is spent self-grooming using the lower incisors and the curved claw on the second digit. Grooming and dust-bathing help rid Rock Hyraxes of ectoparasites (see Predators, Parasites and Diseases). Rock Hyraxes urinate and defaecate in latrines. As they have the habit of urinating in the same place, crystallized calcium carbonate forms deposits that whiten the cliff faces below latrines. The precipitated calcium oxylate where urine soaks through the dung heaps and then crystallizes where it seeps out, was used as medicine (hyraceum) by several South African tribes and by European settlers (Hahn 1934, N. Fairall pers. comm.).

Although Rock Hyraxes are gregarious, low levels of intraspecific aggression play an important role in maintaining colonial life (Olds & Shoshani 1982, Kingdon 1997). Visual communication in the form of flaring of the hairs surrounding the dorsal gland and several typical postures is observed. Piloerection of the dorsal spot can either signal alarm (45°) or threat (90°) (Sale 1970a, Fourie & Perrin 1987a). These clear signals and stereotypical appeasement behaviour limit serious aggressive encounters between individuals (Sale 1970b); agonistic behaviour is mostly observed between ♂♂ during the breeding season (Hoeck *et al.* 1982, Fourie & Perrin 1987a). Olfactory communication functions during reproduction and to establish mother–infant bonds (Sale 1970b). Fourie (1977) recorded 21 vocal and four non-vocal sounds. The multitude of grunts, growls, snarls, spits, snorts and squeals are used in a variety of contexts, but most commonly in showing aggression, appeasement or defensive retreat. The alarm call in the form of a sharp bark is characteristic and differs from that of *Heterohyrax* and *Dendrohyrax*. A repetitious bark appears to function in transmitting territorial and sexual signals. The alarm calls used by the sentinels, especially during group feeding, appear to be specific to the particular threat (Sale 1965b, Davies 1994). Males are more vocal during the breeding season and they also use dorsal gland secretions for signalling (Hoeck 1978a, Kingdon 1997). During this time the dominant territorial ♂♂ monitor urine deposits routinely in search of receptive ♀♀ and most copulations are observed between the territorial ♂♂ and adult ♀♀; peripheral ♂♂ most often mate with subadult ♀♀ (Hoeck *et al.* 1982, Fourie & Perrin 1987a). Fourie & Perrin (1987a) observed receptive ♀♀ approaching the dominant ♂ and signalling their readiness to mate through flaring of the dorsal spot hairs, sniffing the male's anogenital region and presentation of their hind-quarters. Hoeck (1978a) observed ♂♂ initiating mating through a mating call, 'weaving head movements' and dorsal spot flaring, followed by brief copulation. Hoeck *et al.* (1982) reported instances where ♀♀ mated with more than one territorial ♂ and with peripheral ♂♂.

The young are born inside the rock crevices, but despite initial bonding with the mother (Sale 1970a), parental care is minimal

(Olds & Shoshani 1982). Juveniles form nursery groups that often engage in social play; both juveniles and subadults have a much larger area of activity compared with the territorial and peripheral adults (Fourie & Perrin 1987a).

Where Rock and Bush Hyraxes occur together, they huddle in single mixed groups in the early mornings after spending the night in the same holes. They also use the same urination and defaecation sites. Parturition tends to be synchronous (Barry 1994). Newborns are greeted and sniffed intensively by members of both species, and they form a nursery group and play together. Most of their vocalizations are also similar (Hoeck 1982b). However, Bush and Rock Hyraxes do differ in key behavioural patterns. They do not interbreed because their mating behaviour is different and they have a different reproductive anatomy. The male territorial call, which might function as a 'keep out' sign, is also different (Hoeck 1978a).

Reproduction and Population Structure There is a single breeding season per year and for both ♂♂ and ♀♀ this represents a very short period of sexual activity (Millar 1971). In South Africa, Fourie & Perrin (1987b) reported male sexual activity from Feb–May with a peak in Apr; testes mass increased dramatically during this time (with a more than ten-fold difference between active and quiescent ♂♂). Females have a mean oestrous cycle length of 13 days (n = 12; Gombe 1983) and Hoeck *et al.* (1982) observed that ♀♀ may cycle several times over a seven-week period. Gestation is 212–240 days, which is exceptionally long relative to body size. It may represent a plesiomorphic characteristic given the much larger body sizes of some of the ancestral hyrax species or it may reflect the adaptive advantages of producing precocial young (Millar 1971).

Within a family group, the pregnant ♀♀ all give birth within a period of about three weeks; this birth synchrony appears to be a mechanism to avert predation by predators such as eagles (Barry 1994, Barry & Barry 1996). Some authors have suggested that parturition may be linked to rainfall or photoperiod. For example, in most areas in South Africa the breeding season is in late summer (with a peak in Apr), but with a shift from earlier conception in the south-west (Jan–May) relative to the north-east (May–Jul). This results in a shift in births from the end of Aug in the south-west to the end of Mar in the north-east (Millar 1971, Stuart & Stuart 1984, Fourie & Perrin 1987b). In Serengeti N. P., Hoeck (1982a) recorded births from Mar–May while Mendelssohn (1965) found that most of the births in Israel occurred in Apr, supporting photoperiod as one of the proximate causes. Stuart & Stuart (1984) suggested that the timing of the breeding season reflects adaptation to more favourable temperatures for the newborn young. In the arid areas of NW South Africa they reported conception from Sep–Nov and parturition from May–Jul.

The number of young per ♀ varies between 1 and 4 (mean 2.4) in Serengeti N. P. (Hoeck 1982a). In southern Africa, a ♀ with six embryos has been collected, and mean litter-sizes recorded include: 3.3 in the Western Cape (Millar 1971); 2.7 in Mountain Zebra N. P., Eastern Cape (n = 95) (Fourie & Perrin 1987b); 2.4 in the Free State (range 1–5; n = 49) (Van der Merwe & Skinner 1982); and 2.37 in Zimbabwe (n = 19) (Barry 1994). Sale (1969) and Fourie & Perrin (1987b) suggested an increase in litter-size with latitude while several authors demonstrated a relationship between litter-size and female age; first breeders have only 1–2 young (Fairall *et al.* 1986), ♀♀ 2–8 years of age produce the largest litters, followed by a decline in older ♀♀ (Fourie & Perrin 1987b). Millar (1971) indicated that nutritional conditions would affect both litter-size and age at first breeding.

Sketches of *Procavia capensis mackinderi*.

Young are fully developed at birth, fully haired and with eyes open. They are capable of agile movement within a few days and can ingest solid food within a few weeks (Mendelssohn 1965). Birth-weight varies to some degree depending on the number in the litter (mean: 195 g) and the average litter weight (581 g) is high relative to female body weight (Millar 1971). Suckling young assume a strict teat order (Hoeck 1977a). Weaning occurs at 1–5 months (mean: 3 months, Miller 1971), though Young & Evans (1993) mention one individual still suckling at nine months. Both sexes reach sexual maturity at about 16–17 months of age (Millar 1971, Hoeck 1982a). Millar (1971) reported some ♀♀ breeding at five months of age and according to Fourie & Perrin (1987b) ♂♂ reach sexual maturity at 28–29 months. The findings of the latter study suggest that there is a difference between physiological and behavioural sexual maturity in ♂♂; the dominant ♂ effectively prevents mating by sexually mature young ♂♂.

The sex ratio is equal at birth and up to two years of age, whereafter ♀♀ sometimes tend to outnumber ♂♂ (Millar 1971, Hoeck 1982a, Fourie & Perrin 1987a). A high level of juvenile mortality has been recorded by Fourie (1983) and Davies (1994) and appears to be an important factor controlling the dynamics of hyrax populations. Adult ♀♀ live significantly longer than adult ♂♂ and may reach an age of over ten years (Hoeck 1982a, Kingdon 1997); in captivity, animals have lived to more than 14 years (Weigl 2005).

Upon sexual maturity, ♀♀ usually join the adult female group, while ♂♂ disperse before they reach 30 months. Greenwood (1980) proposed that most mammalian species display male-biased dispersal. This was confirmed for Rock Hyraxes by observation in several studies (Hoeck 1982a, 1989, Hoeck *et al.* 1982, Fourie & Perrin 1987a, Davies 1994), and although adult dispersal was also recorded, juvenile male dispersal appeared to be of particular importance. However, preliminary results involving microsatellite

DNA analysis of Rock Hyraxes in Serengeti N. P. showed no sex bias and revealed extremely low levels of genetic variation within and between neighbouring colonies (Gerlach & Hoeck 2001). Prinsloo & Robinson (1992) also suggested, on the basis of the patterns of geographic variation of maternally inherited mtDNA, that there are high levels of historical connectedness among localities in South Africa. Additional molecular markers are needed to reconcile direct and indirect estimates of dispersal and gene flow.

Although occupancy of some habitat islands remains stable over time (Hoeck 1982a, Hoeck *et al.* 1982), most regions are characterized by fluctuations in the numbers of Rock Hyraxes and probably function as metapopulations with local extinction and recolonization events. This is reflected by colonies occupying kopjes in the Serengeti, where low levels of allelic diversity and heterozygosity at eight hyrax-specific microsatellite loci suggest the influence of metapopulation dynamics and population bottlenecks (Gerlach & Hoeck 2001). Davies (1994) suggested that the dynamic population structure of Rock Hyraxes is a consequence of their unpredictable environment.

Predators, Parasites and Diseases Rock Hyraxes form the most important prey in the diet of Verreaux's Eagle *Aquila verreauxii* (Gargett 1990). Because ♂♂ are forced to disperse when mature, one- to two-year-old (as well as old) ♂♂ are particularly at risk of predation. Boshoff *et al.* (1991) found that juveniles constituted 11–33% of the remains of *Procavia* in the nests of Verreaux's Eagles in the Western Cape, and Barry & Barry (1996) found that juveniles comprised 18% of remains in the Matobo N. P., Zimbabwe. Rock Hyraxes also form a main component of the diet of Crowned Hawk-eagles *Stephanoeatus coronatus* (Boshoff *et al.* 1994). Other predators include Martial Eagles *Polemaetus bellicosus* and Tawny Eagles *Aquila rapax*, Leopards *Panthera pardus* (particularly on Mt Kenya), Lions *P. leo*, Caracals *Caracal caracal*, jackals *Canis* spp., Spotted Hyaenas *Crocuta crocuta* and snakes (Coe 1962, Turner & Watson 1965, Palmer & Fairall 1988).

There is an extensive literature on the variety of external parasites (ticks, biting and sucking lice, mites and fleas) that have been collected from Rock Hyraxes (see for example, Hoogstraal & Wassef 1981, Horak & Fourie 1986, and references therein). In the study by Horak & Fourie (1986), only ten individuals of some 10,000 ticks recovered belonged to species that could infest domestic livestock. Dust-bathing probably helps keep parasite burdens relatively low. Rock Hyraxes also harbour a number of internal parasites, including nematodes and cestodes (see, for example, Fourie *et al.* 1987), which could play a role in hyrax mortality in some areas. Fourie *et al.* (1987) investigated the seasonal variation in the densities of both ecto- and endo-parasites on Rock Hyraxes in the Mountain Zebra N. P. in South Africa, and found that ♀♀ showed significant seasonal differences in tick and biting lice densities, with these being highest in the summer (also the time of the highest tick densities). This difference between the sexes was related to social structure (♀♀ being more social than ♂♂, making them vulnerable to increased infestation rates) and to the decreased physiological condition of ♀♀ (see Fourie & Perrin 1985) during the summer. No significant seasonal differences were noted in endoparasite densities.

In Serengeti N. P. the 'sarcoptic mite', which causes mange, is an important cause of mortality for Rock Hyraxes (Hoeck 1982a), and ♀♀ that matched these symptoms have been seen on Mt Kenya (Young & Evans 1993). In Kenya and Ethiopia, Rock Hyraxes might be an important reservoir for the parasitic disease leishmaniasis (Ashford 1970).

Conservation IUCN Category: Least Concern. CITES: Not listed.

Although this species is subject to some localized hunting and may have been extirpated in some small localities, it has a wide distribution on the African continent (and extralimitally). Currently it is present in a number of protected areas across its range, and is generally not believed to be at any risk of extinction in the wild.

Measurements

Procavia capensis johnstoni
HB (♂♂): 484 (395–578) mm, n = 41
HB (♀♀): 496 (439–539) mm, n = 33
HF (sexes combined): 656 mm, n = 63
E (sexes combined): 320 mm, n = 63
WT (♂♂): 3.0 (1.8–4.5) kg, n = 66
WT (♀♀): 3.2 (2.0–5.4) kg, n = 57
WT (subadult): 1.3 (0.2–2.6) kg, n = 86
Serengeti N. P., Tanzania (Hoeck 1982a, H. N Hoeck unpubl.)

Procavia capensis capensis
TL (♂♂): 476 (376–560) mm, n = 26
TL (♀♀): 499 (415–628) mm, n = 31
HF c.u. (♂♂): 67 (56–91) mm, n = 26
HF c.u. (♀♀): 66 (62–78) mm, n = 30
E (♂♂): 32 (28–37) mm, n = 25
E (♀♀): 32 (27–38) mm, n = 28
WT (♂♂): 2.8 (1.5–4.3) kg, n = 10
WT (♀♀): 3.26 (1.8–4.3) kg, n = 12
Former Transvaal, South Africa (Rautenbach 1982)
Smithers & Wilson (1979) give the average weight for ♂♂ as 3.52 kg (range 3.21–4.65, n = 10) and for ♀♀ as 3.09 kg (range 2.47–3.46, n = 10).

Procavia capensis
GLS (♂♂): 92.0 (83.5–104.2) mm, n = 80
GLS (♀♀): 89.8 (79.9–98.6) mm, n = 99
GWS (♂♂): 54.0 (49.1–64.9) mm, n = 86
GWS (♀♀): 51.8 (41.7–60.2) mm, n = 99
Southern Africa (Yom-Tov 1993)

Key References Barry & Mundy 1998, 2002; Bothma 1971; Coe 1962; Fourie & Perrin 1987a, b; Hoeck 1975, 1982a, b, 1989; Millar 1971, 1973; Rübsamen *et al.* 1982; Sale 1965b, 1966, 1970a, b.

Hendrik N. Hoeck & Paulette Bloomer

Superorder TETHYTHERIA

Superorder Tethytheria McKenna, 1975. *In*: Phylogeny of the Primates (Luckett W. P. and Szalay F. S., eds), p. 42.

The superorder Tethytheria contains the extant orders Proboscidea and Sirenia and the extinct order Desmostylia; the extinct order Embrithopoda and family Anthracobunidae are often included as well (Gheerbrant *et al.* 2005a). Like so many other supraordinal nomina currently being used in placental systematics, Tethytheria is not only a phylogenetic hypothesis but also a biogeographic hypothesis, and implies that the origin and early evolution of tethytherians occurred in or around the ancient Tethys Sea, which isolated Afro-Arabia from Europe and Asia throughout the Palaeogene.

A close relationship of proboscideans and sirenians to the exclusion of Hyracoidea has long been suggested by various forms of morphological data and was first articulated by Henri M. D. de Blainville (who used the now-abandoned taxon 'Gravigrades') in 1834. Surprisingly, with the advent of molecular systematics this phylogenetic hypothesis has not come to be similarly well supported by genomic data. Though one large DNA dataset supports Tethytheria (Poulakakis & Stamatakis 2010), another dataset incorporating DNA and amino acids supports a Hyracoidea+Proboscidea clade (Meredith *et al.* 2011). A single retroposon provides support for a hyracoid–sirenian clade, but none have been found that support Tethytheria (Nishihara *et al.* 2005, Poulakakis & Stamatakis 2010). Furthermore, some morphological features that were thought to be tethytherian synapomorphies (Fischer 1990) have been revealed as probable convergences by a better early fossil record of proboscidean and sirenian evolution (Court 1994, Gheerbrant *et al.* 2005b).

The remaining morphological evidence for Tethytheria is nevertheless still compelling, and includes both skeletal and soft-tissue evidence (Fischer & Tassy 1993, Shoshani 1993). There is also some palaeontological evidence for a shared semi-aquatic tethytherian ancestor, as the earliest proboscideans, such as *Eritherium*, *Phosphatherium* and *Daouitherium*, are from marine deposits, and morphological adaptations and isotopic signatures for a semi-aquatic existence are also evident in later genera such as *Moeritherium* (Liu *et al.* 2008). Desmostylians were clearly dedicated to such a life-style. In light of this evidence, a closer relationship of hyracoids and proboscideans, or of hyracoids and sirenians, would imply that 'tethytherian' features are either primitive within Paenungulata, or evolved convergently among early sirenians and proboscideans.

Erik R. Seiffert

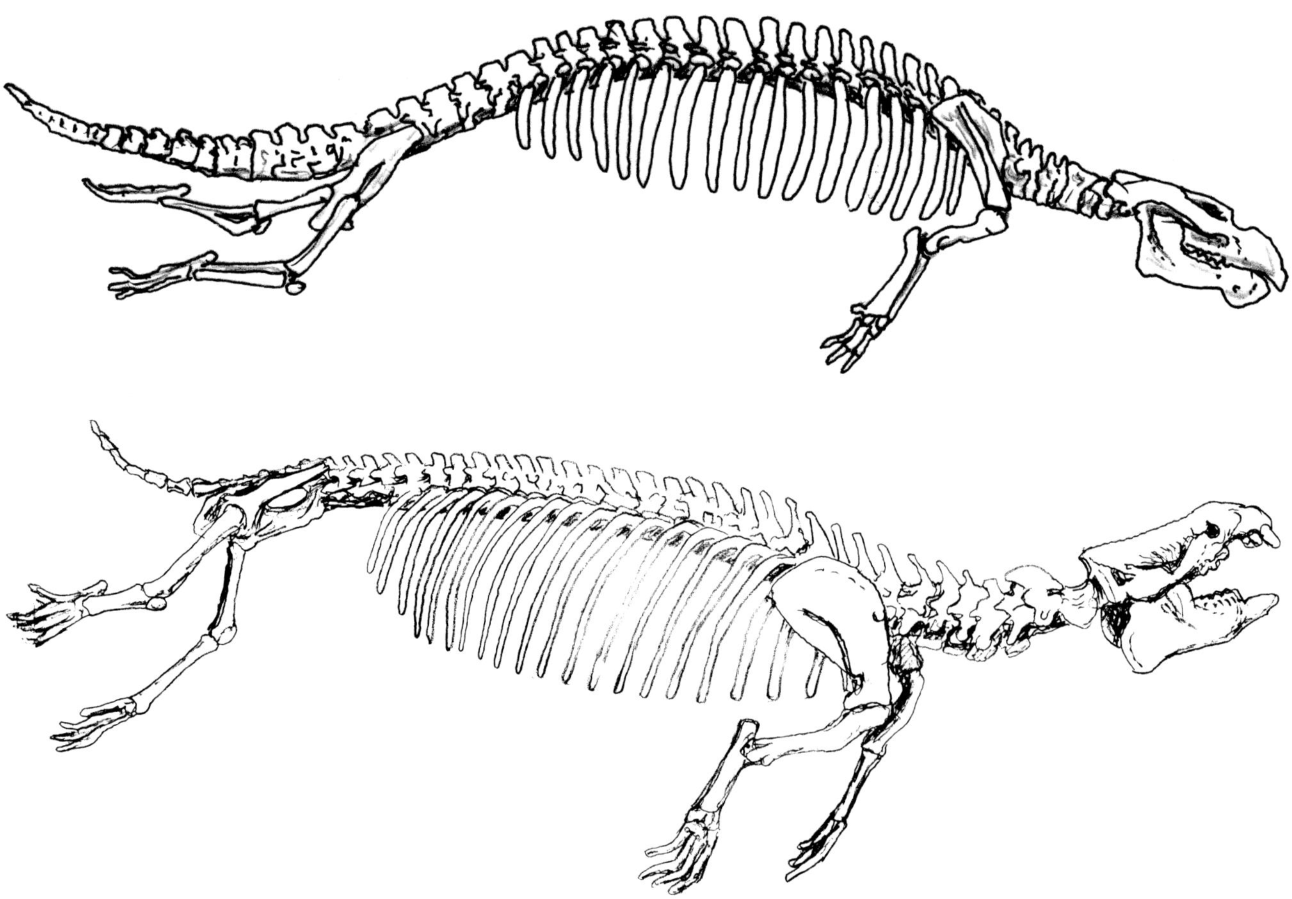

Two early aquatic tethytheres: skeleton of an extinct quadrupedal sirenian, *Pezosiren* (*above*), and the skeleton of a proboscidian, *Moeritherium* (*below*).

Order PROBOSCIDEA – Elephants

Proboscidea Illiger, 1811. Prodromus systematis mammalium, p. 96.

Traditionally (since Illiger 1811) the category or rank of 'order' has been used for Proboscidea. McKenna & Bell (1997), following a cladistic classification, proposed a new category ('parvorder') for Proboscidea. For reasons of stability, we retain the ordinal category here, even though the category parvorder conveys a cladistic message.

The order Proboscidea consists of approximately ten families, 45 genera and 185 species and subspecies. Of these only two or three species are living, all classified in the family Elephantidae. The living elephants include the African and the Asian species. The African group incorporates the Savanna (or African Bush) Elephant *Loxodonta africana* and the Forest Elephant *L. cyclotis*, although this taxonomy is not agreed upon by all scientists.

At the time when Illiger (1811) coined the named Proboscidea, only the living elephants, the extinct American mastodon (named '*Elephas americanus*' by Kerr in 1792) and the woolly mammoth (named '*Elephas primigenius*' by Blumenbach in 1799) were well known. Illiger used the familiar proboscis or trunk as the distinguishing character for this order of mammals ('proboscis' is of Greek origin and it means 'before the mouth'). Today, we recognize some extinct proboscidean species, none of which is believed to have possessed a proboscis, for example *Moeritherium lyonsi*, *Phosphatherium escuilliei*, *Daouitherium rebouli* and *Numidotherium koholense*. It is, therefore, necessary to define the synapomorphies or shared-derived characters for all proboscideans, extinct and extant. Four proboscidean synapomorphies are recognized by Gheerbrant *et al.* (2005): (1) a well-developed zygomatic process of the maxillary bone, which contributes significantly to the ventral of the orbit and the zygomatic arch (in other mammals the maxilla contributes little or none to the ventral border of orbit); (2) internally, the periotic bone has a relatively large size of the mastoid portion (*pars mastoidea*) compared with the *pars cochlearis*; (3) the hypoconulid cuspid of the lower teeth is in the labial position (in other mammals the hypoconulid is either at the centre or on the lingual side of the tooth); and (4) a retracted optical foramen in the orbitotemporal fossa towards the posterior of the cranium. Note that in Gheerbrant *et al.*'s (2005) analysis members of extinct Anthracobunidae are not closely related to Proboscidea.

Classification of the Proboscidea is incomplete mostly because of the paucity of post-cranial elements for early members of this order, and thus there are not enough distinguishing characters assigned to early taxonomic units within the order. Shoshani *et al.* (2001) proposed the name Plesielephantiformes as a sister taxon to Elephantiformes; this suggestion was incorporated in one of the most recent classifications of the Proboscidea (Shoshani & Tassy 2005). In this scheme, the position of Moeritheriidae remains uncertain. The unity of Plesielephantiformes was based on the lophodonty of cheekteeth in members of Numidotheriidae, Barytheriidae and Deinotheriidae, a hypothesis that is not supported for all members of Plesielephantiformes, based on a parsimony analysis by Gheerbrant *et al.* (2005). None the less, E. Gheerbrant (pers. comm.) proposes that lophodonty of cheekteeth may be a synapomorphy for Plesielephantiformes, with some extra homoplasies (with the curently available characters, the most parsimonious solution is that lophodonty is a primitive trait for Proboscidea).

Within the Proboscidea there are shared-characters for the clade of Elephantimorpha, that is, Mammutida (true mastodons) and Elephantida (gomphotheres, stegodontids and elephantids). These characters are typical of the living elephants and they include: forwards displacement of premolars and molars as though they were moving on a slow conveyor belt, columnar arrangement of long bones in limbs (graviportal, in most taxa), and a well-developed proboscis (Shoshani 1996, Tassy 1996). Elephantimorpha probably originated in Africa during the late Oligocene, about 28–27 mya (Kappelman *et al.* 2003, Sanders *et al.* 2004, Shoshani *et al.* 2006b). Forwards displacement of cheekteeth and graviportal adaptation are easily detected on bones and teeth of extinct taxa, yet a soft tissue like the proboscis has to be inferred. The presence of a developed proboscis in extinct forms is conjectured based on the elevated position of the single external naris, the dorsal contact between premaxillae and frontal bones, that the premaxillae make an ascending process for the mesethmoid cartilage in the nasal fossa, and the large infraorbital canals on either side of the face.

Among early members of Proboscidea, Deinotheriidae may have had graviportal limbs (Osborn 1936); however, recent data may lead to different interpretations. A convergent evolution between Deinotheriidae and more advanced proboscideans cannot be definitely ruled out. Deinotheres are also believed to have had a trunk (possibly also evolved in convergence: but note there is no ascending process of the premaxillae in the nasal fossa). Although often depicted in older literature with a developed proboscis (Osborn 1936), Harris (1975), Tassy (1998) and Markov *et al.* (2001) proposed that a shorter trunk than that of the living elephants was probably more likely possessed by deinotheres.

Evolutionary trends of proboscideans from the earliest fossilized member (*Eritherium azzouzorum* that lived in the late Palaeocene

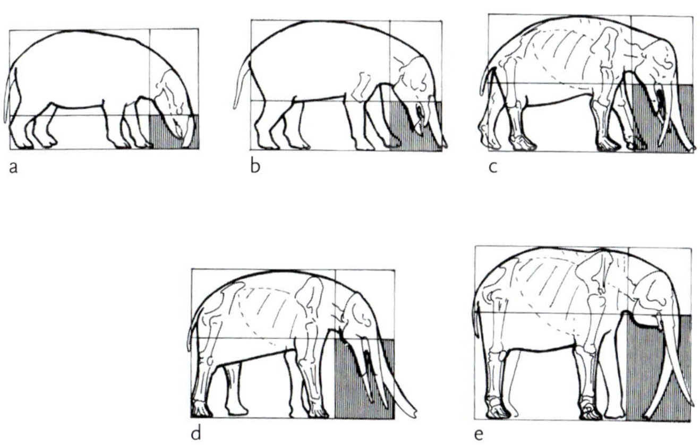

Changes of proportions in the evolution of Proboscidea. Horizontal line indicates rising level of mouth. Texture indicates principal feeding zone. a. Reconstruction of ancestral proboscidean. b. *Phiomia*. c. *Gomphotherium*. d. *Stegotetrabelodon*. e. *Loxodonta africana*.

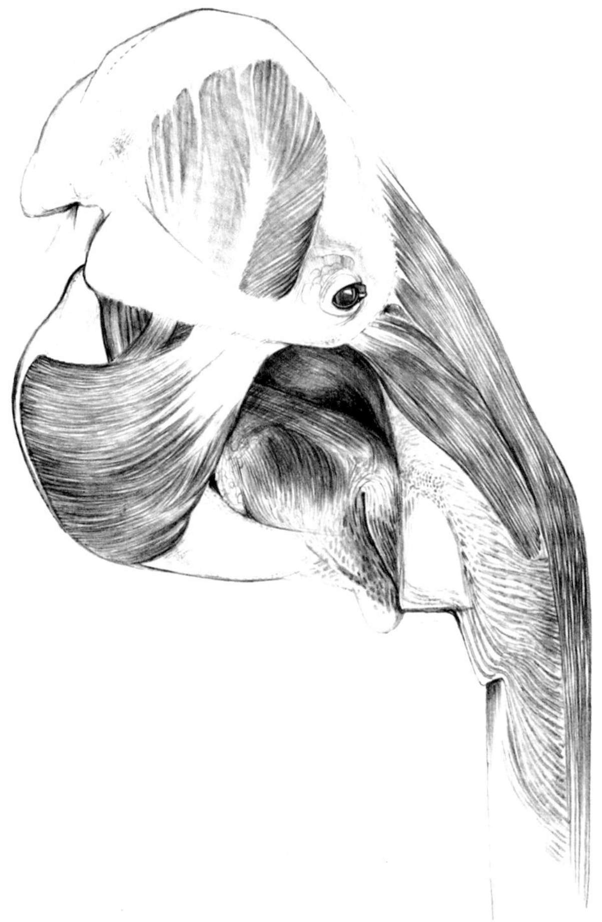

Myology of *Loxodonta africana* head.

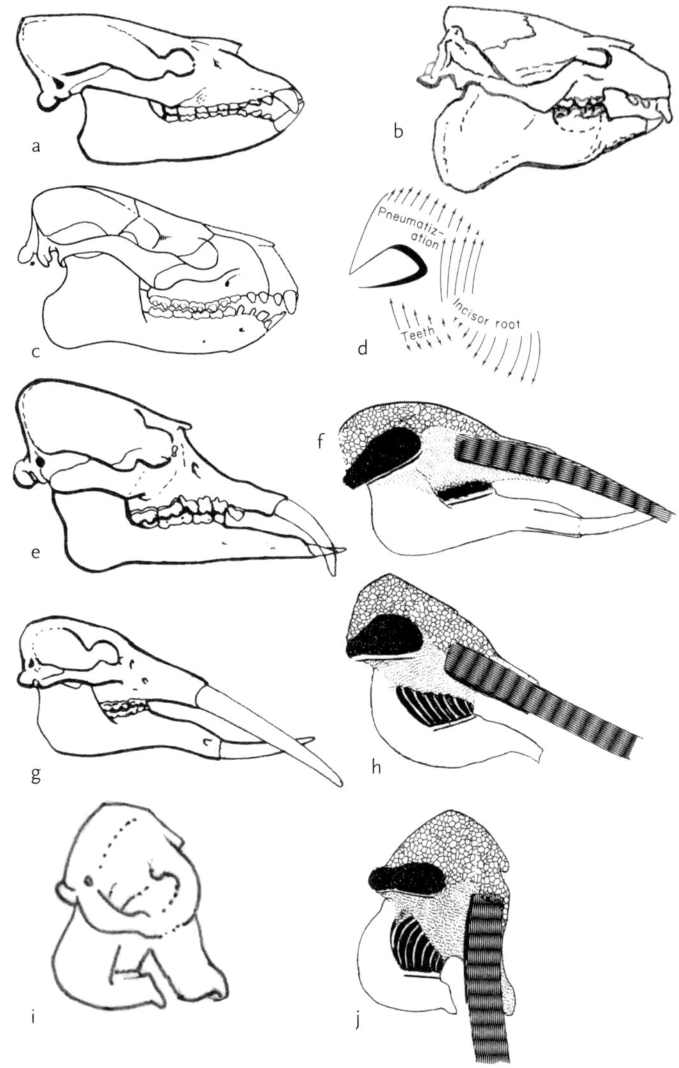

Proboscidean incisors and toothrow in relation to brain and basi-cranial axis. a. Ancestral proboscidean as reconstructed in Kingdon (1979). b. *Moeritherium trigodon*. c. Eocene *Phosphatherium escuilliei*. d. Directions of increases in pneumatization, of changes in orientation of incisors and in size of molars. e. *Phiomia*. f. *Gomphotherium*. g. *Gomphotherium*. h. *Elephas planifrons*. i. *Loxodonta africana*. j. *Loxodonta africana* ('c' after Gheerbrant *et al.* 2005).

epoch, about 60 mya; Gheerbrandt 2009) to the present, include: increase in the number of lophs or plates on premolars and molars; shift from regular tooth replacement (deciduous teeth to permanent teeth) to horizontal displacement of premolars and molars; increase of tusk size (proboscideans have the largest teeth of any mammals, living and extinct); development of a trunk, or proboscis; increase of trunk length; and a ten-fold increase in the encephalization quotient (EQ). All these trends can be summarized as – overall increase in body size. Evolutionary processes incorporate gigantism (over 4 m shoulder height) and dwarfism (only 1 m tall), co-evolution of infrasonic communication and the ability to store water in the pharynx (Shoshani 1998). *Phosphatherium escuilliei* was about the size of a dog (10–15 kg), but it was not a dwarf; it did not have a trunk, tusks, nor horizontal displacement of premolars and molars. Nevertheless, *Phosphatherium* was a proboscidean since it possessed unique proboscidean characters such as a well-developed zygomatic process of the maxillary bone (Gheerbrant *et al.* 2005). The elephant adult brain averages 4783 g, the largest among living and extinct terrestrial mammals; during evolution, EQ has increased by ten-fold, 0.2 for extinct *Moeritherium*, ~2.0 for extant elephants (details in Shoshani *et al.* 2006a).

Despite our increasing knowledge of the early history of the Proboscidea, there is still much uncertainty concerning the place of origin of this group of mammals. Fossils of Proboscidea have been recovered from all continents except for Antarctica, Australia and some oceanic islands. Numidotheres (e.g. *Phosphatherium*, *Daouitherium* and *Numidotherium*), for example, were found in north-west Africa, in the early-middle Eocene. Africa is believed to have been isolated from other continents during most of the Palaeogene (Palaeocene, Eocene and Oligocene), and thus its fauna during these geological epochs was endemic. E. Gheerbrant (pers. comm.) has suggested that Palaeogene proboscideans are representative of the whole African province, and that proboscideans are of African origin (see also Gheerbrant 1997, Gheerbrant *et al.* 1998). Yet, a late Oligocene member of the Elephantimorpha was described in Pakistan by Antoine *et al.* (2003). We thus consider migration and dispersal patterns of the earliest proboscideans from north-west Africa during the Palaeogene to be uncertain. However, one possibility emerges that the northern shores of the Mediterranean Sea (a remnant of the ancient Tethys Sea) might be the place of origin of Proboscidea (for this discussion, members of Anthracobunidae are excluded). North-east Africa (Egypt, Libya) and Arabian Peninsula (Oman) embodied environmental conditions where fossils of *Moeritherium*,

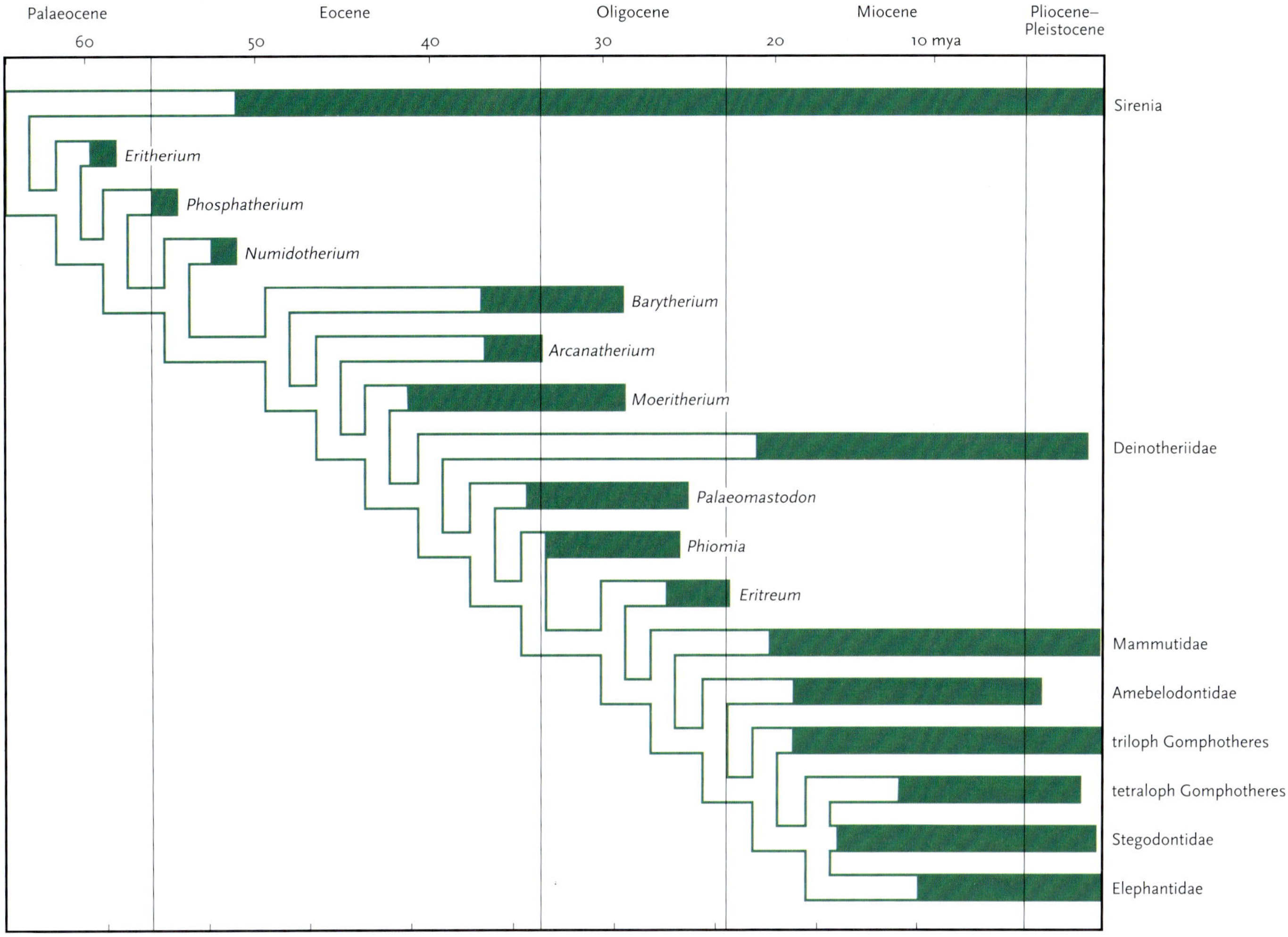

Tentative schematic phylogenetic tree of proboscid evolution. Dark bands indicate known fossil taxa. White tree indicates supposed relationships (after Tassy 1996 & Gheerbrant & Tassy 2009).

Barytherium, *Omanitherium*, *Palaeomastodon* and *Phiomia* have been found in the late Eocene to Oligocene sediments. It seems plausible that north-eastern African proboscideans may have migrated to the Horn of Africa (late Oligocene) and to East Africa (Miocene) where centres of radiations of some proboscideans (including deinotheres and gomphotheres) are believed to have taken place (see Sanders *et al.* 2010).

Another later centre of radiation of extinct gomphotheres is believed to have occurred in Asia. From the Horn of Africa (again following the geological evidence we have thus far), it is suggested that some proboscideans (possibly gomphothere stock) migrated to what is today the Saudi Arabian peninsula (late Oligocene to early Miocene) and from there towards the general area of what is today Pakistan. *Elephas maximus*, originally from East Africa, spread into Asia and Europe during the Pliocene. The living African elephants (*L. africana* and *L. cyclotis*) are believed to have originated in eastern Africa and did not migrate out of the continent. Like the classification and the evolutionary tree, maps of dispersals are subject to constant changes with the discovery of new fossils and/or different interpretations of old material.

The Proboscidea have been grouped together with members of the order Sirenia (manatees and sea cows) in the superorder Tethytheria (which also contains the extinct order Desmostylia; Gheerbrant *et al.* 2005). The Tethytheria and order Hyracoidea form the Paenungulata. However, Tethytheria is not well supported by genomic data; for example, one recent study incorporating DNA and amino acids supports a Hyracoidea+Proboscidea clade (Meredith *et al.* 2011). Refer to the higher-level profiles for further discussion.

Jeheskel Shoshani & Pascal Tassy

Family ELEPHANTIDAE

ELEPHANTS

Elephantidae Gray, 1821. London Med. Repos. 15: 305.

Elephantidae is one of approximately ten families in the order Proboscidea, comprising about 30 species, three of which are extant, one species being found in Asia and two in Africa. Fossil remains of extinct taxa have been recovered from Africa, Asia, Europe and North America, dating to as early as the late Miocene. The fossil record of the living genera (*Loxodonta* and *Elephas*) also goes back to the late Miocene, between 7.3 and 5.4 mya (Maglio 1973, Shoshani & Tassy 1996, Todd & Roth 1996, Tassy 2003).

Maglio (1973) divided the Elephantidae into two subfamilies: Stegotetrabelodontinae (with two genera – *Stegotetrabelodon* and *Stegodibelodon*) and Elephantinae (with four genera – *Primelephas*, *Loxodonta*, *Elephas* and *Mammuthus*). This system has been followed by most other authors (e.g. Coppens *et al.* 1978, Shoshani & Tassy 1996). Kalb *et al.* (1996) included *Stegolophodon* and *Stegodon* in the family Elephantidae. Saegusa (1996) classified the last two genera in the Stegodontidae (and see Tassy 1996 and Shoshani 1996). Shoshani & Tassy (2005) and Shoshani *et al.* (2007) added *Palaeoloxodon* to the Elephantinae, a hypothesis originating with Osborn (1942). Synapomorphies of Elephantidae (excluding stegodontids) include lophes and lophids modified into columnar laminae (Tassy 1996, pp. 41, 44) and hypsodont molars (Shoshani 1996, p. 175).

Relationships among Elephantidae taxa, living and extinct, have been presented by Aguirre (1969), Maglio (1973), Coppens *et al.* (1978), Tassy & Darlu (1987), Froehlich & Kalb (1995), Kalb *et al.* (1996), Shoshani (1996), Shoshani *et al.* (1998) and Tassy (1996), to name a few more recent authors. In general, most authors agree on a cladistic relationship among members of the subfamily Stegotetrabelodontinae and among the basal taxa of Elephantinae (*Primelephas*). Disagreements concerning the relationships among *Loxodonta* (extant), *Elephas* (extant) and *Mammuthus* (extinct) can be summarized in two hypotheses. The first is the traditional, morphology-based hypothesis that *Elephas* and *Mammuthus* are closer to each other than either is to *Loxodonta*, cladistically expressed as (*Loxodonta* (*Mammuthus*, *Elephas*)). This hypothesis was advanced by the morphological studies of Maglio (1973), Tassy & Darlu (1987), Kalb *et al.* (1996), Shoshani (1996), Tassy (1996) and by molecular studies (e.g. Yang *et al.* 1996, Ozawa *et al.* 1997). The second, more recent, molecular-based hypothesis noted that *Loxodonta* and *Mammuthus* are closer to each other than either is to *Elephas*, cladistically expressed as (*Elephas* (*Mammuthus*, *Loxodonta*)), and proposed by Debruyne (2001), Debruyne *et al.* (2003a) and Thomas *et al.* (2000). Most recent studies (Krause *et al.* 2006, Poinar *et al.* 2006, Rogaev *et al.* 2006, Rohland *et al.* 2010) that include larger molecular data for the extinct mammoth corroborate the traditional hypothesis of (*Loxodonta* (*Mammuthus*, *Elephas*)). Shoshani *et al.* (2007) also corroborate the traditional hypothesis based on independent data from the hyoid apparatus.

Molecular studies also contribute to better understanding of species composition within Africa. Until recently, the African elephant was divided into two subspecies, *L. africana africana* and *L. a. cyclotis* (Sikes 1971). However, recent taxonomic revision suggests dividing the African elephant into two species: the Forest Elephant (*L. cyclotis*) and the Savanna Elephant (*L. africana*). However, not all authors agree on this classification (see genus profile for further discussion).

Common features for the living elephants include large size (*Loxodonta* is the largest living terrestrial animal, reaching a height of 4 m at the shoulder and a weight of 6 tonnes); longevity (elephants can live to about 80 years); and presence of trunk or proboscis (not all extinct proboscideans had a trunk). The proboscis is a combination of the upper lip and nose, the two nostrils continue throughout the length of the trunk; the trunk is a multi-purpose prehensile organ and is probably the most important appendage for survival. The skin (dermis and epidermis) can reach 32 mm in thickness, yet it is sensitive, movable and the body is covered with little hair and bristles (most hair is on head and tail, young elephants are more hairy than adult). The

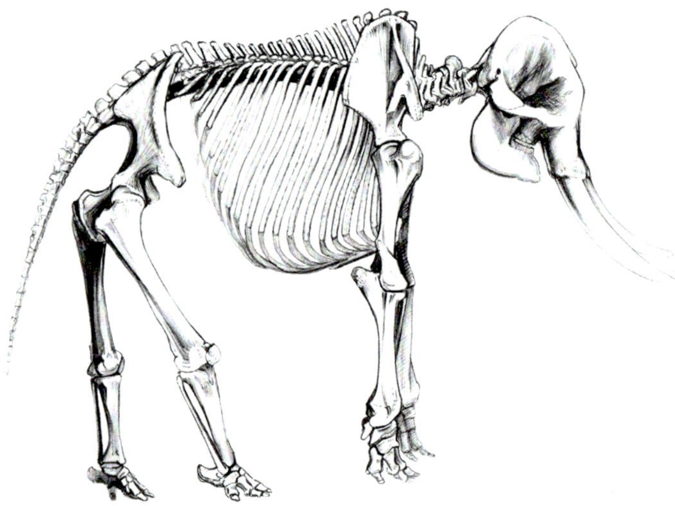

Skeleton of *Loxodonta africana*.

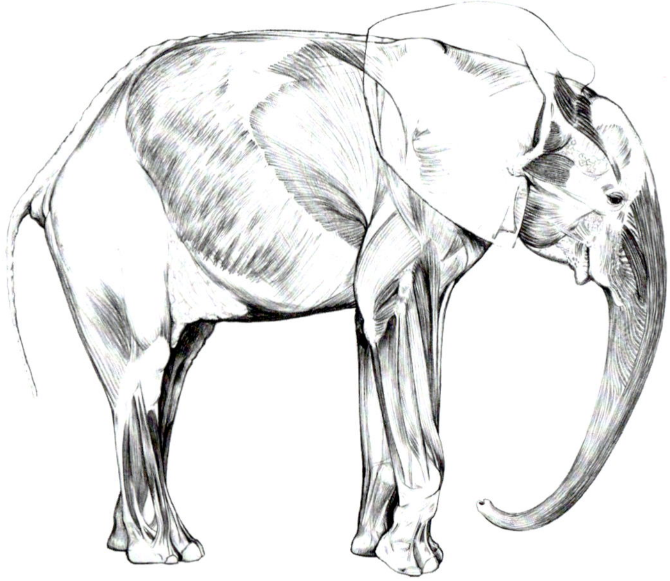

Superficial myology of *Loxodonta africana*.

limbs are columnar with graviportal stance; the sole of the forefoot is larger (used for support) and round, that of the hindfoot (used for support and propulsion) is oval. In addition, newly born and juvenile elephants have relatively 'flat' soleprints, or very few features may be observed. With age, more and more soleprint features are evident and in adults, traditional hunters have long recognized that soleprints are unique for individual elephants (Shoshani *et al.* 2004).

The cranium is pneumatized (filled with air sinuses), thereby reducing the weight of the cranium (Badoux 1961), while providing ample attachment for the nuchal and masticatory muscles. The external nares (where the trunk begins) are elevated. A secondary acoustic meatus is present, as are the alisphenoid canal and the mandibular coronoid canal. The hyoid apparatus consists of five bones. Premolars and molars are composed of plates (lamellae) held together with cementum, the inside of each plate contains dentine, surrounded by enamel, lozenge-shaped in *Loxodonta* (hence the name) and in compressed loops in *Elephas* ('Elephas' means huge arch). Up to 15 plates are present in *Loxodonta* and 29 in *Elephas*. The molars are very large (the last molar can weigh over 5 kg); dental formula is: I $^1/_0$, C $^0/_0$, P $^3/_3$, M $^3/_3$ = 26; chewing teeth (premolars and molars) exhibit forward or horizontal displacement, a feature also shared by non-Elephantidae taxa, but not with all proboscideans); and very large tusks. Tusks are used for defence, offence, display and feeding. Tusks are enlarged incisors (elephants have the longest teeth of any living or extinct mammals), with a record of 345 cm long for an African elephant. Tusks of elephants exhibit a unique and complex pattern of criss-crossing lines called Schreger pattern, also known as 'engine turning' or guillochage (see Fig. 2.7, p. 15 in Shoshani 1996); this pattern is also present in dentine of cheekteeth (summarized from Laursen & Bekoff 1978, Shoshani & Eisenberg 1982, Roth & Shoshani 1988, Shoshani 1996, 2000, Shoshani & Tassy 1996, Shoshani *et al.* 2007).

The vertebral formula is: Cervical 7 (for *Loxodonta* and *Elephas*), Thoracic 20–21 (for *Loxodonta*), 19–20 (for *Elephas*), Lumbar 3–4 (for *Loxodonta*), 3–5 (for *Elephas*), Sacral 4–6 (for *Loxodonta*), 3–5 (for *Elephas*), Caudal 18–33 (for *Loxodonta*), 24–34 (for *Elephas*). The radius and ulna are separate, permanently crossed in pronation position; the tibia and fibula are also separate and the latter articulates with the calcaneum. The ilium is almost vertically expanded laterally, and the acetabular fossa is directed downwards. Both manus and pes are pentadactyl, the manus is larger than the pes, and there is a serial bone arrangement in the carpus, and in the pes the astragalus does not articulate with the cuboid and lacks the astragalar foramen. The long bones generally lack medullary marrow cavities, instead a mesh of cancellous bone allows passage of blood vessels for haematopoetic functions (Shoshani 1996). The epiphyses on long bones fully fuse at about 30 years in females and about 35–40 years for males (Roth 1984). Unlike most mammals, the mandibular fossa is not a convex–concave form, but a double convex part with a double concave cushion of fibrous tissue in between condyle and mandibular fossa (Shoshani *et al.* 1982).

Elephants are adapted to a variety of habitats from desert to mountain-tops, and exhibit nocturnal, diurnal and crepuscular activity patterns to avoid harassment and hunting. Their predators include Lion *Panthera leo*, Tiger *P. tigris* and humans; calves have been preyed upon by crocodiles. Long distance communication between and among herds is achieved by powerful calls containing infrasonic frequencies (5–24 Hz). The degree of development of senses changes with age. Elephants have an excellent sense of hearing, an acute sense of smell, very good sense of touch, unknown sensitivity to taste (seems to be selective) and poor sense of vision, though it is good in dull light. Living elephants migrate long distances in search of food (an adult elephant requires 100–250 kg per day) and water (an adult elephant requires 75–150 l per day) and to reduce or avoid inbreeding. Elephants are keystone and super-keystone species. They are herbivorous (*Loxodonta* is more browser than grazer and *Elephas* is more grazer than browser), social mammals that live in herds led by the matriarch, where the young have the opportunity to be in close contact with the adult and copy or learn the art of survival. Gestation period lasts 18–24 months and the newly born are precocious and able to follow their mothers shortly after birth; they suckle from nipples (not teats) at the axillary mammae. Elephants exhibit complex behaviours, including altruism, tool using and tool making. Both Asian and African elephants have been trained for domestic chores and antics; it has been suggested, however, that the Asian species can perform more complex functions than the African (summarized from Gordon 1966, Sikes 1971, Douglas-Hamilton & Douglas-Hamilton 1975, Moss 1988, Western 1989, Shoshani & Tassy 1996, Meng *et al.* 1997, Payne 1998, Shoshani 2000).

Unlike in other mammals, the vulva is situated between the hindlegs (not under the tail). Having the vulva between the hindlegs may have evolved to accommodate less falling distance for the newborn. The testes are intra-abdominal (a primitive, not derived character). Unique to Elephantidae (*Loxodonta*, *Elephas*, *Mammuthus* and probably other extinct elephantids taxa) is the presence of the temporal gland, located subcutaneously midway between the eye and the ear opening; bull elephants secrete from the temporal glands during a physiological condition called musth during which blood testosterone levels are high and elephants are more aggressive than when not in musth. Also unique to Elephantidae and probably to other proboscideans are the presence of the pharyngeal pouch and the associated function of water storage and sound production. In time of stress (e.g. hot weather and absence of water), Asian and African elephants insert their trunk into their mouths and draw stored water from this pouch and douse themselves; because of their large volume compared with the relatively small surface area, it is difficult for elephants to keep their body cool (Shoshani 1998).

An elephant's heart has a double apex (a feature also common in Sirenia) and paired anterior venae cavae; it can reach 28 kg. Body temperature is 97–99° F (36–37° C) but it can fluctuate about five degrees to conserve energy during thermoregulation. The left lung is slightly smaller than the right, both lungs have several deep fissures and there is little or no plural cavity. The digestive system is not very efficient (about 44% of the food eaten is digested, the rest passes out with the faeces). The soft palate is short and there is no uvula. The stomach is simple (not chambered), intestines reach 30 m, the caecum is sacculated and acts as a fermentation chamber, and the anal flap is present (apparently unique to *Loxodonta*, *Elephas*, *Mammuthus*) (summarized from Todd 1913, Benedict 1936, Short 1962, Laursen & Bekoff 1978, Shoshani & Eisenberg 1982, Roth & Shoshani 1988, Shoshani 1996, 1998, 2000, Shoshani & Tassy 1996, Isaza 2006, Mikota 2006).

Brain weight of a newly born elephant is about 50% of adult brain weight. The elephant brain is the largest among living and extinct terrestrial mammals, with an average weight of 4.78 kg for adults (n = 16). The cerebrum and cerebellum are extremely convoluted (more than that of man), and the temporal lobe is proportionally large and bulges laterally and ventrally. The capacity to store data in elephants is about three times larger than in humans. Average encephalization quotient (EQ, the ratio of actual brain size to expected brain size for a given body size) for mammals = 1 (Jerison 1973). The EQ for elephants ranges from 1.13 to 2.36 (n = 16), with an average of 1.88, and with an average of 2.14 for Asian elephants and 1.67 for African; 2.1 for ♀♀ (n = 4) and 1.3 for ♂♂ (n = 3). Among mammals, those that use and make tools include humans, elephants and chimpanzees; all have a high EQ (details in Shoshani *et al.* 2006a).

Visible, and easily detected, differences between the African and the Asian species include: the shape of the back (which, in profile, is concave in *Loxodonta* and convex or straight in *Elephas*); the ear of the African elephant is large and exceeds the height of the neck and its dorsal end folds medially, whereas in the Asian species, the ear is small, does not exceed the height of the neck and its dorsal end folds laterally.

Living African and Asian elephants are geographically separated and do not interbreed. In captivity, however, animals are artificially placed together, and hybrids among animals that will never meet in the wild are possible. The only known hybrid between a male African elephant and a female Asian elephant was conceived in Chester Zoo, England in 1978 (Howard 1979). 'Motty' lived only 10 days. Immunological experiments confirmed that Motty's tissue behaved like that of a mule, corroborating that it was a hybrid between *L. africana* and *E. maximus* (Lowenstein & Shoshani 1996). These results are not totally surprising since the diploid chromosome number in somatic cells for both elephant species is 56 (Hungerford *et al.* 1966, Norberg 1969).

Of all the living and extinct species and subspecies of Proboscidea (approximately 185), only two extant genera, with three species, are alive today. The vast majority of living elephant populations are continuously decreasing due to shrinking range or habitat fragmentation, and therefore they are listed by CITES as either in Appendix I or Appendix II.

Jeheskel Shoshani & Pascal Tassy

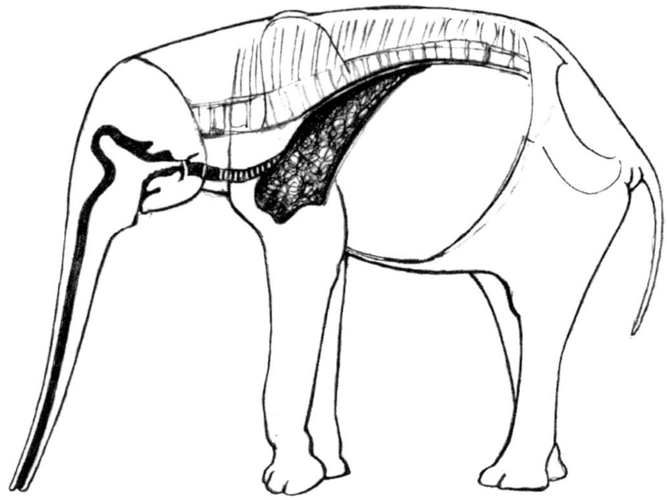

Lungs and air passages of Savanna Elephant *Loxodonta africana.*

GENUS *Loxodonta*
African Elephants

Loxodonta Anonymous, 1827. The Zoological Journal (London) 3: 140–143.

Note: F. Cuvier was the author of the original description of this genus in 1825 (Cuvier 1825 in E. Geoffroy St-Hilaire & F.G. Cuvier, *Hist Nat. Mammifères*, 3 (52): 2) but he used the name 'Loxodonte' in French, which is an invalid format in scientific publications (Article 11(b) of the International Code of Zoological Nomenclature), and thus the emended Latinized form of *Loxodonta* by an anonymous author in 1827 takes precedence.

The current work recognizes two living species of *Loxodonta*: the Savanna Elephant *Loxodonta africana*, characteristic of the savannas of southern and East Africa, and the Forest Elephant *L. cyclotis*, formerly widespread across the equatorial forest of central and western Africa up to Sierra Leone, but now surviving in fragmented populations under threat by poaching and habitat loss (Blake *et al.* 2007). Morphologically, Forest Elephants are smaller, with more slender and straighter tusks, a protuberant mandibular symphysis and smaller ears (Table 7); other morphological differences between the two taxa are given by Grubb *et al.* (2000), who considered *L. cyclotis* to be more primitive than *L. africana.*

The taxonomic validity of two species that were previously considered to be conspecific continues to be debated. During the colonial era there was a profusion of claims for different races of African elephants (Noack 1906, Frade 1931, Dollman 1934, Bourdelle & Petter 1950, Blancou 1962; and see Spinage 1994 for review) but the prevailing view over the past 50 years has been to recognize only two valid subspecies, the forest *L. a. cyclotis* and the savanna *L. a. africana*. None the less, local demes with distinctive features have long been recognized (notably in Namibia and the Rufiji delta) and genetic variation has been demonstrated for several subpopulations (Georgiadis *et al.* 1994, Barriel *et al.* 1999, Comstock *et al.* 2002, Nyakaana *et al.* 2002). Very small forms of *cyclotis*, formerly described in central DR Congo, were once supposed to belong to distinct pygmy forest elephants (described as *Loxodonta pumilio*), but these populations are not now thought to form a distinctive clade on either morphological or molecular grounds (Groves & Grubb 2000a, Debruyne *et al.* 2003b).

Frade (1931, 1955) was among the few earlier authors to recognize two species of African elephants, and several recent authors have presented morphological, ecological and molecular evidence in support of this classification (Groves & Grubb 2000b, Grubb *et al.* 2000, Roca *et al.* 2001, 2004, Rohland *et al.* 2010). Transitional morphological forms between typical savanna and typical forest elephants exist in peripheral zones in Central/East Africa (see Groves & Grubb 2000b, Grubb *et al.* 2000, Roca *et al.* 2004). In their interpretation of hybridism, two schools of thinking have emerged: the first posits that hybridization is ancient and the two forms are indeed separate species (Roca *et al.* 2001, 2004, 2005); the second regards the hybridization as active and the two forms are subspecies with a geographical cline from east to west (Debruyne 2005a, b). Geographical extension of more than two units, however named, is an alternative (Eggert *et al.* 2002).

On the basis of skull differences, Grubb *et al*. (2000) noted that hybrids between the *Loxodonta* forms were 'occasional' and they supposed that the area of sympatry was narrow. Had they recognized a wider area of sympatry or interaction between the *Loxodonta* species or subspecies then they might have drawn different conclusions. The genetic situation is actually much more complicated, with extensive introgression of Forest Elephant haplotypes in populations that are of morphologically Savanna Elephant type now generally acknowledged (e.g. at Garamba, see Roca *et al*. 2004). Nuclear and mtDNA profiles show cytonuclear dissociations that reveal male and female genomes with different evolutionary histories. A nearly continent-wide study showed that *L. africana* bulls are recurrently backcrossing with female hybrids to progressively replace the *cyclotis* nuclear genome over an extremely wide, not narrow, zone of hybridization (Roca *et al*. 2004).

Table 7. Major differences between species of *Loxodonta*.

	L. africana (Savanna Elephant)	*L. cyclotis* (Forest Elephant)
Weight	4000–6500 kg	2000–4500 kg
Height at shoulder (♂♂)	3.2–4.0 m	2.4–3.0 m
Skin	On average lighter	On average darker
Shape and size of ears	Triangular, extend below line of neck	Rounder, usually do not extend below line of neck
Skull, rostrum	More flared	Less flared
Skull, cranium	Much pneumatized	Less pneumatized
Skull, nasal aperture	Narrower	Wider
Skull, anterior end of rostrum	Slight dorsal cavity	Deep dorsal cavity
Skull, occipital plane	Slopes forward	More upright
Skull, mandible	Shorter, taller	Longer, lower
Skull, mandibular condyles	More rounded	Transverse–oval
Tusks	Curved out and forward, thicker	Straighter, downpointing, slender
Cheekteeth	High-crowned	Lower-crowned
Number of nail-like structures ('toes') in adults	Forefeet 4 or 5 Hindfeet 3, 4 or 5	Forefeet 5 Hindfeet 4 or 5

Source: Adapted from Grubb *et al*. (2000)

Fossil remains of *Loxodonta* are restricted to Africa. The earliest known member of the genus has been described from the late Miocene (ca. 5–7 mya) in Kenya and Uganda with the specimens consisting of some primitive isolated teeth showing emergence of the typical loxodont enamel loop (the generic name *Loxodonta* refers to the lozenge shape of the enamel loops on chewing surfaces of the molars). These have been labelled '*Loxodonta* sp. indet. (Lukeino stage)' (Tassy 1986, 1994). Sanders (2007) has described a new loxodont species (*Loxodonta cookei*) dated to the latest Miocene–early Pliocene, ca. 5.0 mya from the Varswater Formation at Langebaanweg, South Africa. This species is distinguished from *L. adaurora* by having anterior and posterior accessory central conules, and from other loxodont elephants by its primitive retention of permanent premolars, lower crown height, fewer molar plates, thicker enamel and lower lamellar frequency. Together with this newly described species, the total number of loxodont fossil taxa is ten. Approximately, from oldest to youngest, they are: '*Loxodonta* sp. indet. (Lukeino stage)', *L. cookei*, *L. adaurora adaurora*, *L. a. kararae*, *L. exoptata*, *L. atlantica angammensis*, *L. a. atlantica*, *L. a. zulu*, *L. cyclotis* and *L. africana*. Beden (1983) employed the name '*Loxodonta schneideri*' from Chad instead of *Stegodibelodon schneideri* described by Coppens (1972); corrected usage of *S. schneideri* was reinstated by Shoshani & Tassy (1996). More common in the latest

Play and other postures in young *Loxodonta africana*.

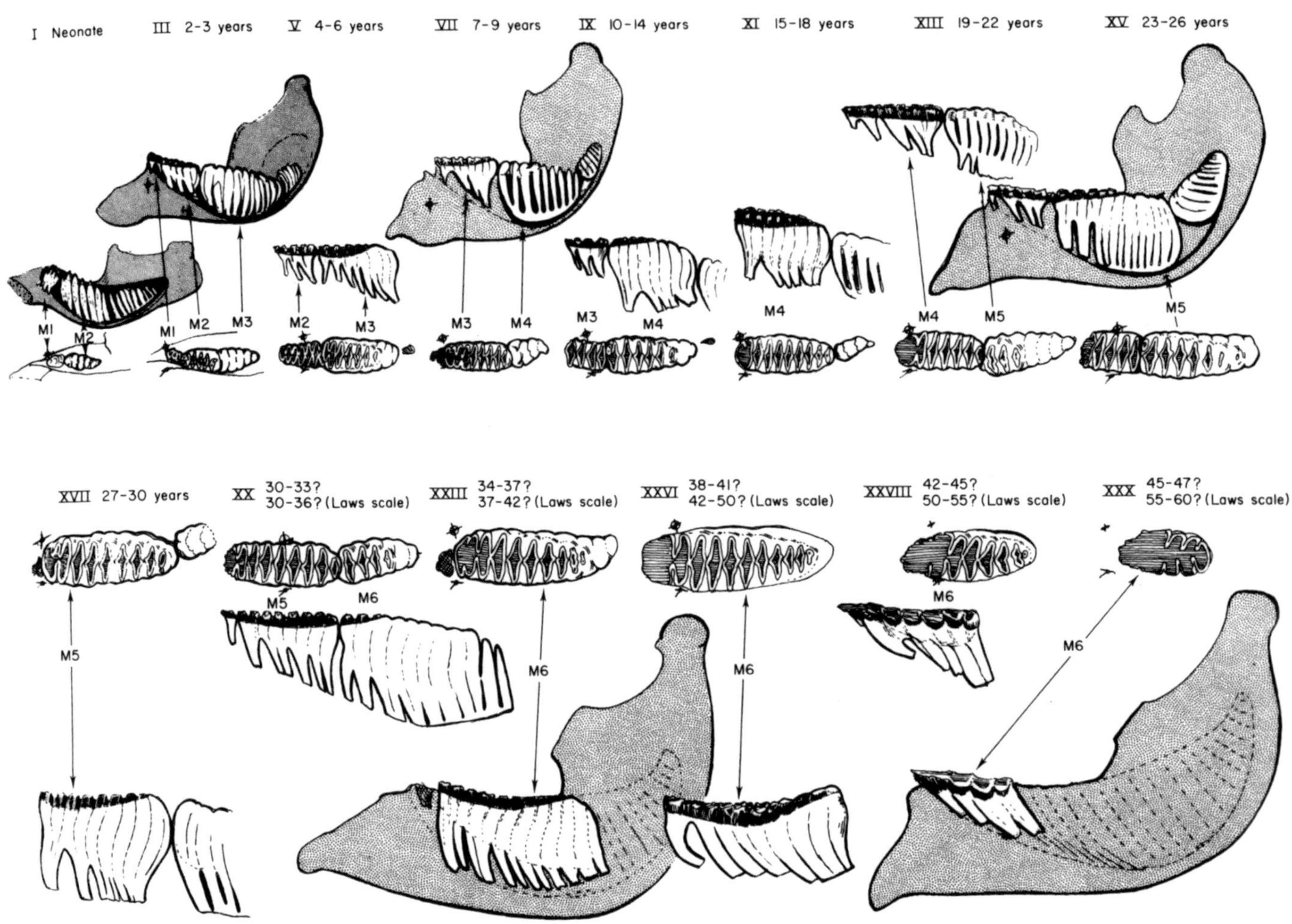

Toothwear and progression in *Loxodonta africana*, following Laws (1966). Roman numerals = 'Laws age scale'. (M1 to M3 equivalent to milk molars in non-proboscids.) Molariform teeth are numbered 1 to 6 – all deciduous.

Miocene and early Pliocene (ca. 5–4 mya) of East Africa are *Loxodonta adaurora* and *L. exoptata* described on the basis of dental and skeletal remains (Maglio 1970, 1973, Beden 1983, 1987). The molars, with a lesser marked lozenge, indicate that *L. adaurora* is probably an early offshoot of the genus. The subspecies *L. adaurora kararae* is a pygmy chronological subspecies described in the latest Pliocene of Kenya at Koobi Fora and the Omo (Shungura) in Ethiopia (Beden 1983). The discovery of *L. a. kararae* has demonstrated that Kingdon's (1979) query whether *L. cyclotis* might not be a dwarfed ecological isolate of *L. adaurora* was not a rhetorical question.

A large species, *Loxodonta atlantica* (larger and with more derived molars than those of *L. africana*), is known from the latest Pliocene of Yayo (Chad) and Omo (Ethiopia) and, in the form of *L. atlantica angammensis*, up to the late Pleistocene (Maglio 1973, Beden 1987). Numerous fossils of *L. atlantica atlantica* show that this was a widespread species, known from northern Africa, to eastern and southern Africa. Although the living species *L. africana* must have been present in Africa together with *L. atlantica*, the former is not known as a fossil before the late Pleistocene of South Africa and Chad.

Although *L. adaurora* had exceptionally large, vertically embedded tusks it shared a typical globular cranium with the extant species (i.e. the profile was rounded compared with the flatter and more elongated frontal profile of gomphotheres and primitive elephantids). Compared with *Elephas* (and *Mammuthus*) the frontoparietal area is concavo-convex. Maglio (1973) observed that *Loxodonta* was the first of the three genera to evolve the short, facially flexed skull that became typical for all later elephants; it evolved more than 2.5 mya before the others. In spite of this early specialization in cranial architecture, 'little change in the dentition occurred in the entire history of the genus and it would seem that this particular combination of cranial and dental specialisations was sufficient for this group' (Maglio 1973).

Six cheekteeth succeed one another in a one-by-one progression. In adult elephants, no more than one and a half teeth are in use. The forward surface of each molar is worn down progressively (while their roots also get resorbed as they wear down). Mandibular tooth rows broadly mirror their antagonists in the upper jaw but the latter travel at a steeper angle. Laws (1966) documented the form and progression of loxodont teeth in the lower jaw of *L. africana*, correlating tooth-wear stages with age-classes. These have provided the basis for population studies of the species ever since, and have been revised by Lee *et al*. (2012).

In terms of behaviour and social structure, *L. africana* and *L. cyclotis* principally differ in that the daughters of Savanna Elephants never leave their mothers. This results in large, stable matriarchal families, while in Forest Elephants daughters leave their mothers once they achieve puberty (see species profiles).

Pascal Tassy & Jeheskel Shoshani

Savanna Elephant *Loxodonta africana*.

Loxodonta africana Savanna Elephant (African Bush Elephant)

Fr. Elephant d'Afrique; Ger. Afrikanischer Elefant

Loxodonta africana Blumenbach, 1797. Handb. Naturgesch., 5th Ed., p. 125. Restricted to the Orange River, South Africa, by Pohle (1926).

Taxonomy The colonial era brought a profusion of claims of different races of African elephants (see genus profile). Ansell (1974) recognized four subspecies within his *africana* section (see also Laursen & Bekoff 1978, Spinage 1994), but their validity is doubtful. Shoshani (2005) recognized a single monotypic species. Synonyms: *angolensis*, *berbericus*, *capensis*, *cavendishi*, *cornaliae*, *hannibali*, *knochenhaueri*, *mocambicus*, *orleansi*, *oxyotis*, *peeli*, *pharaohensis*, *rothschildi*, *selousi*, *toxotis*, *typicus*, *zukowskyi*. Chromosome number: 2n = 56; the normal karyotype has 25 acrocentric/telocentric and two metacentric/submetacentric autosomal pairs. The X chromosome is a large submetacentric; the Y chromosome is small and acrocentric (Hungerford *et al.* 1966, Wallace 1978, Houck *et al.* 2001).

Description The largest of land animals, weighing as much as 6 tonnes, the Savanna Elephant is easily identifiable in having a trunk, tusks, large ears and pillar-like legs. The massive head is relatively larger and broader in older ♂♂ than in younger ♂♂ and ♀♀. The ears are extremely large, as much as 1.2 m across by 2 m vertical. The prehensile trunk, a fusion of nose and upper lip and made up of approximately 150,000 subunits of muscles (Shoshani 1997), contains two nostrils and is equipped with two finger-like tips. The wrinkled skin, which reaches a thickness of 30–40 mm on the legs, forehead, trunk and back, is dark to pale grey or brown, and, in rare instances, depigmented patches can be seen. Newborn elephants may be covered in reddish-brown hair. Adults retain coarse bristles up to 40 mm long on the trunk and chin and as abraded remnants in the crevices of skin over much of body. Tail hair varies in length from stubble to shiny black hairs up to 500 mm long; the tail itself may reach a length of 1.0–1.5 m in adults. The limbs are pillar-like. Elephants walk tip-toed on five digits and one sesamoid predigit, all encased in hoops of tissue, skin and nail-like structures above a cushion of fibro-fatty tissue. As elephants adopted the 'tip-toed' stance 40 mya, the sesamoid bone was co-opted to help support the elephant's weight (Hutchinson *et al.* 2011). 'Toenails' vary from five on both fore- and hindfeet to four on the forefeet and three on the hind. The smooth but cracked soles leave recognizable tracks, with larger circular forefeet and smaller oblong hindfeet. The length of the hindfoot is a reliable indication of shoulder height (Poole 1982, Western *et al.* 1983, Lee & Moss 1995).

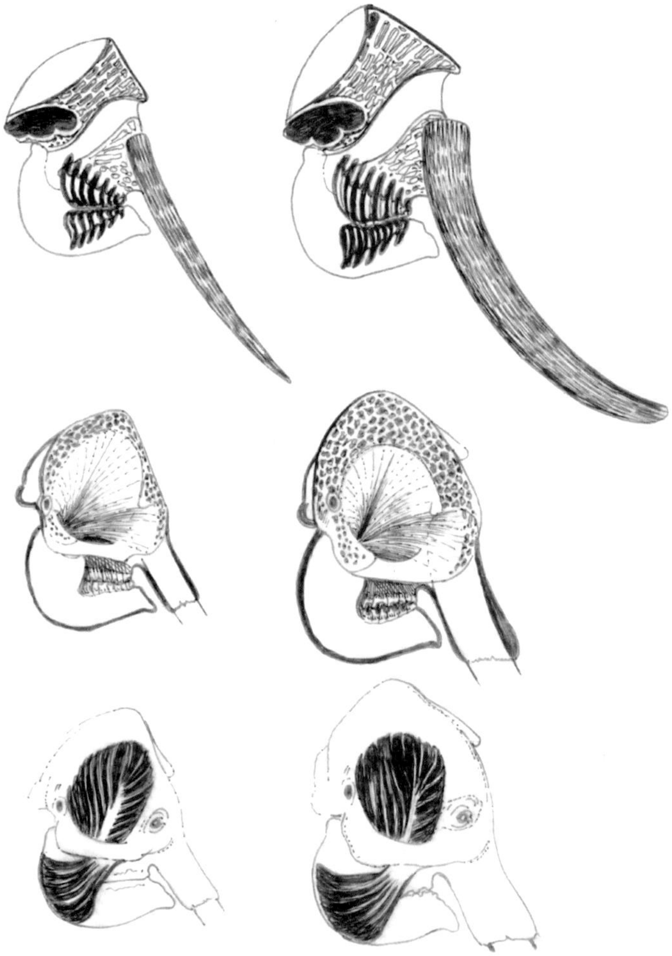

Three aspects of skull structure in female (*left*) and male (*right*) *Loxodonta africana*. (*Above*) Cross section showing brain capsule, honeycomb bone and rooting of incisor and molar. (*Middle*) Cranial buffering with honeycomb revealed by 'shaved' surface. (*Below*) Masseter and temporalis muscles.

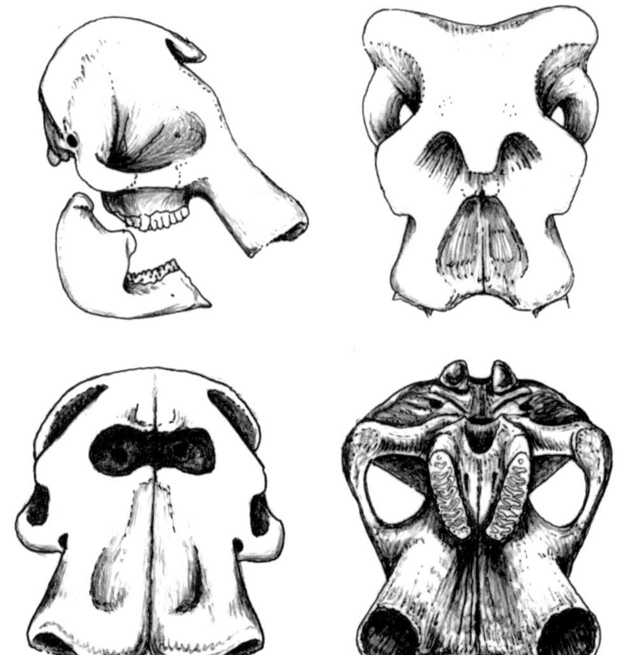

Loxodonta africana skull. Profile and frontal views (*left*). Dorsal and palatal views (*right*).

Sex in elephants can be determined by the shape of the forehead (in profile rounded in ♂♂ and more angular in ♀♀), the presence of mammary glands (two between the forelegs), the thickness and shape of the tusks (thicker and more conical in ♂♂), the slope of the underbelly (sloping downward toward the hindlegs in adult ♂♂; more curved in ♀♀) and the genitals. There is no scrotum, and testes are internal. In nulliparous ♀♀ breast development commences in the fifth month of first pregnancy and is clearly visible by the tenth month (Mutinda 1994). Thereafter, ♀♀ are typically either lactating or pregnant (though inter-calf-intervals may increase substantially and sometimes cease later in life) and breasts are visible (Poole 1989a).

The skull is massive and rounded, the relatively small brain cavity situated low down at the back of the skull. The maxillary and premaxillary bones extend well below the level of the upper toothrow, forming bony supports on either side for the bases of the tusks. In the female skull there is a distinct nuchal eminence not present in that of the male. A small interparietal bone is present at birth, but fuses with the surrounding cranial bones at an early age. The cranium is pneumatized (filled with air sinuses), with a 'honeycombed' structure and, in an adult bull, may be as much as 300–400 mm thick at the frontal region. The orbits are situated towards the front of the skull. The structure of the cranium and mandible is discussed in detail by Van der Merwe *et al.* (1995).

The tusks are modified upper incisors composed of layered dentine (ivory, which shows a cross-grained matrix). The tusk is preceded by a deciduous tooth, which consists of a crown, root and pulpal cavity, the formation of which is completed soon after birth. The deciduous tooth reaches a maximum length of 5 cm but does not erupt through the skin and is later pushed aside and resorbed (Raubenheimer *et al.* 1995). The presence, absence, length, shape and orientation of tusks are subject to much variation both within and between populations. In the Kruger N. P. (South Africa) research has shown that in bulls, tusk growth accelerates with age, while in cows it ceases at about 40 years of age, after which tusk size may even decline probably due to wear (I. Whyte & A. J. Hall-Martin unpubl.). Broken tusks have been observed to grow back rapidly even among older females (J. Poole pers. obs.). Elephants use their tusks for gathering food, digging, defence, offence and display, and it is not unusual for one of the tusks to be subject to greater wear than the other. Both sexes have tusks, although tusklessness occurs naturally (rarely in ♂♂) in low proportions (~2–4%; Poole 1989a) and appears to be increasing (Jachmann *et al.* 1995). In heavily poached areas, tusklessness may reach up to 45% in some female age classes (Poole 1989a). In the Addo Elephant N. P. (South Africa) 98% of the ♀♀ are tuskless (Whitehouse 2002), due to previous heavy exploitation for ivory.

During the course of their lifetime, elephants grow six grinding teeth, known as molars I–VI, which grow and move forward, replacing each other in both upper and lower jaws, one by one, with each successive molar longer and wider than the previous one (Laws 1966). In young and subadult individuals two molars normally occur on each side of the upper and lower jaws and are in use at the same time. Molar displacement and wear has been used as a means of ageing African elephants, based on the following general patterns of eruption: M1 at birth to 0.5 years; M2 partially in wear at 2 years; M3 at 2.5 years, full tooth at 4 years; M4 at 6–8 years, in full wear at 15 years; M5 at 18, full tooth at 25 years; and M6 at 32–34 years, full tooth at ~45 years (Laws 1966, revised by Lee *et al.* 2012). However, there is variation in the rate in which molars come into full wear (e.g. Lindeque 1991).

Hanks (1979) developed a useful visual method for obtaining quick age assessments in the field, based on body size within a family unit; the relative size of footprints and faecal boli may also be useful (Western *et al.* 1983, Jachmann & Bell 1984). Other studies have also investigated patterns of age determination in African elephants, expanding on Laws' earlier pioneering work (Johnson & Buss 1965, Laws 1967, Sikes 1966, 1967, Krumrey & Buss 1968, Fatti *et al.* 1980, Lang 1980, Lark 1984, Jachmann 1985, 1988, Lindeque 1991) in African elephants. Recent work by Shrader *et al.* (2006), based on ten populations in five countries across southern and eastern Africa, confirms the work of earlier studies that the growth patterns of Savanna Elephants are similar across large areas of Africa. Growth curves indicated that shoulder height can provide an accurate estimation of the age of ♀♀ up to 20 years (215 cm) and ♂♂ up to 36 years (290 cm) (Lee & Moss 1995).

Geographic Variation Savanna Elephants can vary dramatically in size across their range, and a number of subspecies have been described based on external features, including the ears (e.g. Lydekker 1907). Regional differences probably reflect a combination of local adaptation (especially in the case of desert populations), founder effects and genetic drift in what sometimes appear to be self-contained gene pools. Savanna Elephant populations show modest levels of phylogeographic subdivision, based on composite microsatellite genotype, an indication of recent population isolation and restricted gene flow between locales (Comstock *et al.* 2002).

Similar Species

L. cyclotis. Smaller body size, with straighter, more downward-pointing tusks and smaller rounder ears. Distributed in the Guinea-Congolian rainforests.

Distribution

Historical Distribution Elephants once inhabited virtually the entire African continent. Neolithic rock paintings from 10,000–12,000 years ago reveal that elephants once existed through much of the Sahara Desert and North Africa (Coulson 2001), but climatic fluctuations might have excluded them from some waterless regions during very arid periods. From ancient historical writings it is clear that elephants occurred from the Mediterranean coast in North Africa (Bryden 1903) and the species is believed to have survived into the first few centuries AD in the Atlas Mts and along the Red Sea coastline and Nubia. At the time of first contact with non-aboriginal people, the distribution of elephants broadly spanned the entire continent south of the Sahara (Mauny 1956, Douglas-Hamilton 1979).

Current Distribution Today, Savanna Elephants are extirpated throughout the region north of the Sahel, and are restricted to south of the Sahara, occupying only 20% of their historic range. Their distribution is patchy and fragmented. The main cause for range decline is habitat loss and poaching for ivory during the last two centuries. In West Africa, where range loss has been most severe, Savanna Elephant populations exist in small, fragmented and isolated enclaves in the Sahelian zone, along forest edge, woodlands and savanna. A relict

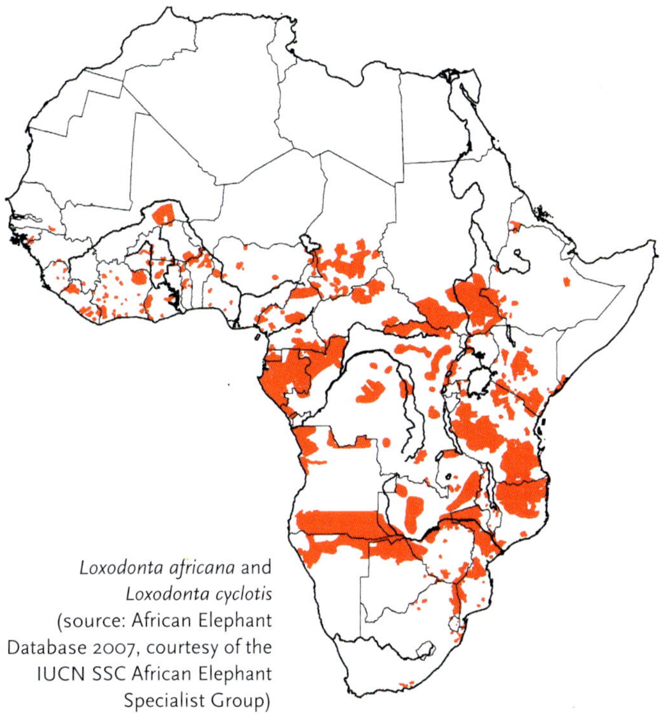

Loxodonta africana and *Loxodonta cyclotis* (source: African Elephant Database 2007, courtesy of the IUCN SSC African Elephant Specialist Group)

population of desert-living elephants occurs in Gourma, Mali (Blake *et al.* 2003, Bouché *et al.* 2009). In central Africa, Savanna Elephants are known to occur in N Central African Republic and N Cameroon, and there is reportedly a narrow zone of hybridization in NE DR Congo (and perhaps in S Central African Republic). In Chad, the only central African range state having only Savanna Elephant populations, elephants occur in Sudanian woodland in the south and in the drier Sahelian Acacia wooded grasslands further north; there are no elephants in the Saharan north of the country (Blanc *et al.* 2007).

In Eastern Africa, Savanna Elephants occupy both forest and savanna habitats. Populations in Uganda, Ethiopia, Eritrea, Somalia and Rwanda are remnant and highly fragmented, while larger ranges and populations still occur in Sudan, Kenya and, particularly, Tanzania. Southern Africa is home to most of Africa's Savanna Elephants where they occur in fragmented ranges, although elephants have been extirpated from Lesotho. The largest remaining population in an unbroken range includes parts of Namibia, N Botswana, Zimbabwe, Zambia and Angola (Blanc *et al.* 2007). Small relict populations of forest-living Savanna Elephants occur along the eastern coast of Africa from Kenya's Arabuko-Sokoke and Shimba Hills forests to the Knysna forest in the Western Cape, South Africa. Many new, but fragmented, populations have been established in South Africa through translocations from Kruger N. P.

Habitat Savanna Elephants occur in virtually every habitat type on the African continent from sub-deserts to swamps, lowland rainforests, gallery and montane forests, upland moors, flood-plains, savannas and various types of woodlands, and range from sea level to as much as 4875 m (Grimshaw *et al.* 1995). Ability to utilize such diverse and contrasting habitats is possible because they are relatively unspecialized herbivores. Preferred habitat such as wooded savanna provides both browse and grass with access to water. Shade and cover for protection and water availability can have an important impact on elephant movements and distribution (Nellemann *et al.* 2002).

Table 8. Estimated numbers of Savanna Elephants, including West Africa.

	Estimate	95% CL ±	Guesstimate From	Guesstimate To	Range covered %	Area (km²)
Estimates from surveys	496,000	36,500			42	883,000
Estimates from low quality surveys and guesses			32,500	45,000	21	448,000
Unassessed range					37	767,000
Total range					100	2,098,000

Source: African Elephant Database 2007, courtesy of the IUCN/SSC African Elephant Specialist Group

In rare cases, elephants have become adapted to desert conditions, for example, in Gourma, Mali (Blake *et al.* 2003), and the Kaokoveld and Damararaland in Namibia (Leggett 2004). The best-studied population is in NW Namibia, where elephants cope with seasonally scarce vegetation and water by moving over large seasonal ranges of approximately 650 km and have home-ranges of up to 12,600 km² (Viljoen 1987, 1989, Viljoen & Bothma 1990, Lindeque & Lindeque 1991, Leggett *et al.* 2003, Leggett 2006a).

Abundance Elephants are least common at high altitudes and in hot deserts where access to water is limited. Densities also tend to be low in dense forest and open grasslands. Mixed woodlands with plenty of grass and browse provide year-round forage and may allow population densities to reach over 5/km² (e.g. Lake Manyara N. P., Tanzania). However, such cases are exceptional and are unlikely to be sustainable over the long term. In most savannas, elephant densities range from 0.5 to 2/km² with densities in forests usually lower. Large spatial and temporal differences in local density are not uncommon due to both environmental and social factors.

A continent-wide assessment of population size for Savanna Elephant (Table 8) is complicated by countries having both Forest and Savanna elephants. None the less, using the results from the African Elephant Status Report (Blanc *et al.* 2007) it is possible to arrive at a general continental population estimate. Such an assessment makes three general assumptions, namely: that central Africa is the only place where Forest Elephants occur; that there are no *cyclotis–africana* hybrids anywhere; and that the only Savanna Elephants that occur in central Africa are those in Chad and N Central African Republic. It is quite likely that all three assumptions are incorrect, and that an unknown number of elephants here allocated to the forest kind are in fact Savanna Elephants (or hybrids) and vice versa. However, given the relatively wide error margins in the estimates, the fact that these figures are rounded, and that there remain considerable amounts of range for which no estimates of abundance are available for either taxon, it is likely that any incorrect allocation would get lost in the general imprecision.

As noted above, determining country-level estimates is complicated, and so we report here only on those countries where available national estimates are very clearly *L. africana* only. For this reason, below we omit discussion of national estimates for all of West Africa (noting only that a recent aerial count in the Gourma region of Mali produced a minimum of 344 elephants; Bouché *et al.* 2009) and several countries elsewhere around the forest rim that likely harbour hybrids (such as Cameroon, Central African Republic and Uganda). In addition, we do not consider national estimates for those countries in which the proportion of surveyed range is so small as to render the national estimate meaningless (e.g. Angola, with population estimates available for only 5% of estimated elephant range in the country). We caution further that increasing levels of ivory poaching since 2007 may render these estimates inaccurate.

The largest number of Savanna Elephants in Africa is found in southern Africa, with a population of nearly 321,000, nearly twice as high as East Africa, mainly in the six countries of Botswana, Zimbabwe, Zambia, South Africa, Namibia and Mozambique (Blanc *et al.* 2007; Table 9). The largest population survives in Botswana, where recent estimates put the population at around 150,000 animals surviving in an area of about 100,000 km², followed by Zimbabwe with approximately 90,000 animals (Blanc *et al.* 2007). Today, Malawi has fewer than 3000 elephants, while Swaziland (where the resident

Table 9. Country and regional elephant numbers for southern Africa.

Country	Elephant numbers: Definite	Probable	Possible	Speculative	Range area (km²)	% of regional range
Angola	818	801	851	60	406,946	31
Botswana	133,829	20,829	20,829	0	100,265	8
Malawi	185	323	632	1,587	7,538	1
Mozambique	14,079	2,396	2,633	6,980	334,786	26
Namibia	12,531	3,276	3,296	0	146,921	11
South Africa	17,847	0	638	22	30,455	2
Swaziland	31	0	0	0	50	0
Zambia	16,562	5,948	5,908	813	201,247	15
Zimbabwe	84,416	7,033	7,367	291	76,931	6
Total	**297,718**	**23,186**	**24,734**	**9,753**	**1,305,140**	**39**

Source: Blanc *et al.* (2007)

Table 10. Country and regional elephant numbers for eastern Africa.

Country	Elephant numbers				Range area (km^2)	% of regional range
	Definite	Probable	Possible	Speculative		
Eritrea	96	0	8	0	5,293	<1
Ethiopia	634	0	920	206	38,365	4
Kenya	23,353	1,316	4,946	2,021	107,113	12
Rwanda	34	0	37	46	1,014	0
Somalia	0	0	0	70	4,526	<1
Sudan	20	0	280	0	318,239	36
Tanzania	108,816	27,937	29,350	900	390,366	44
Uganda	2,337	1,985	1,937	300	15,148	2
Total	**137,485**	**29,043**	**35,124**	**3,543**	**880,063**	**26**

Source: Blanc *et al.* (2007)

population was extirpated around the 1890s; Monadjem 1998) has the smallest population of only 31 individuals.

In eastern Africa, elephant range spans more than 880,000 km^2, of which 80% is accounted for by Tanzania and Sudan (Blanc *et al.* 2007; Table 10). Eastern Africa suffered massive declines of elephants between 1960 and 1990 when 85–98% of the elephants were killed for ivory; some populations have never recovered. Rwanda, Eritrea and Somalia contain only tiny fragmented elephant populations (estimated to number fewer than 300 individuals between them) and even these are thought to be in decline. Elephants were exterminated from Burundi in the 1980s. Sudan boasts a large elephant range, and although Blanc *et al.* (2007) indicated that elephants numbered only 300 individuals, a survey by Fay *et al.* (2007) estimated the population at ~7000. Despite huge losses, Tanzania and Kenya remain strongholds for East African elephants with estimates of 135,000 and 25,000, respectively (Blanc *et al.* 2005, 2007), although current poaching is likely to have eroded these figures. Uganda's elephants (including some Forest Elephant populations) have recovered from a few hundred to nearly 3000 individuals today after a decline of almost 99% (according to the Uganda Wildlife Authority in 2002). Elephant range in some parts of East Africa has been significantly affected by armed conflict and civil disturbances. In other areas it is being reduced by expanding human settlement.

In Chad, the only central African country known to harbour Savanna Elephants only, Blanc *et al.* (2007) estimated the population at nearly 6000 animals, with by far the largest population in Zakouma N. P. However, poaching has had a significant impact on this figure, with the population in Zakouma now at an historical low (Poilecot 2010, Bouche *et al.* 2011).

Adaptations Elephants exhibit some characteristics that have long been interpreted in terms of an aquatic or semi-aquatic ancestry (Andrews 1906, Matsumoto 1923, Osborn 1936, Janis 1988, Gaeth *et al.* 1999), a view initiated by the undoubted relationship between proboscids and sirenians and by the ambiguity of an Eocene fossil, *Moeritherium*, which has been variously allied with one or other order. This has stimulated enquiry into the possibility of retentions from an ancient aquatic ancestry in living elephants. Among the suggestions are the development of a trunk, internal testes and the development of nephrostomes in the mesonephric kidney (Gaeth *et al.* 1999). The latter authors also noted that nephrostomes, briefly present during early stages of foetal development in many mammals, might last longer during a particularly long gestation; as Shoshani & Tassy (2005) have pointed out, this carries no implication of an aquatic phase. Supposedly related to the development of a trunk, two additional adaptations have been noted. Living adult elephants have little or no pleural cavity, and the lungs are fused to the chest wall and diaphragm. This adaptation is thought to be a protection against the high negative intra-thoracic pressures attained when water has to be sucked up the trunk before drinking (Todd 1913, Short 1962). Any connection with an aquatic origin is contradicted by wholly aquatic sirenians having normal lungs with pleural cavities (Rommel & Reynolds 2000).

As elephants ventured into more arid lands in the Miocene, the pharyngeal pouch, a structure unique to elephants, was naturally selected for water storage. Elephants are able to store 4.5 l of water in this pouch located at the base of the tongue. During hot weather elephants can be observed to withdraw water stored there by inserting the trunk up to the pharynx, extracting water therein and spraying it over their backs, under body and ears (Kingdon 1979, Shoshani 1997). While this behaviour has been observed elsewhere, desert-dwelling elephants use this method routinely (Leggett 2004). Several other behavioural adaptations allow elephants to survive very harsh environments, including the ability to go without water for periods of up to five days (Viljoen & Bothma 1990, Leggett 2006b) and young obtaining moisture by eating the moist dung of adults. Another interesting thermoregulatory mechanism involves adult animals urinating on sand and then both the young and old scooping and throwing urine-soaked sands over their bodies (Leggett 2004). Finally, Kahl & Armstrong (2000a) observed juveniles drinking urine excreted by adult elephants during a severe drought in Zimbabwe.

Elephants use their ears in display, in sound localization and as a cooling mechanism. The skin on the backs of the ears is very thin and is laced with numerous blood vessels, which, when fanned, help to reduce the elephant's overall body temperature. Wright & Luck (1984) measured water loss from the skin of a 1200 kg elephant and concluded that it was capable of losing an average of 0.23 l of water/m^2/h (despite the fact that the skin lacks sweat glands). The estimated surface area was 11 m^2, so that some 2.5 l of water would be lost per hour.

Being large bodied, elephants are naturally able to produce low frequency sounds, but several adaptations have allowed them to produce even lower frequencies (down to 5 Hz). First, the trunk itself

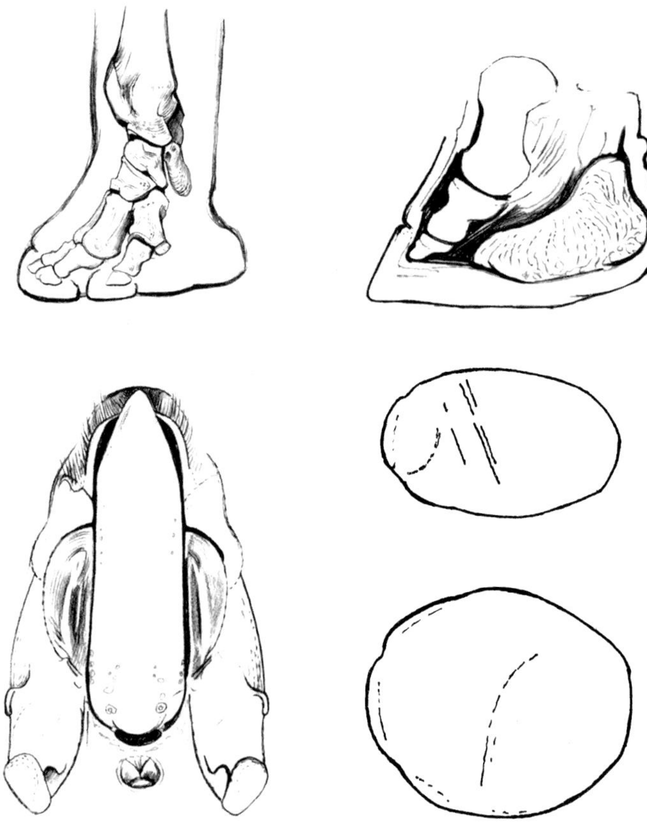

Top left: *Loxodonta africana* carpal area showing toes contained within hoop of tissue, skin and nail.
Top right: Foot profile to show pad of elastic tissue.
Lower right: Footprints of left hindleg (*above*) and left foreleg (*below*).
Lower left: Foetal lower jaw and tongue, showing remarkable length.

adds about 3 m on to the length of the resonating chamber. Second, the hyoid apparatus is attached to the skull by muscles, tendons and ligaments, rather than bones as in most other mammals. This looser arrangement allows for a greater movement and flexibility of the larynx and, therefore, facilitates the production and resonance of low-frequency sounds (Shoshani 1998). Third, the looser arrangement also houses the pharyngeal pouch that permits a further enlargement of the resonating chamber by lowering the loosely attached larynx.

The long ear canal, large tympanic membrane and massive ossicles of elephants reflect special adaptation to low-frequency hearing. The enormous cranium has allowed the evolution of an outer ear canal of about 20 cm in length, thus providing adequate protection for its very large tympanic membrane that in turn allows the high signal-to-noise ratios so important at very low frequencies (Nummela 1995). Massive ossicles seem to occur in animals that receive acoustic information through bodily vibrations (Reuter *et al.* 1998). Elephants and sirenians are unique among modern mammals in having reverted to a reptilian-like cochlear structure (Fischer 1990, O'Connell *et al.* 1998). Since the cochlear structure of reptiles facilitates a keen sensitivity to vibrations it has been suggested that the similar structure in elephants allows them to detect vibrational signals, too (O'Connell *et al.* 1998).

The trunk serves an obvious role in the acquisition of food and water and in tactile communication, but it is also important in chemical communication, for tactile investigation of the environment, olfaction and vomerolfaction (Rasmussen & Schmidt 1998). The tip of the trunk houses two types of vibrissal hairs, small corpuscles and free nerve endings. These features allow the trunk tip to detect vibrations, finely manipulate objects and transfer liquids (Rasmussen & Munger 1996). Within the nasal cavity are seven turbinals specialized for olfaction and detection of hormones emanating from urine, faeces, and from secretions of the trunk, mouth and temporal glands. The tip of an elephant's trunk has densely packed cells known as Pacinian corpuscles that are sensitive to vibrations (Rasmussen & Munger 1996). Elephants often make use of the trunk to throw objects or materials such as mud, soil and vegetation in aggressive encounters and for exploration (Wickler & Seibt 1997), and are reputed to be able to throw with great precision (Chevalier-Skolnikoff & Liska 1993, J. Poole pers. obs.).

Also unique to elephants is the temporal gland, derived from apocrine sweat glands (Estes & Buss 1976), located subcutaneously on the side of the face with its orifice located between the ear and the eye. In Savanna Elephants the gland appears to have different functions in ♂♂ and ♀♀. Females, less frequently young ♂♂ and rarely adult ♂♂ secrete a dark, watery liquid when they are socially aroused, frightened, or subjected to a variety of stresses (Poole & Granli 2003, J. Poole pers. obs.); adult ♂♂ alone secrete a more viscous, chemically distinct substance during their annual period of sexual activity known as musth (Poole 1987).

The brains of *Loxodonta africana* rank among the highest for absolute and relative mass, cortical expansion and complexity, features comparable only to those of some of the Cetacea, man (Cozzi *et al.* 2001) and the great apes. The elephant's brain weighs up to 6 kg (averaging 4.8 kg; Shoshani *et al.* 2006a) and an EQ (encephalization quotient – the ratio between the observed and expected brain weight for a defined body weight; Jerison 1973) of 1.7 to 2.1 (Eisenberg 1981, Shoshani 1998, Shoshani *et al.* 2006a) is comparable with the larger primates. Their large and convoluted neocortex associated with a relatively low density of neurons may indicate that cognitive faculties concentrated in long-term processing and in synthesizing a great diversity of input over time (Hart *et al.* 2008). Hakeem *et al.* (2009) have documented the presence of a type of nerve cell (e.g. Von Economo Neurons) that, in great apes and the Cetacea, has been associated with neural circuitry involved in social awareness and, potentially, Theory of Mind.

The extremely large and convoluted temporal lobes and underlying hippocampus of the elephant's brain likely function in their prodigious memories (for places, individuals and events) and their ability to navigate over long distances (Shoshani 1998, Hakeem *et al.* 2005). A flood of recent publications indicates that elephants are self-aware (Plotnik *et al.* 2006), and capable of a wide range of higher-level cognitive behaviours such as: empathizing with others (Douglas-Hamilton *et al.* 2006, Bates *et al.* 2008a); rudimentary teaching (Bates *et al.* 2010) suffering from the long-term effects of trauma (Bradshaw *et al.* 2005); distinguishing between the many voices (McComb *et al.* 2000) and scents (Bates *et al.* 2008b) of their relatives and companions; making subtle discriminations between predators (Bates *et al.* 2007); recognizing the bones of their own species (McComb *et al.* 2006); using simple tools (Hart *et al.* 2001); social learning (Lee & Moss 1999); and vocal imitation (Poole *et al.* 2005).

Foraging and Food Though primarily browsers, elephants are mixed feeders and can switch to a grass-dominated or browse-

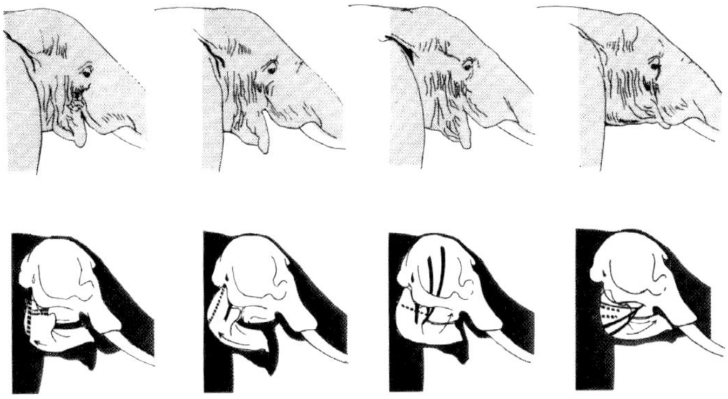

Loxodonta africana chewing cycle. *Above:* stills from film. *Below:* diagrams of a single cycle of mandible action.

dominated diet depending on which provides the highest ratio of protein to fibre; thus, diet changes with season and habitat (e.g. Buss 1961, Laws *et al.* 1975, Williamson 1975). Seasonal movement patterns through different habitats reflect an attempt to achieve the most nutritious diet. Grass predominates in the wet season and woody plants in the dry season. In Kruger N. P., studies involving stable carbon isotope analysis of faeces reveal that elephant populations from the northern part of the park consistently eat higher proportions of grass (40%) during the dry season than do their southern counterparts (10%) (Codron *et al.* 2006; and see Vogel *et al.* 1990). The wet-season diets of elephants from northern and southern parts include similar amounts of grass (50%), because elephants in the south, but not in the north, ate significantly more grass during this time. Codron *et al.* (2006) postulate that the relative homogeneity of woody vegetation in the north of Kruger N. P. (where the variable, and questionable, nutritive value of the dominant Mopane *Colophospermum mopane* may deter browsing) may force elephants to opt for alternative food sources like grass throughout the year. The wider diversity of woody plant species in southern Kruger N. P. allows elephants to exploit this browse more efficiently, especially during the dry season when grass production is low (Codron *et al.* 2006).

Among woody plant resources, forbs and shrubs form the main component of the diet, and elephants will pull down entire trees to consume leaves and bark. The bark of certain trees, such as several Acacia species (*Acacia tortilis*, *A. xanthophloea*, *A. nigrescens*) and Marula *Sclerocarya birrea* is especially favoured; preferred species have strong, pliable bark, which is associated with a high proportion of fibres (Malan & Van Wyk 1993). The bark is loosened with the tusks and then stripped off. Elephants will also take the roots of certain woody species such as *Combretum* spp., *Dalbergia melanoxylon* and *Terminalia* spp.

Due to their large size and digestive system specialized for rapid throughput of course vegetation, elephants satisfy their dietary requirements by consuming considerable quantities of food each day (40–80 kg dry mass, or 1–2% of their body weight; Lindsay 2011). To accomplish this elephants feed for 60% of a 24-hour day and are active for 18–24 hours. While adults usually rest standing up during the day, at night they frequently sleep lying down for a few hours. Foraging is achieved with their incredibly versatile trunk, efficiently selecting individual items of fruit, tugging up tufts of grass, or stripping the bark off tree trunks. The 2–3 pairs of lozenge-shaped molars crush the forage by sliding forwards and backwards. Digestion is inefficient and elephant dung is fibrous. Lactating ♀♀ eat proportionately more than non-lactating ♀♀, and bulls consume greater browse content than cows.

Elephants need to drink in the order of 160 l per day. Dependence on water can force elephants to remain in less preferred, and even degraded, habitats during the dry season, and this may be the driving force behind the seasonal movements and annual migrations (some up to hundreds and even thousands of kilometres) observed in many populations. When freshwater is not available, elephants will drink brackish water, although this has proved fatal if water salinity is unusually high. Weir (1972) found that the largest congregations of elephants in Hwange N. P. gathered at man-made waterholes where the water was rich in sodium (adult elephants require 75–100 g salt

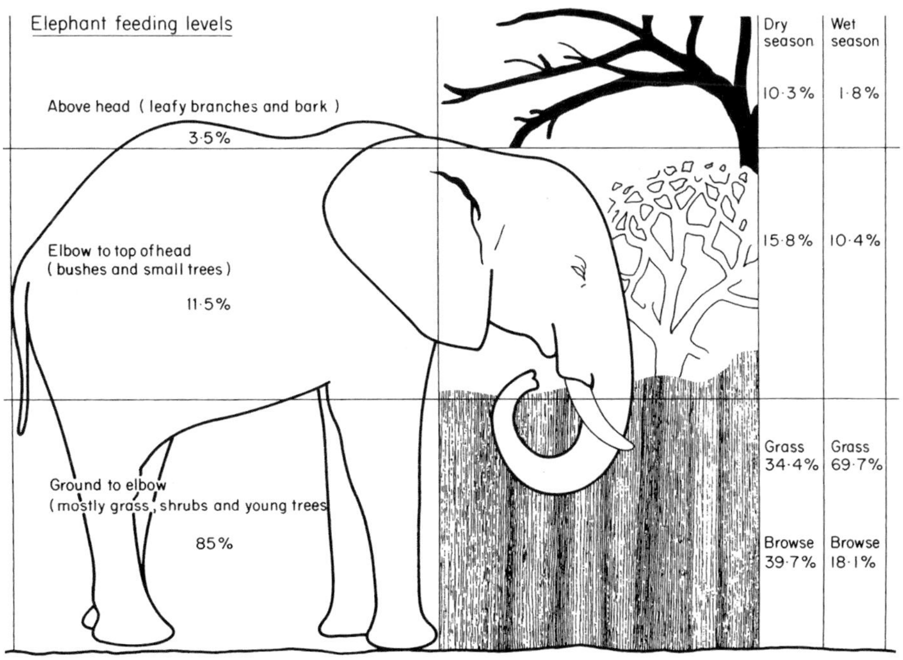

Percentage of food taken by *Loxodonta africana* at Sengwa, classified by level, season and type (browse/grass) (after Guy 1974).

per day); sodium content was usually correlated with carbonate and bicarbonate alkalinity. Elephants also utilized local concentrations of water-soluble sodium in the soil (a practice termed geophagy, the eating of saline or calcareous soils). Ruggiero & Fay (1994) recorded geophagia of enriched termitarium soils, where sodium may be a major attraction, and elephants are well known for making use of the salt licks in the caves of Mt Elgon, where they pry off chunks of the cave walls with their tusks and consume the volcanic agglomerate (Redmond 1982). In Hwange N. P., ♀♀ consume more mouthfuls of soil and spend a greater part of their activity budget feeding on soil than do ♂♂, as they probably have greater requirements due to pregnancy and lactation (Holdo *et al.* 2002). Holdo *et al.* (2002) found that soils consumed by elephants differed from other soils primarily in their high sodium content, and that elephants in non-Kalahari-sand habitats generally do not appear to create or use licks, probably because they are able to obtain their sodium requirements from ubiquitous sodium-rich water supplies (which do not occur naturally in Kalahari-sand habitats).

The activity of elephants can have a dramatic effect on the existing vegetation, and studies of the 'elephant problem' have dominated the literature since the 1970s. In eastern and southern Africa, savanna woodlands have been changed to less wooded grasslands and treeless grasslands (Buechner & Dawkins 1961, Savidge 1968, Laws 1970, Laws *et al.* 1975, Cumming *et al.* 1997). Diverse ecosystem hypotheses have been proposed to explain the elephant habitat interactions from equilibrium theory to stable cycles and multiple stable states. For such long-lived organisms as elephants and trees, quantifying relationships over many decades is virtually impossible as disturbances such as poaching or range contraction become superimposed on the interaction and obscure processes. However, modelling and empirical studies of the Masai Mara ecosystem have provided insights. Dublin *et al.* (1990) proposed that elephants and fire interact in woodlands to drive the change in state to grasslands, while elephants alone can maintain the grassland state. In this ecosystem two equilibria are possible: high tree density and low elephant density, or low tree density and high elephant density. Other states are not stable and reversion to one of the other states will occur. A change in state from high tree density is dependent on an exogenous force, fire, while elephants can maintain the low tree density once the change in state has occurred. Research in other sites suggests that woodlands can be converted to grassland by elephants alone if the density of elephants is high enough. Such changes have occurred in Murchison Falls N. P. in Uganda and Tsavo N. P. in Kenya. Very high local densities, and occupancy by bulls, can accelerate the process. In the Mwaluganje Elephant Sanctuary in the Shimba Hills ecosystem of Kenya, elephants concentrate in a deciduous forest where local densities can exceed 6/km^2 compared with 1.5/km^2 for the entire ecosystem. The killing of trees by debarking and pushing, primarily by bulls, has resulted in woodland decline (Kahumbu 2002). Similar findings in other bull areas have been reported in the Ngorongoro Crater, Tanzania and Nazinga Ranch, Burkina Faso.

In Kruger N. P., concern over excessive impacts of elephants on vegetation and biodiversity led to culling, which aimed to keep the population at around 7000 individuals ($\pm$ 2.5/km^2). This policy was partially successful in minimizing these impacts, but even at this level, elephants induced detectable changes, including elimination of certain species from some landscapes and changes from tree savannas to shrub savannas in others (reviewed by Whyte *et al.* 2003).

It is believed that in the past the elephant–habitat relationship was spatially dynamic, influenced by climatic variation and successional patterns that involved other herbivores and Pleistocene hunter-gatherers and more recent man. Vegetation modification caused by elephants is a normal activity and can contribute to beneficial changes in habitat diversity and biomass turnover. Confinement and concentration of elephants in declining ranges in parts of Africa has caused elevated densities and striking changes that can be detrimental to species diversity and can be a major cause of concern for wildlife management authorities.

Pacing *Loxodonta africana*, note placing of hindfoot on spot vacated by forefoot.

Social and Reproductive Behaviour The social network of Savanna Elephants is unusually large, radiating out from the mother–offspring bond through family, bond group, clan, subpopulation, to independent adult ♂♂, and beyond the population to strangers (Moss & Poole 1983, Archie *et al.* 2005, Wittemyer *et al.* 2005, Moss & Lee 2011). Their fluid society is characterized by group fission and fusion: the membership of groups changes often, forming and splitting along lines that may be predicated on the basis of close social bonds, kinship, home-range, season and, in the case of ♂♂, their reproductive state. Fission–fusion sociality combined with the close and enduring cooperative social relationships as found in elephants is rare in the animal kingdom (Archie *et al.* 2005).

The social lives of adult ♀♀ and ♂♂ are very different. The basic social unit of adult ♀♀ and young elephants is the family group or unit (Douglas-Hamilton 1972). Family size ranges from two, a mother and her youngest calf, to 45 or more individuals and may span four generations of related individuals (Moss 1988, Moss & Lee 2011). Rarely, families may be composed of unrelated individuals (Archie *et al.* 2005). Families vary in their degree of cohesion. Over time families may split to form bond groups or, sometimes, fuse to form new families (Moss & Lee 2011). Bond groups may comprise as many as five family units. Members of a family or a bond group may be distinguished by their pattern of greeting behaviour, coordinated movement and strong affiliative and protective behaviour toward one another. Families using the same dry season home-range have been referred to as belonging to a clan (Moss & Poole 1983). These individuals meet on a regular basis. Elephants are able to discriminate between these more familiar and less familiar elephants on acoustic characteristics alone (McComb *et al.* 2000). Elephants are also able to use scent to discriminate between individual family members and keep track of their specific locations (Bates *et al.* 2008b). In addition, they are able to discriminate between different groups of humans, reacting more fearfully to the scent of those groups who represent a possible threat than to those groups who do not (Bates *et al.* 2007).

During drier periods when food is scarce elephants tend to split up into small groups composed of families or fragments of families. However, when food resources become more plentiful, these smaller

Greetings, trials of strength, digging and other sketches of *Loxodonta africana*.

groups amalgamate and elephants are more likely to be seen in larger aggregations. These groupings may represent entire families, portions of a bond group, several bond groups or even the portions of several clans and independent adult ♂♂ (Poole & Moss 1989).

Group size varies from lone ♂♂ to aggregations of several hundreds or thousands of individuals (although as a consequence of poaching aggregations of more than a few hundred elephants are extremely rare). Group size varies tremendously depending upon the season, habitat type and population. In general, forest or thick bushland tend to support smaller groups of elephants, while large aggregations of elephants are more likely to be observed in open bush or wooded grassland.

Males begin to spend time away from their families at ~ 9 years old and by about 14 years of age they have left their natal families (Lee & Moss 1999). Newly independent ♂♂ usually spend several years in the company of other families near to their natal clan area, either associating with one particular family or moving from family to family, before they embark on their lives as fully adult ♂♂ (Lee *et al.* 2011b). As adult ♂♂ their lives fluctuate between periods of sexual activity and inactivity (Poole 1982, 1987, 1989b).

Elephants are not territorial, but ♀♀ are wary in their use of areas 'belonging' to the families of other clans (J. Poole pers. obs.). Home-range size varies tremendously from 50 km^2 to more than 12,000 km^2. To a large extent the variation can be accounted for by habitat type; elephants living in harsh desert environments typically have the largest home-ranges (see Habitat). Home-range also varies considerably within populations and individual preference, tradition, inter-family relationships, sex, habitat and human activities, all play a role in determining patterns of occupancy (Croze & Moss 2011). In the Kruger N. P. ranges of adult ♀♀ varied in size between 86 km^2 and 2776 km^2, with a mean of 880 km^2 (Whyte 2001b). In N Botswana home-range size rose significantly with increased distance to dry season surface water, and migratory herds travelled up to 200 km in search of water during the dry season; home-range size averaged 1091 km^2 (range 447–3309, n = 9) and 1699 km^2 (range 52–6412, n = 9), using satellite collars and radio collars, respectively (Verlinden & Gavor 1998). In N Kenya, home-ranges ranged from 102 to 5527 km^2 (n = 20) (Thouless 1996), while in the more arid environments of Namibia, they varied between 2136 and 10,738 km^2, with a mean of 5860 km^2 (minimum convex polygon; Lindeque & Lindeque 1991; and see Leggett 2006a).

Female elephants typically utilize their 'clan area' during the dry season and expand beyond this area to meet other clans during periods of food abundance (Moss & Poole 1983, Poole & Moss 1989). Adult males utilize a 'bull' or 'retirement area' during sexually inactive periods, but leave this area during sexually active periods to search over a wide area for receptive females (Moss & Poole 1983, Poole & Moss 1989). At independence young ♂♂ leave their family's clan area and over a period of decades gradually define their own particular 'bull area'. Chiyo *et al.* (2011) found that associations in bull areas were positively correlated with genetic relatedness, suggesting that kinship influences patterns of male associations.

By age 30 the majority of ♂♂ have experienced their first musth (range 17–34; average first musth 28; Poole *et al.* 2011). Musth is a period of heightened sexual and aggressive activity (Poole & Moss 1981, Hall-Martin & Van der Walt 1984), characterized by a distinct posture, the musth walk, swollen and secreting temporal glands, the dribbling of strong smelling urine (Poole & Moss 1981, Poole 1987) and a very low frequency vocalization, the 'musth rumble' (Poole 1987, 1999a, Poole *et al.* 1988). During musth, male elephants experience dramatic surges of circulating testosterone (Hall-Martin & Van der Walt 1984, Poole *et al.* 1984), interact aggressively with other large adult ♂♂, particularly those in musth, and spend much of their time searching for, attempting to gain access to, or guarding oestrous ♀♀ (Poole 1987, 1989b, c, Poole & Moss 1989). The duration of musth is age-related; musth periods are asynchronous and each male is on his own particular cycle (Poole 1987). Musth has a dramatic effect on the relative dominance ranks of ♂♂ (see Poole 1989b); with few exceptions, a musth ♂, whether large or small, ranks above all non-musth ♂♂. Musth ♂♂ interact aggressively whenever they meet, and the presence of an oestrous ♀ is not a prerequisite for a fight (Poole 1989b).

Females come into oestrus about once every 4.5 years (Moss 2001), though in other populations longer intervals have been observed. Behavioural oestrus lasts 4–6 days (Moss 1983, Poole 1989c). Diagnostic behaviours include wariness, the oestrous walk, the chase, mounting, mating, guarding and consort behaviour (Moss 1983, Poole 1989c). During early and late oestrus the ♀ is pursued and sometimes mounted, usually unsuccessfully, by young, non-musth ♂♂; the ♀ is guarded and mated by a musth male(s) during the mid 2–3 days of peak oestrus. Owing to the intense competition from older ♂♂ (Poole 1989b) and strong female preferences for mature individuals (Moss 1983, Poole 1989c), young ♂♂ have very limited mating opportunities during peak oestrus. Once a ♀ goes into consort with a musth ♂ mating occurs relatively infrequently (~ several times per 24 hrs) and lasts about 45 seconds. A mated ♀ rumbles repeatedly with a series of 'post-copulatory' or 'oestrous' rumbles (Poole *et al.* 1988). Her calling is typically joined by loud and excited rumbles, trumpets and roars by nearby elephants (usually members of her family) and is known as the 'mating pandemonium' (Poole *et al.* 1988, Poole 1989c, Poole & Moss 1989).

With the exception of musth, relationships between elephants are, in general, remarkably amiable. Within families and bond groups interactions are, as a rule, highly cooperative and interactions between non-relatives are seldom unfriendly. However, the relative level of agonistic and amiable behaviour differs from population to population depending upon the availability of resources. In areas where water, minerals or high quality food is limited or unevenly distributed, agonistic behaviour is more common (J. Poole pers. obs.). Among sexually inactive ♂♂ relationships are 'gentlemanly', while those between sexually active ♂♂, particularly those in musth, become highly aggressive and combative (Poole 1987). Older ♂♂ are larger and are generally dominant to younger ♂♂ (Poole 1989b).

Agonistic and affiliative displays take many forms (Poole & Granli 2011). Very simply, an elephant may threaten another by attempting to appear larger, approaching closer and using a variety of threatening ear and trunk postures, vocalizations and/or secretions. Submissive behaviour involves attempting to appear smaller, adopting avoidance behaviour and is associated with another set of ear and trunk postures as well as particular vocalizations (Langbauer 2000; for a full description and images see http://www.ElephantVoices.org and Poole & Granli 2011). The survival of calves is, to a great extent, dependent upon the strength of bonds between individual family members and these are maintained by frequent acoustic, chemical and tactile demonstrations of friendly affiliation and solidarity.

Elephants communicate with one another using visual and tactile displays, and vocal and chemical signals. Visuals and tactile displays use a combination of trunk, ear, limb and body postures and movements and signal a broad range of agonistic, defensive and affiliative gestures and complex emotional responses (an extensive description of the different postures and gestures can be viewed on www.ElephantVoices.org; Poole & Granli 2011; see also Kahl & Armstrong 2000b). Sources of chemical signals include fluids from the anogenital tracts and the temporal glands, and saliva, mucus secretions from the eye, ear and interdigital glands may also be involved (Rasmussen & Schmidt 1998). Chemical signals play an essential role in elephant social and reproductive communication and have been the subject of in-depth study (e.g. Rassmussen *et al.* 1996, Rassmussen & Schmidt 1998, Rassmussen & Krishnamurthy 2000, Rassmussen & Wittemyer 2002).

Vocal communication has also been intensively studied. Savanna Elephants produce a broad variety of sounds with components ranging from 9 Hz to over 9000 Hz; Poole 2011), including very low frequency rumbles and higher frequency trumpets, snorts, screams, barks, roars, cries and other calls (Berg 1983, Poole *et al.* 1988, Poole 1994, Langbauer 2000, Soltis *et al.* 2005a, b, Poole 2011). Elephants are able to coordinate their movements by communicating with powerful, very low frequency sounds (Payne *et al.* 1986, Poole *et al.* 1988, Langbauer *et al.* 1991, Garstang *et al.* 1995, Larom *et al.* 1997, McComb *et al.* 2000). Elephants communicate a broad range of complex messages related to interactions of an agonistic, defensive, affiliative, parental care, mating and social nature. Recently, Poole *et al.* (2005) discovered that elephants are capable of vocal production learning (imitation) and posit that this rare talent may have evolved to facilitate social bonding and cohesion in the elephants' highly dynamic fission–fusion society.

Elephants can recognize the specific voices of other elephants up to 2.5 km away, though more usually at distances of 1–1.5 km (McComb *et al.* 2002). Under certain conditions, such as exist in the Etosha N. P. in the dry season, elephants may hear the calls of conspecifics at distances exceeding 10 km (Garstang *et al.* 1995). Furthermore, when an elephant vocalizes with a low frequency rumble an exact replica of this signal propagates separately in the ground (O'Connell *et al.* 1998) and elephants respond appropriately to this signal component (O'Connell *et al.* 2006, 2007).

Comfort behaviour includes: water-bathing; mud-wallowing; dust-bathing; scratching the body by rubbing on trees, logs or anthills; scratching the body (sometimes to remove parasites) with sticks or other objects held in the trunk; swinging the tail, trunk or branches held in the trunk to repel insects; yawning; stretching; exhaling loudly or 'blowing' through the trunk; rubbing the ears with the curved distal portion of the trunk; placing the tip of the trunk over the end of a trunk to reduce the itching; scrunching up the trunk and making a 'squelching' sound to reduce itchiness in the trunk; and ear-flapping to cool the body.

Birthing may take place wherever the ♀ finds herself, but typically occurs in the company of other family members and most often at night. Elephants usually give birth standing with bent hindlegs, but may lie down prior to the birth. Signs of an imminent birth may include: kneeling down or squatting (straining during contraction); lying down and getting up again; kneeling, getting down on all four legs; kicking backwards; beating bushes; vaginal secretions; frequent urination; intense interest by other family members (smelling and vocalizing) in the vaginal orifice and secretions of the pregnant ♀; swollen vulva; and a bulge protruding under the tail (foetus in birth canal) (Moss 1992, Poole 1996, 1999b). The actual birth happens

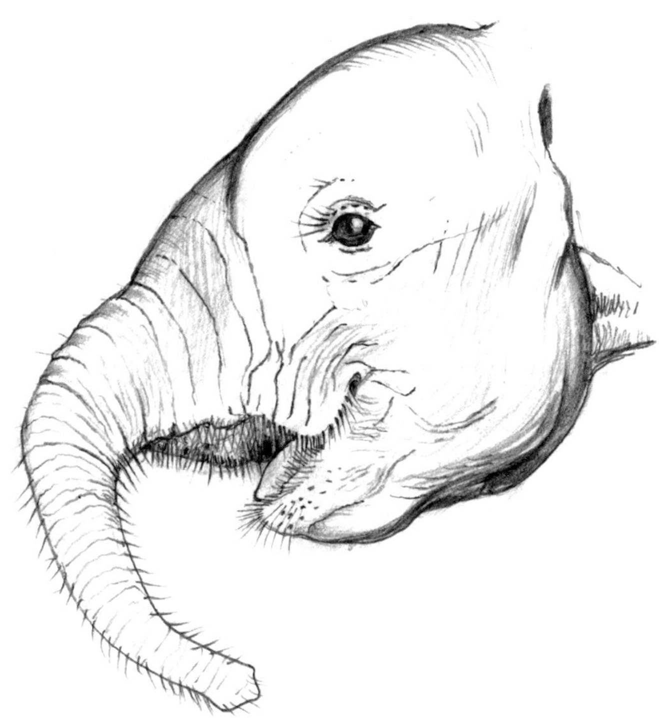

Vibrissae around mouth and trunk of foetal *Loxodonta africana*.

quickly, and the foetus is propelled forward with considerable force. The infant is usually born partially covered by the foetal sac and placenta. Mother and other family members use their tusks to gently free the infant from the birth sac, and show great interest in the placenta, carrying it about, tossing it in the air and swinging it back and forth with their trunks. With assistance from its mother and other adult ♀♀ an infant is usually able to stand within 15 minutes of birth. The birth itself is a momentous occasion for an elephant family and is accompanied by intense rumbling, screaming and trumpeting that may last up to 40 minutes (Poole 1999b, J. Poole pers. obs.). Young ♀♀ between the ages of three and eight years old show intense interest in newborn calves, following the mother and infant around and offering comfort and assistance (Moss 1992).

Young elephants depend on their mothers for milk for two years. After this age, they can support themselves on solid foods, although most calves will carry on sucking for as long as the mother will tolerate it. During the first five years of life, calves develop their social skills and form relationships with members of their family. Sons and daughters are treated differently by mothers and develop sex-specific behaviour early in life (Lee 1986, Lee & Moss 1999). Sons are more exploratory, seek out playmates from unfamiliar families, and are allowed by mothers to suck more frequently than are daughters. Daughters are more socially interactive within the family, and are less demanding of milk from their mothers. The young ♀♀ within a family play a critical role in the survival of calves, shepherding them, protecting them from obstacles or stresses in the environment and ensuring that calves are safe from predators (Lee 1987).

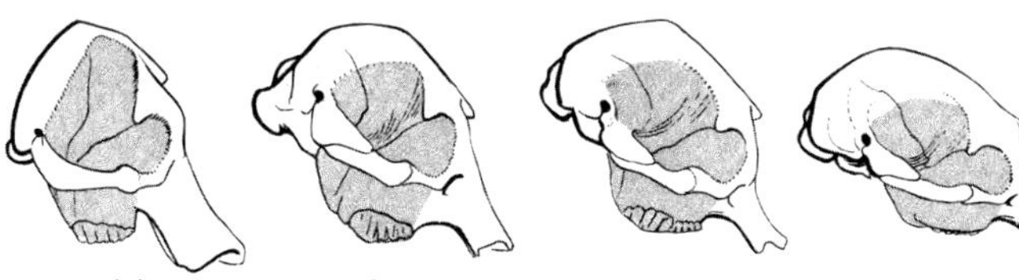

Changes in skull form with age in *Loxodonta africana*.

Elephants may show unusual interest in and/or irritation with other species. Infants and calves enjoy chasing birds, monkeys and other small mammals in play. Larger calves and juveniles, too, will playfully chase a variety of animals including birds, warthogs, antelopes (including buffalo), hyaenas and primates. Adults tend to have a more competitive interaction with other species, displacing other animals at waterholes, putting an end to the cavorting of wildebeests, chasing predators (Lions *Panthera leo*, Spotted Hyaenas *Crocuta crocuta*) and scavengers (jackals, vultures) off kills or away from the vicinity of the family. Elephants are one of the few herbivores that purposefully kill other species, for example, by stepping on birds, hares and other small mammals that have startled them, and goring, and kneeling on, people and their livestock. Young orphaned elephants introduced to wildlife parks have displayed intense aggression towards White Rhinos *Ceratotherium simum* (e.g. Slotow & Van Dijk 2001, Slotow *et al.* 2001; and see Berger & Cunningham 1998). The killing of other animals appears to cause an elephant great distress, resulting in highly aggressive and unusual behaviour, refusal to leave the body, uprooting and trampling of vegetation in the vicinity of the body, and even covering the body in vegetation (J. Poole pers. obs.). Elephants have even been observed to chase Lions and Spotted Hyaenas off a wildebeest kill and then stand over the dead animal (N. Njiraini pers. comm.).

Reproduction and Population Structure The longest study of demography among free-ranging Savanna Elephants comes from the Amboseli population where data have been collected from individuals of known age and identity for 35 years (Moss 2001, Lee *et al.* 2011a). The information that follows comes largely from this study supplemented with additional information from other regions.

The pattern of breeding varies with habitat type and rainfall conditions, though most authors report some seasonality of births coinciding with the months of high rainfall (Hanks 1969, Laws & Parker 1968, Laws 1969, Sherry 1975, Smuts 1975, Craig 1984, Lindeque 1988). In Amboseli where 1030 birth dates were known to within two weeks, births occurred in every month, but 81% occurred between Nov and May (Moss 2001). In general, ♀♀ conceive when they are in good condition during periods of higher food availability. Birth rates show peaks and troughs over time and in some populations (Moss 2001) ♀♀ become synchronized in a 4–5 year cycle of conceptions and births.

Gestation has been calculated as 660 days (Perry 1953, Laws *et al.* 1975, Moss 1983, Poole *et al.* 2011). Twinning is rare. In Amboseli, where the live births of 1192 individuals have been recorded (Moss 2001), one case of twins was observed (0.08%), and Smuts (1975) recorded only two sets of twins out of more than 350 embryos and foetuses examined in Kruger N. P.; however, the incidence was much higher in Etosha N. P. (10%; Lindeque 1988). A female in Kruger N. P. known as MaMerle produced successive sets of twins all of which survived up to weaning age and beyond (I. Whyte pers. obs.). Weight at birth has been estimated at 120 kg (Perry 1953), though they may be lighter (75 kg; Smuts 1975) or heavier. Shoulder heights

Skin folds and lashes around eye of *Loxodonta africana*.

of newborn calves in Amboseli measure 79 cm for ♀♀ and 85 cm for ♂♂. These are somewhat smaller than the minimum measures at Kruger N. P. (85 cm for ♀♀, 92 cm for ♂♂). Growth is most rapid in the first five years of life and growth rates of ♂♂ and ♀♀ diverge after age two. Males are consistently larger than ♀♀ by 5–6 years of age. There are interesting differences between populations in growth rates, which affect adult size. Males reach 2 m in height at seven years in Kruger N. P. and at ten years in Amboseli, while ♀♀ reach this height after nine years in Kruger N. P. and 14 years in Amboseli (Kruger: I. Whyte pers. obs.; Amboseli P. Lee pers. obs.). Growth in body size continues for most of adult life accentuating these differences between the sexes, due to late epiphyseal fusion of long bones, and absent fusion of vertebrae and scapula (Haynes 1991; see also Roth 1984). Full adult height is reached at about 30 years for ♀♀, while ♂♂ continue to grow in height and mass well into their 50s (Poole 1989b, Lee & Moss 1995; and see Lindeque & Van Jaarsveld 1993). Males tend to be over 30% taller and double the weight of ♀♀ (Laws 1966, Lee & Moss 1995).

Elephant calves typically cease to suck when the next calf is born (Lee & Moss 1986), and examination of elephant cows at culls in Kruger N. P. showed that most lactate constantly (Smuts 1975, I. Whyte pers. obs.). As a result 'weaning' is an extended and gentle process (Lee & Moss 1986). There are subtle differences between the sexes of calves in weaning, with sons being allowed to suckle whenever they emit a vocal protest, while mothers are somewhat less tolerant of daughters. Given that the inter-calving period in Kruger N. P. was calculated at 3 years and 8 months (Whyte 2001a), this also represents the average age at weaning. In cases where the inter-calving period is extended, age at weaning would probably be extended accordingly.

Mean inter-birth interval varies from population to population depending upon food availability and other factors, and ranges from 2.9 to 9.1 years (Eltringham 1982; upper range: 9.5, Viljoen 1988). In nutritionally stressed populations, mean inter-birth interval tends to be long (Laws *et al.* 1975). In the Amboseli population inter-birth interval averaged 4.5 years, while in the Addo population inter-birth interval was 3.8 years. In Kruger N. P. the interval was estimated at 3.7 years from a culled sample of 966 adult ♀♀ (Whyte 2001b), comparable with that in Etosha N. P. (3.9; Lindeque 1988). Inter-birth interval is considerably shortened by the death of a calf (Moss 2001). In Amboseli a ♀ has been known to produce a calf as quickly as 22 months after a birth if the first calf dies and 27 months after the birth of a surviving calf (Moss 2001).

Fecundity is fairly constant between the ages of 16 and 40 and then declines slightly, though ♀♀ over 60 can still give birth (Moss 2001). In Kruger N. P., fecundity declines in older cows, but several between 50 and 60 years were pregnant (Freeman *et al.* 2008). Unlike in Addo Elephant N. P. (Whitehouse & Hall-Martin 2000), ♀♀ in Amboseli were able to reproduce until the end of their life-span (Moss 2001). Breast development is a good indication that a ♀ is either pregnant or lactating and thus lack of breast development is an indication that she is neither. In populations that are suffering from high levels of poaching the proportion of ♀♀ of reproductive age with developed breasts can be much lower. For instance, in the Mikumi population in 1989, 31% (n = 35) and 75% (n = 16) of ♀♀ aged 15–19.9 and 20–24.9 years had developed breasts compared with 87% (n = 23) and 92% (n = 26), respectively, in the Amboseli population in the same year (Poole 1989a).

Many studies have estimated birth rate by counting the number of calves less than one-year-old at a given time as a proportion of population size and using this figure as an index of recruitment (Buss & Savidge 1966, Laws 1966, Calef 1988). However, such instantaneous counts can reflect short-term patterns of peaks and troughs and thus produce misleading estimates of population growth rates. Second, these estimates do not take into account the high rate of mortality among calves under one year of age. Thirdly, the method assumes that the estimates of calves' ages are accurate, which is questionable as calves just under or just over 12 months are easily misclassified. Over the course of 25 years in the Amboseli study population, growth rate averaged 2.17%, although lower (–7.42%) and higher (up to 11.3%) were witnessed in any particular year (Moss 2001). Between 1975 and 1994 when the entire Kruger N. P. was fenced and immigration and emigration were thus limited, the observed rate of increase averaged 6.6% per year, though fluctuations of between –2.9% and +14.8% were recorded during this period (Whyte 2001b).

The age of first birth varies across populations (Eltringham 1982). In Amboseli most ♀♀ gave birth for the first time between the ages of 14 and 15 years (median 14.1; Moss 2001). The youngest first birth was 8.9 years and the oldest 21.6 years (Moss 2001). In Kruger N. P. the youngest first birth was 9 years while 93% of 14-year-olds were either pregnant or lactating (Whyte 2001b).

Puberty in males is marked by the first production of sperm at ~ 14 years and sexual maturity, the production of sperm in quantity, by 17 years old (Laws 1969). At this age ♂♂ are only half their full adult weight and cannot compete with older ♂♂ for access to ♀♀ (Poole 1989b). Between the ages of 20 and 25, free-living ♂♂ have begun to show behavioural signs of sexual cycles, and by 29 years of age most ♂♂ have experienced their first musth period (Poole 1987). The duration of musth increases with the age of the ♂ (from days to weeks in younger ♂♂ to 3–4 months in older ♂♂) until approximately 50 years of age, when ♂♂ appear to enter a slow reproductive decline (Poole *et al.* 2011). They are still reproductively active at age 60 (Poole *et al.* 2011). Genetic paternity confirms that older ♂♂ in musth father at least 75% of elephant calves (Hollister-Smith *et al.* 2007). Genetic paternity studies have also shown that despite the roving mating strategy of ♂♂ and the fluid sociality of both sexes, elephants avoid breeding with their close maternal and paternal relatives (Archie *et al.* 2007).

Elephant calves are born at a 1 : 1 sex ratio (Smuts 1975, Whyte 2001b, Moss 2011). Calf mortality in Amboseli averages around 10% in the first year of life (Moss 2001) while average annual natural mortality in Kruger N. P. has been estimated at around 3.2% per annum (Whyte 2001b). Males tend to have higher mortality than ♀♀ throughout much of their lives. While 84% of ♀♀ reach the age of ten years, only 75% of ♂♂ reach this age, and while 82% of ♀♀ survive to the age of first reproduction (14 years), only 39% of ♂♂ survive to the age when they regularly enter musth for the first time and are likely to start obtaining regular matings (Moss 2001).

Males have higher mortality than ♀♀ throughout most of their lives. Life expectancy at birth is 41 years for ♀♀ and 24 years for ♂♂ (Moss 2001). However, in the absence of human-induced mortality, female life expectancy increases to 54 years and males' life expectancy to 39 years. Maximum life-span for Savanna Elephants is unknown, but is at least 65 years for ♀♀ and at least 60 years for ♂♂ (Moss 2001). Since longevity is partly determined by viable molar teeth, animals with exceptionally abrasive diets may, theoretically, have shorter natural life-spans. In captivity, maximum recorded longevity is 53–54 years of age for a wild-born ♀ that was still living (Weigl 2005).

Predators, Parasites and Diseases Non-human predators include Spotted Hyaenas and Lions. Although mortality caused by these predators is probably rare and primarily limited to calves and immature individuals, healthy elephants up to age 15 have been reported killed by Lions (e.g. Joubert 2006, Loveridge *et al.* 2006). The extent to which Spotted Hyaenas and Lions affect mortality patterns varies by season and locality depending upon the elephant's condition, predator population size and their sources of other suitable prey. Elephants do, however, treat both species as a threat and respond to them with aggression.

Defensive formation in *Loxodonta africana*.

Elephants are affected by a number of diseases and ailments. Of all of these, only anthrax *Bacillus anthracis* has been shown to have the potential of limiting an elephant population, and this only in Etosha N. P. (Lindeque & Turnbull 1994). In the Kruger N. P., sporadic outbreaks of anthrax have caused some mortality. In addition, the *Picorna* virus, which causes encephalomyocarditis (EMC) and is transmitted by rodents (Grobler *et al.* 1995), has caused some deaths. No cases of foot-and-mouth disease or bovine tuberculosis have ever been recorded in free-living elephants. Other conditions known to affect elephants include heart diseases, twisted gut, septicaemia and cancerous growths.

A condition known as 'floppy' or 'flaccid trunk syndrome' (FTS), which is caused by a deterioration of the nerves of the trunk, has been recorded at Kariba, Zimbabwe (Kock *et al.* 1994) and Kruger N.P. The condition is progressive, leading to total paralysis of the trunk, which can be fatal as feeding becomes severely impaired. Little is yet known about its causes, but it is suspected that plant toxins and/or pollutants may be involved. In the Amboseli population one or two elephants are seasonally afflicted by a condition known as 'twisting disease', which causes the affected individual to walk with some difficulty and an unnatural gait. Since this condition is only observed during and following the rains and most often in adolescents, plant toxins or a virus are plausible causes. In addition, starvation is not uncommon during periods of severe drought.

Elephants are infested by a host of internal parasites. Condy (1974) and Round (1968) have listed 33 species from 25 genera and 14 families (and see Bauer & Styoe 1985). None of these is known to be fatal, but may contribute to mortality in individuals in poor condition. External parasites include ticks (21 species from four genera), the louse *Haematomyzus elephantis*, ear mites (*Loxonoetus bassoni*, *Otanoetus wetzeli*) and biting flies, including flies, mosquitoes, horse flies, warble, bot and bloodsucking flies (among them tsetse flies) (Zumpt & Wetzel 1970, Sikes 1971, Fain 1983).

Conservation IUCN Category: Vulnerable A2a. CITES: Appendix I in 1989, but the populations of the following Range States have since been transferred back to Appendix II with specific annotations: Botswana (1997), Namibia (1997), South Africa (2000) and Zimbabwe (1997).

Although it is known that earlier humans were capable of killing elephants and mammoths, it is not known what effects they may have had on elephant populations, though it has been speculated that humans may have had a limiting role. However, the activities and practices of modern humans are an extremely important and limiting factor for adult elephants and indirectly, therefore, on immature elephants (Poole 1989a). The Waata people of the Tsavo region (Kenya) were renowned hunters of elephants using bows and poisoned arrows and may even have had a limiting effect on elephant population growth in that area (Ville 1995). In more recent times it is the commercial pursuit of ivory and the expansion of human settlements that has led to the contraction of the range of African elephants (Cumming *et al.* 1990, Milner-Gulland & Beddington 1993).

Although elephant range has undergone considerable contraction, elephant populations occur in a number of protected areas across the continent, with recent evidence of an overall increase in numbers (Blanc *et al.* 2005, 2007). In southern Africa, protected areas known to hold important populations of Savanna Elephants include: Niassa G. R. (Mozambique); Nyika N. P. (Malawi); Etosha N. P. (Namibia); Chobe N. P., Moremi G.R. and Tuli (Botswana); North and South Luangwa, Kafue N. P. (Zambia); Gonarezhou N. P., Hwange N. P. and surrounds, and Matusadona N. P. (Zimbabwe); and Addo Elephant N. P., Kruger N. P., Klaserie G. R., Timbavati G. R., Sabi-Sand G. R., Madikwe G. R. and Tembe Elephant Reserve (South Africa). In East Africa, important protected area include: Omo, Mago and Gambella National Parks (Ethiopia); Aberdare N. P., Mt Kenya N. P., Amboseli, Masai Mara N. R. and Tsavo N. P. (Kenya); Nimule N. P. (Sudan); and Katavi N. P., Mikumi N. P., Ruaha N. P., Rungwa G. R., Selous G. R. and Tarangire N. P. (Tanzania) (Blanc *et al.* 2007). Zakouma N. P. in Chad has been an important stronghold for Savanna Elephants in the central African region but is under intense poaching pressure. The arid-adapted population in Gourma is partly formally protected in the Douentza Elephant Reserve. The 'W' N. P. that straddles the borders of Benin, Niger and Burkina Faso, together with adjacent protected areas, represents the largest elephant range in West Africa (Blanc *et al.* 2007).

Despite their presence in many protected areas, the majority of elephant range is not within protected areas; about 70% and 72% of their range falls outside protected areas, in eastern and southern Africa, respectively (Blanc *et al.* 2007). However, in some countries, such as South Africa, extensive human development precludes the occurrence of elephants outside public and private conservation areas, with the result that about 85% of the range falls within protected areas (Blanc *et al.* 2007). In these confined areas, the impacts of elephants on biodiversity can be dramatic (e.g. Coetzee *et al.* 1979, Moolman & Cowling 1994, Scholes *et al.* 2003) and some argue that such populations can only be constrained by reduction of elephant numbers or limitation of population growth (e.g. via contraception). There are a few options for such management, but all of them pose ethically and morally taxing dilemmas that have been examined by Whyte (2001b) and Whyte & Fayrer-Hosken (2008). It is expected that more and more conservation areas across Africa will be faced with these difficult decisions as human developments expand.

Historically, significant changes in elephant populations and ranges have occurred. In southern Africa, elephants were almost hunted to extinction by the end of the nineteenth century. South Africa contained only a few remnant fragments, which occurred in the Kruger N. P., Addo Elephant N. P., Knysna Forest and Tembe Elephant Park. From these meagre remnants, the national population has increased to nearly 18,000 (Blanc *et al.* 2007), with many new public and privately owned conservation areas being established, most of which have acquired their elephant herds through translocations from Kruger N. P. Genetic inbreeding has occurred in Addo due to the population bottleneck there, though this issue is now being addressed through the introduction of adult bulls from the Kruger N. P. All of these areas are now formally conserved, which has resulted in population increases except in the Knysna area where numbers have declined almost to the point of extirpation.

Poaching continues to pose a significant threat to elephants and has hampered conservation efforts in most parts of Africa. Equally worrying is the growing conflict with humans, which has come about due to human population growth and resulting loss of elephant range and habitat. The 1989 international ban on trade in ivory was a turning point for many populations, though the decision remains highly controversial. A partial re-opening of the ivory trade was approved in 1997 despite the fact that growing demand for ivory in the East continued to threaten elephants in many parts of Africa. Concerns that legal trade would stimulate markets and provide a cover for laundering of illicit ivory led to the implementation in 2000 of a continent-wide monitoring system known as MIKE (Monitoring of Illegal Killing of Elephants). The MIKE was established to determine the effects of CITES decisions on elephants in 50 sites across Africa. The limited and controlled ivory stockpile sales that occurred in 2008 may have contributed positively to elephant conservation in some parts of southern Africa, but it may also have intensified poaching threats in other populations. Indeed, soaring demand for ivory in the Far East (e.g. Vigne & Martin 2011) has caused a marked upsurge in elephant poaching that has the potential to threaten elephant populations across the continent.

Measurements

Loxodonta africana
HF (♂ range 10–65 yr): 360–540 mm
HF (♀ range 10–65 yr): 340–470 mm
Sh. ht (♂): 2.9–3.7 m
Sh. ht (♀): 2.5–3.0 m
Amboseli N. P. (Lee & Moss 1995)
Maximum recorded shoulder height for a ♂ is the Fenykoevi elephant from Angola (housed in the Smithsonian Museum), which stood 4.0 m (Haynes 1991). The largest ♂ measured in Kruger N. P. stood 3.45 m; for females, 2.74 m, but they seldom exceed 2.55 m (I. Whyte pers. obs.).
WT (♂): up to 6048 kg
WT (♀): up to 3232 kg
Moss (1976) recorded that the heaviest bull elephant on record had a mass of 6569 kg.

The heaviest pair of tusks on record, now housed in the collections of the British Museum of Natural History, are from Mt Kilimanjaro with a mass of 102.7 kg (3.11 m long) and 97.0 kg (3.19 m long). In Kruger N. P., which is well known for its large tuskers, the heaviest female tusk weighed was 10.7 kg but they rarely exceed 7.0 kg (I. Whyte & A. J. Hall-Martin unpubl.). Mean tusk mass for bulls exceeding 50 years of age was 40 kg, and the heaviest male tusks recorded in Kruger N. P. were from a bull known as 'Mandleve' and weighed 73.5 kg and 69 kg, respectively. The longest tusk recorded from a ♂ in Kruger N. P. was 2.64 m from a 55-year-old. Several tusks exceeding 50 kg are on display in the elephant museum at Letaba restcamp in Kruger N. P.

Key References Blanc *et al.* 2007; Laws 1966, 1967, 1970; Laws *et al.* 1975; Lee 1986, 1987; Lee & Moss 1986, 1995, 1999; Lee *et al.* 2011a; Lindeque 1988, 1991; Moss 1981, 1988, 2001; Moss & Poole 1983; Moss *et al.* 2011; Poole 1982, 1987, 1989a, b, c; Spinage 1994; Whyte 2001a, b.

Joyce Poole, Paula Kahumbu & Ian Whyte

Forest Elephant *Loxodonta cyclotis*.

Loxodonta cyclotis Forest Elephant

Fr. Eléphant de Forêt; Ger. Waldelefant

Loxodonta cyclotis (Matschie, 1900). Sitzb. Ges. Naturf. Fr. Berlin, p. 194. Yaunde, S Cameroon.

Taxonomy Monotypic. Originally described as a subspecies of the Savanna Elephant *Loxodonta africana* by Matschie (1900), but recent morphological and genetic studies (Grubb *et al.* 2000, Roca *et al.* 2001, 2004, Comstock *et al.* 2002, Rohland *et al.* 2010) confirm the earlier assertions of Frade (1931, 1955) that the Forest Elephant is a distinct species (but see Debruyne 2004, 2005a). Eggert *et al.* (2002) propose a third lineage that includes both Forest and Savanna Elephants of West Africa.

There have been persistent reports of a pygmy elephant (*Loxodonta pumilio*) (Noack 1906) that shares the same habitat as the Forest Elephant (Morrison-Scott 1947). However, there is no genetic evidence for the existence of such a species (see Debruyne *et al.* 2003b). The long-standing belief in a separate race or species was based on sightings of solitary juvenile and subadult Forest Elephants: their small stature and precocial tusk development misled observers into believing they were adults. Behavioural observations make it clear that 'pygmy' elephants are actually immature Forest Elephants (Western 1986, Groves & Grubb 2000a, A. Turkalo pers. obs.), although this view was still disputed until quite recently (Eisentraut & Bohme 1989, Anon 1990, Bohme & Eisentraut 1990). Synonyms: *albertensis*, *cottoni*, *fransseni*, *pumilio*. Chromosome number: 2n = 56 (Hungerford *et al.* 1966).

Description Forest Elephants have many features in common with their Savanna counterparts (most of which are not repeated in this account), but they are markedly smaller in stature, with adult ♀♀ averaging 2 m and ♂♂ 2.4 m at the shoulder. Ear shape has been described as being more rounded and smaller than that of the Savanna Elephant, although there is a wide variation in shape and size. Some Forest Elephants possess ears that are similar to those of Savanna Elephants, with lobes extending well below the chin line. The body colour is generally described as grey, but variations in colour vary from grey to yellow to reddish with the yellow and red colouration usually due to mud wallowing. Newborns possess a noticeable sparse body hair covering (black), but this disappears as the elephant matures. Hair remains present in adults but is sparse and is noticeable in the tail, and in ♀♀ sometimes a 'beard' on the chin is discernible. Tail length varies from half of the rear rump height to longer tails whose hairs touch the ground. The tail is covered in thick skin ending in an asymmetric lobed tip, bordered by long, coarse black hairs (20–200 mm). There is also a difference in the number of 'toenails' or nail-like structures between the Savanna and Forest species. Generally, Forest Elephants have five nails on the front feet and four on the rear, whereas in the Savanna species the front feet have four and the rear three.

Tusks of adult ♂♂ are generally longer and thicker than those of adult ♀♀, this characteristic being easily discerned even in infant and juvenile Forest Elephant ♂♂. Forest Elephant ivory differs from the ivory of the Savanna Elephant in both colour and hardness. Described as 'pink' or 'hot', Forest Elephant ivory is harder, making it the preferred material for carvings. Generally, male Forest Elephants carry thinner tusks than Savanna Elephants, although there is a wide variation in tusk morphology among Forest Elephants.

Geographic Variation None recorded.

Similar Species

L. africana. Larger and taller, with larger more triangular-shaped ears, and heavier, more curved tusks.

Distribution

Historical Distribution Forest Elephants are assumed to have occurred throughout the Guinea-Congolian region of West and central Africa. The limits of past and present distribution remain vague because of intergradations between Forest Elephants and Savanna Elephants along the northern and eastern fringes of the West African and Congo Basin forests (Roth & Douglas-Hamilton 1991). In Garamba N. P., DR Congo, three individual elephants showed genotypes with a combination of Forest and Savanna alleles, suggesting a history of limited hybridization in the ancestors of this population (Roca *et al.* 2001).

Current Distribution Forest Elephants are widespread in the remaining forests of the Guinea-Congolian forests. The majority of elephant range in central Africa is inhabited by Forest Elephants, with Savanna Elephants occurring in N Cameroon, N Central African Republic and Chad (Blanc *et al.* 2007). However, the distribution of elephants in the forests of central Africa is shrinking at an accelerated rate as roads, railways, pipelines and mining and logging companies penetrate forests that were formerly difficult to access, making it easier for hunters to find and kill them. Loss of elephant range in DR Congo has been especially acute over the past several decades as rapidly expanding human populations move into some of the most important elephant landscapes, especially in the eastern forest zone (see Conservation).

In West Africa, elephants are now found in 35 small, isolated forest patches with a median area of 800 km^2 (AfESG 2003).

There are reports of cyclotiform elephants in tongues of riverine forest far into the savanna zone of N Central African Republic and in the Sudanian savanna zone of N Togo (Stalmans & Anderson 1992). Elephants in the mountains of Kenya are Savanna Elephants (Roca *et al.* 2001). See profile of Savanna Elephant (p. 183) for distribution map.

Habitat In central Africa, Forest Elephants inhabit the large contiguous forest tracts, and are also found in areas of overlapping savanna and forest in Lopé in Gabon, Odzala in the Republic of Congo, and in the northern forest savanna mosaic of DR Congo. In West Africa, Forest Elephants are now confined to forest fragments.

Abundance The central African forests, which contain the least disturbed habitats, often support high elephant densities. Although few elephants may be seen close to roads and villages, densities increase with distance from roads to well over 1/km^2 in some remote forests (e.g. Barnes *et al.* 1991, 1997, Hall *et al.* 1997, Thomas *et al.* 2001). Little is known about how the population in central Africa changed during the nineteenth and twentieth centuries when they were under heavy pressure for ivory. The creation of protected areas and the enactment of legislation outlawing commercial hunting for ivory during the past few decades has not been particularly effective due to political instability in the region and the expansion of extractive industries, which has opened up previously inaccessible areas of the rainforest.

Gabon, the only central African country to have been surveyed completely (Barnes *et al.* 1995, 1997, Blanc *et al.* 2007), and the most politically stable country in the region, is thought to have the largest Forest Elephant population, with probably in the order of 25,000 individuals (Blanc *et al.* 2007). The vast forests of DR Congo have the potential to hold many thousands of elephants, possibly as many as 20,000 (Blanc *et al.* 2007, J. Hart pers. comm.), but ivory poaching was intense during the 1970s and 1980s in even the most remote forests (Alers *et al.* 1992). With the recent expansion of the bushmeat trade, elephants are now being killed for meat as well as for ivory (Mubalama & Mapilanga 2001, Blanc *et al.* 2007). Elephants in many remote forests have been nearly or totally eliminated in the years of civil war and political instability spanning the turn of the twenty-first century (J. Hart pers. comm.). Cameroon, Central African Republic and Congo all contain sizeable populations of elephants, but few reliable figures exist. Elephants in Equatorial Guinea may number as many as 1300 individuals, but the majority (~90%) of elephant range in this country remains unsurveyed and this is little more than an informed guess (Blanc *et al.* 2007).

As with Savanna Elephants, a continent-wide assessment of population size for Forest Elephants is complicated by countries having both Forest and Savanna Elephants, and the unclear status of elephants in West Africa. None the less, using the results from the African Elephant Status Report (Blanc *et al.* 2007), and the same general assumptions used to calculate the Savanna Elephant population estimate (see species profile), it is possible to arrive at a general estimate for Forest Elephants in central Africa (here taken to exclude elephant populations from N Central African Republic and Chad) (Table 11).

Table 11. Estimated numbers of Forest Elephants, excluding West Africa.

	Estimate	95% CL ±	Guesstimate		Range covered (%)	Area (km^2)
			From	To		
Estimates from surveys	9500	4200			1	15,500
Estimates from low-quality surveys and guesses			100,000	160,000	29	362,500
Unassessed range					70	860,000
Total range					100	1,238,000

Source: African Elephant Database 2007, courtesy of the IUCN/SSC African Elephant Specialist Group

Few surveys were conducted in West African forests before 2000 and most estimates of population size were guesses that suggested a median population size of 50 and a median density of 0.08 elephants/km^2 (AfESG 1999). Recent surveys show large variations in density, e.g. 0.04 and 0.64/km^2 in two Ghanaian forests (EBM 2000, Danquah *et al.* 2001). Many West African Forest Elephant populations are probably not viable because of their small numbers, unbalanced sex ratios and distorted age structures (Barnes 1999).

The most common method of estimating Forest Elephant numbers is by means of dung counts (e.g. Barnes *et al.* 1997, Hall *et al.* 1997, Walsh & White 1999). Transects are distributed across the study area according to a specified survey design and elephant dung-piles recorded by distance sampling (Buckland *et al.* 2001). The resulting estimates of dung-pile abundance can then be combined with defaecation rate and dung decay rates to give estimates of elephant numbers with confidence limits. The estimates from dung counts are as accurate as those from other methods and give narrower confidence limits than aerial sample surveys (Barnes 2001, 2002). Jachmann (1991) concluded that dung counts should be the preferred method for estimating elephant numbers in all situations.

Work on the use of acoustical methods to monitor and census Forest Elephant was initiated in 2000 at the Dzanga clearing in the Central African Republic (Payne *et al.* 2003). Further work is now in progress to establish the extent to which the rates and patterns of calls of Forest Elephants reflect their numbers, group structure and reproductive health. This information will provide the basis of an acoustical monitoring programme through which much can be learned about the locations, demography and condition of Forest Elephant populations throughout central and West Africa.

Estimates of small populations usually have wide confidence limits that make it difficult to detect changes in animal numbers (Barnes 2002). Advances in DNA technology now provide a reliable method for estimating the numbers of elephants in small forests such as those in West Africa. DNA is extracted from samples of dung and individual elephants are identified by multilocus genotyping. The data from a series of collections can be analysed by mark–recapture methods to give estimates of elephant numbers (Eggert *et al.* 2003).

Adaptations Most of the adaptations described for Savanna Elephants apply also to Forest Elephants. As with their Savanna relatives, both sexes possess temporal glands located on each side of the head at eye level. Temporal glands are most noticeable in adult ♂♂ undergoing musth ('rut'). During this period these glands swell and a secretion (known as temporin) streams down the side of the head. In Forest Elephant ♀♀, temporal gland secretion is seldom observed, and when it does occur is thought to be related to stress (A. Turkalo pers. obs.).

Forest Elephant crania are also exceptionally thick and pneumatized, which effectively reduces the weight of the cranium (Badoux 1961). The honeycomb structure of the cranium serves several primary functions: it allows the angles and areas of attachment for both nuchal and masticatory muscles to achieve optimal mechanical advantage, it allows the tusks to be aligned vertically (at right-angles to the basi-cranial axis), it favours mobility of the head, and also serves as a 'shock absorber' (especially insulating the brain) in those instances when elephants feed by shaking fruits out of trees

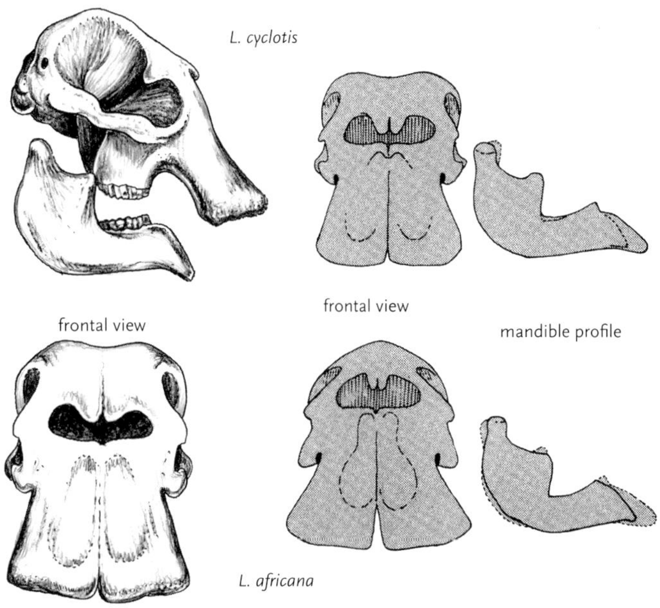

Skulls of *Loxodonta* spp. compared

by ramming their heads against tree trunks (or, during conflicts, against each other) (Maisels *et al.* 2002, A. Turkalo pers. obs.).

Elephant hearing and communication is highly developed and their auditory range extends to low frequency sounds below that of human hearing (>20 Hz). Forest Elephants have been recorded emitting frequencies as low as 5 Hz (S. Gulick pers. comm.). These low frequency sounds are used to communicate over long distances, especially in locating family members. At the Dzanga Clearing in Central African Republic where Forest Elephants have been studied for the last decade, observations in and around the clearing demonstrate that family groups locate one another by means of both audible sounds as well as infrasonic vocalizations. Extensive calling between elephants allows groups to keep track of each other and to coordinate their movements (A. Turkalo pers. obs.).

Foraging and Food The diet of the Forest Elephant consists of a wide variety of plant parts including leaves, fruit, pith, bark, lianas and fruit. Grass, a rare occurrence in the forest, found mostly in and around forest clearings, is also occasionally consumed but forms a very small portion of the diet. Forest Elephants are highly frugivorous and studies in Gabon (White 1994), Congo (Blake

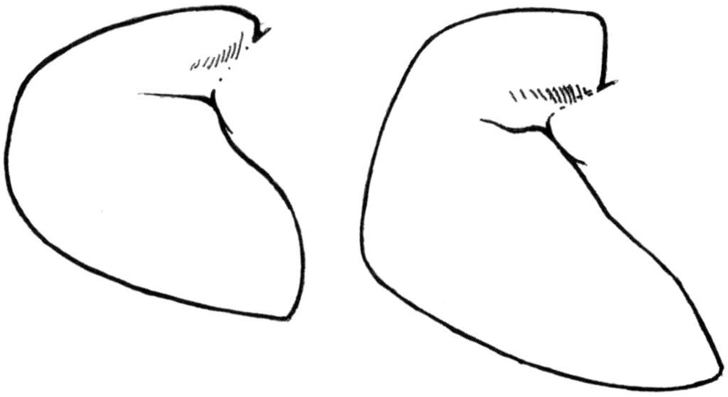
Ear shape in *Loxodonta cyclotis* (*left*) and *L. africana* (*right*).

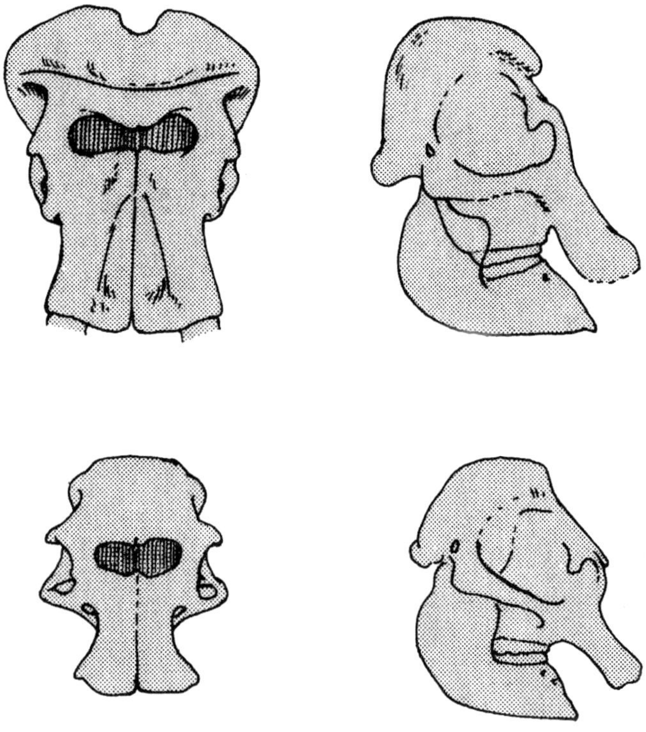

Skull outlines of dwarfed island elephant, *Elephas falconeri* (*below*) with its mainland equivalent, *E. namadicus* (*above*).

2002) and Côte d'Ivoire (Short 1981) showed they were observed to consume between 20 and 70 species of fruit, with favoured fruit species including members of the families Sapotaceae, Tiliaceae, Moraceae and Marantaceae (White *et al.* 1993, Maurois *et al.* 1997, Blake 2002).

Successful germination of many fruit species is linked to elephant digestion, which in turn plays an important role in the species

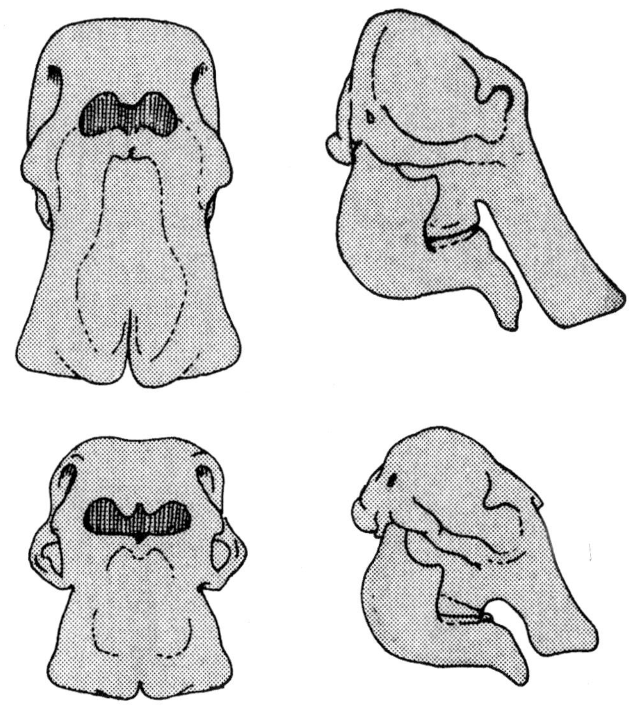

Is *L. cyclotis* (*below*) a dwarfed ecological isolate of extinct *L. adaurora* (*above*)?

composition of the forest (Hawthorne & Parren 2000, Blake *et al.* 2009). The availability of fruits plays an important part in the month-to-month movements of Forest Elephants (Short 1983, White 1994, Theuerkauf *et al.* 2000b). A marked difference in diet occurs between the dry and wet seasons, the wet season diet being particularly rich in fruit in the central African forests, transitioning to leaves and bark in the dry season. In contrast, in West African forests fruit is more abundant in the dry season (Theuerkauf *et al.* 2000b, Danquah & Oppong 2006). The months when fruit is least abundant in West African forest fragments are those when crops are maturing in the farmland outside, so tempting elephants into farmers' fields.

Forest clearings rich in minerals also play an important role in Forest Elephant diet. Elephants have been observed to spend many daylight hours ingesting soils in these clearings where minerals may act as a buffer in their diet in the digestion of secondary plant compounds (Klaus *et al.* 1998, Houston *et al.* 2001). They also use their pillar-like feet and tusks to dig for water that is rich in mineral salts in forest clearings.

From data collected from collared elephants in the Dzanga-Ndoki and Nouabale-Ndoki National Parks, elephants feed most intensively during the day and at night they rest near water points (Blake 2002).

Social and Reproductive Behaviour Forest Elephants are highly social, the basic social unit comprising a family group of several related adult ♀♀ and their young. Initial studies of Forest Elephants showed that they form small family groups averaging three individuals (Merz 1986, White *et al.* 1993, Turkalo 1996, Querouil *et al.* 1999, Theuerkauf *et al.* 2000a), but a long-term study of known individuals at the Dzanga clearing in Central African Republic has shown that Forest Elephants, like their savanna counterparts, also form large, extensive family units that can number up to 20 individuals (A. Turkalo pers. obs.). The lack of natural predators in the forest permits Forest Elephants to form less cohesive family groups unlike in the savanna, where family group members remain close together as a defensive strategy against predation (Western 1986).

Unlike the Savanna Elephant, adult Forest Elephant ♂♂ have not been observed in 'bachelor herds' but only as solitary individuals. Younger ♂♂, juveniles and subadults form loose, non-permanent groups that may last one to several days (A. Turkalo pers. obs., F. Maisels pers. obs.). In forest clearings, which are used extensively by Forest Elephants seeking mineral salts, which are essential to their diet, male hierarchy is tested and established around large holes that are excavated by ♂♂ in their search for high-quality minerals. Aggression is regularly observed in the competition for minerals, but very few injuries are incurred.

From a long-term data set of 15 years, gathered at the Dzanga clearing, Central African Republic, and from four radio-collared elephants in the same area, ranging behaviour varies greatly between individuals (Blake 2002, A. Turkalo pers. obs.). Forest Elephants can have a minimum convex polygon home-range of up to 2000 km^2 and move an average of 7.8 km/day. One individual ranged more than 25 km in less than 48 hours (Blake 2002). By their habitual use of the same paths, they create wide trails in the forest (Vanleeuwe & Gautier-Hion 1998).

Observations at the Dzanga clearing reveal that some ♀♀ are resident, spending more time in the area, whereas ♂♂ noticeably spend less time in the area of the clearing. Larger ♂♂ are more likely to visit the clearing during the dry season during the peak musth season, when they come in search of mineral salts and oestrous ♀♀.

Musth, a phenomenon well documented in male Savanna Elephants, has been observed regularly at the Dzanga clearing. The most obvious musth signs are temporal gland secretion accompanied by a steady urine dribble, resulting in a characteristic pungent body odour. Musth has been observed only in the largest ♂♂ in the Dzanga population. The majority of bulls undergoing musth are observed during the dry seasons, the longest dry season occurring in Dec–Mar, and a shorter dry season occurring during Jul–Aug. The length of musth in ♂♂ is thought to last for at least two months. Musth occurs in individual ♂♂ during the same time period each year (A. Turkalo pers. obs.).

During the guarding of an oestrous ♀ by the ♂, the ♂ stays in close contact with the ♀ allowing her access to (otherwise) unavailable resources such as choice mineral access points otherwise dominated by ♂♂ in the area. The guarding and mating generally lasts for approximately two days during which time the ♂ copulates with the receptive ♀ a couple of times. After the short oestrous period the ♂ leaves the ♀ to go in search of other potential mates (A. Turkalo pers. obs.).

Newborns are immediately mobile after birth and stay in close contact with their mothers and other family members. Allomothering is tolerated by mothers and is performed by older female siblings as well as unrelated subadult and adult ♀♀ (A. Turkalo pers. obs.).

Reproduction and Population Structure Sexual maturity is attained between the ages of 12 and 14 years in ♀♀. Males probably reach sexual maturity at the same age as ♀♀, but become sexually active much later (A. Turkalo pers. obs.). Oestrous occurs during the entire year, but the peak periods are during the two dry seasons (A. Turkalo pers. obs.). Oestrous in free-ranging Forest Elephant ♀♀ is thought to last several days, as in Savanna Elephants (A. Turkalo pers. obs.).

One young is born after a gestation of approximately 660 days (similar to the savanna species). During the 15 years of a long-term study at the Dzanga clearing in Central African Republic, some 675 newborn calves have been recorded to known ♀♀ and twins have never been observed (A. Turkalo pers. obs.). Weaning is dependent on the mother; some calves have been observed suckling until the age of five (A. Turkalo pers. obs.). From the Dzanga clearing dataset, the inter-birth interval has been calculated at four years or less (n = 69) (A. Turkalo pers. obs.) and the sex ratio is 52% female and 48% male (A. Turkalo pers. obs.). Maximum potential life-span in Forest Elephants is thought to approximate that in the Savanna species: 50–60 years.

Predators, Parasites and Diseases Unlike Savanna Elephants, Forest Elephants have no major natural predators. Leopards *Panthera pardus* may prey upon newborn calves (Blake 2004), but since calf mortality is low at 3% (A. Turkalo pers. obs.) this phenomenon is probably rare in the forest environment. Orphaned calves above the age of two years have been observed to survive with siblings or even independently.

Forest Elephants show little sign of disease. Anthrax, which has been associated with mortality in Savanna Elephants, has never been recorded in Forest Elephant populations. Tumorous growths have been observed in a population of Forest Elephants at the Dzanga clearing (A. Turkalo pers. obs.). Infections resulting from wounds incurred from fighting or tree falls during feeding have also been observed but rarely do these infections result in death (A. Turkalo pers. obs.). Although internal parasites have been well documented in Savanna Elephants, little work has been done on the forest species (Chabaud & Rousselot 1956, Kinsella *et al.* 2004). Kinsella *et al.* (2004) collected faecal samples from six African Forest Elephants from the Nouabale-Ndoki N. P., Congo, and the Dzanga-Sangha N. P., which revealed the presence of a schistosome, *Bivitellobilharzia* sp., a tracheal nematode, *Mammomonagamus* sp., and a complex of intestinal strongylids and ciliates. The nematode genera *Decrusia* and *Equinurbia* were also reported for the first time from African elephants, and the ciliate genus *Latteuria* for the first time from wild elephants. One elephant was found freshly dead from natural causes, and 12 species of intestinal parasites (two bot fly larvae, one trematode and nine nematodes) were collected during a complete necropsy.

Conservation IUCN Category: Not listed (included in *Loxodonta africana*, since the view of the African Elephant Specialist Group is that premature allocation into more than one species may leave hybrids in an uncertain conservation status). CITES: Appendix I (see below).

The 1989 CITES Conference of the Parties (COP) voted to transfer African elephants (no distinction being made between Savanna and Forest Elephants) from Appendix II to Appendix I. Although the Savanna Elephant populations of Botswana, Namibia, South Africa and Zimbabwe have since been returned to Appendix II, all Forest Elephant populations remain on Appendix I. Therefore, all commercial trade in Forest Elephant products, such as ivory, meat and hide, between countries is banned. Sport hunting of Forest Elephants is not covered by this ban, and certain countries, such as Cameroon, wish to manage some of their elephants for safari hunting.

Forest Elephants in central Africa roam through vast forests that stretch along the Equator from the Atlantic coast to the Congo–Nile divide. Human densities are sparse over large parts of the central African forest and elephants were numerous in the first half of the twentieth century (Roth & Douglas-Hamilton 1991). During the 1970s and the 1980s, the heavy hunting of central African elephants for ivory was hidden from the outside world by the forest canopy. Comparisons of heavily poached and relatively undisturbed forests suggest large numbers of elephants were killed (Michelmore *et al.* 1994). Poachers used motorized canoes to penetrate the most remote forests of DR Congo (Alers *et al.* 1992). Commercial logging is advancing rapidly, and logging roads open these formerly remote and inaccessible areas for settlement. Elephants avoid roads, and as the road network expands, the area of undisturbed habitat diminishes (Barnes *et al.* 1997). Government wildlife agencies continue to be ineffective as a result of economic and social conditions and the expansion of the bushmeat trade has promoted the market for elephant meat (Blake & Hedges 2004). Poaching for ivory continues, despite the international ban on the ivory trade (Blake *et al.* 2007). Protection is weak in national parks and reserves because of the general lack of resources. Nevertheless, elephant densities remain higher inside the national parks than outside, showing that they still have an important role to play in conservation (Blake *et al.* 2007).

Automatic weapons have proliferated in central Africa because of civil war and strife in Rwanda, DR Congo, Congo and the Central African Republic. As a result of this unrest, refugees have been displaced into areas where elephants were previously undisturbed and it is this loss of their former habitat that is becoming the greatest threat to the last of the great elephant populations of the forest zone (J. Hart pers. comm.).

DR Congo has a human population estimated at 67 million in 2011 (largest national sub-Saharan population after Nigeria and Ethiopia), and despite conflict this population is growing at 2.8% or more annually. Most large forests in DR Congo are visited by people, and even occupied at a light level in relation to fishing, hunting and artisanal mining. Many of these same forests are 'empty' or severely depleted of wildlife, including significant portions of the country's largest forest parks (Salonga, Bili-Uere, Okapi, Maiko and Kahuzi lowlands). Elephants are among the most obvious of the depleted species and have been eliminated, or nearly so, from large areas of their historic range. Range shrinkage continues, and is accelerating in the east (Ituri, Maiko, Kahuzi) (J. Hart pers. comm.). The numbers of Forest Elephants will continue to decline in central Africa unless the expansion of the road network into remote forests can be limited, the trades in ivory and bushmeat can be brought under control, and the remote national parks protected effectively because they are likely to become the last refuges for Forest Elephants (Blake *et al.* 2007).

In West Africa, Forest Elephants are restricted to small forest patches and reserves that are isolated by agriculture from other elephant habitats. In Côte d'Ivoire many of the smaller populations were extirpated during the years of peak poaching in the 1970s and 1980s (Barnes 1999). Elephant numbers have declined significantly in many sites in Côte d'Ivoire even since the ivory trade ban came into effect (Fischer 2005, Hoppe-Dominik *et al.* 2011). While inbreeding, distorted sex ratio and age structures will compromise the long-term viability of the remaining small populations, in the short and medium term it is human–elephant conflict, habitat loss and hunting for meat and ivory that will determine the fate of these elephants. As the human population expands, conflict between elephants and people will intensify because they compete for the same land resources (Parker & Graham 1989). The increasing hectareage of agriculture attracts crop-raiding elephants out of forest reserves, and the suffering of farmers turns rural communities against the concept of conservation (Boafo *et al.* 2004).

Today, Forest Elephants occur in several important protected areas and national parks in the central African region, including: Dja Faunal Reserve, Nki F. R., Abong-Mbang F. R., Boumba-Bek F. R., Lobeke N. P. and Mongokele F. R. (Cameroon); Dzanga-Ndoki N. P., Bangassou F. R. (Central African Republic); Monte Alén N. P. (Equatorial Guinea); Nouabale-Ndoki N. P., Conkouati N. P., Odzala N. P. (Congo); Lopé N. P., Minkebe N. P., Moukalaba N. P., Ivindo N. P., Mpassa N. P., Louango N. P., Mwagne N. P., Waka N. P., Biringou N. P. (Gabon); and Salonga N. P., Kahuzi Biega N. P., Maiko N. P., Okapi N. P. and Garamba N. P. (DR Congo).

Measurements

Loxodonta cyclotis
HB (♂♂): 244.2 (214.0–290.0) cm, n = 103
HB (♀♀): 198.4 (176.0–234.0) cm, n = 170
Dzanga-Ndoki N. P., Central African Republic (A. Turkalo pers. obs.)

Maximum shoulder height measured at the Dzanga clearing was 2.8 m (n = 810) (A. Turkalo pers. obs.). Shoulder height measurements of Forest Elephants from Louango in Gabon were considerably smaller than those obtained at Dzanga, with a maximum measurement of 2.16 m (n = 53) (Morgan & Lee 2003).

Record tusk length in Forest Elephants is unknown because no subspecies are listed for record purposes. The longest estimated tusk length in a ♂ observed at the Dzanga clearing in Central African Republic is 1.8 m (A. Turkalo pers. obs.).

Key References Barnes *et al.* 1997; Blake 2002; Blake *et al.* 2007; Blanc *et al.* 2007; Merz 1986; Querouil *et al.* 1999; Theuerkauf *et al.* 2000a, b.

Andrea Turkalo & Richard Barnes

Order SIRENIA – Dugongs, Manatees

Sirenia Illiger, 1811. Prodromus systematis mammalium et avium … Berlin: C. Salfeld: 140–141.

Dugongidae (1 genus, 1 species)	Dugongs	p. 203
Trichechidae (1 genus, 1 species)	Manatees	p. 209

The order Sirenia comprises four families: two extinct dating to the Eocene (Prorastomidae and Protosirenidae) and two extant: the Dugongidae and the Trichechidae. Sirenians are the only living group of herbivorous marine mammals. The name Sirenia alludes to legends of 'sirens' or mermaids, thought (rightly or wrongly) to have been inspired by manatees or dugongs. *Dugong* is a Filipino Malay name for the animal; *Trichechus*, from Greek *trichos* meaning 'hair', was bestowed on these nearly-hairless mammals to distinguish them from other 'fish' when they were still classified among the latter. The order is represented in Africa by only two species, the Dugong *Dugong dugon* and the West African Manatee *Trichechus senegalensis*.

Except for the earliest forms, which were amphibious quadrupeds (Domning 2001a), sirenians also resemble cetaceans in their fish-like body form, lack of an external ear pinna, short neck, flipper-like forelimbs, loss of hindlimbs, stiffening of the body by a subdermal connective-tissue sheath of helically wound fibres, and propulsion by a horizontally expanded tail fin. The tail fin is triangular and cetacean-like in dugongids, but retains a more primitive paddle shape in manatees. There is never a dorsal fin, and there are two separate nostrils rather than a single blow-hole. The enlarged, prehensile upper lip is a 'muscular hydrostat' formed by variously oriented muscles (Marshall *et al.* 1998, 2003). All body hairs are sinus hairs (specialized tactile organs), especially the elaborately developed facial vibrissae, some of which are also prehensile (Reep *et al.* 2002). The thorax and lungs are elongate and the diaphragm is nearly horizontal, placing the buoyant lungs wholly dorsal to the abdominal organs. The ventricles of the heart form separate lobes. The skeleton typically shows both extensive pachyostosis (increase in bulk of individual elements, especially the ribs) and osteosclerosis (increased proportion of compact to cancellous bony tissue). These conditions provide ballast to neutralize the buoyancy of the lungs, which do not collapse during submergence in these shallow-diving animals (Domning & de Buffrénil 1991). Body sizes have ranged from about 2 m to 9 or 10 m in extinct forms.

The skull is distinguished by enlarged premaxillae that contact the frontals and form a more or less down-turned rostrum, a large mesorostral fossa for the narial passages and musculature, reduction or loss of nasals and lacrimals, lack of paranasal air sinuses, lack of a postorbital bar, a massive zygomatic arch, fusion of parietals and supraoccipital into a thick, dense skullcap (usually with pronounced temporal crests but no sagittal crest), and strongly developed pterygoid processes. There is a ring-shaped ectotympanic, not forming a bulla; the periotic is massive and enclosed in a socket in the squamosal. The mandible is heavy, with a long, laterally compressed, down-turned symphysis that primitively bore parasagittally aligned incisor–canine toothrows. The infraorbital and mental foramina are large, to accommodate the nerves and vessels supplying the fleshy, mobile snout and lips.

The primitive dental formula comprises three incisors, a single canine, five premolars and three molars above and below; whether the five premolars represent a retention of the primitive eutherian condition, or a synapomorphic reversal to that condition, is still unresolved (although optimization of premolar number onto mammalian cladograms derived from morphological or molecular [or both] types of data indicates that the presence of five premolars is apomorphic). Incisors and canines are reduced or (in most forms) absent, except for the single pair of enlarged upper first incisors which form tusks in many species. A pair of horny masticating pads covers the symphyseal parts of the upper and lower jaws.

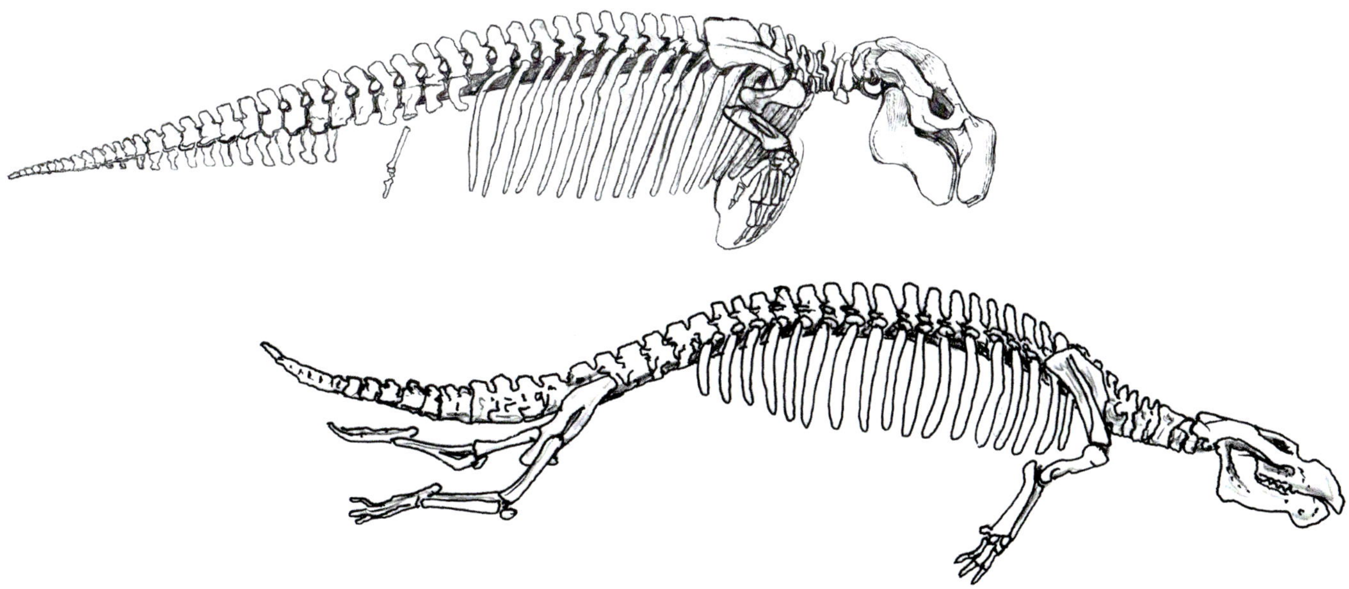

Fossil skeleton of an extinct quadrupedal sirenian, *Pezosiren* (below) compared with the skeleton of the extant Dugong *Dugong dugon*.

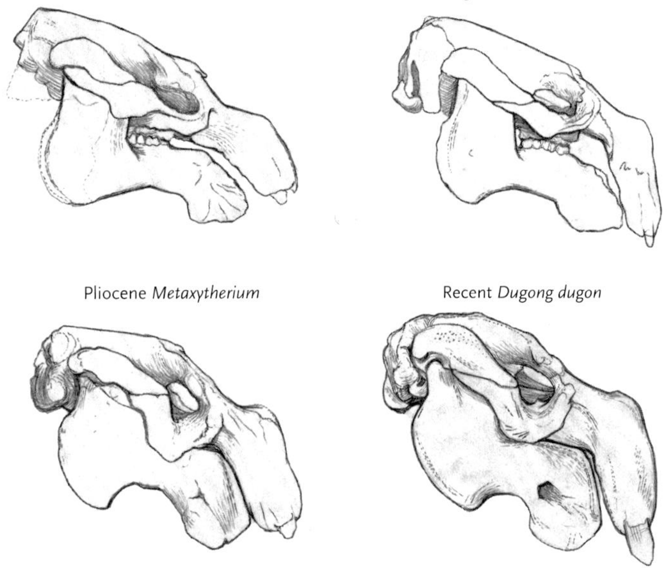

Sirenian evolution.

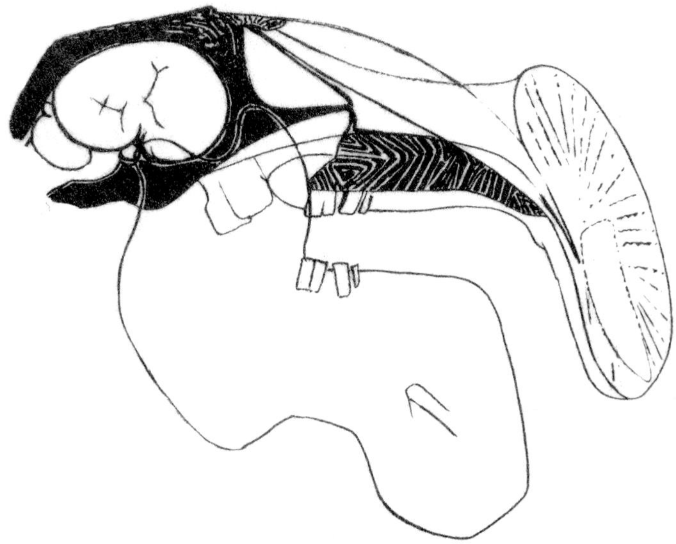

Dugong dugon: cross-section of skull. Note massive bony capsule around brain and incisor embedded in down-turned Os intermaxillare and matching downturn in mandible.

Replacement premolars are reduced in number or lost in post-Eocene sirenians, but at least some deciduous premolars are usually present in immature individuals. Molars are low-crowned, primitively bunodont and more or less bilophodont, with anterior and posterior cingula on uppers and posterior cingula (usually enlarged on M_3 to form a hypoconulid lophule) on lowers. However, in the extant forms, the cheek dentition is radically modified (degenerate molars in the Dugong, supernumerary ones in manatees).

Forelimb elements are less modified than in cetaceans. The shoulder, elbow, wrist and finger joints are freely movable, though radius and ulna are ankylosed and some of the carpals tend to be fused. The deltoid crest and bicipital groove of the humerus are large in most sirenians including the Dugong, but tiny or absent in manatees. Five digits are present; the first is reduced and the fifth enlarged. Hyperphalangy is not normally seen. A clavicle is lacking. Usually no more than three or four of the 14–21 pairs of ribs are connected to the sternum, which in all but protosirenids comprises only one to three elements. There is normally no fusion of vertebrae. Seven cervicals are present, except in *Trichechus*, which has only six (Giffin & Gillett 1996). In most forms the pelvis is attached by a ligament to a single sacral vertebra, but in modern manatees even this connection is no longer as distinct, and the three post-thoracic vertebral regions cannot be clearly distinguished. The vertebral formulae for the extant African species are C 7, T 18–19, L 3, S 1, Ca 26–32 in the Dugong and C 6, T 17–18, L–Ca 25–26 in the West African Manatee.

The fossil record of the Sirenia is substantial and nearly cosmopolitan (Domning 2001b), including important finds in North Africa (Domning *et al.* 2010). The oldest and most primitive sirenians are *Prorastomus* and *Pezosiren* from the early Middle Eocene of Jamaica, ca. 50 mya (Savage *et al.* 1994, Domning 2001a). These pig-sized quadrupeds already displayed aquatic adaptations, and probably spent more time in the water than the modern hippopotamus, feeding as well as resting there. Most later sirenians were marine and consumed mainly sea-grasses. The Dugong's closest fossil relatives are found in the Caribbean; they may have dispersed between there and the Indian Ocean (around the Cape of Good Hope) in just the last 2 million years, so there is a possibility of finding their fossil remains on the coast of West Africa (Domning 1995). The euryhaline *Trichechus* evolved in South America and probably also crossed the equatorial Atlantic to West Africa only in the late Pliocene or Pleistocene (Domning 2005).

The Sirenia are clearly a well-defined, strictly monophyletic group. Morphological evidence links them firmly with proboscideans (elephants) and some extinct groups in a taxon called Tethytheria (McKenna 1975, Shoshani 1993, Gheerbrant *et al.* 2005), indicating that these animals probably share an Old World origin. However, there is little molecular support for Tethytheria – the longest concatenations of mitochondrial and nuclear genes actually reject the clade (Amrine-Madsen *et al.* 2003, and see Meredith *et al.* 2011). Molecular support is stronger (though morphological support is weaker) for the more inclusive taxon Paenungulata (Tethytheria + Hyracoidea). Recent molecular studies have raised the hypothesis of Afrotheria, which associates the Paenungulata with macroscelideans, tubulidentates, chrysochlorids, and tenrecs. This is a radically different view, in that it places sirenians and other paenungulates among the most basal lineages of modern placentals, rather than higher up among the hoofed mammals as in morphology-based trees (e.g. Thewissen & Domning 1992).

Sirenians, like the sea-grasses and mangroves on which they feed, evidently originated in the region of the Tethys Sea, the former east–west seaway separating the northern and southern continents. However, they quickly dispersed throughout the Tethyan realm and have remained pantropical ever since – with the notable exception of the lineage leading to the extinct Steller's Sea-cow (*Hydrodamalis*), which adapted to the temperate and cold waters of the North Pacific. This persistent tropicality very likely accounts for the depleted diversity of the Sirenia, given the global cooling of the last few million years.

A comprehensive annotated bibliography of the order is provided by Domning (1996) and recent reviews of Sirenian evolution are found in Domning (2002) and Domning *et al.* (2010).

Daryl P. Domning

Family DUGONGIDAE

DUGONG

Dugongidae Gray, 1821. London Med. Repos. 15: 309.

Dugong (1 species)	Dugong	p. 203

The family Dugongidae includes a single extant species (Rice 1998, Marsh 2002, Shoshani 2005), the Dugong *Dugong dugon*, which has an extremely large range, being found in coastal waters of at least 48 countries, from East Africa to Vanuatu (Reynolds & Odell 1991, Marsh *et al.* 2002). Despite their wide range, discontinuities in habitat, as well as the impact of localized human activities, cause the species to be sparsely and discontinuously distributed in most regions; in fact, 'throughout much of its range, the Dugong is believed to be represented by relict populations separated by large areas where its numbers have been greatly reduced or where it is already extirpated' (Marsh *et al.* 2002).

Dugongids appeared in the Eocene and were more diverse in the fossil record than is the case today. Most sirenian species that have existed over time were, in fact, members of the family Dugongidae. In addition to the extant Dugong, the family was represented until recently by the Steller's Sea-cow *Hydrodamalis gigas*, which was exterminated by human hunters by 1768 (Reynolds & Odell 1991).

The most thorough, multifaceted and long-term studies of Dugongs have been done in Australia. In the absence of specific information from elsewhere, scientists and managers tend to use information from Australian Dugongs when considering conservation and management in other specific parts of the Dugong's range (see species profile *Dugong dugon*).

Like other sirenians, Dugongs are herbivores. However, unlike other species in the order, Dugongs are obligate bottom feeders, due to their extreme rostral deflection. As a consequence of this adaptation, together with their mostly marine distribution, Dugongs appear to be highly dependent on the integrity of sea-grasses (Marsh 2002, Marsh *et al.* 2002). In fact, Dugong life history attributes (e.g. age at sexual maturation; inter-calf interval) appear to be relatively plastic (see Boyd *et al.* 1999), in part due to changes in availability of sea-grass resources for food. For example, in some locations, ♀♀ may not reach sexual maturity until they are 17 years old, whereas in other locations sexual maturation may be reached as young as six years of age (see species profile). Typical K-strategists, Dugongs live to be at least 73 years old, and the maximum rate of population increase is low. Population models estimate the maximum rate of population increase to be about 5% per year at best (Boyd *et al.* 1999).

The dependence on sea-grasses by Dugongs has been noted. Conversely, Dugongs appear to function as cultivation grazers whose grazing actually increases the productivity of local sea-grass meadows (Preen 1995a). Dugongs possess adaptations associated with herbivory, such as expanded hindgut capacity, the presence of gut microflora for cellulolysis, and a variety of specialized cellular arrangements in the digestive system (see, for example, Marsh *et al.* 1977). In addition, they possess attributes that facilitate an existence in the marine environment: large size (up to approximately 3 m) for thermoregulation; fusiform body shape; a split fluke for locomotion; smooth skin; and general reduction in external protuberances. Although sensory biology of Dugongs has not been well studied, it is likely that they have functional specializations to facilitate communication, navigation and habitat selection. Marsh (2002) and Marsh & Dutton (see species profile) summarize the diagnostic characteristics of Dugongs; Pabst *et al.* (1999) and Marshall (2002) also provide general information regarding Dugong adaptations. Recent volumes by Marsh *et al.* (2011) and Hines *et al.* (in press) provide comprehensive overviews of dugong ecology and conservation.

John E. Reynolds, III

GENUS *Dugong*

Dugong

Dugong Lacépède, 1799. Tab. Div. Subd. Orders Genres Mammiféres 14: 17.

The genus *Dugong* is monotypic, represented by the Dugong *Dugong dugon* (Shoshani 2005), which has a wide distribution in tropical coastal waters from East Africa to Vanuatu within the latitudes of about 26–27° north and south of the equator.

Helene Marsh

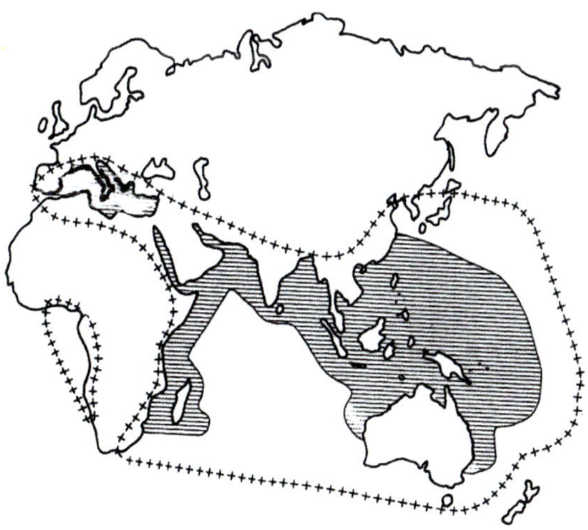

Old World distribution of Dugong past and present (shaded) and Potamogetonaceae (+++++).

Dugong dugon Dugong

Fr. Le Dugong; Ger. Dugong

Dugong dugon (Müller 1776). Linne's Vollstand. Natursyst. Suppl., p. 21. Cape of Good Hope to the Philippines.

Dugong *Dugong dugon* adult male.

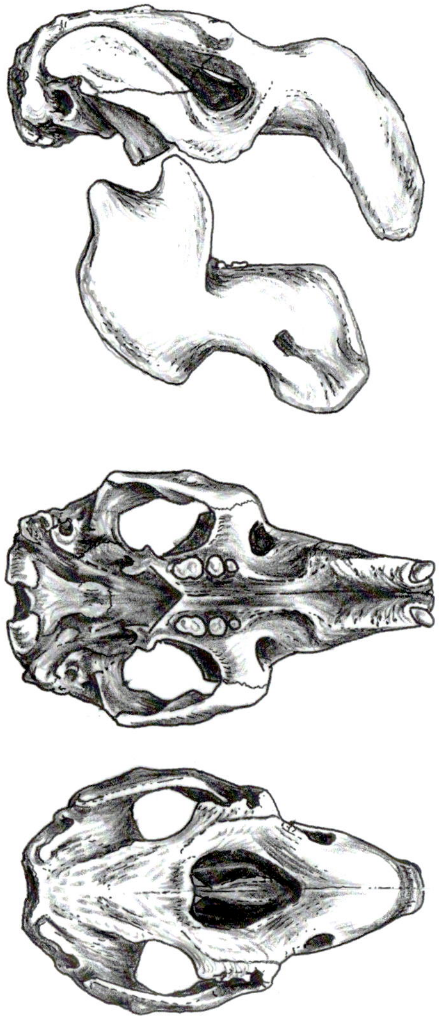

Lateral, palatal and dorsal views of skull of *Dugong dugon*.

Note: there is much information published on Dugongs from outside Africa. The following account, based primarily on data from Africa, is supplemented by relevant information available from other parts of their range.

Taxonomy Monotypic. Synonyms: *australis*, *cetacea*, *dugung*, *hemprichii*, *indicus*, *lottum*, *tabernaculi*. Chromosome number: 2n = 50.

Description Rotund cetacean-like body with flippers and flukes that resemble those of a dolphin without a dorsal fin. Head resembles that of a walrus without the long tusks (Bryden *et al.* 1998). Head distinctive, with mouth opening ventrally beneath broad, flat muzzle. Eyes located on side of head, small and not prominent. Externally, ears consist of small openings on either side of head. Adults grey in colour, but may appear brown from the air or from a boat. Older 'scarback' individuals may have large areas of unpigmented skin on the back above the pectoral fins. Flippers are short (up to 420 mm, average 330 mm; Spain & Heinsohn 1975) and, unlike those of the West African Manatee *Trichechus senegalensis*, lack nails. Hind flippers are absent. Fluke is triangular, like that of a whale, unlike that of manatees (*Trichechus* spp.), which have a paddle-shaped tail. Dugongs exhibit little sexual dimorphism in size. Anal and genital apertures are contiguous in ♀♀. In ♂♂ genital aperture is about halfway between anus and umbilicus. Testes are abdominal (Marsh *et al.* 1984a). Two mammary glands, each opening via a single teat, are situated in the 'armpit' or axilla; mammaries are somewhat reminiscent of the breasts of human ♀♀.

Tusks of mature ♂♂ and some old ♀♀ erupt on either side of the head for up to about 30 mm (Marsh 1980). Molar teeth are simple and peg-like. A Dugong has a total of six cheekteeth (premolar and molar) in each quadrant of its jaw during its life, but these are never all erupted and in wear simultaneously. Three premolars are present at birth; old animals have only two pairs of teeth (molars two and three), and these may grow throughout life (Marsh 1980).

Geographic Variation None recorded.

Similar Species No other extant members of the Family Dugongidae. The Dugong's closest living relatives in the order Sirenia are the manatees of the family Trichechidae, but no members are sympatric.

Dugong dugon

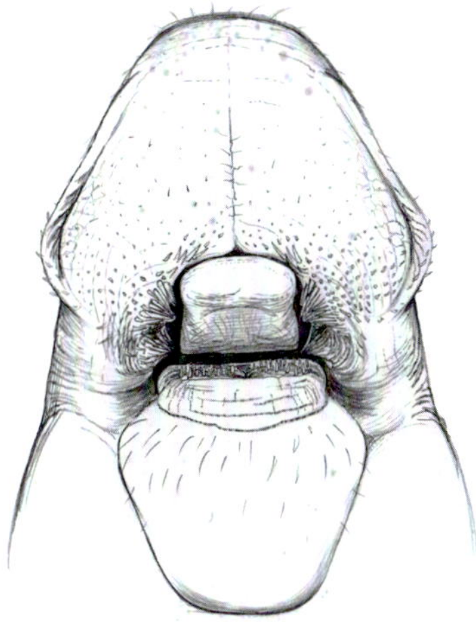

Frontal view of *Dugong dugon* muzzle showing bristles around the oral cavity and intermaxillary rostrum. Embedded in an expansive and flexible complex of oral muscles and skin.

Distribution African range extends from the Red Sea and Gulf of Aden to Mozambique (Marsh *et al.* 2002) and Dugongs have been recorded as vagrants as far south as 50 km north of Durban, KwaZulu–Natal (Skinner & Chimimba 2005). There is little information on current Dugong distribution and abundance along the African Red Sea. Dugongs have been reported off the coasts of Egypt (Vincent 1996, Hanafy *et al.* 2006), Sudan and Djibouti (Robineau & Rose 1982), Somalia (Travis 1967, Cockcroft & Young 1998) and, based on anecdotal reports of live sightings and the location of skulls, are thought to occur in most areas along the coast of Eritrea (Marsh *et al.* 2002). In 2003 and 2004, information was collated through literature reviews, questionnaire surveys and opportunistic sightings, and confirmed the continued existence of small numbers of Dugongs in Kenya, Tanzania, Mozambique, Madagascar, Seychelles and the Comoros Archipelago (Moheli and Mayotte Is.) and established the regional significance of the Bazaroto Archipelago in Mozambique as important Dugong habitat (WWF Eastern African Marine Ecoregion 2004). Dugongs have disappeared from Mauritius and are believed never to have occurred in Réunion.

Extralimital to the African continent and the Red Sea region, the Dugong has a range that spans at least 40 countries and territories in tropical and sub-tropical coastal and island waters including the Arabian Gulf, the coast of India and Sri Lanka, South-East Asia, Taiwan, north to Okinawa in Japan, and east through Papua New Guinea to Vanuatu, between about 26 and 27° north and south of the Equator (Nishiwaki & Marsh 1985, Marsh *et al.* 2002, 2011). Their historic distribution is believed to have been broadly coincident with the tropical Indo-Pacific distribution of its food plants, the phanerogamous sea-grasses of the families Potamogetonaceae and Hydrocharitaceae (Husar 1978) (see illustration on p. 203).

Habitat Dugongs frequent coastal waters. Large concentrations of Dugongs tend to occur in wide shallow protected bays, wide shallow mangrove channels (Hughes & Oxley Oxland 1971) and in the lee of large inshore islands where there are sizeable sea-grass beds (Heinsohn *et al.* 1977). Dugongs are also regularly observed in deeper water further offshore in Australia in areas where the continental shelf is wide, shallow and protected (e.g. Marsh *et al.* 1997), and where deepwater sea-grass beds occur. Dugong feeding scars have been observed at depths of up to 33 m in sea-grass beds in the Great Barrier Reef region (Lee Long & Coles 1997). Around the Lamu archipelago, fishermen reported that Dugong used to be numerous in deeper waters beyond the reef during periods when the sea was calm (Nov–Mar) where they were believed to be feeding on *Syringodium* and *Cymodocea ciliata*. When the sea became rough, Dugongs moved into shallower, more sheltered bays within the reef, at which time they were feeding on *Halophila and Zostera* spp. and *Halodule uninervis*. Similar seasonal movements were reported by fishermen around Mafia I. (Kingdon 1971).

Abundance Anecdotal data suggest that Dugong populations have suffered steep declines in most countries in East Africa since the 1960s (Kenya and Tanzania) and 1980s (Madagascar and Mayotte), while in Mozambique the decline has apparently escalated since the end of the civil war in 1992 (WWF Eastern African Marine Ecoregion 2004). On the basis of the apparent similarity between the distribution of suitable Dugong habitats along the African and Arabian coasts of the Red Sea, Preen (1989a) estimated that the African side could potentially support a population similar in size to that along the Arabian coast, estimated at 4000 in 1986.

The number of Dugongs worldwide is unknown, but there are likely to be more than 100,000 individuals. However, this rather large number is somewhat deceptive if one considers the conservation status of Dugongs in particular countries. Of this estimate there are likely to be some 85,000 animals in Australia (Marsh 2002) and several thousand in the Arabian Gulf; however, in many parts of the species' original range, Dugongs survive only in very small numbers (i.e. a few tens or hundreds of individuals).

Adaptations Dugongs are highly adapted to the marine environment in which they spend their entire lives as evidenced by their relatively large size, streamlined body, whale-like tail, paddle-like forelimbs (flippers) and absence of hindlimbs (Bryden *et al.* 1998). In common with other sirenians, the skeleton of the Dugong is comprised of extremely dense bone that presumably functions to

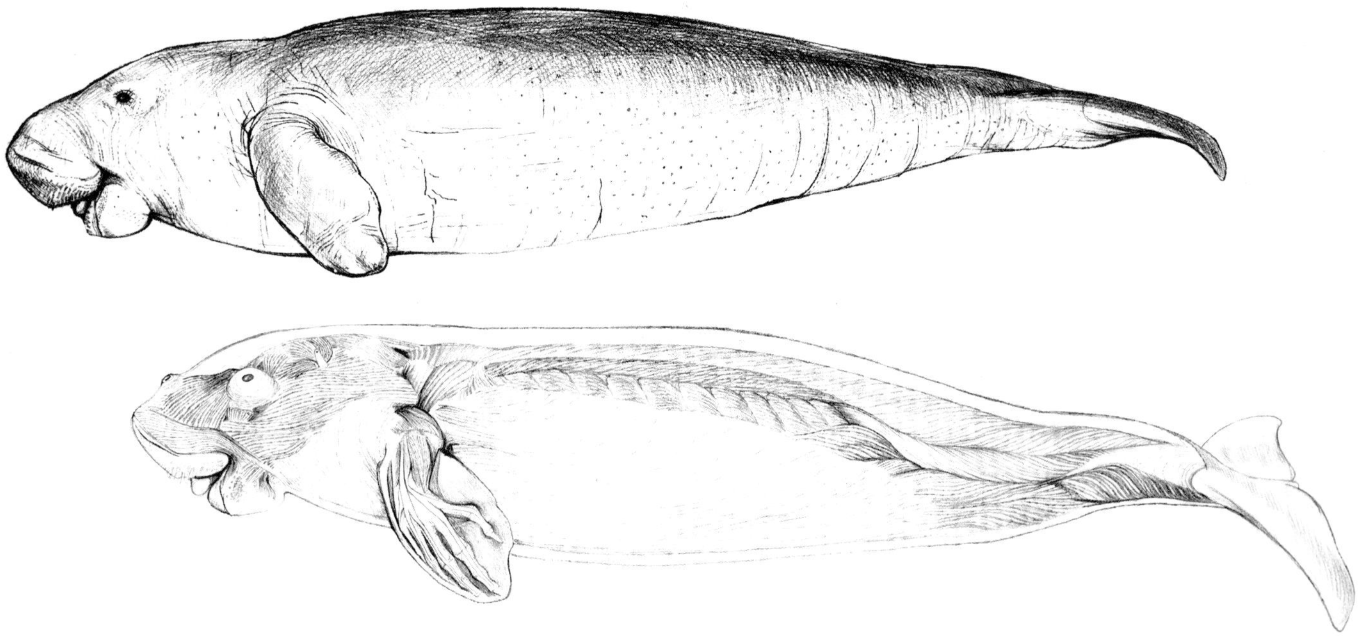

Dugong dugon. The lower image shows the myology. Note that a thick layer of blubber covers all body musculature. The eyes and auditory orifice reach the surface through approximately 25 mm of blubber.

help overcome buoyancy problems while bottom-feeding. Body hair is sparse but the hairs and bristles on the Dugong's head, particularly around the lips and mouth, are highly developed sensory organs that presumably help Dugongs detect their sea-grass food. The Dugong's mouthparts are highly adapted to bottom-feeding on vegetation and the mechanism whereby the muzzle and lips 'rake' food into the mouth has been described and illustrated in Kingdon (1971). The two nostrils are on top of the anterior end of the snout, which allows the Dugong to breathe with very little of the head above the water. The nostrils cannot be closed completely, but contraction of muscles surrounding the nasal ducts closes the ducts. The eyes have no eyelids but are protected by nictitating membranes and are lubricated by oil glands.

Foraging and Food Dugongs are sea-grass specialists, uprooting whole plants when they are accessible leaving long serpentine furrows bare of sea-grass in sea-grass meadows (Anderson & Birtles 1978). When the whole plant cannot be uprooted, they feed only on the leaves (Anderson 1982). In some places, Dugongs prefer 'weedy' or 'pioneer' species of sea-grass, especially species of the genera *Halophila* and *Halodule* (Marsh *et al.* 1982); *Halodule uninervis* is mentioned in most studies that have investigated the diet of the species in the African range (Gohar 1957, Hughes & Oxley-Oxland 1971). These sea-grasses are low in fibre, high in available nutrients and easily digested. Off the Kenya coast, Kingdon (1971) reported seasonal movements between 'pastures' and correlated alterations in Dugong diet. These were attributed to seasonal change in marine flora, which, in turn, could be related to seasonal fluctuations in water temperature and the roughness of the sea. Thus, deep water sea-grasses, notably *Cymodocea and Syringodium*, are allegedly preferred by Dugongs Nov–Mar, when seas are calm and warm, while shallow-water, in-shore species such as *Halodule*, *Halophila* and *Zostera* are apparently grazed Apr–Oct, when the sea, off-shore, is very rough and cold and avoided by Dugongs.

Experiments simulating Dugong grazing indicate that feeding Dugongs alter both the species composition and nutrient qualities of sea-grass communities (Preen 1995a). Dugongs are known to practise coprophagy, at least in captivity, and they have been known to feed on invertebrates in some areas (Kingdon 1971, Preen 1995b). Dugongs have an enlarged hindgut with a rich microflora enabling them to digest cellulose and other fibrous carbohydrates (Nishiwaki & Marsh 1985); they retain low-fibre material for extended periods of time within the long, tubular hindgut thereby digesting all the fibrous material (Lanyon & Marsh 1995).

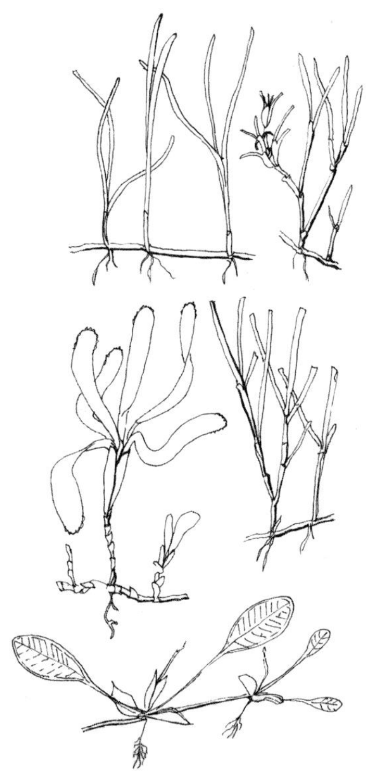

Zostera capensis (*left*) and *Syringodium* sp. (*right*)

Cymodocea ciliata (*left*) and *Halodule uninervis* (*right*)

Halophila sp.

Principal sea-grasses eaten by *Dugong dugon* in African waters.

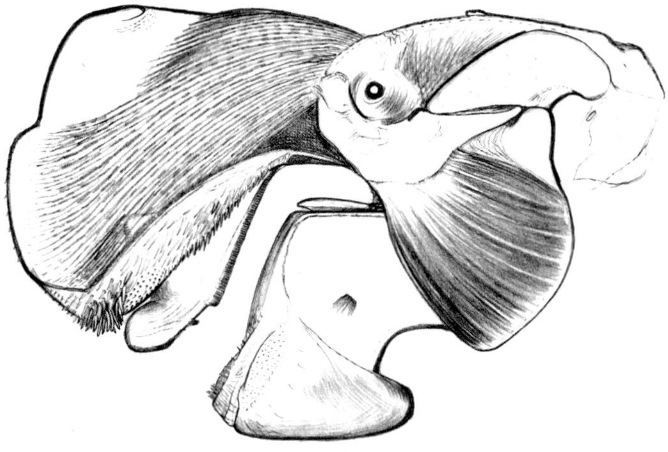

Myology of the facial musculature of *Dugong dugon*. Note the horny bristles around the mouth and embedded in highly developed levator labii muscle.

Social and Reproductive Behaviour Knowledge of the social behaviour of Dugongs is rudimentary. The habits and habitats of Dugongs make them difficult to observe and the lack of distinct size classes or obvious sexual dimorphism limits our capacity to interpret observational data on Dugong behaviour. The only defined long-lasting social unit is the cow and her calf (Anderson 1981); most sightings even in areas of high Dugong density are of one or two animals. Travis (1967) reported a group of more than 500 dugongs off the S Somalia coast, and Kingdon (1971) recorded congregations of over 100 off the Kenyan coast; large aggregations of up to several hundred animals still occur (but outside Africa) (Marsh 2002). Little is known of the structure and function of Dugong herds, but Preen (1995a) reported that grazing helped maintain sea-grass meadows at the stage favoured by Dugongs.

There are no data on long-range Dugong movement patterns in African waters, but such movements can be inferred based on sightings in the Seychelles in 2001 (Marsh *et al.* 2002). Seventy Dugongs fitted with satellite-monitored and GPS transmitters in sub-tropical and tropical waters of Australia exhibited a large range of movements. For example, 26 individuals were relatively sedentary (moved <15 km from the capture site) while 44 moved distances of up to 560 km from their capture sites (Sheppard *et al.* 2006). Regional movements were classified as those greater than 100 km, while local movements were less than 15 km. Local movements were generally tidally driven and represented commuting between and within sea-grass beds. Large-scale movements were rapid and apparently directed at reaching a specific and distant site. Radio-tracked Dugongs that were travelling rarely were far from the coast. Some animals caught close to the high latitude limits of the Dugongs' range on the Australian east coast in winter apparently undertook long distance movements to escape water temperatures less than about 18 °C (Sheppard *et al.* 2006). Preen (2004) sighted large herds of Dugongs in winter in the Arabian Gulf, in contrast to their dispersed distribution in summer. He suggested that Dugongs may aggregate around thermal springs in winter, entailing large-scale movements of approximately 400 km from their summer habitats.

Mating behaviour of Dugongs is variable. Herds of mating Dugongs have been observed at several locations along the N Australian coasts. Presumed groups of ♂♂ engage in bouts of violent fighting before attempting to mount a presumed ♀ (Preen 1989b). In contrast, in South Cove in Shark Bay, Western Australia, presumed male Dugongs defended mutually exclusive small temporary breeding territories or leks to attract ♀♀ (Anderson 1997). There are no reports of Dugong mating behaviour in African waters.

Vocalizations of Dugongs, which sound like bird chirps, span the 3–18 kHz bandwidth (Anderson & Barclay 1995). An audiogram conducted on one captive Dugong in Australia suggested a minimum hearing range of 4–32 kHz (D. Ketten pers. comm.).

Diving behaviour of 15 Dugongs has been documented using time-depth recorders by Chilvers *et al.* (2004). Overall, the Dugongs spent about half of their daily activities within 1.5 m of the sea surface and 72% in less than 3 m. Their mean maximum dive depth was 4.7 m, mean dive duration was 2.6 min, and the average number of dives per hour was 11. The maximum dive depth recorded was 20.5 m, the maximum dive time in water >1.5 m was 12.3 min.

Reproduction and Population Structure Studies in Australia indicate that Dugongs are long-lived with a low reproductive rate, long generation time, and a high investment in each offspring (Marsh 1980, Marsh *et al.* 1984b, Boyd *et al.* 1999, D. Kwan pers. comm.). Females do not bear their first calf until they are at least six years of age and up to 17 years old. Gestation is approximately 13 months (Boyd *et al.* 1999). Usual litter size is one; Dugongs only rarely give birth to twins. Neonates are 1.1–1.25 m in length (n = 18) and weigh 27–35 kg (n = 53) (Marsh 1995). The calf suckles for 18 months or so, and the period between successive births is very variable; estimates range from 2.5 to 7.0 years (Boyd *et al.* 1999, D. Kwan pers. comm.). Dugongs start eating sea-grasses soon after birth, and grow rapidly during the suckling period. Population simulations indicate that a Dugong population is unlikely to be able to increase in size more than 5% per year (Boyd *et al.* 1999). This makes the Dugong highly susceptible to over-exploitation by humans. Body weight at puberty is about 250 kg.

Like the teeth of other marine mammals, Dugong tusks accumulate 'growth layer groups' that are used to estimate age, rather like the growth rings of a tree. The oldest Dugong was estimated to be 73 years old when she died (Boyd *et al.* 1999).

Predators, Parasites and Diseases Given the Dugong's life history, natural mortality rates must generally be low. In Australian waters, there are reports of Dugongs being attacked by sharks (Heithaus 2001), killer whales (Anderson & Prince 1985) and crocodiles. Dugongs are susceptible to a wide range of diseases, some of them infectious or parasitic (Campbell & Ladds 1981, Marsh *et al.* 2000). Blair (1981) describes a range of Dugong parasites. There is no information available from Africa.

Conservation IUCN Category: Vulnerable A2bcd. CITES: Appendix I. CMS: Appendix II.

Dugongs are vulnerable to human impacts because of their life history and their dependence on sea-grasses that are restricted to coastal habitats, and which are often under pressure from human activities. In Africa, Dugongs continue to be threatened by direct human exploitation, habitat loss and destruction, and incidental capture during fishing operations in this region (Marsh *et al.* 2002, WWF Eastern African Marine Ecoregion 2004). From an ecological perspective, the Red Sea is considered to be one of the least damaged bodies of water in the world. None the less, oil spills have occurred in the Red Sea, and

Sketches of *Dugong dugon*. Note the crescent-shaped valvular nostrils, and how the ventral surface follows the same plane as the grazing mouth.

are thought to affect Dugongs through the degradation of sea-grass beds (Preen 1989a). Coastal developments include oil-related industrial centres and tourist resorts. Inadequate environmental standards at oil facilities and the de-ballasting of ships in the Red Sea are all potential threats to sea-grass areas (Marsh *et al.* 2002).

Coastal East African countries are experiencing massive population increases and demographic changes that are resulting in environmental degradation that affects critical near-shore Dugong feeding areas (Cockcroft *et al.* 1994, Wamukoya *et al.* 1996a, b). Sea-grass beds are also being destroyed by trawl nets (Wamukoya *et al.* 1996a, b). Artisanal fisheries mostly operate in inshore waters along the East African coast. These fisheries involve the use of dynamite and/or seine, drift and bottom set gill nets from small traditional boats (Cockcroft *et al.* 1994). The density of gill nets along the coast is high. There are no records of the numbers of Dugongs taken as bycatch in these fisheries; however, surveys, including interviews with fishermen, have revealed that Dugongs are taken in the drift gill net fishery (Guissamulo & Cockcroft 1997). Incidentally captured Dugongs are eaten by fishermen, and eating Dugong is not forbidden in most Muslim communities. In non-Muslim countries the meat may be disguised as pork and sold in restaurants (Dutton 1994). Although legislation now bans Dugong hunting, it is considered to be one of the factors that has caused the decline in Dugong numbers along the south coast of Kenya (Wamukoya *et al.* 1995, Wamukoya, Mirangi *et al.* 1996, Wamukoya *et al.* 1996a, b). Dugongs are afforded legislative protection in most countries in the region. However, most governments have neither the resources nor the capacity to monitor all activities in the marine and coastal zones (WWF Eastern African Marine Ecoregion 2004). The fact that Dugongs range across national borders argues for the development and implementation of regional and international conservation efforts.

Measurements

Dugong dugon
HB (♂♂): 2.4 (2.2–3.31) m; n = 18
WT: 250–570 kg
Australian waters (Spain & Heinsohn 1975, Marsh 1995, Lanyon *et al.* 2010)

Key References Boyd 1999; Kingdon 1971; Marsh *et al.* 2002, 2011; Nishiwaki & Marsh 1985; WWF Eastern African Marine Ecoregion 2004.

Helene Marsh & Paul Dutton

Family TRICHECHIDAE

MANATEES

Trichechidae Gill, 1872. Smithson. Misc. Coll. 11 (1): 14.

Trichechus (1 species)	Manatees	p. 210

The family Trichechidae includes three extant species of manatees (Rice 1998, Reynolds & Powell 2002, Shoshani 2005): the West Indian Manatee *Trichechus manatus*, Amazonian Manatee *T. inunguis* and West African Manatee *T. senegalensis*. Only the latter occurs in Africa. Domning & Hayek (1986) provide evidence that two subspecies of West Indian Manatee exist, namely the Florida Manatee (*T. m. latirostris*) and the Antillean Manatee (*T. m. manatus*). Little genetic data have been published, but Garcia-Rodriguez *et al.* (1998) found 15 haplotypes among West Indian Manatees and suggest three, rather than two lineages.

On current evidence, the first truly manatee-like sirenians (*Potamosiren*) appeared 15 mya, during the Miocene. Extant species are found from the SE USA into Brazil (*T. manatus*), the Amazon R. and tributaries in Brazil, Peru, Ecuador and Colombia (*T. inunguis*), and from S Mauritania to Angola (*T. senegalensis*). Around the mouth of the Amazon R., West Indian and Amazonian manatees may interbreed, as their ranges overlap (Vianna *et al.* 2006).

Manatees are extremely derived anatomically and physiologically (e.g. Pabst *et al.* 1999, Marshall 2002, Reynolds *et al.* 2002, Rommel & Reynolds 2002). The best-studied taxon is the Florida Manatee; in many cases, information available for this subspecies is assumed, in the absence of corroborating data, to apply for other trichechids. Reviews of manatee functional morphology appear in the references cited above.

All living manatees are large animals. The West Indian Manatee is the largest species, with some individuals approaching 1500 kg in weight and 4 m in length. The smallest species is the Amazonian Manatee, which measures less than 3 m and weighs less than 500 kg. Manatees lack pelvic limbs (although vestiges of pelvic bones remain) and have paddle-like pectoral flippers. The body is streamlined (fusiform) with a single, rounded fluke for locomotion. No dorsal fin is present. The skeleton is pachyosteosclerotic; this feature, coupled with the thick, dense integument, provides ballast. The elongated hemidiaphragms and lungs are important in buoyancy control (Rommel & Reynolds 2000). The nostrils are located on the dorsal tip of a somewhat streamlined head. Compared with many other marine mammals, manatees swim slowly and dive only to shallow depths and for short periods of time (< 20 min). The epidermis is somewhat rough in the West Indian Manatee and West African Manatee, but it is much smoother in the Amazonian Manatee. The latter species also tends to possess white-pink areas on the belly. Another external difference in the Amazonian Manatee is the lack of nails at the tips of the pectoral appendages.

Manatees are non-ruminating herbivores (Reynolds & Rommel 1996) that possess a greatly enlarged hindgut (especially the large intestine, which can exceed 20 m in length and 150 kg in weight), a single stomach, a capacious duodenum and an elongated but narrow post-duodenal small intestine. Manatees may consume up to 7% of their body weight per day, but Amazonian Manatees are thought to fast during the dry season for periods of up to 200 days, when vegetation is unavailable to them. Associated with their consumption of large amounts of often-gritty vegetation, manatees possess enlarged, flexible lips equipped with prehensile and tactile vibrissae and moved by a muscular hydrostat; supernumerary molariform teeth that are replaced throughout the lifetime of an individual; and unusual histology of the mucosae of the gastrointestinal tract organs.

The distribution of Florida Manatees is limited by cold weather in winter and perhaps by access to freshwater to drink. The latter limitation may well apply to other taxa. Although manatees possess vascular, counter-current heat exchangers, their inherently low metabolic rates (20–30% lower than expected) influence their ability to withstand cold conditions. Although trichechids are considered totally aquatic, they may voluntarily haul themselves partially out of the water to feed. Amazonian Manatees are confined to freshwater, and there are no known records of the species in salt water, whereas the other species are euryhaline in their distribution, occupying rivers, embayments and coastal marine habitats where sea-grass meadows or other aquatic vegetation is available.

The sensory biology of manatees is not well studied (Glaser & Reynolds 2003). Acoustics is likely to be the most important mode of communication among manatees. Vision is relatively poor, the ability to communicate and select habitat using chemical signals is likely but undemonstrated, and the use of innervated hairs on the body may provide signals similar in nature to those from a lateral line in fishes.

The social structure of manatees is not well understood. In Florida, a loose, fission–fusion society exists (Wells *et al.* 1999). In terms of their life history attributes (Reynolds & Powell 2002), Florida Manatees are K-strategists that may live as long as 60 years, reach sexual maturity as early as two years of age for ♂♂ and 3–4 years for ♀♀, and generally produce a single calf once every 2.5–3 years. Manatees are sperm competitors (Reynolds *et al.* 2004) and display promiscuous breeding. Sexual dimorphism is lacking, although Florida Manatees ♀♀ appear to be heavier, on average, than ♂♂.

The only taxon for which good population data exist is the Florida Manatee, and even for this subspecies, uncertainties in counts and population trends exist. As a general rule, survival of manatees worldwide is jeopardized by a range of current threats (incidental taking in fishing gear, habitat loss, watercraft collisions, hunting and pollution), as well as future threats associated with human population growth and human activities (Reynolds 1999). Recent volumes by Marsh *et al.* (2011) and Hines *et al.* (in press) provide comprehensive overviews of manatee ecology and conservation.

John E. Reynolds, III

GENUS *Trichechus*
Manatees

Trichechus Linnaeus, 1758. Syst. Nat., 10th Edn, 1–34.

The genus *Trichechus* includes three species: West African Manatee *T. senegalensis*, Amazonian Manatee *T. inunguis* and West Indian Manatee *T. manatus* (Shoshani 2005). As their names suggest, only the first occurs in African waters. The generic name originates from the Greek words *trichos* (hair) and *ekhō* (to have) pertaining to the sparse hairs covering manatees. The West Indian and West African species are most similar in anatomy and morphology, although recent genetic analysis suggests closer phylogeny with the Amazonian species (Parr & Duffield 2002, Vianna *et al.* 2006).

James A. Powell

Trichechus senegalensis West African Manatee

Fr. Lamantin; Ger. Seekuh, Manati

Trichechus senegalensis Link, 1795. Beitr. Naturgesch. 1 (2): 109. Senegal.

West African Manatee *Trichechus senegalensis*.

Taxonomy Monotypic. Synonyms: *africanus*, *australis*, *nasutus*, *owenii*, *stroggylonurus*, *vogelii*. Chromosome number: not known.

Description West African Manatees are heavy, adapted to aquatic life with a fusiform shape, broad back and no dorsal fin. Head is small compared with overall body size. Eyes are small and protrude slightly. No external ears. Muzzle blunt and rostrum only slightly deflected downward. Muzzle has massive, movable lip pads that manipulate and grasp vegetation aided by the stiff bristles. Skin colour is grey but can vary from greenish to brown to black depending on growths of epiphytic algae. Manatees in coastal waters may have barnacle growth. The thick skin is tough, slightly rough and underlain by fat; the body is covered by sparse, fine, evenly distributed hairs. Hindlimbs absent, moveable pectoral limbs are paddle-like, used for orienting the body, holding food and 'walking' on the bottom. Each pectoral flipper possesses 3–4 fingernails on the lateral, distal edge. Manatees possess a large, horizontally flattened, rounded tail. Male genital aperture is located near the middle of the ventrum and genital slit for ♀♀ is near the anus just anterior to the peduncle. There are two axillary nipples. Skull is generally broad, with a shortened, slightly deflected snout. Nasal bones variable in form. Cheekteeth brachyodont. Five to seven functional teeth are present at any time in each jaw; teeth are continually replaced by horizontal migration from the back of the jaw to compensate for tooth-wear caused by their abrasive diet of grasses and aquatic plants.

Geographic Variation None recorded.

Similar Species

Dugong dugon. Tail bifurcated similar to Cetacea; muzzle large and disc shaped; peduncle oval in cross-section along dorsal axis (whereas manatees have oval peduncle in cross-section along lateral axis). Strictly marine, feeding on sea-grasses. Indian Ocean coasts only.

Distribution Widely distributed along coastal regions, rivers, estuaries and lagoons of West Africa from S Mauritania to the Cuanza

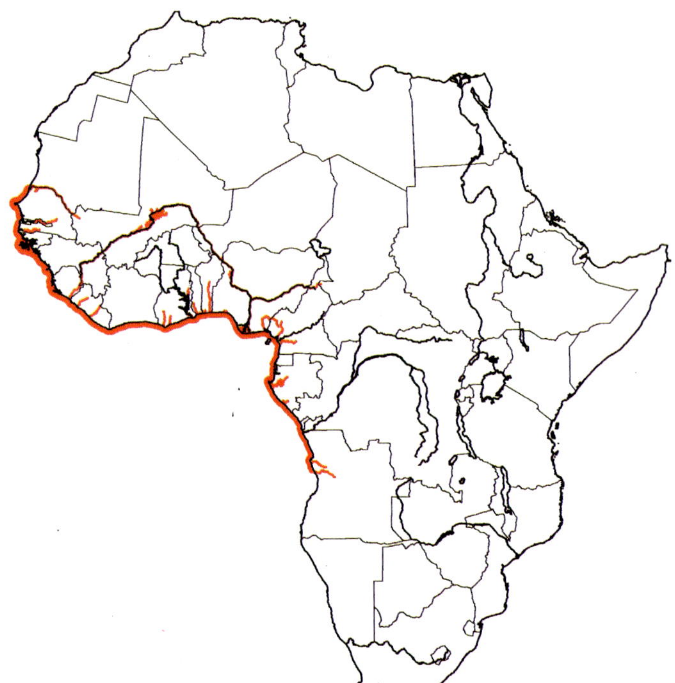

Trichechus senegalensis

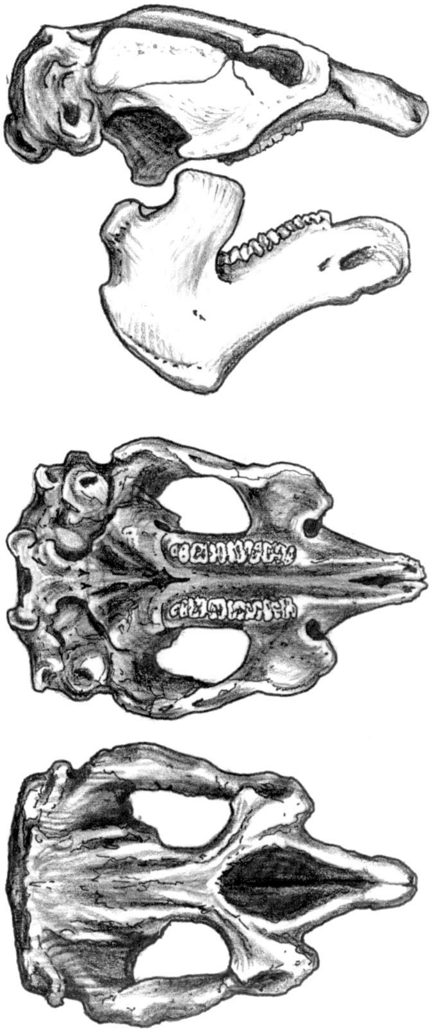

Lateral, palatal and dorsal views of skull of *Trichechus senegalensis*.

and Longa Rivers, Angola (Husar 1978, Nishiwaki 1984, Grigione 1996, Powell 1996, Morais 2006). Ascends many hundreds of kilometres up rivers and can be found offshore in shallow coastal regions such as the Bijagos Archipelago (Guinea Bissau) and Casamance (Senegal) where there are abundant sea-grasses. Isolated populations cut off from the sea are found in L. Volta, Ghana, above the Volta hydroelectric dam and the upper reaches of the Niger R. as far as Guinea. Occur along the entire length of the Gambia R., penetrating into Senegal where there are records as far upstream as Niokolo-Koba N. P. (Grubb *et al.* 1998). In Chad, manatees are present in L. Léré and L. de Tréné along the Mayo-Kebbi, Bahr Keeta and Baningi Rivers; manatees are also reported from the Baningi, Logone and Chari river tributaries of L. Chad. Sightings of manatees reported from L. Chad itself but no confirmed records (Ita 1994). Manatees have penetrated the Niger R. as far upstream as Segou, Mali, which represents a distance of some 2000 km (Hatt 1934).

Habitat Wide variety of marine and freshwater habitats, including shallow coastal areas, estuaries, lagoons, small tidal rivers and hundreds of kilometres up larger freshwater rivers (Husar 1978). Some populations are completely cut off from the sea by natural barriers such as cataracts or by human structures such as hydroelectric and salinity dams. In arid regions where water levels vary considerably between wet and dry seasons, manatees survive drier periods by seeking refuge in deeper pools and seasonally connected lakes, such as Pandam L. in Nigeria or moving downriver (Sikes 1974). Preferred habitats are sheltered estuaries, large, freshwater rivers and shallow coastal regions with freshwater input (Husar 1978). In some areas they co-exist with Common Hippopotamus *Hippopotamus amphibius* (J. Powell pers. obs.).

Abundance Elusive and largely nocturnal, manatees are inconspicuous though present throughout much of their historic range. Recent information and surveys indicate manatees persist in many areas thought previously extirpated (Powell 1996), although numbers appear low throughout. Populations are probably depressed due to their low reproduction rate and human-related mortality preventing populations from building.

Adaptations Manatees are well adapted to their aquatic habitat, being fusiform in shape with a large, rounded tail. Their forelimbs form flippers that tuck in closely to the body when swimming forward. Otherwise, the flippers are used to manoeuvre by sculling and 'walking' along the bottom. Male West Indian Manatees *Trichechus manatus* clasp the ♀ with the flippers during sexual pursuit and cavorting. The muscular lip pads are used to grasp vegetation, aided by stiff bristles that help manipulate food toward the mouth. The body is covered by fine, evenly spaced hairs that, in the West Indian species, have been found to detect subtle pressure differences and vibrations (Reep *et al.* 2002). Internally, manatees have thick, heavy (pachyostotic) bones, which, together with dense skin, help create ballast. The arrangement and orientation of the lungs and diaphragm assist with buoyancy control.

Foraging and Food Food consists of submerged vegetation, emergent plants, overhanging bank growth and floating plants. Fruits and nuts falling into the water are also consumed. Preferred food items include aquatic plants such as *Eichornia crasipes* and *Cymodocea nadosa*, bank sedges such as *Vossia* sp. and *Echinochloa* sp., and leaves of mangroves (*Rhizophora* sp.). The manatees grasp bank grasses with their lips and twist the grass stem at a node until it breaks free. Feeding sign is often easy to observe where there are sections of twisted and broken grass stems mixed among intact bank growth. Food is digested by hindgut fermentation similar to horses and elephants. Husar (1978) estimated that a free-ranging adult could consume as much as 8000 kg of aquatic vegetation in one year. Manatees are considered a pest on rice in Guinea-Bissau and Sierra Leone during the flood season (Reeves *et al.* 1988, Powell 1996).

Fishermen in Senegal and Sierra Leone report manatees 'sucking' the flesh off fish entangled in nets, though this is not confirmed (Reeves *et al.* 1988, Powell 1996). Powell (1978) observed similar behaviour by the West Indian species in Jamaica.

Radio tracking manatees in Côte d'Ivoire indicated that manatees feed almost exclusively at night (Powell 1996). On average, they spent 4–6 hours feeding, sometimes travelling several kilometres between daytime resting areas and feeding areas. Manatees would frequently return to the same areas to feed for several consecutive nights. It is believed that the nocturnal feeding and travelling pattern may be a response to hunting pressure.

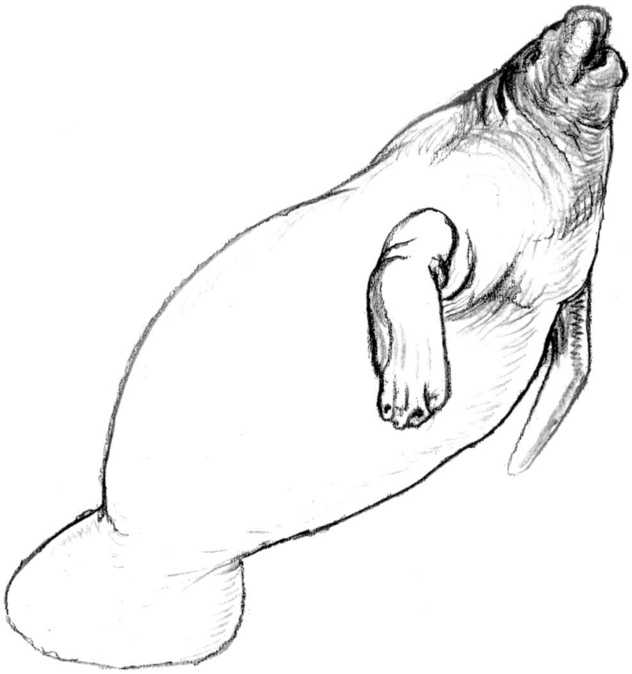

Sketch of *Trichechus senegalensis*.

Social and Reproductive Behaviour Little information is available on the social and reproductive behaviour of West African Manatee. Current knowledge is based on tagging studies and inferences from other manatee species. Males and ♀♀ are largely solitary, but may loosely aggregate at preferred feeding and resting sites; the strongest association is between a ♀ and her calf. Groupings of manatees occur in mating herds, when ♀♀ are in oestrus and followed by a consort of several ♂♂. Radio tagging information from Côte d'Ivoire showed seasonal aggregations occurring just before the rainy season in a certain section of a small tributary (Powell 1996). It is speculated that this aggregation may be related to reproductive behaviour. These observations are consistent with Howell (1968), who reported that manatees breed seasonally in weedy swamps and lagoons at the end of the dry season.

Movements are known to occur in response to changes in seasonal water levels and salinities. In drier periods where river levels fluctuate dramatically, manatees move downstream or, if blocked access, seek deeper pools or lakes connected to main rivers. Where manatees inhabit larger, tidally influenced rivers, they will spend extended time upstream during the dry season in lower salinities (Powell 1996).

Reproduction and Population Structure Typically one young, although, extrapolating from fishermen reports and from what is known of West Indian Manatees, twins may be rare. Gestation is 12–13 months, calving interval probably similar to other manatees at 2.5–3.0 years, with calves remaining with ♀♀ for 1–1.5 years. Newborn calves are about 1 m in length (Cadenat 1957). Howell (1968) reported seasonal breeding/births before the onset of the wet season, although newborn calves have been reported throughout the year (Dekeyser 1952, Cadenat 1957).

Predators, Parasites and Diseases There are no known predators or diseases for West African Manatees, although large crocodiles may occasionally predate calves or injured individuals. Their thick, extremely tough hide may help to protect them from potential predators. Shark predation on West Indian Manatees has been reported (Falcon-Matos *et al.* 2003, J. Powell pers. obs.), but there are no documented cases for West African Manatees.

Only two internal parasites have been recorded for the West African Manatee. The trematode *Chiorchis fabaceus* inhabits the lumen of the caecum and colon (Baylis 1936), and is shared by all three manatees (Beck & Forrester 1988). A new ascaridoid nematode, *Heterocheilus domningi*, was identified from a manatee collected from the lower Congo R. (Sprent 1983) and is very similar to *H. tunicatus* known from the other two species (Sprent 1983, Beck & Forrester 1988). Two ectoparasites have been recorded: *Chelonibia manati* also found on the West Indian Manatee, and *Platylepas hexastylos* (Beck & Forrester 1988).

Conservation IUCN Category: Vulnerable A3cd; C1. CITES: Appendix II. CMS: Appendix II.

Protected by specific laws in all countries where they occur in Africa, although enforcement of laws is minimal. Sanctuaries and reserves established that include protection of manatees and their habitat include: Parc National des Oiseaux de Djoudj, Parc National de Basse Casamance, Delta du Saloum (Senegal), Baboon Island N. P., Kiangs West N. P. (Gambia), Ilhas de Orango N. P., João Vieira e Poilão Marine N. P. (Guinea-Bissau), National Park of Upper Niger (Guinea), Azagny N. P., Iles Eotiles N. P. (Côte d'Ivoire), Digya N. P. (Ghana), 'W' N. P. (Burkina Faso/Niger/Benin), Kainji L. N. P., Pan Dam Wildlife Reserve (Nigeria), Reserve de Faune de Douala-Edea, Korup N. P. (Cameroon), Conkouati N. P. (Congo R.) and five recent national parks in Gabon. Hunting of manatees for their meat, oil and use of body parts for medicinal purposes exists throughout their range. Hunting techniques include harpooning, trapping by a variety of methods, netting in large mesh nets, and catching them on large hooks baited with cassava. Incidental/accidental mortality occurs when manatees become entangled in fishing nets, shark nets, hook lines or stranded when water levels drop (Powell 1996). In Guinea Bissau, Silva & Araújo (2001) reported 209 deaths between 1990 and 1998; of these, 72% died in fishing gear, 13% were hunted, 4% stranded at low tide and for 11% the cause of death was undetermined.

Measurements

Trichechus senegalensis

HB (♂♂): 2.56 (1.77–3.34) m, n = 14

HB (♀♀): 2.59 (2.25–3.03) m, n = 11

Senegal (Cadenat 1957, Dupuy & Maigret 1978, Diop 1984), Sierra Leone (Cole & Okera 1976), Côte d'Ivoire (Powell 1996), Gabon (Peck unpubl.), Nigeria (Dollman 1933) and Gambia (Itoh & Tsuyuki 1974)

Note: there are no known published body weights available for West African Manatee, but it is possible to estimate body weight by extrapolating from a dataset of 2.0–3.35 m long Antillean Manatees (*Trichechus manatus manatus*) from Belize (J. A. Powell *et al.* unpubl.), which weighed 166–640 kg (n = 51).

Key References Husar 1978; Powell 1996.

James A. Powell

Cohort AFROINSECTIPHILLIA

Cohort Afroinsectiphillia Waddell, Kishino & Ota, 2001. Genome Informatics 12: 148.

The afrotherian clade that contains the Aardvark (order Tubulidentata), sengis (order Macroscelidea) and tenrecs and golden-moles (order Afrosoricida) has been named Afroinsectiphillia, which means 'African insect-lovers'. The monophyly of this curious assemblage is now supported by phylogenetic analyses of large genomic datasets (Amrine-Madsen *et al.* 2003, Murata *et al.* 2003, Poulakakis & Stamatakis 2010, Meredith *et al.* 2011), and two types of rare genomic changes (Robinson *et al.* 2004, Nishihara *et al.* 2005, Poulakakis & Stamatakis 2010), but the three major afroinsectiphillian lineages show little superficial morphological similarity. Indeed, Afroinsectiphillia is more notable for its great morphological disparity, in that it contains small-bodied cursors (sengis) that may be descended from herbivorous ancestors; semi-aquatic carnivores (African otter-shrews and Malagasy *Limnogale*); arboreal, spiny and shrew-like tenrecs in Madagascar; and a number of large- and small-bodied fossorial taxa. Digging adaptations are a recurring theme within Afroinsectiphillia, having evolved independently at least three times (in the Aardvark, golden-moles and Malagasy rice tenrecs).

Genetic data suggest that the last common ancestor of the extant afroinsectiphillians lived about 75 mya, in the Late Cretaceous – unfortunately during a huge gap in Afro-Arabia's fossil record of mammalian evolution. However, some of the oldest placental mammal-bearing sites in Africa, which are about 56 million years old (late Palaeocene), have produced possible fossil afrosoricids (Seiffert 2010, Goswami *et al.* 2011), and distant relatives of the living sengis are known from Eocene localities that are 50–45 million years in age. Undoubted fossil aardvarks are not known until about 18 mya, in the early Miocene, but the extinct order Ptolemaiida, known from Eocene and Oligocene deposits in Egypt and the early Miocene of East Africa, might be more closely related to Tubulidentata than to other afrotherian orders (Seiffert 2007).

Perhaps because of their very poor early fossil records, the interrelationships of afroinsectiphillian orders remain unclear. One option, favoured by the largest available DNA datasets, places sengis closest to afrosoricids in the superorder Afroinsectivora (Amrine-Madsen *et al.* 2003, Poulakakis & Stamatakis 2010, Meredith *et al.* 2011); however, rare genomic changes have alternatively provided support for both a sengi–Aardvark clade (Robinson *et al.* 2004) and an Aardvark–afrosoricid clade (Nishihara *et al.* 2005), neither of which have been assigned a higher-level taxonomic name. For now, it is probably best to recognize Afroinsectiphillia and Afrosoricida, and to wait for more data to resolve the position of the Aardvark and sengis relative to afrosoricids.

Erik R. Seiffert

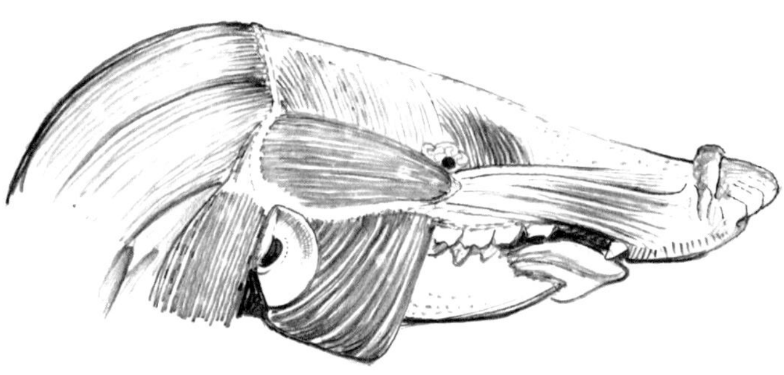

Stuhlman's Golden-mole *Chysochloris stuhlmanni*

Giant Otter-shrew *Potamogale velox*

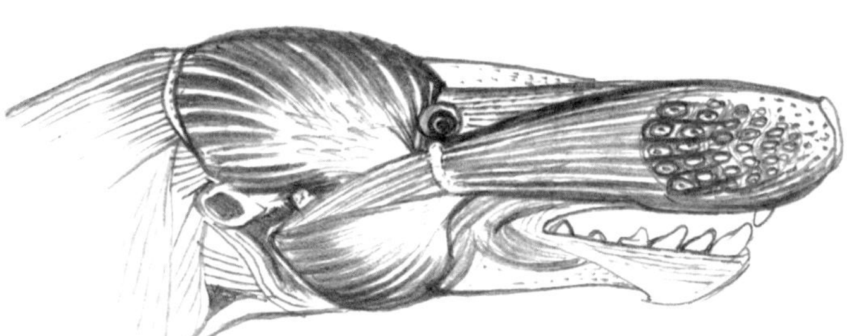

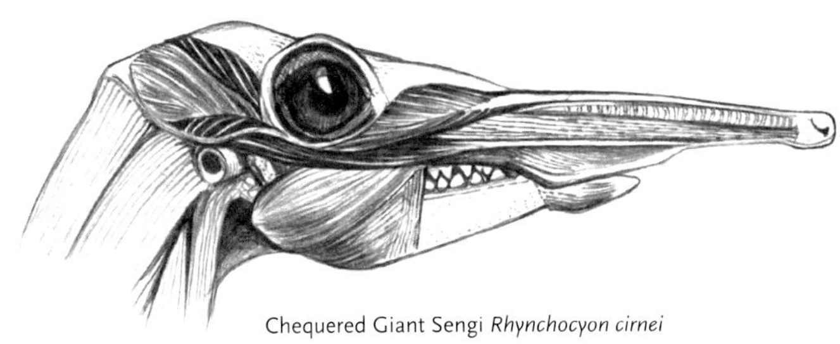

Chequered Giant Sengi *Rhynchocyon cirnei*

Aardvark *Orycteropus afer*

Myology of representatives of the Afroinsectiphillia.

Order AFROSORICIDA – Tenrecs, Otter-shrews, Golden-moles

Afrosoricida Stanhope *et al*. 1998. Proc. Natl Acad. Sci., USA 95: 9967–9972.

Tenrecidae		
Potamogalinae (2 genera, 3 species)	Otter-shrews	p. 216
Chrysochloridae (9 genera, 21 species)	Golden-moles	p. 223

Stanhope *et al*. (1998) erected the Afrosoricida as a new order to name a clade comprising two families: Chrysochloridae or golden-moles (9 genera, 21 species) of sub-Saharan Africa, and Tenrecidae. The Tenrecidae comprise four subfamiles: the tenrecs (3 subfamilies, 8 genera, 27 species) of Madagascar and Comoros Islands (not included in *The Mammals of Africa*), and the otter-shrews (subfamily Potamogalinae; 2 genera, 3 species) of Africa. The African representatives of the Order (Chrysochloridae and Tenrecidae [subfamily Potamogalinae]) comprise 11 genera and 24 species. The Afrosoricida was born out of the final sundering of the Insectivora (also referred to as Lipotyphla), an order that was recognized for many decades, but which recent molecular evidence now shows to be polyphyletic. This profile deals only with the taxa on the continent of Africa; general information on the Tenrecidae of Madagascar may be found in Nowak (1999) and Olson & Goodman (2003). See Bronner & Jenkins (2005) for a taxonomic review of the Afrosoricida.

Biological Characteristics of African Afrosoricida

The two families of the order represented in Africa have very different morphological characteristics. The otter-shrews (family Tenrecidae, subfamily Potamogalinae) are semi-aquatic with an elongated body, long muscular tail (ca 80% of HB), water-repellent pelage, small ears, long vibrissae and short limbs. One species has webbed digits. Otter-shrews are fairly small (mean HB: ca. 140 mm to ca. 320 mm) and are similar in general shape to otters (order Carnivora) although smaller in size. All species live in streams (rather than big waterways) in rainforest and montane forest habitats in West and central Africa. They are generally uncommon and rarely seen. Golden-moles (family Chrysochloridae) are subterranean and highly adapted for life underground. Golden-moles are mostly small in size (mean HB of ca. 80–110 mm, with a range of HB ca. 60 mm to ca. 220 mm, according to the species), body shape is fusiform or cylindrical and the head merges with the body without an obvious neck. External ears are absent; eyes are degenerate and covered with skin. The tail is absent and limbs are very short. Forelimbs are well developed with large digits and nails (especially Digit 3). Most species of golden-moles live in grasslands, semi-deserts and highland areas of southern Africa; two species live in rainforests of central and eastern Africa, and one species in semi-desert habitats of the Horn of Africa.

Species of the Afrosoricida are insectivorous and/or carnivorous. Golden-moles eat insects and earthworms, and otter-shrews eat crabs, aquatic insects, shrimps and fish. Typically, all species have a large set of teeth: 36–40 in golden-moles and 40 in otter-shrews: the incisor and canine teeth are pointed, but the canine is not enlarged as in the Carnivora. Premolar and molar teeth are multi-cusped for crushing and cutting the prey. The social organization of afrosoricids is not well known. As far as is known, otter-shrews and golden-moles are solitary, although the home-ranges of individual otter-shrews may overlap. Reproductive strategies, likewise are not well known; litter-size is small (1–2) and most species have restricted reproductive seasons, the timing of which varies according to the species and the environment; a few species of golden-moles may breed throughout the year.

Populations are often isolated because of restricted distributions and the limited mobility of individuals: otter-shrews are confined to streams and golden-moles are confined to suitable patches of soil. Demes of golden-moles (as for many subterranean mammals) show considerable morphological variation, which has resulted in many named forms (most now considered as synonyms). Movement of individuals to suitable new habitats is difficult for otter-shrews and golden-moles because of their strict environmental requirements and their inability to travel across unsuitable habitats.

Phylogenetic Relationships The Insectivora, as traditionally recognized, included the shrews (Family Soricidae), hedgehogs (Family Erinaceidae), true moles (Family Talpidae), elephant-shrews or sengis (Family Macroscelididae), tree-shrews (Family Tupaiidae), solenodons (Family Solenodontidae), golden-moles (Family Chrysochloridae), tenrecs (Family Tenrecidae) and several extinct families. It was a poorly defined group, essentially a taxonomic wastebin for various 'undifferentiated' insectivorous forms assumed to be the ancestral stock from which other placental mammals radiated. The Insectivora was rendered obsolete, first by the removal of the elephant-shrews to the Order Macroscelidea and the (non-African) tree-shrews to the Order Scandentia, and more recently by the allocation of golden-moles, tenrecs and otter-shrews to the Order Afrosoricida. The remnants of the Insectivora are now allocated to the Order Soricomorpha (shrews, true moles and [non-African] solenodons) and the Erinaceomorpha (hedgehogs).

A close phylogenetic affinity between golden-moles and tenrecs (including otter-shrews) was first proposed more than a century ago, and has been supported by several authors (e.g. Butler 1956), but their recent grouping in the Afrosoricida has evoked controversy. Ironically, this controversy is not about the putative close relationship between the Chrysochloridae and Tenrecidae, which has little morphological support despite overwhelming molecular evidence. Instead, it concerns the name 'Afrosoricida', and the inclusion of this order within the recently erected supraordinal group named 'Afrotheria'.

The name 'Afrosoricida', which literally means 'African shrew-like mammals', is still controversial, and has been rejected by some authors who instead advocate using the names 'Tenrecomorpha' or 'Tenrecoidea' for the clade containing golden-moles and tenrecs. The term 'Afrosoricida' is deemed inappropriate because it suggests a close relationship between shrews and golden-moles and/or tenrecs, and also because it engenders confusion with the shrew subgenus *Afrosorex*. Despite the tendency during the nineteenth and twentieth centuries to include shrews, golden-moles and tenrecs in the same order (Insectivora), or even in the same suborder (Soricomorpha; e.g. Butler 1988), there is now robust morphological and molecular evidence against any close phylogenetic relationship between shrews and the clade containing golden-moles and tenrecs (Asher *et al.* 2003). Consequently, there has been widespread reluctance (especially among shrew biologists) to accept the name 'Afrosoricida'.

Some authors have argued that Tenrecoidea McDowell, 1958 pre-dates Afrosoricida and is the valid name for the clade containing golden-moles and tenrecs (e.g. Malia *et al.* 2002). However, Tenrecoidea Simpson, 1931 was first erected as a superfamily to group several families with 'zalambdodont' cheekteeth (i.e. cheekteeth with three main cusps connected by a V-shaped crest), but excluded chrysochlorids, which were instead assigned to a separate superfamily (Chrysochloridea Broom, 1915). McDowell's (1958) restriction of Tenrecoidea to include only golden-moles and tenrecs radically changed Simpson's grouping concept, violated the principle of reasonable emendation (Simpson 1945: 33) and also his recommendation that superfamily names (ending in -oidea) should be avoided.

A colloquial derivative of 'Tenrecoidea', the term 'tenrecoid', has also been widely misused in the scientific literature for vaguely defined groupings of extant and extinct taxa putatively characterized by zalambdodont cheekteeth. However, it is now known that some of these taxa are not technically zalambdodont, and that zalambdodonty may have arisen independently several times (Broom 1916). This further militates against the nomenclatorial use of 'Tenrecoidea' and 'tenrecoid', regardless of taxonomic rank.

'Tenrecomorpha' Butler, 1972, named as a suborder to accommodate the tenrecs (but not golden-moles), has also been used for the clade containing golden-moles and tenrecs (e.g. Bronner *et al.* 2003). However, 'Tenrecomorpha' has most consistently been used to include only tenrecs (e.g. MacPhee & Novecek 1993) and consequently is better applied at the subordinal level to distinguish between the Tenrecidae and Chrysochloridae, in accordance with both morphological and molecular data suggesting early divergence of these families. The name 'Afrosoricida', although inappropriate, implies the most specific and least confusing phylogenetic grouping concept; it is also a name that is already embedded in the primary scientific literature (Bronner & Jenkins 2005).

Until very recently, the inclusion of Afrosoricida within the supraordinal taxon Afrotheria remained controversial. Numerous studies (employing a variety of sophisticated analytical procedures), which have utilized an impressive suite (>17 kbp) of mtDNA and nuclear gene data by independent research groups, have all found strong probabilistic support for an affinity with the Afrotheria (e.g. Stanhope *et al.* 1998, Murphy *et al.* 2001b). Various molecular synapomorphies uniting all afrotheres (including Afrosoricida) have also been identified; the probability that these arose convergently is infinitesimally small (e.g. Van Dijk *et al.* 2001).

Analyses of morphological data, in contrast, have yielded largely equivocal results regarding an affinity between Afrosoricida and the Soricomorpha. MacPhee & Novacek (1993), however, showed that the only morphological characters that support the grouping of the Afrosoricida with the other families formerly included in the Insectivora are shared primitive characters (symplesiomorphies), which are lacking in the other families, rather than synapomorphies that define and combine them. Monophyly of the Insectivora is only weakly supported by cladistic analyses of morphological data (Asher 1999) and phylogenetic analyses of combined molecular and anatomical data (the latter also from some extinct taxa) strongly support the inclusion of afrosoricids within Afrotheria (e.g. Asher *et al.* 2003). The order Afrosoricida, together with the orders Macroscelidea (sengis) and Tubulidentata (aardvark), has been placed in the Cohort Afroinsectiphillia (Waddell *et al.* 2001, Amrine-Madsen *et al.* 2003) – the 'African insect lovers' – within the Superorder Afrotheria. Further information on Afrotheria is given on p. 143.

Despite the overwhelming molecular evidence for the existence of a clade containing golden-moles and tenrecs, there is surprisingly little morphological support for a close relationship between these families beyond their joint possession of zalambdodont cheekteeth – the homology of which is uncertain given that zalambdodonty has arisen independently several times (Broom 1916). The only non-dental characteristics used previously to defend a grouping of chrysochlorids and tenrecids are either symplesiomorphic or equivocal (MacPhee & Novacek 1993). Golden-moles and tenrecs also diverge markedly in the characters of the male reproductive system, and golden-moles possess many unique adaptations for burrowing (Butler 1988). The absence of morphological synapomorphies to support the monophyly of the Afrosoricida probably reflects the independent evolution of sophisticated phenotypic specializations over the millions of years since the time when golden-moles and tenrecs diverged from each other, thus obscuring any morphological similarities that might offer clues to the ancestry encoded in their genes.

Gary N. Bronner

Family TENRECIDAE

TENRECS, OTTER-SHREWS

Tenrecidae Gray, 1821. London Med. Repos. Rec. 15: 301.

Micropotamogale (2 species)	Pygmy Otter-shrews	p. 216
Potamogale (1 species)	Giant Otter-shrew	p. 220

(African genera only)

The family Tenrecidae is endemic to Madagascar and Africa, and characterized by the presence of a caecum, zalambdodont dentition (molars with a V-shape pattern in most species), a cloaca-like anogenital opening, testes positioned close to the kidney or in an intra-abdominal position, and skull without zygomatic arches. Dental formula: $I\ ^3/_3$, $C\ ^1/_1$, $P\ ^3/_3$, $M\ ^3/_3 = 40$.

The family was formerly allocated to the order Insectivora, suborder Lipotyphla. Based on molecular data, Stanhope *et al.* (1998) evidenced a phylogenetic relationship with Elephants and Hyraxes rather than with other members of the Insectivora, and Emerson *et al.* (1999) confirmed a close relationship with the Chrysochloridae. The Tenrecidae together with the Chrysochloridae (or golden-moles) are now allocated to the order Afrosoricida and separated from other families previously in the Insectivora (i.e. Soricidae, now placed in the order Soricomorpha; Erinaceidae, now in the order Erinaceomorpha).

Of the four subfamilies in the family (see Order profile above), the Potamogalinae is the only subfamily endemic to continental Africa. It contains only the otter-shrews, comprising two genera and three species, all distributed within the Rainforest BZ. They are semi-aquatic mammals looking like small otters with an enlarged flat muzzle, a slender body and a strong tail. They are good swimmers; one species uses the tail and the others use the limbs for propulsion. The characters of the subfamily include: eyes small; ears proportionally narrow; mystacine vibrissae very strong, 3–4 on the cheek, one over the eyes and four (three lateral and one median) on the chin; nose leathery with two opercula for closing the nostrils when diving, without philtrum; pelage shiny, due to the strong overhairs with a large terminal shield (hair apex enlarged like a lance) as in desmans and otters; tail relatively long (75–85% of HB), base enlarged and very strong; and, in one species, digits of fore- and hindfeet webbed. Syndactyly of Digits 2 and 3 of the hindfoot provides a 'comb' for grooming pelage, a character not found in *Limnogale*, the aquatic tenrec of Madagascar. The skull is narrow, without zygomatic arches. One pair of incisors (I^1) in upper jaw and one pair (I_2) in lower jaw are very large for catching prey (and appear caniniform); canine teeth are small. Tooth replacement is rather slow (Kuhn 1964). Skull lacks lacrimal ducts (Butler 1978, Asher 2000). The postcranial skeleton is similar to that of Tenrecs, but without clavicles and with the pubic bones separated.

Otter-shrews were first considered to represent a family (Potamogalidae) within the Order Insectivora. Most authorities now allocate them to the Potamogalinae as a subfamily within the family Tenrecidae (Heim de Balsac & Bourlière 1955, Hutterer 1993, Bronner & Jenkins 2005). Molecular data suggest that otter-shrews belong to the Afrotheria clade (Van Dijk *et al.* 2001), which also contains the Tenrecs (Stanhope *et al.* 1998; see also order profile). In a morphological analysis (Asher 1999), the Potamogalinae clustered with *Limnogale*. In contrast, molecular data (Douady *et al.* 2002) suggest that the Potamogalinae form a sister-group with the tenrecs of Madagascar (Tenrecinae and Oryzorictinae) confirming the conclusion of McDowell (1958), who grouped these two families in the taxon Afrosoricida (=Tenrecoidea).

Two genera of otter-shrews are recognized: *Potamogale*: large size (HB: >170 mm); tail very thick at base and flattened laterally on terminal half; hindfeet not webbed (1 sp.); and *Micropotamogale*: small size (HB: <170 mm); tail not flattened laterally; hindfeet webbed (1 sp.) or not webbed (1 sp.).

Peter Vogel

Subfamily POTAMOGALINAE – Otter-shrews

Potamogalinae Allman, 1865. Proc. Zool. Soc. Lond. 1865: 467.

This subfamily contains two genera and three species, all indigenous to the African continent. Full details of the subfamily are given in the family Tenrecidae profile above.

GENUS *Micropotamogale*

Pygmy Otter-shrews

Micropotamogale Heim de Balsac, 1954. C. R. Acad. Sci. Paris 239: 102. Type species: *Micropotamogale lamottei* Heim de Balsac, 1954.

The genus *Micropotamogale* contains two species of otter-shrews: one lives in West Africa and the other in E DR Congo. Both species have restricted distributions in the Rainforest BZ and montane forests.

The genus is characterized by its adaptations for aquatic life: small elongated body, small size (HB: <170 mm, mean weight ca. 78–100 g) and a tail that is not flattened laterally (cf. *Potamogale*). One species in the genus has webs between the digits of the fore- and

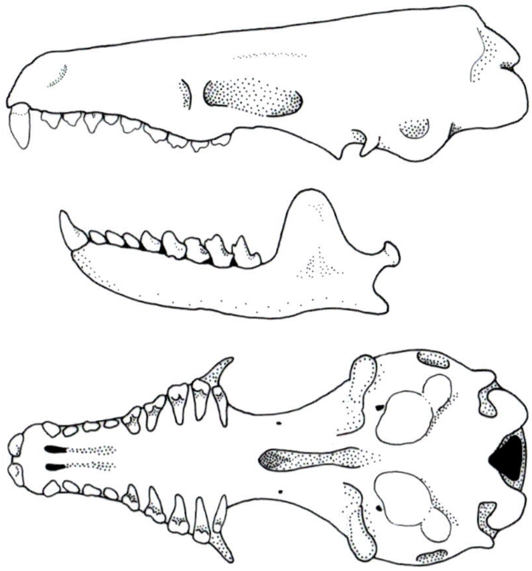

Figure 4. Skull and mandible of *Micropotamogale lamottei* (BMNH 73.170).

Micropotamogale ruwenzorii.

hindfeet (*M. ruwenzori*) and is considered to be highly specialized, and the other species (*M. lamottei*) does not have webs between the digits and is considered less specialized. Skull: elongated and narrow, without zygoma; upper incisors canine-like; canines small, similar to premolars; premolars simple, molars zalambdodont (V-shaped pattern on occlusal surface) (Figure 4). The zalambdodont pattern of teeth is a plesiomorphic character, confirming relationship with the tenrecs. (See also family profile.) Dental formula: $I\ ^3/_3$, $C\ ^1/_1$, $P\ ^3/_3$, $M\ ^3/_3 = 40$.

The two species are distinguished by size, presence/absence of webbing on the feet, and geographical distribution.

Peter Vogel

Micropotamogale lamottei Nimba Otter-shrew (Pygmy Otter-shrew)

Fr. Micropotamogale du Mont Nimba; Ger. Kleine Otterspitzmaus

Micropotamogale lamottei Heim de Balsac, 1954. C. R. Acad. Sci. Paris, 239: 102. Ziela, Mt Nimba, Guinea. 550 m.

Taxonomy Considered by Heim de Balsac (1956) as the most primitive form of the Potamogalinae. Synonyms: none. Chromosome number: 2n = 38 (Vogel *et al.* 1977).

Description Small blackish-brown otter-like mammal, with slender body. Dorsal pelage dense, glossy, brown to blackish-brown; hairs grey to whitish-grey at base, blackish-brown at tip; underfur woolly, grey with brown tips. Ventral pelage similar to dorsal pelage, but when dry slightly speckled by brighter tips to hairs. Head with enlarged snout and numerous stiff vibrissae. Nostrils on upper surface of the leathery rhinarium, closed when diving. Eyes very small. Ears rather short, only little longer than pelage hairs. Fore- and hindlimbs short; narrow digits not webbed and not bordered by stiff hairs. Syndactyly of Digits 2 and 3 of hindfeet form a comb. Tail relatively long (ca. 77% of HB), slightly thickened at base, covered by short dark hairs; not laterally flattened and without visible skin scales. Skull: see genus profile. Nipples: 2 (pectoral) + 1 (abdominal) + 1 (inguinal) = 8.

Geographic Variation None recorded.

Similar Species

M. ruwenzorii. Similar size; ventral pelage grey, pale brown or whitish; feet broad, feet with webs between digits; E DR Congo only.

Potamogale velox. Much larger; ventral pelage white, feet not webbed; tail thick, laterally flattened; Cameroon to E DR Congo.

Micropotamogale lamottei

Distribution Endemic to Africa. Rainforest BZ (Western Region). Known only from Mt Nimba in Liberia, Guinea (Kuhn 1964) and W Côte d'Ivoire (Vogel 1983), and the mountains of the Putu Range, Liberia (Kuhn 1971). Very restricted; most distant localities only 380 km apart.

Habitat Very small creeks, larger streams and small swamps in primary and secondary rainforest in hilly country (Vogel 1983). Also occurs in regions where primary and secondary forest has been replaced by cocoa and coffee plantations, provided dense vegetation remains along streams. May occur in rice cultures (Kuhn 1964).

Abundance Generally uncommon; density usually low; very rare where hills not present. In the Nimba region, drowned animals collected by fishermen from fish bownets suggest a density of about one otter-shrew per 10 km^2. In 1970, otter-shrews were frequently found close to villages; by 1990, rarely found within 3–5 km of villages (P. Vogel, pers. obs.).

Adaptations Efficient swimmer and diver, even though tail and feet are not specially adapted for aquatic life (Guth *et al.* 1959) compared with the other two species of otter-shrews. Nimba Otter-shrews can remain underwater for more than 10 minutes (when escaping from danger) by lowering the metabolic rate. The strong sensitive vibrissae are used for locating underwater prey. The narrower feet and thinner tail (when compared with *M. ruwenzori*) probably allows these otter-shrews to have broader ecological niche compared with other species of otter-shrews.

Foraging and Food Foraging occurs along the river banks and also in open water. The diet is mainly crabs and catfish, with a few insects (Kuhn 1964) and tadpoles (Vogel 1983). Crabs are carried onto land for eating; the crab is then attacked from behind, minimizing the chances of being bitten by the pincers, and the skeleton is crushed at the junction of the cephalothorax and abdomen (Vogel 1983). In captivity, food consumption was about 40 g fish/day.

Social and Reproductive Behaviour Evidence from the rare captures, and from predatory behaviour, suggest that Nimba Otter-shrews are rather solitary and territorial. However, a pair kept over several months in the same enclosure did not exhibit aggressive behaviour (Vogel 1983).

Reproduction and Population Structure Litter size (births and embryo numbers): 2.6 (1–2), mode 2, n = 11 (Kuhn 1971, Vogel 1983, P. Vogel unpubl.). Gestation: >50 days. At birth, young naked, but whiskers present; dorsal pelage present Day 11; eyes open Day 23, first solid food eaten Day 40 (n = 1; P. Vogel unpubl.).

Predators, Parasites and Diseases Endoparasites: *Molineus eburneus* (Nematoda: Trichostrongyloides) recorded from the intestine (Vaucher *et al.* 1979). Ectoparasites: mites *Orycteroxenus liberiensis* and *O. minor* (Fain *et al.* 1974).

Conservation IUCN Category: Endangered.

Suitable habitat is threatened by mining and agricultural development and by increased human activities. Mining activities in Liberia have devastated large regions of suitable habitat (Nicoll & Rathbun 1990), and wetland rice agriculture introduced in 1980 to the Nimba region has resulted in large-scale habitat destruction (P. Vogel unpubl.).

Measurements

Micropotamogale lamottei
HB: 141 (120–155) mm, n = 15
T: 108 (95–134) mm, n = 15
HF: 20 (19–21) mm, n = 15
E: 10 (9–11) mm, n = 13
WT: 78 (32–95) g, n = 6
GLS: 35.9 (32.4–38.1) mm, n = 16
GWS: n. d.
I^1–M^3: 15.4 (14.2–16.2) mm, n = 16
Côte d'Ivoire (P. Vogel unpubl.)
Only adults with complete dentition

Key References Guth *et al.* 1959, 1960; Kuhn 1964; Vogel 1983.

Peter Vogel

Micropotamogale ruwenzorii Rwenzori Otter-shrew

Fr. Micropotamogale du Rwenzori; Ger. Mittel Otterspitzmaus

Micropotamogale ruwenzorii (de Witte and Frechkop, 1955). Bull. Inst. Roy. Sci. Nat. Belg. 31 (84): 9. W slopes of Mt Rwenzori, DR Congo. 1100–1200 m.

Taxonomy Originally described in the genus *Potamogale*, later assigned to a new genus *Mesopotamogale* now considered as a subgenus (Guth *et al.* 1960, Verheyen 1961d, Corbet 1974b). Presence of webbed feet (cf. *Potamogale velox*, *M. lamottei*) suggests that this species could represent a separate genus (P. Vogel, pers. obs.). Subspecies: none. Chromosome number: not known.

Description Small brown otter-like mammal, with a more compact body than the other species of otter-shrews. Dorsal pelage dark brown, woolly underfur grey with brown tips. Ventral pelage grey, pale brown or whitish. Head with a large flat muzzle and numerous stiff vibrissae. Rhinarium leathery. Eyes very small. Ears small but clearly visible, not concealed by pelage. Limbs short; fore- and hindfeet broad and bordered with stiff hairs, webs between digits. Syndactyly of Digits 2 and 3 of hindfeet form a comb. Tail (ca. 84% of HB) bicoloured, with many small scales and well covered with short hairs, dark brown (as dorsal pelage) above, pale brown or cinnamon below; in section, oval at its basal part, then circular towards tip; covered with sparse hairs, scales of the skin present. Tail is stronger than in *M. lamottei*. Skull: see genus profile. Nipples: uncertain; probably the same as *M. lamottei* (i.e. 2 + 1 + 1 = 8).

Geographic Variation None recorded.

Similar Species

Micropotamogale lamottei. Similar size; ventral pelage dark; feet not webbed; tail not laterally flattened; Liberia, Guinea and Côte d'Ivoire.

Potamogale velox. Larger (HB: 224–410 mm); ventral pelage white, feet not webbed; tail thick, laterally flattened.

Distribution Endemic to Africa. Rainforest BZ (East Central Region). Recorded only in Kivu and Rwenzori regions of E DR Congo and W Uganda (Rahm 1966).

Habitat Streams of varying size in Rainforest BZ from 800 to 900 m (n = 4) and in montane forest up to 2200 m (n = 4) (Rahm 1960). Also recorded from streams in gallery forest surrounded by secondary savanna (elephant grass) from 1000 to 1200 m.

Abundance Little quantitative information; locally not rare.

Micropotamogale ruwenzorii

Adaptations The webbed feet suggest a strong adaptation for aquatic life. Both fore- and hindfeet have webs between the digits. The forefeet are relatively large; Digit 1 and Digit 5 are elongated (Verheyen 1961a). Similar morphological adaptations are found in only a few other insectivorous small mammals such as *Limnogale mergulus* (Tenrecidae), *Nectogale elegans* (Soricidae) and *Galemys pyrenaica* (Talpidae) – all of which are separated from related species without webbed feet at the generic level. Swimming is by paddling with the fore- and hindlimbs (as in a dog). The feet may be used synchronously when diving (Rahm 1961) in a similar way to the Pyrenean Desman *Galemys pyrenaica* (Niethammer 1970). The tail does not appear to be used for propulsion. There are several periods of activity each night; grooming occurs after each activity period (Rahm 1961). The anatomy of the species has been studied by Verheyen (1961a, b, c, d).

Foraging and Food Carnivorous and insectivorous. Foraging occurs in water. The preferred prey are crabs with a carapace width <5 cm, aquatic insects, oligochaete worms and fish (*Clarias*, etc.). An otter-shew weighing 135 g eats ca. 80 g of food per night. Manipulation of food is performed only with the mouth, not with the forefeet. Faecal pellets (ca. 2–5 cm long, diameter 0.9 cm) contain fragments of chitin from the crab and insect prey (Rahm 1961).

Social and Reproductive Behaviour No information.

Reproduction and Population Structure Embryo number: 1 (n = 2) or 2 (n = 1) (Rahm 1960, H. Stephan pers. comm.). Young (HB: ca. 80 mm) recorded in Apr in Kivu region, DR Congo (n = 2, labels, RMCA).

Predators, Parasites and Diseases Parasitized by ticks (*Ixodes dawesi*), but only where syntopic with *Potamogale velox* (Arthur 1965, U. Rahm pers. comm.).

Conservation IUCN Category: Near Threatened.

May be threatened by destruction of suitable habitats. Previously assessed as Endangered.

Measurements

Micropotamogale ruwenzorii

HB: 163 (130–200) mm, n = 26

T: 136 (123–150) mm, n = 25

HF: 28 (26–29) mm, n = 10

E: 11 (10–14) mm, n = 12

WT: 98 (75–135) g, n = 10

GLS: 36.8 (33.8–39.4) mm, n = 4

GWS: 16.8 (15.5–17.8) mm, n = 4

I^1–M^3: 17.7, 18.5 mm, n = 2 (adults only)

Central Africa (Rahm 1960, 1961, 1966, U. Rahm unpubl., H. Stephan in litt.)

Key References Rahm 1960, 1961; Verheyen 1961a.

Peter Vogel

GENUS *Potamogale*
Giant Otter-shrew

Potamogale du Chaillu, 1860. Proc. Boston Soc. Nat. Hist. 7: 363. Type species: *Cynogale velox* du Chaillu, 1860.

Potamogale is a monotypic genus, widely distributed in the Rainforest BZ of central Africa. The genus is characterized by its adaptations for aquatic life, large size (mean HB: ca. 320 mm, mean weight ca. 650 g), tail thickened at the base and laterally flattened on the distal half, and without webs between the digits of the fore- and hindfeet. Skull: elongated and narrow, without zygoma; upper incisors canine-like; canines small, similar to premolars; premolars simple, dilambdodont (W-shaped pattern on occlusal surface) (Figure 5). The dilambdodont pattern of the molar teeth is a character interpreted as derived (see also family profile). Dental formula: I $^3/_3$, C $^1/_1$, P $^3/_3$, M $^3/_3$ = 40.

Peter Vogel

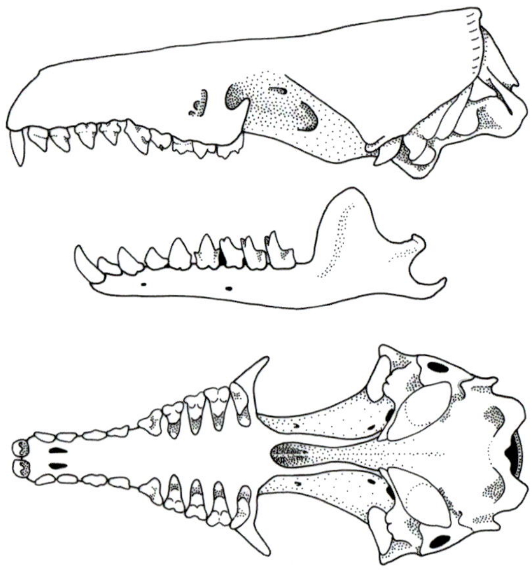

Figure 5. Skull and mandible of *Potamogale velox* (BMNH 26.11.1.64).

Potamogale velox.

Potamogale velox GIANT OTTER-SHREW

Fr. Potamogale; Ger. Große Otterspitzmaus

Potamogale velox du Chaillu, 1860. Proc. Boston Soc. Nat. Hist., 7: 363. Ogowe River, Gabon.

Taxonomy Synonyms: *allmeni*, *argens*. Subspecies: none. Chromosome number: not known.

Description Slim and elongated like an otter, with a flattened head and large muzzle with stiff white vibrissae. Pelage soft, dense and shiny, with thick underfur. Head and dorsal pelage chocolate- or tobacco-brown; hairs white at base, rich chocolate-brown at tip. Dense white underfur; white colour of hairs and underfur does not show through to surface of pelage. Flanks similar, with some white-tipped hairs. Ventral pelage pure white, underfur cream. Colour of dorsal and ventral pelage clearly delineated on lower flanks. Eyes small. Large leathery rhinarium. Ears small but proportionally longer than in otters. Mouth set back from tip of head. Fore- and hindfeet relatively short; five digits on each limb, without webs between digits. Syndactyly of Digits 2 and 3 of hindfeet form a comb. Postaxial border of hindfoot with a wide cutaneous fold. Tail relatively long (ca. 80% of HB), same colour as dorsal pelage; basal part of tail as thick as hind part of body, distal half laterally compressed with sharp dorsal keel. Dense long pelage of body extends on first quarter of the tail, remaining part densely covered with very short adpressed hairs. Skull: see genus profile. Nipples: 2 (pectoral) + 1 (abdominal) +1 (inguinal) = 8 (Dubost 1965).

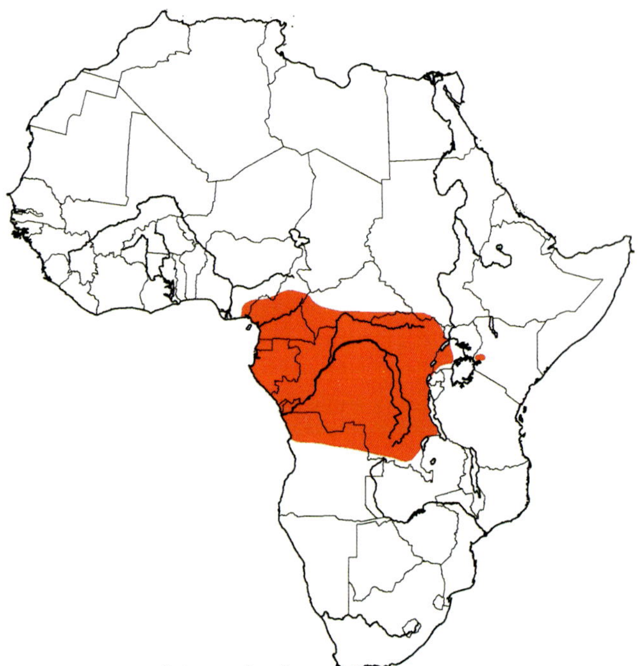

Potamogale velox

Geographic Variation None recorded.

Similar Species

Micropotamogale ruwenzorii. Much smaller; hindfeet with webs between digits; tail not flattened laterally; sympatric in E DR Congo and W Uganda.

Micropotamogale lamottei. Much smaller; hindfeet without webs between digits; tail circular in section; Liberia, Guinea and Côte d'Ivoire.

Distribution Endemic to Africa. Widespread in Rainforest BZ (except for Western Region and West Central Region [Western Nigeria sub-region]). Recorded from Nigeria (east of Cross R.) eastwards to W Uganda and W Kenya, and southwards to C Angola and N Zambia (Corbet 1974a). Sea level to 1800 m. Isolated record from rainforest at Kakamega, Kenya (00° 17' N, 34° 47' E).

Habitat Streams in equatorial rainforests. Lives along rather small slow-flowing forest streams, forest pools and mountain torrents where banks provide habitats for nest chambers (Nicoll & Rathbun 1990). Also along streams bordered by gallery forest. Normally does not occur in large rivers, although one was found in the Ivindo R., Gabon, where the river was several hundred metres wide.

Abundance Mean density in good habitats may be one territorial animal/0.5–1.0 km length of stream; maximum densities of about one individual/100 m of river (Dubost 1965).

Adaptations Very efficient swimmer. Swimming is effected by horizontal undulations of the thick muscular tail, as in fish and crocodiles. These undulations are produced by contractions of the gluteus muscle of the rump, which extends backwards into the tail (an exceptional feature for a tail). Forelimbs are not used for propulsion. This method of swimming is unique amongst aquatic mammals (cf. seals and whales swim by undulations in a vertical plane). Moving on land is rather clumsy. Body pelage is groomed with the comb-like structure formed by fusion of Digits 2 and 3 of the hindfeet. Nocturnal with several bouts of activity each night (Dubost 1965, Nicoll 1985); rests during daytime in burrow in river bank. When disturbed, escapes into the water. Movements over long distances always by water, never by land.

Foraging and Food Forages in water, feeding only on aquatic prey. Hunts by diving, each dive lasting for only several seconds. Prey located using the sensitive vibrissae and odour; eyes apparently not used to locate prey. Analysis of stomach contents and faeces show that diet is mainly fish, crabs, shrimps and water insects; frogs are

Potamogale velox.

rarely eaten (Dubost 1965). In captivity eats 15–20 crabs per night (Durrell 1953).

Social and Reproductive Behaviour Latrines probably used to mark boundaries of territory (Dubost 1965). The den, with the nest chamber, is entered from below or above water level and is usually placed under a tree.

Reproduction and Population Structure Reproduction during wet and dry seasons. Litter size: 1 (n = 3) or 2 (n = 2). Probably two litters/year (Dubost 1965).

Predators, Parasites and Diseases Seven species of Nematoda are known from the digestive system: *Molineus grassei* (Chabaud *et al.* 1966), *Galeiceps cucullus* (Chabaud *et al.* 1967), *Molineus amberti*, *M. brosseti*, *Parastrongyloides callipygus*, *Procamallanus potamogalei* and *Spinitectus menzalei* (Hugot 1979). *Ixodes dawesi*, a tick specific to the Potamogalinae, has been found in several geographic regions (Arthur 1965).

Conservation IUCN Category: Least Concern.

In Cameroon (and probably elsewhere) forest clearance and soil erosion cause murkiness and opaqueness of the water in forest streams, and has resulted in local extinction of some populations (Nicoll 1985). Previously assessed as Endangered.

Measurements

Potamogale velox
HB: 320 (305–337) mm, n = 10
T: 269 (235–290) mm, n = 10
HF: 43 (39–46) mm, n = 10
E: 21 (15–23) mm, n = 10
WT: 656 (517–780) g, n = 7
GLS: 63.7 (60.5–66.3) mm, n = 10
GWS: 26.2 (25.1–27.5) mm, n = 10
I^1–M^3: 30.6 (29.2–32.5) mm, n = 10
Measurements: DR Congo (Mede, Nipa and Kananga; AMNH)
Weight: Several sources

Key References Dubost 1965; Durrell 1953; Nicoll 1985.

Peter Vogel

Family CHRYSOCHLORIDAE

GOLDEN-MOLES

Chrysochloridae Gray, 1825. Ann. Philos., n.s. 10: 335.

Golden-moles are endemic to sub-Saharan Africa, where nine genera and 21 known species (Table 12) inhabit a wide spectrum of altitudinal, climatic and vegetational habitats from deserts to savanna grasslands and montane forests. Most species occur only in southern Africa and many have restricted geographic ranges; only three species occur elsewhere. All species are subterranean, but some forage in litter on the surface.

The family name Chrysochloridae refers not to the pelage colour, which varies from black to pale tawny-yellow, but to the iridescent sheen of coppery green, blue, purple or bronze of their pelage. Morphologically, golden-moles display a high degree of structural conformity comprising both primitive as well as highly specialized mammalian characters (Hickman 1990). Unique characters of chrysochlorids include a hyoid-dentary articulation, extremely enlarged malleus bones in the middle ear of some species that permit great sensitivity to underground vibrations (Mason & Narins 2001), a third bone in the forearm (possibly an ossified tendon), a reduction in the number of phalanges in the fore- and hindfeet, and muscle arrangements not paralleled in the Mammalia (MacPhee & Novacek 1993).

All species have compact fusiform or lozenge-shaped bodies, short and powerful forelimbs and no external tail, although internal caudal vertebrae are present. The dense pelage consists of backward-set guard hairs that are moisture-repellent, and woolly underfur for insulation. The skin is thick and tough, particularly on the head. Eyes are vestigial and covered by skin, and the optic nerve is degenerate. External ear pinnae are absent and ear openings are covered by dense hairs. The muzzle is obtuse and terminates in a leathery nosepad that protects the nostrils, and is used, together with the webbed hindfeet, to shovel soil. In some species, the nostrils also have foliaceous projections that prevent sand from entering the nasal passages during burrowing. The wedge-shaped head is used to push and pack soil loosened by the robust pick-like digging claws of the forelimbs, which are powered by extremely muscular forequarters. There are four digits on the forefoot and five on the hindfoot. The claw on Digit 3 of the forefoot is usually the largest, while those on Digits 1 and 2 are shorter and have more slender claws. Digit 4 is vestigial in all but one species, and bears a minute nail- or knob-like claw. The digits of the hindfeet are linked by small membranes, and each has only a small claw. Nipples: 2 + 2 = 4 in all species.

The skull is conical with a broad braincase, complete zygomatic arches and an elongated snout without any marked interorbital constriction. Cheekteeth are high-crowned and zalambdodont with three main cusps arranged in a triangle. Dental formula is $I\,^3/_3$, $C\,^1/_1$, $P\,^3/_3$, $M\,^{2-3}/_{2-3}$ = 36–40. The first premolars (P1) are either bicuspid and sectorial like the preceding teeth, tricuspid like the other molars, or pseudo-molariform (tricuspid but lacking a posterior crushing basin), but this varies considerably with toothwear. The lower molars of some taxa bear a talonid (low ledge), which occludes with the protocone cusp of the upper molars to provide a small grinding surface, and the cheekteeth interdigitate closely providing a shearing action to slice through prey. The third upper molars (M^3), when present, are either tricuspid (similar to other molars) or minute and peg-like, this to a large extent depending on toothwear (and thus age). The incidence of the M^3 or M_3 molars varies considerably in some taxa (Bronner 1995b). This reflects the almost random sequence of tooth replacement in chrysochlorids, rendering dental formula inadequate for diagnostic purposes. The length of the upper tooth row is measured from canine to upper second molar ($C–M^2$), or canine to upper third molar ($C–M^3$) for those species that retain M^3. In chrysochlorids, the suture between premaxilla and maxilla bone fuses early in life, and hence the canine tooth (which is similar in size to the incisor teeth) may be difficult to discern; it is the fourth tooth from the front of the skull (the first three are incisor teeth).

Several genera of chrysochlorids have an enlarged malleus located in a bony bulla that is often visible externally on the surface of the skull. In *Chrysochloris* and *Cryptochloris* the malleus is elongated and club-shaped, and the caput (head) is housed in a bulla evident as a pronounced temporal bulla (*Chrysochloris*), or smaller subtemporal bulla (*Cryptochloris*). *Eremitalpa* and *Chrysospalax* have an enlarged spherical ('pea-shaped') malleus housed in an anteriolateral expansion of the skull visible as a swelling in the region of the zygomatic arch, rather than a distinct bulla. In *Chlorotalpa* and *Carpitalpa*, the malleus is somewhat inflated and bulbous ('pear-shaped'), but is accommodated entirely within the skull so no external bulla is evident – a condition found also in *Amblysomus* and *Neamblysomus*, which have a small and unspecialized ('normal') malleus similar to those of most other mammals. These specializations of the middle ear are related to their extraordinary sensitivity to vibrations, which are used to detect prey and avoid predators. The enlarged malleus bones in *Eremitalpa* and *Chrysospalax*, which are the heaviest (in relation to body size) among all mammals, may be an adaptation for surface foraging since increased malleus mass enhances sensitivity to higher frequencies of sound than are normally experienced underground (Mayer *et al.* 1995).

Physiologically, chrysochlorids are unusual in mammals in displaying daily and seasonal torpor (Kuyper 1985) or almost complete poikilothermy. They have a high thermal conductance and do not thermoregulate when at rest, thereby considerably reducing energy requirements. Body temperature in the thermal neutral zone is lower than expected for similar-sized mammals, and they are moderate (*Chrysochloris asiatica*, *Amblysomus hottentotus*) to extreme (*Eremitalpa granti*) thermoconformers (Seymour *et al.*

Table 12. Characteristics of the species in the family Chrysochloridae. (n. d. = no data)

Species (subfamily)[a]	Length of claw on Digit 3 of forefoot (range) (mm)[c]	Basal width claw of Digit 3 of forefoot (range) (mm)[c]	HB mean (range) (mm)[c]	GLS mean (range) (mm)[c]	Colour of dorsal pelage	M^3 present (P) or absent (A)	Notes
Amblysomus corriae (A)	13–15	4.5–5.6	119 (108–130) 115 (109–130)	27.4 (25.4–28.9) 26.7 (25.3–28.6)	Smoky-black to reddish-brown	A	South Africa (Western and Eastern Cape Provinces)
Amblysomus hottentotus (A)	13–16	4.3–6.6	123 (114–135) 118 (104–135)	26.5 (25.0–27.7) 25.9 (24.6–27.7)	Chestnut-red to brownish-black	A (usually)	South Africa (coastal and parts of highveld), Swaziland
Amblysomus marleyi (A)	10.5–13	3.7–4.4	102 (96–105) 101 (90–120)	23.9 (23.1–24.6) 23.5 (22.4–24.7)	Dark reddish-brown	A (usually)	South Africa (KwaZulu–Natal)
Amblysomus robustus (A)	14–17	5.3–7.2	131 (118–143) 128 (109–138)	30.8 (29.9–32.0) 28.8 (27.2–29.7)	Dark reddish-brown	A (usually)	South Africa (Mpumalanga Province)
Amblysomus septentrionalis (A)	14–15.7	5.3–6.6	127 (114–135) 125 (105–145)	29.0 (27.2–30.5) 17.9 (16.8–19.5)	Dark reddish-brown	A	South Africa (Mpumalanga Province, ?NE Free State)
Calcochloris leucorhinus (A)	9–11.5	2.3–4	63, 126 (n=2)	21.7 (19.5–22.8) 22.2 (21.0–23.1)	Dark brown to slaty-grey	P (usually)	Angola, DR Congo, Cameroon, Central African Republic
Calcochloris obtusirostris (A)	7–9	<3.5	100 (88–110) 93 (82–110)	22.6 (20.8–24.2) 21.5 (19.0–23.6)	Yellowish-orange to dull reddish-brown	A	South Africa (KwaZulu–Natal), S Mozambique, SE Zimbabwe
Calcochloris tytonis (A)	n. d.	n. d.	n. d.	n. d.	n. d.	P	S Somalia
Carpitalpa arendsi (C)	9–12	3.5–4	124 (115–141) 121 (115–139)	27.7 (26.8–29.0) 27.3 (25.3–28.6)	Blackish-brown	P	E Zimbabwe (Inyanga Highlands)
Chlorotalpa duthieae (C)	7–9	3.5–4	106 (95–125) 102 (96–130)	23.8 (22.9–24.5) 22.8 (22.0–23.9	Reddish-brown	P	South Africa (Eastern and Western Cape Provinces)
Chlorotalpa sclateri (C)	8–9	3–4.5	101 (83–135) 102 (82–123)	23.6 (21.8–27.4) 23.4 (21.2–26.7)	Glossy reddish-brown	P	South Africa (widespread)
Chrysochloris asiatica (C)	10–12	3.5–4.6	105 (100–115) 100 (94–114)	22.9 (22.1–23.6) 22.5 (21.1–25.4)	Blackish to slate-grey	P (usually)	South Africa (western areas)
Chrysochloris stuhlmanni (C)	11	3–4.2	113 (106–123) 107 (103–111)	26.9 (24.8–28.1) 25.9 (24.3–28.3)	Blackish-brown	P	Uganda, Kenya, Tanzania, Cameroon (mostly montane)
Chrysochloris visagiei (C)	n. d.[b]	4.1	106 (n=1)	22.8 (n=1)	Tawny-olive	P	South Africa (Northern Cape Province – type locality only)
Chrysospalax trevelyani (C)	17–19	7	222 (215–235) 216 (208–229)	41.7 (41.0–42.4) 41.6 (39.9–43.2)	Glossy brown	P	South Africa (Eastern Cape Province)
Chrysospalax villosus (C)	15–17	4–6	163 (148–175) 143 (127–160)	33.9, 34.9 (n=2) 33.8 (33.4–34.6)	Yellowish-brown (variable)	P	South Africa (highland areas)
Cryptochloris wintoni (C)	10.3–10.5	4	86, 92 (n=2)	21.3, 22.0 (n=2)	Pale slate-grey	P	South Africa (Northern Cape Province)
Cryptochloris zyli (C)	10	4	82 (n=1)	21.7, 22.2 (n=2)	Lead-grey	P	South Africa (type locality only)
Eremitalpa granti (C)	9–12	3.7–4.9	82 (76–86)	19.5 (18.6–20.4) 16.6 (15.5–18.2)	Tawny-olive to greyish-olive	P (very small)	South Africa (Western Cape Province), S Namibia
Neamblysomus gunningi (A)	12.5–14	4.3–4.9	125 (121–132) 122 (111–132)	28.4 (27.1–29.3) 27.5 (27.0–27.8)	Reddish-brown	P (usually)	South Africa (Drakensberg Mts)
Neamblysomus julianae (A)	8.5–10	3.6–4.3	101 (94–111) 99 (92–111)	23.2 (22.4–23.6) 22.5 (21.8–23.5)	Cinnamon-brown to reddish-brown	A (usually)	South Africa (scattered)

[a] A = subfamily Amblysominae (malleus bone of middle ear typically mammalian in shape and not expanded). C = subfamily Chrysochlorinae (malleus bone of middle ear enlarged to form a spherical or club-like shape).

[b] Probably similiar to *C. asiatica*.

[c] Single line of numerals: sex not recorded. Two lines of numerals: upper line, males; lower line, females.

1998). The low metabolic rate of *C. asiatica* is achieved by lowering the body temperature, whereas in *E. granti* it is also the result of intrinsic metabolic depression. Lowered metabolism and efficient renal function effectively reduce water requirements to the extent that all species are independent of drinking water (Fielden *et al.* 1990a). Far from being 'primitive' characters, such physiological specializations allow chrysochlorids to survive in permanently or seasonally inclement habitats where temperatures are extreme and food is often scarce.

Limited reproductive data available suggest that most species breed throughout the year, perhaps with a peak in the wetter months when prey is more abundant, and may be polyoestrous (Bernard *et al.* 1994). Although these are r-selected characters, other facets of their breeding biology (small litter-sizes [mean = 2] and extended postnatal development) are more typical of a specialized, K-selected reproductive strategy.

Like most subterranean mammals, chrysochlorids are ecological specialists with low vagility. The limited data suggest that all species are opportunistic foragers feeding primarily on earthworms, termites, insect pupae and/or millipedes, although a variety of other prey items, including small vertebrates, are sometimes eaten. Most are solitary and subterranean, constructing semi-permanent burrow systems comprising two tiers: subsurface tunnels (visible from above as raised ridges of soil) for foraging, and deeper tunnels connecting chambers used for resting and raising young, the entrances to which are sometimes marked by small mounds on the ground surface (Roberts 1951, Kuyper 1985). An exception to this is the 'Shark of the Dunes' – the small Namib Golden-mole *Eremitalpa granti namibensis* which 'swims' through the desert sands, often emerging onto the surface to reduce the energetic costs of subterranean travel (Narins *et al.* 1998). The Giant Golden-mole *Chrysospalax trevelyani* is also primarily a surface forager, but in indigenous forests, and is the only species that shows any indication of sociality.

Unlike other mammals, golden-moles have a southern African centre of diversity with only three species occurring outside this region: *Calcochloris leucorhinus* and *Chrysochloris stuhlmanni* in the Rainforest BZ of central Africa; and *Calcochloris tytonis*, in the Somalia–Masai Bushland BZ. The 18 species in southern Africa fall into two broad ecological groups: an arid-adapted group of five species which occur in semi-desert (*Eremitalpa* [1 sp.], *Cryptochloris* [2 spp.]), karoo (*Chrysochloris visagiei*) or moister fynbos habitats (*Chrysochloris asiatica*) along the south-west coast; and a mesic-adapted group of 13 species inhabiting forests, savanna woodland and temperate grassland habitats in the eastern part of the sub-region (*Chrysospalax* [2 spp.], *Chlorotalpa* [2 spp.] *Calcochloris* [1 sp.], *Carpitalpa* [1 sp.], *Neamblysomus* [2 spp.] and *Amblysomus* [5 spp.]). Only two species (*Amblysomus hottentotus* and *Chrysochloris asiatica*) are widespread and show marked geographic variation in size and colour. Populations are often restricted to patches of suitable habitat with friable soils and abundant invertebrate prey, so that the distribution of demes is clumped, even within widespread species. Different species seldom coexist and compete for resources, even though their distributions may appear broadly sympatric. Where two species occur in the same area, they tend to occupy different micro-habitats, probably as a result of ecological displacement. Many of the genera and species in the family are of conservation concern: 11 species (out of 21) and eight genera (out of nine) are placed in various 'Threatened' categories of IUCN because of their rarity, small geographic range and the destruction of their habitats.

Many anatomical specializations shown by extant taxa are evident also in fossils from the early Miocene, and are so numerous and unusual that golden-moles have been described as 'spectacularly autapomorphic' and worthy of ordinal (Chrysochloridea *sensu* Broom 1916) or subordinal (Chrysochloromorpha *sensu* MacPhee & Novacek 1993) status. These proposals have received little attention, however, and until recently chrysochlorids were generally regarded to be only specialized members of the order Insectivora, with possible affinities to the tenrecs of Madagascar based on their joint possession of zalambododont cheekteeth. However, contemporary studies of mitochondrial and nuclear gene sequences have conclusively demonstrated that chrysochlorids and tenrecs form a clade worthy of ordinal status (Liu *et al.* 2001, Madsen *et al.* 2001, Murphy *et al.* 2001a). Stanhope *et al.* (1998) named this clade the Afrosoricida (African shrew-like mammals), a somewhat inappropriate name since it contains no soricids and could engender confusion with *Afrosorex* (a subgenus of *Crocidura* [see Hutterer 2005]). Divergence between golden-moles and tenrecs probably occurred about 50 mya, with the result that they share few morphological synapomorphies, warranting their allocation to distinct suborders (for which the names Tenrecomorpha and Chrysochloridea should be used).

Many previous taxonomic revisions of chrysochlorids were largely intuitive, or based on only elementary statistical evaluation of few specimens, leading to conflicting classifications that obscured rather than resolved inter-specific relationships. Based on phylogenetic analyses of morphometric and cytogenetic variation in three genera, Bronner (1995b) proposed a new classification including a newly described species (Bronner 2000, see also Bronner & Jenkins 2005). Following Simonetta (1968), two subfamilies may be recognized, albeit with differing species allocations: Chrysochlorinae, in which the malleus bone of the middle ear is enlarged with a spherical or club-like shape (*Carpitalpa* [1 sp.], *Chlorotalpa* [2 spp.], *Chrysospalax* [2 spp.], *Chrysochloris* [3 spp.], *Cryptochloris* [2 spp.] and *Eremitalpa* [1 spp.]) and Amblysominae in which the malleus is not expanded and has the typical mammalian shape (*Amblysomus* [5 spp.], *Neamblysomus* [2 spp.] and *Calcochloris* [3 spp.]). Here the genera, and species within each genus, are listed alphabetically (see also Table 12).

In the Descriptions, an indication of the size of each species is recorded comparatively to other species in the family. The size categories, based on mean head and body length (HB), are: Very small: <92 mm; Small: 100–115 mm; Medium-sized: 116–125 mm; Large: ca. 150 mm; Very large: >500 mm.

Gary N. Bronner

GENUS *Amblysomus*
Golden-moles

Amblysomus Pomel, 1848. Arch. Sci. Phys. Nat. Geneve 9: 247. Type species: *Chrysochloris hottentotus* A. Smith, 1829.

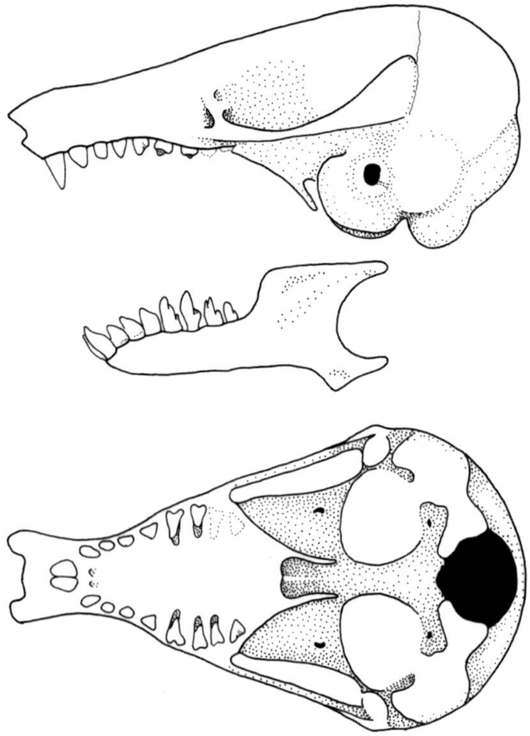

Figure 6. Skull and mandible of *Amblysomus corriae* (BMNH 5.6.5.6).

The most diverse genus of the Chrysochloridae, with five mesic-adapted species endemic to the south-eastern parts of South Africa. Species in the genus occur in habitats ranging from fynbos and coastal afromontane forests of the South-West Cape BZ to grasslands, woodlands and montane scrub in southern African savanna. One species, *A. hottentotus*, is widespread in southern Africa, and the others have restricted geographic distributions. Some previous authors synonymized *Chlorotalpa*, *Neamblysomus* and/or *Calcochloris* under this genus (Roberts 1951; Simonetta 1968; Petter 1981), but phylogenetic analyses of morphometric and cytogenetic data confirm that these are distinct genera (Bronner 1995a, b).

Body fusiform; size ranges considerably from small to medium-sized. Skull elongated (breadth <70% of GLS), zygomatic arches without broad plates sweeping backwards, the malleus of the inner ear relatively small and unspecialized, and the palate relatively narrower (<29% of GLS) compared with that in *Neamblysomus*. M^3 usually absent, molar talonids are well-developed and the first premolars are sectorial (bicuspid) rather than molariform (tricuspid) (Figure 6). The five species are distinguished by body size, colour of pelage, form and size of claw on Digit 3 of forefoot and chromosomal properties (Bronner 1995a). Multivariate analyses of skull measurements may be necessary to discriminate between *A. hottentotus*, *A. septentrionalis* and *A. robustus*.

Gary N. Bronner

Amblysomus corriae FYNBOS GOLDEN-MOLE
Fr. Taupe dorée de fynbos; Ger. Fynbos-Goldmull

Amblysomus corriae Thomas, 1905. Abstracts, Proceedings of the Zoological Society of London 20: 5; Proceedings of the Zoological Society of London 2: 57. Knysna, Western Cape, South Africa.

Taxonomy Traditionally grouped with populations from Zululand in *A. iris*. However, Bronner (1996) showed that the Zululand populations represent only a subspecies (*iris*) of the widespread Hottentot Golden-mole, *A. hottentotus*. The Western Cape populations represent the distinct species *A. corriae*, which includes two subspecies: *A. c. corriae* and *A. c. devilliersi*, which was formerly recognized as a subspecies of *A. hottentotus*. Although distinguishing between these forms using live specimens is difficult, they can be diagnosed with 95% accuracy using a craniometric technique (Bronner 1996). Since *A. corriae* is now considered as a distinct species, the colloquial name 'Zulu Golden-mole' is no longer suitable. Roberts (1951) referred to *A. corriae* as the 'Knysna golden mole', but Broom (1907a) used this name for *Chlorotalpa duthieae*. *Amblysomus corriae* also occurs far from Knysna, and hence this name is not appropriate, and should be replaced with one that more accurately reflects the biogeographic affinity of the species with the fynbos (South-West Cape BZ). Synonyms: *devilliersi*, *swellendamensis*. Subspecies: two. Chromosome number: 2n = 30, FN = 56 (Bronner 1995a).

Description Medium-sized species, similar to *A. hottentotus* in general appearance but with darker colouration. Dorsal pelage smoky-black to reddish-brown. Ventral pelage brownish-grey to pale orange. Hairs 8–9 mm long, dark slaty-grey at base with brownish-grey subterminal bands and iridescent blackish-brown or reddish-brown tips. Muzzle and cheeks somewhat paler than back, sometimes with small buffy eye-spots. Foreclaws marginally more slender than in *A. hottentotus*; claw on Digit 3 largest (13–15 mm long, 4.5–5.6 mm wide at base). Skull: malleus not enlarged; bulla not externally evident; M^3 absent (see also family and and genus profiles). Males on average slightly larger and heavier than ♀♀, but sexes do not differ markedly in cranial proportions.

Geographic Variation

A. c. corriae: Coastal plain below the Outeniqua, Kouga and Baviaanskloof mountain ranges, from George in Western Cape Province northwards along coast to Humansdorp in Eastern Cape Province. Dorsal pelage uniformly smoky-black; ventral pelage

pale brownish-grey without conspicuous reddish-brown tinge seen on flanks and abdomen of *A. hottentotus*.

A. c. devilliersi: Stellenbosch and Paarl in Western Cape Province, eastwards to slopes of Langeberg mountains near Riversdale, and northwards to Hawequas Forest near Worcester, but not into the arid Karoo. Dorsal pelage reddish-black, redder on flanks; ventral pelage reddish-brown to pale orange as in *A. hottentotus*. Considerable variability exists in extent of reddish-brown on dorsal pelage. Some individuals have subtle rufous tinge on flanks, whereas others have reddish-brown colouration extending almost to mid-dorsal region so that blackish colouration is restricted to narrow band along the back. This variability may be related to the stage of moulting.

Similar Species

Chlorotalpa duthieae. Skull smaller (GLS: 22.9–24.5 mm); claw on digit 3: 7–9 mm; M^3 present.

Chrysochloris asiatica. Often smaller (HB: 94–115 mm); skull smaller (GLS: 21.1–25.4 mm); pelage less rufous; M^3 present; malleus enlarged and club-shaped; bulla pronounced and projecting through temporal surface of skull.

Distribution Endemic to Africa. South-West Cape BZ. Recorded from Boland region of Western Cape Province along the coast to Humansdorp in Eastern Cape Province, where it is replaced by *A. h. hottentotus*.

Habitat Sandy soils and soft loams in Mountain Fynbos, Grassy Fynbos and Renosterveld of South-West Cape BZ. Also in afromontane forest and southern African savanna along the southern Cape coast. Coexists with *Chrysochloris asiatica* in Stellenbosch, but favours richer and wetter soils (Broom 1907a). Exists in broad sympatry with *Chlorotalpa duthieae* along the southern Cape coast, but trapping data suggest they select different micro-habitats, *A. corriae* preferring forest fringes and associated fynbos, and *C. duthieae* preferring deeper forest. Thrives in gardens, cultivated lands and livestock paddocks. Present also in exotic plantations, but at lower densities.

Abundance Locally common, but no quantitative data.

Adaptations See profile Family Chrysochloridae.

Foraging and Food Insectivorous. Stomach contents of three individuals comprised only small earthworms. Forages in subsurface tunnels, mainly at night. Sometimes emerges to capture prey on the soil surface, but drags it below ground before eating, often making soft 'chirruping' vocalizations.

Social and Reproductive Behaviour Solitary, except for ♀♀ raising young.

Reproduction and Population Structure Pregnant ♀♀, each with two embryos (1L, 1R) have been captured in Aug, May and Dec, suggesting aseasonal polyoestry. Young altricial and hairless at birth. Gestation period unknown.

Predators, Parasites and Diseases Regularly caught by Barn Owls *Tyto alba* and by domestic cats and dogs. No information on parasites and diseases.

Conservation IUCN Category: Near Threatened.

Conserved in many protected areas in South Africa.

Measurements

Amblysomus corriae
HB (♂♂): 119.7 (108–130) mm, n = 19
HB (♀♀): 115.4 (109–130) mm, n = 24
T: 0 mm
HF (♂♂): 14.7 (12–16) mm, n = 19
HF (♀♀): 13.9 (12–15) mm, n = 24
E: 0 mm
WT (♂♂): 56 (46–62) g, n = 10
WT (♀♀): 47 (46–64) g, n = 13
GLS (♂♂): 27.4 (25.4–28.9) mm, n = 20
GLS (♀♀): 26.7 (25.3–28.6) mm, n = 25
GWS (♂♂): 15.9 (15.1–16.9) mm, n = 20
GWS (♀♀): 15.5 (14.7–16.9) mm, n = 25
$C–M^2$ (♂♂): 6.3 (5.6–6.8) mm, n = 17
$C–M^2$ (♀♀): 6.2 (5.6–7.0) mm, n = 24
A. c. corriae, Knysna/Humansdorp, South Africa (Bronner 1995b)

Key References Bronner 1996; Broom 1907a; Roberts 1951.

Gary N. Bronner

Amblysomus hottentotus HOTTENTOT GOLDEN-MOLE

Fr. Taupe dorée de Hottentote; Ger. Hottentotten-Goldmull

Amblysomus hottentotus (A. Smith, 1829). Zoological Journal 4: 436. 'Interior parts of South Africa' (taken as Grahamstown, Eastern Cape Province, South Africa, following Roberts 1951).

Taxonomy Originally described in the genus *Chrysochloris*. Traditionally taken to include populations that Bronner (1996, 2000) recognized as valid species, namely *A. septentrionalis*, *A. robustus*, *A. marleyi* and *A. corriae* (in part). Synonyms: *affini*, *albifrons*, *albirostris* (*nomen dubium* of Wagner, 1841 and of Broom, 1908), *holosericea*, *iris*, *littoralis*, *longiceps*, *meesteri*, *natalensis*, *pondoliae*, *rutilans*. Subspecies: five. Chromosome number: 2n = 30, FN = 56 (Bronner 1995a).

Description Medium-sized species with considerable geographic variation in size, and marked variability in pelage colour both within and between populations. Dorsal pelage chestnut-red, brownish-black or reddish-black, becoming more rufous on flanks. Ventral pelage bright brown, reddish-brown or dull orange, sometimes with a dark brown mid-ventral line from throat to belly. Extent of reddish tinge is highly variable within populations. Muzzle paler than dorsal pelage with buffy-white to yellowish-orange patches on cheeks, extending laterally to above the subdermal eyes, and often coalescing behind the nosepad. Dorsal surface of hindfeet blackish to light buffy-rufous. Albinistic specimens with dorsal and ventral pelage cinnamon-brown to pale yellow (rarely recorded). Claws of forefeet moderately developed, claw on Digit 3 largest (13–16 mm long and 4.3–6.6 mm at base), claw on Digit 2 much shorter (6–8 mm), claw on Digit 1 about half as long as that on Digit 2. Digit 4 vestigial with a knob-like claw. Skull: malleus not enlarged; bulla not externally evident; zygomatic arches without broad plates sweeping backwards. M^3 are present in less than 2% of individuals and their occurrence often varies between jaw quadrants. If present, they are small and their morphology ranges from triconid (molariform) to peg-like depending on toothwear. Molar talonids well developed, except in specimens from southern Mpumalanga Province. Dental variability is random and localized rather than geographic by nature, and may reflect the irregular sequence of tooth replacement in chrysochlorids (Roberts 1951, Bronner 1996, 2000). Sexual size dimorphism pronounced, ♂♂ larger on average than ♀♀ in external and cranial measurements. Skull: see family and/or genus profiles.

Geographic Variation Size varies clinally in relation to altitude and ambient temperature in the inland part of the geographic range. Five subspecies recognized, but discrimination difficult owing to intergradations in size (Bronner 1996).

A. h. hottentotus: Eastern Cape Province from Van Stadens R. in the south to Great Fish River Valley in north; inland to Elandsberge, Winterhoekberge, Suurberge and Winterberge mountains. Intermediate in size between *A. h. longiceps* and *A. h. iris*; interpterygoid region narrower (<28% of greatest palate width) than in *A. h. pondoliae* (>29%).

A. h. pondoliae: Great Fish River Valley in Eastern Cape Province northwards along KwaZulu–Natal coast to Umdhloti R. near Durban; inland to Drakensberg foothills near Elliot and Maclear, where it is replaced by *A. h. longiceps*. Second largest subspecies.

A. h. iris: Zululand, from Umdloti R. in the south (where it intergrades with *A. h. pondoliae* in the vicinity of Verulam) north-eastwards to L. St Lucia and Cape Vidal (where it is replaced by *Calcochloris obtusirostris*). Pelage darker, reddish-black to dark brown, with a rufous tinge present variably on flanks and ventral pelage. Smallest subspecies with a length of 107–125 mm. Claws gracile, basal width of foreclaw of Digit 3 <5.5 mm, as opposed to >6 mm in *A. h. longiceps*. Mandibular ramus narrower (<60% of palate width) than in *A. h. pondoliae* (60–64%).

A. h. longiceps: Interior of KwaZulu–Natal Province, from Underberg in the south to Van Reenen in the north, and marginally into the Free State to Harrismith and Clarens. Recorded from W Swaziland, but these specimens may be *A. septentrionalis*. Largest subspecies (total length 120–141 mm), similar in size to *A. septentrionalis* and *A. robustus*, from which it can be distinguished craniometrically (Bronner 2000). Claws robust, foreclaw of Digit 3 >6 mm at base, in contrast to <5.5 mm in *A. h. pondoliae*. Mandibular ramus wider (>65% of palate width) than in *A. h. pondoliae*.

A. h. meesteri: Graskop and Mariepskop districts of northern Drakensberg escarpment, southwards to White R. Similar in size to *A. h. pondoliae*, but distinctive in that pelage is reddish-brown with a mid-dorsal band of reddish-black. Molar talonids absent or only weakly developed.

Similar Species

A. septentrionalis and *A. robustus*. Slightly larger; not separable from *A. hottentotus* using any single skull measurement but can be distinguished using a multivariate craniometric technique (Bronner 1996).

A. marleyi. Claw on Digit 3 of forefoot usually shorter (10.5–13 mm) and narrower (3.7–4.4 mm at base). Skull almost always shorter (23.1–24.7 mm).

Chlorotalpa sclateri. On average smaller (HB: 102 [82–135] mm); claw on Digit 3 of forefoot 8–9 mm long; M^3 present.

Chrysospalax trevelyani. Much larger (HB: 208–235 mm, GLS: 39.9–43.2 mm); pelage coarser; malleus enlarged; zygomatic arches with broad plates sweeping backwards.

Chrysospalax villosus. Usually larger (HB: 127–175 mm); skull larger (GLS: 33.4–34.9 mm), pelage coarser, malleus enlarged; zygomatic arches with broad plates sweeping backwards.

Neamblysomus julianae. Usually smaller (HB: 92–111 mm); skull smaller (GLS: 21.8–23.60 mm); pelage less red; subtle chromosomal differences (Bronner 1995a).

Distribution Endemic to Africa. Coastal Forest Mosaic BZ and parts of Highveld BZ. Recorded from Swaziland and South Africa (Eastern Cape and KwaZulu–Natal Provinces, and S Mpumalanga Province below the Drakensberg escarpment, with a marginal intrusion into NE Free State). Marginal intrusion into the South-West Cape BZ (Fynbos and Karoo biomes) in the southern parts of range.

Habitat Wide variety of woodland and grassland habitats, and also in afromontane forests and marshes.

Abundance Locally common in suitable habitats, as well as in gardens, golf courses and cultivated lands. Densities of up to 25 individuals/ha recorded in optimal habitat. Particularly abundant in moist soils near rivers and dams, but also found far from water provided the substrate is friable and not too rocky.

Adaptations Individuals construct two-tier burrow systems: subsurface tunnels used for foraging, and deeper (10–50 cm) tunnels connecting grass- or leaf-lined nests in spherical (15–20 cm diameter) chambers, with two or three spiralling boltholes (29–94 cm deep). Mounds are sometimes thrown up at the junction of the subsurface and deeper tunnel systems. Deeper tunnels are excavated at a rate of up to 9 m daily, and are also used for foraging in dry winter months when invertebrate prey is scarce near the soil surface. Subsurface tunnels are 9.5–240 m long, and may be extended by 4–12 m/day depending on prey abundance and substrate friability, and are used predominantly for foraging during the wet summer months. Burrow systems tend to be more extensive in less fertile habitats.

Activity is for periods of up to two hours by day and night, interspersed with periods of inactivity lasting 3–5 hours. Peak activity is at dawn, sunset and around midnight. Soil temperatures in burrow systems range from 0.8 to 32 °C. Optimal thermal conditions for activity are 23–33 °C, which corresponds with the thermoneutral zone. Individuals enter torpor if ambient temperature drops below 15 °C, or exceeds 30 °C, resulting in a markedly reduced metabolic rate and significant energetic saving. They coexist with Common Mole-rats *Cryptomys hottentotus*, which they tolerate in their burrows, presumably since they do not compete for food; joint occupancy reduces the energetic costs of burrow maintenance. Other species occupying active burrow systems include the Red Toad *Bufo careens*, Wolf-spider *Harpactira* sp., crickets, isopods, scarabid beetles and annelids (Kuyper 1979, 1985, Hickman 1990).

Foraging and Food Carnivorous and insectivorous. The diet is soil-dwelling invertebrates, particularly earthworms. In KwaZulu–Natal, 97% of stomachs contained only earthworms and minute (<0.5%) finely ground vegetable matter, while the contents of the remaining 3% included up to 5% insect remains, especially scarabid beetle larvae (McConnell 1986). Diet sometimes includes slugs, snails, crickets, insect pupae and spiders (Kuyper 1985). Individuals consume up to 90% of their body weight daily (McConnell 1986), but this estimate, based on captive animals, seems extreme and is temperature dependent. In captivity, they have been observed to eat garlic and geophytes, but this was probably coincidental since their digestive tracts are not specialized for herbivory.

Social and Reproductive Behaviour Adults are solitary and fight aggressively with individuals of either sex by wrestling their opponent using the foreclaws, and biting at the abdomen. Fighting is often accompanied by high-pitched squeaks. Individuals actively defend their burrow systems. Some amount of home-range overlap is tolerated, and dominant moles sometimes take over neighbouring burrow systems to increase their home-range. Courtship involves chirruping vocalizations, head-bobbing and foot stamping by ♂, and rasping and squealing vocalizations by ♀ (Kuyper 1985).

Reproduction and Population Structure Reproduction and recruitment of young occur throughout the year, but at low levels, with a peak in the wet summer months. Females show aseasonal polyoestry and a postpartum oestrus, and spermatogenesis in ♂♂ is continuous, reflecting the relatively constant microclimate of their subterranean niche (Bernard *et al.* 1994). Litter-size: 1–3 (mode 2). At birth, young hairless, HB 47 mm, weight ca. 4.5 g. Short, dark-brownish dorsal pelage is evident when 17 g and HB 72 mm, but ventral surface remains naked and teeth unerupted. Pale ventral pelage first appears when HB 80 mm and weight ca. 27 g. Eviction from the maternal burrow occurs when 35–40 g (Kuyper 1979, 1985).

Populations show low levels of genetic diversity, as reported also for other subterranean mammals; this may reflect selection pressures for homozygosity in the largely homogeneous underground environment (Filipucci *et al.* 1991).

Predators, Parasites and Diseases Predators include snakes, owls, white storks, otters, genets, mongooses and jackals. Often also killed by domestic cats and dogs, but seldom consumed. No information on parasites or diseases.

Conservation IUCN Category: Least Concern.

Measurements

Amblysomus hottentotus
HB (♂♂): 123.6 (114–135) mm, n = 20
HB (♀♀): 118.6 (104–135) mm, n = 21
T: 0 mm
HF (♂♂): 14.4 (13–19) mm, n = 21
HF (♀♀): 14.3 (12–18) mm, n = 20
E: 0 mm
WT (♂♂): 56.6 (37–85) g, n = 15
WT (♀♀): 50.1 (37–60) g, n = 19

GLS (♂♂): 26.5 (25.0–27.7) mm, n = 22
GLS (♀♀): 25.9 (24.6–27.7) mm, n = 24
GWS (♂♂): 16.0 (15.1–17.4) mm, n = 24
GWS (♀♀): 17.1 (16.1–19.0) mm, n = 25
C–M^2 (♂♂): 6.7 (6.2–7.2) mm, n = 19
C–M^2 (♀♀): 6.5 (6.0–7.1) mm, n = 24
South Africa (G. Bronner unpubl.)

Key References Bronner 1996; Hickman 1990; Kuyper 1985.

Gary N. Bronner

Amblysomus marleyi MARLEY'S GOLDEN-MOLE

Fr. Taupe dorée de Marley; Ger. Marleys Goldmull

Amblysomus marleyi Roberts, 1931. Ann. Transvaal Mus. 14: 225. Ubombo, KwaZulu–Natal, South Africa.

Taxonomy Traditionally recognized as a subspecies of *A. hottentotus*; now elevated to species rank based on its smaller size, more gracile build and different pelage colour (Bronner 1996). While generally smaller than *A. hottentotus* from further south along the coast, discriminating between these species using external measurements is difficult because of an overlap in size. Multivariate analyses of morphometric data provides unequivocal diagnosis (Bronner 1996). Synonyms: none. Chromosome number: 2n = 30, FN = 56 (Bronner 1995a).

Description Small golden-mole with a delicate build and fine claws on the forefeet. Dorsal pelage very dark reddish-brown. Ventral pelage dullish-orange to greyish-brown. Muzzle and cheeks paler than dorsal pelage often with broad white band behind nosepad that extends to region above the eyes. Throat paler and redder than flanks (Roberts 1931). Guard hairs dark grey distally and yellowish at base, except for uniformly off-white hairs on muzzle. Claw on Digit 3 of forefoot largest (10.5–13 mm long, 3.7–4.4 mm wide at base). Skull: malleus not enlarged; bulla not externally evident; zygomatic arches without broad plates sweeping backwards, M^3 usually absent (see also family and genus profiles). Males on average slightly larger than ♀♀; sexual size dimorphism does not appear to be pronounced, but this may be an artefact of small sample sizes.

Amblysomus marleyi

Geographic Variation None recorded.

Similar Species None within the geographic distribution of the species.

Distribution Endemic to Africa. Coastal Forest Mosaic BZ. Known only from two isolated localities (Ubombo and Ingwavuma) on eastern slopes of Lebombo Mts in KwaZulu–Natal Province. Owl pellet remains from near Weenen, ca. 250 km SW of type locality, appear to represent this species, suggesting that it may be more widespread than currently recognized.

Habitat Moist grasslands and indigenous forests in Natal Lowveld, Bushveld and Lebombo Arid-Mountain Bushveld of southern African savanna. Does not occur below the mountains on the Mozambique plain, where it is replaced by the Yellow Golden-mole *Calcochloris obtusirostris*. Also occurs in gardens.

Abundance Common in localized patches of suitable habitat with soft soils; cryptic and rarely encountered. No quantitative data on abundance.

Adaptations Like other members of this genus, constructs two types of burrows: subsurface tunnels used for foraging and deeper tunnels leading to nesting chambers under rocks and roots of trees. The entrances to deep tunnels are sometimes marked by small mounds of soil.

Remarks Insectivorous, but no quantitative analysis of diet. Solitary.

Predators, Parasites and Diseases No information.

Conservation IUCN Category: Endangered.

Given the restricted range of the species, the fragmented nature of populations and adverse impacts of humans on its habitat, this species is appropriately classified as Endangered.

Measurements

Amblysomus marleyi
HB (♂♂): 102.0 (96–105) mm, n = 4
HB (♀♀): 100.6 (90–120) mm, n = 9
T: 0 mm
HF (♂♂): 12.3 (11–13) mm, n = 4
HF (♀♀): 11.4 (10–13) mm, n = 9

E: 0 mm
WT (♂): n. d.
WT (♀): n. d.
GLS (♂♂): 23.9 (23.1–24.6) mm, n = 4
GLS (♀♀): 23.5 (22.4–24.7) mm, n = 8
GWS (♂♂): 15.3 (14.8–16.0) mm, n = 4
GWS (♀♀): 14.6 (13.9–16.0) mm, n = 8
C–M^2 (♂♂): 5.8 (5.6–6.2) mm, n = 5
C–M^2 (♀♀): 5.6 (5.3–6.1) mm, n = 7
Ubombo/Ingwavuma, KwaZulu–Natal, South Africa.

Key References Bronner 1996; Roberts 1951.

Gary N. Bronner

Amblysomus robustus Robust Golden-mole

Fr. Taupe dorée robuste; Ger. Robuster Goldmull

Amblysomus robustus Bronner, 2000. Mammalia 64: 42. Verloren-Vallei Nature Reserve, Dullstroom, Mpumalanga Province, South Africa.

Taxonomy A monotypic species, with a unique chromosome number amongst chrysochlorids (Bronner 1995a). Easily confused with *A. septentrionalis* (2n = 34; FN = 62) and *A. hottentotus longiceps* (2n = 30; FN = 56), which it resembles in overall size and pelage colour (Bronner 1996), but can be diagnosed craniometrically using a technique outlined in Bronner (2000). Synonyms: none. Chromosome number: 2n = 36, FN = 68.

Description Medium-sized golden-mole, and largest species in genus. Dorsal pelage dark reddish-brown with bright reddish-brown flanks. Ventral pelage orange. Yellowish-orange cheek patches extend laterally to above sub-dermal eyes. Claws on forefeet strongly developed, claw on Digit 3 largest (14–17 mm long and 5.3–7.2 mm wide at base). Juveniles resemble adults but dorsal pelage is greyish-brown, becoming dullish-orange on the flanks. Sexual size dimorphism pronounced, ♂♂ on average larger than ♀♀ in most measurements. Skull: see family and/or genus profiles.

Geographic Variation None recorded.

Similar Species

A. septentrionalis and *A. hottentotus*. On average slightly smaller; not separable from *A. robustus* using any single skull measurement but can be distinguished using a multivariate craniometric technique (Bronner 1996).

Chrysospalax villosus. On average larger (HB: 151 [127–175] mm); skull larger (GLS: 33.4–34.9 mm); pelage finer; malleus enlarged; zygomatic arches with broad plates sweeping backwards.

Neamblysomus julianae. Usually smaller (HB: 92–111 mm); skull smaller (GLS: 21.8–23.60 mm); pelage less red; subtle chromosomal differences (Bronner 1995b).

Distribution Endemic to Africa. Highveld BZ. Recorded only from the Steenkampsberge mountains in the Belfast and Dullstroom districts of E Mpumalanga Province, South Africa.

Habitat Montane grasslands and marshes in Moist Sandy Highveld Grassland (Brendenkamp & van Rooyen 1996). Prefers friable soils, from sands to quite heavy clays. Avoids shallow substrates along rocky ridges (which may act as dispersal barriers) and waterlogged areas.

Abundance Common in suitable natural habitats, also in gardens, orchards and cultivated lands.

Adaptations These golden-moles are robust, with strong claws that enable them to subsist in the heavier soils common on the highveld. In this respect they differ from most other species of *Amblysomus*, which are restricted to sands and soft loams at lower altitudes. They are active mainly in late afternoon and night, and are torpid during the heat of day. Burrow systems are two-tiered comprising subsurface tunnels for foraging and deeper tunnels (20–80 cm) with nest chambers. Burrows sometimes extend upwards into rotting trees, which may provide an abundant source of invertebrate prey and refuge in areas where burrows become waterlogged during heavy rains. Subsurface activity declines during the dry winter season when prey is scarce, and individuals may enter prolonged torpor spanning several days or confine their foraging to deeper tunnels.

Foraging and Food Insectivorous, with a preference for earthworms (n = 2). Golden-moles forage for soil-dwelling invertebrates in subsurface tunnels, mainly at night. In captivity, they consume a wide variety of living invertebrate and vertebrate prey, such as mealworms, kingworms, earthworms and baby mice, but ignore dead prey.

Amblysomus robustus

Social and Reproductive Behaviour No information, except that adults are usually solitary. Individuals confined together fight vigorously, regardless of sex, but this may also be due to stress while in captivity.

Reproduction and Population Structure Pregnancies and lactation recorded in late summer to early autumn (Mar and Apr) suggest that breeding occurs primarily during the wet summer season.

Predators, Parasites and Diseases No information.

Conservation IUCN Category: Vulnerable.

Should be classified as Data Deficient. While the geographic range of the species is limited, and the vegetation type in which it occurs is heavily transformed, insufficient information exists on the status of populations to justify allocation to a threatened category. A persistent population decline seems unlikely since it is common in habitats modified by humans.

Measurements
Amblysomus robustus
HB (♂♂): 131.7 (118–143) mm, n = 6
HB (♀♀): 127.7 (109–138) mm, n = 7
T: 0 mm
HF (♂♂): 18.5 (18–19) mm, n = 6
HF (♀♀): 15.9 (13–18) mm, n = 7
E: 0 mm
WT (♂♂): 72, 98 g, n = 2
WT (♀♀): 61, 73 g, n = 2
GLS (♂♂): 30.8 (29.9–32.0) mm, n = 7
GLS (♀♀): 28.8 (27.2–29.7) mm, n = 14
GWS (♂♂): 19.2 (18.6–20.4) mm, n = 7
GWS (♀♀): 17.6 (16.6–18.2) mm, n = 14
C–M^2 (♂♂): 7.4 (7.1–7.6) mm, n = 6
C–M^2 (♀♀): 6.9 (6.0–7.3) mm, n = 14
Verloren–Vallei/Dullstroom, Mpumalanga Province, South Africa (Bronner 2000)

Key References Bronner 1996, 2000.

Gary N. Bronner

Amblysomus septentrionalis Highveld Golden-mole

Fr. Taupe dorée de montagne; Ger. Highveld-Goldmull

Amblysomus septentrionalis Roberts, 1913. Ann. Transvaal Mus. 4: 73. Wakkerstroom, Mpumalanga Province, South Africa.

Taxonomy Traditionally classified as a subspecies of the Zulu Golden-mole *A. iris*. Now considered as a valid species with a unique chromosome number amongst chrysochlorids; closely related to *A. hottentotus* and *A. robustus*, from which it can be discriminated craniometrically (Bronner 1996, 2000). Synonyms: none. Chromosome number: 2n = 34, FN = 62 (Bronner (1995a).

Description Medium-sized golden-mole similar in size and pelage colour to *A. robustus* from further north in Mpumalanga Province, and *A. h. longiceps* from further south in the interior highlands of KwaZulu–Natal. Dorsal pelage dark reddish-brown, light reddish-brown on flanks. Ventral pelage pale brown to orange. Guard hairs yellowish at base, dark grey subterminally. Pale yellow cheek patches extend laterally to above subdermal eyes. Claws on forefeet well developed, but slightly smaller than in *A. robustus*, claw on Digit 3 largest (14.0–15.7 mm long, 5.3–6.6 mm at base). Skull: malleus not enlarged; bulla not externally evident; zygomatic arches without broad plates sweeping backwards; M^3 absent (see also family and genus profiles). Sexual dimorphism pronounced, ♂♂ on average larger than ♀♀.

Geographic Variation Size varies clinally in relation to altitude and ambient temperature (Bronner 1996). Although consistent craniometric differences exist between populations, sample sizes are too small for a thorough analysis of geographic variation and no subspecies are recognized.

Similar Species
A. hottentotus and *A. robustus*. Not separable from *A. septentrionalis* using any single skull measurement but can be distinguished using a multivariate craniometric technique (Bronner 1996).
Chlorotalpa sclateri. On average smaller (HB: 102 [82–135] mm); claws less robust, claw on Digit 3 of forefoot 8–9 mm; M^3 present.
Chrysospalax villosus. HB on average larger (HB: 151 [127–175] mm); skull larger (GLS: 33.4–34.9 mm); pelage less fine; malleus enlarged; bulla visible externally; zygomatic arches with broad plates sweeping backwards; M^3 present.
Neamblysomus julianae. Usually smaller (HB: 92–111 mm); skull smaller (GLS: 21.8–23.60 mm); pelage less red; subtle chromosomal differences (Bronner 1995a).

Distribution Endemic to Africa. Highveld BZ. Recorded from Mpumalanga Province from Wakkerstroom northwards to Ermelo and Barberton. Populations from NE Free State (Heilbron/Parys) and Swaziland (Piggs Peak/Mbabane) may represent this species.

Habitat Grasslands and edges of marshes in high-altitude southern African savanna. Known only from Moist Sandy Highveld Grassland and Wet Cold Highveld Grassland. Restricted to friable soils in valleys and on mountainsides, where individuals may co-exist with the Rough-haired Golden-mole *Chrysospalax villosus*. Found in thickets of Oldwood trees *Leucosidea sericea* on the banks of streams in valleys, but avoids scrubby vegetation in kloofs and along rocky ridges, where it is replaced by *Chlorotalpa sclateri*.

Abundance Locally common in suitable habitats; also in gardens and cultivated lands.

Amblysomus septentrionalis

Adaptations The sturdily built body and well-developed claws allow these golden-moles to exist in heavier soils than most other species. Individuals construct two-tier burrow systems: subsurface tunnels used for foraging and deeper (15–30 cm) tunnels connecting grass-lined nests. Activity peaks at night, and declines rapidly after dawn when individuals enter daily torpor until late afternoon. Subsurface activity is reduced during dry winter months.

Foraging and Food Soil-dwelling invertebrates, particularly earthworms and pupae.

Social and Reproductive Behaviour No information, except that adults are solitary.

Reproduction and Population Structure Limited data suggest that pregnancies may occur throughout the year, but mostly in the wet summer months (Nov–Mar). Litter-size: 1–2. At birth, young are altricial and raised in grass-lined nests (Roberts 1951, Bernard *et al.* 1994).

Predators, Parasites and Diseases Striped Weasels *Poecilogale albinucha* sometimes prey on these golden-moles. Domestic dogs and cats also kill them regularly, but appear to find them unpalatable. Also recorded from pellets of the Barn Owl *Tyto alba*, indicating that individuals may venture out onto the ground surface, particularly after heavy rains. No information on parasites or diseases.

Conservation IUCN Category: Near Threatened.

While the geographic range of this species is restricted, a persistent population decline seems unlikely since it thrives in habitats modified by humans.

Measurements

Amblysomus septentrionalis
HB (♂♂): 128.6 (114–135) mm, n = 12
HB (♀♀): 125.3 (105–145) mm, n = 25
T: 0 mm
HF (♂♂): 15.3 (13–20) mm, n = 11
HF (♀♀): 14.7 (12–17) mm, n = 21
E: 0 mm
WT: 52–86 g, n = 3
GLS (♂♂): 29.0 (27.2–30.5) mm, n = 16
GLS (♀♀): 28.2 (26.2–30.0) mm, n = 25
GWS (♂♂): 17.9 (16.8–19.5) mm, n = 16
GWS (♀♀): 17.1 (16.1–19.0) mm, n = 25
C–M^2 (♂♂): 7.0 (6.2–7.4) mm, n = 11
C–M^2 (♀♀): 6.9 (6.5–7.2) mm, n = 19
Ermelo/Wakkerstroom, Mpumalanga Province, South Africa (Bronner, 1996)

Key References Bronner 1996; Roberts 1951.

Gary N. Bronner

GENUS *Calcochloris*
Golden-moles

Calcochloris Mivart, 1867. J. Anat. Physiology, London 2: 133. Type species: *Chrysochloris obtusirostris* Peters, 1851.

This genus includes three small species from widely separated parts of Africa. One species (*C. leucorhinus*) lives in primary rainforest and montane forests of central Africa, and another (*C. obtusirostris*) in the sandy soils of eastern South Africa and Mozambique. The third species (*C. tytonis*) is known only from owl pellets in Somalia. See also Family Chrysochloridae.

The diagnostic characters of the genus are a mixture of those in other genera. Body shape fusiform. Claws on forefeet slender and short; claw of Digit 3 largest (7–9 mm long and 2.5–3 mm wide at base), those on Digits 1 and 2 much shorter. Digit 4 vestigial with a short (2 mm) claw. The skull is short and broad (breadth >70% of GLS) with a broad palate (30–35% of GLS). Zygomatic arches without broad plates sweeping backwards. First premolars are tricuspid and lower molars lack talonids (as in *Chrysochloris*). The malleus is small and unspecialized (as in *Amblysomus*), so no external bulla is visible. M^3 are either absent (*C. obtusirostris*) as in *Amblysomus*, or present (*C. leucorhinus* and *C. tytonis*) as in *Chrysochloris* and *Chlorotalpa* (Figure 7).

Considering the mixture of characters within the genus, it is not surprising that previous authors differed greatly in their treatment of the three species, which have been assigned to either *Chrysochloris* (Broom 1907a), *Amblysomus* (Simonetta 1968, Petter 1981) or *Calcochloris* (Meester 1974). In phylogenetic analyses based on multivariate morphometrics, *C. obtusirostris* and *C. leucorhinus*

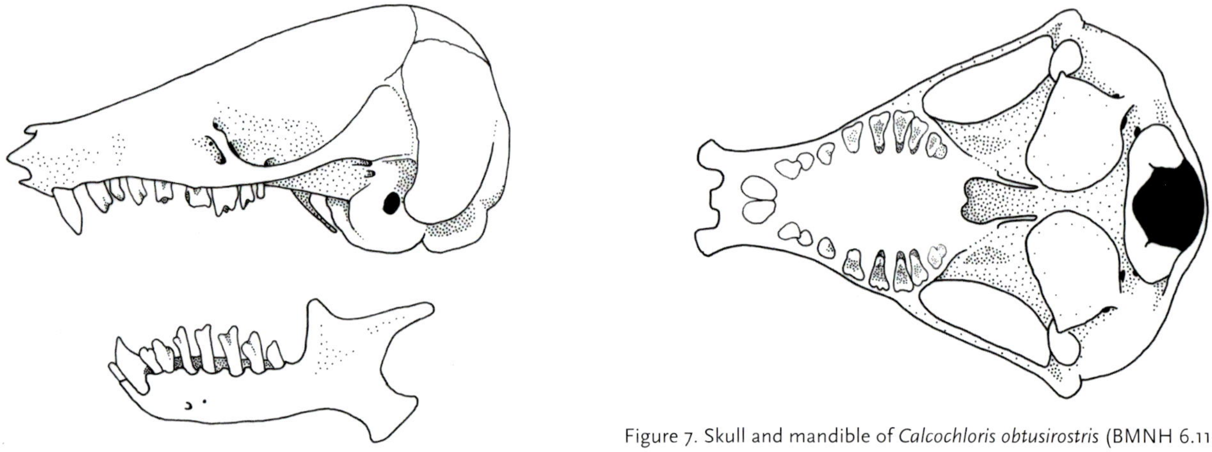

Figure 7. Skull and mandible of *Calcochloris obtusirostris* (BMNH 6.11.5.26).

consistently grouped together in a separate clade, supporting their allocation to a distinct genus. The allocation of *C. tytonis* to this taxon was based only on qualitative characters (hyoid and malleus morphology) since no mensural data were available, and its taxonomic status remains uncertain (Bronner 1995b). The name *Calcochloris* antedates *Chrysotricha*, a name commonly used for this genus until 1953 (Meester 1974). The three species are distinguished by colour and geographic location.

Gary N. Bronner

Calcochloris leucorhinus Congo Golden-mole

Fr. Taupe dorée du Congo; Ger. Kongo-Goldmull

Calcochloris leucorhinus (Huet, 1855). Nouv. Arch. Mus. Hist. Nat., Paris 8: 8. 'Gulf of Guinea Coast, Congo'.

Taxonomy Historically placed in three different genera: included with *Calcochloris obtusirostris* in *Amblysomus* by Simonetta (1968) and Petter (1981), primarily on the basis of malleus morphology ('normal' *cf.* inflated); in *Chrysochloris* on the grounds of dental characters (Allen 1939); and in *Chlorotalpa* (Meester 1974) owing to the absence of a temporal bulla and the presence of M^3. Craniometric analyses indicate, however, that *C. leucorhinus* should provisionally be included in *Calcochloris* (Bronner 1995b). Synonyms: *cahni*, *congicus*, *luluanus*. Subspecies: two. Chromosome number: not known.

Description Small species, similar in size to *C. obtusirostris*. Dorsal pelage darkish-brown to slaty-grey. Ventral pelage similar to dorsal pelage, but paler. Muzzle with creamy-whitish mask extending laterally to almost the ear. Hairs fine, ca. 7 mm on back. Claws on forefeet very slender, claw on Digit 3 largest and narrow (9–11.5 mm long, 2.5–3 mm wide at base). Skull: malleus not inflated, so that there is no distinct subtemporal bulla; zygomatic arches without broad plates sweeping backwards; M^3 usually present, resembling other cheekteeth; first premolar completely molariform but narrower than in *C. obtusirostris* (see also family and genus profiles). Sample sizes too small to evaluate sexual size dimorphism.

Geographic Variation Two subspecies have been described, but their validity is uncertain:

C. l. leucorhinus: DR Congo southwards to N Angola. Lower premolars with well-developed talonids; includes *congicus* and *luluanus* (Meester 1974).

C. l. cahni: Cameroon and Central African Republic. Talonids feeble on $P_{1,2}$, absent on P_3 and M_3. Although the pelage of some *C. l. cahni* is paler than in *C. l. leucorhinus*, one individual from Cameroon has a darker colouration suggesting that colour differences between subspecies are not consistent.

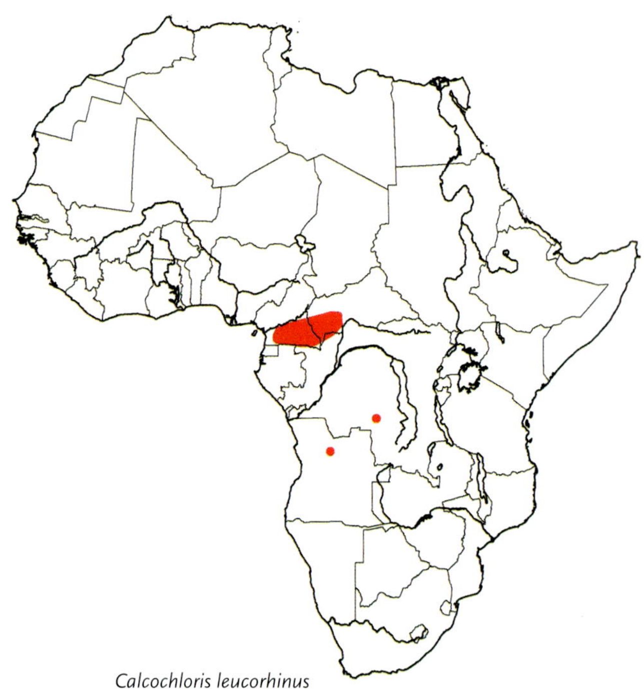

Calcochloris leucorhinus

Similar Species

Chrysochloris stuhlmanni. Skull larger (GLS: 24.3–28.31 mm); malleus enlarged; temporal bulla pronounced and protruding through temporal surface of skull. Might be sympatric but full extent of distribution not known.

Distribution Endemic to Africa. Rainforest BZ (West Central Region [Gabon sub-region] with limited extension into the Northern Rainforest–Savanna Mosaic, and two outliers in Southern Rainforest–Savanna Mosaic). Recorded only from scattered localities in Angola, DR Congo, Cameroon and Central African Republic.

Habitat Lowland equatorial forests and montane forests.

Remarks Apparently no other information available.

Conservation IUCN Category: Data Deficient, previously Least Concern.

Measurements

Calcochloris leucorhinus
HB: 63, 126 mm, n = 2
T: 0 mm
HF: 9, 10 mm, n = 2
E: 0 mm
WT: n. d.
GLS (♂♂): 21.7 (19.5–22.8) mm, n = 4
GLS (♀♀): 22.2 (21.0–23.1) mm, n = 4
GWS (♂♂): 19.2 (18.6–20.4) mm, n = 7
GWS (♀♀): 17.6 (16.6–18.2) mm, n = 14
C–M^3 (♂♂): 5.9 (5.6–6.5) mm, n = 6
C–M^3 (♀♀): 6.7 (4.7–6.4) mm, n = 7
Museum specimens throughout geographic range (Bronner 1995b)

Key References Bronner 1995b; Simonetta 1968.

Gary N. Bronner

Calcochloris obtusirostris Yellow Golden-mole

Fr. Taupe dorée jaune; Ger. Gelber Goldmull

Calcochloris obtusirostris (Peters, 1851). Bericht Verhandl. K. Preuss. Akad. Wiss., Berlin 16: 467. Inhambane, coastal Mozambique.

Taxonomy Originally described in the genus *Chrysochloris*. A polytypic species with three subspecies, all of which were once regarded as valid species. Synonyms: *chrysillus*, *limpopoensis*. Subspecies: three. Chromosome number: 2n = 28, FN = 56 (Bronner 1995a).

Description Small species, distinctive among golden-moles in having underfur that is orange at base, instead of slaty-grey. Dorsal pelage yellowish-orange to dull reddish-brown, owing to a progressive increase in grizzled-brown hair tips. Ventral pelage less variable, yellowish-orange to bright brownish. Muzzle, throat and chin yellowish-white with fewer brown-tipped hairs. Nostrils with unique foliaceous projections. Pelage colour becomes darker and more grizzled with time between moults. Claws on forefeet very slender and flattened. Claw on Digit 3 largest (7–9 mm long, <3.5 mm wide at base), claws on Digits 1 and 2 about same length. Digit 4 small with a short (2 mm) claw. First premolars molariform, canines bicuspid. M^3 absent, molar talonids absent or feeble. Sexual size dimorphism pronounced, ♂♂ on average larger than ♀♀ in most cranial measurements. Skull: see family and/or genus profiles.

Geographic Variation Morphological differences between subspecies are subtle and distinguishing between them is difficult owing to intergradations in size:

C. o. obtusirostris: Inhambane and Gaza districts in Mozambique westwards to Changane and Save river systems and to SE lowlands of Zimbabwe and Northern Province of South Africa.

C. o. limpopoensis: Masiene district of Mozambique southwards along coast to Maputo. Larger than other subspecies (GLS: >22.5 mm), with a wider rostrum (>7.2 mm) than in *C. chrysillus* and more gracile claws than in *C. o. obtusirostris*.

C. o. chrysillus: southern Mozambique Plain from Maputo to Maputaland (north of L. St Lucia) in N KwaZulu–Natal, South Africa; rostrum <7.1 mm.

Similar Species

Carpitalpa arendsi. Larger (HB: 115–141 mm; GLS: 25.3–29.0 mm); M^3 present; pelage darker. Highlands of E Zimbabwe and W Mozambique near Zimbabwe border.

Calcochloris obtusirostris

Distribution Endemic to Africa. Extreme southern part of Zambezian Woodland BZ. Distribution closely follows the limits of the Mozambique sands. Recorded from Inhambane in Mozambique southwards to N KwaZulu–Natal Province, South Africa, and inland to the great escarpment and south-eastern lowveld of Zimbabwe.

Habitat Sandy soils in savanna. In KwaZulu–Natal, *C. o. chrysillus* occurs in Coastal Bushveld–Grassland and Subhumid Lowveld Bushveld. In E Mozambique, *C. o. limpopoensis* occurs in miombo savanna, and in Mozambique and adjacent parts of Zimbabwe (with a marginal intrusion into the Nyadu Sandveld of Kruger N. P.), *C. o. obtusirostris* occurs in *Acacia* and mopane savanna.

Abundance Locally common but no accurate information on densities.

Adaptations The unique foliaceous projections on the nostrils prevent sand entering during burrowing. Nest chambers are located amongst the roots of trees, from where subsurface tunnels lead off for up to 50 m to foraging areas. Deeper tunnels (up to 20 cm below surface) are reused and regularly maintained. If disturbed in subsurface tunnels, these golden-moles burrow deeply downwards in a spiralling pattern. They are particularly active after rains. Watering the soil surface stimulates activity by captive individuals, which sometimes move at a surprising speed in and out of the sand in a porpoise-like fashion (Roberts 1936).

Foraging and Food Mostly insectivorous. Forage in subsurface tunnels, mainly at night. Prey includes tenebrionid larvae, termites, grasshoppers, flies, moths and small lizards. Highly sensitive to vibrations emitted by live prey (Roberts 1936).

Social and Reproductive Behaviour Adults solitary. Emit high-pitched squeaks when agitated or handled.

Reproduction and Population Structure Breeding individuals recorded in Oct (n = 3) and Jan (n = 1), suggesting that reproduction takes place primarily in wet summer months. Litter-size: 2 (n = 1). Gestation period unknown.

Predators, Parasites and Diseases No information.

Conservation IUCN Category: Least Concern.

Occurs in several protected areas. Its cryptic nature allows it to coexist successfully with humans.

Measurements

Calcochloris obtusirostris
HB (♂♂): 100.3 (88–110) mm, n = 17
HB (♀♀): 93.3 (82–110) mm, n = 15
T: 0 mm
HF (♂♂): 11.3 (10–13) mm, n = 17
HF (♀♀): 11.6 (9–14) mm, n = 13
E: 0 mm
WT (♂♂): 28.1 (25–33) g, n = 9
WT (♀♀): 24.5 (15–37) g, n = 11
GLS (♂♂): 22.6 (20.8–24.2) mm, n = 21
GLS (♀♀): 21.5 (19.0–23.6) mm, n = 17
GWS (♂♂): 16.1 (15.4–16.8) mm, n = 23
GWS (♀♀): 15.5 (13.8–16.8) mm, n = 18
C–M^2 (♂♂): 6.1 (5.9–6.3) mm, n = 10
C–M^2 (♀♀): 5.9 (5.6–6.2) mm, n = 10
Museum specimens throughout geographic range (Bronner 1995b)

Key References Bronner 1995b; Roberts 1936

Gary N. Bronner

Calcochloris tytonis Somali Golden-mole

Fr. Taupe dorée de Somalie; Ger. Somalia-Goldmull

Calcochloris tytonis (Simonetta, 1968). Monitore Zoologico Italiano 2 (Suppl.): 31. Giohar (= Villagio Duca degli Abruzzi), southern Somalia.

Taxonomy and Description Originally described in the genus *Amblysomus* on the basis of size and structure of the malleus (Simonetta 1968). Subsequently included in *Chlorotalpa* by Meester (1974) based on the presence of M^3, but included here in the genus *Calcochloris* because the few characters available for this species suggest that it may be closely related to *Calcochloris leucorhinus* (Bronner 1995b). However, *C. tytonis* is known only from a partially complete specimen in an owl pellet, and no measurements are available. It should therefore be regarded as *incertae sedis* until more specimens are available. The incidence of third molars is a primitive character that sheds no light on phylogenetic relationships among chrysochlorids.

Geographic Variation None recorded.

Similar Species None within the known geographic distribution of the species.

Distribution Endemic to Africa. Somalia–Masai Bushland BZ. Known only from the type locality.

Habitat Dense bush and savanna of the Somali Peninsula (Agnelli *et al.* 1988).

Abundance No information.

Remarks The holotype was recovered from the pellet of a Barn Owl *Tyto alba*. Apparently no other information on this species.

Measurements None available.

Conservation IUCN Category: Critically Endangered.

Key References Bronner 1995b; Meester 1974; Simonetta 1968.

Gary N. Bronner

Calcochloris tytonis

GENUS *Carpitalpa*
Arends's Golden-mole

Carpitalpa Lundholm, 1955. Ann. Transvaal Mus. 22: 285. Type species: *Chlorotalpa* (*Carpitalpa*) *arendsi* Lundholm, 1955.

A monotypic genus including only Arends's Golden-mole *Carpitalpa arendsi* from afromontane forests and high-altitude grasslands of E Zimbabwe and adjacent parts of Mozambique. Body shape fusiform. Claws on forefeet gracile, claw of Digit 3 largest (9–12 mm long) but slender (3.5–4.0 mm at base), claws on Digits 1 and 2 much shorter. Digit 4 absent. Skull elongate (breadth <65% of GLS, with a narrow palate 28–29% of GLS). Zygomatic arches without broad plates sweeping backwards. Small, tricuspid M^3 present, upper premolars semi-molariform but crushing basin not fully formed (Bronner 1995b). Definitive characters of this genus are the possession of a club-shaped malleus that, although enlarged, does not project through the temporal region of the skull as a bulla. In this respect it resembles the smaller species of *Chlorotalpa* from South Africa, from which it is distinguished by a relatively narrower skull and absence of well-developed talonids on the lower molars (Figure 8).

Lundholm (1955) described this taxon as a subgenus of *Chlorotalpa*, as well as the subgenus *Kilimatalpa* for *C. stuhlmanni* from East Africa. Simonetta (1968) elevated *Carpitalpa* to generic rank to include *Kilimatalpa*. Meester (1974) assigned *C. arendsi* to *Chlorotalpa* and *C. stuhlmanni* to *Chrysochloris*, commenting that differences between the species of *Chlorotalpa* did not warrant subgeneric separation. Phylogenetic analyses of morphometric data, however, confirm that *C. arendsi* has diverged considerably from the *Chlorotalpa* species and *Chrysochloris (Kilimatalpa) stuhlmanni*, necessitating its allocation to a separate genus (Bronner 1995b).

Gary N. Bronner

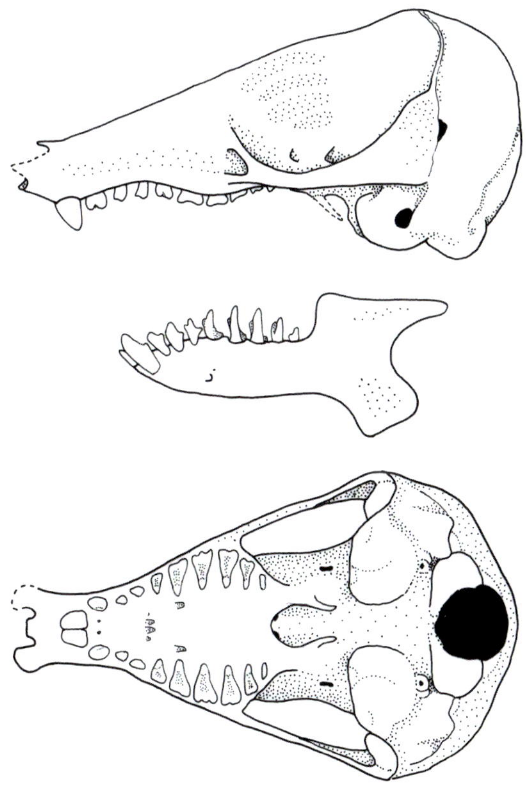

Figure 8. Skull and mandible of *Carpitalpa arendsi* (TM 12965, based on photo by T. Kearney).

Carpitalpa arendsi Arends's Golden-mole

Fr. Taupe dorée d'Arends; Ger. Arends Goldmull

Carpitalpa arendsi (Lundholm, 1955). Ann. Transvaal Mus. 22: 285. Pungwe Falls, Inyanga, Zimbabwe.

Taxonomy Synonyms: none. Chromosome number: not known.

Description Originally described in the genus *Chlorotalpa*. Medium-sized golden-mole with a comparatively strong build, similar in size and cranial proportions to *Neamblysomus gunningi*. Dorsal pelage brownish-black, ventral pelage greyish-brown, with a distinct violet or silver sheen. Underfur dark grey. Muzzle paler; eye patches whitish. Males on average slightly larger than ♀♀ but sexual dimorphism in cranial size is not pronounced (Bronner 1995b). Skull: M^3 present (see family and genus profiles).

Geographic Variation None recorded.

Similar Species

Calcochloris obtusirostris. Smaller (HB: 82–110 mm, GLS: 19.0–24.2 mm); M^3 absent; pelage paler. Lowlands of Zimbabwe and Mozambique.

Distribution Endemic to Africa. Zambezian Woodland BZ. Restricted to the Inyanga Highlands of E Zimbabwe between latitudes 18° S and 20° S, and altitudes of 850–2000 m. Also recorded in W Mozambique in the highlands close to the Mozambique–Zimbabwe border.

Habitat Loamy soils, with detritus, in montane grasslands and forest fringes (Smithers & Wilson 1979). Also recorded in roadside verges and gardens.

Abundance Locally common.

Remarks Arends's Golden-moles construct subsurface tunnels that radiate outwards from roots of trees and clumps of grass into open areas. They are most active in gaps between grass tussocks where they coexist with vlei rats *Otomys irroratus*. After rains, they may forage and travel on the surface of the ground (Smithers & Wilson 1979). In open areas, burrowing may kill the grasses (Lundholm 1955). Insectivorous; also eats earthworms (Smithers & Wilson 1979).

Predators, Parasites and Diseases No information.

Conservation IUCN Category: Vulnerable.

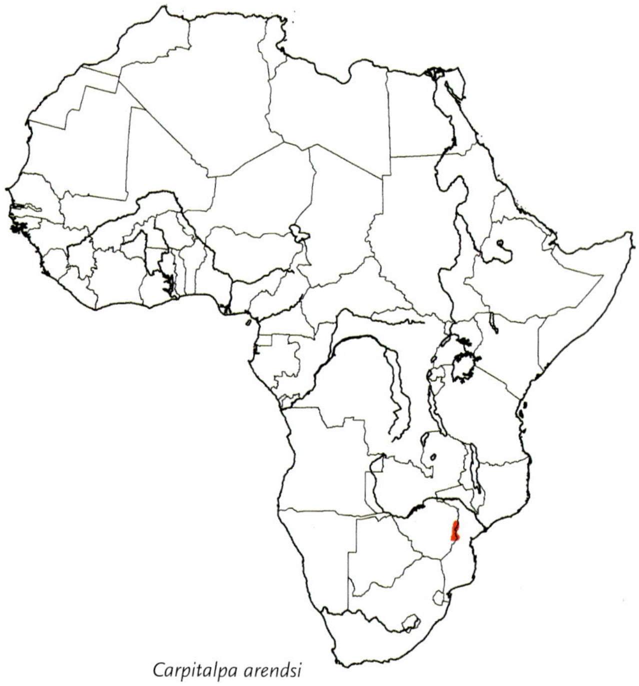

Carpitalpa arendsi

Measurements

Carpitalpa arendsi

HB (♂♂): 123.6 (115–141) mm, n = 11
HB (♀♀): 121.2 (115–139) mm, n = 11
T: 0 mm
HF (♂♂): 14.2 (14–16) mm, n = 9
HF (♀♀): 13.2 (12–16) mm, n = 11
E: 0 mm
WT (♂♂): 53 (43–70) g, n = 6
WT (♀♀): 50 (43–56) g, n = 5
GLS (♂♂): 27.7 (26.8–29.0) mm, n = 7
GLS (♀♀): 27.3 (25.3–28.6) mm, n = 8
GWS (♂♂): 16.8 (16.2–17.5) mm, n = 7
GWS (♀♀): 16.6 (15.8–17.2) mm, n = 8
C–M^3 (♂♂): 7.1 (6.5–7.6) mm, n = 7
C–M^3 (♀♀): 7.3 (6.9–7.7) mm, n = 8
Museum specimens throughout geographic range (Bronner 1995b)

Key References Bronner 1995b; Lundholm 1955; Smithers & Wilson 1979.

Gary N. Bronner

GENUS *Chlorotalpa*
Golden-moles

Chlorotalpa Roberts, 1924. Ann. Transvaal Mus. 10: 64. Type species: *Chrysochloris duthieae* Broom, 1907.

The genus *Chlorotalpa* contains two small species endemic to South Africa. The two species have different habitat requirements and geographic ranges. Sclater's Golden-mole *C. sclateri* lives in a variety of montane grasslands and forested kloofs inland. Duthie's Golden-mole *C. duthieae* lives in sandy habitats of coastal forests.

Body-shape fusiform. Claws on forefoot long and slender, claw on Digit 3 largest (7–9 mm long), claws on Digits 1 and 2 much shorter. Digit 4 minute with a small nail. *Chlorotalpa* resembles *Amblysomus* in general appearance and skull morphology, but is markedly smaller with weaker claws on the forefeet. Skull elongate (breadth 62–66% of GLS) with a narrow palate (<30% of GLS). Zygomatic arches without broad plates sweeping backwards. Small molariform M^3 present, lower cheekteeth have well-developed talonids. The malleus is somewhat inflated and pear-shaped, but does not protrude through the temporal region of the skull to form a bulla; in size and shape, it is intermediate between the club-shaped malleus of *Chrysochloris* and the small, unspecialized malleus of *Amblysomus* (Mayer *et al.* 1995), suggesting that *Chlorotalpa* is transitional between these genera (Figure 9).

Some authors have disputed the use of dental formulae for diagnosing genera amongst golden-moles, and thus have synonymized *Chlorotalpa* under *Amblysomus*. The consistent differences in the morphology of the malleus between these taxa, however, correlate with subtle but complex differences in the configuration of the temporal region of the skull, which must be regarded as taxonomically important. These two genera also differ in chromosomal properties (Bronner 1995b). The two species in the genus are distinguished by pelage colour and habitat requirements.

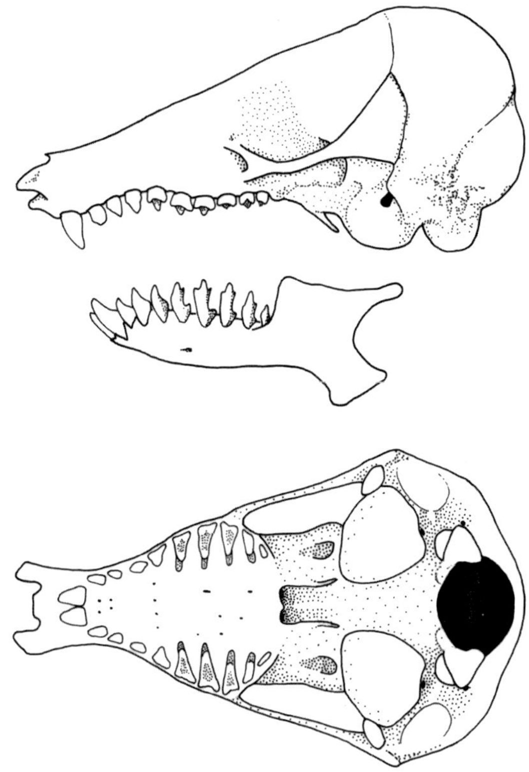

Figure 9. Skull and mandible of *Chlorotalpa duthiae* (BMNH 5.13.10.1).

Gary N. Bronner

Chlorotalpa duthieae Duthie's Golden-mole

Fr. Taupe dorée de Duthie; Ger. Duthies Goldmull

Chlorotalpa duthieae (Broom, 1907). Trans. S. Afr. Phil. Soc. 18: 292. Knysna, Western Cape Province, South Africa.

Taxonomy Originally described in the genus *Chrysochloris*. A monotypic species that some authors (Ellerman *et al.* 1953, Simonetta 1968) treated as a subspecies of *C. sclateri*, but constant differences in pelage colour and chromosomal properties, and their preference for quite dissimilar environments, indicate that they are distinct species (Bronner 1995b). Synonyms: none. Chromosome number: 2n = 30, FN = 60 (Bronner 1995a).

Description Small golden-mole, with a delicate build and slender claws on the forefeet. Dorsal pelage uniformly reddish-black or brownish-black. Ventral pelage with less lustre. Throat and chest greyish. Hairs 8–9 mm long, pale at base, with dark grey subterminally. Muzzle paler than dorsal pelage with buffy-white patches extending to ocular region (Roberts 1951). Nosepad with rounded outer edge, section behind transverse groove narrow (1.5–2 mm). Claw on Digit 3 of forefoot largest (7–9 mm long, 3.5–4.0 mm wide at base). Sexual size dimorphism marked, ♂♂ significantly larger than ♀♀ in most cranial measurements (Bronner 1995b). P1 tricuspid rather than sectorial, P^1 lack protocones. Well-developed talonids invariably present on lower molars. Skull: see also family and/or genus profiles.

Geographic Variation Individuals from Knysna are generally larger than those from further east, but geographic variation is not sufficiently marked to warrant recognition of subspecies.

Similar Species

Amblysomus corriae. Skull larger (GLS: 25.3–28.9 mm); claw on Digit 3 longer (13–15 mm); M^3 absent.

Distribution Endemic to South Africa. South-West Cape BZ. Occurs along the coast from George in Western Cape Province to Port Elizabeth in Eastern Cape Province, South Africa.

Chlorotalpa duthieae

Habitat Alluvial sands and sandy loams in afromontane and coastal forests. Coexists with *Amblysomus corriae* in parts of the range, but trapping data suggest these species select different micro-habitats, *A. c. corriae* apparently favouring fynbos and forest fringes, and *C. duthieae* favouring deeper forest. Thrives in cultivated areas and gardens.

Abundance Locally common but no quantitative data.

Adaptations Constructs predominantly shallow subsurface tunnels that radiate outwards from nests located under the roots of trees.

Foraging and Food Forages in subsurface tunnels and amongst leaf litter, mainly at night. Limited observations suggest the diet comprises mainly earthworms (n = 2 stomachs).

Social and Reproductive Behaviour Adults solitary.

Reproduction and Population Structure Only one pregnancy recorded, in early summer (Nov). Litter-size: 2 (1L, 1R).

Predators, Parasites and Diseases Recorded from pellets of Barn Owls *Tyto alba*, suggesting occasional surface activity. Also caught by domestic cats and dogs.

Conservation IUCN Category: Vulnerable.

Within its limited distribution, adequately conserved in numerous protected areas.

Measurements

Chlorotalpa duthieae
HB (♂♂): 105.7 (95–125) mm, n = 10
HB (♀♀): 101.9 (96–130) mm, n = 11
T: 0 mm
HF (♂♂): 11.6 (9–14) mm, n = 9
HF (♀♀): 11.6 (9–13) mm, n = 9
E: 0 mm
WT (♂♂): 33.0 (26–41) g, n = 3
WT (♀♀): 26.0 (20–27) g, n = 3
GLS (♂♂): 23.8 (22.9–24.5) mm, n = 8
GLS (♀♀): 22.8 (22.0–23.9) mm, n = 10
GWS (♂♂): 14.9 (14.4–15.8) mm, n = 8
GWS (♀♀): 14.3 (13.7–15.2) mm, n = 10
C–M^3 (♂♂): 7.1 (6.5–7.6) mm, n = 7
C–M^3 (♀♀): 7.3 (6.9–7.7) mm, n = 8
Knysna, Nature's Valley and Port Elizabeth, South Africa (Bronner 1995b)

Key References Bronner 1995b; Broom 1907a; Roberts 1951.

Gary N. Bronner

Chlorotalpa sclateri SCLATER'S GOLDEN-MOLE

Fr. Taupe dorée de Sclater; Ger. Sclaters Goldmull

Chlorotalpa sclateri (Broom, 1907). Ann. Mag. Nat. Hist., ser. 7, 19: 263. Matjiesvlei, Beaufort West, Western Cape Province, South Africa.

Taxonomy Originally described in the genus *Chrysochloris*. Differs markedly in colour from *C. duthieae*, although these species are indistinguishable cranially. Synonyms: *guillarmodi*, *montana*, *shortridgei*. Subspecies: four. Chromosome number: 2n = 30, FN = 60 (Bronner 1995a).

Description Small golden-mole with a stocky build and slender claws on the forefeet. Dorsal pelage glossy reddish-brown, darker along mid-line. Ventral pelage dull greyish with rufous tinge. Hairs short, 10–12 mm long and rather coarse. Lips and cheeks creamish-yellow with pale patches extending to above the subdermal eyes and forming a band above the nosepad. Nosepad rounded without obtuse corners as in *Chrysochloris asiatica*. Claw on Digit 3 of forefoot largest (8–9 mm long, 3.0–4.5 mm wide at base). Skull: M^3 present; P1 tricuspid, resembles other cheekteeth, P^1 lacks protocone; lower molars with well-developed talonids (see also family and genus profiles). Males on average slightly larger than ♀♀, but sexual dimorphism not pronounced.

Geographic Variation Distributional limits of the four subspecies are unclear owing to poor geographic sampling, but their known ranges are separated by expanses of seemingly inhospitable habitat suggesting that they may be geographically isolated.

C. s. sclateri: Beaufort West in Western Cape Province north-eastwards along Koueveldberge and Sneeuberge mountains to Graaff–Reinet in Eastern Cape Province. Palate relatively longer (>53% of GLS) than in *C. s. guillarmodi*.

C. s. shortridgei: known from only Sutherland in Northern Cape Province, but may range eastwards along Nieuweveldbeerg mountains toward Beaufort West in Western Cape Province. Slightly larger than other subspecies with wider rostrum (>4.7 mm) than *C. s. sclateri* (<4.5 mm).

C. s. guillarmodi: Herschel and Lady Grey in Eastern Cape Province north-eastwards through Lesotho, with a marginal intrusion into E Free State and KwaZulu–Natal (Taylor *et al.* 1994). Palate <51% of GLS, shorter than in *C. s. sclateri* and *C. s. montana*.

C. s. montana: known from only Wakkerstroom in Mpumalanga Province, but may range southwards along the Low Drakensberg to intergrade with *C. s. guillarmodi* near Harrismith or Clarens. Palate significantly longer (>12.2 mm) than in *C. s. guillarmodi* (<12 mm).

Chlorotalpa sclateri

Similar Species

Amblysomus septentrionalis. On average larger (HB: 126 [105–145] mm); claws more robust, claw on Digit 3 of forefoot 14–15.7 mm; M^3 absent.

Distribution Endemic to Africa. Highveld BZ and South-West Arid BZ. Recorded from Sutherland in Northern Cape Province, South Africa north-eastwards through Eastern Cape Province and Lesotho to E Free State, W KwaZulu–Natal and S Mpumalanga Provinces.

Habitat Restricted to high-altitude grasslands, scrub and forested kloofs in the Karoo and Grassland biomes. *Chlorotalpa s. shortridgei* occurs in Escarpment Mountain Renosterveld. *Chlorotalpa s. sclateri* and *C. s. guillarmodi* in South-eastern Mountain Grassland with marginal intrusion into Wet Cold Highveld Grassland. *Chlorotalpa s. montana* is known from only North-eastern Mountain Grassland where it favours forested kloofs rather than valley grasslands where *A. septentrionalis* is the commoner species.

Abundance Locally common, but no quantitative data.

Adaptations Small size and weak digging claws generally limit the distribution of these golden-moles to light, sandy soils, although they may also invade heavier clays when wet after rains. They construct predominantly shallow subsurface tunnels especially among roots of trees, or in clumps of rocks, where the nesting chambers are located (Roberts 1951). Active mainly at night, but also for short periods during the day, especially after rain.

Foraging and Food Carnivorous and insectivorous. Forages for invertebrates in subsurface tunnels that meander outwards from nesting areas for distances of up to 45 m. Feeds predominantly on earthworms and insect larvae.

Social and Reproductive Behaviour Adults solitary.

Reproduction and Population Structure Pregnancies recorded during the wet summer months (Dec and Jan). Litter-size: 2 (1L, 1R). Young born in grass-lined nests (Roberts 1951). Females have a postpartum oestrus (Bernard *et al.* 1994); this may assist in maximizing breeding success during the short period in summer when rainfall and temperatures are high enough to sustain an abundant supply of food.

Predators, Parasites and Diseases Regularly caught by dogs and cats. No information on wild predators, diseases or parasites.

Conservation IUCN Category: Least Concern, previously Vulnerable.

Although widely distributed, the geographic range is highly fragmented and many of its habitats may have been degraded by overgrazing.

Measurements

Chlorotalpa sclateri
HB (♂♂): 100.9 (83–135) mm, n = 16
HB (♀♀): 102.2 (82–123) mm, n = 13
T: 0 mm
HF (♂♂): 12.7 (10–16) mm, n = 15
HF (♀♀): 11.5 (9–13) mm, n = 13
E: 0 mm
WT (♂♂): 36.4 (22–54) g, n = 11
WT (♀♀): 32.5 (22–48) g, n = 7
GLS (♂♂): 23.6 (21.8–27.4) mm, n = 16
GLS (♀♀): 23.4 (21.2–26.7) mm, n = 13
GWS (♂♂): 15.4 (14.5–17.6) mm, n = 16
GWS (♀♀): 15.1 (14.1–16.9) mm, n = 13
C–M^3 (♂♂): 6.3 (5.8–6.9) mm, n = 15
C–M^3 (♀♀): 6.3 (5.6–6.9) mm, n = 14
Museum specimens throughout geographic range

Key References Bernard *et al.* 1994; Broom 1907b; Roberts 1951.

Gary N. Bronner

GENUS *Chrysochloris*
Golden-moles

Chrysochloris Lacépède, 1799. Tabl. Mamm., p. 7. Type species: *Chrysochloris capensis* Lacépède, 1799 (= *Talpa asiatica* Linnaeus, 1758).

Chrysochloris stuhlmanni.

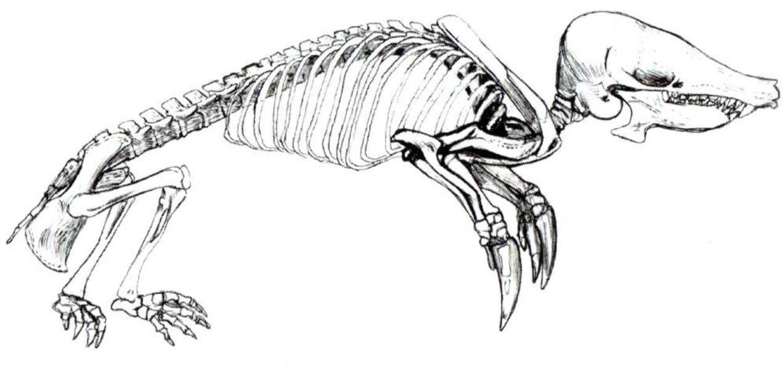

Skeleton of *Chrysochloris stuhlmanni.*

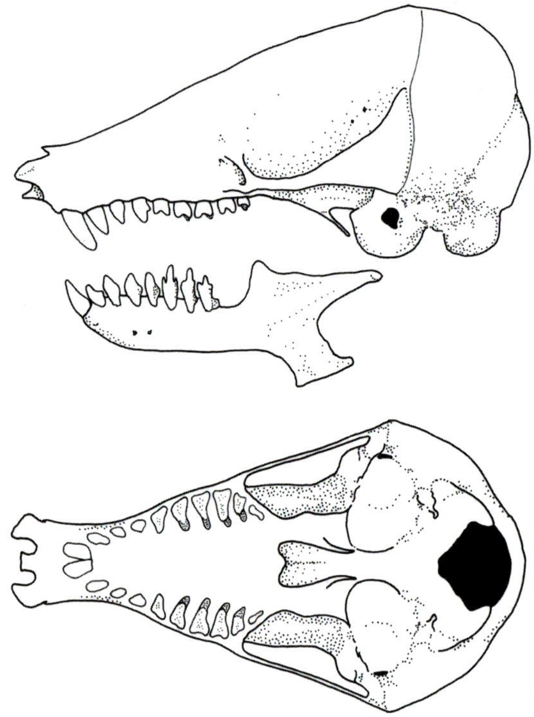

Figure 10. Skull and mandible of *Chrysochloris stuhlmanni* (based on Simonetta 1968).

The genus *Chrysochloris* contains three small species from the geographical extremes of the range of golden-moles in sub-Saharan Africa. The Cape Golden-mole *C. asiatica* (which Linnaeus mistakenly assumed to have come from Siberia) is endemic to sandy soils of SW South Africa, and Stuhlmann's Golden-mole *C. stuhlmanni* occurs on mountains in central and eastern Africa. Visagie's Golden-mole *C. visagiei* is known only from the type locality in Northern Cape Province of South Africa (Meester 1974).

Body shape is fusiform. Claws on the forefeet are slender, the claw of Digit 3 is markedly larger (10–12 mm long) than that on Digit 2 (5–8 mm) and Digit 1 (3–4.5 mm). Digit 4 is vestigial. The most distinctive feature of this genus is the possession of a enlarged, club-shaped malleus orientated vertically and housed in a bony bulla that projects through the temporal surface of the skull (Broom 1907a). Zygomatic arches without broad plates sweeping backwards. In *C. asiatica* and *C. visagiei* the skull is short and broad (breadth >70% of GLS) with a wide palate (30–32% of GLS), though not to the same extent as in *Eremitalpa* and *Cryptochloris* (palate width >35% of GLS). The skull of *C. stuhlmanni* is elongate (breadth <65% of GLS) with a narrow palate (26–28% of GLS), as in the genera *Amblysomus* and *Chlorotalpa*. Pelage colour varies substantially, both within populations (*C. asiatica*) and geographically. Small M^3 present in most specimens (but consistently absent in some populations); morphology ranges from tricuspid to peg-like depending on toothwear. Incidence of molar talonids is variable, being either well-developed (*C. stuhlmanni*) or absent (*C. asiatica* and *C. visagiei*) (Figure 10).

The two species that are comparatively well-known, *C. asiatica* and *C. stuhlmanni*, live in different habitats and have different geographic ranges. *Chrysochloris asiatica* lives in sandy soils in semi-arid habitats of South Africa; *C. stuhlmanni* lives in montane forest soils at higher altitudes in eastern Africa. Both species feed mainly on earthworms.

Gary N. Bronner

Chrysochloris asiatica Cape Golden-mole
Fr. Taupe dorée du Cap; Ger. Kap-Goldmull

Chrysochloris asiatica (Linnaeus, 1758). Syst. Nat., 10th edn, 1: 53. 'In Sibiria', taken as the Cape of Good Hope, South Africa.

Taxonomy Originally described in the genus *Talpa*. A monotypic species in need of revision. Previous authors recognized up to eight subspecies based on considerable geographic variability in size and colour, but this may be largely clinal by nature (Meester 1974). Synonyms: *auratus*, *aurea*, *bayoni*, *capensis*, *calviniae*, *concolor*, *damarensis*, *dixoni*, *elegans*, *inaurata*, *minor*, *namaquensis*, *rubra*, *shortridgei*, *taylori*, *tenuis*, *visserae*. Subspecies: none. Chromosome number: 2n = 30, FN = 56 (Bronner 1995a).

Description Small dark species with pale markings on the muzzle. Dorsal pelage blackish, slaty-grey to brownish-grey, olive-brown, or drab silver. Ventral pelage somewhat paler than dorsal pelage, blackish to pale creamy-grey or greyish-white, tips of hairs often paler on chest and greyer on abdomen. Muzzle and chin creamy-yellow with a profuse admixture of whitish hairs, becoming paler on cheeks, often with whitish eye patches. Underfur dense and slate-grey. Younger animals duller and more greyish. Albino individuals pale to golden-brown (rarely recorded). Nosepad narrow, 9–10 mm long with blunt corners. Claws on forefeet gracile, claw on Digit 3 largest (10–12 mm long, 3.5–4.6 mm wide at base), claw on Digit 2 shorter (5.5–7.2 mm) and claw on Digit 1 very short (3–4.5 mm). Digit 4 reduced to a minute tubercle. Skull broad (breadth 70–78% of GLS) with a relatively wide palate (30–32% of GLS). Malleus enlarged and club-shaped, temporal bulla well-developed. M^3 usually present, though occasionally absent in some individuals from Little Namaqualand; morphology varies from triconid (like other cheekteeth) to peg-like depending on toothwear. Molar talonids absent. Skull: see also family and/or genus profiles. Adult ♂♂ on average larger than ♀♀, but extent of sexual dimorphism has not been evaluated statistically.

Chrysochloris asiatica

Geographic Variation Colour and size vary geographically. Individuals from Cape Peninsula are larger and darker with blackish-grey pelage sometimes suffused with dark brown; those from Little Namaqualand are smaller, with pale pelage, which becomes paler (olive-brown to pale brownish or drab silver) from south to north. Malleus size and shape also vary geographically (Broom 1907a, 1946). These differences may, however, be adequately explained by clinal variation in colour (darker to paler) and size (larger to smaller) from south to north (Meester 1974). Almost complete intergradation between darker and paler colour extremes evident at some localities (Shortridge 1942), suggesting that variation may also have a substantial within-population component.

Similar Species

Amblysomus corriae. Often larger (HB: 108–130 mm), skull larger (GLS: 25.3–28.9 mm); pelage more rufous; M^3 absent, claw on digit 3: 13–15 mm; malleus not enlarged, bulla not visible externally.

Cryptochloris zyli and *Cryptochloris wintoni*. Slightly smaller (HB: < 92 mm); claws on Digits 1, 2 and 4 of forefoot better developed; palate relatively wider (>35% of GLS); malleus enlarged; bulla small.

Eremitalpa granti. Much smaller (HB: 76–86 mm, GLS: 18.6–20.4 mm); body lozenge-shaped, claws on Digits 1, 2 and 4 of forefoot not so markedly shorter than that on Digit 3; skull relatively much broader (breadth > 85% of GLS); malleus pea-shaped, less enlarged; bulla not visible externally.

Distribution Endemic to Africa. Western parts of South-West Arid BZ (Karoo) and South-West Cape BZ. Ranges from the Cape Peninsula (Western Cape Province) eastwards across Cape Flats to Bredasdorp and Swellendam, and northwards along the Atlantic coast to Port Nolloth in Northern Cape Province. Occurs inland to Ceres, Worcester and Stellenbosch in the south, and Calvinia, Garies and Kamieskroon in the north, but not in the arid Karoo. One specimen (type of '*C. damarensis*') recorded from Damaraland in Namibia, but likely incorrect and collected in Little Namaqualand (Meester 1974).

Habitat Sandy soils in a wide variety of vegetation types, most commonly in Renosterveld, Fynbos and Strandveld Succulent Karoo. Common in gardens and cultivated lands; readily invades lawns. Coexists with *A. hottentotus* at Stellenbosch, but prefers drier, sandier soils (Broom 1907a).

Abundance Locally common.

Adaptations Cape Golden-moles make subsurface tunnels, which radiate outwards from bushes or clumps of rocks where nests are located. Occasionally they burrow deeper and throw up small mounds of soil. Activity is mainly at night, particularly after rains. Surface activity declines markedly during the dry summer months (Nov–Mar). Adaptive hypothermia and thermolability reduce energetic requirements and prevent overheating in the closed burrow systems, particularly during dry periods or in arid environments where food resources may be limited. Resting metabolic rate is 68% of that predicted for insectivorous mammals. Individuals may enter torpor if ambient temperature drops below 9 °C, resulting in a five-fold reduction in energy requirements. Assimilation efficiency is high, about 86%, as in other insectivorous and carnivorous mammals (Withers 1978, Bennett & Spinks 1995).

Foraging and Food Insectivorous and carnivorous. Diet includes earthworms, sow-bugs, centipedes, insect larvae and millipedes. Observed to venture onto sandy beaches, presumably to feed on amphipods and isopods associated with rotting kelp (Bennett & Spinks 1995).

Social and Reproductive Behaviour Solitary except for ♀♀ raising young. Aggressive towards conspecifics; agonistic encounters involve 'rolling' and biting.

Reproduction and Population Structure Newborn young recorded in Jun, Jul and Aug, suggesting that breeding is restricted to the wet winter months. Litter-size: 1–3. At birth, young are about 47 mm long, altricial and practically hairless. Postnatal development slow. Pups are suckled by their mother for 2–3 months. Teeth do not erupt until the individual is nearly mature (Broom 1907a).

Predators, Parasites and Diseases Remains are common in pellets of Barn Owls *Tyto alba*, indicating substantial surface activity. Often killed by domestic cats, which seldom eat the carcass, perhaps because of the tough skin and musky secretions of the large anal glands. Disliked by gardeners and green-keepers because their subsurface tunnels are deemed unsightly and often cause plants to wilt and die. No information on parasites and diseases.

Conservation IUCN Category: Least Concern.

Protected in many conservation areas and able to thrive in habitats modified by humans.

Measurements

Chrysochloris asiatica

HB (♂♂): 105.4 (100–115) mm, n = 10
HB (♀♀): 99.6 (94–114) mm, n = 10
T: 0 mm
HF (♂♂): 12.6 (9–14) mm, n = 9
HF (♀♀): 11.9 (11–14) mm, n = 10
E: 0 mm
WT: n. d.
GLS (♂♂): 22.9 (22.1–23.6) mm, n = 9
GLS (♀♀): 22.5 (21.1–25.4) mm, n = 12
GWS (♂♂): 17.6 (16.1–19.0) mm, n = 10
GWS (♀♀): 17.1 (16.1–19.2) mm, n = 11
C–M^3 (♂♂): 6.2 (5.4–6.7) mm, n = 5
C–M^3 (♀♀): 6.4 (6.2–6.5) mm, n = 4
Throughout geographic range and literature records

Key References Bennett & Spinks 1995; Broom 1907a; Meester 1974.

Gary N. Bronner

Chrysochloris stuhlmanni STUHLMANN'S GOLDEN-MOLE

Fr. Taupe dorée de Stuhlmann; Ger. Stuhlmanns Goldmull

Chrysochloris stuhlmanni Matschie, 1884. Sber. Ges. Naturf. Freunde Berlin, p. 123. Ukonjo and Kinyawang, Rwenzori region, Uganda.

Taxonomy A polytypic species subjected to markedly different taxonomic treatments. Lundholm (1955) created the subgenus *Kilimatalpa* within *Chlorotalpa* for this species, whereas Simonetta (1968) placed it in the genus *Carpitalpa*, but assigned one of the subspecies (*tropicalis*) to the genus *Chlorotalpa*. Meester (1974) included *stuhlmanni* in *Chrysochloris* based on the morphology of the malleus, an arrangement followed by Bronner (1995b) who, in addition, assigned it to the subgenus *Kilimatalpa* based on significant divergence from *asiatica* in cranial shape. Synonyms: *balsaci*, *fosteri*, *tropicalis*, *vermiculus*. Subspecies: five but of dubious validity. Chromosome number: not known.

Description Small species, resembling *C. asiatica* in general appearance but with a comparatively narrower and longer skull (breadth <65% of GLS compared to >75%) as in *Amblysomus*. Dorsal pelage blackish-brown, dull brown, greyish-brown or fawn, with silvery sheen. Ventral pelage somewhat paler, sometimes with a russet tint. Pelage shows geographic variation in colour (see below). Muzzle and cheeks often with whitish or russet patches that join across upper lip. Claws on forefeet gracile: claw on Digit 3 of forefoot largest (ca. 11 mm long, 3.0–4.2 mm wide at base), claw on Digit 2 shorter (ca. 8 mm long), claw on Digit 1 shortest (ca. half length of Digit 2). Digit 4 vestigial. Skull elongate (breadth 57–65% of GLS) with a narrow palate (26–27% of GLS). Temporal bulla housing enlarged, club-shaped malleus less developed than in *C. asiatica*. Small, tricuspid M^3 present, molar talonids well-developed but sometimes reduced to feeble ledges by toothwear. See also family and genus profiles. Extent of sexual dimorphism in body size cannot be evaluated because of small samples available for study.

Geographic Variation The validity and geographic limits of subspecies are uncertain owing to the few specimens available.

C. s. stuhlmanni: Rwenzori Mts along border of Uganda and DR Congo, from Mt Rwenzori southwards to Rutshuru and Rukiga district of Rwanda.

C. s. fosteri: Mt Elgon, Uganda and Cherangani Hills, Kenya. Dorsal pelage dull-brown to grey, not so bright-brown as in *stuhlmanni*; ventral pelage more greyish.

C. s. tropicalis: Uluguru mountains and Rungwe district, Tanzania. Dorsal pelage glossy blackish-brown; ventral pelage slightly paler with a rufous tint in inguinal region.

C. s. vermiculus: Yambuya and Kisangani, N DR Congo. Size smaller, similar to *Calcochloris leucorhinus*, but without white facial markings; dorsal pelage pale brown to fawn; ventral pelage slightly paler. Known only from a preserved skin, and therefore *incertae sedis*. Possibly only a geographic variant of *C. leucorhinus*, which occurs 320 km south-west at Bokongu in C DR Congo.

C. s. balsaci: Mt Oku, Cameroon.

Similar Species

Calcochloris leucorhinus. Skull smaller (GLS: 19.5–23.1 mm); malleus small and unspecialized; bulla not visible externally. Might be sympatric but full extent of distribution not known.

Distribution Endemic to Africa. Eastern Rainforest–Savanna Mosaic (Mt Elgon, Uganda and Cherangani Hills, Kenya), Afromontane–Afroalpine BZ (Uganda, E DR Congo), northern edge of Zambezian Woodland BZ (one locality in N Tanzania), Rainforest

Chrysochloris stuhlmanni

BZ (one locality in E DR Congo) and Northern Rainforest–Savanna Mosaic (one locality in Cameroon; Lamotte & Petter 1981). Distribution disjunct and largely restricted to mountain ranges (except *C. s. vermiculus*).

Habitat Well-drained soils at high altitudes, from 1700 to 3500 m. Widespread in montane grasslands, bamboo thickets and ericaceous vegetation, as well as in *Podocarpus* and *Hagenia*/*Hypericum* woodlands. Prefer light soils with sparse vegetation cover.

Abundance Locally common but no quantitative data. Less common in heavily grazed grasslands above 3000 m; avoids exotic plantations (Duncan & Wrangham 1971).

Adaptations Nocturnal. Individuals make subsurface tunnels about 4 cm below the surface but also, in forests, in leaf litter. Food distribution and soil penetrability are important factors governing dispersion. In swampy areas they make branched, interconnecting tunnels with small (6–8 cm diameter) unlined chambers in thick layers (10–15 cm) of sphagnum moss, seldom penetrating the heavy peaty soils below. In better-drained areas, burrowing activity is concentrated in clumps of vegetation that are connected by unbranched tunnels, and small mounds of soil are formed above ground (Duncan & Wrangham 1971, Jarvis 1974). When alarmed, reported to make a 'noise similar to that emitted by the puff-adder' (St Leger 1931), possibly as a threat display.

Foraging and Food Carnivorous and insectivorous. Earthworms and insect (mainly dipteran) larvae form the bulk of the diet, supplemented with myriapods, molluscs and small crustaceans (presumably isopods). Analyses of stomach contents: 100% contained earthworms, 50% contained insect larvae, 12% contained adult insects (n = 8; Duncan & Wrangham 1971, Jarvis 1974).

Social and Reproductive Behaviour Duncan & Wrangham (1971) reported that burrow systems are used by more than one individual, since two golden-moles were caught in one trap in a tunnel that showed signs of renewed use a week later. These two individuals may, however, have been a mother and her near-adult youngster, or the later capture could reflect colonization of the vacant burrow system by another individual.

Reproduction and Population Structure Parous ♀♀ recorded in Jul, each with one embryo in the left uterine horn (n = 3; Duncan & Wrangham 1971). Gestation period not known. At birth, young are altricial and hairless, and remain with the mother for 2–3 months (Kingdon 1974).

Predators, Parasites and Diseases Predation pressure is low and does not seem to affect distribution. Ectoparasites include two species of fleas: *Xiphiopsylla* sp., which also occurs on the mole-rat *Tachyoryctes splendens*; and a species of *Ctenophthalmus* allied to *C. cophurus*, probably transferred to *C. stuhlmanni* by sympatric burrowing rodents (particularly *Lophuromys flavopunctatus*; Duncan & Wrangham 1971). Heligmosomatine nematodes recorded in alimentary tract (Jarvis 1974).

Conservation IUCN Category: Least Concern.

Measurements

Chrysochloris stuhlmanni
HB (♂♂): 112.8 (106–123) mm, n = 5
HB (♀♀): 106.5 (103–111) mm, n = 4
T: 0 mm
HF (♂♂): 12.6 (10–12) mm, n = 5
HF (♀♀): 11.2 (10–12) mm, n = 4
E: 0 mm
WT (♂): 42 g, n = 1
WT (♀): n. d.
GLS (♂♂): 26.9 (24.8–28.1) mm, n = 7
GLS (♀♀): 25.9 (24.3–28.3) mm, n = 6
GWS (♂♂): 16.2 (14.7–17.2) mm, n = 8
GWS (♀♀): 15.8 (14.3–17.2) mm, n = 6
C–M^3 (♂♂): 6.5 (5.8–6.9) mm, n = 7
C–M^3 (♀♀): 6.6 (6.0–7.4) mm, n = 6
Museum specimens throughout geographic range

Key References Duncan & Wrangham 1971; Jarvis 1974; Kingdon 1974.

Gary N. Bronner

Chrysochloris visagiei VISAGIE'S GOLDEN-MOLE

Fr. Taupe dorée de Visagie; Ger. Visagies Goldmull

Chrysochloris visagiei Broom, 1950. Ann. Transvaal Mus. 21: 238. Gouna, 86 km east of Calvinia. Northern Cape Province, South Africa.

Taxonomy Simonetta (1968) treated *visagiei* as a subspecies of *C. asiatica*. The shape of the malleus differs from that of *C. asiatica* (Broom 1950), but also varies geographically within *C. asiatica*, so this character cannot be considered diagnostic. The holotype may be merely an aberrant *C. asiatica* (Meester 1974); more specimens are needed to clarify the status of this taxon. Synonyms: none. Chromosome number: not known.

Description Small species known only from the ♀ holotype specimen. Dorsal pelage pale tawny-olive, hairs grey with fawn tips. Underfur slaty-grey. Ventral pelage somewhat paler, greyish, with fewer tawny hairs. Muzzle and upper lip much paler than back. Forefeet as in *C. asiatica*; claw of Digit 3 largest (4.11 mm wide at base), Digit 4 vestigial with a small nail. Skull with well-developed temporal bullae, malleus club-shaped but shorter and broader (length:breadth <2) than in *C. asiatica* (length:breadth >3) (Broom 1950). Small, tricuspid M^3 present, molar talonids absent. Skull: see also family and/or genus profiles.

Geographic Variation None recorded.

Similar Species None within the geographic distribution of the species.

Distribution Endemic to Africa. South-West Arid BZ (Karoo). Known only from the type locality.

Habitat Bushmanland Nama-Karoo.

Abundance No information.

Remarks Apparently no other information available.

Conservation IUCN Category: Data Deficient, previously Critically Endangered.

A survey (in 2002) of the type locality yielded no signs of golden-moles, and the habitat (gravel plains) is atypical of chrysochlorids. This raises the possibility that the species is either extinct, or that the original specimen was translocated to the type locality accidentally by humans.

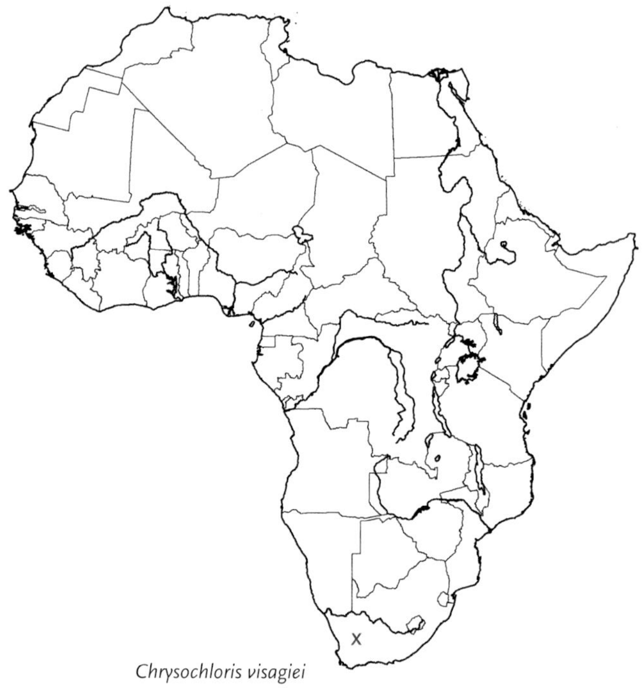

Chrysochloris visagiei

Measurements
Chrysochloris visagiei
HB: 106 mm, n = 1
T: 0 mm, n = 1
HF: 12 mm, n = 1
E: n. d.
WT: n. d.
GLS: 22.8 mm, n = 1
GWS: 17.5 mm, n = 1
C–M^3: 6.5 mm, n = 1
Holotype ♀

Key Reference Broom 1950.

Gary N. Bronner

GENUS *Chrysospalax*
Golden-moles

Chrysospalax Gill, 1883. Standard Nat. Hist. 5 (Mamm.): 137. Type species: *Chrysochloris trevelyani* Günther, 1875.

The genus *Chrysospalax* contains the two largest species of golden-mole, both endemic to eastern parts of South Africa. The Giant Golden-mole *C. trevelyani* lives in indigenous forests, whereas the Rough-haired Golden-mole *C. villosus* lives in densely vegetated mesic grasslands. Both species forage above ground to a greater extent than do other golden-moles. The Giant Golden-mole shows some social behaviour (an unusual character in golden-moles).

Apart from their larger size, these species are distinguished by a long, shaggy pelage with very coarse guard hairs. The claws on the forefeet, although obviously larger than in other species, are long but slender, perhaps reflecting the habit of these golden-moles of constructing only short tunnels and foraging mainly above ground. Skull elongate (breadth: 60–65% of GLS) with a relatively broad palate (30–35% of GLS). The unique feature of the skull is broad plates that develop posteriorly from the zygomatic arches and sweep backwards over the cranium. A sagittal crest for the insertion of strong temporal muscles used in chewing is pronounced, especially in *C. trevelyani* (Figure 11). The malleus is enlarged and pea-shaped, and housed in a large bony bulla (up to 7 mm diameter) that protrudes anteriolaterally at the level of the zygomatic arch to incorporate the glenoid joint, so that the whole side of the skull bulges outwards. This may be an adaptation for surface foraging since increased malleus mass enhances sensitivity to higher frequencies of sound than are normally experienced underground (Mayer *et al.* 1995). Dentition comprises 40 teeth (tricuspid M^3 present), similar in morphology to those of more generalized species (such as *Chrysochloris asiatica*). Larger, lower molars have well-developed talonids. These characters suggest that these are the most primitive of the golden-moles, yet also highly specialized (Roberts 1951). *Chrysospalax villosus* was long referred to the genus *Bematiscus*, but differences between these forms do not warrant more than specific distinction (Meester 1974). The two species are distinguished by the colour and texture of the pelage, size and habitat preference.

Gary N. Bronner

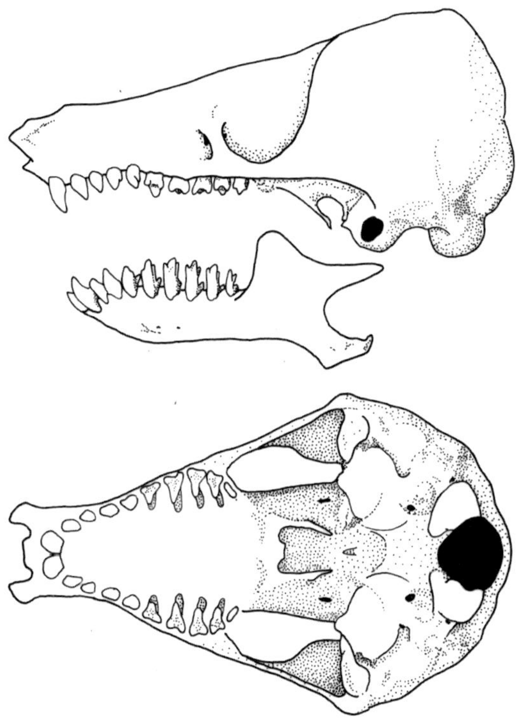

Figure 11. Skull and mandible of *Chrysospalax trevelyani* (based on Simonetta 1968).

Chrysospalax trevelyani Giant Golden-mole

Fr. Taupe dorée géante; Ger. Riesengoldmull

Chrysospalax trevelyani (Günther, 1875). Proc. Zool. Soc. Lond. 1875: 311.
Pirie Forest, King Williams Town, Eastern Cape Province, South Africa.

Taxonomy Synonyms: none. Chromosome number: 2n = 30, FN = 56 (Bronner 1995a).

Description Originally described in the genus *Chrysochloris*. Very large golden-mole, the largest in the family. Dorsal pelage dark glossy-brown, yellowish-brown or greyish-brown. Ventral pelage similar, but paler. Head slightly darker than back, cheeks with yellowish patches above subdermal eyes. Guard hairs long and coarse, up to 20 mm on back, greyish-yellow at base, dark brown with a yellow tinge at tip. Throat, chest and underparts of forelimbs dull yellow; hairs shorter than on dorsal pelage. Underfur dense, buffy-grey. Claws on forelimbs powerful and pick-like; claw on Digit 3 of forefoot largest (17–19 mm long, 7 mm wide at base), claw on Digit 2 slightly smaller, claw on Digit 1 very small. Digit 4 reduced to a small tubercle with a rounded claw. Skull with broad plates that develop posteriorly from the zygomatic arches and sweep backwards over the cranium; M^3 present (see also family and genus profiles). No apparent sexual size dimorphism, but this may be an artefact of small samples available for study.

Geographic Variation There is considerable inter- and intra-population variation in pelage colour; colour variation within demes is as great as between populations and therefore no subspecies are recognized (Broom 1907a).

Similar Species

Amblysomus hottentotus. Much smaller; finer pelage; malleus bones not enlarged; lacks the broad plates sweeping back from the zygomatic arches.

Distribution Endemic to Africa. Coastal Forest Mosaic and Afromontane–Afroalpine BZs. Recorded in Eastern Cape Province from East London northwards along coast to Port St Johns, and inland to Amatola and Kologha Mountains near King Williams Town, Stutterheim and Hogsback. May be locally extinct in many places where it occurred formerly.

Habitat Coastal and Afromontane Forests. Sometimes also penetrates into grasslands adjoining forests (Maddock 1986).

Abundance Locally common but with a clumped dispersion. Greatest abundance in forest patches with soft soils, well-developed undergrowth and deep litter layers. Avoids steep slopes and rocky terrain.

Adaptations Giant Golden-moles make numerous short (1–13 m) tunnels amongst roots of trees; entrances are left open and connected by well-defined surface runways (Maddock 1986). Individuals may move up to 130 m along surface runways during a night, but sometimes forage within only a few metres of the burrow

Chrysospalax trevelyani

from which they emerged. They readily enter water and swim using forefeet as paddles (Ranger 1927). Activity is mainly at night, from shortly after dusk until dawn, although diurnal activity occurs on cool cloudy days. In captivity, they are mainly nocturnal with short peaks of activity from 16:00h to 20:00h, and again from 23:00h to 01:00h (Maddock & Hickman 1985). Activity reduced during dry winter months (Apr–Aug). Torpor is induced if captive individuals are subjected to low temperatures.

Foraging and Food Forages mainly above ground in leaf litter using small runways that meander off from main tracks linking burrow entrances. Foraging areas marked by characteristic signs of rooting where they stir up soil and leaves in search of prey. Diet comprises earthworms, particularly giant earthworms (*Microchaetes* spp.; Roberts 1951), although in parts of their range where giant earthworms do not occur they feed primarily on oniscomorph millipedes (Duckworth & Hickman 1985).

Social and Reproductive Behaviour These are the only chrysochlorids showing any indication of sociality (Hickman 1990). An unconfirmed report of several adult individuals dug out of the same nest in Jul suggests that small groups may overwinter together in burrows.

Reproduction and Population Structure A pregnant ♀ with two embryos was reported by Hickman (1990). Bronner (1992) recorded the birth of a single altricial pup in Oct. Postnatal development is slow, suggesting extended periods of parental care. During copulation, ♂ mounts ♀ from the back in a dog-like fashion.

Predators, Parasites and Diseases Feral dogs, associated with rural subsistence communities, are known predators. At Nqadu Forest, four individuals were killed by a pack of dogs during one evening in 1990, suggesting that they are comparatively defenceless against groups of predators. Their tough, loose skins are difficult to puncture with the result that the dogs seldom consume the carcasses.

Conservation IUCN Category: Endangered.

Suitable forest habitats have been fragmented and degraded by poor agricultural practices, bark-stripping of trees for traditional medicines, collection of firewood, burning, and by the development of commercial plantations, urbanization and coastal resorts.

Measurements

Chrysospalax trevelyani
HB (♂♂): 222.0 (215–235) mm, n = 7
HB (♀♀): 215.8 (208–229) mm, n = 4
T: 0 mm
HF (♂♂): 24.3 (20–28) mm, n = 8
HF (♀♀): 23.4 (21–27) mm, n = 5
E: 0 mm
WT (♂): 470 g, n = 1
WT (♀♀): 410, 500 g, n = 2
GLS (♂♂): 41.7 (41.0–42.4) mm, n = 5
GLS (♀♀): 41.6 (39.9–43.2) mm, n = 7
GWS (♂♂): 26.2 (25.7–27.4) mm, n = 5
GWS (♀♀): 25.9 (24.2–27.4) mm, n = 6
C–M^3 (♂♂): 9.7 (9.4–9.9) mm, n = 5
C–M^3 (♀♀): 9.8 (9.7–10.2) mm, n = 4
Museum specimens throughout geographic range

Key References Hickman 1990; Maddock 1986; Roberts 1951.

Gary N. Bronner

Chrysospalax villosus ROUGH-HAIRED GOLDEN-MOLE

Fr. Taupe dorée à poil dur; Ger. Rauhaar Goldmull

Chrysospalax villosus (A. Smith, 1833). S. Afr. Quart. J. 2: 81. Towards Natal (near Durban), KwaZulu–Natal, South Africa.

Taxonomy Originally described in the genus *Chrysochloris*. Synonyms: *dobsoni*, *leschae*, *rufopallidus*, *rufus*, *transvaalensis*. Subspecies: six, of uncertain validity. Chromosome number: not known.

Description Large golden-mole with extremely coarse pelage. Dorsal pelage yellowish-brown or greyish-brown to dark slaty-grey (see below for geographic variation). Ventral pelage paler, especially on abdomen. Throat dull yellowish-white. Head slightly paler than back, face and muzzle buffy-grey, often with a dark patch around the horny nosepad. Upper lips and chin yellowish-white suffused with brown. Guard hairs glossy and sparsely distributed, 18–21 mm long on back; greyish-brown at base, with broad medial band of yellowish-brown or reddish-brown, and narrow dark-brown tip. Underfur slaty-grey, dense and woolly, about two-thirds length of guard hairs. Claws on forelimbs weaker than in *C. trevelyani*; claw on Digit 3 long (15–17 mm long and 4–6 mm at base), claw on Digit 2 slightly smaller, claw on Digit 1 about one-third the length of Digit 2. Digit 4 reduced to small tubercle with short claw (ca. 1.5 mm long). No apparent sexual dimorphism but this may be an artefact of small samples available for study. Skull with broad plates that develop posteriorly from the zygomatic arches and sweep backwards over the cranium; malleus enlarged; bulla large and protruding laterally; M^3 present (see also family and genus profiles).

Geographic Variation Meester (1974) listed six subspecies distinguished mainly by subtle differences in pelage colour, but the validity of these subspecies is uncertain.

C. v. villosus: 'Near Durban', KwaZulu–Natal. Pelage yellowish-brown faintly brindled with dark brown, flanks and abdomen greyish-brown.

C. v. dobsoni: Pietermaritzburg and KwaZulu–Natal midlands. Probably a synonym of *villosus*; pelage yellowish-brown, brown or brownish-grey.

C. v. leschae: St Cuthberts, Tsolo, to Tabase, Eastern Cape Province. Dorsal pelage uniformly dark brown; ventral pelage slightly paler.

C. v. rufopallidus: Wakkerstroom northwards to Belfast, Mpumalanga Province. Pelage pale yellowish-red; hairs on back comparatively short, 14–16 mm long.

C. v. rufus: Spitzkop, Sabie district, Mpumalanga Province. Pelage rich reddish-brown with dark grey underfur; smaller than *C. v. transvaalensis*, with shorter claws on forefeet (Meester 1953a).

C. v. transvaalensis: Pretoria and Witwatersrand. Pelage pale reddish-brown. (Includes *pratensis.*)

Similar Species

Amblysomus hottentotus, *Amblysomus septentrionalis* and *Amblysomus robustus*. All to some extent smaller in HB (HB lengths: 104–145 mm); skulls smaller (GLS: 24.6–32.0 mm); pelage finer; malleus not enlarged; bulla not evident externally; zygomatic arches without broad plates sweeping backwards; M^3 usually absent.

Distribution Endemic to Africa. Highveld BZ and parts of Afromontane–Afroalpine BZ. Distribution disjunct, recorded only from scattered localities in Eastern Cape, KwaZulu–Natal, Gauteng and Mpumalanga Provinces, South Africa.

Habitat Sandy soils in grasslands, meadows and along edges of marshes. Occurs in Moist Upland Grassland, Moist Cool Highveld Grassland, Rocky Highveld Grassland and North-eastern Mountain Grassland. Recorded from gardens and parklands; also found in dense stands of kikuyu grass *Pennisetum clandestinum* adjoining natural grasslands.

Abundance Extremely rare and secretive. Difficult to detect owing to predilection for dense grasslands with abundant cover.

Chrysospalax villosus

Adaptations These golden-moles construct short burrows, the entrances to which are left open with loose soil piled up at back and sides (like the holes of crabs). Well-defined surface runways lead to foraging areas, which show the characteristic signs of rooting where soil is turned up by the horny nose. Short (30–70 cm) burrows with two entrances are made periodically along surface tracks, where the moles 'dip' into the ground; these probably serve as bolt-holes if alarmed. They have an extraordinary sense of direction, and dash unerringly and rapidly into the nearest burrow entrance at the slightest sign of danger. Sometimes shams death when handled.

Foraging and Food Insectivorous. Roberts (1951) stated that these golden-moles forage above ground at night, and only after rains. Foraging in underground burrows seems probable given the unpredictability of rains and periodic droughts.

Social and Reproductive Behaviour No information.

Reproduction and Population Structure No information, except that ♀♀ give birth to two young.

Predators, Parasites and Diseases No information on predators. Lawrence (1944) recorded a new species of ectoparasitic mite (*Schizoptes*) on this species.

Conservation IUCN Category: Vulnerable.

Only three specimens were collected during the 23-year period 1980–2003. The widespread practice of allowing cattle to graze in marshes and dense grasslands near water during dry winter months leads to trampling and a loss of cover, and this undoubtedly impacts severely on this species. Some areas in which they formerly occurred (eg. Pretoria West) have been completely transformed by urbanization.

Measurements
Chrysospalax villosus
HB (♂♂): 163.0 (148–175) mm, n = 4
HB (♀♀): 142.8 (127–160) mm, n = 6
T: 0 mm
HF (♂♂): 14.3 (14–17) mm, n = 4
HF (♀♀): 17.1 (13–20) mm, n = 6
E: 0 mm
WT (♂♂): 137.7 (108–160) g, n = 3
WT (♀♀): 93, 105 g, n = 2
GLS (♂♂): 33.9, 34.9 mm, n = 2
GLS (♀♀): 33.8 (33.4–34.6) mm, n = 5
GWS (♂♂): 20.5 (19.6–22.2) mm, n = 3
GWS (♀♀): 21.9 (19.6–23.1) mm, n = 5
C–M^3 (♂♂): 7.4 (7.1–7.7) mm, n = 4
C–M^3 (♀♀): 7.2 (6.4–8.1) mm, n = 5
Museum specimens throughout geographic range

Key References Meester 1953; Roberts 1951.

Gary N. Bronner

GENUS *Cryptochloris*
Golden-moles

Cryptochloris Shortridge and Carter, 1938. Ann. S. Afr. Mus. 32: 284. Type species: *Cryptochloris zyli* Shortridge and Carter, 1938.

The genus contains two very small species, *C. zyli* and *C. wintoni*, which differ mainly in colour of the pelage, and are known only from only a few specimens collected at the type localities on the Namaqualand coastal plain in South Africa. Like Grant's Golden-mole (genus *Eremitalpa*), the body is lozenge-shaped, in contrast to the fusiform shape of other genera. The most distinctive external feature is the possession of a large digging pad (possibly supported by a radial sesamoid bone) at the base of Digit 1 of the forefoot, so that the forefoot is much broader than in other species. The forefeet bear long slender claws; claw on Digit 3 long (10–11 mm long, 4.0–4.5 mm wide at base), claws on Digit 1 and 2 only slightly shorter than Digit 3, claw on Digit 4, while moderately developed, is more gracile than in *Eremitalpa*. The skull is short and broad (breadth 70–76% of GLS) with a very wide palate (35–37% of GLS), but slightly narrower than in *Eremitalpa*. Zygomatic arches without broad plates sweeping backwards. The olfactory region is conspicuously inflated and much wider than the anterior part of the cranium (Figure 12). The malleus is club-shaped and enlarged, although not to the extent seen in *Chrysochloris*, with the result that only small subtemporal bullae are present. Tricuspid M^3 present; lower molars lack talonids. These characters suggest that *Cryptochloris* is intermediate between *Eremitalpa* and *Chrysochloris* (Shortridge 1942). Simonetta (1968) regarded *Cryptochloris* as a synonym of *Chrysochloris*. The two species are distinguished by pelage colour, morphology of forelimbs and shape of the malleus.

Gary N. Bronner

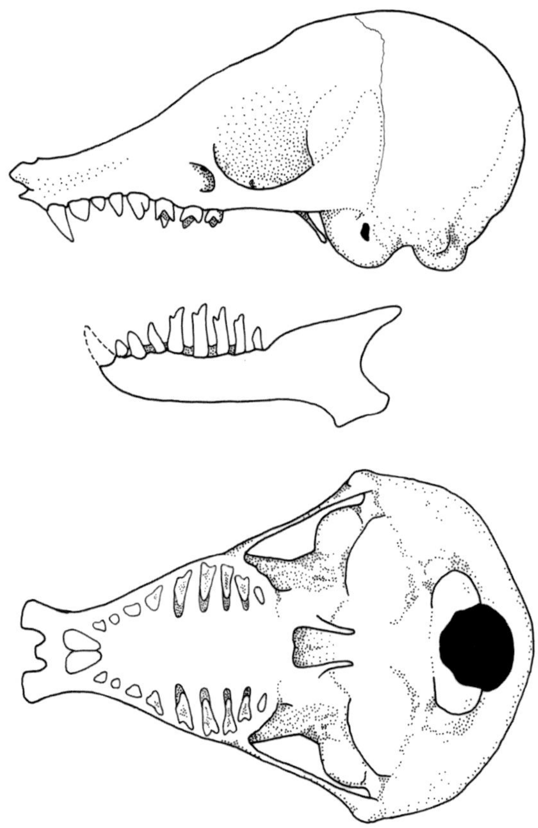

Figure 12. Skull and mandible of *Cryptochloris zyli* (based on Simonetta 1968).

Cryptochloris wintoni DE WINTON'S GOLDEN-MOLE
Fr. Taupe dorée de De Winton; Ger. De Wintons Goldmull

Cryptochloris wintoni (Broom, 1907). Ann. Mag. Nat. Hist., ser. 7, 19: 264.
Port Nolloth, Little Namaqualand, Northern Cape Province, South Africa.

Taxonomy Originally described in the genus *Chrysochloris*. Some authors (e.g. Simonetta 1968) regarded this monotypic species and *C. wintoni* as only subspecifically distinct, but these taxa differ consistently in pelage colour and malleus morphology, indicating that they are not conspecific (Meester 1974). Synonyms: none. Chromosome number: not known.

Description Very small species resembling *Eremitalpa granti* in size and colour. Dorsal pelage pale slaty-grey with yellowish tinge becoming more intense on forehead, cheeks and lips; hairs whitish, grey at base, broad fawn at tips. Ventral surface slightly paler than dorsal pelage, hairs whitish at tips. Body foreshortened and lozenge-shaped. Nosepad 4 mm long, ca. 8 mm wide, corners rounded, anterior digging portion more prominent than in *C. asiatica*. Claw on Digit 3 largest (10.3–10.5 mm long and 4 mm wide at base), claw on Digit 2 shorter (5.0–7.0 mm), claw on Digit 1 also shorter (4.5–4.8 mm). Claw on Digit 4 small (1.5 mm long). Skull: malleus moderately enlarged, club-shaped with head more bulbous than in *C. zyli* (Broom 1907a, b, Roberts 1951); bulla small; M^3 present (see also family and genus profiles). Small sample sizes preclude any evaluation of sexual dimorphism.

Geographic Variation None recorded.

Similar Species

Eremitalpa granti. Smaller (HB: 76–86 mm, GLS: 18.6–20.4 mm); claw on Digit 4 of forefoot longer (3–4 mm); olfactory region of skull not markedly inflated.

Chrysochloris asiatica. Larger (HB: 94–115 mm); claws on Digits 1, 2 and 4 markedly smaller than than on Digit 3; palate relatively narrower (30–32% of GLS); malleus more enlarged so that the temporal bulla is more pronounced.

Distribution Endemic to Africa. South-West Arid BZ (Namib Desert). Recorded only from the type locality at Port Nolloth, Northern Cape Province, South Africa.

Cryptochloris wintoni

Known only from the type locality, situated in a region that is being markedly transformed by alluvial diamond mining.

Habitat Coastal dunes and adjacent sandy areas in Strandveld Succulent Karoo.

Abundance Extremely rare; known from only three specimens and not captured for over 50 years.

Remarks Apparently no other information available.

Conservation IUCN Category: Critically Endangered.

Measurements

Cryptochloris wintoni
HB: 86, 92 mm, n = 2
T: 0 mm
HF: 10.3, 10.5 mm, n = 2
E: 0 mm
WT: n. d.
GLS: 21.3, 22.0 mm, n = 2
GWS: 15.5, 16.2 mm, n = 2
C–M^3: n. d.
Type locality (Broom 1907a, Roberts 1951)

Key References Broom 1907a, b.

Gary N. Bronner

Cryptochloris zyli VAN ZYL'S GOLDEN-MOLE

Fr. Taupe dorée de Van Zyl; Ger. Van Zyls Goldmull

Cryptochloris zyli Shortridge and Carter, 1938. Ann. S. Afr. Mus. 32: 284. Compagnies Drift, Little Namaqualand, Western Cape Province, South Africa.

Taxonomy Some authors (e.g. Simonetta 1968) regarded this monotypic species as only subspecifically distinct from *C. wintoni*. However, these taxa differ consistently in pelage colour and malleus morphology, indicating that they are not conspecific (Meester 1974). Synonyms: none. Chromosome number: not known.

Description Very small species with violet metallic sheen restricted to coarser guard hairs, giving the pelage a spangly appearance. Dorsal pelage dense and short, dark lead-grey; underfur smoky-grey. Ventral pelage similar to dorsal pelage, but drab. Face with well-defined whitish markings (admixed with slightly darker hairs) extending posteriorly to above the subdermal eyes. Claw on Digit 3 long (10.0 mm long, 4 mm wide at base), claws on Digits 1 and 2 of forefoot subequal to Digit 3 so that these three claws meet almost in a point. Claw on Digit 4 small (2.0 mm long) and narrow. Skull: malleus club-shaped but smaller and less bulbous than in *C. wintoni* and *C. asiatica*; M^3 present (see also family and genus profiles). Small sample sizes preclude any evaluation of sexual dimorphism.

Geographic Variation None recorded.

Similar Species

Eremitalpa granti. Dorsal pelage tawny-olive to greyish-yellow, finer in texture. Digit 4 of forefoot with a broad scraper-like claw, 3–4 mm long and 2–3 mm wide at base.

Chrysochloris asiatica. Larger (HB: 94–115 mm); claws on Digits 1, 2 and 4 of forefoot markedly smaller than that on Digit 3; malleus more enlarged so that the temporal bulla is pronounced.

Distribution Endemic to Africa. South-West Arid BZ (Karoo). Recorded only from the type locality near Lamberts Bay, South Africa. Possibly ranges northwards along coast to the Olifants R.

Habitat Coastal dune belt and adjacent sandy areas in Strandveld Succulent Karoo.

Abundance Extremely rare; known from only three specimens (Helgen & Wilson 2001).

Remarks When handled, individuals may emit high-pitched squeaks or sham death (Shortridge 1942).

Predators, Parasites and Diseases No information.

Conservation IUCN Category: Endangered.

Known only from the type locality (situated in a region that is being markedly transformed by alluvial diamond mining) and one other locality.

Cryptochloris zyli

Measurements

Cryptochloris zyli
HB: 82 mm, n = 1
T: 0 mm
HF: 12 mm, n = 1
E: 0 mm
WT: n. d.
GLS: 21.7, 22.2 mm, n = 2
GWS: 15.0, 15.2 mm, n = 2
C–M^3: 8.0 mm, n = 1
South Africa (Shortridge 1942, Helgen & Wilson 2001)

Key References Broom 1946; Helgen & Wilson 2001; Shortridge 1942.

Gary N. Bronner

GENUS *Eremitalpa*
Grant's Golden-mole

Eremitalpa Roberts, 1924. Ann. Transvaal Mus. 10: 63. Type species: *Chrysochloris granti* Broom, 1907.

A monotypic genus endemic to the South-West Arid BZ. Grant's Golden-mole *E. granti* is the smallest of the golden-moles, and inhabits loose sands along the west coast of South Africa and Namibia. Highly specialized for life in the Namib Desert, hence its alternative vernacular name, the Namib Golden-mole.

The body is foreshortened and lozenge-shaped, the claws of the forefeet are long and slender as in *Cryptochloris*. The skull is broader than in any other species (breadth >85% of GLS) with a very wide palate (35–38% of GLS) (Figure 13). Snout short, lacking the marked inflation of the olfactory region seen in *Cryptochloris*. Zygomatic arches without broad plates sweeping backwards. Malleus enlarged and pea-shaped, housed in enlarged epitympanic recess that inflates the internal laminae of the squamosal and temporal bones, but not conspicuous externally. Minute tricuspid M^3 present, but incidence varies among jaw quadrants. Molar talonids absent. Similarities in middle ear structure with *Chrysospalax* probably arose independently in response to a similar mode of foraging predominantly above ground (Mayer *et al.* 1995). Other characters are given in the species profile.

Gary N. Bronner

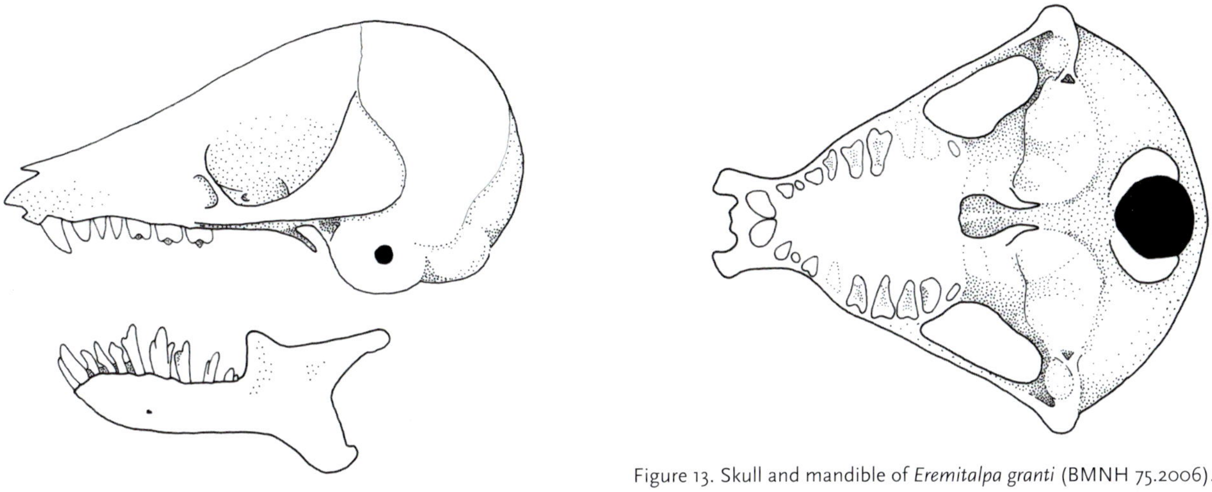

Figure 13. Skull and mandible of *Eremitalpa granti* (BMNH 75.2006).

Eremitalpa granti GRANT'S GOLDEN-MOLE (NAMIB GOLDEN-MOLE)

Fr. Taupe dorée de Grant; Ger. Grants Goldmull

Eremitalpa granti (Broom, 1907). Ann. Mag. Nat. Hist., ser. 7, 19: 265. Garies, Little Namaqualand, Northern Cape Province, South Africa.

Taxonomy Originally described in the genus *Chrysochloris*. Synonyms: *cana*, *namibensis*. Subspecies: two. Chromosome number: not known.

Description Very small golden-mole with a lozenge-shaped body. Pelage silky, 7–12 mm on back, up to 20 mm on flanks. Dorsal pelage tawny-olive to greyish-yellow, flanks paler with intense yellowish tinge; guard hairs grey at base, upper third pale brownish-yellow, darker brown at tip. Ventral pelage greyish-brown to fawn. Forehead and cheeks pale yellow, sometimes with pale pinkish-buff around nosepad. Claw on Digit 3 of forefoot largest (9–12 mm long, 3.7–4.9 mm wide at base), claw on Digit 1 flattened and nearly as long (4–5 mm) as on Digit 2 (5–8 mm) but noticeably narrower. Digit 4 vestigial but with a broad scraper-like claw (3–4 mm long, 2–3 mm wide at base). Hindfoot with prominent thickened pad. Skull: relatively broad (breadth >85% of GLS); malleus pea-shaped, enlarged; bulla not visible externally; M^3 present but minute (see also family and genus profiles).

Geographic Variation

E. g. granti (includes *E. g. cana*): St Helena Bay along coast to Port Nolloth, and inland to Garies. Larger; pelage less yellowish, hairs longer (8–13 mm at mid-back); skull significantly longer (18.6–20.4 mm) and narrower than in *E. g. namibensis* (breadth 79.9–88.3% of GLS).

E. g. namibensis: Orange R. northwards along coast to Walvis Bay, and inland as far as ProNamib. Smaller; pelage with more intense yellowish tinge, hairs shorter (6.5–7 mm at mid-back); skull shorter (16.8–19.4 mm) and relatively broader than in *E. g. granti* (breadth: 85.3–95.7% of GLS).

Similar Species

Cryptochloris zyli. Dorsal pelage lead-grey with violet metallic sheen, coarser in texture. Claw on Digit 4 of forefoot shorter (2.0 mm) and narrow.

Cryptochloris wintoni. Larger (HB: 86–92 mm, GLS: 21.3–22.0 mm, n = 2); claw on Digit 4 of forefoot shorter (1.5 mm); olfactory region of skull markedly inflated.

Chrysochloris asiatica. Much larger (HB: 94–115 mm, GLS: 21.1–25.4 mm); body fusiform; claws on Digits 1, 2 and 4 of forefoot markedly shorter than that on Digit 3; skull relatively less broad (breadth 70–78% of GLS); malleus club-shaped, more enlarged; bulla pronounced and visible externally.

Distribution Endemic to Africa. South-West Arid BZ (Namib Desert and Karoo). Confined to the west coast of southern Africa, from St Helena Bay (Western Cape Province, South Africa) northwards to Walvis Bay (Namibia).

Eremitalpa granti

Habitat Strandveld Succulent Karoo and Namib Desert. Prefers soft, shifting sands of dunes but also present in inter-dune valleys with quite dense vegetation providing sand is not too consolidated.

Abundance Densities are low owing to the arid, energy-sparse conditions of the environment.

Adaptations The broad, hollowed-out claws are adapted for burrowing in loose sand, and the thickened pads on the hindfeet facilitate walking on the sand surface. Primarily nocturnal with occasional bouts of afternoon activity, especially in winter. Activity patterns vary adaptively in relation to environmental conditions, particularly temperature and prey availability. Nocturnal activity related primarily to maintenance of body temperature and water balance rather than predator avoidance (Fielden *et al.* 1992). Diurnal activity involves subsurface sand-swimming, which leaves U-shaped furrows or raised ridges if sand is moist. During the heat of the day, rests under the roots of bushes or dune grasses at depths of up to 50 cm underground, and becomes active in the early afternoon during cool winter days. Surface locomotion at night reduces energetic costs of foraging in an energy-sparse environment, and appears to be more common in sparsely vegetated areas. Some burrows are re-used, presumably those leading to nests where young are raised, or those leading to foraging areas.

The senses are poorly developed, except for touch and sensitivity to vibrations, the latter due to the extraordinary development of the middle ear (Mason & Narins 2001). Energetic requirements are reduced, when necessary, by inactivity, adaptive hypothermia, extreme thermolability and intrinsic metabolic suppression. Assimilation efficiency is high (79%) as in other insectivorous mammals (Seymour *et al.* 1998). Maintenance of water balance is facilitated by a reduction in body temperature and metabolic rate, behaviour (nocturnal activity and surface foraging) and efficient renal function (Fielden *et al.* 1990a).

Foraging and Food Insectivorous and carnivorous. The length of foraging pathways, and foraging tactics, depend on prey dispersion. For example, Grant's Golden-moles may forage over distances of 5–695 m/night (Fielden *et al.* 1990b) and as far as 5800 m (Holm 1969). Average foraging distance is 1412 m (Seymour *et al.* 1998). Surface activity is interspersed with head dips into sand every 3–5 m, presumably to sense seismic cues (Narins *et al.* 1998) or vibrations emitted by prey. Detection of prey is effective over only short distances; search pattern is non-random and is effective at locating localized patches where prey is abundant. Sand-swimming predominates once patches of prey are detected, and when searching beneath the roots of plants. The subspecies *E. g. namibensis* is a highly selective feeder specializing on termites (*Psammotermes allocerus*). Analysis of stomach contents (n = 16): 89% contained termites, 37% beetle larvae, 30% adult beetles and 26% contained root material. Termites comprise 95% (by weight) of stomach contents despite the low biomass of termites relative to other potential invertebrate prey, and their clumped dispersion in subterranean nests under clumps of dune grasses. Ingestion of roots may not be entirely incidental (Fielden *et al.* 1990b). Web-footed geckos (*Palmatogecko rangei*), legless lizards and sand-burrowing skinks may also be eaten (Roberts 1951).

Social and Reproductive Behaviour Adults solitary, but tracks of several individuals have been found in the same area on the same night suggesting that a certain amount of home-range overlap is tolerated. Unlike most other golden-moles, individuals housed together do not fight or display signs of aggression. Whether this is indicative of greater sociality than in other chrysochlorids, or simply an artefact of captivity, is uncertain.

Reproduction and Population Structure Pregnancies recorded in Oct and Nov. Litter-size: 1–2. Placentation and early embryology is typical of eutherian mammals but with some primitive features (Hickman 1990).

Predators, Parasites and Diseases Nocturnal predators include Barn Owls *Tyto alba* and Spotted Eagle-owls *Bubo africanus* and small terrestrial carnivores. Black-backed Jackals *Canis mesomelas* and Striped Polecats *Ictonyx striatus* (and possibly also genets *Genetta genetta*) follow furrows and ridges and then dig for the golden-moles. Diurnal predators include Crows *Corvus alba* and Pale Chanting Goshawks *Melierax canorus* (Fielden *et al.* 1992). Acanthocephalan cystacanths (Origacanthrorynchidae) recorded in the peritoneal cavity (Hickman 1990).

Conservation IUCN Category: Least Concern, previously Vulnerable.

The coastal sands inhabited by *E. g. granti* are mined extensively for alluvial diamonds.

Measurements

Eremitalpa granti granti
HB: 81.5 (76–86) mm, n = 17
T: 0 mm
HF: 10.5 (9–12) mm, n = 7
E: 0 mm
WT: 23 (15–30) g, n = 17
GLS: 19.5 (18.6–20.4) mm, n = 46
GWS: 16.6 (15.5–18.2) mm, n = 46
C–M^3: 5.5 (5.2–5.7) mm, n = 5
Measurements: Port Nolloth, South Africa (Meester 1964)

Key References Fielden *et al.* 1990b, 1992; Narins *et al.* 1998; Seymour *et al.* 1998.

Gary N. Bronner

GENUS *Neamblysomus*
Golden-moles

Neamblysomus Roberts, 1924. Ann. Transvaal Mus. 10: 64. Type species: *Chrysochloris gunningi* Broom, 1908.

This genus contains two species endemic to South Africa. Gunning's Golden-mole *N. gunningi* is found in montane forests of the Drakensberg range, and Juliana's Golden-mole *N. julianae* lives in bushveld habitats of northern South Africa. Body shape fusiform. Claws on forefoot long and slender. Skull elongate (breadth 60–68% of GLS) with narrow palate (<30% of GLS). Zygomatic arches without broad plates sweeping backwards. There is a small, unspecialized malleus bone in the middle ear (so no temporal or lateral bullae is visible externally) as in *Amblysomus* spp. (Figure 14). M^3 present or absent. Lower molars lack well-developed talonids, but a lower tricuspid or peg-like M_3 is present, although variable amongst jaw quadrants (*N. gunningi*) or populations (*N. julianae*). Previous authors synonymized *Neamblysomus* under *Amblysomus* (see Meester 1974). Subtle cranial and dental divergence between these genera has, however, been accompanied by karyotypic differentiation (Bronner 1995b). The two species can be distinguished by size, pelage colouration and skull width.

Gary N. Bronner

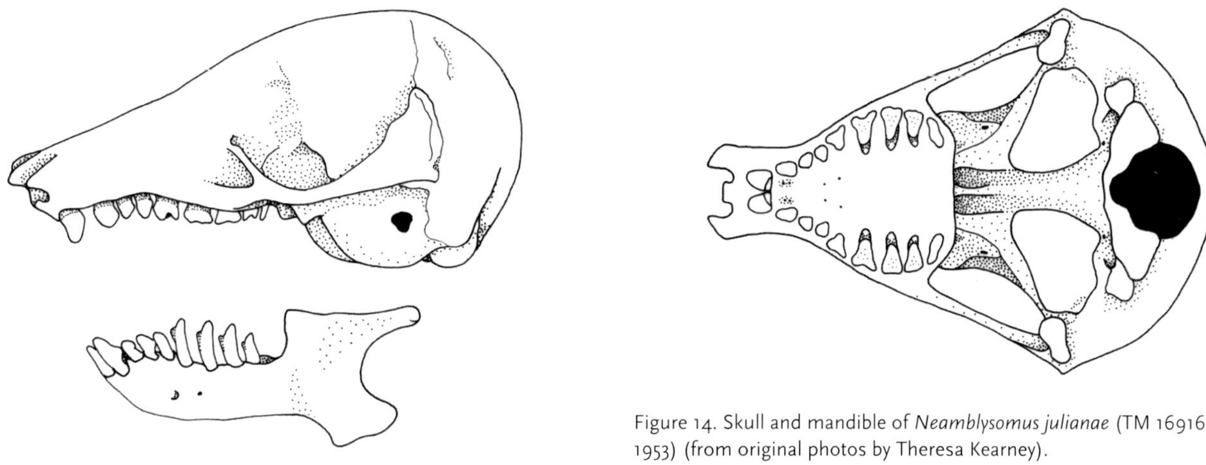

Figure 14. Skull and mandible of *Neamblysomus julianae* (TM 16916 and TM 1953) (from original photos by Theresa Kearney).

Neamblysomus gunningi GUNNING'S GOLDEN-MOLE

Fr. Taupe dorée de Gunning; Ger. Gunnings Goldmull

Neamblysomus gunningi (Broom, 1908). Ann. Transvaal Mus. 1: 14. Woodbush Hill, Soutpansberg district, Limpopo Province, South Africa.

Taxonomy Originally described in the genus *Chrysochloris*. Assigned to *Neamblysomus* by Roberts (1924), Allen (1939), Roberts (1951) and Bronner (1995b) although some authorities, including Ellerman *et al.* (1953) and Meester *et al.* (1986), placed it in *Amblysomus*. Synonyms: none. Chromosome number: 2n = 30, FN = 56 (Bronner 1995a).

Description Medium-sized species, similar in build and appearance to *A. hottentotus*, but with more slender foreclaws. Dorsal pelage dark reddish-brown. Ventral pelage fawn, throat and cheeks yellowish-brown. Claw on Digit 3 of forefoot largest (12.5–14.0 mm long, 4.3–4.9 mm wide at base), claw on Digit 2 much shorter (6.0–7.5 mm), claw on Digit 1 smaller still and narrow. Digit 4 reduced to a small tubercule with a knob-like claw. Skull elongate (breadth 60–63% of GLS) with narrow palate (28–30% of GLS). Molars lack well-developed talonids except in some young individuals. M3 present variably, with a triconid appearance (similar to other molars) in younger individuals, reduced and peg-like in appearance with increasing toothwear. See also family and genus profiles. Males slightly larger on average than ♀♀, but sexes do not differ markedly in cranial size (Bronner 1995b).

Geographic Variation None recorded.

Similar Species

A. septentrionalis and *A. hottentotus*. More robust claws; molar talonids well developed; M3 usually absent; subtle chromosomal differences.

N. julianae. Smaller (HB: 92–111 mm, GLS: 21.8–23.6 mm); skull often narrower (GWS: 14.7–16.7 mm), breadth 67–70% of GLS; lighter (WT: 21–35 g); pelage paler.

Distribution Endemic to Africa. Extreme south of the Zambezian Woodland BZ. Recorded only from the far northern Drakensberg escarpment, between Haenertsburg, New Agatha and Tzaneen, Limpopo Province, South Africa.

Habitat Afromontane forest and adjacent grasslands in North-Eastern Mountain Grassland. Also occurs in cultivated lands and young pine plantations, where it coexists with Hottentot Mole-rats *Cryptomys hottentotus*.

Neamblysomus gunningi

Abundance Locally common; most numerous in moist soils near water-courses and ponds.

Adaptations Burrow systems comprise deep tunnels (15–30 cm) and shallow subsurface tunnels used for foraging at night. Entrances to deep tunnels are marked by mounds of soil. Individuals enter torpor during the day. Particularly active after rains.

Foraging and Food Carnivorous. Forage for invertebrates in subsurface tunnels, and sometimes emerge onto the surface briefly to stir up leaf litter in search of prey. Contents of three stomachs contained only the remains of earthworms. Captive individuals readily eat mealworms (tenebrionid larvae), crickets and baby mice.

Social and Reproductive Behaviour Adults solitary.

Reproduction and Population Structure Pregnancies recorded in Feb, Apr and May suggesting that breeding occurs primarily during the wet summer season.

Predators, Parasites and Diseases No information.

Conservation IUCN Category: Endangered.

The distribution of populations is highly localized and preferred habitats are threatened by commercial forestry operations. In the former Transvaal Province, this species was given the highest regional priority score for mammals (Freitag & van Jaarsveld 1997).

Measurements

Neamblysomus gunningi
HB (♂♂): 125.4 (121–132) mm, n = 9
HB (♀♀): 122.4 (111–128) mm, n = 8
T: 0 mm
HF (♂♂): 15.3 (14–18) mm, n = 9
HF (♀♀): 14.0 (13–17) mm, n = 8
E: 0 mm
WT (♂♂): 62.8 (56–70) g, n = 6
WT (♀♀): 51.0 (39–56) g, n = 4
GLS (♂♂): 28.4 (27.1–29.3) mm, n = 11
GLS (♀♀): 27.5 (27.0–27.8) mm, n = 6
GWS (♂♂): 17.3 (16.1–18.2) mm, n = 11
GWS (♀♀): 16.6 (15.9–16.9) mm, n = 6
C–M^2 (♂♂): 7.1 (6.7–7.3) mm, n = 10
C–M^2 (♀♀): 7.0 (6.8–7.2) mm, n = 7
Throughout geographic range

Key References Broom 1908; Freitag & van Jaarsveld 1997.

Gary N. Bronner

Neamblysomus julianae Juliana's Golden-mole

Fr. Taupe dorée de Juliana; Ger. Julianas Goldmull

Neamblysomus julianae (Meester, 1972). Ann. Transvaal Mus. 28: 35. The Willows, Pretoria, Gauteng Province, South Africa.

Taxonomy Originally described in the genus *Amblysomus* and retained there by Meester *et al.* (1986) but assigned to *Neamblysomus* by Bronner & Jenkins (2005). Synonyms: none. Chromosome number: 2n = 30, FN = 56 (Bronner 1995a).

Description Small species with a gracile build and very slender claws on the forefeet. Dorsal pelage cinnamon-brown to reddish-brown, slightly darker along middle of back; flanks fawn. Ventral pelage fawn to dull reddish-brown. Guard hairs off-white or yellowish at base, slaty-grey medially, reddish-brown or fawn at tip. Muzzle, cheeks and throat speckled with uniformly off-white hairs. Whitish facial patches sometimes present near ocular region. Claw on Digit 3 of forefoot largest (8.5–10.0 mm long, 3.6–4.3 mm wide at base), claw on Digit 2 slightly shorter (5.8–6.7 mm), claw of Digit 1 much shorter barely reaching the insertion of the second claw. Digit 4 vestigial with a knob-like claw. Skull comparatively short with very wide braincase (breadth 67–70% of GLS) and a relatively wide palate (30–32% of GLS). Molar talonids either absent or only weakly developed. M^3 usually absent. M_3 present variably, smaller than other molars with a peg-like appearance. Skull: see also family and/or genus profiles. Males slightly heavier than ♀♀, and slightly (but significantly) larger in six cranial measurements. Sexual size dimorphism thus does not seem to be pronounced, but this may be an artefact of small sample sizes.

Geographic Variation Consistent colour and dental differences exist between populations from geographical extremes of the geographic range. Individuals from two western populations (Pretoria and Nylsvley), located ca. 125 km apart in central South Africa, have a pale cinnamon-brown dorsal pelage and fawn-coloured ventral pelage, and lack lower M_3. Specimens from

Kruger N. P., located ca. 400 km further east, have a dark reddish-brown dorsal pelage and dull reddish-brown ventral pelage, and have minute M_3. Sample sizes are too small for a thorough assessment of intra-specific variation.

Similar Species

N. gunningi. Larger (HB: 111–132 mm, GLS: 27.0–29.3 mm); skull often broader (GWS: 15.9–18.2 mm), breadth 60–63% of GLS; heavier (WT: 39–70 g); pelage darker.

Amblysomus hottentotus, *A. septentrionalis* and *A. robustus*. HB usually larger; skulls larger (GLS: 24.6–32.0 mm); pelage redder; subtle chromosomal differences (Bronner 1995a).

Distribution Endemic to Africa. Extreme south of the Zambezian Woodland BZ. Recorded from three isolated localities in South Africa: The Willows (type locality), Shere and Tierpoort in Pretoria, Gauteng Province; Nylsvley Provincial Nature Reserve in Limpopo Province; and Numbi Gate, Pretoriuskop and Matjulwana districts of Kruger N. P., Mpumalanga Province. Skulls from owl pellets at Witkoppen Cave, ca. 25 km east of Nylsvley Nature Reserve, clearly represent this species, suggesting that it may occur more widely throughout the sandy Sprinkbok Flats. A specimen of this species caught in March 2003 at Malelane in Mpumalanga indicates that it may also be more extensive in the lowveld than previously thought.

Habitat Confined to sandy soils, often pockets of weathered sandstone associated with rocky ridges. The population on Nyl flood-plain occurs in Clay Thorn Bushveld and the population at Pretoria in Rocky Highveld Grassland; in Kruger N. P. occurs in Sour Lowveld Bushveld. Common in well-irrigated gardens. Absent from grasslands on the heavier soils of the Mpumalanga escarpment where the larger-sized *A. septentrionalis* and *A. robustus* occur.

Abundance Locally common, but no quantitative data.

Adaptations The delicate foreclaws and gracile build limit these golden-moles to light, sandy soils. Their small size allows them to burrow through narrow spaces between underground rocks. They enter daily torpor from shortly after dawn until late afternoon. Surface activity increases dramatically after rains. They emit high-pitched squeaks when agitated or handled.

Foraging and Food Forages in shallow subsurface tunnels for invertebrate prey during late afternoon and night. Subsurface foraging activity declines markedly during the dry winter months (May–Sep). However, captive individuals remain active on a daily basis suggesting that the reduction in subsurface activity is the result of foraging being restricted to deeper tunnels, rather than prolonged torpor. No information exists on diets of free-ranging individuals.

Social and Reproductive Behaviour Adults solitary. Fight viciously when confined together, regardless of sex.

Reproduction and Population Structure Breeding individuals have been found in Jul, Aug and Nov–Feb, suggesting aseasonal polyoestry and sustained but low levels of reproduction throughout the year. Litter-size: 2.

Neamblysomus julianae

Predators, Parasites and Diseases Remains have been found in pellets of Barn Owls *Tyto alba*, reflecting the tendency of these golden-moles to forage on the surface. Often captured by domestic cats and dogs.

Conservation IUCN Category: Vulnerable.

The type population is being severely impacted by intensive urbanization and sand mining. In the former Transvaal Province, this species was given the highest regional priority score for mammals (Freitag & van Jaarsveld 1997).

Measurements

Neamblysomus julianae
HB (♂♂): 100.5 (94–111) mm, n = 7
HB (♀♀): 99.2 (92–111) mm, n = 6
T: 0 mm
HF (♂♂): 13.4 (12–14) mm, n = 7
HF (♀♀): 12.0 (9–14) mm, n = 6
E: 0 mm
WT (♂♂): 30 (21–35) g, n = 5
WT (♀♀): 26 (23–29) g, n = 5
GLS (♂♂): 23.2 (22.4–23.6) mm, n = 7
GLS (♀♀): 22.5 (21.8–23.5) mm, n = 8
GWS (♂♂): 16.0 (15.4–16.7) mm, n = 7
GWS (♀♀): 15.4 (14.7–16.2) mm, n = 8
C–M^2 (♂♂): 5.9 (5.7–6.1) mm, n = 6
C–M^2 (♀♀): 5.7 (5.5–5.9) mm, n = 4
Tierpoort, Shere and The Willows districts, Pretoria, South Africa

Key References Bronner 1995a, b; Freitag & van Jaarsveld 1997.

Gary N. Bronner

Order MACROSCELIDEA – Sengis (Elephant-shrews)

Macroscelidea Butler, 1956.

Macroscelididae (4 genera, 15 species)	Sengis	p. 261

The order Macroscelidea contains one family with four genera and 15 species, which are widely distributed in southern, eastern, central and north-west Africa. Representatives of the order live in forests, woodlands, bushlands and semi-arid habitats, but not in sandy deserts. The order, both extant species and fossils, is endemic to Africa.

The taxonomy of the order follows Corbet & Hanks (1968) and Corbet (1974b), and more recently Schlitter (1993, 2005). The order contains a single family, Macroscelididae, with two subfamilies (Table 13). The Rhynchocyoninae (giant sengis) has a single genus *Rhynchocyon* and is comprised of three comparatively large species that live in forests and woodlands of eastern and central Africa. The second subfamily, Macroscelidinae (soft-furred sengis), contains three genera: (a) *Petrodromus*, with a single species, which occurs throughout most of central, eastern and southern Africa; (b) *Macroscelides*, also with a single species, which is confined to the Cape Province of South Africa and W Namibia; and (c) *Elephantulus*, with ten comparatively small species that occur in semi-arid habitats and which, together, have a wide geographic range extending from southern Africa, through eastern Africa, to the Horn of Africa, and a single isolated species in NW Africa (Table 14). Most species of *Elephantulus* have comparatively small geographic ranges.

The most obvious character of sengis (or elephant-shrews) is the elongated, trunk-like and very mobile snout, which ends with a small rhinarium enclosing the nostrils. The snout is often moved up, down and sideways testing the environment for odours. The broad upright ears turn frequently, scanning for sounds, and the large eyes enable the location of small prey and potential predators. The forefeet and legs are of moderate length but the hindlimbs are elongated and slender, facilitating the characteristic rapid quadrupedal bounding (not bipedal hopping) locomotion of sengis; in *Rhynchocyon* both fore- and hindlimbs are especially elongated, as in small antelopes (Evans 1942). *Macroscelides* and *Elephantulus* possess five digits on both fore- and hindfeet, *Petrodromus* has five digits on each forefoot and four on each hindfoot, and *Rhynchocyon* has four digits on both fore- and hindfeet. The relatively long tail is sparsely covered with short hair, which in

Table 13. Principal diagnostic characters of the two subfamilies of Macroscelididae (after Corbet & Hanks 1968).

Character	Rhynchocyoninae (*Rhynchocyon*)	Macroscelidinae (*Elephantulus, Macroscelides, Petrodromus*)
Head and body size	Large (HB: ca. 250 mm)	Medium or small (<200 mm)
Pelage	Slightly coarse, sparse and boldly patterned	Soft, dense, not boldly patterned
Number of digits on forefoot/hindfoot	4/4	5/4 (*Petrodromus*), 5/5 (*Elephantulus, Macroscelides*)
Digit 5 of forefoot	Very short	Long
Nipples	Abdominal only	Nuchal, pectoral and (sometimes) abdominal
Post-anal gland	Present	Absent
Tip of nasal bones of proboscis	Partly ossified	Wholly cartilaginous
Postorbital process of skull	Present	Absent
Bony palate of skull	Not perforated	Perforated
Paraoccipital processes of skull	Well-developed	Rudimentary
Upper incisor teeth	Absent or rudimentary, non-functional	Present, functional
Upper canine teeth	Very large	Small
Shape of pupil of eye	Circular	Vertically elongate

Table 14. Genera in the family Macroscelididae.

Genus	Digits on forefoot	Digits on hindfoot	Head and Body length (mm) [3]	Tail length (mm) [3]	Greatest length of skull (mm) [3]	Nipples [4]	Pectoral gland [6]
Elephantulus (10 spp.)	5	5	114–135	98–157	33.5–37.6	1+1+1 = 6	0 or 2
Macroceledes (1 sp.)	5	5	110 (104–115)	121 (107–134)	33.7 (32.1–34.8)	1+1+1 = 6	0
Petrodromus (1 sp.)	5	4[2]	192 (163–210)	166 (156–187)	56.1 (54.3–58.5)	1+1+0 = 4	1
Rhynchocyon (3 spp.)	4[1]	4[2]	240–270	230–243	67.6–67.6	0+0+2 = 4	0

[1] Digit 1 (= pollex or thumb) absent
[2] Digit 1 (= hallux or 'big toe') absent
[3] Mean and range (*Macroceledes, Petrodromus*). Mean of smallest species – mean of largest species (*Elephantulus, Rhynchocyon*)
[4] antebrachial + pectoral + abdominal/inguinal (on each side) = total both sides
[5] Canine-M3 in *Rhynchocyon*; upper incisors absent or rudimentary (Data for *R. cirnei* only)
[6] 0 = absent. 1 = slightly developed, 2 = well-developed (mostly after Corbet & Hanks 1968)
[7] Gland is postanal, at tail insertion; large and round in shape.

some species forms a small terminal tuft. Scent glands, which are important for chemical communication, are present on several parts of the body (Corbet & Hanks 1968, Rathbun 1979, Faurie 1996); principally these are the pectoral gland (absent in *Rhynchocyon* spp.), subcaudal gland, post-anal gland (*Rhynchocyon* spp. only), pedal gland and sebaceous glands. The presence or absence of each of these glands varies according to species. Testes are internal and thus ♂♂ lack a scrotum, but the prepuse is prominent near the base of the sternum (Woodall 1995a). The dental formula is $I^{0-3}/_{3}$, $C^{1}/_{1}$, $P^{4}/_{4}$, $M^{2}/_{2-3}$ = 42 at maximum (see also genus profiles and Figure 15).

Skull is tall with dome-shaped cranium, and well developed complete zygomatic arches. Orbits are large, with moderate interorbital constriction. Rostrum moderately long. Tympanic region not inflated except in *Macroscelides*. Palate with broad short anterior palatal foramina, and complex posterior palatal foramina (usually four pairs of foramina) in *Elephantulus*, *Macroscelides* and *Petrodromus*; no foramina in *Rhynchocyon*. Toothrow long, without diastema (similar to that in other Afrotheria and in Soricomorpha); toothrow diverges posteriorly so bony palate is considerably wider at level of molar teeth than at level of incisors. Dental formula either $I^{3}/_{3}$, $C^{1}/_{1}$, $P^{4}/_{4}$, $M^{2}/_{2-3}$ = 40 or 42 in *Elephantulus*, *Macroscelides* and *Petrodromus*; or $I^{0}/_{3}$, $C^{1}/_{1}$, $P^{4}/_{4}$, $M^{2}/_{2-3}$ = 34 or 36 in *Rhynchocyon* where upper incisors absent or rudimentary and usually not erupted. Incisors small and subequal (except in *Petrodromus* where I1 is large and caniform). Canine small and incisiform except in *Rhynchocyon* where large and caniform. Upper premolars and molars (four or six in each ramus) large, increasing in size posteriorly, sectorial with well-developed cusps; brachydont or slightly hypsodont; dilambdodont; P^1 usually small and simple; P^4 large and subequal to M^1 (Figure 15). Mandible long and slender, with posteriorly pointing angular process; tall coronoid process (especially in *Elephantulus*, *Macroscelides* and *Petrodromus*).

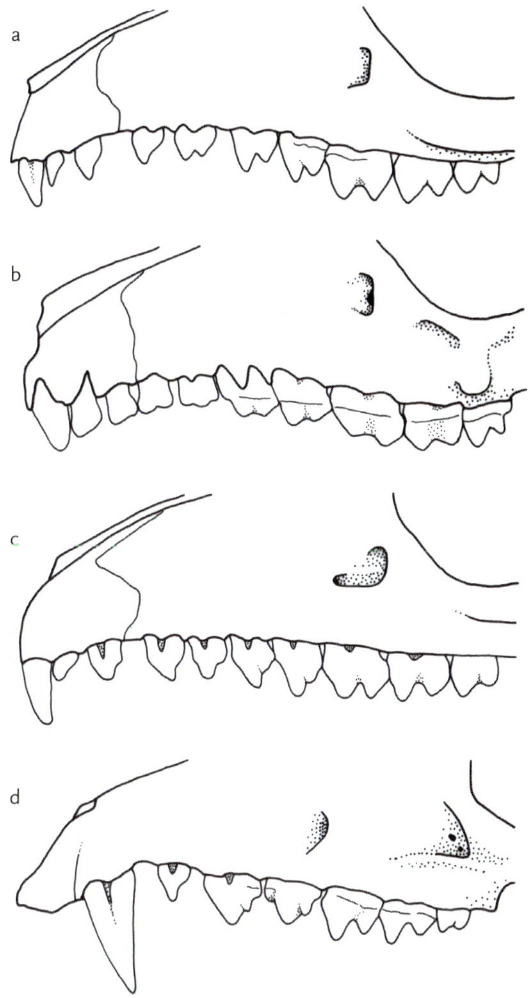

Figure 15. Upper left toothrow of each genus of sengis. (a) *Elephantulus*. (b) *Macroseledes*. (c) *Petrodromus*. (d) *Rhynchocyon*. (BMNH specimens; see Figs 16, 18, 19, 20.)

Sengis are primarily insectivorous, feeding on a wide variety of small invertebrates. The prey are captured by the long tapering tongue that can be extended beyond the tip of the nose, and masticated by the sectorial teeth, in a similar way to other small insectivorous mammals. The digestive tract is of moderate length, as expected for an insectivore, but unlike most insectivores, sengis have a functional caecum (Woodall & Mackie 1987). The caecum perhaps denotes a partly herbivorous diet in some species and it may also reflect the phylogeny of sengis. Most species are socially monogamous and live on home-ranges (sometimes territories) of up to 100 ha (usually less), depending on the species and habitat occupied (Rathbun 1979, Rathbun & Rathbun 2006). Sengis are polycyclic, but primarily crepuscular or diurnal, sheltering in rock crevices, shallow burrows or on comparatively exposed trails at the bases of bushes (Macroscelidinae), while others build leaf nests on the forest floor (Rhynchocyoninae).

Subcaudal gland [6]	Tympanic bullae	I-M^3 (mm) [3]	Notes
1 or 2	Not inflated or moderately inflated	17.0–20.6	Widespread in southern and eastern Africa; one species in NW Africa.
2	Very inflated	15.6 (14.9–16.3)	Namibia and South Africa. Head broad (GWS/GLS = >60%; other genera <60%)
n. d.	Small	29.3 (28.1–30.5)	Widespread in eastern, southern and central Africa. Forelimbs longer relative to hindlimbs than in other genera.
0 [7]	Small or not inflated	28.9 (27.4–29.9) [5]	One species widespread in central and eastern Africa; two species mostly restricted to coastal East Africa.

Reproduction may occur throughout the year in tropical regions, but is seasonal at higher latitudes (Neal 1995). Gestation is comparatively long for such small mammals (4–6 weeks) and is probably related to the precocious state of neonates. There are 2–3 pairs of nipples (antebrachial/pectoral, and abdominal/inguinal) arranged differently according to genus. Litter-size is 1–4 (depending on the species). The eyes are open at birth and the young can walk within a few hours in the Macroscelidinae (Rathbun 1979), but in the Rhynchocyoninae neonates are less developed and nest-bound for about two weeks (Baker *et al.* 2005). Four species of sengis (*Elephantulus edwardii*, *E. brachyrhynchus*, *E. myurus*, *Macrosceledes proboscideus*) are known to have the unusual characteristic of polyovulation (also recorded in several species of mammals belonging to other orders) in which the ovaries produce many more eggs at each ovulation than is required for implantation (11 species examined; Tripp 1971). The reason for polyovulation in these species is uncertain.

The fossil and extant Macroscelidea include six subfamilies (two extant) with 14 genera (four extant) (Holroyd & Mussell 2005). The earliest macroscelid fossils come from the early Eocene (ca. 55 mya), but by the Pliocene (ca. 4 mya), all but the two extant subfamilies have disappeared from the fossil record, and the genera *Elephantulus*, *Macroscelides* and *Rhynchocyon* have first appeared (Butler 1995). *Miorhynchocyon* arose in the early Miocene (ca. 23 mya), and it closely resembled present-day giant sengis in many respects (Novacek 1984). *Petrodromus* is not present in the fossil record until the mid-Pleistocene (ca. 1.0 mya). The dentition from most of the older fossils suggests a herbivorous diet. Indeed, the Pliocene macroscelid genus *Myohyrax* was initially misidentified as a hyracoid because of its highly hypsodont dentition (Patterson 1965).

The taxonomic relationship of sengis with other mammals has been confusing. The historical arrangement was to include them as a family, the Macroscelididae, within the order Insectivora. When the order was split into the two suborders, the Liptotyphla and Menotyphla, the family was assigned with the Tupaiidae (the 'tree-shrews' of India and South-East Asia) in the suborder Menotyphla. Another view, based on morphology and osteology (Novacek 1982, Novacek & Wyss 1986), is that the Macroscelididae are more closely related to the Lagomorpha (hares and rabbits) than to the Insectivora or the Tupaiidae (McKenna 1975). The current view is that the Macroscelididae is the only family in the order Macroscelidea (Butler 1956), and that the Tupaiidae is a separate order, the Scandentia – both families and orders now quite separate from the Insectivora. The vernacular name, 'elephant-shrew', refers to the elongated snout reminiscent of the proboscis of elephants, and their historical relationship to true shrews (order Soricomorpha; formerly Soricidae, order Insectivora); recently, the alternative vernacular name 'sengi' has gained credence because it better reflects their unique phylogeny (Rathbun & Kingdon 2006).

The molecular evidence for sengis belonging to an early African radiation of placental mammals, called the Afrotheria, is overwhelming (see Chapter 7 'Classification: a Mammalian Perspective', p. 101). Afrotheria include the elephants (order Proboscidea), dugongs and manatees (order Sirenia) and hyraxes (order Hyracoidea), which together are referred to as the Paenungulata, the Aardvark (order Tubulidentata), the sengis (order Macroscelidea) and the tenrecs and golden-moles (order Afrosoricida). Although morphological evidence in support of this radiation is less convincing, the Afrotheria hypothesis is focusing considerable interest on fossils (e.g. Zack *et al.* 2005) and morphological traits (e.g. Carter *et al.* 2004), which will undoubtedly result in a better understanding of this extraordinary group of African mammals. The order Macroscelidea, together with the orders Afrosoricida (otter-shrews, golden-moles and the Malagasy tenrecs) and Tubulidentata (Aardvark), has been placed in the Cohort Afroinsectiphillia (Waddell *et al.* 2001, Amrine-Madsen *et al.* 2003) – the 'African insect lovers' – within the Superorder Afrotheria. Further information on Afrotheria is given in Volume 1 of *The Mammals of Africa*.

An extensive bibliography (750 citations) of the scientific literature on sengis was given by Rathbun & Woodall (2002).

Mike Perrin & Galen B. Rathbun

Table 15. Species in the genus *Elephantulus*.

Species	Pectoral gland	Eye-ring	Postorbital patch of dark fur	Canine tooth	Upper incisor teeth
E. brachyrhynchus	Absent	Present	Absent	molariform	I^1 and I^3 larger than I^2
E. edwardii	Absent	Present	Absent	molariform	I^1–I^3 similar in size, or I^2 slightly smaller
E. intufi	Absent	Present	Absent	pointed	I^1–I^3 similar in size
E. myurus	Absent	Present	Absent	small, pointed	I^1 much larger than I^2 and I^3
E. rozeti	Absent	Absent	Absent	molariform	I^2 smaller than I^1 and I^3
E. rupestris	Absent	Indistinct	Absent	slightly molariform	I^2 smaller than I^1 and I^3
E. fuscus	Absent	Present	Absent	molariform	I^1 larger than I^2 and I^3; I^2 and I^3 similar in size
E. fuscipes	Present	Indistinct	Absent	molariform	I^1–I^3 similar in size
E. revoili	Present	Present	Present	molariform	I^1–I^3 similar in size
E. rufescens	Present	Present	Present	slightly molariform	I^2 smaller than I^1 and I^3

[1] Bicoloured = dark above, white or pale below

Family MACROSCELIDIDAE

SENGIS (ELEPHANT-SHREWS)

Macroscelididae Bonaparte, 1838. Nuovi Ann. Sci. Nat. 2: 111.

The characteristics of the family are given in the order profile above. The family contains four genera and 15 species. The genera in the family, and each species in each genus, are described alphabetically. In the Descriptions, an indication of the size of each species is recorded comparatively to other species in the family. The size categories, based on mean head and body length (HB), are: Very small: 90–139 mm; Small: 140–189 mm; Medium-sized: 190–139 mm; Large: 140–189 mm; and Very Large: 190–340 mm.

Mike Perrin & Galen B. Rathbun

GENUS *Elephantulus*

Sengis

Elephantulus Thomas and Schwann, 1906. Abst. Proc. Zool. Soc. Lond. 1906 (33): 10. Type species: *Macroscelides rupestris* A. Smith, 1831.

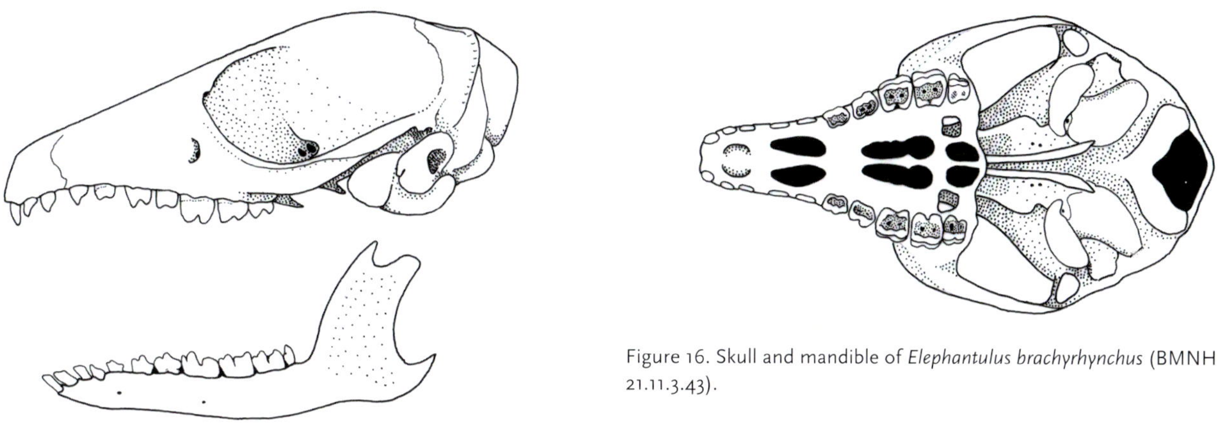

Figure 16. Skull and mandible of *Elephantulus brachyrhynchus* (BMNH 21.11.3.43).

Elephantulus contains ten species (Table 15) distributed throughout southern, eastern and north-west Africa. Five of these species occur only in southern Africa, three only in eastern Africa, one in both eastern and southern Africa, while one occurs only in north-west Africa. Most species have a comparatively restricted geographic range. All species are terrestrial; most live in savanna habitats, but *E. edwardii*, *E. myurus*, *E. rupestris* and *E. rozeti* are rupicolous.

Elephantulus rufescens.

Mean Tail length (as % of HB)	Tail colour; tuft at tip [1]	Distribution and notes
90	Bicoloured, no tuft	Widespread central and eastern Africa. Supratragus simple
130	Bicoloured, small tuft	South Africa
105	Bicoloured (dorsal hairs with black tips), small tuft	Namibia, Botswana, Angola, South Africa. Golden patch behind each ear
110	Bicoloured, no tuft	Southern Africa
110	Bicoloured, small tuft	North-western Africa
115	Bicoloured, well-developed tuft	South Africa, Namibia
90	Bicoloured, no tuft	Malawi, Mozambique, Zambia. Supratragus twisted
67–75	Bicoloured, no tuft	Uganda, NE DR Congo, S Sudan
120	White (dorsal hairs with brown tips), small tuft	Somalia
105	Dark brown, no tuft	Somalia, EC Ethiopia, SE Sudan, Kenya, E Uganda and NC Tanzania

Elephantulus rufescens.

Species in the genus are characterized by small size (HB less than ca. 160 mm), large head and eyes, long ears, long limbs with five digits on each fore- and hindfoot, and three pairs of nipples. Skull characters include upper toothrow (I–M^3) less than 22 mm, upper incisors present, upper canines small, bullae not greatly inflated, I^1 and I^3 not sectorial, and postorbital process absent. The dental formula is $I\,^3/_3$, $C\,^1/_1$, $P\,^4/_4$, $M\,^2/_2 = 40$ in most species, or $I\,^3/_3$, $C\,^1/_1$, $P\,^4/_4$, $M\,^2/_3 = 42$ in *E. brachyrhynchus*, *E. fuscipes* and *E fuscus*. Subcaudal and pectoral glands present in some species. Some of these characters are shared with other genera of sengis. See also family profile and Figures 15 and 16.

Elephantulus spp. are mainly crepuscular with bouts of activity during day and night, suggesting that they are polycyclic (Rathbun 1979; Woodall *et al.* 1989). Some use runways to travel between shelters and foraging areas. None is known to use nesting material. They are primarily insectivorous. Most species that have been well studied are monogamous and the male–female pair defends a territory (Rathbun & Rathbun 2006). Metabolism and thermoregulation has been studied in some species: in *E. intufi*, basal metabolic rate is lower than expected for a species of this size, and T_b can be maintained over a wide range of ambient temperatures (even when T_a is ca. 5 °C); cold temperatures elicit torpor in some species. Those species living in arid environments perhaps have specialized kidneys to aid water conservation. Although they all have long limbs and are extremely agile and can run exceedingly fast, they do not hop on their rear legs. Most species of *Elephantulus* communicate with each other by scent-marking, but also with rapid foot-drumming of the hindfeet on the substrate. Reproductive rate is slow because of the comparatively long gestation, small litter-size (1–4) and limited seasonal breeding especially at higher latitudes (Neal 1995).

Some species of *Elephantulus* show considerable geographic variation in pelage colouration, which has given rise to the description of many taxa (see especially *E. intufi*, *E. rufescens*, *E. rupestris*). The genus was described by Thomas & Schwann (1906) when they split the genus *Macroscelides* into three genera: *Macroscelides* (enlarged tympanic bullae, two lower molars), *Elephantulus* (normal-sized bullae and two lower molars) and *Nasilio* (normal-sized bullae and three lower molars; *brachyrhynchus*, *fuscipes*, *fuscus* only). Allen (1939) and Roberts (1951) also regarded *Nasilio* as a separate genus. In contrast, Ellerman *et al.* (1953) placed it as a subgenus of *Elephantulus*. Currently, *Nasilio* is not recognized as a genus or subgenus (Corbet & Hanks 1968) and the species formerly in this genus are now referred to *Elephantulus*. Although the different species of *Elephantulus* are often difficult to distinguish morphologically, they are genetically distinct with a high degree of divergence in some species (Raman & Perrin 1997); *Elephantulus* is considered to be distinct from the genera *Macroscelides* and *Petrodromus* using genetic criteria. However, recent genetic and morphological evidence suggests that *E. rozeti* in NW Africa is more closely related to *Petrodromus* than *Elephantulus* (Douady *et al.* 2003).

The species are distinguished on the basis of presence/absence of a pectoral gland, the form of the supratragus, the relative size of the tympanic bullae, the shape of the premaxillary suture, the number of molar teeth in the mandible and the presence and form of cusps on premolars. A key to all ten species is given by Koontz & Roeper (1983), and to nine species (excludes *E. fuscus*) by Corbet & Hanks (1969) (Table 15).

A new species, *Elephantulus pilicaudus* from South Africa, was described in 2008 (Smit *et al.* 2008, see also p. 16) after this text was submitted to the publishers.

Mike Perrin & Galen B. Rathbun

Elephantulus brachyrhynchus Short-snouted Sengi (Short-snouted Elephant-shrew)

Fr. Macroscélide à nez court; Ger. Kurznasen-Elefantenspitzmaus

Elephantulus brachyrhynchus (A. Smith, 1836). Report of the Expedition for Exploring Central Africa 1834: 42 [1836]. (The country between L. Latakoo [= Kuruman, Northern Cape Province, South Africa] and the Tropic [= in Botswana]).

Taxonomy Originally described in the genus *Macroscelides* and later assigned to the genus *Nasilio* on the basis of three lower molars rather than two as in most other *Elephantulus* spp. (Corbet & Hanks 1968). Other forms previously considered within this species include *fuscus* and *fuscipes* (both now valid species), *malosae* (now a synonym of *E. fuscus*) and *schinzi* (now a synonym of *E. intufi*) (Corbet 1974b). Molecular and genetic studies suggest low levels of polymorphism in *E. brachyrhynchus*, although higher than in other species of *Elephantulus* (Raman & Perrin 1997). Allozyme and isozyme data link *E. brachyrhynchus* with *E. intufi* and *E. rupestris*, and support the retention of *brachyrhynchus* in the genus *Elephantulus* rather than in the separate genus *Nasilio*, as proposed on the basis of dentition (Corbet 1974b). *Elephantulus brachyrhynchus* probably diverged early within the genus *Elephantulus* (Corbet 1974b). Subspecies: none. Chromosome number: 2n = 26 (Stimson & Goodman 1966, Tolliver *et al.* 1989).

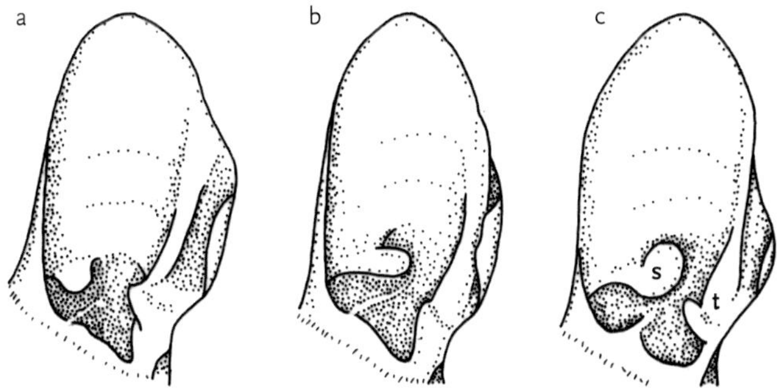

Figure 17. Left ear of (a) *Elephantulus brachyrhynchus*, (b) *E. fuscus* (as *E. brachyrhynchus* – atypical form) and (c) *E. fuscipes* (after Corbet & Hanks 1968). s = supratragus; t = tragus.

Description Small sengi with relatively short snout, eye-ring, and buff patch behind ears. Dorsal pelage reddish- or yellowish-brown, but variable (see below); dorsal hairs ca. 10 mm, dark grey at base, brown at tip; some scattered longer black-tipped hairs. Ventral pelage whitish; ventral hairs grey at base, white at tip. Head large, with short snout, large eyes, indistinct white eye-ring, white upper lip. Ears large, rounded at tip, with small supratragus which is not twisted backwards (Figure 17, cf. *E. fuscus* and *E. fuscipes*). Usually with buff patch behind ears. No dark spot behind eye. Pectoral gland absent, subcaudal gland present. Fore- and hindlimbs whitish. Hindlimbs elongated; five digits; soles brown and naked. Tail relatively long (ca. 90% of HB), but comparatively short for an *Elephantulus* sp.; bicoloured, dark above, paler below; without tuft. Skull: I^1 and I^3 larger than I^2; M_3 present (also in *E. fuscus* and *E. fuscipes*; see profile genus *Elephantulus*). Nipples: 1 + 2 = 6.

Geographic Variation Considerable clinal variation in colour of pelage. Dorsal pelage reddish-yellow or pale buffy-grey (Namibia and Botswana), dark greyish-brown (higher altitudes of Malawi), rufous-grey (Zambia) and grey (East Africa).

Similar Species

E. fuscus. HB and HF similar size; pectoral gland absent; I^1 larger than I^2 and I^3; supratragus twisted; Malawi, Mozambique and Zambia; probably allopatric.

E. fuscipes. HB slightly larger, HF slightly smaller; pectoral gland present, poorly developed; I^1, I^2 and I^3 similar in size; supratragus large and twisted; Uganda, NE DR Congo, S Sudan; marginally parapatric.

E. myurus. HB slightly larger, HF considerably longer (ca. 39 mm); pectoral gland absent; I^1 much larger than I^2 and I^3; M_3 absent; South Africa, Zimbabwe; sympatric in part of range.

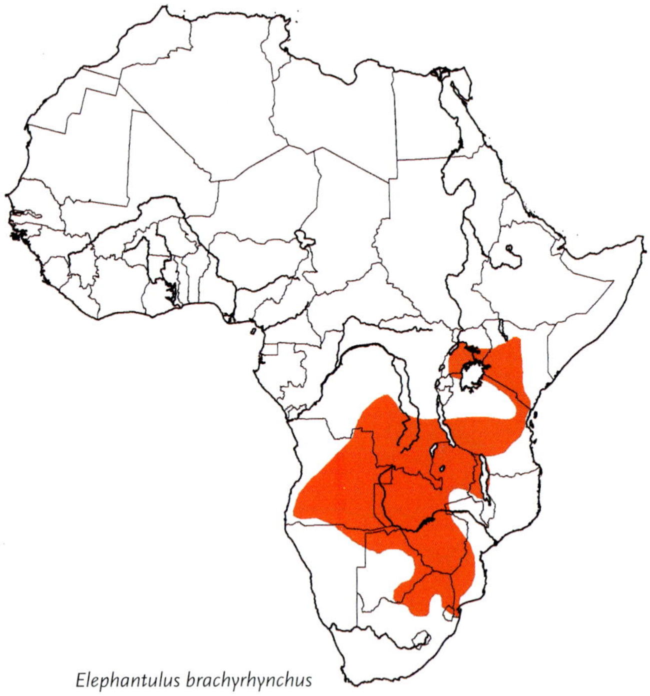

Elephantulus brachyrhynchus

Distribution Endemic to Africa. Zambezian Woodland BZ, adjacent areas of Somalia–Masai Bushland BZ and Eastern and Southern Rainforest–Savanna Mosaics. The most widely distributed species of sengi. Recorded from NE Swaziland, South Africa (N Transvaal), S Mozambique, Zimbabwe, N and E Botswana, NE Namibia, N Malawi, Zambia, S DR Congo, Angola and northwards to Tanzania, Kenya and Uganda. The taxonomic status of populations in Malawi, Mozambique and Zambia are uncertain; they perhaps represent *E. fuscus*.

Habitat Savanna habitats where there is thick cover, long grass and shrubs in riparian woodlands. Lives on sandy ground or on hard substrate but not on rocks (cf. some other *Elephantulus* spp.). Prefers mixed woodland dominated by *Colophospermum mopane*, *Combretum apiculatum*, *C. zeyheri* and *Erythroxylum zambeseacum* in Zimbabwe, where herbs and grasses are abundant and ground cover is good, even during the dry season (Neal 1995). Also occurs in *Brachystegia jullbernardia* mixed woodland (miombo), *Combretum–Terminalia–Eragrostis* low woodland on talus slopes, and *Commiphora–Combretum* wooded bushland thickets (Neal 1995).

Abundance Locally common but usually at low densities. Distances recorded between consecutive captures of individuals in Tanzania are comparatively short (mean ± 1 S.D. = 42.3 ± 38.3 m; median 32 m; n = 65) and not significantly different between the sexes (Leirs *et al.* 1995).

Adaptations Terrestrial; diurnal with most activity at dawn. Runs swiftly, dashing from cover to cover. Shelters in dense cover, but does not dig burrows. Sunbathes (especially in morning) to increase body temperature and to save energy expenditure. In captivity, metabolic rate is 73–94% of that expected (on basis of body size) and T_b is 34.2–37.7 °C (Perrin 1995). When T_a is >35 °C, hyperthermia does not occur because of increased evaporative water loss. Insulation and vasomotor control is also important in controlling body temperature (Perrin 1995). Skin glands on many parts of the body produce secretions used for communication with conspecifics (Faurie 1996). Foot-drumming is common; typically, there are 10–30 drums/bout (with some irregular bouts), drum intervals of 30–40 msec, 100–200 msec between bouts, and <10 bouts per series; the length of a series ranged from 200 msec to 5 sec (Faurie *et al.* 1996). A sharp high-pitched squeal is uttered when fighting, and the alarm call is shrill and penetrating. 'Mew' vocalizations are 545 msec long and have a fundamental of 1.65 kHz (Faurie 1996).

Foraging and Food Insectivorous. In Zimbabwe (Neal 1995), ants and termites are the principal prey, and occasionally plant matter, including leaves and seeds, is eaten (<2% by volume). However, in Tanzania, Leirs *et al.* (1995) recorded the diet (based on analysis of stomach contents) as 47.2 ± 5.3% invertebrates and 51.7 ± 5.2% green plant material (macroscopic analysis), and 40.1% invertebrates and 47.2% green plant material (microscopic analysis). Small amounts of fruits and seeds were also eaten.

Social and Reproductive Behaviour Adults typically occur in monogamous pairs and are territorial (Neal 1995). Their social structure appears to be similar to that described for *E. rufescens* (Rathbun 1979). Males and ♀♀ display aggressive territorial behaviour (Faurie 1996).

Reproduction and Population Structure Females show partial polyovulation, producing a mean of 8.5 ova/ovary (range 0–23, n = 16) at each ovulation (Tripp 1971); this condition is intermediate between those species that do not polyovulate (most sengis) and those that show extreme polyovulation (*E. edwardii*, *E. myurus*, *P. macrosceledes*).

In Zimbabwe, young are born throughout the year (Neal 1995), although the percentage of pregnant ♀♀ varies throughout the year (40% in Feb to 100% in Oct and Nov). Litter frequency is reduced in the cool winter months when insect food is less abundant; average interval between litters is 90 days during the cool dry season and about 60 days during the warmer period of the year (Neal 1995). Mean embryo number varies with season: 1.40 during the cool dry season, 1.74 during the hot dry season and 1.76 in the warm wet season (Neal 1984a). Mean embryo number also varies with the weight of the ♀: smaller ♀♀ (<40 g) have a smaller number of embryos (mean 1.1 [cool dry season] to 1.7 [warm wet season]): and larger ♀♀ (>40 g) have a larger number of embryos (mean 1.6 [cool dry season] to 1.9 [warm wet season] and 2.0 [hot dry season]) (Neal 1995). In Tanzania, the percentages of pregnant ♀♀ each month were mostly lower than in Zimbabwe, and there were several months of the year when the pregnancy rate was 0% (Mar, May, Jun, Jul [no sample], Sep, Nov and Dec) (Leirs *et al.* 1995). Embryo number: 1.4 (1–2), n = 10. At birth; young are precocial, fully furred and the eyes are open; young can run almost immediately after birth, and development is rapid (Faurie 1996). The overall sex ratio in a wild population did not differ from parity (Leirs *et al.* 1995).

Predators, Parasites and Diseases Owls *Tyto alba* and *T. capensis* are known predators (Vernon 1972).

Conservation IUCN Category: Least Concern.

The most widely distributed sengi with no known significant threats.

Measurements

Elephantulus brachyrhynchus
TL (♂♂): 213.8 (202–225) mm, n = 9
TL (♀♀): 212.0 (205–218) mm, n = 4
T (♂♂): 98.9 (91–110) mm, n = 9
T (♀♀): 95.3 (88–101) mm, n = 4
HF (♂♂): 29.9 (29–31) mm, n = 9
HF (♀♀): 28.8 (27–30) mm, n = 6
E (♂♂): 21.1 (20–24) mm, n = 9
E (♀♀): 20.7 (20–21) mm, n = 6
WT: n. d.
GLS (♂♂): 33.5 (32.4–34.3) mm, n = 9
GLS (♀♀): 32.8 (32.2–33.8) mm, n = 5
GWS (♂♂): 18.1(17.6–18.8) mm, n = 9
GWS (♀♀): 17.5 (16.8–18.5) mm, n = 5
I^1–M^3 (♂♂): 17.0 (16.4–17.6) mm, n = 9
I^1–M^3 (♀♀): 17.0 (16.8–17.2) mm, n = 6
Southern Africa (Rautenbach & Schlitter 1977)

Key References Corbet & Hanks 1968; Faurie 1996; Leirs *et al.* 1995; Neal 1995; Perrin 1995.

Mike Perrin

Elephantulus edwardii CAPE SENGI (CAPE ELEPHANT-SHREW)

Fr. Macroscélide du Cap; Ger. Kap-Elefantenspitzmaus

Elephantulus edwardii (A. Smith, 1839). Illustrations of the Zoological Society of South Africa, Mammals: p1. 14. Oliphants River, Western Cape Province, South Africa. Corbet & Hanks (1968) assume the type locality refers to the river that flows into the Atlantic, since the species has subsequently been found in this region, and not to the Oliphants River in the Outdtshoorn District where the species has not been recorded.

Taxonomy Originally described in the genus *Macroscelides*. Ellerman *et al.* (1953) treated *edwardii* as conspecific with *E. myurus* (which they called *E. rupestris*), although there are a number of dental characters that separate the two species (Rautenbach & Nel 1980). The degree of heterozygosity of allozymes and isozymes is very low (Raman & Perrin 1997), suggesting a close relationship with *E. myurus*. Synonyms: *capensis*, *karoensis*, *rupestris*. Subspecies: none. Chromosome number: 2n = 26 (Tolliver *et al.* 1989).

Description Small greyish-coloured sengi, similar to *E. rupestris* and *E. myurus*. Pelage long, soft and silky. Dorsal pelage greyish-brown, turning to ashy-grey tinged with pale tawny on flanks. Ventral pelage greyish-white to white. Head large; snout black above, white laterally. Eyes large with white eye-ring. No spot behind the eye. Ears large, rounded at tip; yellowish-brown patch behind each ear. Tragus and supratragus slightly developed. Pectoral gland absent, subcaudal gland present. Hindlimbs elongated; five digits. Tail long (ca. 130% of HB) with very short hairs, black above, paler below near body but distal third black; hairs longer at tip forming small tuft. First lower premolar single-rooted (cf. double-rooted in *E. myurus*). M_3 absent. Females may be slightly larger than ♂♂ (Stuart *et al.* 2003). Nipples: 1 + 2 = 6.

Geographic Variation There is some variation in pelage colour between populations at extreme ends of the range (Corbet & Hanks 1968).

Similar Species

E. rupestris. Ventral pelage paler; orange-buff rather than yellow-buff behind the ears; tail on average longer, more hairy, distal third not completely black.

E. myurus. Ventral pelage paler; inconspicuous buffy patch behind ears; tail on average longer, less tufted, not black; less truncated supratragus.

Distribution Endemic to Africa. South-West Arid BZ (Karoo) and South-West Cape BZ. Recorded from SW and C Cape in the east to the Little Karoo in the west. Eastern and western parts of range are not disjunct (Rautenbach 1978; C. Stuart pers. comm.), as illustrated in older publications (Corbet & Hanks 1968).

Habitat Rocky areas on mountains, small rocky outcrops and flat rock sills adjacent to vegetated areas of dwarf shrubs, and hard sandy ground with sparse cover.

Abundance Locally common.

Elephantulus edwardii

Adaptations Terrestrial; adapted to rocky habitats. Mainly nocturnal (Woodall *et al.* 1989), but also crepuscular and diurnal (Stuart *et al.* 2003), suggesting they are polycyclic (Leon *et al.* 1983). Sunbathes on cool days. Foot-drumming includes both regular and irregular bouts with enough variation that makes it difficult to distinguish this species from others (Faurie *et al.* 1996). Various vocalizations ('clicks' and 'mews'), and secretions from perianal and pedal scent glands, are used presumably for social interactions and for marking territories (Faurie 1996). Metabolic and water turnover rates are only slightly lower than expected, based on body size (Leon *et al.* 1983).

Foraging and Food Insectivorous. Feeds mainly on ants (Stuart *et al.* 2003) and termites, but also on other arthropods. Stuart *et al.* (2003) observed animals flipping hyrax dung to feed on exposed insects. While foraging on insects attracted to the flowers of *Protea humiflora*, the Cape Sengi acts as a significant pollen vector (Fleming & Nicolson 2002). Captives can be kept on a commercial cereal mixture (Pronutro) plus vitamins and minerals and mealworms (Leon *et al.* 1983). The digestive traits of Cape Sengis are typical of an insectivore, although the assimilation of termites was surprisingly low considering their importance in the diet (Woodall & Currie 1989).

Social and Reproductive Behaviour Probably monogamous and territorial. Occurs singly or in male–female pairs, or as a mother with young (C. Stuart pers. comm.). Captives must be maintained singly or as monogamous pairs due to aggression (Leon *et al.* 1983).

Reproduction and Population Structure Females polyovulate, producing a mean of 44 ova/ovary (range 31–55, n = 3) at each ovulation (Tripp 1971). Young tend to be born in the warmer and wetter months: Sep–Jan in the western part of their range and into February further east (Stuart *et al.* 2003). Litter-size: 1–2 (usually 2). Birth interval: ca. 30 days. Neonates weigh 9–11 g, are highly precocial at birth and gain about 0.6 g per day until they reach about 65% of adult weight at Day 30 (Dempster *et al.* 1992, Stuart *et al.* 2003).

Predators, Parasites and Diseases Ectoparasites include five species of ticks (Fourie *et al.* 1995, Stuart *et al.* 2003) and the louse *Polyplax biseriata* (Stuart *et al.* 2003).

Conservation IUCN Category: Least Concern.

Although this species is endemic to South Africa, it occupies rocky habitats that are comparatively immune to most human disturbances at present.

Measurements

Elephantulus edwardii

TL (♂♂): 249.3 (230–273) mm, n = 9
TL (♀♀): 250.8 (230–275) mm, n = 11
T (♂♂): 131.3 (115–145) mm, n = 9
T (♀♀): 130.7 (116–149) mm, n = 11
HF (♂♂): 35.0 (33–37) mm, n = 9
HF (♀♀): 34.8 (33–39) mm, n = 12
E (♂♂): 29.2 (27–33) mm, n = 9
E (♀♀): 29.5 (26–33) mm, n = 11
WT (♂♂): 46.5, 47.5 g, n = 2
WT (♀♀): 52 (46–67) g, n = 5
GLS (♂♂): 35.3 (33.0–38.0) mm, n = 9
GLS (♀♀): 34.9 (33.5–37.0) mm, n = 12
GWS (♂♂): 19.6 (19.1–20.5) mm, n = 8
GWS (♀♀): 19.3 (18.7–20.3) mm, n = 12
I^1–M^3 (♂♂): 18.2 (17.1–19.9) mm, n = 9
I^1–M^3 (♀♀): 18.3 (17.5–20.8) mm, n = 12
Measurements: southern Africa (Rautenbach & Schlitter 1977)
Weights: South Africa (G. B. Rathbun unpubl.)

Key References Corbet & Hanks 1968; Dempster *et al.* 1992; Leon *et al.* 1983; Raman & Perrin 1997.

Mike Perrin & Galen B. Rathbun

Elephantulus fuscipes Dusky-footed Sengi (Dusky-footed Elephant-shrew)

Fr. Macroscélide à pattes sombres; Ger. Dunkelfuß-Elefantenspitzmaus

Elephantulus fuscipes (Thomas, 1894). Ann. Mag. Nat. Hist., ser. 6, 13: 68. N'doruma, Niam-Niam country, NE DR Congo.

Taxonomy Originally described in the genus *Macroscelides*. Similar to *E. brachyrhynchus*, and probably forms an allopatric pair with *E. fuscus* (Corbet & Hanks 1968). Synonyms: none. Chromosome number: not known.

Description Small dark brown sengi. Dorsal pelage dark brown; hairs dark grey to black, brown at tip; many longer black-tipped hairs especially on dorsum giving pelage a dark saturated appearance, Flanks paler. Ventral pelage whitish-grey; hairs grey at base, white at tip. Head large, without hairy rhinarium (cf. *E. rufescens*). Eyes large, with indistinct white eye-ring and without postorbital dark patch. Ears large, rounded at tip, with swollen supratragus, which is twisted backwards on a constricted stalk (to a greater extent than in *E. fuscus*, cf. *E. brachyrhynchus*; see Figure 17). No buff patch behind ears. Pectoral gland present, but poorly developed and only visible when hairs on chest are parted; subcaudal gland slightly developed. Hindfoot elongated; five digits. Tail relatively short (ca. 67–75% of HB), bicoloured, almost black above; without terminal tuft. M_3 present (also in *E. brachyrhynchus* and *E. fuscus*; see profile genus *Elephantulus*). Nipples: 1 + 2 = 6.

Geographic Variation Pelage colour of individuals at Niangara, DR Congo, at the end of the wet season are brighter and more rufous than those from Faradje several months after the wet season; however, this colour change is not constant and it is uncertain whether such variations are seasonal or geographic (Allen, J. 1922).

Similar Species

E. brachyrhynchus. Dorsal pelage reddish-brown; pectoral gland absent; lacks twisted supratragus.

E. rufescens. Dorsal pelage paler; pectoral gland present; hairy rhinarium; facial pattern includes dark spot behind eye; M_3 absent.
E. fuscus. Pectoral gland absent; similar form of supratragus; Malawi, Mozambique and Zambia only.

Distribution Endemic to Africa. Restricted areas of Northern Rainforest Mosaic and adjacent Eastern Rainforest–Savanna Mosaic. Recorded from SW Sudan, NE DR Congo and Uganda. Marginally sympatric, but not syntopic, with *E. brachyrhynchus* and *E. rufescens*.

Habitat Woodland savanna.

Abundance Uncertain; generally rarely collected and probably uncommon; appears to be common at selected localities in NE DR Congo.

Remarks Apparently no other information available. One pregnant ♀ with one embryo in Mar (Faradje, NE DR Congo; label, RMCA).

Conservation IUCN Category: Data Deficient, previously Least Concern.

Little is known about the status of this species; Nicoll & Rathbun (1990) provisionally assessed it as 'Safe'.

Measurements
Elephantulus fuscipes
HB: 135.2 (126–148) mm, n = 5
T: 91.8 (86–100) mm, n = 6
HF: ca. 27–28 mm
E: n. d.
WT: 59.8 (52–77) g, n = 6

Elephantulus fuscipes

GLS: 32.6 (35.6–36.6) mm, n = 4
GWS: 17.3 (16.9–17.9) mm, n = 3
I^1–M^3: 18.2 (17.8–18.9) mm, n = 4
Body measurements and weight: RMCA
Skull measurements: Uganda, SW Sudan, NE DR Congo (BMNH)

Key Reference Corbet & Hanks 1968.

Mike Perrin

Elephantulus fuscus Dusky Sengi (Dusky Elephant-shrew)

Fr. Macroscélide foncé; Ger. Dunkle-Elefantenspitzmaus

Elephantulus fuscus (Peters, 1852). Reise nach Mossambique, Zoologie 1, Saugethiere, p. 87. Boror, near Quelimane, N Mozambique.

Taxonomy Originally described in the genus *Macroscelides*. Corbet & Hanks (1968) included *fuscus* as a synonym of *E. brachyrhynchus* but Corbet (1974b) points out that the two taxa appear to be sympatric in the vicinity of Tete, Mozambique, and accordingly treats them as separate species. Synonyms: *malosae*. Subspecies: none. Chromosome number: not known.

Description Small sengi very similar to *E. brachyrhynchus*. Dorsal pelage dark brown. Ventral pelage dark grey. Head large. Eyes large, with white eye-ring. Ears large, rounded at tip, with swollen supratragus, which is twisted backwards on a constricted stalk (but to a lesser extent than in *E. fuscipes*, cf. *E. brachyrhynchus*; see Figure 17). Pectoral gland absent; subcaudal gland present. Hindfeet elongated; five digits; soles dark and naked. Tail relatively long (ca. 90% of HB), black above in mid-dorsal line, paler below. Skull: infraorbital foramen is diagonal in lateral view and triangular when viewed anteriorly. M_3 present (also in *E. brachyrhynchus* and *E. fuscipes*; see profile Genus *Elephantulus*). Nipples: 1 + 2 = 6.

Geographic Variation None recorded.

Similar Species
E. brachyrhynchus. Very similar; supratagus swollen and not twisted backwards; infraorbital foramen oval in shape; not sympatric.

Distribution Endemic to Africa. Zambezian Woodland BZ. Recorded from Mozambique (Lower Zambezi Valley in the Vila Pery, Beira and Tete districts), S Malawi and SE Zambia (Meester *et al.* 1986).

Habitat Grassland with scattered bushes and trees on a sandy substrate (Smithers & Lobão Tello 1976).

Remarks Little is known about this species, although its biology is presumed to be similar to *E. brachyrhynchus*.

Conservation IUCN Category: Data Deficient, previously Least Concern.

Elephantulus fuscus

Little is known about the status of this species; Nicoll & Rathbun (1990) provisionally assessed it as 'Safe'.

Measurements
Elephantulus fuscus
HB: 114.0 (105–125) mm, n = 5
T: 102.8 (86–116) mm, n = 5
HF: 28.6 (28–30) mm, n = 5
E: 19.9 (19.5–20) mm, n = 5
WT: 45, 46 g, n = 2
GLS: 33.9 (32.7–35.5) mm, n = 7
GWS: 17.8 (17.2–18.4) mm, n = 5
I^1–M^3: 17.3 (16.3–17.9) mm, n = 7
Zambia and Malawi (BMNH, D.C. D. Happold unpubl.)

Key References Corbet 1974b; Corbet & Hanks 1968; Meester *et al.* 1986.

Mike Perrin

Elephantulus intufi Bushveld Sengi (Bushveld Elephant-shrew)

Fr. Macroscélide des régions sèches; Ger. Trockenland-Elefantenspitzmaus

Elephantulus intufi (A. Smith, 1836). Report of the Expedition for Exploring Central Africa 1834: 42 [1836]. Flats beyond Kurrichane, Marico district, North West Province, South Africa.

Taxonomy Originally described in the genus *Macroscelides*. Ellerman *et al.* (1953) included all the forms now considered as *E. rupestris* (Corbet & Hanks 1968) within *E. intufi*. Genetic studies suggest that this species is most closely related to *E. rupestris* (Raman & Perrin 1997). Morphological variation within and between populations of *E. intufi* is low (Matson & Blood 1997). Synonyms: *alexandri*, *schinzi*, *kalaharicus*, *mossamedensis*, *namibensis*, *campbelli*, *mchughi*, *omahekensis*, *canescens*. Subspecies: none. Chromosome number: 2n = 26 (Tolliver *et al.* 1989).

Description Small pale-coloured sengi, with white eye-ring. Pelage long and dense; hairs ca. 12 mm on mid-dorsal line. Dorsal pelage yellowish-buff; hairs dark grey, with yellowish-buff tip; long black-tipped guard hairs (especially on the rump). Flanks paler. Ventral pelage white or greyish-white; hairs white, grey at base. Head large. Eyes with white eye-ring. Ears large, rounded at tip, buffy, with conspicuous long white hairs on inner margins. Golden-buffy patches behind each ear. Pectoral gland absent, subcaudal gland possibly present. Hindlimbs elongated; five digits; soles of hindfeet naked, pale brown. Tail relatively long (ca. 105% of HB), dense short white hairs (with black tips) above, white below; no tuft at tip. M_3 absent. No evidence of sexual dimorphism (Matson & Blood 1997). Nipples: 1 + 2 = 6.

Geographic Variation Individuals from the eastern and southern parts of the geographic distribution tend to be paler than those from N Namibia (Corbet & Hanks 1968).

Similar Species
E. myurus. Almost always larger in size; soles of feet black; NE South Africa and Zimbabwe.
E. rupestris. On average larger in size; dorsal pelage brownish; eye-ring less conspicuous; tail tufted with long black tuft at tip; tail relatively longer; Eastern Cape Province, South Africa, and W Namibia.

This species and *E. rupestris* are often difficult to distinguish in the field, especially where the two occur in close proximity where bushveld habitat meets rock outcrops (Tolliver *et al.* 1989).

Distribution Endemic to Africa. South-West Arid BZ (Kalahari Desert and parts of Namib Desert). Recorded from Namibia, SW Angola, S Botswana and South Africa (adjacent to the borders with Namibia and Botswana).

Habitat Arid habitats with low shrubs and scattered grasses, usually on compact sandy soils.

Abundance Widespread, but usually not abundant or dense and often rare (G. B. Rathbun pers. obs.).

Adaptations The behavioural ecology of this species is similar to *E. rufescens* of eastern Africa (Rathbun 1979). In Namibia (Rathbun & Rathbun 2006, G. B. Rathbun pers. obs.), Bushveld Sengis normally shelter at comparatively exposed sites at the bases of bushes or under

Elephantulus intufi

large boulders, and may temporarily use rodent burrows to escape disturbances. Their activity is polycyclic, with crepuscular peaks. Like many *Elephantulus*, they build and maintain paths through the surface litter with sweeping actions of the forefeet; the paths connect basking, sheltering and feeding areas within the home-range.

Bushveld Sengis are well adapted to their arid environment in aspects of their thermal biology and metabolism. Basic metabolic rate is 73–94% of that expected in relation to size; T_b is maintained at 33.5–37.7 °C over a wide range of T_a (5–30 °C), even during cold nights (Perrin 1995). Hyperthermia is prevented at high temperatures by evaporative water loss, which may be as high as 156% of the normal rate for short periods, and by behavioural means (habitat selection, inactivity, etc.). When T_a is cool, T_b is maintained by resting in (warmer) burrows, vasoconstriction and good insulation. Torpor has not been reported.

Communication with conspecifics probably includes foot-drumming, scent-marking and vocalizations (high-pitched alarm squeak and soft 'purrs' [Faurie 1996, G. B. Rathbun pers. obs.]). Foot-drumming is highly variable (Faurie *et al.* 1996) and probably is a graded response to stressful situations (G. B. Rathbun pers. obs.).

Foraging and Food Insectivorous. Feeds primarily on ants. Also catches larger prey, which are carried to the safety of cover for ingestion. Sometimes eats some plant material.

Social and Reproductive Behaviour In Namibia (Rathbun & Rathbun 2006), opposite sex pairs share highly overlapping home-ranges that are exclusive of neighbours; mate-guarding by ♂♂ results in social monogamy. Average territory area of ♂ is 0.47 ha compared to 0.32 ha for ♀♀. Pairs shelter separately and interact infrequently and, except for brief polygamous associations, remain paired for life.

Reproduction and Population Structure Reproduction may be seasonal in the southern part of the range, corresponding with the warm and wet months of Aug–Mar (Skinner & Smithers 1990), but is nearly year-round in Namibia (Rathbun & Rathbun 2006). Litter-size: 1–2. Mean embryo number in Botswana: 1.9 (n = 9; Smithers 1971). Young are precocious at birth; the mother only visits neonates infrequently to nurse, and there is no direct or indirect paternal investment (Rathbun & Rathbun 2006).

Predators, Parasites and Diseases Barn Owls *Tyto alba* are a known predator (Vernon 1972) and the Horned Adder *Bitis caudalis* also preys on Bushveld Sengis (G. B. Rathbun pers. obs.). Ectoparasites include one species of ixodid tick, *Rhipicephalus parvus* (Fourie *et al.* 1995).

Conservation IUCN Category: Least Concern.

The unspecialized habitat requirements and comparatively wide distribution of the species in southern Africa suggest that it is not threatened at present. Nicoll & Rathbun (1990) provisionally assessed it as 'Safe'.

Measurements

Elephantulus intufi

TL (♂♂): 237.7 (218–246) mm, n = 7
TL (♀♀): 244.3 (238–254) mm, n = 4
T (♂♂): 118.3 (102–128) mm, n = 7
T (♀♀): 121.8 (117–131) mm, n = 4
HF (♂♂): 32.7 (32–39) mm, n = 10
HF (♀♀): 34.0 (32–36) mm, n = 6
E (♂♂): 23.7 (22–25) mm, n = 10
E (♀♀): 24.0 (23–26) mm, n = 6
WT (♂♂): 45.9 (43–54) g, n = 12
WT (♀♀): 45.7 (35–62) g, n = 15
GLS (♂♂): 33.8 (32.7–35.0) mm, n = 10
GLS (♀♀): 34.2 (33.0–35.0) mm, n = 6
GWS (♂♂): 19.3 (18.9–19.8) mm, n = 8
GWS (♀♀): 19.6 (19.0–20.0) mm, n = 5
I^1–M^3 (♂♂): 17.1 (16.3–17.7) mm, n = 10
I^1–M^3 (♀♀): 17.3 (16.7–13.2) mm, n = 6
Measurements: southern Africa (Rautenbach & Schlitter 1977)
Weight: Omaruru, Namibia (G. B. Rathbun unpubl.)

Key References Corbet & Hanks 1968; Ellerman *et al.* 1953; Meester *et al.* 1986; Rathbun & Rathbun 2006; Smithers 1971.

Mike Perrin & Galen B. Rathbun

Elephantulus myurus EASTERN ROCK SENGI (EASTERN ROCK ELEPHANT-SHREW)

Fr. Macroscélide oriental; Ger. Östliche Klippen-Elefantenspitzmaus

Elephantulus myurus Thomas and Schwann, 1906. Proceedings of the Zoological Society, London 1906: 586. Woodbush, Limpopo Province (formerly NE Transvaal), South Africa.

Taxonomy Originally described as a subspecies of *E. rupestris*, but recognized as specifically distinct by Roberts (1935). *Elephantulus myurus* is closely related to *E. rupestris* on the basis of allozyme and isozyme comparisons, although it also shows affinities with *E. edwardii* (Raman & Perrin 1997). Synonyms: *jamesoni*, *mapogonensis*, *centralis*, *fitzsimonsi*. Subspecies: none, although *fitzsimonsi* may represent a valid subspecies (race) (Corbet & Hanks 1968). Chromosome number: 2n = 30 (Ford & Hamerton 1956).

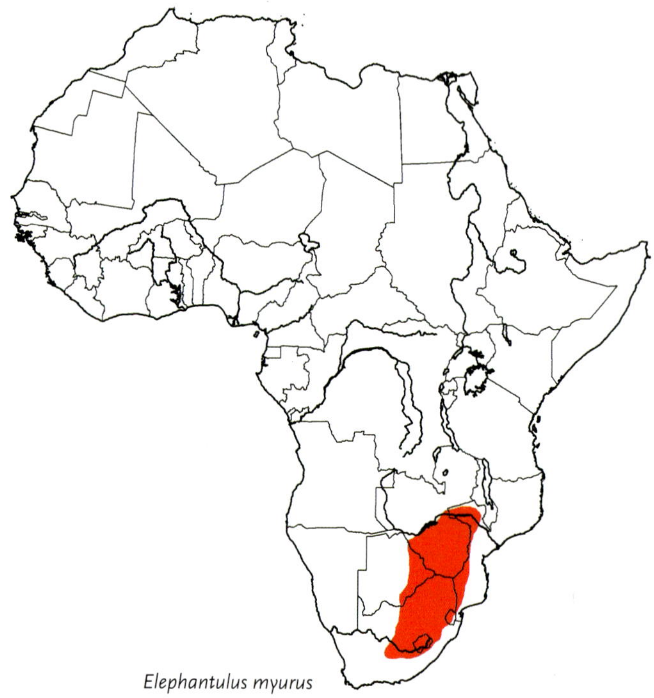

Elephantulus myurus

Description Small greyish sengi with white eye-ring. Dorsal pelage buffy-grey, soft and woolly, becoming greyer towards the rump. Flanks paler and greyer than dorsal pelage. Ventral pelage white or pale grey. Pectoral gland absent; subcaudal gland present. Head large, with large eyes and conspicuous white eye-ring. Ears large, brown, rounded at tip, with fringe of white hairs in inner margin. Buffy patch behind ears not conspicuous. Hindlimbs long; five digits; soles of hindfeet black. Tail relatively long (ca. 110% of HB), with sparse short hairs, dark above, paler below, without tuft at tip. M_3 absent. Nipples: 1 + 2 = 6.

Geographic Variation Very little variation, although the length of the upper toothrow decreases clinally to the north (Corbet & Hanks 1968).

Similar Species

E. rupestris. Dorsal pelage greyish-brown; comparatively conspicuous patch behind ears with rufous-yellow hair at the base of the ears; tail hairy with black tuft at tip; South Africa and W Namibia; only marginally sympatric.

Distribution Endemic to Africa. Zambezian Woodland BZ and parts of the Highveld BZ. Recorded from South Africa (NE Eastern Cape Province northwards to KwaZulu–Natal and former Transvaal), Lesotho, Zimbabwe, SE Botswana and SW Mozambique. Not recorded from Swaziland (Monadjem 1998).

Habitat Rocky outcrops on hill slopes, kopjes or piles of boulders in semi-arid environments. Holes and crannies in rocks that are used for shelter are essential for survival. Partially sympatric with *E. brachyrhynchus*, which lives in adjacent sandy, flat ground.

Abundance Locally common with marked interannual changes in density depending on the amount of rainfall. Densities may be high (88/ha) when insect abundance peaks during periods of average rainfall, but decline quickly to small isolated groups (even in prime habitats) when rainfall is low (Du Toit 1993). Tends to be commoner where there are large boulders and rock debris, and less common in isolated rock outcrops and unbroken hill slopes.

Adaptations Terrestrial, in rocky habitats. Predominantly diurnal, with a peak of activity at dawn, but also active at night (Woodall *et al.* 1989). Shelters in crevices and under rocks, without building any nest (Du Toit 1993). Extremely agile, capable of jumping from rock to rock. Eastern Rock Sengis are well adapted to semi-arid environments: the structure of the kidney shows characters for conserving water (Downs 1996); when T_a is cold, these sengis become hypothermic and enter torpor (Perrin 1995). Deep torpor occurs when T_b is 5 °C; during deep torpor (which does not exceed 24 h in length) oxygen consumption drops to just 2% of basal metabolic rate (Lovegrove *et al.* 2000a, b).

As in many other species of *Elephantulus*, communication is probably by foot-drumming, scent-marking and vocalizations, which includes a 'mew' with a fundamental of 1.47 kHz that averages 468 msec (Faurie 1996). Foot-drumming is performed during social encounters and also to indicate alarm; some drumming traits are probably specific to Eastern Rock Sengis (Faurie *et al.* 1996).

Foraging and Food Insectivorous. The diet is mostly small insects, especially termites and ants. Forages from cover, with quick hops to seize prey; large prey (the size of a large grasshopper) is carried into cover for ingestion. The caecum is small, but definitely plays a role in the digestion of protein (Woodall & Mackie 1987).

Social and Reproductive Behaviour Mate-guarding by ♂♂ results in social monogamy (Ribble & Perrin, 2005). Pair bonding is weak, and unattended ♀♀ may be temporarily attended by an adjacent monogamous ♂ (Ribble & Perrin 2005), similar to *E. rufescens* (Rathbun 1979) and *E. intufi* (Rathbun & Rathbun 2006).

Reproduction and Population Structure The reproductive biology of this species is documented in a large series of papers by Van der Horst & Gillman (e.g. 1941) and Van der Horst (e.g. 1946, 1954); ♀♀ polyovulate, producing a mean of 49 ova/ovary (range 25–89, n = 50) at each ovulation (Tripp 1971). Young are born during the warm, wet summer (Sep–Mar). Gestation: ca. eight weeks. Embryo number: 1.7 (1–2) in Botswana (Smithers 1971); 1.9 (1–2) in South Africa (Tripp 1971). Young are precocial at birth and independent at a young age; ♀♀ become sexually mature at five or six weeks of age. Males have relatively small, abdominal testes, which are reproductively active throughout the year, but with a significant reduction in sperm numbers and viability during winter (Apr–Jun), when reproduction stops (Woodall & Skinner 1989).

Predators, Parasites and Diseases Known predators include Barn Owls *Tyto alba* (Vernon 1972), Cape Eagle-owls *Bubo capensis mackinderi* (Gargett & Grobler 1976) and Rhombic Skaapsteker *Psammophylax rhombeatus rhombeatusi* (Van Wyk, 1988). Parasites include eight species of ticks and two non-tick arthropods (Fourie *et al.* 1995).

Conservation IUCN Category: Least Concern.

This species occupies rocky habitats that are in little danger of modification from human activities and the species is comparatively widespread in southern Africa.

Measurements
Elephantulus myurus
TL (♂♂): 275.1 (239–288) mm, n = 21
TL (♀♀): 276.3 (264–292) mm, n = 13
T (♂♂): 147.3 (135–160) mm, n = 21
T (♀♀): 147.2 (137–159) mm, n = 13
HF (♂♂): 39.5 (38–42) mm, n = 24
HF (♀♀): 39.3 (37–41) mm, n = 13
E (♂♂): 26.6 (25–28) mm, n = 24
E (♀♀): 26.5 (25–28) mm, n = 13
WT n. d.
GLS (♂♂): 38.5 mm, n = 22*
GLS (♀♀): 38.1 mm, n = 6*
GWS (♂♂): 20.4 (19.5–21.2) mm, n = 20
GWS (♀♀): 20.1 (19.4–20.4) mm, n = 7
I^1–M^3 (♂♂): 20.6 (19.9–21.2) mm, n = 25
I^1–M^3 (♀♀): 20.6 (20.0–21.3) mm, n = 12
Southern Africa (Rautenbach & Schlitter 1977)
*Minimum and maximum not available

Key References Corbet & Hanks 1968; Faurie *et al.* 1996.

Mike Perrin & Galen B. Rathbun

Elephantulus revoili Somali Sengi (Somali Elephant-shrew)

Fr. Macroscélide de Somalie; Ger. Somali-Elefantenspitzmaus

Elephantulus revoili (Huet, 1881). Bull. Soc. Philomath. Paris, ser. 7, 5: 96. Medjourtine, NE Somalia.

Taxonomy Originally described in the genus *Macroscelides*. A clearly defined but little-known species, closely related only to *E. rufescens* (Corbet & Hanks 1968). Synonyms: none. Subspecies: none. Chromosome number: not known.

Description Small sengi with distinctive facial pattern, and fairly long hairy tail ending in a small tuft. Dorsal pelage pale brown to pinkish-grey; hairs grey or black at base, brown or pinkish-grey at tip; black-tipped guard hairs. Flanks paler. Ventral pelage whitish-grey; hairs black on basal third, white on terminal two-thirds. Head large; rhinarium hairy below. Eyes large, with narrow white eye-ring and dark postorbital patch that extends back to below ear. Ears grey-brown; dark buffy patch behind ears. Pectoral gland present, marked by dense fringes of short white hair. Subcaudal gland present. Hindfeet elongated; five digits. Tail relatively long (ca. 120% of HB); hairs white with brown tips; longer at terminal end forming small tuft at tip. Dorsal pelage of juveniles paler than in adults, with more yellow and less grey. M_3 absent. Nipples: 1 + 2 = 6.

Geographic Variation Individuals from the south of the geographic distribution tend to be paler, with hairs grey at base (rather than black) and fewer black-tipped guard hairs, than individuals from the north.

Similar Species
E. rufescens. On average slightly smaller, darker in colour; tail less hairy; distribution mostly further south in East Africa.

Distribution Endemic to Africa. Somalia–Masai Bushland BZ. Recorded from N coast of Somalia between 44° 00' E and 44° 20' E; isolated specimens from Run, Garoe, Somalia (48° 17' N, 48° 20' E).

Elephantulus revoili

Possibly sympatric with *E. rufescens* but in different habitats. Occurs above 1300 m in northern part of range.

Habitat Stony ground in montane habitats.

Abundance Possibly rare; very few specimens have been collected.

Remarks Apparently no other information available.

Conservation IUCN Category: Data Deficient.

Little is know about the status of this species or its habitat. The small geographic range may be cause for concern.

Measurements
Elephantulus revoili
HB: 132 (122–148) mm, n = 6
T: 157 (144–167) mm, n = 6
HF: 37.3 (34–39) mm, n = 6
E: 24.3 (23–25) mm, n = 5
WT: n. d.
GLS: 37.6 (36.4–38.7) mm, n = 6
GWS: 20.8 (18.6–21.9) mm, n = 6
I^1–M^3: 19.2 (17.3–21.0) mm, n = 5
Somalia (Corbet & Hanks 1968, BMNH)

Key Reference Corbet & Hanks 1968.

Mike Perrin

Elephantulus rozeti North African Sengi (North African Elephant-shrew)

Fr. Macroscélide de Rozet; Ger. Nordafrikanische Elefantenspitzmaus

Elephantulus rozeti (Duvernoy, 1833). Mem. Soc. Hist. Nat. Strasbourg 1 (2) art. 4: 18. Near Orana, Algeria.

Taxonomy Originally described in the genus *Macroscelides*. This species is not conspecific with *E. rufescens* (Corbet 1974b), as suggested by Ellerman *et al.* (1953). Based on molecular genetics and penis morphology *E. rozeti* is more closely related to *Petrodromus* than to other *Elephantulus* (Douady *et al.* 2003). Synonyms: *atlantis*, *clivorum*, *deserti*, *moratus*. Subspecies: two. Chromosome number: 2n = 28 (Matthey 1954).

Description Small sengi. Pelage long, soft and silky; hairs 12–14 mm on mid-dorsal line. Dorsal pelage yellowish-brown to pale buffy-brown; hairs black at base, brown or buff at tip; some longer hairs with black tips. Flanks paler. Ventral pelage greyish-white; hairs black at base, white at tip. Rhinarium naked. Eyes large, but smaller than other *Elephantulus*, without eye-ring. Ears large, rounded at tip, with large tragus; buffy patch behind ears (sometimes inconspicuous). Postorbital dark spot absent. Pectoral gland absent; subcaudal gland present and especially well developed. Hindfeet elongated; five digits. The only species of *Elephantulus* with very rugose interdigital pads. Tail relatively long (ca. 110% of HB), hairy, dark above, paler below, with small pencil at tip. M_3 absent. Nipples: 1 + 2 = 6.

Geographic Variation Corbet & Hanks (1968) recognize two subspecies:

E. r. rozeti: Morocco and Algeria north of the Atlas Mts. Dorsal pelage dark; hairs with brown tip ca. 2 mm; I^1–M^2 on average longer (mean 17.7 [17.0–18.8] mm).

E. r. deserti: Tunisia and Morocco south of the Atlas Mts (but not into the desert). Dorsal pelage pale greyish-buff, brown tips 3–4 mm; I^1–M^2 on average shorter (mean 16.9 [16.5–17.6] mm).

Similar Species This is the only species of sengi in North Africa. No close physical resemblance with any other *Elephantulus* species.

Distribution Endemic to Africa. Mediterranean Coastal BZ of NW Africa from SW Morocco/Western Sahara to Tunisia and W Libya. Altitudinal range: sea level to 1600 m, and occasionally to 2725 m (Cuzin & Séguignes 1990). The Atlas Mts divide the western part of the range into a coastal region with typical Mediterranean climate, and a drier southern region continuous with the Sahara Desert. In NE Algeria and Tunisia, the range is more likely to be continuous from the coast to the edge of the desert (Corbet & Hanks 1968).

Habitat Semi-arid rocky areas, including outcrops, mountainsides and scree slopes; often adjacent to bush-scrub and grassland habitats (Cabrera 1932, Séguignes 1988).

Elephantulus rozeti

Abundance Widespread and locally common, but not abundant. Can maintain high population densities in optimal habitat (M. R. Perrin unpubl.).

Adaptations Terrestrial. Mainly crepuscular and nocturnal, although may sunbathe for short periods of time (Aulagnier & Thévenot 1986). Shelters among rocks and boulders where T_a remains comparatively constant and is neither too hot nor too cold, or in burrows made by other species. During summer, they are mainly active at dusk and from 02:00h to 04:00h, but in winter, activity is mainly during the day from 10:00h to 16:00h (Séguignes 1983). Prominent subcaudal gland probably used for scent-marking (Cabrera 1932). Locomotion is similar to other sengis; bipedal hopping has not been observed (Cabrera 1932).

T_b varies from 32 °C in the shade to 37 °C in the sun, and individuals actively move from shade to sun as a means of controlling T_b. In experimental conditions, when T_a is about 20 °C, enters torpor and is capable of arousing spontaneously. When T_a is 10–19.5 °C, T_b is the same as T_a, and individuals are incapable of arousal until T_a increases to above 20 °C (Séguignes 1983). Sandbathes by rubbing head and body in sandy soil accompanied by vigorous kicking of the hindlimbs, similar to other *Elephantulus* species.

Foraging and Food Insectivorous. Forages at base of rock scree and boulders for insects.

Social and Reproductive Behaviour No data available.

Reproduction and Population Structure Mating occurs Jan–Apr, and young are born Mar–Jun. At higher altitudes in Morocco, young are born later (Apr). Gestation: ca. 75 days (longer than for other small sengis). May produce two litters/year at low altitudes, but only one litter/year at higher altitudes. Embryo number: 2.4 (1–4), n = 48; mode 2. In 48 ♀♀, the number of embryos/female was 1 (n = 2), 2 (n = 28), 3 (n = 15) and 4 (n = 3). Females producing two young/litter contributed 49% of young. Young are born with eyes open and with hair, and can eat insects while still suckling (Séguignes 1989). Reproduction is constrained by the comparatively short period of warm weather (spring, summer) and the comparatively long gestation. The mean embryo number and maximum embryo number is greater than in other sengis.

Predators, Parasites and Diseases Remains of these sengis are commonly found in owl pellets in Algeria (Kowalski & Rzebik-Kowalski 1991). They are also preyed upon by rock-dwelling snakes (M. R. Perrin unpubl.). A remarkable number of endoparasites have been described (Seurat 1913, 1914, Delanoe 1931, Fain & Lukoschus 1976, Quentin 1978, 1979, Quentin & Seureau 1978, Kolebinova 1979).

Conservation IUCN Category: Least Concern.

Little is known about the status of this species; Nicoll & Rathbun (1990) provisionally assessed it as 'Safe' because the rocky habitats where it lives are rarely disturbed by humans.

Measurements

Elephantulus rozeti
HB: 108.1 (100–120) mm, n = 14
T: 113.4 (105–137) mm, n = 14
HF: 31.8 (30–34) mm, n = 13
E: 24.6 (21–27) mm, n = 14
WT: 45.3 ± 1.4 g, n = 7
GLS: 33.8 (32.7–34.8) mm, n = 8
GWS: 19.9 (19.2–20.5) mm, n = 8
I^1–M^3: 17.5 (16.8–18.8) mm, n = 8
Measurements: Morocco and Algeria (BMNH)
Weight: Morocco and Algeria (M. R. Perrin unpubl.)

Key References Aulagnier & Thévenot 1986; Cabrera 1932; Corbet & Hanks 1968; Séguignes 1983, 1988, 1989.

Mike Perrin & Galen B. Rathbun

Elephantulus rufescens Rufous Sengi (Rufous Elephant-shrew)

Fr. Macroscélide roux; Ger. Rotbraune Elefantenspitzmaus

Elephantulus rufescens (Peters, 1878). Monatsb. K. Akad. Wiss. Berlin 1878: 198. Ndi, Taita, Kenya.

Taxonomy Originally described in the genus *Macroscelides*. Corbet & Hanks (1968) do not recognize any subspecies, although Corbet (1974b) recognizes six subspecies (see below). This is a clearly defined species not closely resembling any other except *E. revoili*. Review of taxonomy and biology given by Koontz & Roeper (1983). Synonyms: *boranus*, *delicatulus*, *dundasi*, *hoogstraali*, *mariakanae*, *ocularis*, *peasi*, *phaeus*, *pulcher*, *rendilis*, *renatus*, *somalicus*. Subspecies: six. Chromosome number: 2n = 34 (Robbins & Baker 1978).

Description Small sengi with distinctive facial pattern. Pelage dense, hairs ca. 10 mm on mid-dorsal line. Dorsal pelage rufous, grey, buff or yellowish (according to locality – see below); hairs dark grey, grey-buff on terminal quarter. Ventral pelage pure white. Head large, with hairy rhinarium (a character shared only with *E. revoili*). Eyes large, with conspicuous white eye-ring that enlarges to form a large white eyebrow; dark postorbital patch extends to below ear. Ears large, rounded at tip; buff or whitish-buff postauricular area present. Pectoral gland present, surrounded by short woolly white hairs. Subcaudal gland present, but rudimentary. Hindlimbs elongated; five digits. Feet white in adults (brown in juveniles), with dark claws. Tail relatively long (ca. 105% of HB), dark brown, sparsely haired, without tuft at tip. M_3 absent. Nipples: 1 + 2 = 6.

Geographic Variation Six subspecies are recognized by Corbet (1974b):

E. r. boranus: S Ethiopia.
E. r. dundasi: N Kenya, Uganda and Sudan.

E. r. peasi: E Ethiopia.
E. r. pulcher: N Tanzania.
E. r. rufescens: SE Kenya.
E. r. somalicus: N Somalia.

Colour variation may be related to the type and colour of the soil. Kingdon (1974: plate opp. p. 65) refers to a 'red population' in SE Kenya (= *rufescens*?), a 'grey population' in NW Tanzania (= *pulcher*?) and a 'yellow population' in C Tanzania. Intermediate forms are also present. Within a single population, there may be individual variation in respect of some skull characters (supernumerary teeth, number of accessary cups on premolars, length of palatal foramina). However, the level of variation is low, lower than in other *Elephantulus* spp. (Matson *et al.* 1984).

Similar Species

E. revoili. On average slightly larger; dorsal pelage pale brown to pinkish-grey; tail with tuft at tip; Somalia only. Both species have a hairy rhinarium and similar distinctive facial pattern.

Distribution Endemic to Africa. Somalia–Masai Bushland BZ. Recorded from Somalia, EC Ethiopia, extreme SE Sudan, Kenya, E Uganda and N Tanzania (southwards at least to Ruaha R.).

Habitat Dry woodland and bushland; open wooded steppe and grassland, where there are thickets that can be used for shelter.

Abundance Comparatively common in suitable habitats.

Adaptations Terrestrial. Active at all times of the day and night (polycyclic); in SE Kenya peaks of activity were crepuscular (06:00 to 09:00h, and 16:00 to 19:00h) with the least activity at mid-day (Rathbun 1979). Typical daylight activity budgets for an individual ♀ (and ♂): 30% (13%) foraging, 21% (40%) trail clearing, 24% (30%) locomotion, 13% (13%) social interactions, 6% (0%) inter-specific interactions, 2% (1%) grooming and 3% (1%) combined activities (Rathbun 1979). When active, individuals travel between sheltering places (just a simple depression under a bush) along well-defined trails, which are maintained by 'sweeping' with the forefeet, by biting grass and small twigs, and pushing larger debris out of the way with the forefeet and forehead. Burrows generally are not used (Rathbun 1979). Body temperature is maintained at 34–38 °C when T_a is 10–35 °C (Perrin 1995). Sunbathing occurs frequently, especially in the early morning (Rathbun 1979). Basal metabolic rate is 82% of expected for weight (Thompson & Nicoll 1986). Both sexes scent-mark with pectoral and pedal glands (Rathbun & Redford 1981, Koontz 1984). Foot-drumming probably functions in both intra- and inter-specific communication, the latter perhaps as a predator invitation signal (Roeper 1981).

Elephantulus rufescens

Foraging and Food Insectivorous. Foraging takes place on trails or on nearby (1–2 m) barren areas for a few seconds to a few minutes; foraging often alternates with trail clearing. When foraging in leaf litter, the nose is used as a probe while the forefeet clean away obstructing leaves and twigs. Prey is captured by rapid in-and-out flicking of the very long tongue (Rathbun 1979). The principal invertebrate prey (decreasing frequency of stomach content) in SE Kenya are harvester termites (*Odontotermes*), black ants, beetles, crickets and grasshoppers, spiders and silverfish. During the wet season, the fruits of a shrub (*Premma resinosa*) form a small part of the diet along with some green plant matter. Rufous Sengis are selective feeders; termites and ants form ca. 75% of the diet, but comprise only about 40% of the available invertebrates (Rathbun 1979). The diet in Meru National Park (C Kenya) is similar: arthropods (mainly termites and ants) form 93 to >99% of the diet, depending on the season; 0–7% is stem and leaf matter (Neal 1984b).

Social and Reproductive Behaviour Socially monogamous. In SE Kenya, home-ranges of ♂♂ and ♀♀ are congruent and vary in area from 0.16 ha to 0.52 ha. Where the habitat was not saturated, the home-ranges of most pairs were contiguous with those of neighbouring pairs (Rathbun 1979). Pairs defend their home-ranges (same-sex specifically). The pair association is stable and may last for as long as the individuals live, perhaps 2–3 years. Pair-bonding behaviour is infrequent and the ♀ is dominant over the ♂ (Rathbun 1979). Sexual interactions are brief, facilitated by familiarity between the ♂ and ♀ (Lumpkin & Koontz 1986). Vaginal marking by the ♀ occurs during oestrus, which is nearly the only time when pairs are inseparable (Rathbun 1979, Lumpkin *et al.* 1982). Male mate-guarding, rather than indirect paternal investment (trail maintenance), is probably responsible for their social monogamy (Rathbun & Rathbun 2006).

Reproduction and Population Structure Reproduction occurs throughout the year (Rathbun 1979, Neal 1982). In Meru N. P., Kenya, pregnancy rate was 65–100% (monthly n = 2–17), and was highest in wet season and lowest in the dry season. Mean embryo number varies with the weight of the ♀: smaller ♀♀ (<40 g) have only single embryos; larger ♀♀ (>40 g) have a mean embryo number of 1.51 (range: 1–2; n = 37) (Neal 1982). However, there were no seasonal changes in embryo number or recruitment of young into the population (Neal 1982). Females do not superovulate. Placenta: haemochorial (Oduor-Okelo *et al.* 1980). Gestation: 57 days. Birth wt: ca. 10 g. Average daily gain ca. 1.0 g (Rathbun *et al.* 1981). Birth-sites are at comparatively exposed locations on the surface at the base of a shrub; no nesting material is used. At birth, young are highly precocial and remain at the birth-site for the first 1–2 days. Mean birth interval is 56–65 days (n = 7). Juveniles are weaned by Day 30 and both parents aggressively drive the young from their territory when the next litter is born. Juveniles either disperse to a vacant territory, try to establish a new one, or fall prey to predators. Mortality of young is high: survival to 100 days of age is ca. 40% (Rathbun 1979). Annual potential productivity is 8.3 young/annum (Neal 1982).

Predators, Parasites and Diseases In Kenya, only Speckled Sand Snakes *Psammophis punctulatus* are known predators (Rathbun 1979). In Serengeti N. P., Tanzania, 0.8% of owl pellets (n = 345) contained sengi bones (Laurie 1970). Three species of *Rhipicephalus* ticks, one species of sucking louse (Fourie *et al.* 1995) and a mite (Fain & Lukoschus 1976) have been recorded. Several endoparasties are also known (Hoogstraal *et al.* 1950, Kolebinova 1981).

Conservation IUCN Category: Least Concern.

Little is known about the status of this species; Nicoll & Rathbun (1990) provisionally assessed it as 'Safe'.

Measurements
Elephantulus rufescens
HB: 128.3 (102–199) mm, n = 22
T: 134.5 (111–163) mm, n = 22
HF: 33.8 (30–54) mm, n = 22
E: 25.0 (22–39) mm, n = 22
WT: 57.3 (47.1–70.2) g, n = 16
GLS: 36.1 (35.2–37.2) mm, n = 12
GWS: 20.1 (19.4–20.8) mm, n = 12
I^1–M^3: 17.6 (17.2–18.1) mm, n = 12
Body measurements and weight: Kibwezi, Kenya (G. B. Rathbun unpubl.)
Skull measurements: Cherangani Mts, Kenya (Matson *et al.* 1984)

References Corbet 1974b; Neal 1982, 1984b; Rathbun 1979.

Mike Perrin & Galen B. Rathbun

Elephantulus rupestris Western Rock Sengi (Western Rock Elephant-shrew)

Fr. Macroscélide occidental; Ger. Westliche Klippen-Elefantenspitzmaus

Elephantulus rupestris (A. Smith, 1831). Proc. Zool. Soc. Lond. 1830–1831: 11.
Mountains near the mouth of the Orange River (= Little Namaqualand [Shortridge 1934]).

Taxonomy Originally described in the genus *Macroscelides.* According to Corbet & Hanks (1968), 'Ellerman *et al.* (1953) included this species in *E. intufi* and used the name *rupestris* for the species that we call *E. myurus* and *E. edwardii.*' This error was caused by confusion over type specimens. Meester *et al.* (1986) recognized *rupestris* as a valid species, as do Corbet & Hanks (1968). Morphological analysis shows that there is little variation within and between populations (Matson & Blood 1997). Allozyme and chromosomal studies suggest that *E. rupestris* is closely related to *E. intufi* (Tolliver *et al.* 1989), and allozyme and isozyme data generally show a close relationship to *E. myurus* (Raman & Perrin 1997). Synonyms: *barlowi*, *gordoniensis*, *kobosensis*, *montanus*, *okombahensis*, *tarri*, *typus*, *vandami*. Subspecies: none. Chromosome number: 2n = 26 (Wenhold & Robinson 1987).

Description Small sengi with broad patch of rufous or yellowish-brown hair behind ears. Dorsal pelage greyish-brown or rufous-brown, becoming grey on flanks; hairs dark grey, with rufous-brown at tip; some longer hairs with black tips especially on mid-dorsal line. Ventral pelage greyish-white; hairs grey at base, tip white. Head large with elongated snout, narrow black line dorsally. Eyes large; eye-ring indistinct. Ears large, rounded at tip, with distinct patch of rufous hair from base to nape of neck. Pectoral gland absent; subcaudal gland present. No dark spot behind the eye. Hindlimbs elongated; five digits. Soles of hindfeet naked, black. Tail relatively very long (ca. 115% of HB), black above, paler below, with black tuft at tip. M_3 absent. Mean measurements of ♀♀ slightly larger than for ♂♂. Nipples: 1 + 2 = 6.

Geographic Variation All named forms show variation in pelage colour.

Similar Species
E. myurus. Ventral pelage greyer; eye-ring conspicuous; tail without tuft at tip; South Africa, Zimbabwe.
E. intufi. Ventral pelage greyish-white; eye-ring present; tail usually shorter, paler, with small tuft at tip; Namibia, S Botswana.
E. edwardii. Ventral pelage greyish-white; eye-ring present; tail black above, distal one-third completely black with small tuft at tip; western parts of South Africa.

Elephantulus rupestris and *E. intufi* are often difficult to distinguish in the field, especially where the two occur in close proximity where bushveld habitat meets rock outcrops (Tolliver *et al.* 1989).

Distribution Endemic to Africa. South-West Arid BZ (Namib Desert and Karoo) and parts of South-West Cape BZ. Recorded from South Africa (Eastern Cape Province to Little Namaqualand, where isolated populations occur) and W Namibia (Kaokoveld to just south of the border with Angola). Absent from NE Namibia and Western Cape Province of South Africa.

Habitat Rocky kopjes and outcrops or piles of boulders in arid and semi-arid areas.

Abundance No information.

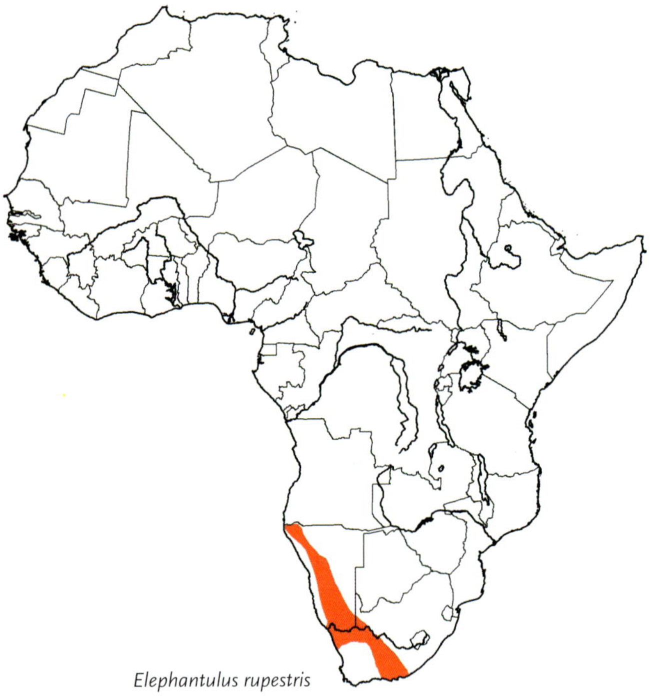

Elephantulus rupestris

Adaptations Little is known about this species. Terrestrial, and adapted to living among rocks. Predominantly crepuscular. Locomotion quadripedal, often characterized by habitual use of specific routes across rock surfaces. Vocalizations ('mews' and 'clicks'), foot-drumming and scent marking are used, probably for communication (Faurie 1996). Foot-drumming is characterized by 30–50 drums/bout (more than in other *Elephantulus* spp.), drum intervals of 15–25 msec, and one bout/series; the length of a series is 1–1.5 sec (Faurie *et al*. 1996). This species is more active and less aggressive than other related species (Faurie 1996).

Foraging and Food Insectivorous. Principal prey are ants and termites.

Social and Reproductive Behaviour Probably similar to *E. myurus*. Often trapped as male–female pairs (Withers 1979), and thus probably monogamous and territorial, as in other sengis (Rathbun & Rathbun 2006).

Reproduction and Population Structure Little information. In Namibia, ♀♀ were pregnant in Sep. Embryo number: 1–2 (Shortridge 1934). Females may produce several litters each year.

Predators, Parasites and Diseases Barn Owls *Tyto alba* are known predators (Vernon 1972). Ectoparasites include ten species of fleas, 11 species of ixodod ticks and one species of mite (Fourie *et al.* 1995).

Conservation IUCN Category: Least Concern.

The rocky habitats of this species are not subject to human interference, and are comparatively widely distributed in southern Africa.

Measurements

Elephantulus rupestris
TL (♂♂): 269.0 (239–292) mm, n = 24
TL (♀♀): 274.5 (255–292) mm, n = 19
T (♂♂): 145.5 (122–166) mm, n = 24
T (♀♀): 149.7 (137–163) mm, n = 19
HF (♂♂): 36.5 (34–38) mm, n = 26
HF (♀♀): 37.1 (35–39) mm, n = 20
E (♂♂): 26.0 (22–29) mm, n = 26
E (♀♀): 26.4 (24–38) mm, n = 20
WT: n. d.
GLS (♂♂): 36.5 (35.0–37.9) mm, n = 21
GLS (♀♀): 36.9 (35.6–38.6) mm, n = 17
GWS (♂♂): 20.0 (19.2–20.9) mm, n = 24
GWS (♀♀): 20.0 (19.1–20.9) mm, n = 16
I^1–M^3 (♂♂): 19.1 (18.2–20.1) mm, n = 26
I^1–M^3 (♀♀): 19.1 (18.4–19.7) mm, n = 20
Southern Africa (Rautenbach & Schlitter 1977)

Key References Corbet & Hanks 1968; Matson & Blood 1997; Meester *et al.* 1986.

Mike Perrin

GENUS *Macroscelides*

Round-eared Sengi

Macroscelides A. Smith, 1829. Zoological Journal, London 4: 435. Type species: *Macroscelides typus* A. Smith, 1829.

Macroscelides proboscideus.

The genus is monotypic (Table 13) and occurs only in semi-arid and arid environments of Cape Province, South Africa, W Namibia and SW Botswana. The genus is characterized by small size (HB usually less than 120 mm), five digits on fore- and hindfeet, three pairs of nipples, very inflated tympanic bullae (visible dorsally on the skull) and two molars on each mandible (Figure 18). Further information is given in the species profile. The single species is *Macroscelides proboscideus*.

Mike Perrin

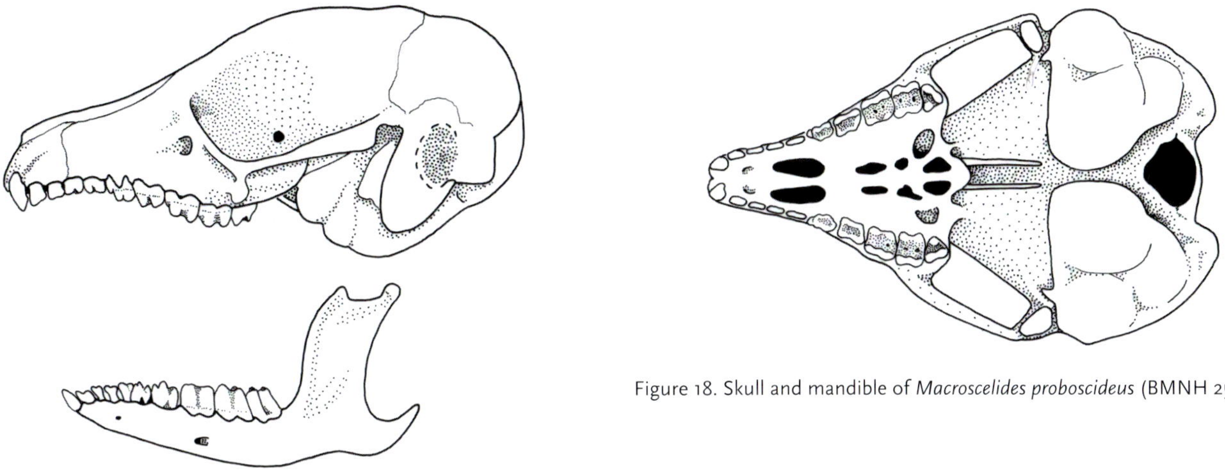

Figure 18. Skull and mandible of *Macroscelides proboscideus* (BMNH 25.1.2.11).

Macroscelides proboscideus ROUND-EARED SENGI (ROUND-EARED ELEPHANT-SHREW)

Fr. Macroscélide à oreilles courtes; Ger. Kurzohr-Rüsselspringer

Macroscelides proboscideus (Shaw, 1800). Gen. Zool. Syst. Nat. Hist. 1 (2), Mammalia, p. 536. Cape of Good Hope (specified as Roodewal, Oudthoorn, Cape Province [now Western Cape Province] by Roberts 1951).

Taxonomy Originally described in the genus *Sorex*. The only species in the genus *Macroscelides*. Synonyms: *ausensis*, *brandvleiensis*, *calvinensis*, *chiversi*, *flavicaudatus*, *harei*, *hewetti*, *isabellinus*, *jaculus*, *langi*, *melanotis*, *typicus*, *typus*. Subspecies: uncertain. Roberts (1951) recognized ten subspecies, based on pelage colouration, although only *flavicaudatus* in the north of the range may be valid, perhaps even at the species level (Corbet & Hanks 1968). The status of the form *melanotis* is uncertain: it is now considered to be conspecific with *M. proboscideus* because the specimen from Benguella, Angola, was a misidentified *E. intufi*, and the location of the holotype from Namibia was supposedly from South Africa (Corbet & Hanks 1968). The genetic distance between *M. proboscideus* and *Petrodromus tetradactylus* is comparatively low (0.323), implying a close relationship between these two species (Tolliver *et al.* 1989, Corbet 1995). Chromosome number: 2n = 26 (Wenhold & Robinson 1987).

Description A small round-bodied sengi with a round face and round ears. Pelage long (up to 17 mm), soft and silky and rather fluffy-looking. Dorsal pelage buffy-grey to greyish-brown; hairs black at base with buffy-grey tip. Flanks yellowish-brown. Ventral pelage greyish-white; hairs black at base with white tip. Head large with long thin snout. Eyes moderate (smaller than in most *Elephantulus*), without white eye-ring. Ears broad and rounded (shorter than in most *Elephantulus*), white on inner edges and without buffy patch behind ears; tragus large; supratragus well-developed, but not twisted. Pectoral gland absent; subcaudal gland present. Hindlimbs elongated; five digits. Tail long (ca. 105% of HB), proximal half dark above, paler below; distal half all black with longer hairs forming tuft at tip. Skull with very enlarged tympanic bullae (which can be felt on the dorsal posterior part of the head). Nipples 1 + 1 + 1 = 6.

Geographic Variation Individuals from the Cape Province, South Africa, have darker pelage than those from Namibia, which are pinkish in the Namib Desert. According to Corbet & Hanks (1968) this variation is clinal, but darker specimens recently trapped near the type location for *melanotis*, east of the Namib Desert in Namibia (M. Griffin pers. comm.), suggest that this is not always the case.

Similar Species

Elephantulus spp. (*E. edwardii*, *E. rupestris*, *E. intufi*). Large ears not rounded at tip; supratragus small; tympanic bulla not enlarged; white eye-ring present; buffy colour behind ears.

Distribution Endemic to Africa. South-West Arid BZ (Namib Desert and Karoo) and South-West Cape BZ. Recorded from

Northern, Western and Eastern Cape Provinces, South Africa (as far west as Grahamstown), W Namibia and extreme SW Botswana.

Habitat Desert and semi-arid regions. In the Namib Desert, associated with boulders and small rocky outcrops on gravel plains, where vegetation is exceedingly sparse. In the Karoo, found where gravel and sandy plains are well vegetated with low bushes and scrub. Livestock grazing reduces population numbers of this sengi (Eccard *et al.* 2000).

Abundance Locally common but not abundant. Numbers and densities may fluctuate over time, but quantitative data are lacking.

Adaptations Terrestrial. Highly cursorial (ca. 20 km/h), especially when running on trails (which are sometimes hundreds of metres in length) (Sauer 1973). Mainly nocturnal and crepuscular (Woodall *et al.* 1989, Roxburgh & Perrin 1994). Shelters under bushes, rocks, boulders, or scree (depending on habitat). May dig their own burrows (Sauer 1973) or use existing burrows.

The digestive tract is well adapted to an omnivorous diet (Woodall 1987, Kerley 1995, Spinks & Perrin 1995). The stomach is a simple unilocular glandular sac, the intestines are long and there is a functional caecum containing a diverse microflora (as in many purely herbivorous mammals). The characters of the digestive system change seasonally in response of the seasonal changes in diet (Woodall 1987).

Temperature regulation is atypical for a small mammal adapted to arid habitats (Roxburgh & Perrin 1994). T_b remains fairly constant (35–37.8 °C) over a wide range of T_a (5–38 °C) and is regulated behaviourally by basking, sheltering, and altering activity patterns. Vasodilation is also used when T_a is high. Evaporative water loss is comparatively low. Metabolic rate is close to the weight-specific prediction. Individuals enter torpor in cold weather (Roxburgh & Perrin 1994) and when deprived of food (Lovegrove *et al.* 1999).

The kidneys show adaptations to conserve water, with the pelvic region enlarged and elongated for efficient reabsorption of fluid. However, the kidneys are not as efficient at concentrating urine as those of *Elephantulus*. Captive sengis can maintain weight without drinking water when insects are available (Downs 1996). It is likely that in the wild, these sengis obtain all their water from their food. In these respects, they exhibit many similarities to arid-adapted small rodents.

Foraging and Food Omnivorous, feeding mainly on insects and herbage. In the Karoo of South Africa, the diet was 63.0% insects, 36.7% herbage, and a trace of seeds (Kerley 1995). At other locations, the proportion of insects was 46–88%, with considerable individual and seasonal variation. The consumption of insects was highest in Oct–Nov (77.4%, n = 8) and lowest in Jun–Jul (45.5%, n = 4). In captivity, Round-eared Sengis are also omnivorous (Unger & Schratter 2000).

Social and Reproductive Behaviour In the Namib Desert, home-ranges are up to 1.0 km², with varying degrees of overlap between loosely associated male–female pairs (Sauer 1973) that result in various degrees of social monogamy, depending on habitat and the population density (Rathbun & Rathbun 2006). Well-defined straight trails are maintained through gravel and substrate litter that connect shelters and feeding areas (Sauer 1973). Nesting material is not used in nests, and ♀♀ visit the highly precocial neonates infrequently (once a day?) to suckle. At about Day 5, the mother begins provisioning young with invertebrates collected in her cheeks (Sauer 1973). There is no direct paternal investment. Scent-glands (Faurie 1996) and foot-drumming (Faurie *et al.* 1996) are probably important in communication.

Reproduction and Population Structure Females polyovulate, producing 21 ova/ovary (n = 2) at each ovulation (Tripp 1971). In the Karoo of South Africa (Bernard *et al.* 1996), spermatogenesis and births occur throughout the year, but most pregnancies occur in the summer months (Sep–Feb), although a few occur in early winter. Oestrous cycle: ca. 10 weeks; gestation: ca. 56 days. Litter-size: 1–2 (Trautmann & Carbone 1991). At birth, young are highly precocial (Unger 2000), as in other Macroscelidinae; they are fully haired, and the eyes are open. Female suckles young for ca. 2 weeks, stops provisioning (see above) at ca. Week 3 and the young are independent at ca. Week 6 (Sauer 1973).

Predators, Parasites and Diseases Barn Owls *Tyto alba* are known predators (Vernon 1972). Ectoparasites include three species of fleas (Fourie *et al.* 1995).

Conservation IUCN Category: Least Concern.

Measurements
Macroscelides proboscideus
HB: 110.1 (104–115) mm, n = 13
T: 121.4 (107–134) mm, n = 13
HF: 34.1 (32–36) mm, n = 13
E: 22.2 (20–25) mm., n = 13
WT (♂♂): 38.0 (32–47) g, n = 4
WT (♀♀): 38.4 (31–47) g, n = 5
GLS: 33.7 (32.1–34.8) mm, n = 13
GWS: 21.0 (20.2–21.8) mm., n = 13
I^1–M^3: 15.6 (15.0–16.3) mm, n = 13
Measurements: South Africa (BMNH)
Weight: South Africa (Smithers 1983)

Key References Corbet & Hanks 1968; Raman & Perrin 1997; Roxburgh & Perrin 1994; Sauer 1973; Woodall 1987.

Mike Perrin & Galen B. Rathbun

GENUS *Petrodromus*
Four-toed Sengi

Petrodromus Peters, 1846. Bericht Verhandl. K. Preuss. Akad. Wiss. Berlin 11: 258. Type species: *Petrodromus tetradactylus* Peters, 1846.

Petrodromus tetradactylus.

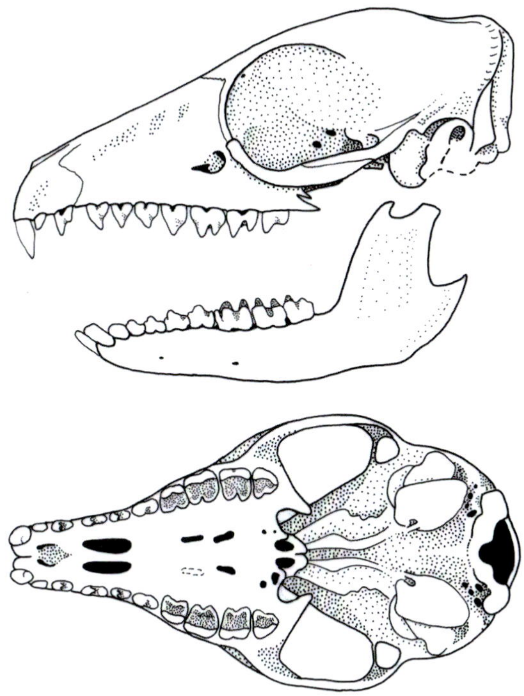

Figure 19. Skull and mandible of *Petrodromus tetradactylus* (BMNH 20.10.10.19).

This monotypic genus (Table 13) is a comparatively unspecialized representative of the family. Characters of the genus include the narrow skull (proportionally the narrowest in the family); smallest braincase, smallest bullae and olfactory chamber; and the most constricted interorbital region (Figure 19). It also has proportionally longer forelimbs than hindlimbs compared with other sengis (Evans 1942). Based on molecular techniques and morphology, *Petrodromus* forms a distinct clade with the monospecific genus *Macroscelides* (Tolliver *et al.* 1989, Corbet 1995, Raman & Perrin 1997). However, more recent morphological and molecular data indicate that *Elephantulus rozeti* is more closely related to *Petrodromus* than to other *Elephantulus* (Douady *et al.* 2003). *Petrodromus* is the largest member of the subfamily Macroscelidinae, and is distinguished from other genera by five digits on forefoot, four digits on hindfoot (Digit 1 absent), and two pairs of nipples. In size, it is intermediate between the small to medium-sized *Macroscelides* and *Elephantulus* and the very large *Rhynchocyon* (Corbet & Hanks 1968). Dental formula is $I\ ^{3}/_{3}$, $C\ ^{1}/_{1}$, $P\ ^{4}/_{4}$, $M\ ^{2}/_{2} = 40$. The first upper incisors are slightly recurved and over twice as long as the other incisors and canines. *Petrodromus* does not polyovulate (Tripp 1971). The 'pseudo-ungulate' habit and behaviour is less developed than in the three species of *Rhynchocyon*.

The single species is *Petrodromus tetradactylus*.

Galen B. Rathbun

Petrodromus tetradactylus Four-toed Sengi (Four-toed Elephant-shrew)

Fr. Pétrodrome à quatre orteils; Ger. Vierzehen-Elefantenspitzmaus

Petrodromus tetradactylus Peters, 1846. Bericht Verhandl. K. Preuss. Akad. Wiss., Berlin 11: 258. Tette, Mozambique.

Taxonomy A varied species with one or two forms that may prove to justify specific rank (Corbet & Hanks 1968). Synonyms: *beirae*, *matschie*, *mossambicus*, *nigriseta*, *occidentalis*, *robustus*, *rovumae*, *sangi*, *schwanni*, *sultani*, *swynnertoni*, *tordayi*, *tumbanus*, *venustus*, *warreni*, *zanzibaricus*. Subspecies: nine (Corbet 1974b). Chromosome number: 2n = 28, with eight pairs of metacentric and submetacentric chromosomes. Distinguished from all other Macroscelidea by presence of a small pair of subtelocentric autosomes on chromosomes 4 and 13 (Tolliver *et al.* 1989).

Description Medium-sized sengi with four digits on hindfoot and conspicuous facial markings. Pelage soft and dense. Dorsal pelage varied: rusty-red, buffy-grey, or dark brown to grey. Dark wide mid-dorsal stripe is present in some forms; naked patch on rump at base of tail. Flanks buff or orange or pale grey. Ventral pelage white. Head large with moderately long snout; dark brown or black patch from behind eye to base of ear. Eye large, with distinctive white eye-ring. Ears broad and upright, rusty-brown to buffy-yellow. Limbs elongated and slender; forefeet with five digits, hindfeet with four digits (Digit 1 absent). Tail relatively long (ca. 85% of HB), slender, black above, paler below, sparsely haired; stiff bristles (see Adaptations) on the underside of the distal third in some forms. Pectoral gland slightly developed; subcaudal gland absent. Little or no sexual dimorphism. Nipples: 2 (one antebrachial, one pectoral) + 0 (inguinal) = 4.

Petrodromus tetradactylus.

Geographic Variation Colour and pattern variation in this species is extensive, and complex, and not well understood (see incomplete descriptions in Corbet & Hanks 1968). Some of the nine subspecies are provisional because further information may indicate clinal variation (Corbet & Hanks 1968, Corbet 1974b).

P. t. beirae: C Mozambique, south of Zambezi R.
P. t. rovumae: E Tanzania and N Mozambique.
P. t. schwanni: coast of S Mozambique.
P. t. sultani: coast of SE Kenya and NE Tanzania.
P. t. swynnertoni: Chimanimani Mts, E Zimbabwe.
P. t. tetradactylus: Malawi, Zambia, E DR Congo, W Tanzania, Ruanda.
P. t. tordayi: C and W DR Congo, NE Angola.
P. t. warreni: N KwaZulu–Natal, South Africa and Mozambique.
P. t. zanzibaricus: Mafia and Zanzibar Is.

Similar Species

Elephantulus rufescens. On average smaller, with distinct pectoral gland; similarly distinct white eye-ring with a dark posterior patch on cheek; nearly sympatric with *P. tetradactylus* in Kenya.

Distribution Endemic to Africa. Zambezian Woodland BZ, Coastal Forest Mosaic BZ and Southern Rainforest–Savanna Mosaic where rainfall exceeds about 700 mm/year, and from sea level to ca. 1400 m (Corbet & Hanks 1968). One subspecies (*tordayi*) in Rainforest BZ of DR Congo. Recorded from SE Kenya, S Uganda, Tanzania, Mozambique, Zambia, Malawi, SE Zimbabwe, DR Congo, E Congo, NE Angola and NE South Africa (KwaZulu–Natal Province). Also Mafia and Zanzibar Is.

Habitat Dense woody thickets in forests, closed canopy woodlands, rocky outcrops and riparian areas. In Kenya, sometimes sympatric and syntopic with *Rhynchocyon chrysopygus*, but the two species avoid competition because Four-toed Sengis prefer denser cover, prey on different invertebrates and are crepuscular rather than diurnal (FitzGibbon 1995).

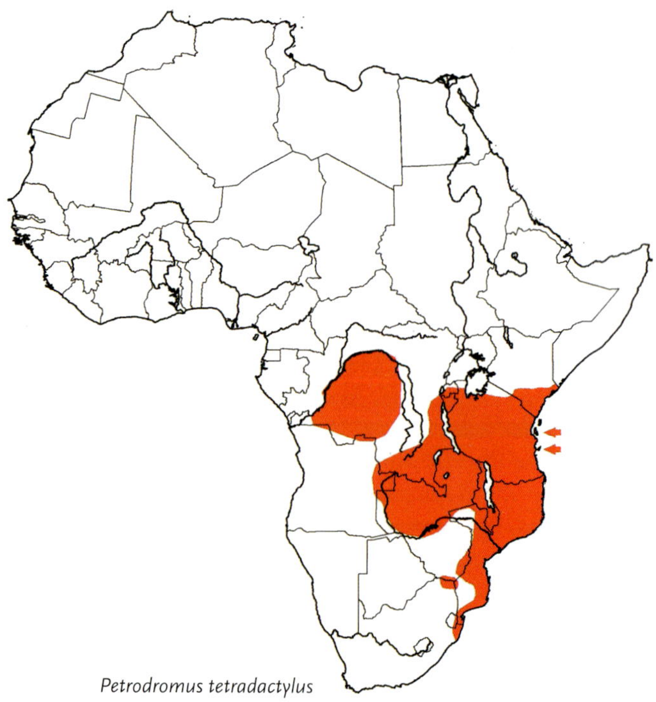

Petrodromus tetradactylus

Abundance Widely distributed. In parts of the Arabuko–Sokoke Forest, Kenya, abundance varies according to habitat. Most abundant in habitats dominated by *Cynometra* trees, moderately abundant in *Afzelia* habitats and least abundant in *Brachystegia* habitats. In *Afzelia* habitats, the estimated density was 2.1 animals/ha (FitzGibbon 1995). Although comparatively common locally, these sengis are not easy to see because they are crepuscular and secretive. The most obvious evidence of their presence is the loud sound that they make by foot-drumming on the substrate (especially when disturbed), and their network of narrow paths through the leaf litter (Ansell & Ansell 1969, Kingdon 1974, Rathbun 1979).

Adaptations Terrestrial. Four-toed Sengis are highly cursorial and very alert. They are active at all times of the day, mainly at dawn and dusk, and usually rest in the middle of the day. The level of activity at night is uncertain, but may continue to at least 22:25h (FitzGibbon 1995). Movement is normally by walking and running; not by ricochets as reported in many accounts. A network of narrow paths is made through the leaf litter, and is maintained by sweeping motions with the front feet. Paths are also used to reach feeding areas, and to flee disturbances (Kingdon 1974, Rathbun 1979). These sengis do not build or use nests, but instead rest and groom on favoured sections of path. To escape predators, they sometimes retreat into hollow tree trunks or logs, and into holes in the ground (Ansell & Ansell 1969).

T_b is 33–37.5 °C, generally 1–2 °C lower than in smaller *Elephantulus* spp. Because of the larger size of this species, oxygen consumption/g body weight and evaporative water loss/g body weight is lower than in *Elephantulus* (Perrin 1995). The kidneys are unable to produce concentrated urine (cf. the arid-adapted sengis) so, when water is scarce, temperature regulation is effected by vasodilation and selection of comparatively cool micro-climates. When fed mealworms in captivity in the absence of drinking water, body weight can be maintained (Downs 1996). The stiff bristles along the underside of the tail probably function to spread sweat and sebaceous gland products on the substrate during scent marking (Jennings & Rathbun 2001). Other forms of communication include foot-drumming (similar to most Macroscelidinae), and soft purrs and chirps. Captured animals sometime scream loudly (Jennings & Rathbun 2001).

Foraging and Food Insectivorous and omnivorous. Prey is located by scuffing and disturbing leaf litter with the hindfeet and long mobile nose. The tongue is very long and used to glean exposed prey (Kingdon 1974). In coastal Kenya, prey (in decreasing order of importance) consists of beetles, termites, plant matter, centipedes, ants, crickets, millipedes and spiders (Rathbun 1979, FitzGibbon 1995). In other areas, ants and termites are the principal prey, together with some green plant material, seeds and fruits (Ansell & Ansell 1969).

Social and Reproductive Behaviour Appears to be solitary or to live in pairs (Brown 1964). Home-ranges in the Arabuko–Sokoke Forest, Kenya, dominated by *Afzelia* trees, averaged 1.2 ± 0.2 ha (FitzGibbon 1995). Although it is suspected that Four-toed Sengis form monogamous pairs for life, little is actually known of their social behaviour (Rathbun 1979, FitzGibbon 1995). Captive animals are best maintained as opposite sex pairs to reduce intra-specific aggression, which suggests that they may be territorial and monogamous in the wild (Jennings & Rathbun 2001).

Reproduction and Population Structure Birth of young appears to vary according to location and climate. Young have been recorded for most months of the year in eastern Africa (Jennings & Rathbun 2001). In Zambia, foetuses have been recorded in Jan, Jul and Oct (Ansell 1960). In southern Africa, breeding occurs mainly during the wet months of Aug–Oct (Smithers 1983). Litter-size: 1, occasionally 2. At birth, young weigh about 31.5 g, are highly precocial and are able to walk within hours of birth (Tripp 1971, Rathbun 1979). Four-toed Sengis have not been successfully bred in captivity (Nicoll & Rathbun 1990).

Predators, Parasites and Diseases Predators include the Gabon Viper (*Bitis gabonica*), domestic cats and probably raptors and native carnivores (Ansell & Ansell 1969, Jennings & Rathbun 2001). Ectoparasites include ticks (9 spp.), lice (1 sp.), mites (2 spp.) and fleas (2 spp.) (Fourie *et al.* 1995). Blood parasites include *Trypanosoma petrodromi*, *Plasmodium brodeni* and filarial worms (Jennings & Rathbun 2001).

Conservation IUCN Category: Least Concern.

Most subspecies are widely distributed and thus in little danger of extinction. The subspecies *sangi* (included in *P. t. sultani* by Corbet 1974b) is restricted to the Taita Hills in Kenya, and may be endangered or already extinct due to loss of forest habitat (Nicoll & Rathbun 1990). The Giriama people of Kenya trap Four-toed Sengis for meat with snares and deadfall traps. FitzGibbon *et al.* (1995) found that about 15 individuals/km^2/year were being harvested, a rate that they believe is sustainable. The subspecies *beirae* is considered 'rare' based on its restricted distribution in KwaZulu–Natal, South Africa (Jennings & Rathbun 2001).

Measurements

Petrodromus tetradactylus
HB: 192.9 (163–210) mm, n = 33
T: 166.4 (156–187) mm, n = 33
HF: 54.8 (51–58) mm, n = 33
E: 35.9 (34–39) mm, n = 33
WT: 198.3 (129–250) g, n = 11
GLS: 56.1 (54.3–58.5) mm, n = 12
GWS: 29.4 (28.1–30.5) mm, n = 12
I^1–M^3: 29.3 (28.1–30.5) mm, n = 12
Upper canine: 3.98 (3.53–4.90) mm, n = 12
Body measurements and weight: Arabuko–Sokoke Forest, Kenya (G. B. Rathbun unpubl.)
Skull measurements: Kenya (BMNH)

Key References Corbet & Hanks 1968; FitzGibbon 1995; Jennings & Rathbun 2001; Perrin 1995; Rathbun 1979.

Galen B. Rathbun

GENUS *Rhynchocyon*
Giant Sengis

Rhynchocyon Peters, 1847. Bericht Verhandl. K. Preuss. Akad. Wiss., Berlin 12: 36. Type species: *Rhynchocyon cirnei* Peters, 1847.

Rhynchocyon cirnei.

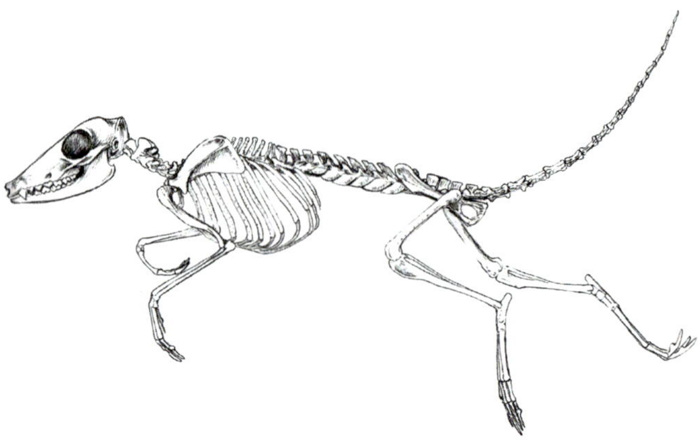

Skeleton of *Rhynchocyon petersi.*

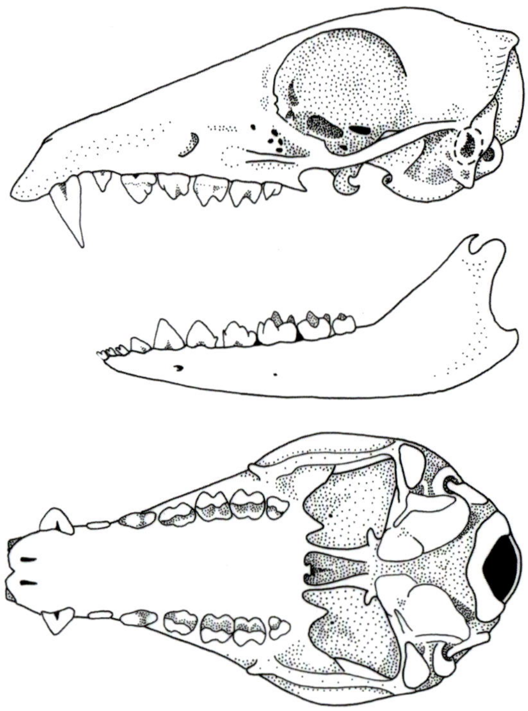

Figure 20. Skull and mandible of *Rhynchocyon cirnei* (BMNH [4287], RMCA 90-042-M-200).

The genus *Rhynchocyon* contains three species that live in woodlands and forests of central and eastern Africa. The species are allopatric: one (*R. cirnei*) is widespread and the other two (*R. chrysopygus* and *R. petersi*) have restricted distributions. The genus is placed in a separate subfamily (Rhynchocyoninae, see Tables 13 and 14) to the other genera of sengis (see order Macroscelidea).

The three species in the genus have very similar and somewhat bizarre morphology that gives the impression of a cross between a miniature antelope, anteater and rodent. The genus is characterized by large size (the largest of all sengis), narrow body (ungulate-like), long and spindly legs, long rodent-like tail and sparse, slightly coarse, brightly coloured and patterned pelage. The feet are digitigrade, with four digits on forefoot (a unique character of this genus), four digits on hindfoot and long claws (especially on three well-developed front digits). The sole of the metatarsal has hair, and there is no carpal pad. The eyes are large and dark, with a round pupil. The pinnae are hairless, moderate in size and carried upright. The nose is extremely elongated and flexible (trunk-like rather than rigid), with the nostrils at the tip, and the small mouth is set well behind the nostrils; the tip of the nasal bones is ossified (cf. other genera of sengis). Pectoral gland absent, post-anal gland well developed, large and round. Females have two pairs of nipples, both abdominal/inguinal. The dental formula is $I\,^{0-1}/_3$, $C\,^1/_1$, $P\,^4/_4$, $M\,^2/_2 = 34–38$. The presence/absence of upper incisors varies according to species and individuals. If present, upper incisors are rudimentary and may not even be visible; the single upper incisor may be present on both sides or on only one side of the upper jaw (see Corbet & Hanks 1968) (cf. *Elephantulus*, *Macrosceledes* and *Petrodromus*, which have well-developed upper incisors). Other characteristics of the teeth include notched lower incisors, large upper canines, and progressively complex cheekteeth ending with hypsodont and dilambdondont molars (Figure 20). Palatal foramina are absent. The many differences between *Rhynchocyon* and the other genera of sengis are listed by Corbet & Hanks (1968).

These large sengis are terrestrial and exclusively diurnal. When not active, they shelter in large terrestrial nests made of leaves. They feed exclusively on arthropods and other small invertebrates, which they gather by excavating in the soil and litter with the claws of the forefeet. Many of their characters, such as their alert, high-strung nature and swift cursorial half-bounding gait, are reminiscent of a small ungulate (Evans 1942). Their morphological and behavioural characters are quite unlike those usually exhibited by other small insect-eating mammals (Kingdon 1974, Rathbun 1979).

The genus is the sole extant representative in the subfamily Rhynchocyoninae (Corbet & Hanks 1968), which dates from at least the early Miocene (Butler 1995). The three species are most easily distinguished by the colour and pattern of the pelage.

A distinctively coloured *Rhynchocyon* was camera-trapped in the NE Udzungwa Mts of Tanzania (Rovero & Rathbun 2006) and was described as a new species (*Rhynchocyon udzungwensis*) in 2008 (Rovero *et al.* 2008, see also p. 16), after this text was submitted to the publishers.

Galen B. Rathbun

Rhynchocyon chrysopygus Golden-rumped Giant Sengi (Golden-rumped Elephant-shrew)

Fr. Macroscélide à croupe dorée; Ger. Goldrücken Rüsselhündchen

Rhynchocyon chrysopygus Günther, 1881. Proc. Zool. Soc. Lond. 1881: 164.
River Mombaça, corrected by Moreau *et al.* (1946) to Mombasa, Kenya.

Rhynchocyon chrysopygus.

Rhynchocyon chrysopygus

Taxonomy Generally considered a full species (Corbet & Hanks 1968), although Kingdon (1974) once treated it as an 'incipient species' of *R. cirnei*. This view was based mainly on indistinct dark stripes at the leading edge of the rump patch of some adults and most neonates, which are reminiscent of the prominent dorsal stripes of *R. cirnei*. Synonyms: none. Subspecies: none. Chromosome number: not known.

Description Large sengi with prominent yellow rump. Dorsal pelage maroon or russet-brown to dark rufous with bright yellow rump, which has slightly longer hairs than rest of body. Ventral pelage rufous. Head grizzled yellowish-brown; neck tending to maroon. Snout very long, eyes large. Ears broad, upright, naked and black. Fore- and hindlimbs black and long, four digits on forefoot, four digits on hindfoot, all with well-developed claws. Tail long (ca. 85% of HB), thick at base tapering to tip, with short, sparse hairs, black above, paler below; tip of tail with irregular white patch. Some individuals have indistinct dark central stripes at the anterior edge of the rump patch, reminiscent of the pattern in *R. cirnei*. Pectoral gland absent; post-anal gland well developed. No sexual dimorphism in body size, but ♂♂ have longer canines and thicker rump skin than ♀♀. Nipples: 0 + 2 = 4.

Geographic Variation None recorded.

Similar Species This is the only sengi with a yellow patch on the rump.

Distribution Endemic to Africa. Coastal Forest Mosaic BZ. Occurs only in small and fragmented forests in the coastal region of Kenya from south of the Tana R. southwards, through Arabuko–Sokoke Forest, to Rabai near Mombasa (FitzGibbon 1994).

Habitat Coastal semi-deciduous forest, woodlands with a more-or-less closed canopy, coral rag scrub, and abandoned and overgrown agricultural lands with a closed canopy. Thick leaf litter is characteristic of all habitats.

Abundance Mostly a rare species with a restricted distribution, although may be locally common in a few favoured habitats. In Arabuko–Sokoke Forest maximum densities are about 75 individuals/km^2 in comparatively undisturbed forest habitats dominated by *Afzelia* trees. Densities are lower in less desirable habitats, including forest edges and near human settlements. Total population in Arabuko–Sokoke Forest is estimated at about 22,000 (FitzGibbon 1994).

Adaptations Terrestrial and diurnal. Golden-rumped Sengis build nests of leaves where they rest, singly, during the night. When building a nest, a cup-shaped hole about 8 cm deep is dug in the soil and dead leaves are arranged to form a lining; additional leaf material is dragged and piled on top of the nest forming a pile about 15 cm high and 50 cm in diameter. There is no proper entrance; the sengi just burrows in at the side and the leaves settle down again. In time, the leaves weather and the nest is inconspicuous. Usually each member of a pair (see below) builds and uses several nests (Rathbun 1979).

These sengis leave their nests just before dawn and return just before sunset; they are active most of the day except for a mid-day rest period (FitzGibbon 1995). During the day, they spend, on average, nearly 80% of their time foraging, 12% walking, 4% resting and a little time building nests and interacting with conspecifics (Rathbun 1979). Walking is the main form of locomotion; when disturbed, an individual either freezes, or runs for about 10 m before stopping. When in danger, they either (a) walk away slowly while slapping the tail on the leaf litter, (b) run away while stotting in a similar way to an antelope, or (c) flee in the typical half-bounding gait. When necessary, they can run extremely swiftly (up to ca. 27 kph) (Rathbun 1979). Unlike *Petrodromus* and some *Elephantulus* spp., this sengi does not build trails.

The distinctively coloured rump-patch, with its associated thick dermal shield, may serve to attract bites from pursuing conspecifics during agonistic encounters; the thickness of the dermal shield probably gives protection against serious bites from a conspecific. Similarly, the distinct rump patch may serve as a predator invitation signal by enticing would-be predators to prematurely expose themselves to sengis foraging on the forest floor (Rathbun 1978). The senses of smell, sight and sound are acute. There is no information on metabolism, temperature regulation and water conservation in this species.

Foraging and Food Insectivorous. Prey is located by probing in dense leaf litter with the long flexible snout, and small prey and food fragments are flicked up with the long extensible tongue (Rathbun 1979). Earthworms are excavated from the soil with the strong claws of the forefeet, leaving distinctive 3 cm-deep conical holes. Large prey, such as earthworms and centipedes, are pinned to the ground with a forefoot, and then awkwardly ingested. The principal prey (in decreasing order of importance) are beetles, centipedes, termites, crickets, ants, spiders and earthworms (Rathbun 1979). The density of this sengi is positively correlated to the abundance of spiders (FitzGibbon 1995), although spiders comprise only a small proportion of the diet.

Social and Reproductive Behaviour Facultatively monogamous and territorial. Territories are 1.5–5.0 ha, territories of ♂♂ being slightly larger than those of ♀♀. Both sexes mark their territories with secretions from a post-anal gland (Rathbun 1979, FitzGibbon 1995, 1997). The territories of a monogamous pair overlap to large extent, but there is little overlap with the territories of neighbouring pairs. Males intrude on neighbouring territories more frequently than do ♀♀. Territorial defence is sex-specific, male–male aggression being most common. The ♀ of a pair is closely followed by her mate during oestrus for 1–2 days, otherwise the pair rarely interact. Male mate guarding is probably responsible for their monogamous social structure (Rathbun & Rathbun 2006). Individuals spend each night, and rest intervals during the day, in separate leaf-litter nests on the forest floor. Different nests are used every few nights (FitzGibbon & Rathbun 1994). Vocalizations include very soft chattering between conspecifics and a loud distress scream when captured. The tail is noisily slapped on the dry leaf litter every 1–2 seconds in conflict situations.

Reproduction and Population Structure Golden-rumped Sengis are reproductively active throughout the year, with an average interval between births of 82 days. Gestation: ca. 42 days. Litter-size: 1. Weight at birth: ca. 80 g. The precocial neonate remains in the nest until about Day 14, when it is weaned. Young remain on the parental territory indefinitely, eventually finding a vacant territory or becoming prey. Longevity is about 3–4 years (Rathbun 1979).

The structure of the penis is unique among sengis, being composed mostly of connective tissue rather than vascular bodies (Woodall 1995b). The spermatozoa are also distinct from other sengis in that they have the shortest spermatozoon and the fewest gyres (Woodall 1995b, Woodall & FitzGibbon 1995). Whether the placenta is chorioallantoic or endotheliochorial is not clear (Cutler *et al.* 1998). Golden-rumped Sengis have rarely been maintained, and have never bred, in captivity.

Predators, Parasites and Diseases A wide variety of large snakes, raptors and mammalian carnivores are known or suspected predators (Rathbun 1979). Ectoparasites include two species of fleas and one species of tick (Fourie *et al.* 1995).

Conservation IUCN Category: Endangered.

The decline in numbers is associated with reduction in area, and in the extent and quality of habitats. FitzGibbon *et al.* (1995) estimate that about eight individuals/km^2 are harvested for meat by local people in the Arabuko–Sokoke Forest, a harvesting rate that they believe is sustainable. However, forest destruction for agricultural and urban development, and logging for building materials, wood carving and charcoal production, is a serious threat to their habitat (Rathbun & Kyalo 2000).

Measurements

Rhynchocyon chrysopygus
HB: 277.9 (218–304) mm, n = 80
T: 240.5 (213–270) mm, n = 80
HF 74.0 (68–79) mm, n = 80
E: 33.5 (30–38) mm, n = 80
WT: 534.8 (410–690) g, n = 40
GLS: 67.6 mm, n = 1
GWS: 36.2 mm, n = 1
C–M^3: 27.8 mm, n = 1
Canine length (♂♂): 5.0 ± S.E. 0.2 mm, n = 53
Canine length (♀♀): 3.1 ± S.E. 0.1 mm, n = 44
Arabuko–Sokoke Forest, Kenya.
Body measurements and weight: G.B. Rathbun unpubl.
Canine lengths: FitzGibbon 1995
Skull measurements: BMNH

Key References Corbet & Hanks 1968; FitzGibbon 1994, 1995, 1997; Rathbun 1978, 1979; Rathbun & Kyalo 2000; Woodall & FitzGibbon 1995.

Galen B. Rathbun

Rhynchocyon cirnei Chequered Giant Sengi (Chequered Elephant-shrew)

Fr. Macroscélide de Cirne; Ger. Geflecktes-Rüsselhündchen

Rhynchocyon cirnei Peters, 1847. Bericht Verhandl. K. Preuss. Akad. Wiss., Berlin 12: 37. Quelimane, Bororo District, Mozambique.

Taxonomy Corbet & Hanks (1968) suggest that six allopatric forms can be recognized and that further collecting may demonstrate clinal variation linking some of these forms; they also suggest that allopatric *R. c. stuhlmanni* from DR Congo and Uganda may be a separate species. Synonyms: *claudei*, *hendersoni*, *macrurus*, *melanurus*, *nudicaudata*, *reichardi*, *shirensis*, *stuhlmanni*, *swynnertoni*. Subspecies: six. Chromosome number: not known.

Description Large sengi with distinctive pattern of lines and spots on back and flanks. Dorsal pelage very variable (see below), buff to dark brown; three longitudinal stripes from mid-back to base of tail on either side of mid-dorsal line; two central lines are nearly continuous, while second and third lines on each side often broken into pale and dark spots. In some races, the darker background obscures the stripes so only a faint pattern is visible. Ventral pelage whitish. Head with long snout, grizzled yellow or cream. Ears upright, naked. Fore- and hindlimbs elongated, buff; four digits on forefoot, four digits on hindfoot, all with well-developed claws. Tail long (ca. 90% of HB), nearly hairless, white at tip. Pectoral gland absent; post-anal gland well developed. No sexual dimorphism except for longer and wider canines in ♂♂. Nipples: 0 + 2 = 4.

Geographic Variation Six subspecies (or races) distinguished by sometimes subtle variation in pelage colouration and pattern (Corbet 1974b, Kingdon 1974 especially plate p. 42).

R. c. cirnei: Mozambique. Dorsal spots pale chestnut.
R. c. hendersoni: Livingstonia, N Malawi. As *R. c. reichardi*, but entire pelage very dark.
R. c. macrurus: SE Tanzania and perhaps N Mozambique. Dorsal pelage rufous, chequered pattern prominent on inland specimens but obscured by dark pelage from those near coast.
R. c. reichardi: N Malawi, NE Zambia, SE DR Congo, SW Tanzania. Dorsal pattern of dark and pale spots.
R. c. shirensis: S Malawi. Dorsal spots blackish-chestnut, white spots absent.
R. c. stuhlmanni: NE DR Congo, Uganda (and perhaps near Bangui, Central African Republic). Similar to *reichardi*, but dorsal pelage very dark so that the spots and stripes may be obscured; tail nearly white.

Similar Species

R. chrysopygus. Large yellow patch on rump; some individuals have indistinct stripes on back.
R. petersi. Mid-back to rump jet black and face grizzled rufous-orange; pale orange tail and orange-brown ears; some individuals with indistinct stripes at anterior margin of the black colouration.

No other species of sengi has distinctive stripes on the back and flanks.

Rhynchocyon cirnei

Distribution Endemic to Africa. Rainforest BZ (East Central Region) and restricted parts of Zambezian Woodland and Somalia–Masai Bushland BZs. Recorded from Mozambique north of the Zambezi R.; highlands associated with the Rift Valley in Malawi, Zambia and Tanzania; W Uganda; S Tanzania; DR Congo between the Congo and Ubangi rivers (Corbet & Hanks 1968).

Habitat Montane and lowland forests, closed-canopy woodlands and riparian thickets where substrate is usually covered with dense leaf litter.

Abundance Widespread, but no detailed information.

Adaptations Terrestrial and diurnal. Builds leaf-litter nests on the forest floor (Allen & Loveridge 1933). Highly cursorial (Lawrence & Loveridge 1953, Ansell & Ansell 1973). Based on numerous short accounts of fleeting observations, the natural history of this species is probably similar to the better-known Golden-rumped Sengi. However, for such a widespread species, very little is known about its biology.

Foraging and Food Stomachs of specimens from montane region of NE Zambia contained beetles, bees or wasps, fly larvae, and bugs (Ansell & Ansell 1973).

Social and Reproductive Behaviour In NE Zambia, all sightings were of solitary individuals, except for one pair (Ansell & Ansell 1973).

Reproduction and Population Structure Litter-size usually one, with twin foetuses reported from Zambia, Malawi and Uganda (Lawrence & Loveridge 1953, Brown 1964, Ansell & Ansell 1973).

Predators, Parasites and Diseases The chequered pattern of the pelage is probably related to camouflage from predators (Kingdon 1974). Forest Cobras *Naja melanoleuca*, Robust Chimpanzees *Pan troglodytes* and African Golden Cats *Profelis aurata* are known predators. Mbuti Pygmies and local people near Kisangani, DR Congo, harvest Chequered Sengis for meat. Ectoparasites include one species of flea, a species of dipteran fly and several species of *Ixodes* ticks (Fourie *et al.* 1995). The blood protozoan *Sarcocystis* sp. has also been found (Keymer 1971).

Conservation IUCN Category: Near Threatened.

Of special concern are two subspecies: *R. c. hendersoni* is restricted to an isolated montane forest in Malawi, and may be threatened with extinction from habitat destruction (Nicoll & Rathbun 1990); and *R. c. cirnei* (known only from the holotype) in coastal Mozambique. Habitat loss may also threaten the other subspecies with restricted distributions. Chequered Sengis are not known to have been successfully bred in captivity (Nicoll & Rathbun 1990), although it is rumoured that they have been successfully maintained in private collections.

Measurements
Rhynchocyon cirnei
HB: 272.7 (242–303) mm, n = 67
T: 243.4 (220–265) mm, n = 66
HF: 85.7 (81–91) mm, n = 67
E: 31.0 (29–34) mm, n = 67
WT: 352.0 (320–420) g, n = 10
GLS (♂♂): 67.6 (62.2–70.8) mm, n = 39
GWS (♂♂): 36.1 (33.5–38.0) mm, n = 39
C–M^3: 28.9 (27.4–29.9) mm, n = 12
Canine length (♂♂): 3.4 (3.3–3.5) mm, n = 5
Canine length (♀♀): 2.2 (1.7–2.3) mm, n = 5
Body and skull measurements: *R. c. stuhlmanni* from Niapu, DR Congo (Allen, J. 1922)
Weight and canine lengths: *R. c. reichardi* from NE Zambia (Ansell & Ansell 1973)
C–M^3: BMNH
Note: *R. c. reichardi* (mean values: HB = 242.0 mm, T = 213.8 mm, HF = 66.8 mm, E = 29.5 mm, n = 10; Ansell & Ansell 1973) is on average smaller than *R. c. stuhlmanni*.

Key Reference Corbet & Hanks 1968.

Galen B. Rathbun

Rhynchocyon petersi Black-and-rufous Giant Sengi (Black-and-rufous Elephant-shrew)

Fr. Macroscélide de Peters; Ger. Schwarzbraunes Rüsselhündchen

Rhynchocyon petersi Bocage, 1880. Jorn. Sci. Math. Phys. Nat., Lisboa 1 (7): 159. Mainland Tanganyika (Tanzania), opposite Zanzibar I.

Rhynchocyon petersi.

Taxonomy Corbet & Hanks (1968) suggested that there may be a cline between this species and *R. cirnei*. In contrast, Kingdon (1974) suggested a hybrid zone between the two taxa, because some individuals of *R. petersi* have indistinct dorsal stripes reminiscent of the pattern in *R. cirnei*; on this basis *R. petersi* may be considered as an 'incipient species' of *R. cirnei*. Synonyms: *adersi*, *fischeri*, *usambarae*. Subspecies: two. Chromosome number: not known.

Description Large black and orange-rufous coloured sengi. Pelage bright and shiny. Dorsal pelage (from shoulder blades to rump and thighs) black; upper back and flanks orange to rufous-orange or dull maroon (see Geographic Variation). Ventral pelage orange to rufous-orange. Head with well-developed snout, orange-brown; forehead grizzled, tinged with rufous. Ears upright, orange-brown. Fore- and hindlimbs long, orange-brown; four digits on

Rhynchocyon petersi

forefoot, four digits on hindfoot, all with well-developed claws. Tail pale orange, nearly hairless, with faint irregular white area near tip. Pectoral gland absent; post-anal gland well developed. Nipples: 0 + 2 = 4.

Geographic Variation

R. p. petersi: Tanzania (mainland) and Kenya. Pelage of shoulders, flanks and ventrum orange-rufous; head yellowish.

R. p. adersi: Zanzibar and Mafia Is. Pelage of shoulders, flanks and ventrum dull maroon; head rufous.

Similar Species No other species of sengi has the distinctive black and orange-rufous dorsal pelage and grizzled rufous face, nor the orange-coloured skin on tail and ears.

Distribution Endemic to Africa. Coastal Forest Mosaic BZ and marginally in adjacent BZs. Recorded from Tanzania and Kenya (in the 'Eastern Arc' mountains and coastal forests from the Rabai Hills near Mombasa, Kenya, to just south of Dar es Salaam, Tanzania). Also Zanzibar and Mafia Is.

Habitat Evergreen and semi-deciduous forests, dense woodlands, coral rag scrub and abandoned and overgrown agricultural lands with closed canopies where there is a thick covering of leaf litter.

Abundance Mostly a rare species with a fragmented and restricted distribution. Maximum densities, estimated from transect surveys of nests, between 19/km^2 and 79.3/km^2 (Hanna & Anderson 1994, Coster & Ribble 2005).

Adaptations Little is known about this species. Terrestrial and diurnal. Very keen senses, highly cursorial (Allen & Loveridge 1927).

Foraging and Food Insectivorous, but no details except that soldier ants (*Dorylus* spp.) are eaten in Nkuka Forest, Tanzania (Allen & Loveridge 1933). Captive animals successfully maintained on a diet of dry cat food supplemented with crickets and mealworms (Baker *et al.* 2005).

Social and Reproductive Behaviour Generally observed as solitary individuals, and presumed to be facultatively monogamous (as for Golden-rumped Sengi). In captivity, kept as monogamous pairs. Multi-sex groups show greater tolerance between unrelated adult ♀♀ than between ♂♂ (Baker *et al.* 2005).

Reproduction and Population Structure Based on captive animals (Baker *et al.* 2005), gestation ca. 40 days, litter-size 1–3. Young not as precocial as in species of Macroscelidinae. Neonates remain in a leaf nest for about 3 weeks and are visited only by the mother once each day. The ♂ does participate in rearing the young. Interval between births about 80 days.

Predators, Parasites and Diseases Ectoparasites include three species of fleas and three species of ixodid ticks (refs in Fourie *et al.* 1995).

Conservation IUCN Category: Vulnerable.

The geographic range is small, and there is continuing decline in area and quality of habitats. Fragmentation of habitats is due to urban and agricultural expansion into forests, many of which are already small and isolated (Nicoll & Rathbun 1990). Extraction of timber for woodcarving, firewood and charcoal production are also threats to habitats (Hanna & Anderson 1994). In 2000, Black-and-rufous Sengis from Tanzania were imported to North American zoos, where they have bred successfully (Baker *et al.* 2005).

Measurements

Rhynchocyon petersi
HB (♂): 324 mm, n = 1
HB (♀♀): 275, 270 mm, n = 2
T (♂): 230 mm, n = 1
T (♀♀): 240, 213 mm, n = 2
HF (♂): 71 mm, n = 1
HF (♀♀): 67, 83 mm, n = 2
E (♂): 31 mm, n = 1
WT: n. d.
GLS: n. d.
GWS: n. d.
C–M^3: n. d.
Measurements (♂♂): Allen & Loveridge 1927
Measurements (♀♀): Hollister 1918, Loveridge 1922

Key References Corbet & Hanks 1968; Hanna & Anderson 1994; Nicoll & Rathbun 1990.

Galen B. Rathbun

Order TUBULIDENTATA – Aardvark

Tubulidentata Huxley, 1872. A Manual of the Anatomy of Vertebrated Animals. New York: D. Appleton, pp. 288–289.

Orycteropodidae (1 genus, 1 species)	Aardvark	p. 289

The order Tubulidentata consists of a single family, the Orycteropodidae, and one extant species, *Orycteropus afer*. Tubulidentata is the only order of mammal to be represented by a single extant species. The name Tubulidentata derives from the unique microstructure of the orycteropodid teeth: each tooth consists of hundreds of tubes of dentine agglomerated and surrounded by cement. The name Orycteropodidae and *Orycteropus* derives from Greek '*orukter*' and '*pous*' meaning 'digging foot', referring to the Aardvark's adaptations for digging.

All Tubulidentata, fossil and extant, share the same following osteological characters. The skull is long, roughly tubular, particularly elongated on the snout and widest at the level of the jugals. Maxilla, premaxilla, frontal and nasal bones are well developed. The nasal bones are triangular in shape, broaden caudally but are never fused. The lacrimal bone presents a rostral development and the lacrimal foramen is situated on the border of the orbit. The frontal bones bulge dorsally in front of the orbit as a result of the highly developed nasal chamber where the olfactory bulbs are contained. The position of the anterior border of the orbit in relation to the upper toothrow differs among living and extinct species. The parietals are fused in one unit at maturity. In Tubulidentata, a part of the parietal bone joins the alisphenoid so that frontal and squamosal are not in contact. There is no sagittal crest and only faint temporal ones. The lambdoid crest, between the parietal and occipital bones, can be straight or V-shaped (in dorsal view) according to species. In juvenile Aardvarks an interparietal bone is present between the parietal and the occipital bones, but this disappears (fuses) by adulthood. The zygomatic arch is complete but slender and the orbit is only separated from the temporal fossa by a postorbital process (i.e. there is no postorbital bar). The palate is long and narrow. The palatine bone is elongated, presents two post-palatine foramina and ends caudally in a strong post-palatine torus. The tympanic cavity is simple: the tympanic bone is annular, and there is no auditory bulla. The mastoid bone is visible laterally and caudally. Moreover, post-temporal and mastoid foramens are present on its caudal aspect. The body of the mandible is very slender and broadens at the level of the molars. The mandibular symphysis is rarely complete. On the ascending ramus, the coronoid process is long and projects over the condylar process, which can be flat or concave according to species. Likewise, the mandibular fossa on the cranium is flat or has a tubercle.

The humerus shows a very broad distal epiphysis, possesses a deltoid tuberosity and a well-developed deltoid crest (except in two extinct species). The vertebral formula for the extant Aardvark is Cervical 7, Thoracic 13, Lumbar 8, Sacral 6, Caudal 25–28. In some fossil forms (e.g. *Amphiorycteropus abundulafus* from Chad), and at the juvenile stage of the extant form, the number of sacral vertebrae is only five. The pelvis is large and characterized by a dorso-caudal extension of the iliac bone. The sacrum does not enter into contact with the ischium. The pubic symphysis is unreduced in comparison with other digging mammals (MacPhee 1994). The proximal epiphyses of the tibia and fibula are fused, and the diaphysis of the tibia is bent. Scaphoid and centrum are fused (Clark & Sonntag 1926) and the pollex (Digit 1) has disappeared, so that the hand has only four digits. The tarsus is serial (the talus does not articulate with the cuboid); the talus retains an astragalar foramen and its distal articulation surface is ball-like, supported by a distinct neck. The hindfoot, which can be plantigrade when the animal is digging, possesses five toes. A detailed anatomical description was originally given in a monograph of *Orycteropus afer*, divided in a series of three papers (Sonntag 1925, Sonntag & Woolard 1925, Clark & Sonntag 1926).

The oldest known unquestionable members of Tubulidentata are *Myorycteropus minutus* (Pickford 1975) and *M. africanus* (MacInnes 1956) from the early Miocene (more precisely, between 20 and 18 mya) of Kenya. However, they already show the peculiar tubulidentate tooth structure, and are probably dedicated diggers. Therefore, the Tubulidentata must have diverged from the eutherian mammal lineage earlier, during the Palaeocene or, more likely, the Cretaceous. The phylogenetic position of the Tubulidentata has been heavily discussed since the first description of the extant Aardvark in 1766, but even more during the last 20 years. The presupposed (but incorrect) absence of teeth, and most of all the myrmecophagous diet, led many authors to include the Tubulidentata in the (now obsolete) order 'Edentata', along with pangolins, armadillos, sloths and South American anteaters. The extensive work by Sonntag (Sonntag 1925, Sonntag & Woolard 1925, Clark & Sonntag 1926) on the anatomy of the extant Aardvark demonstrated that this placental mammal must be placed in an order on its own. Resemblances between Aardvarks and other ant-eating mammals (e.g. Xenarthra, pangolins) are in fact likely due to convergent evolution. Since then, mammalian phylogenies, based on morphological characters and fossil record, supposed that Tubulidentata belonged to the higher-level taxon Ungulata (Thewissen 1985, Novacek 1989, Shoshani & McKenna 1998). However, over the past two decades, molecular analyses have led to the description of a supraordinal clade of mammals, the Afrotheria, which has regrouped Proboscidea, Hyracoidea, Sirenia, Macroscelidea, Afrosoricida and the Tubulidentata (see, for example, Springer *et al.* 1997, Stanhope *et al.* 1998, Asher *et al.* 2003). Afrotheria is considered to be a group of placental mammals that originated in Africa in the late Cretaceous. Among the Afrotheria, the position of Tubulidentata needs further clarification (see Robinson & Seiffert 2004).

According to the fossil record, the earliest Eurasian fossil aardvarks have been found in the early middle Miocene of Turkey (Van der Made 2003). Therefore, they must have dispersed from Africa into Eurasia before or around that period like numerous other mammals (see Rögl 1999). The widest distribution area of the order spans Africa, Europe (France, Italy, Moldavia, Greece, Turkey) and Asia (Iran, Pakistan). However, Tubulidentata became extinct from Eurasia before the end of the Pliocene, thus restricting their distribution to the African continent. Fossil records (Romer 1938), as well as archaeological clues (Keimer 1944, Frechkop 1946, Manlius 2002), suggest that Aardvarks were, until recently, present north of the Sahara. The extant species lives now only in sub-Saharan Africa.

Thomas Lehmann

Family ORYCTEROPODIDAE

AARDVARK

Orycteropodidae Gray, 1821. London Medical Repository 15 (1): 305.

Orycteropus (1 species)	Aardvark	p. 289

The only family in the order Tubulidentata, comprising a single extant species, the Aardvark *Orycteropus afer*. Patterson (1975) subdivided the Orycteropodidae into two subfamilies. The Plesiorycteropodinae are represented by a single genus, *Plesiorycteropus*, including sub-fossil taxa that lived around 1000 years ago on Madagascar, and represented by two species: *P. madagascariensis* and *P. germainepetterae*. However, MacPhee (1994) argued that this genus does not belong to the Tubulidentata, but rather represents an order on its own: Bibymalagasia. More recent analyses (Lehmann 2004) tend to confirm that there is only one subfamily of Tubulidentata, namely the Orycteropodinae. The subfamily comprises at least four genera (Lehmann 2009). Two of these, *Myorycteropus*, from the early Miocene of Rusinga and Mfangano in Kenya, and *Leptorycteropus*, from the late Miocene of Lothagam also in Kenya, are monospecific and extinct. Pickford (1975, 2004) proposed that the genera *Myorycteropus* and *Leptorycteropus* were synonyms of a third genus, the extant *Orycteropus*, but their validity is borne out by other authors (Van der Made 2003, Lehmann *et al.* 2005, Lehmann 2009). A recent revision of the family (Lehmann 2009) split *Orycteropus* in two based on a cladistic analysis of morphological characters, and in so doing created the genus *Amphiorycteropus*. Consequently, whereas the genus *Orycteropus* was until recently known by 12 extinct species and only one living species, the latest classification recognizes two Plio-Pleistocene and one extant African species in *Orycteropus*, and includes five species known from middle Miocene to Pliocene in Africa and Eurasia in *Amphiorycteropus* (Lehmann 2009).

The following features characterize the extant Aardvark and the members of the genus *Orycteropus*: large species with an elongated snout, absence of incisors or canines, rectangular and 8-shaped outline of the molars, large deltoid crest, broad distal epiphysis of the humerus, button-like bicipital tuberosity of the radius, articular axis of the sigmoid notch of the ulna perpendicular to the diaphysis, six sacral vertebrae, presence of a falciform process on the tibia, and talus as long as broad with a concave cotylar fossa.

However, in the smaller extinct *Leptorycteropus*, the canines are preserved, the deltoid crest is not strongly developed, the distal epiphysis of the humerus is reduced, the pubis is oriented medio-ventrally, five vertebrae form the sacrum, there is no falciform process on the tibia and the tibial crest is short. In contrast to *Orycteropus*, *Myorycteropus*, which is 50% smaller than the extant Aardvark, shows upper teeth that stand vertical in their alveoli, a low mandibular angle, an articular axis of the sigmoid notch of the ulna oblique to the diaphysis, a *caput femoris* not oriented medio-laterally, no falciform process on the tibia, a short tibial crest with a real cnemial tuberosity and a vertical cotyloid facet on the talus. Finally, *Amphiorycteropus* can be distinguished from the other genera by a V-shaped nuchal line, the anterior border of the orbit situated at the level of the M^2, molars trapezoidal in shape, a pointed bicipital tuberosity on the radius, a proximo-distally elongated talus, and a mandibular angle superior to 73°.

Thomas Lehmann

GENUS *Orycteropus*

Aardvark

Orycteropus E. Geoffroy St. Hilaire, 1796. Extrait d'un mémoire sur le *Myrmecophaga capensis*, Gmellin. Bulletin de la Société Philomathique de Paris, no. 50, pp. 1–2 [see Lehmann 2007 for discussion about the various incorrect former attributions of authority and date to this genus].

Orycteropus is a monotypic genus, represented by a single extant species, the Aardvark *Orycteropus afer*, distributed throughout sub-Saharan Africa in a wide range of habitats including grassland, all savanna types, semi-arid Karoo (South Africa) and some forests but not desert.

Patterson (1975) included five extinct species of *Orycteropus* in the genus with confidence, noting that others required more complete specimens for clarification. Since then, a new species has been described, some additional material has been discovered, and the genus has been split in two, with the description of *Amphiorycteropus* (see Lehmann *et al.* 2004, Lehmann 2009). Thus, including fossil forms, the genus currently comprises three described species, all known only from Africa so far: *O. afer* (extant, Africa), *O. crassidens* (Kenya) and *O. djourabensis* (Chad). Following Patterson (1975) and Lehmann (2009), three forms previously included in *Orycteropus* – cf. *Myorycteropus minutus*, cf. *Amphiorycteropus pottieri* and cf. *A. seni* – are considered with reservation pending the availability of further material. The species *O. pilgrimi* has been shown to be a synonym of *A. browni*.

The oldest established record for the extant Aardvark is from the Palaeolithic of Algeria (Romer 1938) or perhaps the early Pliocene of South Africa (Langebaanweg; Pickford 2005). Historically, the genus appeared in East Africa some 7 mya. It soon spread over the whole continent, reaching Chad by 4 mya, and replaced all former taxa (Lehmann 2008, 2009).

Thomas Lehmann & Andrew Taylor

Aardvark *Orycteropus afer.*

Orycteropus afer AARDVARK (ANTBEAR)

Fr. Oryctérope; Ger. Erdferkel

Orycteropus afer Pallas, 1766. Miscellanea Zoologica 64. Cape of Good Hope, Western Cape, South Africa.

Taxonomy There are 18 subspecies recognized, based primarily on colour, size and degree of frontal inflation, but this number is almost certainly too high, and the scarcity of material makes it difficult to determine the status or the limits of distribution of subspecies (Meester 1971). The name Aardvark derives from Dutch and means 'earth pig'. Synonyms: *adametzi*, *aethiopicus*, *albicaudus*, *angolensis*, *erikssoni*, *faradjius*, *haussanus*, *kordofanicus*, *lademanni*, *leptodon*, *matschiei*, *observandus*, *ruvanensis*, *senegalensis*, *somalicus*, *wardi*, *wertheri*. Chromosome number: 2n = 20 (Benirschke *et al.* 1970, Yang *et al.* 2003).

Description Medium- to large-sized species with heavy arched body and greyish-brown appearance. Head appears small relative to body, particularly in large specimens. Long tubular muzzle ends in a soft, swollen snout, with numerous bristle-like hairs inside mobile nostrils. Dark-brown eyes, small for the size of the animal, with vibrissae below lower eye-lashes. Ears long and tubular, commonly held upright. Thick neck, wider than head. Dorsal pelage grey and sparse with irregular bare patches. Flank pelage less sparse, becoming thicker and longer on legs and rump; pelage on legs black. Body colour may be influenced by colour of local soils. Legs stocky and powerful, with four toes on each forefoot and five on each hindfoot. Strong nails present on all toes; those on forefeet longer and more robust than those on hindfeet. All toes are united by webbing to varying degrees. Webbing on the forefeet is greatest between Digits 1 and 2; on the hindfeet webbing is greatest between Digits 2 and 3. Only digital pads are present; plantar and carpal pads are missing. Tail thick at base, tapering to a narrow tip. There is no sexual dimorphism. One abdominal pair and one inguinal pair of nipples. Scent glands resembling scrotal sacks occur in the groin area of both sexes. The orifices of these are long slits opening on either side of the vulva in ♀♀ and just behind the prepuce in ♂♂. The sacs are short and wide and filled with a yellow secretion smelling like the anal glands of mustelids. In ♂♂, the penis is short and shaped like a truncated cone. The genital orifice of the ♀ is a long cleft behind the centre of the genital eminence and is preceded by a large cordate posteriorly bilobed plate, the clitoris.

The dentition is heterodont and diphyodont, with a highly variable permanent dental formula of $I\ ^0/_0$, $C\ ^0/_0$, $P\ ^{2-4}/_{2-4}$ (usually $^3/_2$), $M\ ^3/_3$ = 20–28. In contrast with numerous mammals, the third molars of the Tubulidentata erupt at an early stage of the ontogenesis and are thus not a good criterion for age determination. The tubulidentate teeth have no enamel and grow continuously from open roots. A digitation of the pulp runs inside each tube and produces centripetal layers of dentine until closure of the tube upon the surface. The premolars are peg-like whereas the molars are

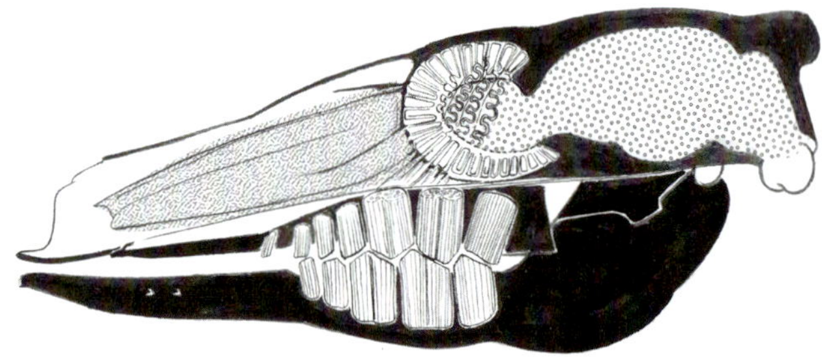

Mid-line section of *Orycteropus afer* skull.

bicolumnar (8-shaped in occlusal view) with longitudinal grooves on each side.

Geographic Variation Numerous subspecies have been described, but the validity of many of these is in doubt (see Taxonomy). For example, Smithers (1971) synonymized the form *albicaudus* from Botswana, a form stated to have a very short tail and to be pure white, in the nominate *afer*. The form *erikssoni* was described as a large forest form from the Congo, characterized by short hairs, short ears and large claws, with the skull large, the anterior base of the zygoma narrow and the mandible slender, and was said to differ unmistakably in many features from East African and southern African animals (Hatt 1934).

Similar Species The Aardvark is unlikely to be confused with any other species.

Distribution Endemic to Africa. Widespread, but localized, south of the Sahara from Senegal to Ethiopia and south to South Africa, being absent from the Sahara and Namib Deserts (Kingdon 1971, Shoshani *et al.* 1988, Skinner & Chimimba 2005). Not recorded from Lesotho (Lynch 1994), but they almost certainly do occur. Although sometimes shown as absent from the forests of the Congo Basin (e.g. Shoshani *et al.* 1998), Aardvarks are in fact widely distributed throughout the region (Hatt 1934, Schouteden 1948, Rahm 1966, Pagès 1970, Carpaneto & Germi 1989, J. Hart pers. comm.); for example, in Congo they occur in savannas and forests from at least the Bateke Plateaux northwards to the border with Cameroon and the Central African Republic (F. Maisels pers. comm.), while in Gabon they have been camera-trapped in the Ivindo N. P. (and a skull was found in Minkebe N. P.) (P. Henschel pers. comm.). Distribution in West African rainforests is poorly known (Grubb *et al.* 1998). The distribution of the Aardvark is largely determined by the distribution of suitable ant and termite species (see Foraging and Food).

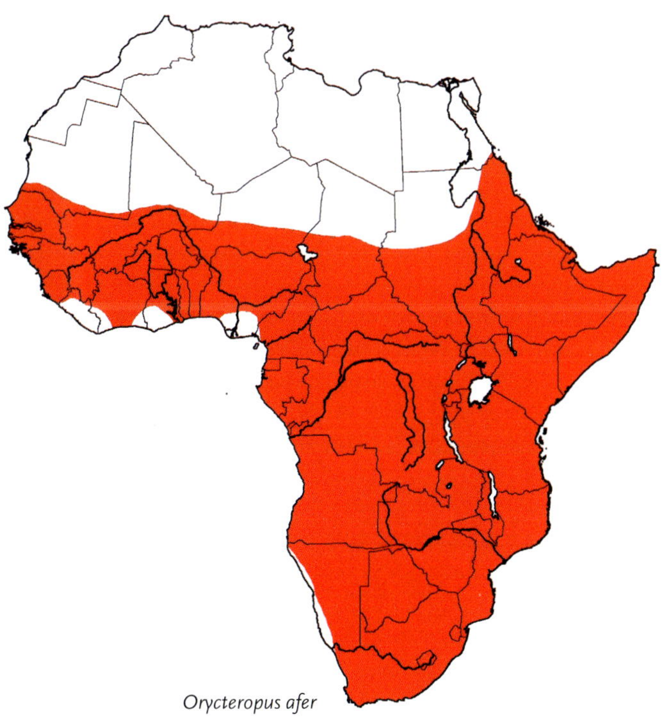

Orycteropus afer

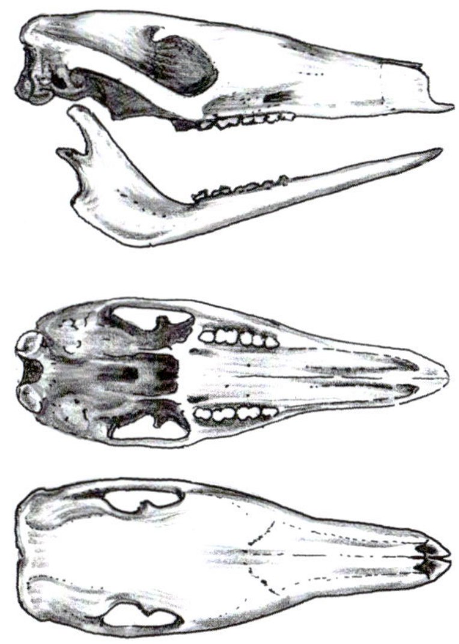

Lateral, palatal and dorsal views of skull of *Orycteropus afer.*

In Pre-Dynastic Egypt, the Aardvark occurred in the Nile Delta as has been proved by faithful representations on jugs of that period, and in paintings in tombs (Keimer 1944, Manlius 2002).

Habitat Present in a wide range of habitats, including grassland, all savanna types, woodlands and thickets, semi-arid Karoo (South Africa) and the transition zones (savanna-forest) of West Africa; also known from all the forested zones of central Africa except, seemingly, swamp forests (F. Maisels pers. comm.). Recorded in submontane forests of the Udzungwa Mts (Rovero & De Luca 2007) and to elevations of 3200 m in the Bale Mts of Ethiopia (Yalden *et al.* 1996). Their preferred habitat is flat or gently sloping ground that is not too rocky, which facilitates the digging of burrows and excavation of ant nests. Steep slopes are sometimes utilized, mainly in traversing areas. Presence of ants and termites is vital. The presence of free-standing water is not essential.

Abundance Locally common, but rarely seen due to nocturnal and evasive behaviour. Densities in the Nama Karoo of South Africa were estimated at 1–2/km^2 (Taylor & Skinner 2003). Densities vary according to abundance of prey. Burrows are often the only indication of presence; however, high burrow densities do not imply high Aardvark densities because many burrows are abandoned.

Adaptations Many adaptations of the Aardvark are associated with their myrmecophagous diet and digging specialization. Aardvarks are nocturnal and find both food and refuge underground. Prey is located by a highly developed sense of smell. The macrosmatic brain, and the fact that Aardvarks have very large olfactory lobes, with more olfactory bulbs than any other eutherian mammal, suggest that olfaction is probably the most developed sense (Shoshani *et al.* 1988). These are contained within the turbinate bones of the nasal cavities (Sonntag & Woollard 1925). The enlarged olfactory lobes result in a slight swelling of the skull just in front of the eyes. When foraging for food, bristle-like hairs inside the nostrils help prevent soil particles from being inhaled (Pocock 1924). The tongue is long (ca. 30 cm), vermiform and specialized for penetration of narrow tunnels of ant and termite nests. Large salivary glands provide

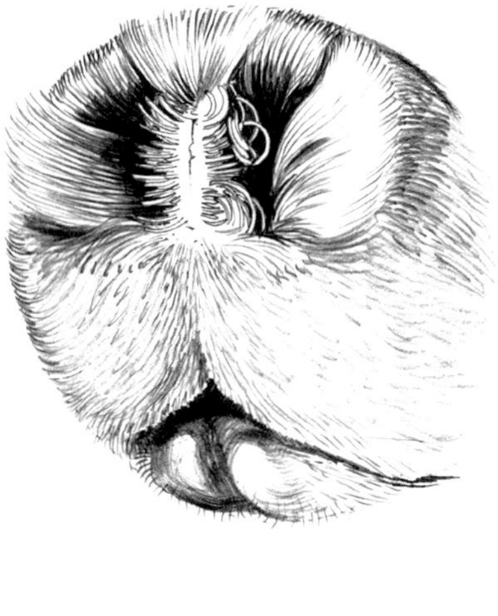

Frontal view of mouth and nostrils of *Orycteropus afer*.

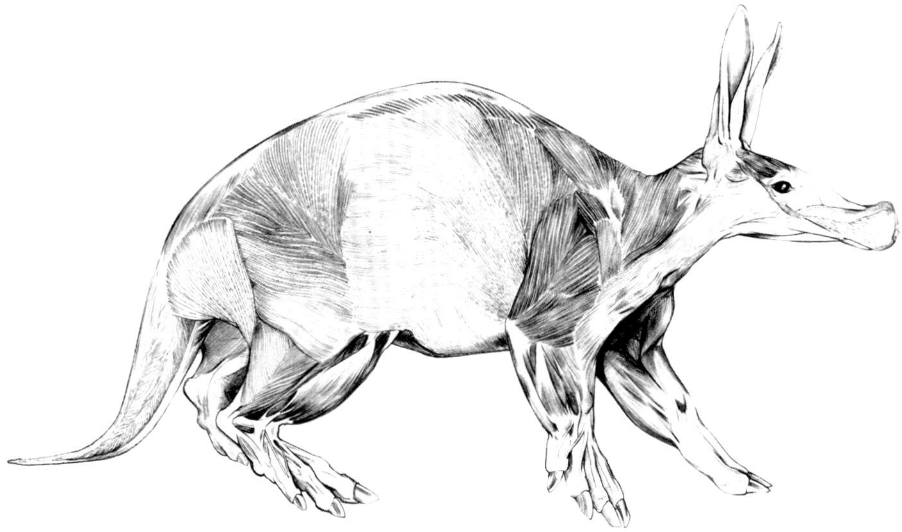

Myology of *Orycteropus afer*. Note short, thick muscles on limbs and large, hard nails.

copious amounts of saliva, facilitating the adherence of small prey but also soil. Cheekteeth are present even though most food is swallowed without chewing. The muscular stomach usually suffices for digestion. However, Aardvarks do occasionally chew specific types of prey, with the ant species *Messor capensis* being one example. They also have a large caecum, unusual for a myrmecophagous mammal. Hearing is acute, with the middle ear having a large tympanic membrane (Hunt & Korth 1980) and the detection of ambient noises facilitated by the long mobile ears; the ears are not generally used for prey location. Ears are kept erect mainly by the stiffness of the inflected margins of the lower half of the pinna, rather than being strengthened by additional cartilage. They can be folded flat during burrowing. Eyesight is poor, dark-adapted with no colour vision, and there is no tapetum (Franz 1908, Sonntag & Woollard 1925). The outer layer of the cornea is greatly keratinized to defend against the bites of ants. The skin shows a strong development of the dermis reticular layer as well as a thick epidermis (Sokolov *et al.* 1995) that helps protect against penetration of biting mandibles of the prey. Sensory hairs on the face below the eyes serve to alert the individual to obstacles and help prevent damage to the eyes during burrowing and foraging.

Aardvarks are digitigrade and are well known for their prodigious digging activities. The musculature of the forelimb is highly developed (see Thewissen & Badoux 1986) and the joints are strengthened. For instance, the distal epiphysis of the metacarpal bones presents a strong vertical crest that prevents uncontrolled latero-medial movement of the digit. The radius is shorter than the humerus. The brachial index (length of radius as a percentage of the length of humerus) is always less than 75% (Lehmann 2004), a condition that, according to MacPhee (1994), characterizes fossorial mammals. The powerful forelimbs and robust nails allow excavation of heavy soils and hard epigeal termitaries. The hindlimbs provide a firm base while digging, and are used to shovel soil backwards along passages and out of burrows. The sturdy tail provides additional support while digging. Nails could potentially inflict serious wounds on small carnivores, but are ineffective against the largest predators. Aardvarks are fast runners, attaining speeds of up to 40 km/h and escape predation by entering burrows. In the Karoo, burrows generally occur in clusters; in one extreme case 58 burrows occurred in an area 40 m × 200 m. These burrows were not interconnected and most were abandoned (A. Taylor pers. obs.). Densities of up to 15 burrows/ha have been recorded in the Rwenzori N. P. in Uganda (Melton 1976). Aardvarks regularly change burrows. New burrows may be used for only one night, or for more than a month, but the average length of tenancy in South Africa is between five and nine days (Taylor & Skinner 2003). When changing burrows, old ones are often renovated rather than new ones excavated (though this may depend on the substrate).

Burrows normally have one entrance, although occasionally they have more. Out of 18 burrows excavated in Uganda, 13 had one entrance, two had two entrances, two had three entrances and one had five (Melton 1976). In the Karoo, Aardvarks under observation always exited burrows from the same entrance they entered, implying that each had only one exit/entrance hole (Taylor & Skinner 2003). Burrows descend steeply before levelling out, may turn in any direction and often extend over 10 m in length. With diameters of 350–450 mm, tunnels are generally just broad enough for Aardvarks to move through. They often fork and terminate with an enlarged chamber used for sleeping and giving birth. Depths of 3 m in soft soil are easily achieved. In some areas Aardvarks have to contend with freezing temperatures at night. Their sparse hair and lack of body fat is not adapted for this, but the use of deep burrows buffers them from the cold. Burrow temperatures vary less than ambient temperatures, being warmer in cold weather and cooler in hot weather. As an Aardvark pushes its way through a burrow, soil is displaced behind it, creating an additional barrier to temperature extremes as well as predators. Aardvark body temperatures vary between 34° C when they are inactive inside burrows, and 37° C when they are active on the surface (Taylor & Skinner 2004).

Aardvarks become active soon after dusk. They are normally active all night in summer, for periods of up to eight hours or more, and return to their burrows before dawn. In winter they become active earlier and in cold areas such as the Karoo sometimes emerge in the afternoon (Taylor & Skinner 2003). They then often return to their burrows before midnight at ambient temperatures of approximately 2° C. Activity patterns become shorter, lasting up to seven hours. Nearer the Equator, activity periods are more consistent.

Foraging and Food Aardvarks predominantly prey on ants and termites, digging them out of the ground or from epigeal termitaries. Geographical variation in ant and termite faunas leads to variation in the prey species eaten. Ant genera so far identified as food items include *Aenictus*, *Alaopone*, *Anoplolepis*, *Camponotus*, *Crematogaster*, *Dorylus*, *Messor*, *Monomorium*, *Pheidole*, *Solenopsis*, *Tetramorium* and *Typhlophone*; termite genera include *Allodontermes*, *Basidentitermes*, *Cubitermes*, *Hodotermes*, *Macrotermes*, *Microhodotermes*, *Odontotermes*, *Pseudacanthotermes* and *Trinervitermes*.

As well as ants and termites, Aardvarks are also known to eat the pupae of dung beetles (Scarabaeidae). Adult scarab beetles lay their eggs in dung and store them up to 40 cm below the surface, from where Aardvarks dig them up. Evidence for this comes from stomach contents, direct observation of diggings for the larvae, spoor and dung in eastern and southern Africa (Kingdon 1971, P. Lindsey pers. comm.).

The most detailed studies of feeding ecology have been carried out at Tussen die Riviere N. R. in the Free State, South Africa (Taylor *et al.* 2002). Here more ants are eaten than termites, with a dietary

Aardvark *Orycteropus afer* sketches.

ratio of four ants to one termite. These proportions have not been determined elsewhere. The ant *Anoplolepis custodiens*, which is an abundant species, comprises about 70% of the total number of prey eaten. The termite *Trinervitermes trinervoides*, with its many epigeal termitaries, makes up about 20% of the diet. There are 13 other prey species known from this site, but they constitute a small proportion (about 10%) of the diet.

Seasonal changes in diet have been recorded, but reports are conflicting. In Uganda, Aardvarks are reported to eat fewer termites in the dry season than in the wet because the termites become quiescent and are harder to obtain (Melton 1976). In the Karoo, the termite *T. trinervoides* becomes quiescent in winter (May–Aug), but they confine themselves in termitaries where they are highly concentrated and more easily extracted. Aardvarks regularly target termitaries at these times and consume large quantities of termites. They do not feed from termitaries between Oct and Mar (Taylor *et al.* 2002).

Aardvarks move slowly when foraging, keeping their nose close to the ground and can be heard sniffing continuously. When a nest is located, they push their snouts flat against the ground while continuing to sniff. They then either start digging frantically to reach the prey or move on foraging. All this time the ears are held erect, indicating that hearing is probably only used for predator detection. Mammals that use sound to locate prey, such as the Bat-eared Fox *Otocyon megalotis*, always cup their ears forwards and downwards in the direction of the prey.

Aardvarks feed in discreet bouts of short duration, moving from one ant or termite nest to the next. Most feeding bouts vary between 10 sec and 2 min, but feeds from termitaries may last over 30 min. On average, Aardvarks make about 25 separate feeds per hour and may feed from over 200 nests in one night. Foraging speeds vary between 0.5 and 1.0 km/h (Taylor 1998). There are almost always some ants or termites left active on the surface at the end of a feed because the Aardvark's tongue is not adapted to lapping them off the surface.

Aardvarks feed from subterranean nests that vary in depth from shallow scratches at the surface to depths of over 1 m. Digs of approximately 200 mm are the norm. Very deep excavations are normally restricted to deep-living termite species such as *Hodotermes mossambicus* and such feeds may last over 30 min. In the case of *Trinervitermes* termitaries, Aardvarks dig into the centre of the mound and below ground level where large numbers of workers and larvae are concentrated.

Mechanical and chemical defences of ants and termites often play an important role in the feeding behaviour of myrmecophagous mammals. When the number of soldiers gets too high, other myrmecophagous mammals are forced to stop feeding, but this is not generally so for Aardvarks. Some ants, such as *Dorylus helvolus*, do bite hard enough to cause discomfort, but the attempts of most species are ineffective. The chemical defences of termites such as *Trinervitermes* spp., which deter many potential predators, do not stop Aardvarks either. On the contrary, these species are eaten in large numbers and many soldiers are ingested.

While active, Aardvarks spend the majority of their time searching for food. The small size of their prey requires them to consume hundreds of thousands of ants and termites per night, and this necessitates them spending all their time foraging to satisfy their energy requirements. Nightly foraging distances are governed by ant densities. In the Karoo, it is not unusual for them to travel 4 km or more per night (Taylor 1998), although Melton (1976) recorded distances of up to 14 km per night in Uganda (though these latter estimates were based on spoor and may not be reliable).

There have been reports of Aardvarks eating the fruits of a geocarpic plant, the Aardvark Cucumber *Cucumis humifructus* (Meeuse 1958). Seeds from this species have been found in faeces and the plant has been found growing at the entrance of burrows where droppings are often deposited. Due to its unique habit of forming fruit underground, fruits of this plant need to be dug up for dispersal. It has been suggested that the Aardvark devours the fruit for its moisture content and the consumed seeds are then dispersed by the Aardvark with the added bonus of being deposited in manure, but this has yet to be observed directly. Aardvarks, however, fulfil their water requirements from ants and termites, so this argument seems implausible. In addition, Aardvarks lack the mouthparts necessary to break open the tough rind of the fruit. An alternative hypothesis is that seed harvesting ants store the seeds of the Aardvark Cucumber in their nests and then Aardvarks consume these seeds coincidentally when eating the ants. Kingdon (1971) reports that fungus gardens of termitaries are rejected in the wet season, but may be eaten in the dry season.

Commensal feeding associations have been recorded. The Aardwolf *Proteles cristatus*, which is unable to open termitaries itself, exploits the ability of the Aardvark to do so. During winter in South Africa *Trinervitermes trinervoides* remain within their termitaries where Aardwolves cannot reach them. This considerably reduces the amount of food available and Aardwolves lose condition. At these times they often follow Aardvarks and feed on freshly exposed termites (Taylor & Skinner 2000). When Aardvarks become active during the day, Southern Anteater-chats *Myrmecocichla formicivora* hang around and feed on ants from freshly opened nests (Taylor & Skinner 2001). Similar associations are known with Clapper Lark *Mirafra apiata* (Vernon & Dean 1988) and baboons (J. Kingdon pers. comm.).

Social and Reproductive Behaviour Aardvarks are solitary with very limited contact between conspecifics. When a ♂ and ♀ are in close proximity, they detect each other by sound and smell. On approach they sniff each other vigorously, especially around the base of the tail. On contact they occasionally rear up on their hindlegs as part of the investigative process. Interactions are short, usually lasting less than 10 min. If sexual interest is shown, the interaction may last longer.

Home-range areas are 200–300 ha in the Karoo (Taylor & Skinner 2003), but could be larger in areas with lower prey abundance. No seasonal variation is recorded, although non-cyclic spatial shifts in home-ranges do occur. The degree of territoriality is unknown. Meetings between adult ♂♂ have not been observed, so the degree of antagonism is unrecorded. However, one young habituated ♂ behaved very nervously when a large adult ♂ was close by, and repeatedly hid inside burrows for short periods. This young animal appeared to perceive the adult by scent. Although densities are low, home-ranges are not mutually exclusive. Limited overlap occurs between ♂♂ and ♀♀, ♂♂ and ♂♂, and ♀♀ and ♀♀. Scent glands produce a viscous, strong-smelling liquid that is regularly used by both sexes to scent-mark throughout their home-ranges. In ♀♀ the gland opens on either side of the vulva and in ♂♂ they open just behind the prepuce. Scent-marking is achieved by wiping the gland over freshly excavated soil at feeding sites, burrow entrances and faecal scrapes.

During three consecutive years in the Karoo, mating occurred in Oct and Nov; one observation was also made in May. During copulation, the ♂ hangs on tightly to the ♀ with his forefeet, the claws inflicting numerous scratches on the back and flank of the ♀. Several attempted mountings may occur in a short space of time, with each lasting up to 15 sec. Behaviour of mothers with young is unknown in the wild. In captivity, young first follow their mother out of the nesting chamber at about 18 days. Females consume faeces and urine excreted by babies and have been observed to scent-mark them. It is not known how long young remain in the burrow with their mother, nor the age when the young disperse and become totally independent.

Although Aardvarks do not interact directly with other species, their burrows are utilized by many other species. Unused burrows provide sleeping shelter for warthogs, porcupines, Aardwolves, pangolins, jackals, genets, Black-footed Cats *Felis nigripes*, mongooses and any mammals small enough to enter (Smithers 1971). Hyaenas and African Wild Dogs *Lycaon pictus* may use them to shelter their young. They are even known to provide roosts for bats, notably *Nycteris* spp. Snakes and lizards are also important users of burrows, as are some birds such as the Southern Anteater-chat and Blue Swallow *Hirundo atrocaerulea*, which make nesting chambers in the roof of the burrow entrance. Excavated termitaries also provide shelter for many snakes and lizards as well as nests for small mammals such as the Southern African Pygmy Mouse *Mus minuotoides* and Lesser Dwarf Shrew *Suncus varilla* (Smithers 1971, A. Taylor pers. obs., S. Cilliers pers. comm.). Aardvarks never share burrows with conspecifics, but have been known to share a burrow with porcupines, probably in a separate chamber (A. Taylor pers. obs.).

Aardvarks do not vocalize, although young animals may make a bleating sound when stressed.

Reproduction and Population Structure Information about reproduction is available mostly from captive animals, supplemented by limited observations from the wild. Births in the wild have been reported from May to Jun in Ethiopia, early Nov in Uganda (Kingdon 1971), Oct and Nov in DR Congo (Shoshani *et al.* 1988) and May to Aug in southern Africa (Smithers 1983). In the Karoo, mating in Oct and Nov extrapolates to births in Jul (A. Taylor pers. obs.). One young, occasionally two, are born after an average gestation of 35 weeks (range 33.5–37.0) (Chicago Zoological Park records). Babies are born alert and active, weighing 1.8 kg (range 1.40–1.95 kg) (Chicago Zoological Park records). Captive animals grow quickly, reaching 10 kg after seven weeks, and 40 kg after just seven months; wild animals grow much more slowly. One captive ♂ gained only 9 kg in one year (23 kg to 32 kg). Population structure is poorly known, but meagre evidence from the Karoo suggests it is 1 ♂ to 1 ♀ (Nel *et al.* 2000). One captive specimen lived to nearly 30 years (Weigl 2005).

Predators, Parasites and Diseases Lions *Panthera pardus*, Leopards *Panthera leo*, Spotted Hyaenas *Crocuta crocuta* and African Rock Pythons *Python sebae* are the main predators of the Aardvark, although smaller predators may take young. Shoshani *et al.* (1988), reviewing the literature available at the time, list a number of ectoparasites including ticks, such as *Haemaphysalis muhsami*, *Hyalomma impressum* and a number of species of the genus *Rhiphicephalus*; sucking lice (*Haematopinus notophallus* and *Hybophthirus notophallus*), a flea (*Echidnophaga larina*) and various flies. Endoparasites include flagellates (*Trichomonas* sp. and *Trypanosoma* spp.); an amoeba (*Entamoeba* sp.); a thorny headed worm (*Nephridiacanthus longissimus*); various roundworms and a pentastome (*Armillifer armillatus*). Diseases are unknown.

Conservation IUCN Category: Least Concern. CITES: Not Listed.

Although reduced in areas where their habitat has been altered as a result of human activities, the Aardvark has a wide, nearly pan-African distribution, south of the Sahara, occurs in many well-managed protected areas, and is not currently believed to be threatened. In central and West Africa, numbers may be declining as a result of the expansion of human populations, habitat destruction and hunting for meat. Hatt (1934) recorded indigenous hunters in the Congo killing Aardvarks trapped in burrows, and Mbuti pygmies in the Ituri Forest in DR Congo are reported smoking them out of their burrows (Carpaneto & Germi 1989). The meat is prized, including the skin; many parts of the Aardvark, such as the claws and teeth, are used to make bracelets, as charms and curios, and for some medicinal purposes (Carpaneto & Germi 1989).

Measurements

TL (♂♂): 1600 (1490–1750) mm, n = 15 TL (♀♀): 1580 (1400–1730) mm, n = 16
T (♂♂): 544 (443–620) mm, n = 15
T (♀♀): 539 (464–630) mm, n = 16
HF c.u. (♂♂): 256 (240–268) mm, n = 15
HF c.u. (♀♀): 247 (225–280) mm, n = 16
E (♂♂): 180 (167–210) mm, n = 15
E (♀♀): 177 (165–185) mm, n = 16
WT (♂♂): 53.3 (41.3–64.5) kg, n = 15
WT (♀♀): 51.4 (40.4–57.7) kg, n = 16
Zimbabwe (Smithers & Wilson 1979)

TL (♂♂): 1721 (1640–1785) mm, n = 4
TL (♀♀): 1770 (1610–1850) mm, n = 6
T (♂♂): 637 (610–720) mm, n = 4
T (♀♀): 644 (580–700) mm, n = 6
HF c.u. (♂♂): 276 (270–290) mm, n = 4
HF c.u. (♀♀): 270 (263–275) mm, n = 6
E (♂♂): 169 (162–175) mm, n = 15
E (♀♀): 171 (158–180) mm, n = 16
GLS (♂♂): 253 (252–254) mm, n = 3
GLS (♀♀): 247 (240–255) mm, n = 6
GWS (♂♂): 92 (84–98) mm, n = 4
GWS (♀♀): 92 (85–96) mm, n = 6
NE DR Congo (Hatt 1934)
Hatt (1934) recorded a ♂ specimen from the forests of Congo (which he attributed to the form *O. a. erikssoni*) with measurements: TL: 1980 mm; T: 760 mm; and HF c.u.: 300 mm.

Key References Kingdon 1971; Melton 1976; Shoshani *et al.* 1988; Skinner & Chimimba 2005; Taylor 1998; Taylor & Skinner 2000, 2001, 2003, 2004; Taylor *et al.* 2002.

Andrew Taylor

Glossary

abbrev. = abbreviation
adj. = adjective
cf. = *confer*, compare with; as opposed to
Lat. = Latin
n. = noun
pl. = plural
q.v. = *quod vide*, 'which see'
v. = verb

acetabulum: the concave socket (fossa) in the pelvic bone in which the head of the femur articulates.

acrocentric: describes a chromosome which has the centromere (*q.v.*) very near one end and which therefore appears to have only one arm (= telocentric (*q.v.*) for practical purposes).

ad libitum: (*Lat.*) as much as one likes; having unrestricted access to a resource (e.g. water or food).

aestivate: state of torpor (*q.v.*) induced by cold or drought; usually associated with a reduced metabolic rate and inactivity.

afroalpine: describes habitats and/or vegetation occurring above the treeline on African mountains. Includes montane grassland and heathlands.

afromontane: refers to mountainous regions in Africa, e.g. afromontane forests and afromontane grasslands.

Albertine Rift Valley: *see* Rift Valley (*q.v.*).

allele: an alternative form of a gene. A diploid organism carries two alleles (which may be same or different) for each gene locus. At any one locus, there may be several possible alleles (although only two are present in a single organism).

allelomimetic behaviour: behaviour in social animals in which each animal does the same thing as those nearby.

Allen's Rule: A rule that states that structures in endotherms such as limbs (which are more prone to heat loss) are reduced in size by means of natural selection over time in cooler climates (to reduce heat loss).

allogrooming: grooming behaviour directed at another individual. *cf.* autogroom (*q.v.*).

allomothering: non-parental mothering; caring for young by individuals (male or female) that are not the parents of the young.

allopatry (*adj.* **allopatric**): the situation where populations of the same or different species have non-overlapping geographic ranges; refers also to populations of the same, or different, species that are geographically separated. *cf.* sympatry (*q.v.*); syntopy (*q.v.*).

allozyme: one of a number of forms of the same enzyme having different electrophoretic properties and which are encoded by alternate alleles at the same genetic locus.

altimontane: collective term for the belts of ericaceous and afroalpine vegetation on the high mountains of tropical East Africa (White 1983).

altricial: describes young born in an undeveloped state. *cf.* precocious (*q.v.*).

altruism: behaviour which enhances the reproductive and genetic fitness of another individual at the expense of its own.

alveolus (*pl.* **alveoli**, *adj.* **alveolar**): small cavity; socket that houses the root of a tooth.

amastoidy: a condition characteristic of paenungulate mammals in which the mastoid process is concealed by the expansion and overlap of the squamosal.

angular process: process at the posterior lower corner of the mandible; situated ventral to the coronoid process (*q.v.*).

ante-orbital: in front of the orbit (*q.v.*).

antebrachial: anterior to the arm (forelimb).

anterior palatal foramina: the two foramina (*q.v.*) on the ventral part of the skull.

apomorphy (*adj.* **apomorphic**): situation in which a novel character evolves from a pre-existing character. In cladistics (*q.v.*), an apomorphic character shared among two or more species (synapomorphy [*q.v*]) indicates shared descent from a common ancestor and hence monophyly (*q.v.*). *cf.* plesiomorphy (*q.v*).

arboreal: living above the ground (in trees and shrubs). *cf.* scansorial (*q.v.*); terrestrial (*q.v.*).

auditory bulla: *see* tympanic bulla.

auditory meatus (*pl.* **auditory meati**): the external opening of the ear; the passage leading from the tympanic membrane (ear drum) to the external ear.

autapomorphy: derived trait uniquely characteristic of a taxon.

autogroom: grooming behaviour in which an individual grooms itself. *cf.* allogroom (*q.v.*).

autosomal: pertaining to any chromosome other than the sex chromosomes.

bachelor herd: a herd comprised entirely of males, usually mature, but of mixed age.

baculum (*pl.* **bacula,** *adj.* **bacular**): the os penis, or penis bone, which supports the penis in some mammals.

bai (*pl.* **bais**): an opening or clearing.

basal metabolic rate: metabolic rate required for survival in the thermal neutral zone (*q.v.*); a state that requires the lowest expenditure of energy when at rest.

basicranium: the base of the skull.

basisphenoid: cranial bone in middle of base of skull; the median posterior part of the sphenoid bone, forming part of the floor of the braincase.

Bergmann's Rule: The theory that the size of a warm-blooded animal in a single, closely related, evolutionary line, increases along a gradient from warm to cold temperatures.

bicuspid: having two points or cusps (particularly of teeth).

bifid: divided by a shallow notch.

bilophodont; describes cheekteeth having two transverse ridges.

bipedal: body supported by the two hindlimbs; movement not using the forelimbs.

blastula: a hollow ball of undifferentiated cells (derived from a fertilized ovum by cell division), which represents one of the earliest stages of embryonic development.

BP: (*abbrev.*) before the present.

brachydont: describes a premolar or molar tooth with low crowns. *cf.* hypsodont (*q.v*).

braincase (= cranium): that part of the skull housing the brain; the part of the skull posterior to the front line of the orbits. *cf.* rostrum (*q.v.*).

buccal: On the cheek side of the mouth or teeth or penetrating to the cheek or sometimes used broadly as pertaining to the cavity of the mouth.

bulla: *see* tympanic bulla.

bunodont: describes molar teeth, entirely covered by enamel, that have low, rounded, hill-like cusps (as opposed to sharp, pointed cusps). (*cf.* hypsodont, lophodont).

bushmeat: meat for human consumption derived from non-domesticated mammals, birds and reptiles taken from their natural habitats and domiciles.

bushveld: savanna vegetation type characterized by a grassy ground layer and a moderately dense upper layer of shrubs and scattered trees.

BZ: (*abbrev.*) Biotic Zone.

C or **c**: (*abbrev.*) canine tooth; upper case denotes adult dentition, lower case denotes deciduous dentition (milk teeth). *See also* canine.

C–M^1, C–M^2, C–M^3: in golden-moles, the length of the upper toothrow, measured from the most anterior part of the canine to the most posterior part of the most posterior molar. For golden-moles, the most posterior molar is M^2 or M^3

c.u.: (*abbrev.*) (*Lat. cum unguis* = with nail) measurement of the hindfoot when length of the nail on the claw is included in the measurement. Usually hindfoot is measured without the claw because claws may be broken or worn. When length of claw is included, it is conventional to record as 'HF c.u.'. *cf.* s.u. (*q.v.*)

caecum (*pl.* **caeca**): a blind-ending pouch in the alimentary canal (often enlarged as a fermentation chamber) located at the junction of the small and large intestines.

canine: the most anterior tooth on the maxilla bone and in a similar position on the mandible; situated immediately posterior to the incisors; if incisors are absent, the most anterior tooth in the jaw. Unicuspid; tall and pointed in most mammals. Never more than one canine on each side of each upper and lower jaw; absent in some taxa.

caniniform: having shape and appearance of a canine tooth.

carotid: pertaining to the carotid artery located in the front of the neck though which blood from the heart flows to the brain.

caudal: pertaining to the tail; in the direction of the tail.

Cenozoic (= Cenozoic Era): geological era, ca. 65 mya to today, comprising the Quaternary and Tertiary Periods: the Age of Mammals.

central Africa: Cameroon (south of the Sanaga R.), Central African Republic (but only south of ca. 7° N), Equatorial Guinea, Gabon, DR Congo (except SE). Mainly rainforest habitats and rainforest–savanna mosaics.

centromere: the part of a chromosome where sister chromatids are linked together during mitosis.

cerebellum: the part of the hindbrain that controls and coordinates motor movements, posture, balance and muscle tone.

cerebrum (= cerebral hemispheres): the anterior part of the brain that is involved in voluntary movements, processing sensory information, olfaction, learning, memory, communication and other functions.

cervical: pertaining to the neck.

cf. (in general usage): compare or compare with. In the context of descriptions, implies a difference or contrast: e.g. 'In *Elephatulus edwardii*, first lower premolar single-rooted (*cf. E. myurus* in which the first lower premolar is double-rooted).'

cf. (in taxonomy): precedes the specific name if there is uncertainty in the assignment.

cheekteeth: the premolar (*q.v.*) and molar (*q.v.*) teeth combined.

choana (*pl.* **choanae**): the openings of the internal nostrils on the skull, situated immediately posterior to the bony palate.

chromosome: one of the thread-like bodies within the nucleus of a cell, which carry the genes (genetic material) in linear order; each chromosome is composed of one long molecule of DNA (and two long molecules at cell division). Chromosomes occur in pairs (one from each parent) and are visible as rod-like bodies in cells that are dividing. The total number of chromosomes in a cell is expressed as the diploid number (2n).

cingulum (*pl.* **cingula**): ridge around the base of the crown of a tooth.

CITES: (*abbrev.*) Convention on International Trade of Endangered Species of Wild Fauna and Flora; an international treaty set up to ensure that international trade in wild animals and plants does not threaten the survival of species in the wild. It accords varying degrees of protection to more than 33,000 species of animals and plants. Appendix I lists species that are the most endangered among CITES-listed animals and plants. Appendix II lists species that are not necessarily now threatened with extinction but that may become so unless trade is closely controlled. Appendix III is a list of species included at the request of a Party that already regulates trade in the species and that needs the cooperation of other countries to prevent unsustainable or illegal exploitation.

clade: branch of a phylogenetic tree containing the set of all organisms descended from a common ancestor.

cladistic (analysis): a methodology that provides a classification in which organisms are grouped in terms of the time when they had a common ancestor.

cline (*adj.* **clinal**): in context of geographic variation, a gradual and sequential change of a character(s) without a significant break such as would justify division into separate subspecies or species.

cloaca: the single common opening for faeces, urine and genital products; not present in mammals except that a cloaca-like structure is present in some species of Tenrecidae.

cochlea (*pl.* **cochleae,** *adj.* **cochlear**): a hollow structure, spirally coiled like a snail's shell, situated in the skull and containing the internal organ of hearing.

comparatively: in profiles of Afrosoricida and Macroscelidea, used in the context of describing the size of one character compared with the size of the same character in a different species. Sizes described as small, medium or large (if range is divided into three) or very small, small, medium, large, very large (if range is divided into five). *cf.* relatively (*q.v.*).

competitive exclusion: the principle that two different species cannot indefinitely occupy the same ecological niche.

concatenation: a chain of linked elements.

concave: having a curvature that curves inwards; having an outline or a surface curved like the interior of a circle or sphere. *cf.* convex (*q.v.*).

concavity: a concave depression in an outline or surface.

conductance: in thermal biology, the rate at which heat passes across a temperature gradient, e.g. the density and thickness of the pelage affects the rate at which body heat passes from the body to the outside. Thick pelage, which traps and holds air, results in low thermal conductance.

condylar process: process at the posterior upper corner of the mandible, which forms the lower hinge of the jaw articulation; fits into the glenoid fossa (*q.v.*) of the skull.

condylarth (*adj*): as in the Condylarthra, an extinct order of mammals.

condyle: a rounded process on a bone, which articulates with a socket-like concavity in another bone.

congeneric: belonging to the same genus.

conspecific: belonging the same species. *cf.* heterospecific (*q.v.*).

contiguous: touching; sharing a boundary (as in geographic ranges).

convex: having a curvature that bulges outwards; having an outline or a surface curved like the exterior of a circle or sphere. *cf.* concave (*q.v.*).

coprophagy: the eating of faeces. Includes the eating of an individual's own faeces as they are voided from the anus.

coronoid canal: a foramen (canal) in the coronoid process (*q.v.*) of the mandible.

coronoid process: angular pointed process on the upper margin of the mandible, situated anteriorly to the condylar process (*q.v.*); does not participate in the jaw articulation.

corpus luteum (*pl.* **corpora lutea**): a glandular mass of tissue on the surface of an ovary, that develops after the extrusion of an ovum from a Graafian follicle (*q.v.*); secretes the hormone progesterone.

cotype: originally synonymous with syntype but now used as synonym of paratype (*q.v.*).

CR: (*abbrev.*) *see* crown–rump length.

cranial profile: the shape of the cranium (that part of the skull which surrounds the brain) when viewed from the side.

craniodental: pertaining to the skull and teeth.

cranium: that part of the skull housing the brain. Also called braincase.

crepuscular: at, active in, twilight, when light intensity is higher than at night but lower than during the day. *cf.* diurnal (*q.v.*); nocturnal (*q.v.*).

Cretaceous Period: period (within the Mesozoic Era); 146–65 mya.

crown: (1) top of head; (2) exposed part of a tooth (visible above gum), especially the grinding surface.

crown–rump length (CR): distance from the crown of head to the rump of a foetus (i.e. maximum length of a foetus in its natural form).

cursorial: pertaining to running.

cusp (*adj.* **cuspidate**): a prominence or sharp point, such as on the occlusal surface of some teeth. *See also* t.

cutaneous: (*adj.*) pertaining to the skin.

Cyrenaica: a region of North-East Libya. Includes the Cyrenaican Plateau and that part of the Mediterranean Coastal Biotic Zone between the plateau and the sea, as well as drier terrain south of the plateau.

cytochrome *b*: a protein involved in electron exchange in the mitochondria. It is the product of a gene in the mitochondrial genome. The sequence of this gene is often compared between species in phylogenetic studies to infer relatedness.

cytogenetics (*adj.* **cytogenetic**): the study of the microscopic structure of chromosomes, especially the mapping of genes.

cytonuclear (*adj.*): pertaining to the nucleus and the cytoplasm of a cell.

Dahomey Gap: the geographic region where savanna habitat extends southwards to the West African coast in E Ghana, Togo, Benin (formerly Dahomey) and extreme SW Nigeria. The presence of savanna forms a break (or gap) in the extensive Rainforest Biotic Zone, which extends along the West Africa coast from Sierra Leone to Cameroon. The Dahomey Gap is an important biogeographical barrier separating the faunas to the east and west of the Gap.

deciduous teeth: *see* milk teeth (*q.v.*).

delayed implantation: a means of lengthening the interval between copulation and parturition by delaying the implantation of the blastula (*q.v.*), so that both copulation and parturition can occur in the most optimal seasons. Development to blastula stage is followed by a period of halted development lasting several weeks or months; then the blastula implants and embryonic development proceeds normally, usually without any further interruption, until the young is born.

deme: a unit of population which is interbreeding and which is separate from any other such population.

dental formula: a simple numerical method of denoting the number of incisor (I), canine (C), premolar (P) and molar (M) teeth on one side of the upper jaw and lower jaw, and the total number of teeth. For example, the dental formula of a primitive mammal is $I\ ^3/_3$, $C\ ^1/_1$, $P\ ^4/_4$, $M\ ^3/_3 = 44$, which means there are three incisors, one canine, four premolars and three molars on each side of the upper jaw and also the lower jaw, making a total of 44 teeth. The formula may also be expressed in the form $^{3143}/_{3143} = 44$. Each incisor, premolar and molar is numbered according to its position in the tooth row; superscript numbers indicate upper jaw, subscript numbers indicate lower jaw (mandible) e.g. P^4 (upper fourth premolar), M_2 (lower second molar).

dentine: the substance, also known as ivory, comprising tusks (*q.v.*) and the interior hard part of vertebrate teeth. Lies under the enamel of teeth (but may be exposed if the enamel wears) and surrounds the pulp chamber and root canals.

diastema: space in the mouth between the incisor teeth and cheekteeth in those mammals that feed on grasses, herbs etc. (e.g. rodents, hares, rabbits, ruminants, etc.).

dichromatism: condition in which members of a species show one of only two distinct colours or colour-patterns.

dilambdodont: molar tooth with W-shaped ridges. *cf.* zalambdodont (*q.v.*).

dimorphism: *see* sexual dimorphism.

diphyly: the derivation of a taxon from two separate lines of descent. *cf.* monophyly (*q.v.*).

diploid number (2n): total number of chromosomes (including sex chromosomes) in a somatic cell of an organism.

distal: the end of any structure furthest away from the mid-line of the body or furthest from the point of its attachment. *cf.* proximal (*q.v.*).

diurnal: at, active in, daytime; when light intensity is high. *cf.* crepuscular (*q.v.*); nocturnal (*q.v.*).

DNA: (*abbrev.*) deoxyribonucleic acid; the very large self-replicating molecule that carries the genetic information of a chromosome; each molecule is composed of two complementary chains of DNA.

DNA hybridization: technique of comparing the similarity between two DNA molecules by reassociating single strands from each molecule and determining the extent of double-helix formation. In phylogenetics, this technique is used to determine the relatedness of two or more taxa.

dorsoventral (dorsoventrally): from dorsal to ventral surface; from back to belly of an animal.

E: (*abbrev.*) length of external (outer) ear (= pinna), measured from tip of ear to the posterior point of the ear conch. Length and shape usually affected by preservation.

East Africa: Kenya, Uganda, Rwanda, Burundi and Tanzania.

eastern Africa: SE Sudan, Ethiopia, Eritrea, Djibouti, Somalia, Kenya, Uganda, Tanzania, Malawi (but only south of L. Malawi and east of the Shire R. Valley) and Mozambique (but only east of Malawi and north of the Zambezi R.).

ecotype: a genetically distinct geographic variety or population within a species, which is adapted to specific environmental conditions.

ectoparasite: a parasite that lives on the exterior of an organism (e.g., ticks, fleas, lice). *cf.* endoparasite (*q.v.*).

ectotympanic: a bony element within the middle ear that supports the tympanic membrane or eardrum.

edaphic: influenced by conditions of soil or substratum.

emargination: a distinct notch or indentation.

embryo number: number of foetuses within the uterus or uteri of the female (as assessed by autopsy). Expressed as mean number (with range from minimum to maximum, and sample size). *cf.* litter-size (*q.v.*).

enamel: hard material that forms a cap over the dentine component of a tooth; usually the most visable part of a tooth.

encephalization quotient (EQ): a measure of comparative brain size or weight defined as the ratio of the actual brain weight to the expected brain weight predicted for an animal of a given body weight.

endemic: restricted to, peculiar to, or prevailing in, a specified country or region.

endoparasite: a parasite that lives in the interior of an organism (e.g., nematodes, cestodes, blood parasites). *cf.* ectoparasite (*q.v.*).

entotympanic: an independent ossification found in the floor of the tympanic cavity in various extant and extinct eutherian groups, including, for example, Scandentia, Chiroptera, Dermoptera, Hyracoidea, Pholidota, Xenarthra, Carnivora, and Macroscelidea.

Eocene: geological Epoch (within the Tertiary Period); 55–38 mya.

epiphysis (*pl.* epiphyses): any part of a long bone that is formed from a different centre of ossification and which later fuses with the bone to form its terminal part.

epitympanic recess: a hollow located on the roof of the middle ear.

erg: a large, relatively flat area of desert covered with wind-swept sand with little or no vegetation cover (sometimes referred to as a dune sea).

evaporative water loss: the loss of water from the body through the skin and/or the lungs. A mechanism used by mammals to reduce T_b (*q.v.*) when T_a (*q.v.*) is high. Excessive evaporative water loss may lead to dehydration if free (drinking) water is unavailable.

exfoliating: shedding flakes (e.g. of bark), or breaking into relatively thin slabs (e.g. of granitic rock).

exoccipital condyles: a pair of projections from the occipital bone on either side of the foramen magnum (*q.v.*) which articulate with the first of the spinal vertebrae.

extant: living at the present time. *cf.* extinct.

F. R.: (*abbrev.*) Forest Reserve.

facultative: having the capacity to switch from one mode of life or action to another depending on conditions or circumstances. *cf.* obligate (*q.v.*).

female-defence polygyny: a mating system in which males control access to females directly, usually by virtue of female gregariousness (Emlen & Oring 1977).

fenestra (*pl.* fenestrae): opening in a bone, or between two bones.

flank: the side of the body of a mammal.

FN: (*abbrev.*) *see* fundamental number.

foliaceous: (*adj.*) resembling the leaf of a plant.

folivore (*adj.* folivorous): an animal that eats leaves.

foot-drumming: activity of banging/drumming the soles of the hindfeet on the substrate to produce a noise used for communication between individuals; footdrums vary in length, intensity and form according to species; a frequent means of communication in sengis.

foramen (*pl.* foramina): an aperture (which is usually small, round or elliptical) in a bone, or between bones, for the passage of a nerve, blood vessel or muscle.

foramen magnum: the large opening at the posterior end of the skull through which the spinal cord passes.

forest island: *see* relict forest.

fossorial: adapted for digging; burrowing. *cf.* subterranean (*q.v.*).

founder effect: the loss of genetic diversity that occurs when a new isolated population is derived from a very small number of individuals.

fovea: small pit or depression.

frontal bone: one of a pair of bones forming the anterior part of the braincase.

frugivorous: fruit-eating.

fundamental number (FN): an ambiguous term sometimes defined as (1) the total number of chromosomal arms in the full chromosomal complement of an organism (i.e. including the sex chromosomes), or (2) the total number of chromosomal arms found in the autosomal chromosomes only (i.e. excluding the sex chromosomes). When only the autosomal chromosomes are included, some authors (but not all) use aFN instead of FN to avoid ambiguity.

fusiform: elongated and tapering at both ends.

fynbos: the heath shrublands characteristic of the Cape Floristic Kingdom (within the South-West Cape Biotic Zone) of South Africa. Dominant plants are sclerophyllous, evergreen, low (<3 m), bushy and fine-leafed, but there are also scattered taller bushes and, less often, very widely spaced trees. Contains an exceptionally high number of endemic species of plants. The three main components on nutrient-poor sandy soils are species of Ericaceae, Restionaceae and Proteaceae. Also includes 'renosterveld' (*q.v.*) (dominated by species of Asteracea) on nutrient-richer silt or clay soils.

G. R.: (*abbrev.*) Game Reserve.

gallery forest: type of forest outlier (in a savanna region) found in narrow sheltered valleys and ravines on hillsides, where soils are moist enough, and conditions humid enough, to support rainforest trees (Rosevear 1953).

genetic drift: in population genetics, refers to changes in the frequencies of genes in small, isolated populations that occur with time, as a result of random sampling events as opposed to natural selection.

genotype: genetical term to describe the genetic constitution of an individual inherited from its parents. *cf.* phenotype (*q.v.*).

genus (*pl.* genera): taxonomic division superior to species and subordinate to family.

geocarpic: having fruit mature underground.

geophagy (= geophagia): the eating of saline or calcareous soil.

gestation: the development of embryo/foetus, which takes place in the uterus; the period during which this development takes place. The gestation period is defined as the interval between conception and parturition (birth). Strictly speaking, the gestation period is not the interval between copulation (mating) and parturition (birth), although many authors take it to be this interval. The gestation period (*sensu stricto*) may be lengthened by reproductive delays, such as delayed implantation (*q.v.*).

glans penis: the bulbous tip of the penis.

glenoid (= glenoid fossa): the cavity (fossa) in the squamosal bone of the skull for the articulation of the condyle of the mandible (lower jaw) in mammals; visible on lateral and/or ventral views of skull depending on the taxon.

GLS: (*abbrev.*) greatest length of skull, measured from anterior end of incisor teeth or nasal bone (whichever is most anterior) to the posterior end of the skull (occiput, occipital condyles or auditory bullae, whichever is most posterior).

Gondwana (= Gondwanaland): the southernmost of the two Mesozoic (*q.v.*) supercontinents that later fragmented (as a result of continental drift) into the landmasses of Africa, Madagascar, Antarctica, Australia, South America, the Indian subcontinent and the Arabian peninsula.

Graafian follicle: the structure in the mammalian ovary which contains the developing ovum, and from which the ovum is released at the time of ovulation.

gracile: lightly built. *cf.* robust.

graminivore (*adj.* graminivorous): feeding on grasses.

granivore (*adj.* granivorous): an animal that eats grains and seeds.

graviportal: having limbs adapted to bearing great weights (as in elephants).

gregarious: living together in groups, flocks, herds.

Gregorian Rift Valley: *see* Rift Valley.

guard hair: long thin bristle-like hairs, mainly on the back and flanks, which project beyond the soft hairs of the pelage; when present, conspicuous but never as numerous as soft hairs; probably tactile in function.

gular: pertaining to the upper part of the throat (the gula) as in gular gland, gular pouch, gular region.

GWS: (*abbrev.*) greatest width of skull, usually measured across the widest point of the zygomatic arches (*q.v.*).

haematopoietic (*n.* **haematopoensis**): pertaining to the formation of blood cells from stem cells.

hallux: the first digit of the hindlimb.

hammada: flat plain covered with pebbles of various sizes on a hard substrate in an arid environment. Much of the Sahara Desert is formed of hammada.

haplotype: the set of alleles borne on one of a pair of homologous chromosomes; a group of genes that is inherited together from a single parent.

HB: (*abbrev.*) length of head and body (measured from the tip of the nose to the most posterior point of the pelvis (anterior to the first tail vertebra).

hemidiaphragm: half of the diaphragm, the muscle that separates the chest cavity from the abdomen and that serves as the main muscle of respiration.

hermaphrodite (*adj.* hermaphroditic): an individual having both male and female reproductive organs.

heterospecific: (*adj.*) belonging to different species. *cf.* conspecific (*q.v.*)

heterothermic: condition when the body temperature fluctuates in relation to the ambient temperature T_a (*q.v.*). *cf.* homeothermic (*q.v.*).

heterozygosity: a measure of the genetic variation in a population, expressed through the presence of different alleles (*q.v.*) at one or more loci on homologous chromosomes.

HF: (*abbrev.*) length of hindfoot (measured from the 'ankle bone' to the tip of the longest digit usually without including the claw). *See also* c.u.; s.u.

high forest: rainforest that has matured, stabilized and reached the climax stage of succession.

highveld: high plateaux characteristic of inland southern Africa, dominated by grasses.

Holocene: geological Epoch (within the Quaternary Period) (*q.v.*) following the Pleistocene Epoch (*q.v.*); ca.11,000 mya to today. Sometimes referred to as the 'Recent' Epoch.

holotype (= type): the single specimen designated or indicated by the original author of the original description of a new species or subspecies, to be the standard reference to the essential characters of the new taxon. *See also* cotype (*q.v.*), lectotype (*q.v.*), neoparatype (*q.v.*), neotype (*q.v.*), paratype (*q.v.*), syntype (*q.v.*), topotype (*q.v.*) and type locality (*q.v.*).

home-range: the area (expressed in square metres, or square kilometres) routinely used by an animal for its day-to-day activities and requirements, and which contains the resources required for survival and reproduction. Within the home range there may be a 'core area' or 'centre of activity', which is utilized more frequently than other parts (e.g. 80% or 90% of known time-based observations). *cf.* territory (*q.v.*).

homeothermic: describes an organism having a body temperature that is maintained at a constant level (within limits), independently of the ambient temperature T_a (*q.v.*). For the limits, see upper and lower critical temperatures [*q.v.*]). *cf.* heterothermic (*q.v.*).

homoplasy: similarity between different organisms or taxa resulting from evolution along similar lines (e.g. convergent evolution) rather than descent from a common ancestor.
homozygous (*n.* homozygosity): describes a diploid organism that has inherited the same allele of any particular gene from each of its parents. *cf.* heterozygous.
hyoid: a small bone or bones in the throat located at the base of the tongue and supporting the muscles of the tongue.
hyperthermia: elevation of body temperature above normal limits due to increase in Ta (*q.v.*) or increase in metabolic rate. *cf.* hypothermia (*q.v.*).
hypoconulid: the most posterior cusp on the lingual (inner) side of a lower molar tooth.
hypothermia: drop in body temperature below normal limits; occurs in some small mammals when Ta (*q.v.*) falls, and is a means of conserving energy in cold weather. *cf.* hyperthermia (*q.v.*).
hypsodont: describes a premolar or molar tooth with high crowns; has short roots. *cf.* bunodont, lophodont.
I or i: (*abbrev.*) incisor (*q.v.*) tooth or teeth; upper case denotes adult dentition, lower case denotes deciduous dentition (milk teeth). *See also* incisor.
I. (*pl.* Is.): (*abbrev.*) Island.
I^1-M^3: (*abbrev.*) the length of the upper toothrow from the anterior end of the first upper incisor to the posterior end of the last molar. In taxa where the last molar is M^1 or M^2 (instead of M^3), the measurement is I^1-M^1or I^1-M^2.
implantation: the embedding of a blastula (*q.v.*) in the lining of the uterus.
incertae sedis: (*Lat.*) refers to a taxon of uncertain status and that is pehaps taxonomically invalid.
incisive foramen (*pl.* foramina): paired foramina located at the anterior end of the bony palate, just posterior to the incisor teeth.
incisor: tooth rooted in the premaxillary bone (most anterior bone of upper jaw) and in an equivalent position in the mandibular bone (lower jaw); always anterior to the canine teeth, if present. The number and form of the incisor teeth varies in different species. *See also* dental formula (q.v.).
infraorbital foramen (*pl.* foramina): foramen that connects the orbit (eye-socket) with the rostrum (premaxilla and maxilla bones); this foramen contains the masseter medialis muscle in some taxa. Also referred to as ante-orbital foramen and antorbital foramen.
infrasonic: describes sounds with frequencies below the range that can be heard by humans.
infundibulum: a funnel-shaped cavity or depression.
inguinal: situated in the groin, that is the area between the lower lateral part of the abdomen and the thigh.
inner: nearest to mid-line of body, e.g. inner incisor, inner margin of ear.
insectivore (*adj.* insectivorous): an animal that eats insects.
inselberg: isolated rocky hill; term used mainly in western Africa. *See also* jebel (*q.v.*); kopje (*q.v.*), krantz (*q.v.*).
inter-aural: between the ears.
inter-digital: between the digits (e.g. interdigital gland).
inter-orbit (= inter-orbital constriction): narrowest part of the skull between the orbits when viewed dorsally at the level of the frontal bones.
interparietal: bone (not paired) on dorsal surface of the skull, lying between and partly posterior to the parietal bones.
interpterygoid region: the region between the two plate-like pterygoid bones (*q.v.*) that form part of the roof of the mouth.
inter-specific: between different species.
intra-specific: within one species; between members of the same species.
intromission: insertion of the penis into the vagina during copulation.
Isoptera: taxonomic order containing the termites (within the phylum Insecta).
isozyme: any one of several different forms of an enzyme encoded by the same or different genes and which each differ in properties such as optimum pH or isoelectric point.
iteroparous: repeated reproduction throughout a season or a lifetime (*cf.* semelparous – reproduces once and then dies).
ivory: *see* dentine.
jebel (= gebel): isolated rocky hill in savanna and arid habitats; term used mainly in northern Africa. *See also* inselberg (*q.v.*); kopje (*q.v.*).
jugal: bone of the skull which forms the middle part of the zygomatic arch (*q.v.*); anteriorly joins the maxillary process (*q.v.*) of the maxilla bone and posteriorly joins the squamosal process of the squamosal bone.
K-selection: selection for characteristics suitable in stable environments (slow development of young, small litter size, relatively long time to maturity, relatively long life span) *cf.* r-selection (*q.v.*).
Kaokoveld: area of sandstone and lava hills of the Namib Desert (Biotic Zone 11b) in NW Namibia, inland from the Skeleton Coast.
Karoo: southern African term for grassy dwarf shrubland on the semi-arid central plateau of the western half of South Africa at altitudes of 500–2000 m.
karyogram: a photographic representation of a karyotype (*q.v.*) as it appears at mitosis. Individual chromosomes are arranged in pairs from largest to smallest.
karyological: pertaining to the nucleus and chromosomes.
karyotype: the chromosomes in the cell of an animal. *See also* karyogram.
kloof: a deep, narrow valley or gorge.
kopje: rocky hill with boulders, rock crevices and sparse vegetation; term used mainly in southern Africa. *See also* inselberg (*q.v.*); jebel (*q.v.*); kranz (*q.v.*).
krantz (*pl.* **krantzes**): a wall of rock encircling a mountain or summit; hence any precipitous or overhanging wall of rocks.
L.: (*abbrev.*) Lake.
labial: pertaining to the lips; situated near the lips; on the outer side of the teeth within the mouth. *cf.* lingual (*q.v.*).
labially: in the direction of the lips. *cf.* lingually (*q.v.*).
lachrymal bone: small bone in skull situated at the upper anterior margin of the orbit, near the lachrymal (tear) gland.
lambdoid crest: *see* supraoccipital crest.
lamina (*pl.* laminae): in the context of premolar and molar teeth, a transverse row of cusps which form a transverse ridge of bone and dentine across the tooth.
larynx (= voice box): the structure in the throat of mammals that produces sound.

Laurasia: the northernmost of the two Mesozoic (*q.v.*) supercontinents that later fragmented (as a result of continental drift) into the landmasses of Europe, Asia (excluding the Indian subcontinent) and North America.

lectotype: a specimen chosen from syntypes (*q.v.*) to stand as 'the type' of a species or subspecies when no holotype (*q.v.*) was designated by the author who described the new species or subspecies.

lek: a traditional site where males gather for the sole purpose of attracting and courting females, and to which females come for the sole purpose of mating with the male of their choice.

limiting factor: the principle that the growth or functioning of an organism is limited when any essential factor (or resource) is lacking or in short supply, regardless of the quantity available of any other factor (or resource). The principle may also be applied to the proliferation and/or distribution of a population or species.

lingual: pertaining to the tongue; situated near the tongue; on the inner (tongue) side of the mouth. *cf.* labial (*q.v.*).

lingually: in the direction of the tongue. *cf.* labially (*q.v.*).

Lipotyphla (*adj.* lipotyphlan): a suborder of the order Insectivora comprised of shrews, moles, hedgehogs, tenrecs and golden-moles; now considered to be polyphyletic and therefore not currently recognised as a valid taxon.

litter-size: number of young born to a female. Expressed as mean number (with range from minimum to maximum, and sample size). *cf.* embryo number (*q.v.*).

localized movements: movements of an animal within part or all of its home-range within a single day or within a limited period of time. *cf.* migration (*q.v.*); nomadic movements (*q.v.*).

longitudinal: lengthwise; running in a head to tail direction. *cf.* transverse (*q.v.*).

lophodont: molar teeth having the crowns formed in transverse crests or ridges (*cf.* bunodont).

lordosis: dorsi-flexion of the spine, often performed by females prior to copulation.

lower critical temperature: the lowest ambient temperatures at which the animal must increase its metabolic rate to maintain a constant body temperature. If ambient temperature decreases further, and the animal is unable to increase its metabolic rate sufficiently to maintain a constant body temperature, it will enter hypothermia and may eventually die. *cf.* upper critical temperature.

lowveld: savanna at lower altitudes below the Great Escarpment of South Africa, with vegetation consisting of a grassy lower layer and a woody upper layer of shrubs and/or trees, at either high density (woodland) or intermediate density (bushveld, *q.v.*).

lumbar: describes vertebrae between the thoracic vertebrae and the sacrum; vertebrae of the lower back in humans.

M: (*abbrev.*) molar tooth or teeth. *See also* molar (*q.v.*).

Maghreb: *see* North-West Africa (*q.v.*).

malleus: the most external bone of the three ossicles of the middle ear, the handle of which is joined to the ear drum.

mandible: lower jaw.

mandibular ramus: one of the two branches (sides) of the mandible (*q.v.*); the two rami are joined at the mandibular symphysis (*q.v.*).

mandibular symphysis: the junction, marked by a small vertical ridge, where the left and right sides of the mandible are fused.

masseter muscle: one of the muscles which raises the lower jaw, e.g. during chewing. Particularly well developed and powerful in herbivores.

mastoid: one of a pair of bones, often with a prominent process, situated near the posterior end of the skull behind the auditory meatus (*q.v.*).

matriarchal; describes social groups dominated and/or led by a female (often one of the oldest females in the group).

maxilla (*pl.* maxillae): one of the pair of bones in the skull that forms that part of each upper jaw in which the canine (*q.v.*) (if present) and cheekteeth (premolars and molars) (*q.v.*) are rooted.

maxillary process: projection of bone from the maxilla, which forms the anterior portion of the zygomatic arch (*q.v.*); usually oriented vertically to the anterior–posterior line of the skull.

maxillary: pertaining to the maxilla (*q.v.*).

meatus: a passage or channel; the opening of a passage.

medial: situated in the middle.

Mesozoic (= Mesozoic Era): geological era; ca. 250–ca. 65 mya, comprising the Triassic, Jurassic and Cretaceous Periods: the Age of Reptiles.

metabolic rate: *see* basal metabolic rate (*q.v.*).

metabolic water: water produced by oxidative processes within the body; an important source of water for arid-adapted mammals when free (drinking) water and water within the food is in short supply or unavailable.

metacarpals: the long bones of the hand, situated between the carpal bones of the wrist and the proximal phalanges of the fingers.

metacentric: describes a chromosome with the centromere (*q.v.*) at or very near the middle of its length, so there are two arms of equal or almost equal length (ratio not greater than 1 : 1.1). *cf.* submetacentric (*q.v.*).

metacromion: a process projecting backward and downward from the acromion (the lateral triangular projection of the spine of the scapula that forms the point of the shoulder and articulates with the clavicle of the scapula) of some mammals.

metapopulation: in population dynamics and ecology, a group of spatially separated populations of the same species which interact at some level.

metatarsals: the long bones of the foot, situated between the tarsal bones of the ankle and the most proximal phalanges of the toes.

microsatellite locus (*pl.* microsatellite loci): a region within a DNA sequence where short sequences of one, two, three or four of the nucleotides (A [adenine], T [thiamine], G [guanine] and C [cytosine]) are repeated consecutively; e.g. ACACAC, CAACAA. The number of times a sequence is repeated varies between individuals, within populations of a species, or between species which makes microsatellite loci useful tools in systematics (q.v.).

migration: movements of species that travel, predictably and more-or-less directly, from one habitat to another (and back again), along predetermined routes, in response to seasonal changes in climate, food supply or any other resource. *cf.* localized movements (*q.v.*); nomadic movements (*q.v.*).

milk teeth: teeth occurring in newborn mammals, or appearing soon after birth, and preceding the permanent teeth of the adult animal. Also called deciduous teeth.

Miocene: geological Epoch (within the Tertiary Period); ca. 23–5 mya.

miombo: a vernacular name applied to trees in the genus *Brachystegia*; a type of savanna woodland in the Zambezian region where *Brachystegia* spp. are the commonest trees or one of the commonest trees.

mitochondrial DNA: the small amount of DNA contained within the mitochondria of a cell.

molar: grinding or cutting tooth rooted in the maxilla bone and the mandible; there are usually one, two or three molar teeth in each ramus of the jaw. Together with the premolars, if present, they form the 'cheekteeth'. The number and form of the molar teeth varies in different species. Not preceded by deciduous (milk) teeth.

molariform: similar in form to a molar tooth; used to describe the form of the premolar teeth in some taxa.

monoestrous: Generally, having one oestrus cycle (*q.v.*) during a single reproductive season. In some taxa, having one litter per year. *cf.* polyoestrous (*q.v.*).

monogamous: having only one mate, usually for the whole of an animal's lifetime.

monogamy (*adj.* monogamous): a mating system in which one male mates with one female. Neither sex has the opportunity of monopolizing additional members of the opposite sex. Fitness often maximized through shared parental care (Emlen & Oring 1977). *cf.* polygyny (*q.v.*).

monophyletic: describes a taxonomic group descended from a common ancestor that was itself a member of that taxonomic group, and including *all* the descendants of that ancestor (Groves 2001). *cf.* paraphyletic (*q.v.*); polyphyletic (*q.v.*).

monophyly: derivation of taxa from a common ancestor. *cf.* diphyly (*q.v.*), polyphyly (*q.v.*).

monospecific: describes a genus containing only one known species. *See also* monotypic (q.v.).

monotypy (*adj.* monotypic): describes a taxon containing only one immediately subordinate taxonomic unit, e.g. a monotypic family contains only one genus; a monotypic genus contains only one species. *cf.* polytypy (*q.v.*).

mopane: a vernacular name applied to the tree *Colophospermum mopane*; a type of savanna woodland in the Zambezian region in which *C. mopane* is the commonest species of tree.

Mt: (*abbrev.*) Mount.

mtDNA: (*abbrev.*) mitochondrial DNA (*q.v.*).

Mts: (*abbrev.*) Mountains.

multiparous: describes either the birth of more than one offspring at once, or a female that has given birth two or more times.

muzzle: the snout; the nose and jaws of a mammal.

mya: (*abbrev.*) millions of years ago.

myrmecophagy (*adj.* myrmecophagous): the eating of ants and/or termites.

mystacine vibrissae: vibrissae (*q.v.*) resembling a moustache.

n. d.: (*abbrev.*) no data.

N. P.: (*abbrev.*) National Park.

N. R.: (*abbrev.*) Nature Reserve (not National Reserve).

naris (*pl.* **nares**): a nostril.

Neogene Period: geological period and system; 23.03 ± 0.05 to 2.5 mya, comprising the second part of the Cenozoic Era.

neonate: a newly born animal.

neoparatype: any specimen described at the same time as the neotype (*q.v.*).

neotype: a specimen selected as the type in cases where the primary types are definitely known to be lost or destroyed.

nipple: external opening of mammary gland. Nipple number and position vary according to the taxon, but are consistent within a species. Nipples arranged in pairs, one of the pair on each side of body. Nipple number (if given) is: the number and the position of the nipples on one side of the body, and the total number for both sides. For example, in golden-moles, nipple number is: 1 (abominal) + 1 (inguinal *q.v.*) = 4.

nocturnal: at, active in, the night; when light intensity is at its lowest; between sunset and sunrise. *cf.* crepuscular (*q.v.*); diurnal (*q.v.*).

nomadic movements: irregular and unpredictable movements, from one locality to another, made by species living in unpredictable habitats. *cf.* localized movements (*q.v.*); migration (*q.v.*).

nomen dubium: when the available evidence is not sufficient to permit the identification of a species, its name is considered to be a *nomen dubium* and therefore not available for taxonomic purposes.

nomen nudum: a name which is not valid because, when it was originally published, the organism to which it referred was not adequately described, defined or sketched. The name is therefore invalid because it is impossible to associate it indisputably with any specific organism.

nominate subspecies: the subspecies that bears the name of the species to which it belongs (e.g. *Amblysomus corriae corriae*), only brought into existence by the creation of one or more other subspecies in the same species (e.g. *Amblysomus corriae devilliersi*).

North Africa: those parts of Mauritania, Morocco, Algeria, Tunisia, Libya and Egypt that are north of the Sahara Desert.

North-Central Africa: Southern Chad and southern Sudan, west of the Nile R. and south of the Sahara. Mainly savanna habitats.

North-West Africa (= Mahgreb): those parts of Mauritania, Morocco, Algeria, Tunisia and NW Libya which are north of the Sahara Desert.

nosepad: area of naked, thickened dry skin which surrounds and protects the nostrils in Golden-moles; used for pushing and moving soil during tunnel excavation.

nuchal: pertaining to the nape of the neck.

nulliparous: not having given birth.

obligate: obligatory; limited to one mode of life or action irrespective of conditions or circumstances. *cf.* facultative (*q.v.*).

occipital condyles: the pair of smooth, rounded processes of the occipital bone at the posterior end of the skull on either side of the occipital foramen which acts as a hinge between the head and the neck.

occiput: the posterior part of the skull, above the foramen magnum (*q.v.*).

occlusal: pertaining to the biting surface of a tooth.

oestrous cycle: the reproductive cycle in most non-pregnant female mammals, comprising recurring physiological changes induced by reproductive hormones.

oestrus (*adj.* oestrous): The period when a female is sexually receptive.

Oligocene: geological Epoch (within the Tertiary period), ca. 38–23 mya.

omnivore (*adj.* omnivorous): an animal that eats a wide range of foods.

orbit: bony cavity (eye-socket) in which the eye is situated.
oscillogram: a visual representation (graph) of sound waves in which amplitude (sound wave pressure) is plotted on the vertical axis, and time on the horizontal axis (*cf*. sonogram *q.v.*).
osteophagia: the chewing on bones engaged in by herbivorous animals suffering from a deficiency of phosphorus and calcium in their diet.
outer: furthest from the mid-line of the body.
ovulation: the release of female gametes (ova, eggs) from the ovary.
ovum: the female gamete or egg cell.
owl pellets: waste material, normally in an egg-like shape, regurgitated by owls, which contains undigested fragments of bone, hair, feathers and scales etc. from the prey; for mammalogists, useful in determining indirectly the species of small mammals in a habitat.
P or **p**: (*abbrev.*) premolar tooth or teeth; upper case denotes adult dentition, lower case denotes deciduous dentition (milk teeth). *See also* premolar.
pachyosteosclerosis: a thickening of the bones of the ribs and vertebrae, resulting in unusually solid bone structure with little to no marrow.
Paenungulata (*adj*. **paenungulate**): a taxon (belonging to the Supercohort Afrotheria) that includes the elephants (Proboscidea), manatees and dugongs (Sirenia), hyraxes (Hyracoidea) and two extinct orders.
Palaearctic: a zoogeographic region comprising Europe, North Africa and Asia (except for southern portion of the Arabian Peninsula and tropical and sub-tropical parts of the Oriental Region.
Palaeocene: geological Epoch (within the Tertiary Period); 65–55 mya.
palatal foramen (*pl*. **palatal foramina**): a foramen (*q.v.*) in the palatine bone.
palate: the roof of the mouth. The anterior part (hard palate) comprises the bony palate (formed by the premaxillae, maxillae and palatine bones), and a covering mucous membrane. The posterior part (soft palate) is composed only of muscular tissue covered by a mucous membrane.
Paleogene: geological period and system; 65.5 ± 0.3 to 23.03 ± 0.05 mya, comprising the first part of the Cenozoic Era.
papilla (*pl*. **papillae**, *adj*. **papillate**): a small projection or protuberance.
paraoccipital process: narrow strut-like bone which projects ventrally from the paraoccipital bone on the posterior part of the skull.
parapatry (*adj*. **parapatric**): the situation where two or more species have geographic ranges that are contiguous (*q.v.*) but do not overlap. This abutting may be along a line of habitat discontinuity, ecotone, or altitudinal/climatic contour, but may also arise from competitive exclusion of one (or both) by the other.
paraphyletic: describes a taxon containing units that have evolved from a single ancestral taxon but which do not contain all of the descendents of the most recent common ancestor.
paratype: a specimen collected at the same time and place as the holotype (*q.v.*) and designated by the original authority as such. There may be one or more paratypes.
parietal: one of the pair of bones forming the vault of the braincase, situated between the frontal and the occipital bones.
parous: having given birth.
parturition: the act of giving birth.
pectoral: pertaining to or situated on the chest.
pedal glands: glands which are situated in the feet opening between the hooves.
pelage: the hairy, woolly or furry covering of the body in mammals. The pelage variously consists of hairs, guard hairs and underfur according to the species.
petrophily (*adj*. **petrophilic**): occupying habitats dominated by large boulders.
phalanx (*pl*. **phalanges**): one of the bones in a finger.
pharynx (*adj*. **pharyngeal**): the passage leading from the oral and nasal cavities in the head to the oesophagus and larynx (*q.v.*).
phenotype: the visible characters of an individual resulting from the interaction between the genotype (*q.v.*) and the environment.
philopatry: the tendency to remain in or consistently return to the place where one was born.
philtrum: a medial vertical cleft or groove in the rhinarium (*q.v.*) extending from between the nostrils to the upperlip.
phylogenetics: pertaining to the line of descent of a taxon; a method of classification that attempts to show the evolutionary relatedness of organisms.
phylogeny (*adj*. **phylogenetic**): the evolutionary history and line of descent of a species or higher taxonomic unit.
piloerection: the erection of hairs which occurs as a means of conserving heat (*see* temperature regulation) and, in some mammals, as a threat display.
pinna (*pl*. **pinnae**): the external (outer) ear.
placental scars: scars on the inner surface of the uterus of a mammal; the site where a placenta was attached during pregnancy. The number of placental scars can give an indication of the number of litters that a female has had (when the average number of young/litter is known).
placentation: the formation, type and structure of the placenta; the arrangement of placentae.
plantar: of the sole of the foot.
plantigrade: describes locomotion during which the entire sole of each foot touches the ground.
Pleistocene: geological Epoch (within the Quaternary Period); ca. 1.7 mya to 10,000 year ago.
plesiomorphy: (*adj*. **plesiomorphic**): in cladistics (*q.v.*), describes a pre-existing character state across a wide taxonomic grouping. Sharing of that state (symplesiomorphy [*q.v.*]) amongst a subset of taxa within the wider grouping is not indicative of monophyly (*q.v.*). *c.f.* apomorphy (*q.v.*).
pleural cavity: the body cavity occupied by the lungs.
Pliocene: geological Epoch (within the Tertiary period), ca. 5–2 mya.
poikilothermic: the condition when the body temperature varies with the surrounding ambient temperature.
pollex: digit 1 ('thumb') of forefoot ('hand').
polygyny (*adj*. **polygynous**): a mating system in which one male mates with several females. *See* resource-defence polygyny; female-defence polygyny. *cf*. monogamy (*q.v.*).
polymorphism (*adj*. **polymorphic**): the existence, within a species or population, of individuals having different forms (colour, size, shape etc.).

polyoestrous: Generally, having two or more oestrous cycles (*q.v.*) during a single reproductive season. In some taxa, having more than one litter/year. *cf.* monoestrous (*q.v.*).

polyovulation: production of many ova (and hence many corpora lutea) during a single oestrus; sometimes called 'superovulation'; recorded in some species of sengis.

polyphyletic: describes a taxon derived from two or more ancestral sources; not of a single, immediate line of descent (Mayr *et al.* 1953). *cf.* monophyletic (*q.v.*).

polyphyly: derivation of a taxon from two or more ancestral sources. *cf.* monophyly (*q.v.*).

polytypy (*adj.* polytypic): a taxon that has several taxa in the next lower taxonomic category, e.g. a polytypic family contains two or more genera, a polytypic genus has two or more species. *cf.* monotypy (*q.v.*).

postauricular: behind the external ear (pinna).

postorbital: behind (posterior to) the orbit.

postorbital process: the bony projection arising from upper rim of orbit and projecting outwards and downwards around the posterior of the orbit.

postpartum oestrus: an oestrus cycle immediately after (or very soon after) parturition and while lactating.

precocious (= precocial): describes young born in a well developed state enabling them to move around and forage soon after birth. *cf.* altricial (*q.v.*).

premaxilla (*pl.* premaxillae): one of a pair of bones at the anterior end of the skull which bears the incisor teeth.

premolar: tooth on the maxilla bone situated immediately anterior to the molar teeth; usually preceded in time by deciduous (milk) tooth; multi-cusped. Structure and function of premolar teeth vary according to species.

preorbital: anterior to the eye.

prepuce (*adj.* preputial): the skin surrounding and protecting the head of the penis; foreskin.

preputial gland: a gland situated adjacent to the penis or vaginal opening.

primiparous: a female that has given birth only once. *cf.* multiparous (*q.v.*).

promiscuity: a mating system in which each male mates with several females and each female mates with several males.

protocone: The main inner (lingual) cusp of an upper molar tooth.

protuberant: projecting forward (bulging) beyond the surrounding surface.

proximal: nearest to the body or to the mid-line of the body; nearest to the point of attachment. *cf.* distal (*q.v.*).

proximate cause: a cause that produces an effect directly without any intervening agency.

pterygoid process: one of a pair of narrow, ventrally projecting processes of the pterygoid bones situated immediately posterior to the bony palate and forming the walls of the mesopterygoid fossa (*q.v.*).

pterygoid: one of a pair of cranial bones forming part of the roof of the mouth.

pubic bone: anterior inferior part of the pelvis, articulating with its opposite number in the anterior midline at the pubic symphysis.

Quaternary Period: period within the Cenozoic Era; 2 mya to today, comprising two epochs: the Pleistocene and Holocene (Recent) (*q.v.*).

r-selection: selection for characteristics suitable in unstable fluctuating environments (rapid development of young, large litter size, relatively short time to maturity, relatively short life span) *cf.* K-selection (*q.v.*).

R.: (*abbrev.*) River.

ramus: one half (left or right) of the lower jaw or mandible.

relatively: in profiles of Afrosoricida and Macroscelidea, used in the context of describing the size of one character relative to the size of a different character in the same species. Usually expressed as a percentage. e.g. Tail 80–90 % of HB. *cf.* comparatively (*q.v.*).

relict forest: a forest that persists where local conditions are favourable after the disappearance of forest from the surrounding area as a result of climate change or human activity. Relict forests include those at the base of inselbergs that are watered by rainwater running off the inselberg, and forests growing in graveyards and sacred sites that are protected. Sometimes known as forest islands.

relict population: one that persists where local conditions are favourable after the extinction of the species from at least part of its former range.

Renosterveld: a dominant vegetation type in the Cape Floristic Region (*q.v.*), and characterized by the dominance of members of the Daisy Family (Asteraceae), specifically one species - Renosterbos *Elytropappus rhinocerotis*, from which the vegetation type gets its name.

reproductive capacity: number of young produced by a female during the breeding season.

reproductive chronology: the timing and duration of events, such as spermatogenesis, copulation, ovulation, gestation, parturition, lactation and reproductive inactivity, throughout the year.

reproductive strategy: the strategy adopted to maximise reproductive success, determined primarily by litter-size and reproductive chronology (*q.v.*).

resource-defence polygyny: a mating system in which a male controls access to several females indirectly, by monopolizing critical resources (Emlen & Oring 1977).

retarded embryonic development: a means of lengthening the interval between copulation and parturition so that both events can occur in the most optimal seasons. The implanted embryo enters a period of retarded (slowed) growth which may last 4–8 months, after which development proceeds normally.

reticulation (*adj.* reticulated): having a net-like pattern.

rhinarium: area of naked moist skin surrounding the nostrils.

ridge (= commissure): in teeth, a ridge connecting two cusps.

Rift Valley: deep valley extending from the Red Sea through Ethiopia and East Africa to Malawi; formed ca. 12 mya by subsidence of the valley floor and uplifting of the edges to form mountains and highlands. Four parts: (1) Ethiopian Rift Valley dividing the Ethiopian plateau into two parts; (2) Albertine Rift Valley (Western Rift Valley) in Uganda, E DR Congo and W Tanzania; (3) Gregorian Rift Valley (Eastern Rift Valley) in N Kenya, C Kenya and N Tanzania; (4) Malawian Rift Valley – the extension of the

Albertine Rift Valley in Malawi. The Rift Valleys are noted for their many deep and beautiful lakes.

riparian: growing on or living on the banks of streams or rivers.

riverine forest: forest growing along the banks of a river or stream where conditions are moister than in the surrounding area. Sometimes referred to as 'fringing forest'.

Robertsonian fusion: a chromosomal event involving the apparent fusion of non-homologous single armed (telocentric *q.v.*) chromosomes to form a bi-armed (metacentric *q.v.* or submetacentric *q.v.*) chromosome. Modern studies have revealed that all chromosomes have two arms, even if the smaller one is not detectable by light microscopy. Consequently, the term Robertsonian translocation is becoming more commonly used than Robertsonian fusion.

rostral: pertaining to the rostrum.

rostrum: that portion of the skull anterior to the front line of the orbits and supporting the upper part of the muzzle, comprised of the nasals, premaxillae and maxillae bones.

rupiculous: rock-living.

sacrum (*adj*. sacral): the fused vertebrae to which the pelvic girdle is attached.

sagittal crest: longitudinal crest of raised bone on the mid-dorsal line of the cranium.

scansorial: climbs or scrambles over logs and in low vegetation close to the ground. *cf.* terrestrial (*q.v.*); arboreal (*q.v.*).

sclerophyllous: describes vegetation having hard leaves that are resistant to drought.

scrotal: pertaining to, or within, the scrotum (*q.v.*).

scrotum: an external sac containing the testes and epididymides in male mammals.

sebkha: A geological feature, in North Africa, which is a smooth, flat, plain usually high in salt (also sabkha).

sectorial: adapted for cutting.

selenodont: molar teeth having longitudinal crescent-shaped ridges.

Senegambia: Senegal and Gambia.

sensu lato: (*Lat.*) in a broad sense.

sensu stricto: (*Lat.*) in a strict sense.

sensu: (*Lat.*): in the sense of.

sex ratio: the number of males to the number of females, usually expressed as a proportion to one male, e.g. 1 : 1 (equal numbers of males and females), 1 : 0.5 (= twice as many males as females), 1 : 2 (= twice as many females as males).

sexual dimorphism: observable (phenotypic) difference(s) (e.g. in colour, size or form) between the males and females of a species or higher taxon.

Sh. Ht.: (*abbrev.*) shoulder height.

sibling species: pairs or groups of true species that are reproductively isolated, but genetically closely related and so similar in appearance that they are difficult to separate solely on the basis of morphological characters.

side-stripe: longitudinal stripe(s) of contrasting colour on each flank, usually from shoulder to rump or upper part of hindlimbs. May be bordered by additional side-stripe above and below.

singleton: a neonate which is born singly as opposed to being one of a larger litter.

sister species: species that are thought to have arisen from a single dichotomous splitting event.

sonogram: a visual representation (graph) of sound in which frequency is plotted on the vertical axis, and time on the horizontal axis; can be analysed manually or with computer programs (*cf.* oscillogram *q.v.*).

South-Central Africa: Angola, SE DR Congo, Zambia and Malawi (but only west of L. Malawi and the Shire R. Valley).

southern Africa: south of the Cunene and Zambezi Rivers, i.e. Namibia, Botswana, Zimbabwe, southern Mozambique and South Africa (after Smithers 1983).

species: populations(s) of closely related and similar organisms, which are capable of interbreeding freely with one another, and cannot or typically do not interbreed with members of other species.

sperm: any male gamete; the male cell which fuses with a female gamete (ovum, egg cell) to produce a fertilized egg or zygote from which an embryo will develop.

spermatogenesis: the formation of sperm in the testes.

s.u.: (*abbrev.*) (*Lat. sans unguis* = without claw) sometimes added as a suffix to the hindfoot measurement to emphasize that HF has been measured without the claw. However, since this is the standard method of measurement, most authors write 'HF', not 'HF s.u.'. *cf.* c.u. (*q.v.*).

sub, sub-: prefix meaning under, signifying beneath or ventral to (as in anatomical features) or south of (as in sub-Saharan); less than (as in subsonic); not quite, nearly, almost, somewhat (as in subequal, subtriangular). In taxonomy, indicates a group just below the status of the taxa immediately following it (e.g. a genus may contain two or more subgenera).

subauricular: below the ear.

subcaudal: below the tail.

subdermal: just below the skin.

submetacentric: describes a chromosome with the centromere (*q.v.*) somewhat nearer one end than the other, so there are two arms of somewhat unequal length (ratio 1 : 1.2–1.9). *cf.* metacentric (*q.v.*); subtelocentric (*q.v.*).

subspecies: a geographically localized and isolated subdivision of a species, which differs genetically, morphologically and taxonomically from other subdivisions of the species.

subtelocentric: describes a chromosome with the centromere (*q.v.*) much nearer one end than the other, so there are two arms of very unequal length (ratio 1 : >2).

subterminal: just below the end or tip.

subterranean: living permanently below the ground; *cf.* fossorial (*q.v.*).

suckling: the act of a mother giving milk directly from her breast (mammary glands) to her young. Mothers suckle; their young suck.

sulcus (*pl.* sulci): a groove, fissure or furrow.

superovulation: *see* **polyovulation** (q.v.).

supinate: to turn or rotate the hand or forearm, or the hindlimb and foot.

supracaudal: above the tail.

supraoccipital crest: ridge of bone, oriented transversely across the back of the skull, at the junction of the parietal and/or supraoccipital bones and the occipital bone. Sometimes referred to as the lambdoid crest.

supraorbital ridge: ridge of bone along upper rim of orbit (eye-socket).

supraorbital: above (dorsal to) the orbit.
supraordinal: describes a taxon above the level of the order.
supratragus: in some species of sengis, a fleshy lobe on inner lower surface of outer ear situated above the tragus.
sympatry (*adj.* **sympatric**): the situation where populations of two or more different species have overlapping geographic ranges; refers also to populations of two or more species whose geographic ranges are partly or wholly overlapping. They may or may not interact. *cf.* allopatry (*q.v.*); syntopy (*q.v.*).
symplesiomorphy: a primitive or ancestral character shared by two or more groups, which is inherited from ancestors older than the last common ancestor.
synapomorphy (*adj.* **synapomorphic**): situation in which a homologous character is present in two or more taxa and is thought to have originated in their most recent common ancestor. *See also* apomorphy.
syndactyly: of digits; whole of part fusion of two or more digits (e.g. Digits 2 and 3 of the hindfoot in otter-shrews).
synonym: one or more different names for the same taxonomic unit. A synonym may be a 'senior synonym' (the oldest name), or a 'junior synonym' (a more recent name), which is no longer considered as valid. May be used to refer to all names that have been associated, at some time in the past, with the taxonomic unit as currently understood.
syntopy (*adj.* **syntopic**): describes the situation where two or more species use the same or similar habitats and activity times. They may or may not interact. *cf.* allopatry (*q.v.*); sympatry (*q.v.*).
syntype: any specimen, or one of a series of specimens, used to designate a species when a holotype (*q.v.*) and paratype(s) (*q.v.*) have either not been selected, or have been lost or destroyed.
systematics: the science of arranging organisms in a way that reflects their evolutionary relationships; such relationships may be expressed as a phylogeny (*q.v.*). Often defined (somewhat incorrectly) as a synonym of taxonomy (*q.v.*).
T: (*abbrev.*) length of tail, measured from anterior of the first caudal vertebra to the posterior end of the last caudal vertebra (excluding any tufts, bristles etc. at tip of tail).
T_a: (*abbrev.*) ambient temperature; the temperature in which an animal is living. *cf.* Tb (*q.v.*).
talonid: heel at the posterior end of a lower molar tooth.
tapetum lucidum: light-reflecting layer behind or in the retina of the eyes of some vertebrates which reflects light back through the retina thereby increasing the sensitivity of the eye to dim light.
taxon (*pl.* **taxa**): any defined unit (e.g. family, genus, species, subspecies) in the classification of organisms.
taxonomy: the science of biological nomenclature; the study of the rules, principles and practice of naming and classifying species and other taxa. Sometimes considered as an integral part (and near synonym) of systematics (*q.v*).
T_b: (*abbrev.*) body temperature; the temperature of the core (central) part of an animal. *cf.* Ta (*q.v.*).
telocentric: describes a chromosome that appears to have a terminal centromere (*q.v.*) and therefore only one arm. Modern studies have revealed that all chromosomes have two arms but the smaller arm of telocentric chromosomes is not visible under a light microscope.
temporalis: a broad radiating muscle arising from the coronoid process (*q.v.*) of the lower jaw and attaching to the upper part of the skull.
termitarium (*pl.* **termitaria**): a place where termites (Insecta: Isopoda) live. Often a large mound of modified hard soil. The shape and size of a termitarium is unique to each species of termite.
terrestrial: living on the ground. *cf.* arboreal (*q.v.*); scansorial (*q.v.*).
territory: an area defended by an individual against certain other members of the species, usually by overt aggression or advertisement; territory is marked by the urine, faeces or glandular secretions of the territory's owner. *cf.* home-range (*q.v.*).
Tertiary Period: geological period, 65–2 mya, comprising five epochs: Palaeocene, Eocene, Oligocene, Miocene and Pliocene (*q.v.*); followed by the Quaternary Period (*q.v.*).
testes: the male gonads, or testicles, in which spermatozoa are formed and in which the male hormone is produced.
Tethys Sea: the sea separating the two supercontinents, Gondwana (*q.v.*) and Laurasia (*q.v.*) during much of the Mesozoic Era before the opening of the Indian and Atlantic oceans during the Cretaceous Period (*q.v.*).
thermal conductance: a measure of the ability of substances (including pelage) to transfer heat.
thermolability (*adj*, **thermolabile**): the ability of a homeotherm (e.g. camel) to allow its body temperature to vary over a 24-hour period, without either hibernating or aestivating.
thermoneutral zone: the range of body temperatures within which an animal does not have to increase its metabolic rate to increase T_b (*q.v.*) (when T_a (*q.v.*) is low) and reduce T_b (when T_a is high).
thermoregulation: regulation of body temperature, either by metabolic or behavioural means (or both simultaneously) so that T_b (*q.v*) is kept more or less constant.
thoracic: pertaining to, or situated upon, the chest.
through-put time: time taken for food to pass through the digestive tract.
tibia (*pl.* **tibiae**): one of the two bones forming the lower leg (the shin bone); part of hindlimb between knee and ankle.
TL: (*abbrev.*) total length from tip of snout to posterior end of tail. Equivalent to the head and body length and tail length added together. *See also* HB (*q.v.*). and T (*q.v.*).
toothrow: Generally, the row of teeth from the most anterior incisor tooth to the most posterior molar. In golden moles, the row of teeth from the canine to the most posterior molar. Sometimes used in contexts of specific types of teeth, e.g. premolar toothrow, molar toothrow.
topotype: any specimen from the type locality (*q.v.*), i.e. the same locality as that from which the holotype (*q.v.*) was taken.
topotypical: pertaining to the type locality (e.g. a topotypical population is one found at the type locality).
torpor (*adj.* **torpid**): a state in which there is a (usually short-term) reduction of metabolic rate and a lowering of T_b (*q.v.*) when T_a (*q.v.*) declines; arousal from torpor occurs when T_a increases and without high energy costs to the individual. Torpor is associated with a state of inactivity and reduced responsiveness to stimuli. Torpor lasts for only short periods of time (hours or days). *cf.* hibernation.

tragus: a cartilaginous structure, usually small, projecting from the inner side of the external ear just anterior to the auditory meatus (*q.v.*).
transverse: in a direction across the body from side to side. *cf.* longitudinal (*q.v.*).
Triassic Period: period (within the Mesozoic Era); 248–208 mya. The first mammals appeared in this period.
triconid: describes a molariform tooth having three cusps.
tricuspid: having three points or cusps (particularly of teeth).
trifid: divided into three by two emarginations (*q.v.*).
tubercle: a small rounded protuberance.
tusks: long, continuously growing incisor or canine teeth that protrude (usually in pairs) beyond the mouth in some mammals including elephants (in which the tusks are incisors), and warthogs and other pigs (in which the tusks are canines); comprised of dentine (ivory). Some mammals, e.g. hyraxes, have 'tusk-like' incisors.
tympanic bulla (*pl.* tympanic bullae): one of a pair of usually rounded bony capsules, on underside of skull (one on each side), housing structures of the middle and inner ear in many mammals. Also called auditory bulla (*q.v.*).
type description: the original description of a species; the original description of the holotype (and paratype[s] if included).
type locality: the locality from which a holotype (*q.v.*), lectotype (*q.v.*) or neotype (*q.v.*) was collected. Also called topotypical locality.
type population: the population from which the holotype was selected.
type series: the holotype and all specimens collected at the same place and time and used, together with the holotype, to describe a new species.
type species: usually the species that was the first to be described under the name of a new genus. Not all genera had a designated type species when they were first created; in such cases, other rules determine which species will be the type species.
type specimen: *see* holotype.
umbraculum, a membrane that shades the pupil of the eye allowing basking hyraxes to stare into the sun without harm.
underfur: dense and often woolly layer of the pelage, situated close to the skin and below the soft hairs and guard hairs; usually short and present in those species which experience lower Ta.
unicuspid: having one cusp or point (particularly of teeth).
upper critical temperature: the highest ambient temperatures at which the animal must increase its metabolic rate to maintain a constant body temperature. If the ambient temperature increases above the upper critical temperature and the animal is unable to cool itself, it will enter hyperthermia and may eventually die. *cf.* lower critical temperature.
uvula: the conical projection from the posterior edge of the soft palate that plays a role in the articulation of sounds and the closing the nasopharynx during swallowing.
vagility: the ability to move about, disperse or migrate.
vagrant: an individual that has been found well outside the normal geographic range of its species, e.g. a bat or bird that has been wind-borne, or an animal that has been transported as a stowaway on a ship, to a distant locality.
vascularized: infiltrated with capillaries.
vasoconstriction: constriction of the capillaries of the blood system near the surface of the skin in order to reduce the rate of heat loss through the skin; a mechanism used by many mammals to conserve heat when T_a (*q.v.*) is low. *cf.* vasodilation (*q.v.*).
vasodilation: the dilation (or opening) of the capillaries of the blood system near the surface of the skin in order to increase the rate of heat loss through the skin; a mechanism used by many mammals to cool themselves when T_a (*q.v.*) is high. *cf.* vasoconstriction (*q.v.*).
veld: Afrikaans word, used mainly by southern African biologists, to refer to a wide variety of grassland vegetation types typically used for grazing. *See also* bushveld, highveld, lowveld.
vertebra (*pl.* vertebrae): any of the bones that make up the backbone.
vertebral formula: the number of vertebrae in each part of the spine, from anterior to posterior: the parts are cervical (C), thoracic (T), lumbar (L), sacral (S), caudal (Ca).
vestigial: small and imperfectly developed; a structure having a smaller and more simple form than the corresponding structure in an ancestral species.
vibrissa (*pl.* vibrissae): long stiff hairs on the face, especially around nostrils and lips; often associated with the perception of tactile sensation; 'whiskers'.
vlei: southern African term for a marsh or swamp, either permanent or seasonal.
wadi: a desert valley, usually dry at the surface except after heavy rainfall.
water turnover: the rate at which water (fluids) is utilized and replaced in the body per unit time (normally expressed as ml/kg body weight/day); the amount of water an animal processes through its body each day. Water turnover is related to water availability, the urine concentrating ability of the kidney, amount of protein in the diet and Ta (*q.v.*). Water turnover rates are characteristically low in arid-adapted mammals when compared with non arid-adapted mammals.
West Africa: ca. south of 18° N from Senegal to the Sanaga R. in Cameroon, and Bioko I. (Equatorial Guinea) (Rosevear 1965).
WT: (*abbrev.*) weight (mass) of an individual, usually expressed in grams (g) or kilograms (kg).
xiphisternum: The posterior segment, or extremity, of the sternum (sometimes called the xiphoid process).
zalambdodont: cheekteeth with three main cusps connected by crests (ectolophs) forming a V-shape; largest cusp is at the apex of the V (on the lingual or tongue side of the tooth); assumed to be derived from the primitive tribosphenic teeth found in some extinct early mammals. *cf.* dilambdodont (*q.v.*).
ZW: (*abbrev.*) *see* zygomatic width.
zygomatic arch: one of a pair of cheekbones, formed of the maxillary process anteriorly, jugal bone medially and squamosal bone posteriorly. Ranges from massive, broad, widely flared and bony, to frail, slender and cartilaginous. When present, provides protection to the eyes and orbits. Also called zygoma.
zygomatic width (ZW): greatest width between the outer aspect of one zygomatic arch to the equivalent position on the opposite zygomatic arch. *See also* GWS.

Bibliography

Chapter 1. *Mammals of Africa*: An Introduction and Guide

Gunther, P. 2003. *Mammifères du monde. Inventaire des noms scientifiques français et anglais*. Editions Cade, Paris, 377 pp.

Wilson, D. R. & Cole, F. R. 2000. *Common Names of Mammals of the World*. Simithsonian Institution Press, Washington, DC, 204 pp.

Wilson, D. E. & Reeder, D. M. (eds) 2005. *Mammal Species of the World: A Taxonomic and Geographic Reference* (3rd edn). Johns Hopkins Press, Baltimore, 2142 pp.

Chapter 2. Thinking Mammals: An Introduction to African Mammals in Science, Natural History and Culture

Darwin, C. 1859. *On the Origin of Species by Means of Natural Selection*. John Murray, London, 502 pp.

Eiseley, L. 1946. *The Immense Journey*. Random House Inc., New York, 210 pp.

Hamilton, W. D. 1996. *Narrow Roads of Gene Land*. W. H. Freeman, Oxford, 552 pp.

Namer, E. 1980. la vie et l'Oevre de J. C. Vanini. Paris

Orlando, L., Metcalf, J. L., Alberdi, M. T., Telles-Antunes, M., Bonjean, D., Otte, M., Martin, F., Eisenmann, V., Mashkour, M., Morello, F., Prado, J. L., Salas-Gismondi, R., Shockey, B. J., Wrinn, P. J., Vasil'ev, S. K., Ovodov, N. D., Cherry, M. I., Hopwood, B., Male, D., Austin, J. J., Hänni, C. & Cooper, A. 2009. Revising the recent evolutionary history of equids using ancient DNA. *Proceedings of the National Academy of Sciences of the United States of America* 106: 21754–21759.

Roosevelt, T. 1909. *African Game Trails*. Syndicate Publishing, New York.

Selous, F. C. 1881. *A Hunter's Wanderings in Africa*. Bentley & Son, London.

Selous, F. C. 1893. *Travel and Adventure in South-east Africa*. Rowland Ward, London.

Selous, F. C. 1896. *African Nature Notes and Reminiscences*. Rowland Ward, London.

Wilson, O. E. 1994. *Naturalist*. Island Press, Washington, DC.

Chapter 3. The Evolution of a Continent: Geography and Geology

Adams, W. M. A., Goudie, A. S. & Orme, A. R. 1996. *The Physical Geography of Africa*. Oxford University Press, Oxford, 456 pp.

Adamson, D. A., Gasse, F., Street, F. A. & Williams, M. A. J. 1980. Late Quaternary history of the Nile. *Nature* 287: 50–55.

Aitchison, J., Ali, J. R. & Davis, A. M. 2005. A revised model for India–Asia collision, based on new observations from Tibet. *Abstracts for Geological Society of America Annual Meeting*. Session 171, October 2005.

Archibald, J. D. 2003. Timing and biogeography of the eutherian radiation: fossils and molecules compared. *Molecular Phylogenetics and Evolution* 28: 350–359.

Asher, R. J., Meng, J., Wible, J. R., Mckenna, M. C., Rougier, G. W., Hashzeveg, D. & Novacek, M. J. 2005. Stem Lagomorpha and the antiquity of Glires. *Science* 307: 1091–1094.

Barker, P. F. & Burrell, J. 1977. The opening of the Drake passage. *Marine Geology* 25: 15–34.

Beadle, L. C. 1974. *The Inland Waters of Tropical Africa: An Introduction to Tropical Limnology*. Longmans, New York, 350 pp.

Benton, M. J. 2000. *Vertebrate Palaeontology* (2nd edn) Blackwell Science, Oxford, xii + 452 pp.

Bininda-Emonds, O. R. P. (ed.) 2004. *Phylogenetic Supertrees: Combining Information to Find the Tree of Life*. Springer-Verlag, New York, 564 pp.

Bininda-Emonds, O. R. P., Cardillo, M., Jones, K. E., MacPhee, R. D. E., Beck, R. M. D., Grenyer, R., Price, S. A., Vos, R. A., Gittleman, J. L. & Purvis, A. 2007. The delayed rise of present-day mammals. *Nature* 446: 507–512.

Bonis, L. de, Jaeger, J. J., Coiffait, B. & Coiffait, P.-E. 1988. Decouverte du plus ancien primate catarrhinien connu dans l'Eocene superieur d'Afrique du Nord. *Comptes Rendus de l'Academie des Sciences (Paris)*, Série IIa 306: 929–934.

Brunet, M. *et al*. 2002. A new hominid from the Upper Miocene of Chad, Central Africa. *Nature* 418: 145–151.

Buch, M. W. 1997. Etosha pan: the third largest lake in the world? *Madoqua* 20: 49–64.

Burgess, N. D., Butynski, T. M., Cordeiro, N. J., Doggart, N. H., Fjeldsa, J., Howell, K. M., Kitahama, F. B., Loader, N. H., Lovett, J. C., Mbilinye, B., Menegon, M., Moyer, D. C., Nashanda, E., Perkin, A., Rovero, F., Stanley, W. T. & Stuart, S. N. 2007. The biological importance of the Eastern Arc Mountains of Tanzania and Kenya. *Biological Conservation* 134: 209–231.

Burke, K. 1996. The African Plate. *South African Journal of Geology* 99: 341–409.

Burke, K. & Wells, G. L. 1989. Trans-African drainage system of the Sahara: was it the Nile? *Geology* 17: 743–747.

Butynski, T. M. & Ehardt, C. L. 2003. Notes on ten restricted-range birds in the Udzungwa Mts, Tanzania. *Scopus* 23: 13–28.

Butzer, K. W. & Cooke, H. B. S. 1982. *The Palaeoecology of the African Continent* (*Cambridge History of Africa*, Vol. I). Cambridge University Press, Cambridge.

Butzer, K. W. & Hansen, C. L. 1968. *Desert and River in Nubia: Geomorphology and Prehistoric Environments at the Aswan Reservoir*. University of Wisconsin Press, Madison, 560 pp.

Cane, M. A. & Molnar, P. 2001. Closing of the Indonesian seaway as a precursor to East African aridification around 3–4 Ma ago. *Nature* 411: 157–159.

Clauzon, G., Suc, J., Gautier, F., Berger, A. & Loutre, M. 1996. Alternate interpretation of the Messinian Salinity Crisis: controversy resolved? *Geology* 24: 363–366.

Clayton, D. 1976. The chorology of African mountain grasses. *Kew Bulletin* 31: 273–288.

Coe, M. J. & Skinner, J. D. 1993. Connections, disjunctions and endemism in eastern and southern African mammal faunas. *Transactions of the Royal Society of South Africa* 48: 233–255.

Coiffait, P. E., Coiffait, B., Jaeger, J.-J. & Mahboubbi, M. 1984. Un nouveau gisement à mammifères fossiles d'age Eocene superieur sur le versant sud des Nementcha (Algerie orientale). *Comptes Rendus des Seances de l'Academie des Sciences (Paris)*, Série IIa 299: 893–898.

Commission for the Geological Map of the World. 1985–1990. *International geological map of Africa, 1 : 5,000,000*. UNESCO, Paris.

Cooke, H. B. S. 1968. The evolution of mammals on southern continents, II. The fossil mammals of Africa. *Quarterly Review of Biology* 43: 234–264.

Cooke, H. B. S. 1972. The fossil mammal fauna of Africa. In: *Evolution, Mammals and Southern Continents (eds A. Keast, F. C. Erk & B. Glass)*. State University of New York Press, Albany, pp. 89–139.

Cooke, H. B. S. 1978. Suid evolution and correlation of African hominid localities: an alternative taxonomy. *Science* 201: 460–463.

Cox, C. B. & Moore, P. D. 2000. *Biogeography: An Ecological and Evolutionary Approach. 6th edn.* Blackwell Scientific, Oxford, 298 pp.

Cox, K. G. 1989. The role of mantle plumes in the development of continental drainage patterns. *Nature* 242: 873–877.

Crochet, J. Y., Thomas, H., Sen, S., Roger, J., Gheerbrant, E. & Al-Sulaimani, Z. 1992. Decouverte d'un Peradectide (Marsupialia) dans l'Oligocene inferieur du Sultanat d'Oman; nouvelles données sur la palaeobiogeographie des Marsupiaux de la plaque arabo-africaine. *Comptes Rendus de l'Academie des Sciences*, Série IIa 314: 539–545.

Darwin, Charles. 1859. *On the Origin of Species by Means of Natural Selection*. John Murray, London, 502 pp.

Denton, G. H. 1999. Cenozoic climate change. In: *African Biogeography, Climate Change and Human Evolution (T. G.* Bromage & F. Schrenk). Oxford University Press, New York, pp 94–114.

Dercourt, J., Gaetani, M., Vrielynck, B., Barrier, E., Biju-Duval, B., Brunel, M. F., Cadet, J. P., Crasquim, B. & Sandulescu, M. (eds) 2000. *Peri-Tethys Palaeogeographical Atlas.* Commission for the Geological Map of the World, Paris.

DeWit, M., Jeffrey, M., Bergh, H. & Nicolaysen, L. 1999. Gondwana reconstruction and dispersion. *Search and Discovery Article 30001*. American Association of Petroleum Geologists

De Wit, M. J. C., Marshall, T. R. & Partridge, T. C. 2000. Fluvial deposits and drainage evolution. In: *The Cenozoic of Southern Africa* (eds T. C. Partridge & R. R. Maud). Oxford Monographs on Geology and Geophysics 40 : 55–72.

Duggen, S., Hoernle, K., Van den Bogaard, P., Ruepke, L. & Morgan, J. P. 2003. Deep roots of the Messinian salinity crisis. *Nature* 422: 602–606.

Ebinger, C. J. & Sleep, H. N. 1998. Cenozoic magmatism throughout East Africa resulting from impact of one large plume. *Nature* 395: 788–791.

Eizirik, E., Murphy, W. J. & O'Brien, S. J. 2001. Molecular dating and biogeography of the early placental mammal radiation. *Journal of Heredity* 92: 212–219.

Fage, J. D. 1963. *An Atlas of African History.* Edward Arnold, London, 64 pp.

Falconer, J. D. 1911. *The Geology and Geography of North Nigeria*. Macmillan, London.

Garrels, R. M. & MacKenzie, F. T. 1971. *Evolution of Sedimentary Rocks.* W. W. Norton, New York, xvi + 397 pp.

Gheerbrant, E. & Rage, J.-C. 2006. Palaeobiogeography of Africa: How distinct from Gondwana and Laurasia? *Palaeogeography, Palaeoclimatology, Palaeoecology* 241: 224–246.

Gilchrist, A. R. & Summerfield, M. A. 1990. Differential denudation and flexural isostasy in formation of rifted margin upwarps. *Nature* 346: 739–742.

Goudie, A. S. 2002. *Great Warm Deserts of the World: Landscapes and Evolution*. Oxford University Press, Oxford, 476 pp.

Goudie, A. S. 2005. The drainage of Africa since the Cretaceous. *Geomorphology* 67: 437–456.

Gradstein, F. M., Ogg, J. G. & Smith, A. G. 2004. *A Geologic Time Scale.* Cambridge University Press, Cambridge, xix + 589 pp.

Griffin, D. L. 2002. Aridity and humidity: two aspects of the late Miocene climate of North Africa and the Mediterranean. *Palaeogeography, Palaeoclimatology, Palaeoecology* 182: 65–91.

Grove, A. T. 1983. Evolution of physical geography of East African Rift Valley region. In: *Evolution, Time & Space: The Emergence of the Biosphere* (eds R. W. Sims, J. H. Price & P. E. S. Whalley). Academic Press, London.

Grubb, P. 1999. Evolutionary processess implicit in distribution patterns of modern African mammals. In: *African Biogeography, Climatic Change and Human Evolution* (eds T. G. Bromage & F. Schrenk). Oxford University Press, New York, pp. 150–164.

Grubb, P., Sandrock, O., Kullmer, O., Kaiser, T. M. & Schrenk, F. 1999. Relationships between eastern and southern African mammal faunas. In: *African Biogeography, Climatic Change and Human Evolution* (eds. T. G. Bromage & F. Schrenk). Oxford University Press, Oxford, pp. 150–164.

Hamilton, A. C. 1982. *Environmental History of East Africa*: a study of the Quaternary. Academic Press, London, 328 pp.

Harris, J. M. 1991. Family Giraffidae. In: *Koobi Fora Research Project*, Vol. 3 (ed. J. M. Harris). Clarendon Press, Oxford, pp. 93–138.

Harvey, T. J. 1976. The palaeolimnology of Lake Mobutu Sese Seko, Uganda–Zaire: the last 28,000 years. Doctoral Dissertation, Duke University, USA.

Hedges, S. B. 2001. Afrotheria: plate tectonics meets genomics. *Proceedings of the National Academy of Sciences of the United States of America* 98: 1–2.

Hempton, M. R. 1987. Constraints on Arabian Plate motion and extensional history of the Red Sea. *Tectonics* 6: 687–705.

Holmes, A. 1965. *Principles of Physical Geology* (2nd edn). Ronald Press, New York, xv +1288 pp.

Hopwood, A. T. & Hollyfield, J. P. 1954. *An Annotated Bibliography of the Fossil Mammals of Africa (1742–1950).* Fossil Mammals of Africa No. 8, British Museum (Natural History), London, 194 pp.

Hu, Y., Meng, J., Wang, Y. & Li, C. 2005. Large Mesozoic mammals fed on young dinosaurs. *Nature* 433: 149–152.

Jacobs, L. L., Congleton, J. D., Brunet, M., Dejax, J., Flynn, L. J., Hell, J. V. & Mouchelin, G. 1988. Mammal teeth from the Cretaceous of Africa. *Nature* 336: 158–160.

Jaeger, J.-J. 2003. Mammalian evolution: isolationist tendencies. *Nature* 426: 509–511.

Jaeger, J.-J. & Marivaux, L. 2005. Shaking the earliest branches of anthropoid primate evolution. *Science* 310: 244–245.

Janis, C. M. 1993. Tertiary mammal evolution in the context of changing climates, vegetation, and tectonic events. *Annual Review of Ecology and Systematics* 24: 467–500.

Ji, Q., Luo, Z.-X., Yuan, C.-X., Wible, J. R., Zhang, J.-P. & Georgi, J. A. 2002. The earliest known eutherian mammal. *Nature* 416: 816–822.

Ji, Q., Luo, Z.-X., Yuan, C.-X. & Tabrum, A. E. 2006. A swimming Mammaliaform from the Middle Jurassic and ecomorphological diversification of early mammals. *Science* 311: 1123–1127.

Johnson, T. C., Kelts, K. & Odada, E. 2000. The Holocene history of Lake Victoria. *Ambio* 29: 2–11.

Johnson, W. E., Eizirik, E., Pecon-Slatery, J., Murphy, W. J., Antunes, A., Teeling, E. & O'Brien, S. J. 2006. Felid phylogeny. *Science* 311: 73.

Kappelman, J., Tasmussen, D. T., Sanders, W. J., Feseha, M., Bown, T., Copeland, P., Crabaugh, J., Fleagle, J., Glantz, M., Gordon, A., Jacobs, B., Maga, M., Muldoon, K., Pan, A., Pyne, L., Richmond, B., Ryan, T., Seiffert, E. R., Sen, S., Todd, L., Weimann, M. C. & Winkler, A. 2003. Oligocene mammals from Ethiopia and faunal exchange between Afro-Arabia and Eurasia. *Nature* 426: 549–552.

Kaser, G. & Osmaston, H. 2002. *Tropical Glaciers. International Hydrology Series.* Cambridge University Press, xx +207 pp.

Kemp, T. S. 1982. *Mammal-like Reptiles and the Origin of Mammals.* Academic Press, London, xiv + 363 pp.

Kendall, R. L. 1969. An ecological history of the Lake Victoria basin. *Ecological Monographs* 39: 121–176.

Kielan-Jaworowska, Z., Cifelli, R. L. & Luo, Z.-X. 2004. *Mammals from the Age of Dinosaurs.* Columbia University Press, New York, 648 pp.

Kilham, P. 1972. Biogeochemistry of African lakes and rivers. PhD dissertation, Duke University, USA, ix + 199 pp.

Kilham, P. 1990. Mechanisms controlling the chemical composition of lakes and rivers: data from Africa. *Limnology and Oceanography* 35: 80–83.

King, L. C. 1962. *The Morphology of the Earth.* Oliver & Boyd, New York & London. Hafner, New York, xii + 699 pp.

King, S. D. & Ritsema, J. 2000. African hotspot volcanism: small-scale convection in the upper mantle beneath cratons. *Science* 290: 1137–1140.

Kingdon, J. 1971 *East African Mammals: An Atlas of Evolution in Africa*. Vol. 1: *Primates*. Academic Press, London.

Kingdon, J. 1990. *Island Africa.* William Collins, London, 287 pp.

Kingdon, J. 1997. *The Kingdon Field Guide to African Mammals.* Academic Press, London, 464 pp.

Kingdon, J. 2003. *Lowly Origin.* Princeton University Press, Princeton, 396 pp.

Kowalski, K. & Rzebik-Kowalska, B. R. 1991. *Mammals of Algeria.* Ossolineum, Warsaw, 370 pp.

Krause, D.W., Prasad, G.V. R., Koenigswald, W. von, Sahni, A. & Grine, F. E. 1997. Cosmopolitanism among Late Cretaceous mammals. *Nature* 390: 504–507.

Krause, D. W., Gottfried, M. D. & O'Connor, P. M. 2003. A cretaceous mammal from Tanzania. *Acta Palaeontologica Polonica* 48: 321–330.

Krijgsman, W., Hilgen, F. J., Raffi, I., Sierro, F. J. & Wilson, D. S. 1999. Chronology, causes and progression in the Messinian salinity crisis. *Nature* 400: 652–655.

Kumar, S. & Hedges, S. B. 1998. Plate tectonics meets genomics. *Nature* 98: 188–193.

Livingstone, D. A. 1963. *Chemical composition of rivers and lakes: data of geochemistry* (6th edn). US Geological Survey Professional Paper 440–G, 64 pp.

Luo, Z.-X. & Wible, J. R. 2005. A Late Jurassic digging mammal and early mammalian diversification. *Science* 308: 103–107.

Luo, Z.-X., Ji, Q., Wible, J. R. & Yuan, C.-X. 2003. An early Cretaceous tribosphenic mammal and metatherian evolution. *Science* 302: 1934–1940.

Madsen, O., Scally, M., Douady, C. J., Kao, D. J., DeBry, R. W., Adkins, R., Amrine, H. M., Stanhope, M. J., de Jong, W. W. & Springer, M. S. 2001. Parallel adaptive radiations in two major clades of placental mammals. *Nature* 409: 610–614.

Maglio, V. J. & Cooke, H. B. S. 1978. *Evolution of African Mammals*. Harvard University Press, Cambridge, MA, 656 pp.

Marivaux, L., Welcomme J. L., Antoine, P. O., Metais., G., Baloch, I. M., Bennami, M., Chaimanee, Y., Ducrocq, S. & Jaeger, J. J. 2001. A fossil lemur from the Oligocene of Pakistan. *Science* 294: 587–591.

McBrearty, S. & Jablonski, N. G. 2005. First fossil chimpanzee. *Nature* 437: 105–108.

McCarthy, T. S. 1983. Evidence for the former existence of a major, south-flowing river in Griqualand west. *Transactions of the Geological Society of South Africa* 86: 37–49.

McCauley, J. F., Schaber, G. G., Breed, C. S., Grolier, M. J., Haynes, C. V., Issawi, B., Elachi, C. & Blom, R. 1982. Subsurface valleys and geoarchaeology of the eastern Sahara revealed by shuttle radar. *Science* 218: 1004–1020.

Mercer, J. M. & Roth, V. L. 2003. The effects of Cenozoic global change on squirrel phylogeny. *Science* 299: 1568–1570.

Moore, A. & Blenkinsop, T. 2002. The role of mantle plumes in the development of continental-scale drainage patterns: the southern African example: revisited continental scale drainage patterns. *South African Journal of Geology* 105: 353–360.

Moore, A. & Larkin, P. A. 2001. Drainage evolution in south-central Africa since the break-up of Gondwana. *South African Journal of Geology* 104: 47–68.

Mote, P. W. & Kaser, G. 2007. The shrinking glaciers of Kilimanjaro: Can Global Warming be blamed? *American Scientist* 95: 318–325.

Murphy, W. J., Eizirik, E., Johnson, W. E., Zhang, Y. P., Ryder, O. A. & O'Brien, S. J. 2001a. Molecular phylogenetics and the origins of placental mammals. *Nature* 409: 614–618.

Murphy, W. J., Eizirik, E., O'Brien, S. J., Madsen, O., Scally, M., Douady, C. J., Teeling, E., Ryder, O. A., Stanhope, M. J., de Jong, W. W. & Springer, M. S. 2001b. Resolution of the early placental mammal radiation using Bayesian phylogenetics. *Science* 294: 2348–2351.

Murphy, W. J., Eizirik, E., Johnson, W. E. & Zhang, Y. P. 2007. Molecular phylogenetics and the origins of placental mammals. *Nature* 409: 614–618.

Osmaston, H. A. 1967. The sequence of glaciations in the Ruwenzori and their correlation with glaciations of other mountains in East Africa and Ethiopia. *Palaeoecology of Africa* 2: 26–28.

Osmaston, H. A. 1989. Glaciers, glaciations and equilibrium line altitudes on the Ruwenzori. In: *Quaternary and Environmental Research on East African Mountains* (ed. W. C. Mahaney). Balkema, Rotterdam/Brookfield, pp. 31–104.

Partridge, T. C. & Maud, R. R. 1987. Geomorphic evolution of Southern Africa since the Mesozoic. *South African Journal of Geology* 90: 179–208.

Penny, D. & Phillips, M. J. 2007. Mass survivals. *Nature* 446: 501–502.

Pik, R., Marty, B., Cariguan, J. & Lave, J. 2003. Stability of the Upper Nile drainage network (Ethiopia) deduced from (U-Th) therma chronometry: implications for uplifts and erosion of the Afar plume dome. *Earth and Planetary Science Letters* 215: 73–88.

Pritchard, J. M. 1979. *Landform and Landscapes in Africa*. Arnold, London, 160 pp.

Rasmussen, D. T., Tilden, C. D. & Simons, E. L. 1989. New specimens of the giant creodont *Megistortherium* (Hyaenodontidae) from Moghara, Egypt. *Journal of Mammalogy* 70: 442–447.

Rich, T. H., Vickers-Rich, P., Constantine, A., Flannery, T. F., Kool, L. & Van Klaveren, N. 1997. A Tribosphenic mammal fom the Mesozoic of Australia. *Science* 275: 1438–1442.

Richardson, J. L., Harvey, T. J. & Holdship, S. A. 1978. Diatoms in the history of shallow East African lakes. *Potskie Archiwum Hydrobiologii* 25: 341–353.

Robinson, T. J., Yang, F. & Harrison, W. R. 2002. Chromosome painting refines the history of genome evolution in hares and rabbits. *Cytogenetic and Genome Research* 96: 223–227.

Rogers, N., Macdonald, R., Fitton, J. G., George, R., Smith, M. & Barreiro, B. 2000. Two mantle plumes beneath the East African Rift System. Sr. Nd. & Pb isotope evidence from Kenya Rift basalts. *Earth and PlanetaryScience Letters* 176: 387–400.

Rook, L., Gallai, G. & Torre, D. 2006. Lands and endemic mammals in the late Miocene of Italy: constraints for palaeogeographic outlines of Tyrrhennian area. *Palaeogeography, Palaeoclimatology, Palaeoecology* 238 (1–4): 263–269.

Rose, K. D. 2006. *The beginnings of the Age of Mammals*. Johns Hopkins University Press, Baltimore, 448 pp.

Rosendahl, B. R. (ed.) 1988. *Seismic Atlas of Lake Tanganyika, East Africa*. Project Probe Geophysical Atlas Series, Vol. 1. Duke University, North Carolina, ii + 82 pp.

Rovero, F. & Rathbun, G. B. 2005. A potentially new Giant Sengi (Elephant Shrew) from the Udzungwa Mts, Tanzania. *Journal of East African Natural History* 95: 111–115.

Rovero, F., Jones, T. & Sanderson, J. 2005. Notes on Abbot's duiker *Cephalophus spadix* and other forest antelopes of Mwanihana forest, Udzungwa Mts, Tanzania. *Tropical Zoology* 18: 13–23.

Said, R. 1981. The Geological Evolution of the River Nile. Springer-Verlag, New York, 151 pp.

Salama, R. B. 1987. The evolution of the Nile. The buried saline rift lakes in Sudan al Arab Rift, the Sudd buried saline lake. *Journal of African Earth Sciences* 6: 899–913.

Schlueter, T. 2006. *Geological Atlas of Africa*, xi + 272 pp.

Scholz, C. A., Johnson, T. C., Cohen, A. S., King, J. W., Peck, J. A., Overpeck, J. T., Talbot, M. R., Brown, E. T., Kalindekafe, L., Amoako, P.Y. O., Lyons, R. P., Shanahan, T. M., Castaneda, I. S., Heil., C. W., Forman, S. L., McHargue, L. R., Beuning, K. R., Gomez, J. & Pierson, J. 2007. East African megadroughts between 135 and 75 thousand years ago and bearing on early-modern human origins. *Proceedings of the National Academy of Sciences of the United States of America* 104: 16416–16421.

Schuster, M., Duringer, P., Ghienne, J.-F., Vigneaud, P., Mackaye, H. T., Likius, A. & Brunet, M. 2006. The age of the Sahara Desert. *Science* 311: 821.

Seiffert, E. R., Simons, E. L. & Simons, C. V. M. 2004. Phylogenetic, biogeographic and adaptive implications of new fossil evidence bearing on crown anthropoid origins and early stem catarrhine evolution. In: *Anthropoid Origins: New Visions* (eds C. F. Ross & R. F. Kay). Academic Press, New York, pp. 147–171.

Seiffert, E. R., Simons, E. L., Clyde, W. C., Rossie, J. B., Attia, Y., Bown, T. M., Chatrath, P. & Mathison, M. E. 2005. Basal anthropoids from Egypt and the antiquity of Africa's higher primate radiation. *Science* 310: 300–304.

Siesser, W. G. 1980. Late Miocene origin of the Benguela upwelling system off northern Namibia. *Science* 208: 283–285.

Sige, B., Jaeger, J. J., Suore, J. & Vianey-Ligaud, M. 1990. *Altiaatlasius coulchii* new genus new species an omomyid primate from the late palaeocene of Morocco and the origins of the Euprimates. *Palaeontographia, Abteilung A, Palaeozoologie, Stratigraphie* 214: 31–56.

Smith, A. B. & Peterson, K. J. 2002. Dating the time of origin of major clades: molecular clocks and the fossil record. *Annual Review of Earth and Planetary Sciences* 30: 65–88.

Smith, A. G., Smith, D. G. & Funnell, B. M. 1994. *Atlas of Mesozoic and Cenozoic Coastlines*. Cambridge University Press, Cambridge, ix + 99 pp.

Springer, M. S., Stanhope, M. J., Madsen, O. & de Jong, W. W. 1997. Endemic African mammals shake the phylogenetic tree. *Nature* 388: 61–64.

Springer, M. S., Murphy, W. J., Eizirik, E. & O'Brien, S. J. 2003. Placental mammal diversification and the Cretaceous–Tertiary boundary. *Proceedings of the National Academy of Sciences of the United States of America* 100: 1056–1061.

Springer, M. S., Stanhope, M. J., Madsen, O. & de Jong, W. W. 2004. Molecules consolidate the placental mammal tree. *Trends in Ecology and Evolution* 19: 430–438.

Stager, J. C., Reinthal, P. H. & Livingstone, D. A. 1986. A 25,000-year history for Lake Victoria East Africa and some comments on its significance for the evolution of cichlid fishes. *Freshwater Biology* 16: 15–20.

Stampfli, G. M., Borel, G. D., Cavazza, W., Mosar, J. & Ziegler, P. A. 2001. Palaeotectonic and palaeogeographic evolution of the western Tethys & peri-Tethyan domain. *GCP Project* 369.

Stanhope, M. J., Waddell, V. G., Madsen, O., de Jong, W., Hedges, S. B., Cleven, G. C., Kao, D. & Springer, M. S. 1998. Molecular evidence for multiple origins of Insectivora and for a new order of endemic African insectivore mammals. *Proceedings of the National Academy of Sciences of the United States of America* 95: 9967–9972.

Stewart, C. B. & Disotell, T. R. 1998. Eurasian ape as great ape ancestor. *Current Biology* 8: R582–R588.

Suwa, G., Kono, R. T., Katoh, S., Asfaw, B. & Beyene, Y. 2007. A new species of great ape from the late Miocene epoch in Ethiopia. *Nature* 448: 921–924.

Tabuce, R. & Marivaux, L. 2004. Mammalian interchanges between Africa and Eurasia: an analysis of temporal constraints on plausible anthropoid dispersals during the Palaeogene. *Anthropological Science* 113: 27–32.

Tabuce, R., Adnet, S., Capetta, H., Noubhani, A. & Quillevere, F. 2005. Aznag (Ouarazate basin, Morocco), a new African Middle Eocene (Lutetian) vertebrate-bearing locality with selachians and mammals. *Bulletin de la Société Geologique de France* 176: 381–400.

Tabuce, R., Marivaux, L., Adaci, M., Bensalah, M., Hartenberger, J.-L., Mahboubi, M., Mebrouk, F., Tafforeaux, P. & Jaeger, J.-J. 2007. Early tertiary mammals from North Africa reinforce the molecular Afrotheria clade. *Proceedings of the Royal Society B* 274: 1159–1166.

Talbot, M. 1980. Environmental responses to climatic change in the West African Sahel over the past 20,000 years. In: *The Sahara and the Nile* (eds M. A. J. Williams and H. Faure). Balkema, Rotterdam, pp. 37–82.

Talbot, M. R., Williams, M. A. J. & Adamson, D. A. 2000. Strontium isotope evidence for late Pleistocene reestablishment of an integrated Nile drainage network. *Geology* 28: 343–346.

Thomas, D. S. G. & Shaw, P. 1991. *The Kalahari Environment*. Cambridge University Press, Cambridge, 298 pp.

Thomas, H. 1979. Les Bovidae miocene des rifts est-africains; implications biogeographiques. *Bulletin de la Société Geologique de France* XXI (3): 395–399.

Thomas, H., Bernor, R. & Jaeger, J.-J. 1982. Origines du Peuplement mammalien en Afrique du Nord durant le Miocene terminal. *Geobios* 15: 283–297.

Turner, A. & Anton, M. 2004. *Evolving Eden. An illustrated guide to the evolution of the African large-mammal fauna*. Columbia University Press, New York, 269 pp.

Uenzelman-Nebel, G. 1998. Neogene sedimentation history of the Congo fan. *Marine and Petroleum Geology* 15: 635–650.

UNESCO 1971. *Tectonics of Africa*. UNESCO, Paris.

Urvoy, Y. 1942. Les basins du Niger. *Memoires de l'Institut Fondamental d'Afrique Noire* 4: 1–139.

Werdelin, L. & Sanders, W. J. 2010. *Cenozoic Mammals of Africa*. University of California Press, Berkeley, 1008 pp.

White, T. D. & Harris, T. D. 1977. Suid evolution and correlation of African hominid localities. *Science* 198: 13–21.

Wible, J. R., Rougier, G. W., Novacek, M. J. & Asher, R. J. 2007. Cretaceous eutherians and Laurasian origin for placental mammals near the K/T boundary. *Nature* 447: 1003–1007.

Yoder, A. D., Burns, M. M., Zehr, S., Deleosse, T., Veron, G., Goodman, S. M. & Flynn, J. F. 2005. Single origin of Malagasy carnivora from an African ancestor. *Nature* 421: 734–737.

Zack, S. P., Penkrot, T. A., Bloch, J. I. & Rose, K. D. 2005. Affinities of 'hyopsodontids' to elephant shrews and a Holarctic origin of Afrotheria. *Nature* 434: 497–501.

Chapter 4. Africa's Environmental and Climatic Past

Abreu, V. S. & Anderson, J. B. 1998. Glacial eustacy during the Cenozoic: sequence stratigraphic implications. *American Association of Petroleum Geologists Bulletin* 82: 1385–1400.

Adegoke, O. S., Jan du Chene, R. E., Agumanu, A. K. & Ajayi, P. O. 1978. Palynology and age of the Kerri–Kerri Formation, Nigeria. *Revista Espanola de Micropalaeontologia* 10: 276–83.

Adkins, R. M., Gelke, E. L., Rowe, D. & Honeycutt, R. L. 2001. Molecular phylogeny and divergence time estimates for major rodent groups: evidence from multiple genes. *Molecular Biology and Evolution* 18: 777–791.

Alvarez, L. W., Alvarez, W., Asaro, F. & Michel, H. V. 1980. Extraterrestrial cause for Cretaceous–Tertiary extinction. *Science* 208: 1095–1108.

Andrews, P. & Van Couvering, J. A. H. 1975. Palaeoenvironments in the East African Miocene. In: *Approaches to Primate Paleobiology* (ed. F. S. Szalay). Contributions to Primate Biology 5. Karger, Basle, pp. 62–103.

Anhuf, D., Ledru, M. P., Behling, H., Da Cruz Jr, F. W., Cordeiro, R. C., Van der Hammen, T., Karmann, I., Marengo, J. A., De Oliveira, P. E., Pessenda, L., Siffedine, A., Albuquerque, A. L. & Da Silva Dias, P. L. 2006. Paleo-environmental change in Amazonian and African rainforest during the LGM. Palaeogeography, Palaeoclimatology, Palaeoecology 239(3–4): 510-527. 2006. Paleo-environmental change in Amazonian and African rainforests during the LGM. *Paleogeography, Paleoclimatology, Paleoecology* 239: 510–527.

Aubreville, A. 1949. *Climats, forêts et desertification de l'Afrique tropicale*. Societe D'editions Geographiques Maritimes et Coloniales, Paris, 351 pp.

Awad, M. Z. 1994. Stratigraphic, palynological and palaeoecological studies in the east-central Sudan (Khartoum and Kosti basins) late Jurassic–mid-Tertiary. *Berliner Geowissenschaftlische Abhandlungen* 161.

Awad, M. Z. & Brier, F. A. 1993. Oligo-Miocene to Quaternary palaeo-environment in Gezira area, Central Sudan. In: *Geoscientific Research in Northeast Africa* (eds U. Thorweihe and H. Schandelmeier). A. A. Balkema, Rotterdam, pp. 465–470.

Axelrod, D. I. & Raven, P. H. 1978. Late Cretaceous and Tertiary vegetation history of Africa. In: *Biogeography and Ecology of Southern Africa* (eds M. J. A. Werger & A. C. van Bruggen). W. Junk Publishers, The Hague, pp. 79–130.

Baker, B. H., Mohr, P. A. & Williams, L. A. J. 1972. *Geology of the Eastern Rift System of Africa*. Geological Society of America Special Publication 136.

Bancroft, H. 1935. Some fossil dicotyledonous woods from Mount Elgon, East Africa. *American Journal of Botany* 22: 164–182.

Barker, P. F. & Burrell, J. 1977. The opening of the Drake passage. *Marine Geology* 25: 15–34.

Barron, E. J. & Washington, W. M. 1982. Cretaceous climate: a comparison of atmospheric simulations with the geologic record. *Palaeoclimatology, Palaeogeography, Palaeoecology* 40: 103–133.

Barron, E. J. & Washington, W. M. 1985. Warm Cretaceous climates: high atmospheric CO_2 as a plausible mechanism. In: *The Carbon Cycle and Atmospheric CO_2: Natural Variations, Archaean to the Present* (eds E. T. Sundquist & W. S. Broecker). American Geophysical Union, Washington, DC, pp. 546–553.

Beard, K. C. 2004. *The Hunt for the Dawn Monkey: Unearthing the Origins of Monkeys, Apes and Humans*. University of California Press, Berkeley, 348 pp.

Bininda-Emonds, O. R. P., Cardillo, M., Jones, K. E., MacPhee, R. D. E., Beck, R. M. D., Grenyer, R., Price, S. A., Vos, R. A., Gittleman, J. L. & Purvis, A. 2007. The delayed rise of present-day mammals. *Nature* 446: 507–512.

Boltenhagen, E. 1976. Pollen et spores Senoniens du Gabon. *Cahiers du Micropalaeontologie* 3: 3–21.

Bonnefille, R. 1984. Cenozoic vegetation and environments of early hominids in east Africa. In: *The Evolution of the East Asian Environment* (ed. R. O. Whyte). University of Hong Kong, Centre of Asian Studies 2, pp. 579–612.

Bonnefille, R. 2006. Rain forest responses to past climatic changes in tropical Africa. In: *Tropical rain forest responses to climatic change* (eds M. B. Bush & J. R. Flenley). Praxis, Chichester, pp. 117–170.

Bonnefille, R. & Riollet, G. 1988. The Kashiru Pollen Sequence (Burundi). Palaeoclimatic implications for the last 40,000 yr B. P. in Tropical Africa. *Quaternary Research* 30: 19–35.

Boucot, A. J., Chen, X. U. & Scotese, C. R. with contributions by Morley, R. J. 2008. Preliminary compilation of Cambrian through Miocene climatically sensitive deposits. *SEPM Special Publication Series*.

Boureau, E., Cheboldaeff-Salard, M., Koeniguer, J.-C. & Louvet, P. 1983. Evolution des flores et de la vegetation Tertiaires en Afrique, au nord de l'Equateur. *Bothalia* 14: 355–367.

Brenac, P. 1988. Evolution de la vegetation et du climat dans l'Ouest Cameroun entre 25000 et 11000 ans BP. Actes Xeme Symposium Association Palynologiques Langues Francais. *Travaux de la Section des Sciences & Techniques de l'Institut Francais de Pondicherry* 25: 91–103.

Brenan, J. M. P. 1978. Some aspects of the phytogeography of tropical Africa. *Annals of the Missouri Botanical Garden* 65: 437–478.

Bush, M. B. & Silman, M. R. 2004. Observations on Late Pleistocene cooling and precipitation in the lowland Neotropics. *Journal of Quaternary Science* 19: 677–684.

Caratini, C., Tissot, C., Kar, R. K., Venkatachala, B. S. & Sarr, R. 1991. Palaeocene palynoflora Senegal from Walalane borehole, Senegal. *Palaeoecology of Africa* 22: 123–133.

Cerling, T. E., Hay, R. L. & O'Neil, J. R. 1977. Isotopic evidence for dramatic climatic changes in East Africa during the Pleistocene. *Nature* 267: 137–138.

Cerling, T. E., Harris, J. M., Ambrose, S. H., Leakey, M. G. & Solounias, N. 1997. Dietary and environmental reconstruction with stable isotope analyses of herbivore tooth enamel from the Miocene locality of Fort Ternan, Kenya. *Journal of Human Evolution* 33: 635–650.

Chesters, K. I. M. 1955. Some plant remains from the Upper Cretaceous and Tertiary of West Africa. *Annals and Magazine of Natural History* 12: 498–504.

Chesters, K. I. M. 1957. The Miocene flora of Rusinga Island, Lake Victoria, Kenya. *Palaeontographica* 101: 30–67.

Chiarugi, A. 1933. Legni fossili della Somalia Italiana. *Palaeontographica Italica* 32 (Suppl. 1): 97–167.

Coe, M. J. & Skinner, J. D. 1993. Connections, disjunctions and endemism in the eastern and southern African mammal faunas. *Transactions of the Royal Society of South Africa* 48: 233–256.

Coetzee, J. A. 1967. Pollen analytical studies in East and Southern Africa. *Palaeocology of Africa* 3: 1–146.

Coetzee, J. A. 1978. Climatic and biological changes in south-western Africa during the late Cainozoic. *Palaeoecology of Africa* 10: 13–29.

Coetzee, J. A. 1993. African flora since the Terminal Jurassic. In: *Biological Relationships between Africa and South America* (ed. P. Goldblatt). Yale University Press, New Haven, pp. 37–59.

Coetzee, J. A. & Muller, J. 1984. The phytogeographic significance of some extinct Gondwana pollen types from the Tertiary of the southwestern Cape (South Africa). *Annals of the Missouri Botanical Garden* 71: 1088–1099.

Colinvaux, P., De Oliviera, P. E., Moreno, J. E., Miller, M. C. & Bush, M. B. 1996. A long pollen record from lowland Amazonia: forest and cooling in glacial times. *Science* 274: 85–88.

Colinvaux, P., De Oliveira, P. E., Moreno, J. E. & Patino, J. 1999. *Amazon Pollen Manual and Atlas/Manual e atlas palinologico da Amazonia.* Harwood Academia Press, New York, 344 pp.

Colyn, M. 1991. *L'importance zoogeographique du basin du fleuve Zaire pour la speciation: le cas des primates simians*. Belgique Annales

Sciences Zoologiques, Musee Royal de l'Afrique Central Tervuren 264, 1–250..

Corner, E. J. H. 1949. The durian theory and the origin of the modern tree. *Annals of Botany* 13: 367–414.

Corner, E. J. H. 1964. *The Life of Plants*. Wiedenfeld & Nicolson, London, 325 pp.

Crane, P. R. & Lidgard, S. 1990. Angiosperm radiation and patterns of Cretaceous palynological diversity. In: *Major Evolutionary Radiations* (eds P. D. Taylor & G. P. Larwood). Systematics Association Special Volume 42, pp. 377–407.

Davis, C. C., Webb, C. O., Wurdack, K. J., Jaramillo, C. A. & Donoghue, M. J. 2005. Explosive radiation of Malpighiales supports a Mid-Cretaceous Origin of Modern Tropical Rain Forests. *American Naturalist* 165: E36–E65.

Denton, G. H. 1999. Cenozoic climate change. In: *African Biogeography, Climate Change and Human Evolution (T. G.* Bromage & F. Schrenk). Oxford University Press, New York, pp 94–114.

Doyle, J. A., Biens, P., Doerenkamp, A. & Jardine, S. 1977. Angiosperm pollen from the pre-Albian Cretaceous of equatorial Africa. *Bulletin du Centres de Recherches Exploration–Production Elf-Aquitaine* 1: 451–473.

Doyle, J. A., Jardine, S. & Doerenkamp, A. 1982. *Afropollis*, a new genus of early angiosperm pollen, with notes on the Cretaceous palynostratigraphy and palaeoenvironments of northern Gondwana. *Bulletin du Centres Recherches Exploration–Production Elf-Aquitaine* 6: 39–117.

Duperon-Laudoueneix, M. 1991. Importance of fossil woods (Conifers and Angiosperms) discovered in continental Mesozoic sediments of Northern Equatorial Africa. *Journal of African Earth Sciences* 12: 391–396.

Dupont, L. M. & Wienelt, M. 1996. Vegetation history of the savanna corridor between the Guinean and the Congolian rainforest during the last 150,000 years. *Vegetation History and Archaeobotany* 5: 273–292.

Dutrillaux, B., Couturier, J., Muleris, M., Lombard, M. & Chauvier, G. 1982. Chromosomal evolution of 42 spp or sspp of cercopithecoids (Primates, Catarrhini). *Annales de Génétique* 25: 96–109.

Fjeldsa, J. & Lovett, J. C. 1996. Geographical patterns of old and young species in Africa. *Biodiversity and Conservation* 6: 315–346.

Friis, E. M. 1883. Upper Cretaceous (Senonian) floral structures of juglandalean affinity containing normapolles pollen. *Review of Palaeobotany and Palynology* 39: 161–183.

Germeraad, J. H., Hopping, C. A. & Muller, J. 1968. Palynology of Tertiary sediments for tropical areas. *Review of Palaeobotany and Palynology* 6: 189–348.

Grubb, P., Sandrock, O., Kullmer, O., Kaiser, T. M. & Schrenk, F. 1999. Relationships between Eastern and Southern African mammalian faunas. In: *African Biogeography, Climate Change and Human Evolution* (eds T. G. Bromage & F. Schrenk). Oxford University Press, Oxford, pp 253–267.

Guiraud, R. & Maurin, J. C. 1991. Le rifting en Afrique au Cretace Inferieur synthese structurale, mise en evidence de deux etapes dans la genese des bassins, relations avec les ouvertures oceaniques peri-africaines. *Bulletin de la Société Geologique de France* 162: 811–823.

Hamilton, A. C. 1976. The significance of patterns of distribution shown by forest plants and animals in tropical Africa for the reconstructions of Upper Pleistocene palaeoenvironments: a review. *Palaeoecology of Africa and of the surrounding islands and Antarctica* 9: 63–97. Hamilton, A. C. 1982. *Environmental History of East Africa*: a study of the Quaternary. Academic Press, London, 328 pp.

Hardenbol, J., Thierry, J., Farley, M. B., Jacquin, T., Graciansky, P.-C. de & Vail, P. R. 1998. Mesozoic and Cenozoic sequence chronostratigraphic framework of European Basins. In: *Mesozoic and Cenozoic sequence stratigraphy of European Basins (Eds* P.-C. de Graciansky, J. Hardenbol, T. Jacquin & P. R. Vail). SEPM Special Publication 60: 3–13.

Harris, J. M. & White, T. D. 1979. Evolution of the Plio-Pleistocene African Suidae. *Transactions of the American Philosophical Society* 69 (2): 1–128.

Hedges, S. B., Parker, P. H., Sibley, C. G. & Kumar, S. 1996. *Nature* 381: 226–229.

Heizemann, P. J. & Begun, D. R. 2001. The oldest Eurasian hominoid. *Journal of Human Evolution* 41: 463–481.

Herngreen, G. F. W. & Duenas-Jimenez, H. 1990. Dating of the Cretaceous Une Formation, Colombia, and the relationship with the Albian–Cenomanian African–South American microfloral province. *Review of Palaeobotany and Palynology* 66: 345–359.

Herngreen, G. F. W., Kedves, M., Rovina, L. V. & Smirnova, S. B. 1996. Cretaceous palynofloral provinces: a review. In: *Palynology: Principles and Applications (Eds J.* Jansonius & D. C. McGregor). American Association of Stratigraphic Palynologists Foundation, pp. 1157–1188.

Hildebrand, A. R., Penfield, G. T., King, D. A., Pilkington, M., Camargo, Z. A., Jacobsen, S. B. & Boynton, W. V. 1991. Chicxulub Crater: a possible Cretaceous/Tertiary boundary impact crater on the Yucatan Peninsula, Mexico. *Geology* 19: 867–871.

Hoogheimstra, H. & Agwu, C. O. C. 1988. Changes in the vegetation and trade winds in Equatorial Northwest Africa 140,000–70,000 yr BP as deduced from two marine pollen records. *Palaeogeography, Palaeoclimatology, Palaeoecology* 66: 173–213.

Hopping, C. A. 1967. Palynology and the oil industry. *Review of Palaeobotany and Palynology* 6: 23–48.

Jacobs, B. F. 1999. Estimation of rainfall variables from leaf characters in tropical Africa. *Palaeogeography, Palaeoclimatology, Palaeoecology* 145: 231–250.

Jacobs, B. F. 2002 Estimation of low latitude palaeoclimates using fossil angiosperm leaves: examples from the Miocene Tugen Hills, Kenya. *Paleobiology* 28: 399–421.

Jacobs, B. F. & Deino, A. L. 1996. Test of climate-leaf physiognomy regression models, their application to two Miocene floras from Kenya, and 40Ar/39Ar dating of the Late Miocene Kapturo site. *Palaeogeography, Palaeoclimatology, Palaeoecology* 123: 259–271.

Jacobs, B. F. & Herendeen, P. S. 2004. Eocene dry climate and woodland vegetation in tropical Africa reconstructed from fossil leaves from northern Tanzania. *Palaeogeography, Palaeoclimatology, Palaeoecology* 213: 115–123.

Jaeger, J. J., Denys, C. & Coiffait, B. 1985. New Phiomorpha and Anomaluridae from the late Eocene of N. W. Africa; phylogenetic implications. In: *Evolutionary Relationships among Rodents: A Multidisciplinary Analysis* (eds W. P. Luckett & J. L. Hartenberger). Plenum Press, New York, pp. 685–712.

Janis, C. 1997. Ungulate teeth, diets, and climatic changes at the Eocene/Oligocene boundary. *Zoology* 100: 203–220.

Janis, C. 2003. Tectonics, climate change, and the evolution of mammalian ecosystems. *Evolution on Planet Earth* 17: 319–338.

Jaramillo, C., Pardo-Trujillo, A., Rueda, M., Harrington, G., Bayona, G., Torres, V. & Mora, G. 2007. Palynology of the Upper Paleocene Cerrejon Formation, Northern Colombia. *Palynology* 31: 153–189.

Jolivet, L. and Facenna, C. 2000. Mediterranean extension in the Africa-Eurasia collision. *Tectonics* 19: 1994–1106.

Kappelman, J., Rasmussen, D. T., Sanders, W. J., Mulugeta Feseha, Bown, T., Copeland, P., Crabaugh, J., Fleagle, J., Glantz, M., Gordon, A., Jacobs, B., Murat Maga, Muldoon, M., Pan, A., Pyne, L., Richmond, B., Ryan, T., Seiffert, E. R., Sevket Sen, Todd, L., Wiemann, M. C. & Winkler, A. 2004. Oligocene mammals from Ethiopia and faunal exchange between Afro-Arabia and Eurasia. *Nature* 426: 549–552.

Kedves, M. 1971. Presence de types sporomorphs importants dans les sediments prequaternaires Egyptiens. *Acta Botanica Academiae Scientarum Hungaricae* 17: 371–378.

Kedves, M. 1986. Etudes palynologiques sur les sediments prequaternaires de l'Egypt, Eocene. *Revista Espanola de Micropalaeontologia* 18: 5–26.

King, L. C. 1962. *The morphology of the Earth: A Study and Synthesis of World Scenery*. Hafner, New York, 699 pp.

Kingdon, J. 1971. *East African Mammals: An Atlas of Evolution in Africa*, Vol. 1. Academic Press, London.

Kingdon, J. 1980. The role of visual signals and face patterns in African forest monkeys (guenons) of the genus *Cercopithecus*. *Transactions of the Zoological Society of London* 35: 425–475.

Kingdon, J. 1981. Where have the colonists come from? A zoogeographical examination of some mammalian isolates in eastern Africa. *African Journal of Ecology* 19: 115–124.

Kingdon, J. 1982. *East African Mammals: An Atlas of Evolution in Africa*, Vol. 3. Academic Press, London.

Kingdon, J. 1988. What are face patterns and do they contribute to reproductive isolation in guenons? In: *A Primate Radiation: Evolutionary Biology of the African Guenons* (eds A. Gautier-Hion, F. Bourlière, J.-P. Gautier & J. Kingdon). Cambridge University Press, Cambridge, pp. 227–245.

Kingdon, J. 1990. *Island Africa*. Collins, London, 287 pp.

Kingdon, J. 1997. *The Kingdon Field Guide to African Mammals*. Academic Press, London, 464 pp.

Kingdon, J. 2003. *Lowly Origin*. Princeton University Press, Princeton, 396 pp.

Kingdon, J. 2007. Primate visual signals in noisy environments. In: Soligo, C. *Primate Evolution and the Environment. Folia Primatologica* 78(5–6): 389–404.

Kingston, J. D., Marino, B. D. & Hill, A. P. 1994. Isotopic evidence for Neogene hominid palaeoenvironments in the Kenya Rift Valley. *Science* 264: 955–959.

Klak, C., Reeves, G. & Hedderson, T. 2004. Unmatched tempo of evolution in Southern African semi-desert ice plants. *Nature* 427: 63–65.

Knaap, W. A. 1972. A montane pollen species from the Upper Tertiary of the Niger Delta. *Journal of the Mining, Geological and Metallurgical Society, Nigeria* 6: 23–29.

Krutzsch, W. 1989. Palaeogeography and historical phytogeography (palaeochorology) in the Neophyticum. *Plant Systematics and Evolution* 162: 5–61.

Kubitski, K. 2005. Major evolutionary advances in the history of green plants. *Acta Phytotaxonomica et Geobotanica* 56: 1–11.

Legoux, O. 1978. Quelques espèces de pollen caracteristiques du Neogene du Nigeria. *Bulletin du Centres des Recherches Exploration–Production Elf-Aquitaine* 2: 265–317.

Leroy, S. & Dupont, L. 1994. Development of vegetation and continental aridity in northwestern Africa during the Late Pliocene: the pollen record of ODP Site 658. *Palaeogeography, Palaeoclimatology, Palaeoecology* 109: 295–316.

Lezine, A. M. & Vergnaud-Grazzini, C. 1993. Evidence of forest extension in West Africa since 2200 BP: a pollen record from eastern tropical Atlantic. *Quaternary Science Reviews* 12: 203–220.

Linder, H. P. 2003. The radiation of the Cape flora, southern Africa. *Biological Reviews of the Cambridge Philosophical Society* 78: 597–638.

Livingstone, D. A. 1967. Postglacial vegetation of the Ruwenzori Mountains of Equatorial Africa. *Ecological Monographs* 37: 25–52.

Lonnberg, E. 1929. The development and distribution of the African fauna in connection with and depending upon climatic changes. *Archiv for Zoologi* 21A (4): 133.

Lopatin, A. V. 2006. Early Paleogene insectivore mammals of Asia and establishment of the major groups of Insectivora. *Palaeontological Journal* 40: S205–405.

Louchart, A., Mourer-Chauviré, C., Vignaud, P., Mackaye, H. T. & Brunet, M. 2005. A finfoot from the Late Miocene of Toros Menalla (Chad, Africa): Palaeobiogeographical and palaeoecological implications. *Palaeogeography, Palaeoclimatology, Palaeoecology* 222: 1–9.

Louvet, P. 1971. Sur l'evolution des flores tertiare de l'Afrique nord-equatoriale. PhD thesis, CNRS Paris, 497 pp.

Lydekker, R. 1908. *The Game Animals of Africa*. Rowland Ward, London.

Maley, J. 1987. Fragmentation de la fôret dense humide africaine et extension des biotypes montagnards au Quaternaire recent: nouvelles donnees polliniques et chronologiques. Implications palaeoclimatiques et biogeographiques. *Palaeoecology of Africa* 18: 307–344.

Maley, J. 1989. Late Quaternary climatic changes in the African rainforest: forest refugia and the major role of sea surface temperature variations.

In: *Palaeoclimatology and Palaeometeorology: Modern and Past Patterns of Global Atmospheric Transport* (eds M. Leinen & M. Sarnthein). Kluwer Academic, Leiden, pp. 585–616.

Maley, J. 1991. The African rainforest vegetation and palaeoenvironments during the Late Quaternary. *Climatic Change* 19: 79–98.

Maley, J. 1996. The African Rainforest – main characteristics of changes in vegetation and climate from the Upper Cretaceous to the Quaternary. In: *Essays on the Ecology of the Guinea–Congo Rainforest* (eds I. J. Alexander, M. D. Swaine & R. Watling). *Proceedings of the Royal Society of Edinburgh, B* 104: 31–73.

Maley, J. & Livingstone, D. A. 1983. Extension d'une élément montagnarde dans le sud du Ghana (Afrique de l'Ouest) au Pleistocene superieur et à l'Holocene inferieur: premières données palynologiques. *Comptes Rendus de l'Academie des Sciences (Paris)* Série II, 296: 1287–1292.

Mayr, E. & O'Hara, R. J. 1986. The biogeographic evidence supporting the Pleistocene refuge hypothesis. *Evolution* 40: 55–67.

Medus, J., Popoff, M., Fourtanier, E. & Sowunmi, M. A. 1988. Sedimentology, pollen, spores and diatoms of a 148 m deep Miocene drill hole from Oku Lake, East Central Nigeria. *Palaeogeography, Palaeoclimatology, Palaeoecology* 68: 79–94.

Meon, H. 1990. Palynologic studies of the Cretaceous–Tertiary boundary interval at El Kef outcrop, northwestern Tunisia: palaeogeographic implications. *Review of Palaeobotany and Palynology* 65: 85–94.

Misonne, X. 1969. African and Indo-Australian Muridae. Evolutionary trends. *Annals Museum of Royal Africain Centrale* 8: 1–219.

Montgelard, C., Bentz, S., Tirard, C., Verneau, O. & Catzeflis, F. M. 2002. Molecular systematics of Sciurognathi (Rodentia): the mitochondrial cytochrome b and 12s rRNA genes support the Anomaluroidea (Pedetidae and Anomaluridae). *Molecular Phylogenetics and Evolution* 22: 220–233.

Moodley, Y. & Bruford, M. W. 2007. Molecular biogeography: towards an integrated framework for conserving pan-African biodiversity. *PloS ONE* 2007 (2): e454.

Moreau, R. E. 1952. Africa since the Mesozoic with particular reference to certain biological problems. *Proceedings of the Zoological Society London* 121: 869–913.

Moreau, R. E. 1963a. Vicissitudes of the African biomes in the late Pliestocene. *Proceedings of the Zoological Society London* 141: 395–421.

Moreau, R. E. 1963b. The distribution of tropical African birds as an indicator of past climatic changes. In: *African Ecology and Human Evolution* (eds F. C. Howell & F. Bourliere). Aldine Press, Chicago, pp. 28–42.

Moreau, R. E. 1966. *The Bird Faunas of Africa and Its Islands*. Academic Press, London, 424 pp.

Moreau, R. E. 1969. Climatic changes and the distribution of forest vertebrates in West Africa. *Journal of Zoology* (London) 158: 39–61.

Morley, R. J. 1986. New approaches to stratigraphic and palaeoenvironmental modelling in Neogene deltaics with emphasis on the Niger Delta. *Proceedings of the 3rd Annual Conference of the Nigerian Association of Petroleum Explorationists*.

Morley, R. J. 2000. *Geological Evolution of Tropical Rainforests*. Wiley & Sons, London, 362 pp.

Morley, R. J. 2003. Interplate dispersal routes for megathermal angiosperms. *Perspectives in Plant Ecology, Evolution and Systematics* 6: 5–20.

Morley, R. J. & Richards, K. 1993. Gramineae cuticle: a key indicator of late Cenozoic climatic change in the Niger Delta. *Review of Palaeobotany and Palynology* 77: 119–127.

Morley, R. J. & Rosen, R. 1996. High resolution biostratigraphic zonation and palaeoclimatic record for the Offshore Niger Delta. *Proceedings of the IXth IPC Meeting, Houston, Texas, USA*, June 1996, pp. 111–112 (abstract).

Morley, R. J., Morley, H. P. & Restrepo-Pace, P. 2003. Unravelling the tectonically controlled stratigraphy of the West Natuna Basin by means of palaeo-derived Mid Tertiary climate changes. *IPA Proceedings 29* (Proceedings, Indonesian Petroleum Association 29th Annual Convention and Exhibition), Vol. 1.

Murphy, W. J., Eizirik, E., Johnson, W. E. & Zhang, Y. P. 2007. Molecular phylogenetics and the origins of placental mammals. *Nature* 409: 614–618.

Niklas, K. J., Tiffney, B. H. & Knoll, A. 1980. Apparent changes in the diversity of fossil plants. *Evolutionary Biology* 12: 1–89.

Niklas, K. J., Midgley, J. J. & Rand, R. H. 2003. Tree size frequency distributions, plant density, age and community disturbance. *Ecology Letters* 6: 405–411.

Pancost, P. D., Steart, D. S., Handley, L., Collinson, M. E., Hooker, J. E., Scott, A. E., Grassineau, N. V. & Glasspool, I. J. 2007. Increased terrestrial methane cycling at the Palaeocene–Eocene thermal maximum. *Nature* 449: 332–335.

Parrish, J. T. 1987. Global palaeogeography and palaeoclimate of the Late Cretaceous and Early Tertiary. In: *The Origins of Angiosperms and Their Biological Consequences* (eds E. M. Friis, W. G. Chaloner & P. R Crane). Cambridge University Press, Cambridge, pp. 51–73.

Parrish, J. T. & Barron, E. J. 1986. *Paleoclimates and Economic Geology*. Society of Economic Palaeontologists and Mineralogists Short Course no. 18, 162 pp.

Parrish, J. T., Ziegler, A. M. & Scotese, C. R. 1982. Rainfall patterns and the distribution of coals and evaporites in the Mesozoic and Cenozoic. *Palaeogeography, Palaeoclimatology, Palaeoecology* 40: 67–101.

Pearson, P. H., Ditchfield, P. W., Singano, J., Harcourt-Brown, J. C., Nicholas, C. J., Olsson, K. R., Shackleton, N. J. & Hall, M. A. 2001. Warm tropical sea surface temperatures in the Late Cretaceous and Eocene epochs. *Nature* 413: 481–487.

Pickford, M. 1986. A revision of the Miocene Suidae and Tayassuidae (Artiodactyla, Mammalia) of Africa. *Tertiary Research Special Paper* No. 7, 82 pp.

Pickford, M. 1993. Old World suid systematics, phylogeny, biogeography and biostratigraphy. *Paleontologia y Evolucion* 26–27: 237–269.

Poumot, C. 1987. Palynological evidence for eustatic events in the tropical Neogene. *Bulletin du Centres Recherches Exploration–Production Elf-Aquitaine* 13: 347–353.

Prasad, V., Strömberg, C. A. E., Alimohammadian, H. & Sahni, A. 2005. Dinosaur Coprolites and the Early Evolution of Grasses and Grazers. *Science* 310: 1177–1180.

Prothero, D. R. 1994. *The Eocene–Oligocene Transition: Paradise Lost (Critical Moments in Paleobiology and Earth History)*. Columbia University Press, New York, 310 pp.

Raup, D. M. and Sepowski Jr, J. J. 1984. Periodicity of extinctions in the geological past. *Proceedings of the National Academy of Sciences of the United States of America* 81: 801–805.

Retallack, G. J. 1992a. Comment on the paleoenvironment of *Kenyapithecus* at Fort Ternan. *Journal of Human Evolution* 23, 363–369.

Retallack, G. J. 1992b. Middle Miocene fossil plants from Fort Ternan (Kenya) and evolution of African grasslands. *Paleobiology* 18: 383–400.

Retallack, G. J., Dugas, D. P. & Bestland, E. A. 1990. Fossil soils and grasses of a Middle Miocene East African grassland. *Science* 247: 1325–1328.

Reyment, R. A. 1965. *Aspects of the Geology of Nigeria*. University Press, Ibadan, 145 pp.

Richard, D., Pancost, R. D., Steart, D. S., Handley, L., Collinson, M. E., Hooker, J. J., Scott, A. S., Grassineau, N. V. & Glasspool, I. J. 2007. Increased terrestrial methane cycling at the Palaeocene–Eocene thermal maximum. *Nature* 449: 332–335.

Richardson, J. E., Weitz, F. M., Fay, M. F., Cronk, Q. C. B., Linder, H. P., Reeves, G. & Chase, M. W. 2001. Rapid and recent origin of species richness in the Cape flora of South Africa. *Nature* 412: 181–183.

Salard-Cheboldaeff, M. 1990. Intertropical African palynostratigraphy from Cretaceous to Late Quaternary times. *Journal of African Earth Sciences* 11: 1–24.

Schiotz, A. 1975. *The Tree Frogs of Eastern Africa*. Steenstrupia, Copenhagen, 232 pp.

Scholtz, A. 1985. The palynology of the upper lacustrine sediments of the Arnot Pipe, Banke, Namaqueland. *Annals of the South African Museum* 95: 1–109.

Seiffert, E. R. 2006. Revised age estimates for the later Paleogene mammal faunas of Egypt and Oman. *Proceedings of the National Academy of Sciences of the United States of America* 103: 5000–5005.

Sepulchre, P., Ramstein, G., Fluteau, F., Schuster, M., Tiercelin, J.-J. & Brunet, M. 2006. Tectonic uplift and Eastern Africa aridification. *Science* 313: 1419–1423.

Seward, A. C. 1935. *Leaves of Dicotyledons from the Nubian Sandstones of Egypt*. Geological Survey of Egypt. Survey Department, Ministry of Finance, Cairo, 21 pp. + plates.

Smith, A. G., Smith, D. G. & Funnell, B. M. 1994. *Atlas of Mesozoic and Cenozoic coastlines*. Cambridge University Press, Cambridge, 99 pp.

Sowunmi, M. A. 1991. Late Quaternary environments in equatorial Africa: palynological evidence (Lake Mobutu Sese Seko). *Palaeoecology of Africa* 22: 213–238.

Springer, M. S., Murphy, W. J., Eizirik, E. & O'Brien, S. J. 2003. Placental mammal diversification and the Cretaceous–Tertiary boundary. *Proceedings of the National Academy of Sciences of the United States of America* 100: 1056–1061.

Stanhope, M. J., Waddell, V. G., Madsen, O., de Jong, W., Hedges, S. B., Cleven, G. C., Kao, D. & Springer, M. 1998. Molecular evidence for multiple origins of Insectivora and for a new order of endemic African insectivore mammals. *Proceedings of the National Academy of Sciences of the United States of America* 95: 9967–9972.

Steppan, S. J., Adkins, R. M. & Anderson, J. 2004. Phylogeny and divergence – date estimates of rapid radiation in muroid rodents based on multiple nuclear genes. *Systematic Biology* 53: 533–553.

Stewart, C.-B. & Disotell, T. 1998. Eurasian ape as great ape ancestor. *Current Biology* 8: R582–R588.

Street-Perrott, F. A. & Grove, A. T. 1976. Holocene vegetation and late Quaternary lake levels. *Nature* 261: 335–390.

Tosi, A. J., Detwiler, K. M. & Disotell, T. R. 2005. X-chromosomal window into the evolutionary history of the guenons (Primates: Cercopithecini). *Molecular Phylogenetics and Evolution* 36: 58–66.

Trauth, M. H., Maslin, M. A., Deino, A. & Strecker, M. R. 2005. Late Cenozoic Moisture History of East Africa. *Science 309: 2051–2053.*

Turner, A. 1995. Plio-Pleistocene correlations between climatic change and evolution in terrestrial mammals: the 2. 5 Ma event in Africa and Europe. *Acta Zoologica Cracovensia* 38: 45–58.

Vail, P. R. & Wornardt, W. W. 1990. Well log-seismic sequence stratigraphy: an integrated tool for the 90s. *GCSSEPM Foundation 11th Annual Conference*, December 1990, 379–388.

Vakhrameev, V. A. 1992. *Jurassic and Cretaceous Floras and Climates of the Earth* (translated by Ju. V. Litvinov). Cambridge University Press, Cambridge, 318 pp.

Van Zinderen Bakker, E. M. 1975. The origin and palaeoenvironment of the Namib Desert biome. *Journal of Biogeography* 2: 65–73.

Van Zinderen Bakker, E. M. & Coetzee, J. Q. 1988. A review of late Quaternary pollen studies in East, Central and Southern Africa. *Review of Palaeobotany and Palynology* 55: 155–174.

Van Zinderen Bakker, E. M. & Mercer, J. H. 1986. Major late Caenozoic climatic events and palaeo-environmental changes in Africa viewed in a world-wide context. *Palaeogeography. Palaeoclimatology, Palaeoecology* 56: 217–235.

Wayland, E. J. 1940. Desert versus forest in East Africa. *The Geographical Journal* 96: 329–340.

White, F. 1983. *The Vegetation of Africa: A Descriptive Memoir to Accompany the UNESCO/AETFAT/UNSC Vegetation Map of Africa*. UNESCO, Paris.

White, F. 1983. *The Vegetation of Africa: A Descriptive Memoir to Accompany the UNESCO/AETFAT/UNSO Vegetation Map of Africa*. UNESCO, Paris, 356 pp + map.

Wible, J. R., Rougier, G. W., Novacek, M. J. & Asher, R. J. 2007. Cretaceous eutherians and Laurasian origin for placental mammals near the KT boundary. *Nature* 447: 1003–1008.

Zachos, J. C., Pagini, M., Sloan, L., Thomas, E. & Billups, K. 2001. Trends, rhythms and aberrations in global climate 65 Ma to Present. *Science* 292: 686–693.

Zachos, J. C., Wara, W. M., Bohaty, S., Delaney, M. L., Pettrizzo, M. R., Brill, A., Bralower, T. J. & Premoli-Silva, I. 2003. A transient rise in tropical sea surface temperature during the Palaeocene–Eocene thermal maximum. *Science 302: 1551–1554.*

Chapter 5. The Biotic Zones of Africa

Acocks, J. P. H. 1975. Veld types of South Africa (2nd edn). *Memoir of the Botanical Survey of South Africa* 40: 1–128.

Clayton, W. D. 1963. The Vegetation of Katsina Province, Nigeria. *Journal of Ecology* 51: 345–351.

Dowsett-Lemaire, F. & Dowsett, R. J. 1998. Parallels between F. White's phytochoria and Avian Zoochoria in Tropical Africa: an Analysis of the Forest Elements. In: *Chorology, Taxonomy and Ecology of the Flora of Africa and Madagascar* (eds C. R. Huxley, J. M. Lock & D. F. Cutler). Royal Botanic Gardens, Kew, pp. 87–96.

Greenway, P. J. 1965. The vegetation and flora of Mt Kilimanjaro. *Tanganyika Notes & Records* 64: 97–107.

Greenway, P. J. 1973. A classification of the vegetation of East Africa. *Kirkia* 9: 1–68.

Hall, J. B. & Swaine, M. D. 1976. Classification and ecology of closed-canopy forest in Ghana. *Journal of Ecology* 64: 913–951.

Hall, J. B. & Swaine, M. D. 1981. *Distribution and Ecology of Vascular Plants in a Tropical Rain Forest. Forest Vegetation in Ghana*. Junk, The Hague, 392 pp.

Happold, D. C. D. 1996. Mammals of the Guinea-Congo rainforests. In: *Essays on the Ecology of the Guinea-Congo Rainforest* (eds I. J. Alexander, M. D. Swaine & R. Watling). *Proceedings of the Royal Society of Edinburgh B* 104: 243–284.

Hedberg, O. 1964. Features of afroalpine plant ecology. *Acta Phytogeographica Suecica* 19: 1–144.

Howell, P., Lock, J. M. & Cobb, S. (eds) 1988. *The Jonglei Canal: Impact and Opportunity*. Cambridge University Press, Cambridge, 450 pp.

Jackson, G. 1974. Cryptogeal germination and other seedling adaptations to the burning of vegetation in savanna regions: the origin of the pyrophytic habit. *New Phytologist* 73: 771–780.

Keay, R. W. J. 1959a. *Vegetation Map of Africa South of the Tropic of Cancer*. Explanatory Notes. Oxford University Press, Oxford, 24 pp + map.

Keay, R. W. J. 1959b. *An Outline of Nigerian Vegetation* (3rd edn). Government Printer, Lagos, 52 pp.

Lock, J. M. 1998. Aspects of fire in Tropical African vegetation. In: *Chorology, Taxonomy and Ecology of the Flora of Africa and Madagascar* (eds C. R. Huxley, J. M. Lock & D. F. Cutler). Royal Botanic Gardens, Kew, pp. 181–189.

Lock, J. M. & Milburn, T. R. 1971. The seed biology of *Themeda triandra* Forsk in relation to fire. *Symposium of the British Ecological Society* 11: 337–349.

Rosevear, D. R. 1953. *Checklist and Atlas of Nigerian Mammals*. Government Printer, Lagos, 131 pp. + maps.

Rutherford, M. C. & Westfall, R. H. 1986. Biomes of Southern Africa – an Objective Categorization. *Memoirs of the Botanical Survey of South Africa* 54, 98 pp.

Sinclair, A. R. E. & Norton-Griffiths, M. (eds) 1979. *Serengeti: Dynamics of an Ecosystem*. University of Chicago Press, Chicago, 397 pp.

Swaine, M. D. & Hall, J. B. 1986. Forest structure and dynamics. In: *Plant Ecology in West Africa* (ed. G. W. Lawson). John Wiley & Sons, Chichester, pp. 47–93.

Swaine, M. D., Hall, J. B. & Lock, J. M. 1976. The forest–savanna boundary in west-central Ghana. *Ghana Journal of Science* 16: 35–52.

Tutin, C. E. G. 1998. Gorillas and their food plants in the Lopé Reserve, Gabon. In: *Chorology, Taxonomy and Ecology of the Flora of Africa and Madagascar* (eds C. R. Huxley, J. M. Lock & D. F. Cutler). Royal Botanic Gardens, Kew, pp. 227–243.

Walter, H. 1979. *Vegetation of the Earth and Ecological Systems of the Geo-biosphere* (2nd edn) (Translated from 3rd German edn by Joy Wieser). Springer-Verlag, Berlin, 558 pp.

Walter, H. & Lieth, H. 1967. *Klimmadiagramm-Weltatlas*. VEB Gustav Fischer-Verlag, Jena.

White, F. 1983. *The Vegetation of Africa: A Descriptive Memoir to Accompany the UNESCO/AETFAT/UNSC Vegetation Map of Africa*. UNESCO, Paris, 356 pp + map.

Chapter 6. Mammalian Evolution in Africa

Adkins, R. M., Gelke, E. L., Rowe, D. & Honeycutt, R. L. 2001. Molecular phylogeny and divergence time estimates for major rodent groups: evidence from multiple genes. *Molecular Biology and Evolution* 18: 777–791.

Adkins, R. M., Walton, A. H. & Honeycutt, R. L. 2003. Higher-level systematics of rodents and divergence time estimates on two congruent nuclear genes. *Molecular Phylogeny and Evolution* 26: 409–420.

Archibald, J. D. 2003. Timing and biogeography of the Eutherian radiation: fossils and molecules compared. *Molecular Phylogeny and Evolution* 28: 350–359.

Beard, K. C. 2004. *The Hunt for the Dawn Monkey: Unearthing the Origins of Monkeys, Apes and Humans*. University of California Press, Berkeley, 348 pp.

Bininda-Emonds, O. R. P., Cardillo, M., Jones, K. E., Ross, D., MacPhee, E., Beck, R. M. D., Grenyer, R., Price, S. A., Vos, R. A., Gittleman, J. L. & Purvis, A. 2007. The delayed rise of present day mammals. *Nature* 446: 507–512.

Booth, A. H. 1960. *The Small Mammals of West Africa*. Longmans, London, 68 pp.

Bronner, G. N., Hoffmann, M., Taylor, P. J., Chimimba, C. T., Best, P. B., Matthee, C. A. & Robinson, T. J. 2003. A revised systematic checklist of the extant mammals of the southern African subregion. *Durban Museum Novitates* 28: 56–106.

Burgess, N. & Clarke, G. P. 2000. *Coastal Forests of Eastern Africa*. IUCN Forest Conservation Programme, Gland, 443 pp + maps.

Burgess, N. D., Butynski, T. M., Cordeiro, N. J., Doggart, N. H., Fjeldsa, J., Howell, K. M., Kitahama, F . B., Loader, N. H., Lovett, J. C., Mbilinye, B., Menegon, M., Moyer, D. C., Nashanda, E., Perkin, A., Rovero, F., Stanley, W. T. & Stuart, S. N. 2007. The biological importance of the Eastern Arc Mountains of Tanzania and Kenya. *Biological Conservation* 134: 209–231.

Cartmill, M. 1974. Rethinking primate origins. *Science* 184: 436–443.

Clutton-Brock, T. H., O'Riain, M. J., Brotherton, P. N. M., Gaynor, D., Kansky, R., Griffin, A. S. & Manser, M. 1999. Selfish sentinels in cooperative mammals. *Science* 284: 1640–1644.

Coe, M. J. & Skinner, J. D. 1993. Connections, disjunctions and endemism in the eastern and southern African mammal faunas. *Transactions of the Royal Society of South Africa* 48: 233–256.

Corbet, G. B. & Hanks, J. 1968. A revision of the elephant-shrew family (Macroscelididae). *Bulletin of the British Museum (Natural History)* 16 (2): 5–111.

Creel, S. R. 1996. Behavioral endocrinology and social organization in dwarf mongooses. In: *Carnivore Behavior, Ecology, and Evolution*, Vol. 2 (ed. J. L. Gittleman). Cornell University Press, New York, pp. 46–77.

Dawkins, R. 1986. *The Blind Watchmaker*. W. W. Norton, New York.

Douady, E., Teeling, O. A., Ryder, M. J., Stanhope, M. J. & de Jong, A. 2003. Molecular and genomic data identify the closest living relatives of primates. *Molecular Biology and Evolution* 15: 1647–1660.

Dutrillaux, B., Couturier, J., Muleris, M., Lombard, M. & Chauvier, G. 1982. Chromosomal phylogeny of forty-two species of subspecies of cercopithecoids (Primates, Catarrhini). *Annales de Génétique* 25: 96-109.

Dutrillaux, B., Muleris, M. & Couturier, J. 1988. Chromosome evolution of Cercopithecinae. In: *A Primate Radiation. Evolutionary Biology of African guenons* (eds A. Gautier-Hion, F. Bourliere, J.P. Gautier & J. Kingdon). Cambridge University Press, Cambridge, pp. 150-159.

Eiseley, L. 1946. *The Immense Journey*. Random House Inc., New York.

Ewer, R. F. 1973. *The Carnivores*. Cornell University Press, New York, 494 pp.

Gheerbrandt, E., Sudre, J., Tassy, P., Amaghzaz, M., Bouya, B. & Larochene, M. 2005. Nouvelles données sur *Phosphatherium escuilliei* (Mammalia, Proboscidea) de l'Éocène inférieur du Maroc, apports à la phylogénie des Proboscidea et des ongulés lophodontes. *Geodiversitas* 27: 239–333.

Hamilton, A.C. 1982. Environmental history of East Africa: a study of the Quaternary. Academic Press, London, 328 pp.

Hayman, R. W. 1936. A note on *Dologale dybowskii* Pousargues. *Annals and Magazine of Natural History* (10) 18: 626–630.

Hempton, M. R. 1987. Constraints on Arabian Plate motion and extensional history of the Red Sea. *Tectonics* 6: 687–706.

Huchon, D., Madsen, O., Sibbald, M. J. J. B., Ament, K., Stanhope, M. J., Catzeflis, F., de Jong, W. W. & Douzery, E. J. P. 2002. Rodent phylogeny and a timescale for the evolution of Glires: evidence from an extensive taxon sampling using three nuclear genes. *Molecular Biology and Evolution* 19: 1053–1065.

Hutchinson, G. E. 1959. Homage to Santa Rosalia, or why are there so many kinds of animals? *American Naturalist* 93: 145–159.

Jansa, S. A. & Weksler, M. 2004. Phylogeny of muroid rodents: relationships within and among major lineages as determined by IRBP sequences. *Molecular Phylogenetics and Evolution* 31: 256–276.

Jarvis, J. U. M., O'Riain, M. J., Bennett, N. C. & Sherman, P. W. 1994. Mammalian eusociality: a family affair. *Trends in Ecology and Evolution* 9: 47–51.

Ji, Q., Luo, Z.-X., Yuan, C.-X., Wible, J. R., Zhang, J.-P. & Georgi, J. A. 2002. The earliest known eutherian mammal. *Nature* 416: 816–822.

Kielan-Jaworowska, Z. & Dashzeveg, D. 1989. Eutherian mammals from the Early Cretaceous of Mongolia. *Zoologica Scripta* 18: 347–355.

Kingdon, J. 1971. *East African Mammals: An Atlas of Evolution in Africa*. Vol. I: *Primates*. Academic Press, London.

Kingdon, J. 1974. *East African Mammals: An Atlas of Evolution in Africa*. Vol. IIA: *Insectivores and Bats*. Academic Press, London.

Kingdon, J. 1977. *East African Mammals: An Atlas of Evolution in Africa*. Vol. IIIA: *Carnivores*. Academic Press, London.

Kingdon, J. 1982. *East African Mammals: An Atlas of Evolution in Africa*. Vol. III: *Bovids*. Academic Press, London.

Kingdon, J. 1990. *Arabian Mammals: A Natural History*. Academic Press, London, 279 pp.

Kingdon, J. 1997. *The Kingdon Field Guide to African Mammals*. Academic Press, London, 464 pp.

Kingdon, J. 2003. *Lowly Origin*. Princeton University Press, Princeton, 396 pp.

Kingdon, J., Agwanda, B., Kinnaird, M., O'Brien, T., Holland, C., Gheysens, T., Boulet-Audet, M. & Vollrath, F. 2011. A poisonous surprise under the coat of the African Crested Rat. *Proceedings of the Royal Society B* 1–6.

Lonnberg, E. 1929. The development and distribution of the African fauna in connection with and depending upon climatic changes. *Archiv fur Zoologi* 21A No. 4: 1–33.

Lovett, J. C. & Waser, S. 1993. *Biogeography and Ecology of the Rain Forests of Eastern Africa*. Cambridge University Press, Cambridge, UK, 351 pp.

MacDonald, J. T. & Nel, J. A. J. 1986. Comparative diets of four sympatric small carnivores. *South African Journal of Wildlife Research* 16: 115–121.

Matthew, W. D. 1904. The arboreal ancestry of the Mammalia. *American Naturalist* 38: 811–818.

Matthee, C. A. & Davis, S. K. 2001. Molecular insights into the evolution of the family Bovidae: A nuclear DNA perspective. *Molecular Biology and Evolution* 18: 1220–1230.

Mayr, E. 1970. *Populations, Species and Evolution*. Harvard University Press, Cambridge, MA.

Meredith, R. W., Janečka, J. E., Gatesy, J., Ryder, O. A., Fisher, C. A., Teeling, E. C., Goodbla, A., Eizirik, E., Simão, T. L. L., Stadler, T., Rabosky, D. L., Honeycutt, R. L., Flynn, J. J., Ingram, C. M., Steiner, C., Williams, T. L., Robinson, T. J., Burk-Herrick, A., Westerman, M., Ayoub, N. A., Springer, M. S. & Murphy, W. J. 2011. Impacts of the Cretaceous Terrestrial Revolution and KPg extinction on mammal diversification. *Science* 334: 521–524.

Misonne, X. 1969. African and Indo-Australian Muridae. Evolutionary trends. *Annals Museum of Royal Africain Centrale* 8: 1–219.

Mohres, F. P. 1953. Ueber die Ultraschallorientierung der Hufeisennasen. *Zeitschrift für Vergleishende Physiologie* 34: 547–588.

Montgelard, C., Bentz, S., Tirard, C., Verneau, O. & Catzeflis, F. M. 2002. Molecular systematics of Sciurognathi (Rodentia): the mitochondrial cytochrome b and 12s rRNA genes support the Anomaluroidea (Pedetidae and Anomaluridae). *Molecular Phylogenetics and Evolution* 22: 220–233.

Moodley, Y. & Bruford, M. W. 2007. Molecular biogeography: towards an integrated framework for conserving pan-African biodiversity. *PLoS ONE* 2007 (2): e454.

Moreau, R. E. 1952. Africa since the Mesozoic with particular reference to certain biological problems. *Proceedings of the Zoological Society of London* 121: 869–913.

Moreau, R. E. 1963. Vicissitudes of the African biomes in the late Pleistocene. *Proceedings of the Zoological Society of London* 141: 395–421.

Moreau, R. E. 1969. Climatic change and the distribution of forest vertebrates in West Africa. *Journal of Zoology (London)* 158: 39–61.

Murphy, W. J., Eizirik, E., Johnson, W. E., Zhang, Y. P., Ryder, O. A. & O'Brien, S. J. 2001a. Molecular phylogenetics and the origins of placental mammals. *Nature* 409: 614–618.

Murphy, W. J., Eizirik, E., O'Brien, S. J., Madsen, O., Scally, M., Douady, C. J., Teeling, E., Ryder, O. A., Stanhope, M. J., de Jong, W. W. & Springer, M. S. 2001b. Resolution of the early placental mammal radiation using Bayesian phylogenetics. *Science* 294: 2348–2351.

Murphy, W. J., Eizirik, E., Johnson, W. E. & Zhang, Y. P. 2007. Molecular phylogenetics and the origins of placental mammals. *Nature* 409: 614–618.

Rasa, O. A. E. 1977. The ethology and sociology of the dwarf mongoose (*Helogale undulata rufula*). *Zeitschrift für Tierpsychologie* 43: 337–406.

Rathbun, G. 1979a. *The Social Structure and Ecology of Elephant Shrews*. Verlag P. Parey, Berlin, 76 pp.

Rathbun, G. 1979b. *Rhynchocyon chrysopygus*. *Mammalian Species 117: 1–4*.

Rich, T. H., Vicers-Rich, P., Constantine, A., Flannery, T. F., Kool, L. & Klaveren, N. von. 1997. A tribosphenic mammal from the Mesozoic of Australia. *Science* 278: 1438–1442.

Robinson, T. J. & Seiffert, E. R. 2004. Afrotherian origins and inter-relationships: new views and future prospects. *Current Topics in Developmental Biology* 63: 37–60.

Romer, A. S. 1945. *Vertebrate Paleontology*. University of Chicago Press, Chicago, viii + 687 pp.

Rood, J. P. 1975. Population dynamics and food habits of the banded mongoose. *East African Wildlife Journal* 13: 89–111.

Rood, J. P. 1983. The social system of the dwarf mongoose. In: *Advances in the Study of Mammalian Behavior* (eds J. F. Eisenberg & D. G. Kleinman). American Society of Mammalogists, Special Publication 7. Pittsburgh, Pennsylvania, pp. 454–488.

Rood, J. P. 1986. Ecology and social evolution in the mongooses. In: *Ecological Aspects of Social Eevolution* (eds D. Rubenstein & R. Wrangham). Princeton University Press, Princeton, pp. 131–152.

Rood, J. P. 1990. Group size, survival, reproduction, and routes to breeding in dwarf mongooses. *Animal Behaviour* 39: 566–572.

Rose, K. D. 2006. *The Beginning of the Age of Mammals*. Johns Hopkins University Press, Baltimore, 448 pp.

Sige, B., Jaeger, J. J., Sudre, J. & Vianey-Liaud, M. 1990. *Altiatlasius koulchii* n. gem. et sp., primate omomyide du Paleocene superior du Maroc, et les origins des Euprimates. *Palaeontographica Abteilung* A 214: 31–56.

Simpson, G. G. 1937. The beginning of the Age of Mammals. *Biological Reviews* 12: 1–47.

Springer, M. S., Cleven, G. C., Madsen, O., de Jong, W. W., Waddell, V. G., Amrine, H. M. & Stanhope, M. J. 1997. Endemic African mammals shake the phylogenetic tree. *Nature* 388: 61–64.

Springer, M. S., Murphy, W. J., Eizirik, E. & O'Brien, S. J. 2003. Placental mammal diversification and the Cretaceous–Tertiary boundary. *Proceedings of the National Academy of Sciences of the United States of America* 100: 1056–1061.

Springer, M. S., Stanhope, M. J., Madsen, O. & de Jong, W. W. 2004. Molecules consolidate the placental mammal tree. *Trends in Ecology and Evolution* 19: 430–438.

Steppan, S. J., Adkins, R. M. & Anderson, J. 2004. Phylogeny and divergence – date estimates of rapid radiation in muroid rodents based on multiple nuclear genes. *Systematic Biology* 53: 533–553.

Stewart, C.-B. & Disotell, T. 1998. Eurasian ape as great ape ancestor. *Current Biology* 8: R582–R588.

Tabuce, R., Marivaux, L., Adaci, M., Bensalah, M., Hartenberger, J. L., Mahboubi, M., Mebrouk, F., Tafforeau, P. & Jaeger, J.-J. 2007. Early Tertiary mammals from North Africa reinforce the molecular Afrotheria clade. *Proceedings of the Royal Society B* 274: 1159–1166.

Veron, G., Colyn, C., Dunham, A. E., Taylor, P. J. & Gaubert, P. 2004. Molecular phylogeny of the mongooses (Carnivora, Herpestidae) and origin of the sociality in this family. *Molecular Phylogenetics and Evolution* 30: 582–598.

Vorontsov, N.N. 1967. Evolutsia Pishchevaritelnoy Systemy Mysheobraznykh Gryzunov "Nauka", Novosibirsk, 239 pp. (in Russian). English translation, 1979. Evolution of the Alimentary System in Myomorph Rodents. Smithsonian Institution, Washigton D.C. & National Science Foundation (translated by the Indian Documentation Centre, New Dahli), v. I: 346 pp.

Waddell, P. J., Kishino, H. & Ota, R. 2001. A phylogenetic foundation for comparative mammalian genomics. *Genome Informatics* 12: 141–154.

White, M. J. D. 1978. *Modes of Speciation*. W. H. Freeman, San Francisco.

Wible, J. R., Rougier, G. W., Novacek, M. J. & Asher, R. J. 2007. Cretaceous eutherians and Laurasian origin for placental mammals near the KT boundary. *Nature* 447: 1003–1008.

Willows-Munro, S., Robinson, T. J. & Mathee, C. A. 2005. Utility of nuclear DNA intron markers at lower taxonomic levels: phylogenetic resolution among nine *Tragelaphus* species. *Molecular Phylogenetics and Evolution* 35: 624–636.

Chapter 7. Classification: A Mammalian Perspective

Allen, G. M. 1939 (and reprint 1945). A checklist of African mammals. *Bulletin of the Museum of Comparative Zoology at Harvard College* 83: 1–763.

Ansell, W. F. H. 1989. *African Mammals, 1938–1988*. The Trendrine Press, Cornwall, 77 pp.

Asher, R. J., Novacek, M. J. & Geisler, J. H. 2003. Relationships of endemic African mammals and their fossil relatives based on morphological and molecular evidence. *Journal of Mammalian Evolution* 10: 131–194.

Avise, J. C. 1994. *Molecular Markers, Natural History, and Evolution*. Chapman & Hall, London, 511 pp.

Avise, J. C. & Johns, G. C. 1999. Proposal for a standardized temporal scheme of biological classification for extant species. *Proceedings of the National Academy of Sciences of the United States of America* 96: 7358–7363.

Estes, R. D. 1991. *The Behavior Guide to African Mammals*. University of California Press, Berkeley, 411 pp.

Fabricius, C., Lowry, D. & van den Berg, P. 1988. Fecund black wildebeest x blue wildebeest hybrids. *South African Journal ofWildlife Research* 18: 35–37.

Flagstad, O., Syvertsen, P. O., Stenseth, N. C. & Jakobsen, K. S. 2001. Environmental change and rates of evolution: the phylogeographic pattern within the hartebeest complex as related to climatic variation. *Proceedings of the Royal Society B* 268: 667–677.

Gentry, A., Clutton-Brock, J. & Groves, C. P. 2004. The naming of wild animals species and their domestic derivatives. *Journal of Archaeological Science* 31: 645–651.

Goodman, M., Porter, C. A., Czelusniak, J., Page, S. L., Schneider, H., Shoshani, J., Gunnell, G. & Groves, C. P. 1998. Toward a phylogenetic classification of Primates based on DNA evidence complemented by fossil evidence. *Molecular Phylogeny and Evolution* 9: 585–598.

Groves, C. P. 2000. What are the elephants of West Africa? *Elephant* 2(4): 7–8.

Groves, C. P. 2001. *Primate Taxonomy*. Smithsonian Institution Press, Washington, DC, 350 pp.

Groves, C. P. & Grubb, P. 2000. Do *Loxodonta cyclotis* and *L. africana* interbreed? *Elephant* 2(4): 4–7.

Grubb, P. 1985. Geographical variation in the bushbuck of eastern Africa (*Tragelaphus scriptus*, Bovidae). In: *African Vertebrates: Systematics, Phylogeny and Evolutionary Ecology*. (ed. K.-L. Schumann). Zoologisches Forschungsinstitut und Museum Alexander Koenig, Bonn, Germany. pp. 11–27.

Grubb, P., Groves, C. P., Dudley, J. P. & Shoshani, J. 2000. Living African elephants belong to two species, *Loxodonta africana* (Blumenbach, 1797) and *Loxodonta cyclotis* (Matschie, 1900). *Elephant* 2(4): 1–4.

Hoffmann, M., Grubb, P., Groves, C. P., Hutterer, R., Van der Straeten, E., Simmons, N. & Bergmans, W. 2009. A synthesis of African and western Indian Ocean Island mammal taxa (Class: Mammalia) described between 1988 and 2008: an update to Allen (1939) and Ansell (1989). *Zootaxa* 2205: 1–36.

International Commission on Zoological Nomenclature. 1999. *International Code of Zoological Nomenclature*, 4th edn. International Trust for Zoological Nomenclature, London.

Jolly, C. J. 1993. Species, subspecies, and baboon systematics. In: *Species, Species Concepts, and Primate Evolution* (eds W. H. Kimbel & L. B. Martin). Plenum Press, New York, pp. 67–107.

Jolly, C. J., Woolley-Barker, T., Beyene, S., Disotell, T. R. & Phillips-Conroy, J. E. 1997. Intergeneric hybrid baboons. *International Journal of Primatology* 18: 597–627.

Mayr, E. 1963. *Animal Species and Evolution*. Harvard University Press, Cambridge, MA, 797 pp.

Moodley, Y. & Bruford, M. W. 2007. Molecular biogeography: towards an integrated framework for conserving pan-African biodiversity. *PLoS One* 2 (5): e454.

Pettigrew, J. D. 1986. Flying primates? Megabats have the advanced pathway from eye to midbrain. *Science* 231: 1304–1306.

Pettigrew, J. D. 1991. Wings or brain? Convergent evolution in the origins of bats. *Systematic Zoology* 40: 199–216.

Roca, A. L., Georgiadis, N., Pecon-Slattery, J. & O'Brien, S. J. 2001. Genetic evidence for two species of elephant in Africa. *Science* 293: 1473–1477.

Rohland, N., Reich, D., Mallick, S., Meyer, M., Green, R. E., Georgiadis, N. J., Roca, A. L. & Hofreiter, M. 2010. Genomic DNA sequences from Mastodon and Woolly Mammoth reveal deep speciation of forest and savanna elephants. *PLoS Biology* 8(12): e1000564.

Ruxton, A. E. & Schwarz, E. 1929. On hybrid hartebeests and on the distribution of the *Alcelaphus buselaphus* group. *Proceedings of the Zoological Society of London* 1929: 567–583.

Selous, F. C. 1893. Exhibition of, and remarks upon, the skull of an antelope believed to be a hybrid between the sassaby and the hartebeest. *Proceedings of the Zoological Society of London* 1893: 1–2.

Simpson, G. G. 1961. *Principles of Animal Taxonomy*. Oxford University Press, Oxford, 247 pp.

Stanhope, M. J., Waddell, V. G., Madsen, O., de Jong, W., Blair Hedges, S., Cleven, G. C., Kao, D. & Springer, M. 1998. Molecular evidence for multiple origins of Insectivora and for a new order of endemic African insectivore mammals. *Proceedings of the National Academy of Sciences of the United States of America* 95: 9967–9972.

Wible, J. R., Rougier, G. W., Novacek, M. J. & Asher, R. J. 2007. Cretaceous eutherians and Laurasian origin for placental mammals near the K/T boundary. *Nature* 447: 1003–1006.

Wilson, D. E. & Cole, F. R. 2000. *Common Names of Mammals of the World*. Smithsonian Institution Press, 204 pp.

Wilson, D. E. & Reeder, D.-A. (eds) 1993. *Mammal Species of the World*, 2nd edn. Smithsonian Institution Press, Washington, DC, 1206 pp.

Wilson, D. E. & Reeder D.-A. (eds) 2005. *Mammal Species of the World*, 3rd edn (2 vols). Boston University Press, Baltimore, 2142 pp.

Chapter 8. Behaviour and Morphology

Bertram, B. 1973. Lion population regulation. *East African Wildlife Journal* 11: 215–225.

Bertram, B. 1975. Social factors influencing reproduction in wild lions. *Journal of Zoology (London)* 177: 463–482.

Coe, M. J. 1967. 'Necking' behaviour in the giraffe. *Journal of Zoology (London)* 151: 313–321.

Cronin, J. E. & Sarich, V. M. 1976. Molecular evidence for dual origin of mangabeys among Old World monkeys. *Nature* 260: 700–702.

Currie, P. 2004. Muscling in on hominid evolution. *Nature* 428: 373–374.

Darwin, C. 1859. *On the Origin of Species by Means of Natural Selection*. John Murray, London, 502 pp.

Davenport, T., Stanley, W., Sargis, E., De Luca, D., Mpunga, N., Machaga, S. & Olson, L. 2006. A new genus of African monkey, *Rungwecebus*: morphology, ecology, and molecular phylogenetics. *Science* 312: 1378–1381.

Dutrillaux, B., Couturier, J., Muleris, M., Lombard, M. & Chauvier, G. 1982. Chromosomal phylogeny of forty-two species or subspecies of Cercopithecoids (Primates, Catarrhini). *Annales de Génétique* 25: 96–109.

Dwyer, P. 1968. The little bent-wing bat: evolution in progress. *Australian Natural History* 16: 55–58.

Fada, C. & Liers, H. 2009. The role of growth stop as a morphogenetic factor in *Mastomys natalensis* (Rodentia: Muridae). *Biological Journal of the Linnean Society* 97: 791–800.

Gautier-Hion, A. 1978. Food niches and coexistence in sympatric primates in Gabon. In: *Recent Advances in Primatology* (eds D. J. Chivers & J. Herbert), Academic Press, London, pp. 269–286.

Goethe, J. W. von. 1790. *Die Metamorphose der Pflanzen*. Carl Wilhelm Ettinger, Gotha.

Horowitz, S. L. & Schapiro, H. 1951. Modification of mandibular architecture following removal of temporalis muscle in the rat. *Journal of Dental Research* 30: 276–280.

Jolly, C. J. 1970. The Seed-Eaters: A new model of hominoid differentiation based on a baboon analogy. *Man* 5: 5–26.

Kingdon, J. 1971. *East African Mammals: An Atlas of Evolution in Africa*. Vol. I: *Primates*. Academic Press, London.

Kingdon, J. 1974. *East African Mammals: An Atlas of Evolution in Africa*. Vol. IIA: *Insectivores and Bats*. Academic Press, London.

Kingdon, J. 1977. *East African Mammals: An Atlas of Evolution in Africa*. Vol. IIIA: *Carnivores*. Academic Press, London.

Kingdon, J. 1979. *East African Mammals: An Atlas of Evolution in Africa*. Vol. IIIB: *Large Mammals*. Academic Press, London.

Kingdon, J. 1980. The role of visual signals and face patterns in African forest monkeys (guenons) of the genus *Cercopithecus*. *Transactions of the Zoological Society of London* 35: 425–475.

Kingdon, J. 1982. *East African Mammals: An Atlas of Evolution in Africa*. Vol. III: *Bovids*. Academic Press, London.

Kingdon, J. 1983. *African Mammal Drawings: the Welcome Volume* (of facsimile reproductions). Pangolin Prints, Islip, UK. R

Kingdon, J. 1988. What are face patterns and do they contribute to reproductive isolation in guenons? In: *A Primate Radiation: Evolutionary Biology of the African Guenons* (eds A. Gautier-Hion, F. Bourlière, J.-P. Gautier & J. Kingdon). Cambridge University Press, Cambridge, pp. 227–245.

Kingdon, J. 1990. *Arabian Mammals: A Natural History*. Academic Press, London, 279 pp.

Kingdon, J. 1997. *The Kingdon Field Guide to African Mammals*. Academic Press, London, 464 pp.

Kingdon, J. 2003. *Lowly Origin*. Princeton University Press, Princeton, 396 pp.

Kuhme, W. 1964. Uber die soziale Bindung innerhalb eines Hyanenhundes Rudels. *Naturwissenschaften* 51.

Kuhme, W. 1965. Freilandstudien zur Soziologie des Hyanenhundes *Lycaon pictus lupinus* Thomas 1902 *Zeitschrift für Tierpsychologie*. 22 (5), 495–541.

Le Gros Clark, W. E. 1959. *The Antecedents of Man: an introduction to the evolution of the primates.* Edinburgh University Press, Edinburgh, 374 pp.

Liers, H., Stuyck, J., Verhagen, R. & Verheyen, W. 1990. Seasonal variation in growth of *Mastomys natalensis* (Rodentia: Muridae) in Morogoro, Tanzania. *African Journal of Ecology* 28: 298–306.

Lovejoy, C. O., Suwa, G., Simpson, S. W., Matternes, J. H. & White, T. D. 2009. The Great Divides: *Ardipithecus ramidus* Reveals the Postcrania of Our Last Common Ancestor with African Apes. *Science* 326: 72–86.

Mitchell, W. J. T. 1994 *Picture Theory: Essays on Verbal and Visual Representation.* University of Chicago Press, Chicago, 462 pp.

Monfort, A. & Monfort, N. 1974. Notes sur l'ecologie et le comportement des oribis (*Ourebia ourebi* Zimmermann 1783). *Terre et Vie* 28: 169–208.

Moore, W. J. 1965. Masticatory function and skull growth. *Journal of Zoology (London)* 146: 123–131.

Penfield, W. & Rasmussen, T. 1950. *The Cerebral Cortex of Man: a clinical study of localization of function.* Macmillan, London, 248 pp.

Schaller, G. 1972. *The Serengeti Lion: A Study of Predator–Prey Relations.* Chicago University Press, Chicago, 504 pp.

Schwartz, E. 1928. Notes on the classification of the African monkeys of the genus *Cercopithecus*. *Annals and Magazine of Natural History* 10: 1.

Simmons, R. E. & Scheepers, L. 1996. Winning by a neck: sexual selection in the evolution of the giraffe. *American Naturalist* 148: 771–786.

Stedman, H. H., Kozyak, B. W., Nelson, A., Thesier, D. M., Su, L. T., Low, D. W., Bridges, C. R., Schrager, J. B., Minugh-Purvis, N. & Mitchell, M. A. 2004. Myosin gene mutation correlates with anatomical changes in the human lineage. *Nature* 428: 415–418.

Struhsaker, T. T. & Leland, L. 1979. Socioecology of five sympatric monkey species in the Kibale Forest, Uganda. *Advances in the Study of Behaviour* 9: 159–228.

Van den Brink, F. H. 1967. *Field Guide to the Mammals of Britain and Europe.* Collins, London, 221 pp.

Wainwright, S. A. 1988. Form and function in organisms. *American Zoologist* 28: 671–680.

Wentworth Thompson, D. 1911. *On Growth and Form.* Cambridge University Press, Cambridge, 1116 pp.

Young, J. Z. 1957. *The Life of Mammals: their anatomy and physiology.* Oxford University Press, Oxford, 192.

Class Mammalia

Alroy, J. 1999. The fossil record of North American mammals. Evidence for a Palaeocene evolutionary radiation. *Systematic Biology* 48: 107–118.

Benton, M. J. 1999. Early origins of modern birds and mammals; molecules vs. morphology. *Bioessays* 21: 1043–1051.

Bininda-Emonds, O. R. P., Cardillo, M., Jones, K. E., MacPhee, R. D. E., Beck, R. M. D., Grenyer, R., Price, S. A., Vos, R. A., Gittleman, J. L. & Purvis, A. 2007. The delayed rise of present-day mammals. *Nature* 446: 507–512.

Bown, T. M. & Kraus, M. J. 1979. Origin of the tribosphenic molar and metatherian and eutherian dental formulae. In: *Mesozoic Mammals* (eds J. A. Lillegraven, Z. Kielan-Jaworowska & W. A. Clemens). University of California Press, Berkeley, CA.

Carroll, R. L. 1988. *Vertebrate Palaeontology and Evolution.* Freeman & Co., New York, 698 pp.

Cheney, D. L. & Seyfarth, R. M. 1990. *How Monkeys See the World: inside the mind of another species.* University of Chicago Press, Chicago, 388 pp.

Cheney, D. L. & Seyfarth, R. M. 2007. *Baboon Metaphysics: the evolution of a Social Mind.* University of Chicago Press, Chicago, 358 pp.

Cifelli, R. L. 2001. Early mammalian radiations. *Journal of Palaeontology* 75: 1214–1226.

Cohen, M. & Larsson, E. 1988. Human endogenous retroviruses. *Bioessays* 9: 191–196.

Crook, J. H. & Gartlan, J. S. 1966. Evolution of primate societies. *Nature* 2001: 1200–1203.

Darwin, C. 1859. *On the Origin of Species by Means of Natural Selection.* John Murray, London, 502 pp.

Dawkins, R. 1986. *The Blind Watchmaker.* W. W. Norton, New York.

Douady, C. J. & Douzery, E. J. P. 2003. Molecular estimation of Eulipotyphla divergence times and the evolution of 'Insectivora'. *Molecular Phylogenetics and Evolution* 28: 285–296.

Dunbar, R. I. M. 1988. *Primate Social Systems.* Cornell University Press, Ithaca, New York, 373 pp.

Foote, M., Hunter, J. P., Janis, C. M. & Sepkowski, J. J. 1999. Evolutionary and preservational constraints on origins of biologic groups: divergence times of eutherian mammals. *Science* 283: 1310–1314.

Fox, D. 1999. Why we don't lay eggs. *New Scientist* 2190: 26.

Hillson, S. 1986. *Teeth.* Cambridge University Press, Cambridge, 388 pp.

Hopson, J. A. 1994. Synapsid evolution and the radiation of non-eutherian mammals. In: *Major Features of Vertebrate Evolution* (eds D. R. Prothero & R. M. Schoch). Palaeontological Society, Lawrence, Kansas.

Hopson, J. A. & Crompton, A. W. 1969. Origin of mammals. *Evolutionary Biology* 3: 15–72.

Jarvis, J. U. M. 1993. Eusociality has evolved independently in two genera of bathyergid mole-rats. *Behavioral Ecology and Sociobiology* 33: 353–360.

Ji, Q., Luo, Z.-X., Yuan, C.-X., Wible, J. R., Zhang, J.-P. & Georgi, J. A. 2002. The earliest known eutherian mammal. *Nature* 416: 816–822.

Kemp, T. S. 1983. The relationships of mammals. *Zoological Journal of the Linnean Society* 77: 353–384.

Kemp, T. S. 2005. *The Origin and Evolution of Mammals.* Oxford University Press, Oxford. 342pp.

Kumar, S. & Hedges, S. B. 1998. A molecular timescale for vertebrate evolution. *Nature* 392: 917–920.

Larsson, E., Andersson, A. C. & Nilsson, B. O. 1994. Expression of an endogenous retrovirus (erv3 HERV-R) in human reproductive and embryonic tissues – evidence for a function for envelope gene products. *Uppsala Journal of Medical Science* 99: 113–120.

Luo, Z.-X., Crompton, A. W. & Sun, A.-L. 2001. A new mammaliaform from the Early Jurassic and evolution of mammalian characteristics. *Science* 292: 1535–1540.

Luo, Z.-X., Kielan-Jaworowska, Z. & Cifelli, R. L. 2002. In quest for a phylogeny of Mesozoic mammals. *Acta Palaeontologica Polonica* 47: 1–78.

Mallet, F., Bouton, O., Prudhomme, S., Cheynet, V., Oriol, G., Bonnaud, B., Lucotte, G., Duret, L. & Mandrand, B. 2004. The endogenous retroviral locus ERVWE1 is a bona fide gene involved in hominoid placental physiology. *Proceedings of the National Academy of Sciences of the United States of America* 101: 1731–1736.

Martin, R. D. 1990. *Primate Origins and Evolution. A Phylogenetic Reconstruction.* Princeton University Press, Princeton, New Jersey.

McKenna, M. C. 1975. Towards a phylogenetic classification of the Mammalia. In: *Phylogeny of the Primates* (eds W. P. Luckett & F. S. Szalay). Plenum Press, New York.

Novacek, M. J. 1986. The primitive eutherian dental formula. *Journal of Vertebrate Palaeontology* 6: 191–196.

Oftedal, O. T. 2002. The mammary gland and its origin during synapsid evolution. *Journal of Mammary Gland Biology and Neoplasia* 7: 225–252.

Romer, A. S. 1945. *Vertebrate Palaeontology.* University of Chicago Press, Chicago.

Rose, K. D. 1996. On the origin of the order Artiodactyla. *Proceedings of the National Academy of Sciences of the United States of America* 93: 1705–1709.

Rose, K. D. 2006. *The Beginning of the Age of Mammals.* Johns Hopkins University Press, Baltimore, 448 pp.

Sarich, V. M. & Wilson, A. C. 1967. Immunological time scale for hominid evolution. *Science* 158: 1200–1203.

Seyfarth, R. M. & Cheney, D. L. 1984. Grooming, alliances, and reciprocal altruism in vervet monkeys. *Nature* 308: 541–543.

Sherman, P.W., Jarvis, J. U. M. & Braude, S. H. 1992. Naked mole-rats. *Scientific American* 267: 42–48.

Springer, M. S., Stanhope, M. J., Madsen, O. & de Jong, W. W. 1997. Endemic African mammals shake the phylogenetic tree. *Nature* 388: 61–64.

Springer, M. S., Murphy, W. J., Eizirik, E. & O'Brien, S. J. 2003. Placental mammal diversification and the Cretaceous–Tertiary boundary. *Proceedings of the National Academy of Sciences of the United States of America* 100: 1056–1061.

Springer, M. S., Stanhope, M. J., Madsen, O. & de Jong, W. W. 2004. Molecules consolidate the placental mammal tree. *Trends in Ecology and Evolution* 19: 430–438.

Vorbach, C., Capecchi, M. R. & Penninger, J. M. 2006. Evolution of the mammary gland from the innate immune system? *Bioessays* 28: 606–616.

Wible, J. R., Rougier, G. W., Novacek, M. J. & Asher, R. J. 2007. Cretaceous eutherians and Laurasian origin for placental mammals near the KT boundary. *Nature* 447: 1003–1008.

Supercohort Afrotheria

Amrine-Madsen, H., Koepfli, K-P., Wayne, R. K. & Springer, M. S. 2003. A new phylogenetic marker, apolipoprotein B, provides compelling evidence for eutherian relationships. *Molecular Phylogenetics and Evolution* 28: 225–240.

Andrews, C. W. 1906. *A descriptive catalogue of the Tertiary Vertebrata of the Fayum, Egypt; based on the collection of the Egyptian Government in the Geological Museum, Cairo, and on the collection in the British Museum (Natural History).* Trustees of the British Museum (Natural History).

Archibald, J. D. 2003. Timing and biogeography of the eutherian radiation: fossils and molecules compared. *Molecular Phylogenetics and Evolution* 28: 350–359.

Asher, R. J. & Seiffert, E. R. 2010. Systematics of endemic African mammals. In: *Cenozoic Mammals of Africa* (eds. L. Werdelin & W. J. Sanders). University of California Press, Berkeley. pp. 903–920.

Bronner, G. N. & Jenkins, P. D. 2005. Order Afrosoricida. In: *Mammal Species of the World: A Taxonomic and Geographic Reference* (Third Edition) (eds. D. E. Wilson & D. M. Reeder). Johns Hopkins University Press, Baltimore. pp. 70–81.

Butler, P. M. 1956. The skull of *Ictops* and the classification of the Insectivora. *Proceedings of the Zoological Society of London* 126: 453–481.

Butler, P. M. 1972. The problem of insectivore classification. In: *Studies in Vertebrate Evolution* (eds. K. A. Joysey & T. S. Kemp). Oliver and Boyd, Edinburgh. pp. 253–265.

Cote, S., Werdelin, L., Seiffert, E. R. & Barry, J. C. 2007. Additional material of the early Miocene mammal Kelba and its relationship to the order Ptolemaiida. *Proceedings of the National Academy of Sciences of the United States of America* 104: 5510–5515.

De Jong, W. W., Leunissen, J. A. M. & Wistow, G. J. 1993. Eye lens crystallins and the phylogeny of placental orders: evidence for a macroscelid-paenungulate clade? In: *Mammal Phylogeny: Placentals* (eds. F. S. Szalay, M. J. Novacek & M. C. McKenna). Springer-Verlag, Berlin. pp. 5–12.

De Jong, W. W., Zweers, A. & Goodman, M. 1981. Relationship of aardvark to elephants, hyraxes and sea cows from alpha-crystallin sequences. *Nature* 292: 538–540.

Fischer, M. S. 1989. Hyracoids, the sister-group of perissodactyls. In: *The Evolution of Perissodactyls* (eds. D. R. Prothero & R. M. Schoch). Oxford University Press, New York. pp. 37–56.

Gheerbrant, E. 1992. Les mammifères paléocenes du Bassin d'Ouarzazate (Maroc). I. Introduction général et palaeoryctidae. *Palaeontographica* Abt. A 224: 67–132.

Gheerbrant, E. 1994. The Paleocene mammals from the Ouarzazate Basin (Morocco) II. Todralestidae (Proteutheria, Eutheria). *Palaeontographica* Abt. A 231: 133–188.

Gheerbrant, E. 1995. Les mammifères paléocenes du Bassin d'Ouarzazate (Maroc). III. Adapisoriculidae et autres mammifères (Carnivora, ?Creodonta, Condylarthra, ?Ungulata et *incertae sedis*). *Palaeontographica*, Abt. A 237: 39–132.

Gheerbrant, E. 2009. Paleocene emergence of elephant relatives and the rapid radiation of African ungulates. *Proceedings of the National Academy of Sciences of the United States of America* 106: 10717–10721.

Gheerbrant, E., Sudre, J., Capetta, H., Iarochène, M., Amaghzaz, M. & Bouya, B. 2002. A new large mammal from the Ypresian of Morocco: evidence of surprising diversity of early proboscideans. *Acta Palaeontologica Polonica* 47: 493–506.

Gheerbrant, E., Sudre, J., Sen, S., Abrial, C., Marandat, B., Sigé, B. & Vianey-Liaud, M. 1998. Nouvelles données sur les mammifères du Thanétien et de l'Yprésian du Bassin d'Ouarzazate (Maroc) et leur contexte stratigraphique. *Palaeovertebrata* 27: 155–202.

Gheerbrant, E., Sudre, J., Tassy, P., Amaghzaz, M., Bouya, B. & Iarochene, M. 2005. Nouvelles données sur Phosphatherium escuilliei (Mammalia, Proboscidea) de l'Éocène inférieur du Maroc, apports à la phylogénie des Proboscidea et des ongulés lophodontes. *Geodiversitas* 27: 239–333.

Gill, T. N. 1872. Arrangement of the families of mammals with analytical tables. *Smithsonian Miscellaneous Collections* 11: 1–98.

Goswami, A., Prasad, G. V. R., Upchurch, P., Boyer, D. M., Seiffert, E. R., Verma, O., Gheerbrant, E. & Flynn, J. J. 2011. A radiation of arboreal basal eutherian mammals beginning in the Late Cretaceous of India. *Proceedings of the National Academy of Sciences of the United States of America* 108: 16333–16338.

Hartenberger, J-L. 1986. Hypothèse paléontologique sur l'origine des Macroscelidea (Mammalia). *Comptes Rendus de l'Academie des Sciences,* Serie IIa 302: 247–249.

Kappelman, J., Rasmussen, D. T., Sanders, W. J., Feseha, M., Bown, T., Copeland, P., Crabaugh, J., Fleagle, J., Glantz, M., Gordon, A., Jacobs, B., Maga, M., Muldoon, K., Pan, A., Pyne, L., Richmond, B., Ryan, T., Seiffert, E. R., Sen, S., Todd, L., Wiemann, M. C. & Winkler, A. 2003. Oligocene mammals from Ethiopia and faunal exchange between Afro-Arabia and Eurasia. *Nature* 426: 549–552.

Kingdon, J. 1979. *East African Mammals*. Volume IIIB: *Large Mammals*. Academic Press, London.

Lavergne, A., Douzery, E., Stichler, T., Catzeflis, F. M. & Springer, M. S. 1996. Interordinal mammalian relationships: evidence for paenungulate monophyly is provided by complete mitochondrial 12S rRNA sequences. *Molecular Phylogenetics and Evolution* 6: 245–258.

MacInnes, D. G. 1956. Fossil Tubulidentata from East Africa. *Fossil Mammals of Africa* 10: 1–52.

Madsen, O., Scally, M., Douady, C. J., Kao, D. J., DeBry, R. W., Adkins, R., Amrine, H. M., Stanhope, M. J., de Jong, W. W. & Springer, M. S. 2001. Parallel adaptive radiations in two major clades of placental mammals. *Nature* 409: 610–614.

Marivaux, L., Adaci, M., Bensalah, M., Rodrigues, H. G., Hautier, L., Mahboubi, M. h., Mebrouk, F., Tabuce, R. & Vianey-Liaud, M. in press. Zegdoumyidae (Rodentia, Mammalia), stem anomaluroid rodents from the Early to Middle Eocene of Algeria (Gour Lazib, Western Sahara): new dental evidence. *Journal of Systematic Palaeontology*

McKenna, M. 1975. Toward a phylogenetic classification of the Mammalia. In: *Phylogeny of the Primates* (eds. W. P. Luckett & F. S. Szalay). Plenum Press, New York. pp. 21–46.

Meredith, R. W., Janečka, J. E., Gatesy, J., Ryder, O. A., Fisher, C. A., Teeling, E. C., Goodbla, A., Eizirik, E., Simão, T. L. L., Stadler, T., Rabosky, D. L., Honeycutt, R. L., Flynn, J. J., Ingram, C. M., Steiner, C., Williams, T. L., Robinson, T. J., Burk-Herrick, A., Westerman, M., Ayoub, N. A., Springer, M. S. & Murphy, W. J. 2011. Impacts of the Cretaceous Terrestrial Revolution and KPg extinction on mammal diversification. *Science* 334: 521–524.

Mess, A. & Carter, A. M. 2006. Evolutionary transformations of fetal membrane characters in Eutheria with special reference to Afrotheria. *Journal of Experimental Zoology. B. Molecular and Developmental Evolution* 306: 140–163.

Murata, Y., Nikaido, M., Sasaki, T., Cao, Y., Fukumoto, Y., Hasegawa, M. & Okada, N. 2003. Afrotherian phylogeny inferred from complete mitochondrial genomes. *Molecular Phylogenetics and Evolution* 28: 253–260.

Murphy, W. J., Eizirik, E., Johnson, W. E., Zhang, Y. P., Ryder, O. A. & O'Brien, S. J. 2001a. Molecular phylogenetics and the origins of placental mammals. *Nature* 409: 614–618.

Murphy, W. J., Eizirik, E., O'Brien, S. J., Madsen, O., Scally, M., Douady, C. J., Teeling, E., Ryder, O. A., Stanhope, M. J., de Jong, W. W. & Springer, M. S. 2001b. Resolution of the early placental mammal radiation using Bayesian phylogenetics. *Science* 294: 2348–2351.

Nessov, L. A., Zhegallo, V. I. & Averianov, A. O. 1998. A new locality of Late Cretaceous snakes, mammals and other vertebrates in Africa (western Libya). *Annales de Paléontologie* 84: 265–274.

Nikaido, M., Nishihara, H., Hukumoto, Y. & Okada, N. 2003. Ancient SINEs from African endemic mammals. *Molecular Biology and Evolution* 20: 522–527.

Nishihara, H., Satta, Y., Nikaido, M., Thewissen, J. G. M., Stanhope, M. J. & Okada, N. 2005. A retroposon analysis of afrotherian phylogeny. *Molecular Biology and Evolution* 22: 1823–1833.

Novacek, M. J. 1986. The skull of leptictid insectivorans and the higher-level classification of eutherian mammals. *Bulletin of the American Museum of Natural History* 183: 1–112.

Novacek, M. J. 1992. Mammalian phylogeny: shaking the tree. *Nature* 356: 121–125.

Patterson, B. 1975. The fossil aardvarks (Mammalia: Tubulidentata). *Bulletin of the Museum of Comparative Zoology at Harvard College* 147: 185–237.

Prothero, D. R., Manning, E. M. & Fischer, M. 1988. The phylogeny of the ungulates. In: *Phylogeny and Classification of the Tetrapods* (ed. M. J. Benton). Clarendon Press, Oxford. pp. 201–234.

Rasmussen, D. T. & Gutierrez, M. 2009. A mammalian fauna from the late Oligocene of northwestern Kenya. *Palaeontographica Abteilung A* 288: 1–52.

Rasmussen, D. T. & Simons, E. L. 2000. Ecomorphological diversity among Paleogene hyracoids (Mammalia): a new cursorial browser from the Fayum, Egypt. *Journal ofVertebrate Paleontology* 20: 167–176.

Robinson, T. J., Fu, B., Ferguson-Smith, M. A. & Yang, F. 2004. Cross-species chromosome painting in the golden mole and elephant-shrew: support for the mammalian clades Afrotheria and Afroinsectiphillia but not Afroinsectivora. *Proceedings of the Royal Society of London B* 271: 1477–1484.

Robinson, T. J. & Seiffert, E. R. 2004. Afrotherian origins and interrelationships: new views and future prospects. *Currents Topics in Developmental Biology* 63: 37–60.

Sánchez-Villagra, M. R., Narita, Y. & Kuratani, S. 2007. Thoracolumbar vertebral number: the first skeletal synapomorphy for afrotherian mammals. *Systematics and Biodiversity* 5: 1–7.

Scally, M., Madsen, O., Douady, C. J., de Jong, W. W., Stanhope, M. J. & Springer, M. S. 2001. Molecular evidence for the major clades of placental mammals. *Journal of Mammalian Evolution* 8: 239–277.

Seiffert, E. R. 2002. The reality of afrotherian monophyly, and some of its implications for the evolution and conservation of Afro-Arabia's endemic placental mammals. *Afrotherian Conservation* 1: 3–6.

Seiffert, E. R. 2003. A phylogenetic analysis of living and extinct Afrotherian placentals. Ph.D. thesis, Duke University, USA.

Seiffert, E. R. 2007. A new estimate of afrotherian phylogeny based on simultaneous analysis of genomic, morphological, and fossil evidence. *BMC Evolutionary Biology* 7: 224.

Shoshani, J. 1986. Mammalian phylogeny: comparison of morphological and molecular results. *Molecular Biology and Evolution* 3: 222–242.

Shoshani, J. 1992. The controversy continues: an overview of the evidence for Hyracoidea–Tethytheria affinity. *Israel Journal of Zoology* 38: 233–244.

Sigé, B., Jaeger, J-J., Sudre, J. & Vianey-Liaud, M. 1990. *Altiatlasius koulchii* n. gen. n. sp., primate omomyidé du Paléocène du Maroc, et les origines des euprimates. *Palaeontographica* Abt. A 214: 31–56.

Simons, E. L., Holroyd, P. A. & Bown, T. M. 1991. Early Tertiary elephant shrews from Egypt and the origin of the Macroscelidea. *Proceedings of the National Academy of Sciences of the United States of America* 58: 9734–9737.

Simpson, G. G. 1945. The principles of classification and a classification of mammals. *Bulletin of the American Museum of Natural History* 85: 1–350.

Solé, F., Gheerbrant, E., Amaghzaz, M. & Bouya, B. 2009. Further evidence of the African antiquity of hyaenodontid ('Creodonta', Mammalia) evolution. *Zoological Journal of the Linnean Society* 156: 827–846.

Springer, M. S., Amrine, H. M., Burk, A. & Stanhope, M. J. 1999. Additional support for Afrotheria and Paenungulata, the performance of mitochondrial versus nuclear genes, and the impact of data partitions with heterogeneous base compounds. *Systematic Biology* 48: 65–75.

Springer, M. S., Cleven, G. C., Madsen, O., de Jong, W. W., Waddell, V. G., Amrine, H. M. & Stanhope, M. J. 1997. Endemic African mammals shake the phylogenetic tree. *Nature* 388: 61–64.

Springer, M. S., Murphy, W. J., Eizirik, E. & O'Brien, S. J. 2003. Placental mammal diversification and the Cretaceous-Tertiary boundary. *Proceedings of the National Academy of Sciences of the United States of America* 100: 1056–1061.

Springer, M. S., Stanhope, M. J., Madsen, O. & de Jong, W. W. 2004. Molecules consolidate the placental mammal tree. *Trends in Ecology and Evolution* 19: 430–438.

Stanhope, M. J., Madsen, O., Waddell, V. G., Cleven, G. C., de Jong, W. W. & Springer, M. S. 1998. Highly congruent molecular support for a diverse superordinal clade of endemic African mammals. *Molecular Phylogenetics and Evolution* 9: 501–508.

Szalay, F. S. 1977. Phylogenetic relationships and a classification of the eutherian Mammalia. In: *Major Patterns of Vertebrate Evolution* (eds. M. K. Hecht, P. C. Goody & B. M. Hecht). Plenum Press, New York. pp. 375–394.

Tabuce, R., Marivaux, L., Adaci, M., Bensalah, M., Hartenberger, J-L., Mahboubi, M., Mebrouk, F., Tafforeau, P. & Jaeger, J.-J. 2007. Early Tertiary mammals from North Africa reinforce the molecular Afrotheria clade. *Proceedings of the Royal Societyof London B* 274: 1159–1166.

Van Dijk, M. A. M., Madsen, O., Catzeflis, F., Stanhope, M. J., de Jong, W. W. & Pagel, M. 2001. Protein sequence signatures support the African clade of mammals. *Proceedings of the National Academy of Sciences of the United States of America* 98: 188–193.

Vianey-Liaud, M., Jaeger, J-J., Hartenberger, J-L. & Mahboubi, M. 1994. Les rongeurs de l'Eocène d'Afrique nord-occidentale [Glib Zegdou (Algérie) et Chambi (Tunisie)] et l'origine des Anomaluridae. *Palaeovertebrata* 93–118.

Waddell, P. J., Kishino, H. & Ota, R. 2001. A phylogenetic foundation for comparative mammalian genomics. *Genome Informatics* 12: 141–154.

Waddell, P. J. & Shelley, S. 2003. Evaluating placental inter-ordinal phylogenies with novel sequences including RAG1, gamma-fibrinogen, ND6, and mt-tRNA, plus MCMC-driven nucleotide, amino acid, and codon models. *Molecular Phylogenetics and Evolution* 28: 197–224.

Wagner, J. A. 1855. *Die Säugethiere in Abbildungen nach der Natur.* Weiger, Leipzig.

Werdelin, L. & Nilsonne, Å. 1999. The evolution of the scrotum and testicular descent in mammals: a phylogenetic view. *Journal of Theoretical Biology* 196: 61–72.

Wible, J. R., Rougier, G. W., Novacek, M. J. & Asher, R. J. 2007. Cretaceous eutherians and Laurasian origin for placental mammals near the K/T boundary. *Nature* 447: 1003–1006.

Zack, S. P., Penkrot, T. A., Bloch, J. I. & Rose, K. D. 2005. Affinities of 'hyopsodontids' to elephant shrews and a Holarctic origin of Afrotheria. *Nature* 434: 497–501.

Cohort Paenungulata

Andrews, C. W. 1906. *A descriptive catalogue of the Tertiary Vertebrata of the Fayum, Egypt; based on the collection of the Egyptian Government in the Geological Museum, Cairo, and on the collection in the British Museum (Natural History).* Trustees of the British Museum (Natural History), 324 pp.

Asher, R. J., Novacek, M. J. & Geisler, J. H. 2003. Relationships of endemic African mammals and their fossil relatives based on morphological and molecular evidence. *Journal of Mammalian Evolution* 10: 131–194.

Gheerbrant, E. 2009. Paleocene emergence of elephant relatives and the rapid radiation of African ungulates. *Proceedings of the National Academy of Sciences of the United States of America* 106: 10717–10721.

Gheerbrant, E., Sudre, J., Sen, S., Abrial, C., Marandat, B., Sigé, B. & Vianey-Liaud, M. 1998. Nouvelles données sur les mammifères du Thanétien et de l'Yprésian du Bassin d'Ouarzazate (Maroc) et leur contexte stratigraphique. *Palaeovertebrata* 27: 155–202.

Gheerbrant, E., Sudre, J., Capetta, H., Mourer-Chauviré, C., Bourdon, E., Iarochene, M., Amaghzaz, M. & Bouya, B. 2003. Les localités à mammifères des carrières de Grand Daoui, bassin des Ouled Abdoun, Maroc, Yprésien: premier état des lieux. *Bulletin de la Société Geologique de France* 174: 279–293.

Gheerbrant, E., Domning, D. P. & Tassy, P. 2005a. Paenungulata (Sirenia, Proboscidea, Hyracoidea, and relatives). In: *The Rise of Placental Mammals: Origins and Relationships of the Major Extant Clades* (eds K. D. Rose & J. D. Archibald). Johns Hopkins University Press, Baltimore. pp. 84–105.

Gheerbrant, E., Sudre, J., Tassy, P., Amaghzaz, M., Bouya, B. & Iarochene, M. 2005b. Nouvelles données sur *Phosphatherium escuilliei* (Mammalia, Proboscidea) de l'Éocène inférieur du Maroc, apports à la phylogénie des Proboscidea et des ongulés lophodontes. *Geodiversitas* 27: 239–333.

Gregory, W. K. 1910. The orders of mammals. *Bulletin of the American Museum of Natural History* 27: 1–525.

Lucas, S. G. 1993. Pantodonts, tillodonts, uintatheres, and pyrotheres are not ungulates. In: *Mammal Phylogeny: Placentals* (eds F. S. Szalay, M. J. Novacek & M. C. McKenna). Springer-Verlag, Berlin. pp. 182–194.

Rasmussen, D. T., Gagnon, M. & Simons, E. L. 1990. Taxeopody in the carpus and tarsus of Oligocene Pliohyracidae (Mammalia: Hyracoidea) and the phyletic position of hyraxes. *Proceedings of the National Academy of Sciences of the United States of America* 87: 4688–4691.

Savage, R. J. G., Domning, D. P. & Thewissen, J. G. M. 1994. Fossil Sirenia of the west Atlantic and Caribbean region. V. The most primitive known sirenian, *Prorastomus sirenoides* Owen, 1855. *Journal of Vertebrate Paleontology* 14: 427–449.

Shoshani, J. 1993. Hyracoidea–Tethytheria affinity based on myological data. In: *Mammal Phylogeny: Placentals* (eds. F. S. Szalay, M. J. Novacek & M. C. McKenna). Springer Verlag, New York. pp. 235–256.

Springer, M. S., Stanhope, M. J., Madsen, O. & de Jong, W. W. 2004. Molecules consolidate the placental mammal tree. *Trends in Ecology and Evolution* 19: 430–438.

Order Hyracoidea

Allen, G. M. 1939 (and reprint 1945). A checklist of African mammals. *Bulletin of the Museum of Comparative Zoology at Harvard College* 83: 1–763.

Ansell, W. F. H. 1978. *The Mammals of Zambia*. The National Parks and Wildlife Service, Chilanga, Zambia, 126 pp.

Ansell, W. F. H. & Dowsett, R. J. 1988. *Mammals of Malawi*. Trendrine Press, Zennor, Cornwall, 170 pp.

Ashford, R. W. 1970. A possible reservoir for Leishmania tropica in Ethiopia. *Transactions of the Royal Society of Tropical Medicine and Hygiene* 64: 936–937.

Ashford, R. W., Bray, M. A., Hutchinson, M. P. & Bray, R. S. 1973. The epidemiology of cutaneous leishmaniasis in Ethiopia. *Transactions of the Royal Society of Tropical Medicine and Hygiene* 67: 568–601.

Aumann, T. & Chiweshe, N. 1995. Diet notes for the black eagle *Aquila verreauxii* in areas of different land use in the Matobo Hills, Zimbabwe. *Journal of African Raptor Biology* 10: 8–11.

Barry, R. E. 1994. Synchronous parturition of *Procavia capensis* and *Heterohyrax brucei* during drought in Zimbabwe. *South African Journal of Wildlife Research* 24: 1–5.

Barry, R. E. & Barry, L. M. 1996. Species composition and age structure of remains of hyraxes (Hyracoidea: Procaviidae) at nests of black eagles. *Journal of Mammalogy* 77: 702–707.

Barry, R. E. & Mundy, P. J. 1998. Population dynamics of two species of hyraxes in the Matobo National Park, Zimbabwe. *African Journal of Ecology* 36: 221–233.

Barry, R. E. & Mundy, P. J. 2002. Seasonal variation in the degree of heterospecific association of two syntopic hyraxes (*Heterohyrax brucei* and *Procavia capensis*) exhibiting synchronous parturition. *Behavioral Ecology and Sociobiology* 52: 177–181.

Barry, R. & Shoshani, J. 2000. *Heterohyrax brucei*. *Mammalian Species* 645: 1–7.

Bartholomew, G. A. & Rainy, M. 1971. Regulation of body temperature in the rock hyrax, *Heterohyrax brucei*. *Journal of Mammalogy* 52: 81–95.

Bjornhag, G., Becker, G., Buchholz, C. & Engelhardt, W. von 1994. The gastrointestinal tract of the Rock Hyrax (*Procavia habessinica*). 1. Morphology and motility patterns of the tract. *Comparative Biochemistry and Physiology A – Physiology* 109: 649–653.

Boshoff, A. F., Palmer, N. G., Avery, G., Davies, R. A. G. & Jarvis, M. F. J. 1991. Biographical and topographical variation in the prey of the black eagle in the Cape Province, South Africa. *Ostrich* 62: 59–72.

Boshoff, A. F , Palmer, N. G., Vernon, C. J. & Avery, G. 1994. Comparison of the diet of crowned eagles in the Savanna and Forest Biomes of south-eastern South Africa. *South African Journal of Wildlife Research* 24: 26–31.

Bothma, J. du P. 1964. A taxonomic revision of the extant Hyracoidea of southern Africa. MSc thesis, University of Pretoria, South Africa.

Bothma, J. du P. 1966. Colour variation in Hyracoidea from southern Africa. *Journal of Mammalogy* 47: 687–693.

Bothma, J. du P. 1967. Recent Hyracoidea (Mammalia) of southern Africa. *Annals of the Transvaal Museum* 25: 117–152.

Bothma, J., du P. 1971. Order Hyracoidea. In: *The Mammals of Africa: An Identification Manual* (eds J. Meester & H. W. Setzer). Smithsonian Institution Press, Washington, DC. Part 12: 1–8.

Brown, K. J. & Downs, C. T. 2005. Seasonal behavioural patterns of free-living rock hyrax (*Procavia capensis*). *Journal of Zoology, London* 265: 311–326.

Burgess, N. D., Kock, D., Cockle, A., FitzGibbon, C., Jenkins, P. & Honess, P. 2000. Mammals. In: *Coastal Forests of Eastern Africa* (eds N. D. Burgess & G. P. Clarke). IUCN Forest Conservation Programme. IUCN, Gland, Switzerland and Cambridge, UK, pp. 173–190.

Caro, T. M. & Alawi, R. M. 1985. Comparative aspects of behavioural development in two species of free-living hyrax. *Behaviour* 95: 87–109.

Carroll, R. L. 1988. *Vertebrate Paleontology and Evolution*. W. H. Freeman and Company, New York, 698 pp.

Castley, J. G. & Kerley, G. I. H. 1993. Trends in indigenous forest patch size in the North-Eastern Cape. In: *The Relevance of Island Biogeographic Theory in Commercial Forestry* (ed. D. A. Everard). Environmental Forum Report, FRD, Pretoria, pp. 148–149.

Chen, G.-F. 2003. A new genus of Pliohyracinae (Hyracoidea, Mammalia) from the Late Pliocene of Dege, Sichuan, China. *Vertebrata Palasiatica* 41: 240–248.

Churcher, C. S. 1956. The fossil Hyracoidea of the Transvaal and Taungs deposits. *Annals of the Transvaal Museum* 22: 477–501.

Clemens, E. T. 1977. Sites of organic acid production and patterns of digesta movement in the gastrointestinal tract of the rock hyrax. *Journal of Nutrition* 107: 1954–1961.

Clemens, E. T. & Maloiy, G. M. O. 1978. Cyclic changes in the osmolality and electrolyte composition in the gastrointestinal tract of the rock hyrax. *Journal of Nutrition* 108: 988–993.

Coe, M. J. 1962. Notes on the habits of the Mount Kenya hyrax (*Procavia johnstoni mackinderi* Thomas). *Proceedings of the Zoological Society of London* 138: 639–644.

Coetzee, C. G. 1966. The relative position of the penis in southern African dassies (Hydracoidea) as a character of taxonomic importance. *Zoologica Africana* 2: 223–224.

Corbet, G. B. 1978. *The Mammals of the Palaearctic Region: A Taxonomic Review*. British Museum (Natural History), London. Cornell University Press, Ithaca, NY.

Corbet, G. B. 1979. The taxonomy of *Procavia capensis* in Ethiopia, with special reference to the aberrant tusks of *P. c. capillosa* Brauer (Mammalia, Hyracoidea). *Bulletin of the British Museum (Natural History) Zoology* 36: 251–259.

Cordeiro, N. J., Seddon, N., Capper, D. R., Ekstrom, J. M. M., Howell, K. M., Isherwood, I. S., Msuya, C. A. M., Mushi, J. T., Perkin, A. W., Pople, R. G. & Stanley, W. T. 2005. Notes on the ecology and status of some forest mammals in four Eastern Arc Mountains, Tanzania. *Journal of the East Africa Natural History Society* 94: 175–189.

Court, N. & Hartenberger, J.-L. 1992. A new species of the hyracoid mammal *Titanohyrax* from the Eocene of Tunisia. *Palaeontology* 35: 309–317.

Court, N. & Mahboubi, M. 1993. Reassessment of Lower Eocene *Seggeurius amourensis*: aspects of primitive dental morphology in the mammalian order Hyracoidea. *Journal of Paleontology* 67: 889–893.

Crawford-Cabral, J. & Veríssimo, L. N. 2005. *The Ungulate Fauna of Angola: Systematic List, Distribution Maps, Database Report*. Instituto de Investigação Científica Tropical, Lisboa, 277 pp.

Davies, R. A. G. 1994. Black eagle *Aquila verreauxii* predation on Rock Hyrax *Procavia capensis* and other prey in the Karoo. PhD thesis, University of Pretoria, South Africa.

De Barros Machado, A. 1969. *Mamiferos de Angola ainda naô citados ou ponco conhecidos*. Museo do Dundo, Lisboa.

De Smet, K. 1989. Studie van de verspreiding en biotoopkeuze van de grote mammalia in Algerije in het kader van het naturbehoud. PhD thesis, Rijskuniversiteit Gent, Belgium. [Distribution and habitat choice of larger mammals in Algeria, with special reference to nature protection; English translation from Dutch by the World Conservation Monitoring Centre, Cambridge.]

DeBlieux, D. D., Baumrind, M. R., Simons, E. L., Chatrath, P. S., Meyer, G. E. & Attia, Y. S. 2006. Sexual dimorphism of the internal mandibular chamber in Fayum Pliohyracidae (Mammalia). *Journal of Vertebrate Paleontology* 26: 160–169.

Dekeyser, P. 1955. *Les mammiferes de l'Afrique Noire Française*. Institut Français d'Afrique Noire, Dakar, 426 pp.

DeNiro, M. J. & Epstein, S. 1978. Carbon isotopic evidence for different feeding patterns in two hyrax species occupying the same habitat. *Science* 201: 906–908.

Dobson, G. E. 1876. On peculiar structures in the feet of certain mammals which enable them to walk on smooth perpendicular surfaces. *Proceedings of the Zoological Society of London* 1876: 526–534.

Dor, M. 1987. *Zoological Lexicon: Vertebrata*. Dvir Publishers, Tel-Aviv (Israel), 2 vols.

Dorst, J. & Dandelot, P. 1970. *A Field Guide to the Larger Mammals of Africa*. Collins, London, 287 pp.

Dubrovo, I. A. 1978. New data on fossil Hyracoidea. *Paleontological Journal* 12: 375–383.

Eley, R. M. 1994. The hyrax: a most mysterious mammal. *Biologist* 41: 141–144.

Ellerman, J. R., Morrison-Scott, T. C. S. & Hayman, R. W. 1953. *Southern African Mammals 1758 to 1951: A Reclassification*. British Museum (Natural History), London.

Eloff, A. K. & Van Hoven, W. 1985. Volatile fatty acid production in the hindgut of *Procavia capensis*. *Comparative Biochemistry and Physiology* A 80: 291–295.

Eltringham, S. K., Morley, R. J., Kingdon, J., Coe, M. J. & McWilliam, N. C. 1998. Checklist: mammals of Mkomazi. In: *Mkomazi: The Ecology, Biodiversity and Conservation of a Tanzanian Savanna* (eds M. Coe, N. McWilliam, G. Stone & M. Packer). Royal Geographical Society (with The Institute of British Geographers), London, pp. 505–510.

Engelhart, W. von, Wolter, S., Lawrenz, H. & Homsley, J. A. 1978. Production of methane in two non-ruminant herbivores. *Comparative Biochemistry and Physiology A* 60: 309–311.

Estes, R. D. 1991. *The Behavior Guide to African Mammals: Including Hoofed Mammals, Carnivores and Primates*. University of California Press, Berkeley and Los Angeles, 601 pp.

Fa, J. E., Yuste, J. E. G. & Castelo, R. 2000. Bushmeat markets on Bioko Island as a measure of hunting pressure. *Conservation Biology* 14: 1602–1613.

Fain, A. & Lukoschus, F. S. 1981. A new Psoroptidae (Acari: Astigmata) from *Dendrohyrax dorsalis* in Zaire. *International Journal of Acarology* 7: 143–146.

Fairall, N., Vermeulen, P. J. & Van der Merwe, M. 1986. A general model of population growth in the hyrax *Procavia capensis*. *Ecological Modelling* 34: 115–132.

Fischer, M. S. 1989. Hyracoids, the sister-group of perissodactyls. In: *The Evolution of Perissodactyls* (eds D. R. Prothero & R. M. Schoch). Oxford University Press, New York & London, pp. 37–56.

Fischer, M. S. 1992. *Hyracoidea. Handbuch der Zoologie. Band VIII Mammalia*. Walter de Gruyter, Berlin & New York.

Flower, W. H. & Lydekker, R. 1891. *An Introduction to the Study of Mammals Living and Extinct*. Adam and Charles Black, London, 763 pp.

Fourie, L. J. 1983. The population dynamics of the rock hyrax *Procavia capensis* (Pallas 1766) in the Mountain Zebra National Park. PhD thesis, Rhodes University, South Africa.

Fourie, L. J. & Perrin, M. R. 1985. Seasonal change in the body fat of the hyrax *Procavia capensis* (Pallas 1766) using a body fat rating index. *Koedoe* 28: 61–67.

Fourie, L. J. & Perrin, M. R. 1987a. Social behaviour and spatial relationships of the Rock Hyrax. *South African Journal of Wildlife Research* 17: 91–98.

Fourie, L. J. & Perrin, M. R. 1987b. Some new data on the reproductive biology of the Rock Hyrax. *South African Journal of Wildlife Research* 17: 118–122.

Fourie, L. J. & Perrin, M. R. 1989. Quantitative and qualitative aspects of the diet of the rock hyrax (*Procavia capensis* Pallas, 1766) in the Mountain Zebra National Park. *Journal of African Zoology* 103: 361–370.

Fourie, L. J., Horak, I. G. & Visser, E. 1987. Quantitative and qualitative aspects of the parasite burdens of rock dassies (*Procavia capensis* Pallas, 1766) in the Mountain Zebra National Park. *South African Journal of Zoology* 22: 101–106.

Fourie, P. B. 1977. Acoustic communication in the rock hyrax, *Procavia capensis*. *Zeitschrift für Tierpsychologie* 44: 194–219.

Gargett, V. 1990. *The Black Eagle: A Study*. Acorn Books, Randburg, South Africa, 279 pp.

Gargett, V., Gargett, E. & Damania, D. 1995. The influence of rainfall on black eagle breeding over 31 years in the Matobo hills, Zimbabwe. *Ostrich* 66: 114–121.

Gasc, J.-P. 1967. *Squelette hyobranchial*. In: *Traité de Zoologie* (ed. P.-P. Grasse) 16 (1): 550–583, 1103–1106. Masson et Cie Editeurs, Paris.

Gaylard, A. 1994. The tree hyrax (*Dendrohyrax arboreus*) as a rare forest species: habitat and diet. MSc thesis, University of Port Elizabeth, South Africa.

Gaylard, A. & Kerley, G. I. H. 1997. Diet of tree hyraxes *Dendrohyrax arboreus* (Hyracoidea: Procaviidae) in the Eastern Cape, South Africa. *Journal of Mammalogy* 78: 213–221.

Gaylard, A. & Kerley, G. I. H. 2001. Habitat assessment for a rare, arboreal forest mammal, the tree hyrax *Dendrohyrax arboreus*. *African Journal of Ecology* 39: 205–212.

George, W. & Crowther, G. 1981. Space partitioning between two small mammals in a rocky desert. *Biological Journal of the Linnean Society* 15: 195–200.

Gerlach, G. & Hoeck, H. N. 2001. Islands on the plains: metapopulation dynamics and female biased dispersal in hyraxes (Hyracoidea) in the Serengeti National Park. *Molecular Ecology* 10: 2307–2317.

Gheerbrant, E., Sudre, J., Capetta, H., Mourer-Chauviré, C., Bourdon, E., Iarochene, M., Amaghzaz, M. & Bouya, B. 2003. Les localités à mammifères des carrières de Grand Daoui, bassin des Ouled Abdoun, Maroc, Yprésien: premier état des lieux. *Bulletin de la Société Géologique de France* 174: 279–293.

Glover, T. D. & Sale, J. B. 1968. The reproductive system of male rock hyrax (*Procavia* and *Heterohyrax*). *Journal of Zoology, London* 156: 351–362.

Gombe, S. 1983. Reproductive cycle of the Rock Hyrax (*Procavia capensis*). *African Journal of Ecology* 21: 129–133.

Gray, J. E. 1868. Revision of the species of *Hyrax*, founded on the specimens in the British Museum. *Annals and Magazine of Natural History* 4 (1): 35–51.

Greenwood, P. J. 1980. Mating systems, philopatry and dispersal in birds and mammals. *Animal Behaviour* 28: 1140–1162.

Grimshaw, J. M., Cordeiro, N. J. & Foley, C. A. H. 1995. The mammals of Kilimanjaro. *Journal of East African Natural History* 84: 105–139.

Grobler, J. H. & Wilson, V. J. 1972. Food of the leopard *Panthera pardus* (Linn.) in the Rhodes Matopos National Park, Rhodesia, as determined by faecal analysis. *Arnoldia Rhodesia* 5 (35): 1–10.

Grubb, P., Jones, T. S., Davies, A. G., Edberg, E., Starin, E. D. & Hill, J. E. 1998. *Mammals of Ghana, Sierra Leone and the Gambia*. Trendrine Press, Zennor, Cornwall, 265 pp.

Hahn, H. 1934. Die Familie der Procaviidae. *Zeitschrifte für Säugetierkundliche* 9: 207–358.

Harrison, D. L. & Bates, P. J. J. 1991. *The Mammals of Arabia*. Harrison Zoological Museum, Sevenoaks, Kent, 354 pp.

Hatt, R. T. 1933. An annotated catalogue of the Hyracoidea in the American Museum of Natural History, with a description of a new species from the Lower Congo. *American Museum Novitates* 594: 1–13.

Hatt, R. T. 1936. The hyraxes collected by the American Museum Congo Expedition. *Bulletin of the American Museum of Natural History* 72: 117–141.

Hayman, R. W. 1963. Mammals from Angola, mainly from the Lunda district. *Companhia de Diamantes de Angola, Lisboa, Publicaçoes Culturais Museu do Dundo* 66: 84–139.

Hirata, S., Yamakoshi, G., Fujita, S., Ohashi, G. & Matsuzawa, T. 2001. Capturing and toying with hyraxes (*Dendrohyrax dorsalis*) by wild chimpanzees (*Pan troglodytes*) at Bossou, Guinea. *American Journal of Primatology* 53: 93–97.

Hoeck, H. N. 1975. Differential feeding behaviour of the sympatric hyrax *Procavia johnstoni* and *Heterohyrax brucei*. *Oecologia* 22: 15–47.

Hoeck, H. N. 1977a. 'Teat order' in hyrax (*Procavia johnstoni* and *Heterohyrax brucei*). *Zeitschrift für Säugetierkunde* 42: 112–115.

Hoeck, H. N. 1977b. Heterohyrax brucei *(Procaviidae) – Fortbewegung im geast und nahrungsaufnahme (freilandaufnahmen). Film E 2266 des Institut fur den Wissenschaftlichen Film, Gottingen*. Publikation von H. N. Hoeck, Publikationen zu Wissenschaftlichen Filmen, Sektion Biologie, Serie 10, Nummer 60–E 2266. Gottingen, Germany.

Hoeck, H. N. 1978a. Systematics of the Hyracoidea: toward a clarification. In: *Ecology and Taxonomy of African Small Mammals* (ed. D. A. Schlitter). *Bulletin of the Carnegie Museum of Natural History* No. 6, Pittsburgh, pp. 146–151.

Hoeck, H. N. 1978b. Heterohyrax brucei *(Procaviidae) – young playing with each other and with young* Procavia johnstoni *(open air shots)*. Film E 2473 des Institute fur den Wissenschaftlichen Film, Gottingen 1978. Publikationen von H. N. Hoeck, Publikationen zu Wissenschaftlichen Filmen, Sektion Biologie, Serie 11, Nummer 29–E 2473. Gottingen, Germany.

Hoeck, H. N. 1978c. Heterohyrax brucei *(Procaviidae) – mating behaviour (open air shots)*. Film E 2474 des Institute fur den Wissenschaftlichen Film, Gottingen 1978. Publikation von H. N. Hoeck, Publikationen zu Wissenschaftlichen Filmen, Sektion Biologie, Serie 11, Nummer 30–E 2474. Gottingen, Germany.

Hoeck, H. N. 1982a. Population dynamics, dispersal and genetic isolation in two species of hyrax (*Heterohyrax brucei* and *Procavia johnstoni*) on habitat islands in the Serengeti. *Zeitschrift für Tierpsychologie* 59: 177–210.

Hoeck, H. N. 1982b. *Ethologie von busch – und klippschliefer*. Film 1338 des Institut fur den Wissenschaftlichen Film, Gottingen 1982. Publikationen von H. N. Hoeck, Publikationen zu Wissenschaftlichen Filmen, Sektion Biologie, Serie 15, Nummer 31–D 1338. Gottingen, Germany.

Hoeck, H. N. 1982c. *Nahrungsokologie bei busch-und klippschliefer. Sympatrische Lebensweise*. Film D 1371 des Institut fur den Wissenschaftlichen Film, Gottingen 1980. Publikationen von H. N. Hoeck, Publikationen zu Wissenschaftlichen Filmen, Sektion Biologie, Serie 15, Nummer 32–D 1371. Gottingen, Germany.

Hoeck, H. N. 1989. Demography and composition in hyrax: a 17-year study. *Oecologia* 79: 353–360.

Hoeck, H. N., Klein, H. & Hoeck, P. 1982. Flexible social organization in hyrax. *Zeitschrift für Tierpsychologie* 59: 265–298.

Hoffmann, M., Hoeck, H. N. & De Smet, K. 2008. Does the Bush Hyrax *Heterohyrax brucei* occur in North Africa? *Afrotheria Conservation* 6: 6–8.

Hoogstraal, H. & Wassef, H. Y. 1981. Notes on African *Haemaphysalis* ticks. XIII. Identity of *H. (Rhipistoma) cooleyi*, a parasite of the rock hyrax in South Africa (Acarina: Ixodidae). *Onderstepoort Journal of Veterinary Research* 48: 135–140.

Hoppe-Dominik, B. 1984. Etude du spectre des proies de la panthère, *Panthera pardus*, dans le Parc National de Taï en Côte d'Ivoire. *Mammalia* 48: 477–487.

Horak, I. G. & Fourie, L. J. 1986. Parasites of domestic and wild animals in South Africa. XIX. Ixodid ticks and fleas on rock dassies (*Procavia capensis*) in the Mountain Zebra National Park. *Onderstepoort Journal of Veterinary Research* 53: 123–126.

Hufnagl, E. 1972. *Libyan Mammals*. The Oleander Press, Cambridge, 85 pp.

Hungerford, D. A. & Snyder, R. L. 1969. Chromosomes of the rock hyrax, *Procavia capensis* (Pallas, 1767). *Experientia* 25: 870.

Jones, C. 1978. *Dendrohyrax dorsalis*. *Mammalian Species* 113: 1–4.

Jones, C. 1984. Tubulidentates, proboscideans, and hyracoideans. In: *Orders and Families of Recent Mammals of the World* (eds S. Anderson & J. K. Jones, Jr). John Wiley & Sons, New York, pp. 523–535.

Kappelman, J., Rasmussen, D. T., Sanders, W. J., Feseha, M., Bown, T., Copeland, P., Crabaugh, J., Fleagle, J., Glantz, M., Gordon, A., Jacobs, B., Maga, M., Muldoon, K., Pan, A., Pyne, L., Richmond, B., Ryan, T., Seiffert, E. R., Sen, S., Todd, L., Wiemann, M. C. & Winkler, A. 2003. Oligocene mammals from Ethiopia and faunal exchange between Afro-Arabia and Eurasia. *Nature* 426: 549–552.

Kingdon, J. 1971. *East African Mammals: An Atlas of Evolution in Africa*. Vol. I: *Primates, Hyraxes, Pangolins, Protoungulates, Sirenians*. Academic Press, London, 446 pp.

Kingdon, J. 1997. *The Kingdon Field Guide to African Mammals*. Academic Press, San Diego, 465 pp.

Kingdon, J. & Howell, K. M. 1993. Mammals in the forest of eastern Africa. In: *Biogeography and Ecology of the Rain Forests of Eastern Africa* (eds J. C. Lovett & S. K. Wasser). Cambridge University Press, Cambridge, UK, pp. 229–241.

Kitching, J. W. 1965. A new giant hyracoid from the Limeworks Quarry, Makapansgat, Potgietersrus. *Palaeontologia Africana* 9: 91–96.

Klein, R. G. & Cruz-Uribe, K. 1996. Size variation in the Rock Hyrax (*Procavia capensis*) and Late Quaternary climatic change in South Africa. *Quaternary Research* 46: 193–207.

Kock, D. & Stanley, W. T. 2009. Mammals of Mafia Island, Tanzania. *Mammalia* 73: 339–352.

Kolbe, F. F. 1967. A plea for dassies. *African Wildlife* 21: 237–248.

Kotler, B. P., Brown, J. S. & Knight, M. H. 1999. Habitat and patch use by hyraxes: there's no place like home? *Ecology Letters* 2: 82–88.

Kowalski, K. & Rzebik-Kowalska, B. 1991. *Mammals of Algeria*. Zaklad Narodowy Imienia Ossolinskich Wydawnictwo Polskiej Akademii Nauk Wroclaw, Poland, 370 pp.

Kryštufek, B. & Vohralík, V. 2001. *Mammals of Turkey and Cyprus*. Vol. 1: *Introduction, Checklist, Insectivora*. Zgodovinsko drustvo za juzno Primorsko, Koper, 140 pp.

Kundaeli, J. N. 1976a. Commercial exploitation of tree hyrax on Mt Kilimanjaro. *Tanzania Notes and Records* 78 and 80: 57–64.

Kundaeli, J. N. 1976b. Distribution of tree hyrax (*Dendrohyrax validus validus*. True) on Mount Kilimanjaro, Tanzania. *East African Wildlife Journal* 14: 253–264.

Lawes, M. J., Mealin, P. E. & Piper, S. E. 2000. Patch occupancy and potential metapopulation dynamics of three forest mammals in fragmented Afromontane forest in South Africa. *Conservation Biology* 14: 1088–1098.

Lensing, J. E. 1983. Feeding strategy of the rock hyrax and its relation to the rock hyrax problem in southern South West Africa. *Madoqua* 13: 177–196.

Leon, B. 1980. Fermentation and production of volatile fatty acids in the alimentary tract of the rock hyrax, *Procavia capensis*. *Comparative Biochemistry and Physiology A* 65: 411–420.

Louw, E., Louw, G. N. & Retief, C. P. 1972. Thermoability, heat tolerance and renal function in the dassie or hyrax, *Procavia capensis*. *Zoologica Africana* 7: 451–469.

Malbrant, R. & Maclatchy, A. 1949. *Faune de l'Equateur Africain Français. Tome II. Mammifères.* Paul Lechevalier, Paris, 323 pp.

Maloiy, G. M. O. & Eley, R. M. 1992. *The Hyrax.* Regal Press, Nairobi, 122 pp.

McKenna, M. C. 1975. Towards a phylogenetic classification of the Mammalia. In: *Phylogeny of the Primates* (eds W. P. Luckett & F. S. Szalay). Plenum, New York, pp. 21–46.

McNairn, I. S. & Fairall, N. 1984. Metabolic rate and body temperature of adult and juvenile hyrax (*Procavia capensis*). *Comparative Biochemistry & Physiology A* 79: 539–545.

Meester, J. A. J., Rautenbach, I. L., Dippenaar, N. J. & Baker, C. M. 1986. Classification of Southern African Mammals. *Transvaal Museum Monograph* 5: 1–359.

Meltzer, A. 1967. The rock hyrax. MSc thesis, Tel Aviv University, Israel.

Meltzer, A. 1973. Heat balance and water economy of the rock hyrax (*Procavia capensis syriaca* Schreber 1784). PhD thesis, Tel-Aviv University, Israel.

Mendelssohn, H. 1965. Breeding the Syrian hyrax. *International Zoo Yearbook* 5: 116–125.

Meyer, G. E. 1978. Hyracoidea. In: *Evolution of African Mammals* (eds V. J. Maglio & H. B. S. Cooke). Harvard University Press, Cambridge, MA, pp. 284–314.

Millar, R. P. 1971. Reproduction in the rock hyrax (*Procavia capensis*). *Zoologica Africana* 6: 243–261.

Millar, R. P. 1972. Reproduction in the rock hyrax (*Procavia capensis*) with special reference to seasonal sexual activity in the male. DPhil thesis, University of Liverpool, UK.

Millar, R. P. 1973. An unusual light-shielding structure in the eye of the dassie, *Procavia capensis* Pallas (Mammalia: Hyracoidea). *Annals of the Transvaal Museum* 28: 203–205.

Milner, J. M. 1994. Relationships between the forest dwelling people of South-west Mau and tree hyrax, *Dendrohyrax arboreus*. *Journal of East African Natural History* 83: 17–29.

Milner, J. M. & Harris, S. 1999a. Activity patterns and feeding behaviour of the tree hyrax, *Dendrohyrax arboreus*, in the Parc National des Volcans, Rwanda. *African Journal of Ecology* 37: 267–280.

Milner, J. M. & Harris, S. 1999b. Habitat use and ranging behaviour of tree hyrax, *Dendrohyrax arboreus*, in the Virunga Volcanoes, Rwanda. *Afrioan Journal of Ecology* 37: 281–294.

Mollaret, H. 1962. Naissance de damans en captivité. *Mammalia* 25: 529–532.

O'Donoghue, P. N. 1963. Reproduction in the female hyrax (Dendrohyrax arboreal ruwenzorii). *Proceedings of the Zoological Society of London* 141: 207–237.

Olds, N. & Shoshani, J. 1982. *Procavia capensis. Mammalian Species* 171: 1–7.

Olubayo, R. & Karstad, L. H. 1983. Fatal toxoplasmosis in a tree hyrax (*Dendrohyrax arboreus*). *Bulletin of Animal Health and Production in Africa* 29: 263–264.

Osborn, H. F. 1907. *Evolution of the Mammalian Molar Teeth*. The Macmillan Company, London, 250 pp.

Osborne, T. O. 1987. Rock Hyrax *Procavia capensis* new to Zambia. *Mammalia* 51: 327–327.

Pakenham, R. H. W. 1984. *The Mammals of Zanzibar and Pemba Islands*. Harpenden: printed privately.

Palmer, R. & Fairall, N. 1988. Caracal and African wild cat diet in the Karoo National Park and the implications thereof for hyrax. *South African Journal of Wildlife Research* 18: 30–34.

Pickford, M. 1996. Pliohyracids (Mammalia, Hyracoidea) from the upper Middle Miocene at Berg Aukas, Namibia. *Comptes Rendus de l'Académie des Sciences de Paris* 322: 501–505.

Pickford, M. 2003. Giant dassie (Hyracoidea, Mammalia) from the middle Miocene of South Africa. *South African Journal of Science* 99: 366–367.

Pickford, M. 2004. Revision of the Early Miocene (Hyracoidea) (Mammalia) of east Africa. *Comptes Rendus Palevol* 3: 675–690.

Prinsloo, P. 1993. Molecular and chromosomal phylogeny of the Hyracoidea. PhD thesis, University of Pretoria, South Africa.

Prinsloo, P. & Robinson, T. J. 1991. Comparative cytogenetics of the Hyracoidea: chromosomes of two *Hyrax* species from South Africa. *Zeitschrift für Säugetierkunde* 56: 331–338.

Prinsloo, P. & Robinson, T. J. 1992. Geographic mitochondrial DNA variation in the Rock Hyrax, *Procavia capensis*. *Molecular Biology and Evolution* 9: 447–456.

Rahm, S. & Frewein, J. 1980. On the anatomy of the gastrointestinal tract in *Procavia, Heterohyrax* and *Dendrohyrax* Hyracoidea. *Zentralblatt für Veterinarmedizin C* 9: 307–320.

Rahm, U. 1966. Les mammifères de la forêt équatoriale de l'est du Congo. *Annales de Musée Royal de l'Afrique Centrale, Sciences Zoologiques* 149: 39–121.

Rahm, U. 1969. Notes sur le cri du *Dendrohyrax dorsalis* (Hyracoidea). *Mammalia* 33: 68–79.

Rasmussen, D. T. & Simons, E. L. 1988. New Oligocene hyracoids from Egypt. *Journal of Vertebrate Paleontology* 8: 67–83.

Rasmussen, D. T., Pickford, M., Mein, P., Senut, B. & Conroy, G. C. 1996. Earliest known procaviid hyracoid from the late Miocene of Namibia. *Journal of Mammalogy* 77: 745–754.

Rautenbach, I. L. 1982. The mammals of the Transvaal. *Ecoplan Monograph* 1: 1–211.

Richard, B. P. 1964. Notes sur la biologie du daman des arbres *Dendrohyrax dorsalis*. *Biologica Gabonica* 1: 73–84.

Rio, B. & Galat, G. 1982. Arboreal locomotion of a *Dendrohyrax dorsalis* (Temminck, 1853). *Mammalia* 46: 449–456.

Roberts, A. 1951 (and reprint 1954). *The Mammals of South Africa*. Trustees of 'The Mammals of South Africa' Book Fund, Johannesburg.

Roberts, D. 1999. A preliminary investigation of the taxonomic implications of the loud calls of *Dendrohyrax* (Hyracoidea, Procavidae) in Africa. MSc thesis, Oxford Brookes University, UK.

Roberts, D. 2001. Geographical variation in the loudcalls of Tree Hyrax – *Dendrohyrax validus* (True 1890) – in the Eastern Arc Mountains, East Africa: taxonomic and conservation implications. MSc thesis, University of Reading, UK.

Roche, J. 1962. Nouvelles données sur la reproduction des Hyracoides. *Mammalia* 26: 517–529.

Roche, J. 1972. Systematique du genre *Procavia* et des damans en general. *Mammalia* 36: 22–49.

Rübsamen, K., Hume, I. D. & Engelhardt, W. V. 1982. Physiology of the Rock Hyrax. *Comparative Biochemistry and Physiology A* 72: 271–277.

Rudnai, J. 1984a. Suckling behaviour in captive *Dendrohyrax arboreus* (Mammalia: Hyracoidea). *South African Journal of Zoology* 19: 121–123.

Rudnai, J. 1984b. Activity cycle and space utilization in captive *Dendrohyrax arboreus*. *South African Journal of Zoology* 19: 124–128.

Rudnai, J. 1992. Dendrohyrax arboreus sheltering in termitaria. *East African Natural History Society Bulletin* 22: 45.

Rudnai, J. 1998. Reproduction in a group of captive southern tree hyrax *Dendrohyrax arboreus*. *East Africa Natural History Society Bulletin* 28: 5–8.

Rudnai, J. & Frere, P. 2000. Vocalisation in the southern tree hyrax *Dendrohyrax arboreus*. *East Africa Natural History Society Bulletin* 30: 13–20.

Sale, J. B. 1965a. Observations on parturition and related phenomena in the hyrax (Procaviidae). *Acta Tropica* 22: 37–54.

Sale, J. B. 1965b. The feeding behaviour of rock hyraxes (genera *Procavia* and *Heterohyrax*) in Kenya. *East African Wildlife Journal* 3: 1–18.

Sale, J. B. 1965c. Hyrax feeding on a poisonous plant. *East African Wildlife Journal* 3: 127.

Sale, J. B. 1966. The habitat of the rock hyrax. *Journal of the East African Natural History Society* 25: 205–214.

Sale, J. B. 1969. Breeding season and litter size in Hyracoidea. *Journal of Reproduction and Fertility* (Suppl.) 6: 249–263.

Sale, J. B. 1970a. Unusual external adaptations in the rock hyrax. *Zoologica Africana* 5: 101–113.

Sale, J. B. 1970b. The behaviour of the resting rock hyrax in relation to its environment. *Zoologica Africana* 5: 87–99.

Schlitter, D. A. 1993. Order Hyracoidea. In: *Mammal Species of the World: A Taxonomic and Geographic Reference* (eds D. E. Wilson & D. M. Reeder) (2nd edn). Smithsonian Institution Press, Washington, DC, pp. 373–374.

Sclater, W. L. 1900. *The Mammals of South Africa*. R. H. Porter, London.

Seibt, V. U., Hoeck, H. N. & Wickler, W. 1977. *Dendrohyrax validus* True, 1890 in Kenia. *Zeitschrift für Säugetierkunde* 42: 115–118.

Seiffert, E. R. 2003. A phylogenetic analysis of living and extinct Afrotherian placentals. PhD thesis, Duke University, USA.

Shortridge, G. C. 1934. *The Mammals of South West Africa*, Vols I & II. Heinemann, London, 437 pp.

Shoshani, J. 1992. Ecology and taxonomy: a new approach using *Heterohyrax* vs. *Procavia* as a case study. *Israel Journal of Zoology* 38: 399–416.

Shoshani, J. 2005. Order Hyracoidea. In: *Mammal Species of the World: A Taxonomic and Geographic Reference* (eds D. E. Wilson & D. M. Reeder) (3rd edn). Johns Hopkins University Press, Baltimore, pp. 87–89.

Shultz, S. 2002. Population density, breeding chronology and diet of Crowned Eagles *Stephanoaetus coronatus* in Taï National Park, Ivory Coast. *Ibis* 144: 135–138.

Skinner, J. D. & Chimimba, C. T. (eds) 2005. *The Mammals of the Southern African Subregion* (3rd edn). Cambridge University Press, Cambridge, UK, 814 pp.

Smith, R. M. 1977. Movement patterns and feeding behaviour of leopard in the Rhodes Matopos National Park, Rhodesia. *Arnoldia Rhodesia* 8 (13): 1–16.

Smithers, R. H. N. 1971. *The Mammals of Botswana*. Museum Memoir No. 4. Trustees of the National Museums and Monuments of Rhodesia, Salisbury, 340 pp.

Smithers, R. H. N. & Lobão Tello, J. L. P. 1976. *Checklist and Atlas of the Mammals of Mocambique*. Museum Memoir No. 8. Trustees of the National Museums and Monuments of Rhodesia. Salisbury, Rhodesia, 184 pp.

Smithers, R. H. N. & Wilson, V. J. 1979. *Checklist and Atlas of the Mammals of Zimbabwe Rhodesia*. Museum Memoir No. 9. Trustees of the National Museums and Monuments of Rhodesia, Salisbury, 147 pp.

Sokolov, V. E. & Sale, D. 1981. The structure of skin in the Hyracoidea (Mammalia). Part 2. *Zoologicheskii Zhurnal* 60: 1849–1860.

Storr, G. C. C. 1780. *Prodromus methodi Mammalium ... inauguralem disputationem propositus*. F. Wolffer, Tubigen, 43 pp.

Stuart, C. T. & Stuart, T. D. 1984. The breeding season of the rock hyrax, *Procavia capensis*, a lesson in adaptation. *The Naturalist* 28 (3): 13–15.

Sudre, J. 1979. Nouveaux mammifères éocènes du Sahara occidental. *Palaeovertebrata* 9: 83–115.

Swynnerton, G. H. & Hayman, R. W. 1950. A checklist of the land-mammals of the Tanganyika Territory and the Zanzibar Protectorate. *Journal of the East Africa Natural History Society* 20: 274–392.

Tabuce, R., Mahboubi, M. & Sudre, J. 2001. Reassessment of the Algerian Eocene hyracoid *Microhyrax*. Consequences on the early diversity and basal phylogeny of the Order Hyracoidea (Mammalia). *Eclogae Geologicae Helvetiae* 94: 537–545.

Taylor, C. R. & Sale, J. G. 1969. Temperature regulation in the hyrax. *Comparative Biochemistry and Physiology* 31: 903–907.

Taylor, P. J. 1998. *The Smaller Mammals of KwaZulu–Natal*. University of Natal Press, Pietermaritzberg, 139 pp.

Thomas, O. 1892. On the species of the Hyracoidea. *Proceedings of the Zoological Society of London* 1892: 50–76.

Topp-Jørgensen, J. E. & Pedersen, U. B. 2001. Mammalian abundance in response to different hunting pressures in the Udzungwa Mountains, Tanzania. MSc thesis, University of Copenhagen, Denmark.

Topp-Jørgensen, J. E., Marshall, A. R., Brink, H. & Pedersen, U. B. 2008. Quantifying the response of tree hyraxes (*Dendrohyrax validus*) to human disturbance in the Udzungwa Mountains, Tanzania. *Tropical Conservation Science* 1: 63–74.

True, F. W. 1890. Mammals from Mt Kilima-Njaro. *Proceedings of the United States National Museum* XIII: 228–229.

Turner, M. I. M. & Watson R. M. 1965. An introductory study on the ecology of hyrax (*Dendrohyrax brucei* and *Procavia johnstoni*) in the Serengeti National Park. *East African Wildlife Journal* 3: 49–60.

Van der Merwe, M. & Skinner, J. D. 1982. Annual reproductive pattern in the dassie *Procavia capensis*. *South African Journal of Zoology* 17: 130–135.

Walker, A., Hoeck, H. N. & Perez, L. 1978. Microwear of mammalian teeth as an indicator of diet. *Science* 201: 908–910.

Weigl, R. 2005. *Longevity of Mammals in Captivity; from the Living Collections of the World*. Kleine Senckenberg-Reihe 48: 1–214.

Wilson, V. J. 1969. The large mammals of the Matopos National Park. *Arnoldia Rhodesia* 4 (13): 1–32.

Yom-Tov, Y. 1993. Does the Rock Hyrax, *Procavia capensis*, conform with Bergmanns Rule? *Zoological Journal of the Linnean Society* 108: 171–177.

Yom-Tov, Y. & Tchernov, E. (eds) (1988). *The Zoogeography of Israel: The Distribution and Abundance at a Zoogeographical Crossroad*. Dr W. Junk Publishers, Dordrecht, 600 pp.

Young, T. P. & Evans, M. R. 1993. Alpine vertebrates on Mount Kenya, with particular notes on the rock hyrax. *Journal of the East Africa Natural History Society* 82: 55–79.

Ziegler, S., Nikolaus, G. & Hutterer, R. 2002. High mammalian diversity in the newly established National Park of Upper Niger, Republic of Guinea. *Oryx* 36: 73–80.

Superorder Tethytheria

Court, N. 1994. The periotic of *Moeritherium* (Mammalia, Proboscidea): homology or homoplasy in the ear region of Tethytheria McKenna, 1975? *Zoological Journal of the Linnean Society* 112: 13–28.

Fischer, M. S. 1990. Un trait unique de l'oreille des éléphants et des siréniens (Mammalia): un paradoxe phylogénétique. *Comptes Rendus de l'Academie des Sciences (Paris)*, Série III, 311: 157–162.

Fischer, M. S. & Tassy, P. 1993. The interrelations between Proboscidea, Sirenia, Hyracoidea, and Mesaxonia: the morphological evidence. In: *Mammal Phylogeny: Placentals* (eds. F. S. Szalay, M. J. Novacek & M. C. McKenna). Springer Verlag, New York. pp. 217–243.

Gheerbrant, E., Domning, D .P. & Tassy, P. 2005a. Paenungulata (Sirenia, Proboscidea, Hyracoidea, and relatives). In: *The Rise of Placental Mammals: Origins and Relationships of the Major Extant Clades* (eds. K. D. Rose & J. D. Archibald). Johns Hopkins University Press, Baltimore. pp. 84–105.

Gheerbrant, E., Sudre, J., Tassy, P., Amaghzaz, M., Bouya, B. & Iarochene, M. 2005b. Nouvelles données sur Phosphatherium escuilliei (Mammalia, Proboscidea) de l'Éocène inférieur du Maroc, apports à la phylogénie des Proboscidea et des ongulés lophodontes. *Geodiversitas* 27: 239–333.

Liu, A. G. S. C., Seiffert, E. R. & Simons, E. L. 2008. Stable isotope evidence for an amphibious phase in early proboscidean evolution. *Proceedings of the National Academy of Sciences of the United States of America* 105: 5786–5791.

Meredith, R. W., Janečka, J. E., Gatesy, J., Ryder, O. A., Fisher, C. A., Teeling, E. C., Goodbla, A., Eizirik, E., Simão, T. L. L., Stadler, T., Rabosky, D. L., Honeycutt, R. L., Flynn, J. J., Ingram, C. M., Steiner, C., Williams, T. L., Robinson, T. J., Burk-Herrick, A., Westerman, M., Ayoub, N. A., Springer, M. S. & Murphy, W. J. 2011. Impacts of the Cretaceous Terrestrial Revolution and KPg extinction on mammal diversification. *Science* 334: 521–524.

Nishihara, H., Satta, Y., Nikaido, M., Thewissen, J. G. M., Stanhope, M. J. & Okada, N. 2005. A retroposon analysis of afrotherian phylogeny. *Molecular Biology and Evolution* 22: 1823–1833.

Poulakakis, N. & Stamatakis, A. 2010. Recapitulating the evolution of Afrotheria: 57 genes and rare genomic changes (RGCs) consolidate their history. *Systematics and Biodiversity* 8: 395–408.

Shoshani, J. 1993. Hyracoidea–Tethytheria affinity based on myological data. In: *Mammal Phylogeny: Placentals* (eds. F. S. Szalay, M. J. Novacek & M. C. McKenna). Springer Verlag, New York. pp. 235–256.

Order Proboscidea

AfESG (African Elephant Specialist Group). 1999. Strategy for the conservation of West African elephants. IUCN/SSC African Elephant Specialist Group, Nairobi.

AfESG (African Elephant Specialist Group). 2003. Strategy for the conservation of West African elephants. Unpublished report, African Elephant Specialist Group, Ouagadougou, 46 pp.

Aguirre, E. E. 1969. Evolutionary history of the elephant. *Science* 164: 1366–1376.

Alers, M. P. T., Blom, A., Sikubwabo Kiyengo, C., Masunda, T. & Barnes, R. F. W. 1992. Preliminary assessment of the forest elephant in Zaire. *African Journal of Ecology* 30: 279–291.

Andrews, C. W. 1906. *A Descriptive Catalogue of the Tertiary Vertebrate of the Fayum, Egypt*. Natural History Museum, London.

Anon. 1990. New evidence supports existence of pygmy elephant. *International Society of Cryptozoology Newsletter* 9: 1–6.

Ansell, W. F. H. 1974. Order Proboscidea. Part II. In: *The Mammals of Africa: An Identification Manual* (eds J. Meester & H. W. Setzer). Smithsonian Institution Press, Washington, DC, pp. 1–5.

Antoine, P. O., Welcome, J.-L., Marivaux, L., Baloch, I., Benammi, M. & Tassy, P. 2003. First record of Paleogene Elephantoidea (Mammalia, Proboscidea) from the Bugti Hills of Pakistan. *Journal of Vertebrate Paleontology* 23: 977–980.

Archie, E. A., Hollister-Smith, J. A., Poole, J. H., Lee, P. C., Moss, C. J., Maldonado, J. E., Fleischer, R. C. & Alberts, S. C. 2007. Behavioural inbreeding avoidance in wild African elephants. *Molecular Ecology* 16: 4128–4148.

Archie, E. A., Moss, C. J. & Alberts, S. C. 2005. The ties that bind: genetic relatedness predicts the fission and fusion of social groups in wild African elephants. *Proceedings of the Royal Society of London B* 1–10.

Badoux, D. M. 1961. A mechanical interpretation of the shape and cyrtocephaly in the skull of the Indian elephant, *Elephas maximus*. *Säugetierkunde Mitteilungen* 9: 97–105.

Barnes, R. F. W. 1993. Indirect methods for counting elephants in forest. *Pachyderm* 16: 24–30.

Barnes, R. F. W. 1999. Is there a future for elephants in West Africa? *Mammalia* 29: 175–199.

Barnes, R. F. W. 2001. How reliable are dung counts for estimating elephant numbers? *African Journal of Ecology* 39: 1–9.

Barnes, R. F. W. 2002. The problem of precision and trend detection posed by small elephant populations in West Africa. *African Journal of Ecology* 40: 179–185.

Barnes, R. F. W., Barnes, K. L., Alers, M. P. T. & Blom, A. 1991. Man determines the distribution of elephants in the rain forests of northeastern Gabon. *African Journal of Ecology* 29: 54–63.

Barnes, R. F., Blom, A. & Alers, M. P. T. 1995. A review of the status of Forest Elephants, *Loxodonta africana* in Central Africa. *Biological Conservation* 71: 125–132.

Barnes, R. F. W., Beardsley, K., Michelmore, F., Barnes, K. L., Alers, M. P. T. & Blom, A. 1997. Estimating forest elephant numbers with dung counts and a geographic information system. *Journal of Wildlife Management* 61: 1384–1393.

Barriel, V., Thuet, E. & Tassy, P. 1999. Molecular phylogeny of Elephantidae. Extreme divergence of the extant forest elephant. *Comptes Rendus de l'Académie des Sciences (Paris)*, serie III, 322: 447–454.

Bates, L. A., Sayialel, C. N., Njiraini, N. W., Poole, J. H., Moss, C. J. & Byrne, R. W. 2007. Elephants classify human ethnic groups by odour and garment colour. *Current Biology* 17: 1938–1942.

Bates, L. A., Lee, P. C., Njiraini, N., Poole, J. H., Sayialel, K, Sayialel, S., Moss, C. J. & Bryne, R. W. 2008a. Do elephants show empathy? *Journal of Consciousness Studies* 15: 204–225.

Bates, L. A., Sayialel, C. N., Njiraini, N. W., Poole, J. H., Moss, C. J. & Byrne, R. W. 2008b. African elephants have expectations about the locations of out-of-sight family members. *Biology Letters* 23: 34–36

Bates, L. A., Handford, R., Lee, P. C., Njiraini, N., Poole, J. H., Sayialel, K., Sayialel, S., Moss, C. J. & Byrne, R. W. 2010. Why do African elephants (*Loxodonta africana*) simulate oestrus? An analysis of longitudinal data. *PLoS ONE* 5: e10052.

Bauer, C. & Stoye, M. 1985. Endoparasiten beim afrikanischen und asiatischen Elefanten. *Der Praktische Tierarzt* 66: 55–60.

Beden, M. 1983. Family Elephantidae. In: *Koobi for a Research Project*. Vol. 2: *The Fossil Ungulates: Proboscidea, Perissodactyla, and Suidae* (ed. J. M. Harris). Clarendon Press, Oxford, pp. i–xvii, 1–321.

Beden M. 1987. *Les Faunes Plio-pléistocènes de la Basse Vallée de l'Omo (Ethiopie). Tome 2: Les Eléphantidés (Mammalia–Proboscidea)*. Editions du CNRS Cahiers de Paléontologie, Paris, 162 pp.

Benedict, F. G. 1936. *The Physiology of the Elephant*. Carnegie Institution of Washington (Publication No. 474), Washington, DC, 302 pp.

Berg, J. K. 1983. Vocalizations and associated behaviours of the African elephant (*Loxodonta africana*) in captivity. *Zeitschrift für Tierpsychologie* 63: 63–79.

Berger, J. & Cunningham, C. 1998. Behavioural ecology in managed reserves: gender-based asymmetries in interspecific dominance in African elephants and rhinos. *Animal Conservation* 1: 33–38.

Blake, S. 2002. The ecology of Forest Elephant distribution, ranging and habitat use in the Ndoki Forest, Central Africa. PhD thesis, University of Edinburgh, UK.

Blake, S. 2004. Do leopards kill forest elephants? Evidence from northern Congo. *Mammalia* 68: 225–227.

Blake, S. & Hedges, S. 2004. Sinking the flagship: the case of forest elephants in Asia and Africa. *Conservation Biology* 18: 1191–1202.

Blake, S., Bouche, P., Rasmussen, H. B., Orlando, A. & Douglas-Hamilton, I. 2003. The last sahelian elephants: ranging behaviour, population status and recent history of the desert elephants of Mali. Unpublished report. Save the Elephants, Nairobi, 48 pp.

Blake, S., Strinberg, S., Boudjan, P., Makombo, C., Bila-Isia, I., Ilambru, O., Grossmann, F., Bene-Bene, L., de Semboli, B., Mbenzo, V., S'hwa, D., Bayongo, R., Williamson, L., Fay, M., Hart, J. & Maisels, F. 2007. Forest elephant crisis in the Congo Basin. *PLoS Biology* 5: 945–953.

Blake, S., Deem, S. L., Mossimbo, E., Maisels, F. & Walsh, P. 2009. Forest Elephants: tree planters of the Congo. *Biotropica* 41: 459–468.

Blanc, J. J., Barnes R. F. W., Craig, G. C., Douglas-Hamilton, I., Dublin, H. T., Hart, J. A. & Thouless, C. R. 2005. Changes in elephant numbers in major savanna populations in eastern and southern Africa. *Pachyderm* 38: 19–28.

Blanc, J. J., Barnes, R. F., Craig, G. C., Dublin, H. T., Thouless, C. R., Douglas-Hamilton, I. & Hart, J. A. 2007. *African Elephant Status Report 2007: An Update from the African Elephant Database*. Occasional Paper Series of the IUCN Species Survival Commission, No. 33. IUCN/SSC African Elephant Specialist Group. IUCN, Gland, vi + 276 pp.

Blancou, L. 1962. A propos des formesnaines de l'éléphant d'Afrique. *Mammalia* 26: 343–361.

Boafo, Y., Dubiure, U.-F., Danquah, E., Manford, M., Nandjui, A., Hema, E. M., Barnes, R. F. W. & Bailey, B. 2004. Long-term management of crop raiding by elephants around Kakum Conservation Area in southern Ghana. *Pachyderm* 37: 68–72.

Bohme, W. & Eisentraut, M. 1990. Zur weiteneren Dokumentation des Zwergelefanten (*Loxodonta pumilio* NOACK, 1906). *Zeitschrift des Kolner Zoo Heft* 4. 33 Jahrgang, 153–158.

Bouché, P., Doamba, B., Sissoko, B. & Bouju, S. 2009. Les éléphants du Gourma, Mali: Statut et menaces pour leur conservation. *Pachyderm* 45: 47–56.

Bouché, P., Douglas-Hamilton, I., Wittemyer, G., Nianogo, A. J., Doucet, J.-L., Lejeune, P. & Vermeulen, C. 2011. Will elephants soon disappear from West African savannahs? *PLoS ONE* 6: e20619.

Bourdelle, E. & Petter, F. 1950. Note relative à un éléphant nain du Gabon. *Mammalia* 14: 4.

Bradshaw, I. G. A., Schore, A. N., Brown, J. L., Poole, J. H. & Moss, C. J. 2005. Elephant breakdown. Social trauma: early disruption of attachment can affect the physiology, behaviour and culture of animals and humans over generations. *Nature* 433: 807.

Bryden, H. A. 1903. The decline and fall of the South African elephant. *Fortnightly Review* 79: 100–108.

Buckland, S. T., Anderson, D. R., Burnham, K. P., Laake, J. L., Borchers, D. L. & Thomas, L. 2001. *Introduction to Distance Sampling: Estimating Abundance of Biological Populations*. Oxford University Press, Oxford.

Buechner, H. K. & Dawkins, H. C. 1961. Vegetation change induced by elephants and fire in the Murchison Falls National Park, Uganda. *Ecology* 42: 752–766.

Buss, I. O. 1961. Some observations on food habits and behaviour of the African elephant. *Journal of Wildlife Management* 25: 131–148.

Buss, I. O. & Savidge, J. M. 1966. Changes in population numbers and reproductive rate of elephants in Uganda. *Journal of Wildlife Management* 30: 791–809.

Calef, G. W. 1988. Maximum rate of increase in the African elephant. *African Journal of Ecology* 26: 323–327.

Chabaud, A. G. & Rousselot, R. 1956. Nématodes parasites d'un éléphant du moyen Congo. *Annales de Parasitologie* 31: 579–597.

Chevalier-Skolnikoff, S. & Liska, J. 1993. Tool use by wild and captive elephants. *Animal Behaviour* 46: 209–219.

Chiyo, P. I., Archie, E. A., Hollister-Smith, J. A., Lee, P. C., Poole, J. H., Moss, C. M. & Alberts, S. C. 2011. Association patterns of African elephants in all-male groups: the role of age and genetic relatedness. *Animal Behaviour* 81: 1093–1099.

Codron, J., Lee-Thorp, J. A., Sponheimer, M., Codron, D., Grant, R. C. & de Ruiter, D. J. 2006. Elephant (*Loxodonta africana*) diets in Kruger National Park, South Africa: spatial and landscape differences. *Journal of Mammalogy* 87: 27–34.

Coetzee, B. J., Engelbrecht, A. H., Joubert, S. C. J. & Retief, P. F. 1979. Elephant impact on *Sclerocarya caffra* trees in *Acacia nigrescens* tropical plains thornveld of the Kruger National Park. *Koedoe* 22: 39–60.

Comstock, K. E., Georgiadis, N., Pecon-Slattery, J., Roca, A. L., Ostrander, E. A., O'Brien, S. J. & Wasser, S. K. 2002. Patterns of molecular genetic variation among African elephant populations. *Molecular Ecology* 11: 2489–2498.

Condy, J. B. 1974. Observations on internal parasites in Rhodesian elephants. *Proceedings and Transactions of the Rhodesia Scientific Association* 55: 67–99.

Coppens, Y. 1972. Un nouveau proboscidien de Pliocène du Tchad. *Stegodibelodon schneideri* nov. gen. nov. sp., et le phylum des Stegotetrabelodontinae. *Comptes Rendus de l'Académie des Sciences (Paris)*, serie D 274: 2962–2965.

Coppens, Y., Maglio, V. J., Madden, C. T. & Beden, M. 1978. Proboscidea. In: *Evolution of African Mammals* (eds V. J. Maglio & H. B. S. Cooke). Harvard University Press, Cambridge, MA, pp. 336–367.

Coulson, D. 2001. *African Rock Art: Painting and Engraving on Stone*. Harry N Abrams Inc., New York, 256 pp.

Cozzi, B., Spagnoli, S. & Bruno, L. 2001. An overview of the central nervous system of the elephant through a critical appraisal of the literature published in the XIX and XX centuries. *Brain Research Bulletin* 54: 219–227.

Craig, G. C. 1984. Foetal mass and date of conception in African elephants: a revised formula. *South African Journal of Science* 80: 512–516.

Croze, H. & Moss, C. J. 2011. Patterns of occupancy in time and space. In: *The Amboseli Elephants: A Long-Term Perspective on a Long-Lived Mammal* (eds C. J. Moss, H. J. Croze & P. C. Lee). University of Chicago Press, Chicago, pp. 89–105.

Cumming, D. H. M., Du Toit, R. H. & Stuart, S. N. 1990. *African Elephants and Rhinos: Status Survey and Conservation Action Plan*. IUCN/SSC African Elephant and Rhino Specialist Group. IUCN, Gland, 73 pp.

Cumming, D. H. M., Fenton, M. B., Rautenbach, I. L., Taylor, R. D., Cumming, G. S., Cumming, M. S., Dunlop, J. M., Ford, A. G., Hovorka, M. D., Johnston, D. S., Kalcounis, M., Mahlangu, Z. & Portfors, C. V. R. 1997. Elephants, woodlands and biodiversity in southern Africa. *South African Journal of Science* 93: 231–236.

Cuvier, F. 1825. Eléphant d'Afrique. In: *Histoire naturelle des Mammifères, avec des figures originales, coloriées, dessinées d'après des animaux vivants; publiée sous l'autorité de l'administration du Museum d'Histoire Naturelle* (eds E. Geoffroy Saint-Hilaire & F. Cuvier, for the years of 1824–1847, 7 vols), pp. 1–2 + 1 plt. Vol. 3. A. Belin, Paris.

Danquah, E. & Oppong, S. K. 2006. Food plants of forest elephants and their availability in the Kakum Conservation Area, Ghana. *Pachyderm* 40: 51–59.

Danquah, E., Boafo, Y., Dubiure, U. F., Awo, N., Héma, E. M. & Appiah, M. A. 2001. Elephant census in the Ankasa Conservation Area in south-western Ghana. *Pachyderm* 31: 63–69.

Debruyne, R. 2001. New mitochondrial data demonstrating a close relationship between *Mammuthus primigenius* (Blumenbach, 1799) and *Loxodonta africana* (Blumenbach, 1797). In: *Proceedings of the 1st International Congress of 'La Terra degli Elefanti': The World of Elephants* (eds G. Cavarretta, P. Gioia, M. Mussi & M. R. Palombo). Consiglio Nazionale delle Ricerche, Rome, pp. 630–632.

Debruyne, R. 2004. Contribution of molecular phylogeny and morphometrics to the systematics of African elephants. *Journal de la Société de Biologie* 198: 335–342.

Debruyne, R. 2005a. A case study of apparent conflict between molecular phylogenies: the interrelationships of African elephants. *Cladistics* 21: 31–50.

Debruyne, R. 2005b. Phylogeography of forest and savannah African elephants. In: *2nd World of Elephants Congress, Short Papers and Abstracts* (eds L. D. Agenbroad & R. L. Symington). Hot Springs, Mammoth Site Publication, pp. i–vii, 1–216.

Debruyne, R., Barriel, V. & Tassy, P. 2003a. Mitochondrial cytochrome b of the Lyakhov mammoth (Proboscidea, Mammalia): new data and phylogenetic analyses of Elephantidae. *Molecular Phylogenetics and Evolution* 26: 421–434.

Debruyne, R., Van Holt, A., Barriel, V. & Tassy, P. 2003b. Status of the so-called African pygmy elephant (*Loxodonta pumilio* (Noack 1906)): phylogeny of cytochrome b and mitochondrial control region sequences. *Comptes Rendus Biologies* 326: 687–697.

Deraniyagala, P. E. P. 1955. *Some Extinct Elephants, Their Relatives and the Two Living Species*. Ceylon National Museums Administration, Colombo, 161 pp.

Dollman, G. 1934. Pygmy elephants. *Natural History* 4: 266–271.

Douglas-Hamilton, I. 1972. On the ecology and behaviour of the African elephant. PhD thesis, Oxford University, UK.

Douglas-Hamilton, I. 1979. African Elephant Ivory Trade – Final Report to the US Fish and Wildlife Service. Typescript.

Douglas-Hamilton, I. & Douglas-Hamilton, O. 1975. *Among the Elephants*. Viking Press, New York, 285 pp.

Douglas-Hamilton, I., Bhalla, S., Wittemyer, G. & Vollrath, F. 2006. Behavioural reactions of elephants towards a dying and deceased matriarch. *Applied Animal Behaviour Science* 100: 87–102.

Dublin, H. T., Sinclair, A. R. E. & Mcglade, J. 1990. Elephants and fire as causes of multiple stable states in the Serengeti–Mara woodlands. *Journal of Animal Ecology* 59: 1147–1164.

EBM. 2000. Preliminary report on estimates of elephant numbers from dung counts in Kakum Conservation Area. Unpublished report, Elephant Biology & Management Project, Conservation International, Cape Coast, Ghana, 4 pp.

Eggert, L. S., Rasner, C. A. & Woodruff, D. S. 2002. The evolution and phylogeography of the African elephant inferred from mitochondrial DNA sequence and nuclear microsatellite markers. *Proceedings of the Royal Society of London B* 269: 1993–2006.

Eggert, L. S., Eggert, J. A. & Woodruff, D. S. 2003. Estimating population sizes for elusive animals: the forest elephants of Kakum National Park, Ghana. *Molecular Ecology* 12: 1389–1402.

Eisenberg, J. F. 1981. *The Mammalian Radiations. An Analysis of Trends in Evolution, Adaption, and Behavior*. The University of Chicago Press, Chicago, xx + 610 pp.

Eisentraut, M. & Bohme, W. 1989. Gibt es zwei Elefantenarten in Afrika? *Zeitschrift des Kolner Zoo Heft* 2. 32 Jahrgang, 61–68.

Eltringham, S. K. 1982. *Elephants*. Blandford Press, Poole, Dorset, 262 pp.

Estes, J. A. & Buss, I. O. 1976. Microanatomical structure and development of the African elephant's temporal gland. *Mammalia* 40: 429–436.

Fain, A. 1983. Notes sur les genres *Loxonaetus* Fain, 1970 et *Otanoetus* Fain et Zumpt, 1974 (Acari, Anoetidae). *Revue de Zoologie Africaine* 97: 846–851.

Fatti, L. P., Smuts, G. L., Starfield, A. M. & Spurdle, A. A. 1980. Age determination in African elephants. *Journal of Mammalogy* 61: 547–551.

Fay, M., Elkan, P., Marjan, M. & Grossman, F. 2007. *Aerial Surveys of Wildlife, Livestock, and Human Activity in and around Existing and Proposed Protected Areas of Southern Sudan, Dry Season 2007*. WCS – Southern Sudan Technical Report. 150 pp.

Fischer, F. 2005. Elephants in Cote d'Ivoire – a warning for West African conservation. *Pachyderm* 38: 64–75.

Fischer, M. S. 1990. The unique ear of elephants and manatees (Mammalia): a phylogenetic paradox. *Comptes Rendus de l'Académie des Sciences, Série III* 311: 157–162.

Frade, F. 1931. Sur l'existence en Afrique de deux espèces d'éléphants. *Bulletin de la Société Portugaise des Sciences Naturelles, Lisbonne* 9: 135.

Frade, F. 1955. Sous-famille des Loxodontinae Osborn 1918. In: *Traite de Zoologie* (ed. P. P. Grasse). 17 (1): 774–780.

Freeman, E. W., Whyte, I. J. & Brown, J. L. 2008. Reproductive evaluation of elephants culled in Kruger National Park, South Africa between 1975 and 1995. *African Journal of Ecology* 47: 192–201.

Froehlich, D. & Kalb, J. E. 1995. Internal reconstruction of elephantid molars: applications for functional anatomy and systematics. *Paleobiology* 21 (3): 379–392.

Gaeth, A. P., Short, R. V. & Renfree, M. B. 1999. The developing renal, reproductive, and respiratory systems of the African elephant suggest an aquatic ancestry. *Proceedings of the National Academy of Sciences of the United States of America* 96: 5555–5558.

Garstang, M., Larom, D., Raspet, R. & Lindeque, M. 1995. Atmospheric controls on elephant communication. *Journal of Experimental Biology* 198: 939–951.

Georgiadis, N., Bischof, L., Templeton, A., Patton, J., Karesh, W. & Western, D. 1994. Structure and history of African Elephant populations. 1. Eastern and Southern Africa. *Journal of Heredity* 85: 100–104.

Gheerbrant, E. 1997. The oldest known proboscidean and the role of Africa in the radiation of modern orders of placentals. *Bulletin of the Geological Society of Denmark* 44: 181–185.

Gheerbrant, E. 2009. Paleocene emergence of elephants relatives and the rapid radiation of African ungulates. *Proceedings of the National Academy of Sciences of the United States of America* 106: 10717–10721.

Gheerbrant, E. & Bardet, N. 1999. Les vertébrés des phosphates. In: *Maroc, Mémoire de la Terre*. Editions du Muséum National d'Histoire Naturelle, Paris, pp. 160–163.

Gheerbant, E. & Tassy, P. 2009. L'origine et l'évolution des elephants. *Comptes Rendus Palevol* 8: 281–294.

Gheerbrant, E., Sudre, J., Sen, S., Abrial, C., Marandat, B., Sige, B. & Vianey-Liaud, M. 1998. Nouvelles données sur les mammifères du Thanétien et de l'Yprésien du bassin d'Ouarzazate (Maroc) et leur contexte stratigraphique. *Palaeovertebrata* 27: 155–202.

Gheerbrant, E., Sudre, J., Tassy, P., Amaghzaz, M., Bouya, B. & Iarochene, M. 2005. Nouvelles données sur *Phosphatherium escuilliei* de l'Eocène inférieur du Maroc, apports à la phylogénie des Proboscidea et des ongulés lophodontes. *Geodiversitas* 27 (2): 239–333.

Gordon, J. A. 1966. Elephants do think. *African Wildlife* 20 (1): 75–79.

Grimshaw, J. M., Cordeiro, N. J. & Foley, C. A. H. 1995. The mammals of Kilimanjaro. *Journal of East African Natural History* 84: 105–139.

Grobler, D. G., Raath, J. P., Braack, L. E. O., Keet, D. F., Gerdes, G. H., Barnard, B. J. H., Kriek, N. P. J., Jardine, J. & Swanepoel, R. 1995. An outbreak of encephalomyocarditis-virus infection in free-living elephants in the Kruger National Park. *Onderstepoort Journal of Veterinary Research* 62: 97–108.

Groves, C. P. & Grubb, P. 2000a. Are there pygmy elephants? *Elephant* 2 (4): 8–10.

Groves, C. P. & Grubb, P. 2000b. Do *Loxodonta cyclotis* and *L. africana* interbreed? *Elephant* 2 (4): 4–7.

Grubb, P., Groves, C. P., Dudley, J. P. & Shoshani, J. 2000. Living African elephants belong to two species: *Loxodonta africana* (Blumenbach, 1797) and *Loxodonta cyclotis* (Matschie, 1900). *Elephant* 2 (4): 1–4.

Guy, P. 1974. Feeding behaviour of the African elephant in the Sengwa Research Area of Rhodesia. M.Sc. thesis, University of Rhodesia, Rhodesia.

Hakeem, A. Y., Hof, P. R., Sherwood, C. C., Switzer, R. C., Rasmussen L. R. L. & Allman, J. M. 2005. Brain of the African elephant (*Loxodonta africana*): neuroanatomy from magnetic resonance images. *Anatomical Record* 287A: 1117–1127.

Hakeem, A. Y., Sherwood, C. C., Bonar, C. J., Butti, C., Hof, P. R. & Allman, J. M. 2009. Von Economo Neurons in the elephant brain. *Anatomical Record* 292: 242–248.

Hall, J. S., Inogwabini, B.-I., Williamson, E. A., Omari, I., Sikubwabo, C. & White, L. J. T. 1997. A survey of elephants (*Loxodonta africana*) in the Kahuzi–Biega National Park lowland sector and adjacent forest in eastern Zaire. *African Journal of Ecology* 35: 213–233.

Hall-Martin, A. J. & Van der Walt, L. A. 1984. Plasma testosterone levels in relation to musth in the male African elephant. *Koedoe* 27: 147–149.

Hanks, J. 1969. Seasonal breeding of the African elephant in Zambia. *East African Wildlife Journal* 7: 167.

Hanks, J. 1979. *A Struggle for Survival: The Elephant Problem*. Struik, Cape Town, 176 pp.

Harris, J. M. 1975. Evolution of feeding mechanisms in the family Deinotheriidae (Mammalia: Proboscidea). *Zoological Journal of the Linnean Society* 56: 331–362.

Hart, B. L., Hart, L. A., McCoy, M. & Sarath, C. R. 2001. Cognitive behaviour in Asian elephants: use and modification of branches for fly switching. *Animal Behaviour* 62: 839–847.

Hart, B. L., Hart, L. A. & Pinter-Williams, N. 2008. Large brains and cognition: Where do elephants fit in? *Neuroscience and Biobehavioral Reviews* 32: 86–98.

Hawthorne, W. D. & Parren, M. P. E. 2000. How important are forest elephants to the survival of woody plant species in Upper Guinean forests? *Journal of Tropical Ecology* 16: 133–150.

Haynes, G. 1991. *Mammoths, Mastodonts & Elephants: Biology, Behavior, and the Fossil Record*. Cambridge University Press, Cambridge, UK, 413 pp.

Heffner, R. & Heffner, H. 1980. Hearing in the elephant. *Science* 208: 518–520.

Holdo, R. M., Dudley, J. P. & Mcdowell, L. R. 2002. Geophagy in the African Elephant in relation to availability of dietary sodium. *Journal of Mammalogy* 83: 652–664.

Hollister-Smith, J. A., Poole, J. H., Archie, E. A., Vance, E. A., Georgiadis, N. J., Moss, C. J. & Alberts, S. C. 2007. Age, musth and paternity success in wild male African elephants, *Loxodonta africana*. *Animal Behaviour* 74: 287–296.

Hoppe-Dominik, B., Kühl, H. S., Radl, G. & Fischer, F. 2011. Long-term monitoring of large rainforest mammals in the Biosphere Reserve of Taï National Park, Côte d'Ivoire. *African Journal of Ecology* 49: 450-458.

Houck, M. L., Kumamoto, A. T., Gallagher, D. S. & Benirschke, K. 2001. Comparative cytogenetics of the African elephant (*Loxodonta africana*) and Asiatic elephant (*Elephas maximus*). *Cytogenetics and Cell Genetics* 93: 249–252.

Houston, D. C., Gilardi, J. D. & Hall, A. J. 2001. Soil consumption by elephants might help minimize toxic effects of plant secondary compounds in forest browse. *Mammal Review* 3: 249–254.

Howard, A. L. 1979. 'Motty' – birth of an African/Asian elephant at Chester Zoo. *Elephant* 1 (3): 36–41.

Hungerford, D. A., Chandra, H. S., Snyder, R. L. & Olmer, F. A., Jr 1966. Chromosomes of three elephants, two Asian (*Elephas maximus*) and one African (*Loxodonta africana*). *Cytogenetics* 5 (3): 243–246.

Hutchinson, J. R., Delmer, C., Miller, C. E., Hildebrandt, T., Pitsillides, A. A. & Boyde, A. 2011. From flat foot to fat foot: structure, ontogeny, function, and evolution of elephant "sixth toes". *Science* 334: 1699–1703.

Illiger, C. D. 1811. Prodromus systematis mammalium et avium additis terminis zoographicis uttriusque classis. Salfeld, Berlin, 301 pp.

Isaza, R. 2006. Respiratory system. In: *Biology, Medicine, and Surgery of Elephant* (eds M. E. Fowler & S. K. Mikota). Blackwell Publishing, Ames, IA, pp. 291–298.

Jachmann, H. 1985. Estimating age in African elephants: a revision of Law's molar evaluation technique. *African Journal of Ecology* 23: 199–202.

Jachmann, H. 1988. Estimating age in African elephants: a revision of Laws' molar evaluation technique. *African Journal of Ecology* 26: 51–56.

Jachmann, H. 1991. Evaluation of four survey methods for estimating elephant densities. *African Journal of Ecology* 29: 188–195.

Jachmann, H. & Bell, R. H. V. 1984. The use of elephant droppings in assessing numbers, occupancy and age structure: a refinement of the method. *African Journal of Ecology* 22: 127–141.

Jachmann, H., Berry, P. S. M. & Imae, H. 1995. Tusklessness in African elephants: a future trend. *African Journal of Ecology* 33: 230–235.

Janis, C. M. 1988. New ideas in ungulate phylogeny and evolution. *Trends in Ecology and Evolution* 3 (11): 291–297.

Jerison, H. J. 1973. *Evolution of the Brain and Intelligence*. Academic Press, London and New York, 482 pp.

Johnson, C. F. 1998. Vulnerability, irreplaceability and reserve selection of the elephant-impacted flora of the Addo Elephant National Park, Eastern Cape, South Africa: MSc thesis, University of Rhodes, South Africa.

Johnson, C. F., Cowling, R. M. & Phillipson, P. B. 1999. The flora of the Addo Elephant National Park, South Africa: are threatened species vulnerable to elephant damage? *Biodiversity and Conservation* 8: 1447–1456.

Johnson, O. W. & Buss, I. O. 1965. Molariform teeth of male African elephants in relation to age, body dimensions, and growth. *Journal of Mammalogy* 46: 373–384.

Joubert, D. 2006. Hunting behaviour of lions (*Panthera leo*) on elephants (*Loxodonta africana*) in the Chobe National Park, Botswana. *African Journal of Ecology* 44: 279–281.

Kahl, M. P. & Armstrong, B. D. 2000a. Observations on African elephants (*Loxodonta africana*), including possible urine drinking, during a severe drought in Zimbabwe. *Elephant* 2 (4): 21–25.

Kahl, M. P. & Armstrong, B. D. 2000b. Visual and tactile displays in African elephants, *Loxodonta africana*: a progress report (1991–1997). *Elephant* 2 (4): 19–21

Kahumbu, P. 2002. Elephant habitat interactions in the Shimba Hills, Kenya. PhD dissertation, Princeton University, USA.

Kalb, J. E., Froehlich, D. J. & Bell, G. L. 1996. Phylogeny of African and Eurasian Elephantoidea of the late Neogene. In: *The Proboscidea: Evolution and Palaeoecology of Elephants and their Relatives* (eds J. Shoshani & P. Tassy). Oxford University Press, Oxford, pp. 101–116.

Kappelman, J., Rasmussen, D. T., Sanders, W. J., Feseha M., Bown, T., Copeland, P., Crabaugh, J., Fleagle, J., Glantz, M., Gordon, A., Jacobs, B., Maga, M., Muldoon, K., Pan, A., Pyne, L., Richmond, B., Ryan, T., Seiffert, E. R., Sen, S., Todd, L., Wiemann, M. C. & Winkler, A. 2003. Oligocene mammals from Ethiopia and faunal exchange between Afro-Arabia and Eurasia. *Nature* 426: 549–552.

Kingdon, J. 1979. *East African Mammals: An Atlas of Evolution in Africa*. Vol. III Part B: *Large Mammals*. Academic Press, London, v + 436 pp.

Kinsella, J. M., Deem, S. L., Blake, S. & Freeman, A. S. 2004. Endoparasites of African forest elephants (*Loxodonta africana cyclotis*) from the Republic of Congo and Central African Republic. *Comparative Parasitology* 71: 104–110.

Klaus, G., Klaus-Hugi, C. & Schmid, B. 1998. Geophagy by large mammal at natural licks in the rain forest of the Dzanga National Park, Central African Republic. *Journal of Tropical Ecology* 14: 829–839.

Kock, N. D., Goedegebuure, S. A., Lane, E. P., Tyrrell, D. & Kock, M. D. 1994. Flaccid trunk paralysis in free ranging elephants in Zimbabwe. *Journal of Wildlife Diseases* 30: 432–435.

Krause, J., Dear, P. H., Pollack, J. L., Slatkin, M., Spriggs, H., Barnes, I., Lister, A. M., Ebersberger, I., Paabo, S. & Hofreiter, M. 2006. Multiplex amplification of the mammoth mitochondrial genome and the evolution of Elephantidae. *Nature* 439: 724–727.

Krumrey, W. A. & Buss, I. O. 1968. Age estimation, growth, and relationships between body dimensions of the female African elephant. *Journal of Mammalogy* 49: 22–31.

Lang, E. M. 1980. Observations on growth and molar change in the African elephant. *African Journal of Ecology* 18: 217–234.

Langbauer, W. R. Jr 2000. Elephant communication. *Zoo Biology* 19: 425–445.

Langbauer, W. R. Jr, Payne, K. B., Charif, R. A., Rapaport, L. & Osborn. F. 1991. African elephants respond to distant playbacks of low-frequency conspecific calls. *Journal of Experimental Biology* 157: 35–46.

Lark, R. M. 1984. A comparison between techniques for estimating the ages of African elephants (*Loxodonta africana*). *African Journal of Ecology* 22: 69–71.

Larom, D., Garstang, M., Payne, K., Raspet, R. & Lindeque, M. 1997. The influence of surface atmospheric conditions on the range and area reached by animal vocalizations. *Journal of Experimental Biology* 200: 421–431.

Laursen, L. & Bekoff, M. 1978. *Loxodonta africana*. *Mammalian Species* 92: 1–8.

Laws, R. M. 1966. Age criteria for the African elephant (*Loxodonta africana*). *East African Wildlife Journal* 4: 1–37.

Laws, R. M. 1967. Eye lens weight and age in African elephants. *East African Wildlife Journal* 5: 46–52.

Laws, R. M. 1969. The Tsavo Research Project. *Journal of Reproduction and Fertility* (Suppl.) 6: 495–531.

Laws, R. M. 1970. Biology of African elephants. *Scientific Progress* 58: 251–262.

Laws, R. M. & Parker, I. S. C. 1968. Recent studies on elephant populations in East Africa. In: Comparative Nutrition of Wild Animals. Proceedings of a Symposium Held at the Zoological Society of London, 10–11 November, 1966 (ed. M. A. Crawford). *Symposia of the Zoological Society of London* 20: 319–359.

Laws, R. M., Parker, I. S. C. & Johnstone, R. C. B. 1975. *Elephants and their Habitats: The Ecology of Elephants in North Bunyoro, Uganda*. Clarendon Press, Oxford, 376 pp.

Lee, P. C. 1986. Early social development among African elephant calves. *National Geographic Research* 2: 388–400.

Lee, P. C. 1987. Allomothering among African elephants. *Animal Behaviour* 35: 278–291.

Lee, P. C. & Moss, C. J. 1986. Early maternal investment in male and female elephant calves. *Behavioral Ecology & Sociobiology* 18: 353–361.

Lee, P. C. & Moss, C. J. 1995. Statural growth in known-age elephants (*Loxodonta africana*). *Journal of Zoology, London* 236: 29–41.

Lee, P. C. & Moss, C. J. 1999. The social context for learning and behavioural development among wild African elephants. In: *Mammalian Social Learning* (eds H. O. Box & K. R. Gibson). Cambridge University Press, Cambridge, pp. 102–125.

Lee, P. C., Lindsay, W. K. & Moss, C. J. 2011a. Ecological patterns of variability in demographic rates. In: *The Amboseli Elephants: A Long-Term Perspective on a Long-Lived Mammal* (eds C. J. Moss, H. J. Croze & P. C. Lee). University of Chicago Press, Chicago, pp. 74–88.

Lee, P. C., Poole, J. H., Njiraini, N. & Moss, C. J. 2011b. Male social dynamics: independence and beyond. In: *The Amboseli Elephants: A Long-Term Perspective on a Long-Lived Mammal* (eds C. J. Moss, H. J. Croze & P. C. Lee). University of Chicago Press Chicago, pp. 260–271.

Lee, P. C., Sayialel, S., Lindsay, W. K. & Moss, C. J. 2012. African elephant age determination from teeth: validation from known individuals. *African Journal of Ecology* 50: 9–20.

Leggett, K. E. A. 2004. Coprophagy and unusual thermoregulatory behaviour in desert-dwelling elephants of north-western Namibia. *Pachyderm* 36: 113–115.

Leggett, K. E. A. 2006a. Home range and seasonal movement of elephants in the Kunene Region, Northwest Namibia. *Journal of African Zoology* 41: 17–36.

Leggett, K. E. A. 2006b. Effect of artificial water points on the movement and behaviour of desert-dwelling elephants of north-western Namibia. *Pachyderm* 40: 24–34.

Leggett, K. E. A., Fennessy, J. & Schneider, S. 2003. Seasonal distributions and social dynamics of elephants in the Hoanib River catchment, northwestern Namibia. *Journal of African Zoology* 38: 305–316.

Lindeque, M. 1988. Population dynamics of elephants in Etosha National Park, S.W.A. /Namibia. PhD thesis, University of Stellenbosch, South Africa.

Lindeque, M. 1991. Dentition and age estimation of elephants in the Etosha National Park, Namibia. *Madoqua* 18: 17–25.

Lindeque, M. & Lindeque, P. M. 1991. Satellite tracking of elephants in northwestern Namibia. *African Journal of Ecology* 29: 196–206.

Lindeque, M. & Van Jaarsveld, A. S. 1993. Post-natal growth of elephants *Loxodonta africana* in Etosha National Park, Namibia. *Journal of Zoology, London* 229: 319–330.

Lindeque, P. M. & Turnbull, P. C. B. 1994. Ecology and epidemiology of anthrax in the Etosha National Park, Namibia. *Onderstepoort Journal of Veterinary Research* 61: 71–83.

Lindsay, W. K. 2011. Habitat use, diet choice, and nutritional status in female and male Amboseli elephants. In: *The Amboseli Elephants: A Long-Term Perspective on a Long-Lived Mammal* (eds C. J. Moss, H. J. Croze & P. C. Lee). University of Chicago Press, Chicago, pp. 51–73.

Loveridge, A. J., Hunt, J. E., Murindagomo, F. & Macdonald, D. W. 2006. Influence of drought on predation of elephant (*Loxodonta africana*) calves by lions (*Panthera leo*) in an African wooded savannah. *Journal of Zoology, London* 270: 523–530.

Lowenstein, J. M. & Shoshani, J. 1996. Proboscidean relationships based on immunological data. In: *The Proboscidea: Evolution and Palaeoecology of Elephants and Their Relatives* (eds J. Shoshani & P. Tassy). Oxford University Press, Oxford, pp. 49–54.

Lydekker, R. 1907. The ears as a race-character in the African elephant. *Proceedings of the Zoological Society of London* 1: 380–403.

Maglio, V. J. 1970. Early Elephantidae of Africa and a tentative correlation of African Plio-Pleistocene deposits. *Nature* 225 (5230): 328–332.

Maglio, V. J. 1973. Origin and evolution of the Elephantidae. *Transactions of the American Philosophical Society of Philadelphia, New Series* 63 (3): 1–149.

Maisels, F., Blake, S. & Turkalo, A. K. 2002. Wild forest elephants shake down fruit and leaves from trees. *Pachyderm* 33: 88–90.

Malan, J. W. & Van Wyk, A. E. 1993. Bark structure and preferential bark utilisation by the African Elephant. *IAWA Journal* 14: 173–185.

Markov, G. N., Spassov, N. & Simeonovski, V. 2001. A reconstruction of the facial morphology and feeding behaviour of the deinotheres. In: *Proceedings of the 1st International Congress of 'La Terra degli Elefanti', The World of Elephants* (eds G. Cavarretta, P. Gioia, M. Mussi & M. R. Palombo). Consiglio Nazionale delle Ricerche, Rome, pp. 652–655.

Matschie, P. 1900. Uber geographische Albarten des Afrikanischen elephantens. *Sitzungsberichte Gesellschaft naturforschunde Freunde Berlin* 8: 189–197.

Matsumoto, H. 1923. A contribution to the knowledge of Moeritherium. *Bulletin of the American Museum of Natural History* 48: 97–140.

Mauny, R. A. 1956. Repartition de la grande faune Ethiopienne du nord-ouest africaine du paleolithique à nos jours. *Proceedings of the 3rd Pan-African Congress of Prehistory*, pp. 102–105.

Maurois, C., Chamberlan, C. & Marechal, C. 1997. An outline of the diet of forest elephants, *Loxodonta africana cyclotis*, in the Parc National d'Odzala, Republic of the Congo. *Mammalia* 61: 127–130.

McComb, K., Moss, C., Sayialel, S. & Baker, L. 2000. Unusually extensive networks of vocal recognition in African elephants. *Animal Behaviour* 59: 1103–1109.

McComb, K., Reby, D., Baker, L., Moss, C. & Sayialel, S. 2002. Long-distance communication of social identity in African elephants. *Animal Behaviour* 65: 317–329.

McComb, K., Baker, L. & Moss, C. 2006. Elephants show high level of interest in the skulls and ivory of their own species. *Biology Letters* 2: 26–28.

McKenna, M. C. & Bell, S. K. (eds) 1997: *Classification of Mammals: Above the Species Level*. Columbia University Press, New York, 631 pp.

Meng, J., Shoshani, J. & Ketten, D. 1997. Evolutionary evidence for infrasonic sound and hearing in proboscideans. *Journal of Vertebrate Paleontology*, Abstracts of Papers Volume 17, Supplement to Number 3: 64A.

Meredith, R. W., Janečka, J. E., Gatesy, J., Ryder, O. A., Fisher, C. A., Teeling, E. C., Goodbla, A., Eizirik, E., Simão, T. L. L., Stadler, T., Rabosky, D. L., Honeycutt, R. L., Flynn, J. J., Ingram, C. M., Steiner, C., Williams, T. L., Robinson, T. J., Burk-Herrick, A., Westerman, M., Ayoub, N. A., Springer, M. S. & Murphy, W. J. 2011. Impacts of the Cretaceous Terrestrial Revolution and KPg extinction on mammal diversification. *Science* 334: 521–524.

Merz, G. 1986. Movement patterns and group size of the African Forest Elephant *Loxodonta africana cyclotis* in the Tai National Park, Ivory Coast. *African Journal of Ecology* 24: 133–136.

Michelmore, F., Beardsley, K., Barnes, R. F. W. & Douglas-Hamilton, I. 1994. A model illustrating the changes in forest elephant numbers caused by poaching. *African Journal of Ecology* 32: 89–99.

Mikota, S. K. 2006. Preventative health care and physical examination. In: *Biology, Medicine, and Surgery of Elephant* (eds M. E. Fowler & S. K. Mikota). Blackwell Publishing, Ames, IA, pp. 67–73.

Milner-Gulland, E. J. & Beddington, J. R. 1993. The exploitation of elephants for the ivory trade: an historical perspective. *Proceedings of the Royal Society of London B* 252: 29–37.

Monadjem, A. 1998. *Mammals of Swaziland*. The Conservation Trust of Swaziland and Big Game Parks, 154 pp.

Moolman, H. J. & Cowling, R. M. 1994. The impact of elephant and goat grazing on the endemic flora of South African succulent thicket. *Biological Conservation* 68: 53–61.

Morgan, B. J. & Lee, P. C. 2003. Forest elephant (*Loxodonta africana cyclotis*) stature in the Reserve de Faune du Petit Loango, Gabon. *Journal of Zoology, London* 259: 337–344.

Morrison-Scott, T. C. S. 1947. A revision of our knowledge of African elephants' teeth, with notes on forest and 'pygmy' elephants. *Proceedings of the Zoological Society of London* 117: 505–527.

Moss C. 1976. *Portraits in the Wild: Animal Behaviour in East Africa*. Hamish Hamilton, London, 363 pp.

Moss, C. 1988. *Elephant Memories: Thirteen Years in the Life of an Elephant Family*. William Morrow and Company Inc., New York, 336 pp.

Moss, C. J. 1981. Social Circles. *Wild News* 16: 2–7.

Moss, C. J. 1983. Oestrous behaviour and female choice in the African elephant. *Behaviour* 86: 167–196.

Moss, C. 1992. *Echo of the Elephants*. BBC Books, London.

Moss, C. J. 2001. The demography of an African elephant (*Loxodonta africana*) population in Amboseli, Kenya. *Journal of Zoology, London* 255: 145–156.

Moss, C. J. & Lee, P. C. 2011. Female social dynamics: fidelity and flexibility. In: *The Amboseli Elephants: A Long-Term Perspective on a Long-Lived Mammal* (eds C. J. Moss, H. J. Croze & P. C. Lee). University of Chicago Press, Chicago, pp. 203–223.

Moss, C. J. & Poole, J. H. 1983. Relationships and social structure in African elephants. In: *Primate Social Relationships: An Integrated Approach* (ed. R. A. Hinde). Blackwell Scientific Publications, Oxford, pp. 315–325.

Moss, C. J., Croze, H. & Lee, P. C. (eds) 2011. *The Amboseli Elephants: A Long-Term Perspective on a Long-Lived Mammal*. University of Chicago Press, Chicago, 376 pp.

Mubalama, L. & Mapilanga, J. J. 2001. Less elephant slaughter in the Okapi Faunal reserve, Democratic Republic of Congo, with Operation Tango. *Pachyderm* 31: 36–41.

Mutinda, H. S. 1994. Assessment of reproductive status of free ranging female African elephants (*Loxodonta africana*) using non-invasive techniques. MSc thesis, University of Nairobi, Kenya.

Nellemann, C., Moe, S. R. & Rutina, L. P. 2002. Links between terrain characteristics and forage patterns of elephants (*Loxodonta africana*) in northern Botswana. *Journal of Tropical Ecology* 18: 835–844.

Noack, T. 1906. A dwarf form of the African elephant. *Annals and Magazine of Natural History* 17: 501–503.

Norberg, H. S. 1969. The chromosomes of the Indian female elephant (*Elephas indicus* syn. *E. maximus* L.). *Hereditas* 63 (1): 279–281.

Nummela, S. 1995. Scaling of the mammalian middle ear. *Hearing Research* 85: 18–30.

Nyakaana, S., Arctander, P. & Siegismund, H. R. 2002. Population structure of African savannah elephant inferred from mitochondrial control region sequences and nuclear microsattelite loci. *Heredity* 85: 90–98.

O'Connell, C., Hart, L. & Arnason, B. T. 1998. Comments on 'Elephant hearing'. *Journal of the Acoustical Society of America* 105: 2051–2052.

O'Connell-Rodwell, C., Arnason, B. T. & Hart, L. A. 2000. Seismic properties of Asian elephant (*Elephas maximus*) vocalizations and locomotion. *Journal of the Acoustical Society of America* 108: 3066–3072.

O'Connell-Rodwell, C. E., Wood, J. D., Rodwell, T. C., Puria, S., Partan, S. R., Keefe, R., Shriver, D., Arnason, B. T. & Hart, L. A. 2006. Wild elephant (*Loxodonta africana*) breeding herds respond to artificially transmitted seismic stimuli. *Behavioral Ecology & Sociobiology* 59: 842–850.

O'Connell-Rodwell, C., Wood, J. D., Kinzley, C., Rodwell, T. C., Poole, J. & Puria, S. 2007. Wild African elephants (*Loxodonta africana*) discriminate between familiar and unfamiliar conspecific seismic alarm calls. *Journal of the Acoustic Society of America* 122: 823–830.

Osborn, H. F. 1936. *Proboscidea: a monograph on the discovery, evolution, migration and extinction of the mastodonts and elephants of the world. Vol. 1: Moerithierioidea, Deinotherioidea, Mastodontoidea*. The American Museum Press, New York, xi + 802 pp.

Osborn, H. F. 1942. *Proboscidea: a monograph of the discovery, evolution, migration and extinction of the mastodonts and elephants of the world. Vol. II: Stegodontoidea, Elephantoidea*. The American Museum Press, New York, xxvii + 805–1675 pp. (plus one errata page).

Ozawa, T., Hayashi, S. & Mikhelson, V. M. 1997. Phylogenetic position of mammoth and Steller's sea cow within Tethytheria demonstrated by mitochondrial DNA sequences. *Journal of Molecular Evolution* 44 (4): 406–413.

Parker, I. S. C. & Graham, A. D. 1989. Men, elephants and competition. *Symposia of the Zoological Society of London* 61: 241–252.

Payne, K., Thompson, M. & Kramer, L. 2003. Elephant calling patterns as indicators of group size and composition: the basis of an acoustical monitoring system. *African Journal of Ecology* 42: 99–107.

Payne, K. B. 1998. *Silent Thunder: In the Presence of Elephants*. Simon & Schuster, New York, 288 pp.

Payne, K. B., Langbauer, Jr W. R. & Thomas, E. M. 1986. Infrasonic calls of the Asian elephant (*Elephas maximus*). *Behavioral Ecology & Sociobiology* 102: 283–316.

Perry, J. S. 1953. The reproduction of the African elephant *Loxodonta africana*. *Philosophical Transactions of the Royal Society B* 237: 93–149.

Plotnik, J. M., de Waal, F. B. M. & Reiss, D. 2006. Self-recognition in an Asian elephant. *Proceedings of the National Academy of Sciences of the United States of America* 103: 17053–17057.

Pohle, H. 1926. Notizen uber africanische Elephanten. *Zeitschrift für Säugetierkunde* 1: 58–64.

Poilecot, P. 2010. Le braconnage et la population d'e´le´phants au Parc National de Zakouma (Tchad). *Bois et Forêts des Tropiques* 303: 93–102.

Poinar, H. N., Schwarz, C., Ji Qi, Shapiro, B., MacPhee, R. D. E., Buigues, B., Tikhonov, A., Huson, D. H., Tomsho, L. P., Auch, A., Rampp, M., Miller, W. & Schuster, S. C. 2006. Metagenomics to paleogenomics: large-scale sequencing of Mammoth DNA. *Science* 311: 392–394.

Poole, J. H. 1982. Musth and male–male competition in the African elephant. PhD thesis, University of Cambridge, UK.

Poole, J. H. 1987. Rutting behavior in African elephants: the phenomenon of musth. *Behavior* 102: 283–316.

Poole, J. H. 1989a. The effects of poaching on the age structures and social and reproductive patterns of selected East African elephant populations. In: *The Ivory Trade and the Future of the African Elephant*. Vol. II: *Technical Reports*. By The Ivory Trade Review Group. Prepared for the 7th CITES Conference of the Parties.

Poole, J. H. 1989b. Announcing intent: the aggressive state of musth in African elephants. *Animal Behaviour* 37: 140–152.

Poole, J. H. 1989c. Mate guarding, reproductive success and female choice in African elephants. *Animal Behaviour* 37: 842–849.

Poole, J. H. 1994. Sex differences in the behavior of African elephants. In: *The Differences Between the Sexes* (ed R. Short & E. Balaban). Cambridge University Press, Cambridge, UK, pp. 341–346.

Poole, J. H. 1996. Coming of Age with Elephants: a Memoir. Hyperion Books, New York, 287 pp.

Poole, J. H. 1999a. Signals and assessment in African Elephants: evidence from playback experiments. *Animal Behaviour* 58: 185–193.

Poole, J. H. 1999b. Ella's Easter Baby. *Care for the Wild Magazine*.

Poole, J. H. 2011. The behavioral context of African elephant acoustic communication. In: *The Amboseli Elephants: A Long-Term Perspective on a Long-Lived Mammal* (eds C. J. Moss, H. J. Croze & P. C. Lee). University of Chicago Press, Chicago, pp. 125–161.

Poole, J.H. & Granli, P.K. 2011. Signals, gestures and behaviors of African elephants. In: *The Amboseli Elephants: A Long-Term Perspective on a Long-Lived Mammal* (eds C. J. Moss, H. J. Croze & P. C. Lee). University of Chicago Press, Chicago, pp. 109–124.

Poole, J. H. & Moss, C. J. 1981. Musth in the African elephant, *Loxodonta africana*. *Nature* 292: 830–831.

Poole, J. H. & Moss, C. J. 1989. Elephant mate searching: Group dynamics and vocal and olfactory communication. In: *The Biology of Large African Mammals in Their Environment* (eds P. A. Jewell & G. M. O. Maloiy). Clarendon Press, Oxford. *Symposia of the Zoological Society of London* 61: 111–125.

Poole, J. H., Kasman, L. H., Ramsay, E. C. & Lasley, B. L. 1984. Musth and urinary testosterone concentrations in the African elephant, *Loxodonta Africana*. *Journal of Reproduction and Fertility* 70: 255–260.

Poole, J. H., Payne, K. B., Langbauer, W. Jr & Moss, C. J. 1988. The social contexts of some very low frequency calls of African elephants. *Behavioral Ecology & Sociobiology* 22: 385–392.

Poole, J. H., Tyack, P. L., Stoeger-Horwath, A. S. & Watwood, S. L. 2005. Elephants are capable of vocal learning. *Nature* 434: 455–456.

Poole, J. H., Lee, P. C., Njiraini, N. & Moss, C. J. 2011. Longevity, competition and musth: a long-term perspective on male reproductive strategies. In: *The Amboseli Elephants: A Long-Term Perspective on a Long-Lived Mammal* (eds C. J. Moss, H. J. Croze & P. C. Lee). University of Chicago Press, Chicago, pp. 272–286.

Querouil, S., Magliocca, F. & Gautier-Hion, A. 1999. Structure of population, grouping patterns and density of forest elephants in northwest Congo. *African Journal of Ecology* 37: 161–167.

Rasmussen, L. E. L. & Krishnamurthy, V. 2000. How chemical signals integrate Asian elephant society: the known and the unknown. *Zoo Biology* 19: 405–423.

Rasmussen, L. E. L. & Munger, B. 1996. The sensorimotor specializations of the trunk tip of the Asian elephant, *Elephas maximus*. *The Anatomical Record* 246: 127–134.

Rasmussen, L. E. L. & Schmidt, B. A. 1998. Chemical signals in the reproduction of Asian (*Elephas maximus*) and African (*Loxodonta africana*) elephants. *Animal Reproduction Science* 53: 19–34.

Rasmussen, L. E. L. & Wittemyer, G. 2002. Chemosignaling of musth by individual wild African elephants (*Loxodonta africana*): implications for conservation and management. *Proceedings of the Royal Society of London B* 269: 853–860.

Rasmussen, L. E. L., Hall-Martin, A. & Hess, D. L. 1996. Chemical profiles of African bull elephants (*Loxodonta africana*): physiological and ecological implications. *Journal of Mammalogy* 77: 422–439.

Raubenheimer, E. J., Vanheerden, W. F. P., Van Niekerk, P. J., De Vos, V. & Turner, M. J. 1995. Morphology of the deciduous Tusk (Tush) of the African Elephant (*Loxodonta africana*). *Archives of Oral Biology* 40: 571–576.

Redmond, I. 1982. Salt-mining elephants of Mount Elgon. *Swara* 5: 28–31.

Rench, B. 1956. The intelligence of elephants. *Scientific American* 196: 44–49.

Reuter, T., Nummela, S. & Hemila, S. 1998. Elephant hearing. *Journal of the Acoustical Society of America* 104: 1122–1123.

Roca, A. L., Georgiadis, N., Pecon-Slattery, J. & O'Brien, S. J. 2001. Genetic evidence for two species of elephant in Africa. *Science* 293: 1473–1477.

Roca, A. L., Georgiadis, N. & O'Brien, S. J. 2004. Cytonuclear genomic dissociation in African elephant species. *Nature Genetics* 37: 96–100.

Roca, A. L., Georgiadis, N. & O'Brien, S. J. 2005 Cyto-nuclear genomic dissociation in African elephant species. In: *2nd World of Elephants Congress, Short Papers and Abstracts* (eds L. D. Agenbroad & R. L. Symington). Hot Springs, Mammoth Site Publication, pp. i–vii, 1–216.

Rogaev, E. I., Moliaka, Y. K., Malyarchuk, B. A., Kondrashov, F. A., Derenko, M. V., Chumakov, I. & Grigorenko, A. P. 2006. Complete mitochondrial genome and phylogeny of Pleistocene Mammoth *Mammuthus primigenius*. *PLoS Biology* 4 (3): e73.

Rohland, N., Reich, D., Mallick, S., Meyer, M., Green, R. E., Georgiadis, N. J., Roca, A. L. & Hofreiter, M. 2010. Genomic DNA sequences from Mastodon and Woolly Mammoth reveal deep speciation of forest and savanna elephants. *PLoS Biology* 8(12): e1000564.

Rommel, S. A. & Reynolds, J. E., III. 2000. Diaphragm structure and function in the Florida manatee (*Trichechus manatus latirostris*). *Anatomical Record* 259: 41–51.

Roth, H. H. & Douglas-Hamilton, I. 1991. Distribution and status of elephants in West Africa. *Mammalia* 55: 489–527.

Roth, V. L. 1984. How elephants grow: heterochrony and the calibration of developmental stages in some living and fossil species. *Journal of Vertebrate Paleontology* 4 (1): 126–145.

Roth, V. L. & Shoshani, J. 1988. Dental identification and age determination in *Elephas maximus*. *Journal of Zoology, London* 214 (4): 567–588.

Round, M. C. 1968. *Checklist of the Helminth Parasites of African Mammals*. Technical Communication No. 38. Commonwealth Institute of Helminthology, St Albans, UK..

Ruggiero, R. G. & Fay, J. M. 1994. Utilization of termitarium soils by elephants and its ecological implications. *African Journal of Ecology* 32: 222–232.

Saegusa, H. 1996. Stegodontidae: evolutionary relationships. In: *The Proboscidea: Evolution and Palaeoecology of Elephants and their Relatives* (eds J. Shoshani & P. Tassy). Oxford University Press, Oxford, pp. 178–190.

Sanders, W. J. 2007. Taxonomic review of fossil Proboscidea (Mammalia) from Langebaanweg, South Africa. *Transactions of the Royal Society of South Africa* 62: 1–16.

Sanders, W. J., Kappelman, J. & Rasmussen, D. T. 2004. New large-bodied mammals from the late Oligocene site of Chilga, Ethiopia. *Acta Palaeontologica Polonica* 49 (3): 365–392.

Sanders, W. J., Gheerbrant, E., Harris J. M., Saegusa H. & Delmer C. 2010. Proboscidea. In: *Cenozoic Mammals of Africa* (eds L. Werdelin & W. J. Sanders). University of California Press, Berkeley. pp. 161–251.

Savidge, J. M. 1968. Elephants in the Ruaha National Park, Tanzania: a management problem. *East African Agricultural and Forestry Journal* 33 (Special issue): 191–196.

Scholes, R. J., Bond, W. J. & Eckhardt, H. C. 2003. Vegetation dynamics in the Kruger National Park ecosystem. In: *The Kruger Experience – Ecology and Management of Savanna Heterogeneity* (eds J. du Toit, K. H. Rogers & H. C. Biggs). Island Press, Washington, D.C. pp. 242–262.

Sherry, B. Y. 1975. Reproduction of elephant in Gonarezhou, south-eastern Rhodesia. *Arnoldia Rhodesia* 7 (29): 1–16.

Short, J. 1981. Diet and feeding behaviour of the forest elephant. *Mammalia* 45: 177–186.

Short, J. 1983. Density and seasonal movements of forest elephant (*Loxodonta africana cyclotis*, Matschie) in Bia National Park, Ghana. *African Journal of Ecology* 21: 175–184.

Short, R. V. 1962. The peculiar lungs of the elephant. *New Scientist* 16 (316): 570–572.

Shoshani, J. 1992. Comparing the living elephants. In: *Elephants: Majestic Creatures of the Wild* (consult. ed. J. Shoshani). Rodale Press, Emmaus, PA, pp. 36–51.

Shoshani, J. 1993. Elephants: the super keystone species. *Swara* 16 (2): 25–29.

Shoshani, J. 1996. Para- or monophyly of the gomphotheres and their position within Proboscidea. In: *The Proboscidea: Evolution and Palaeoecology of Elephants and Their Relatives* (eds J. Shoshani & P. Tassy). Oxford University Press, Oxford, pp. 149–177.

Shoshani, J. 1997. It's a nose! It's a hand! It's an elephant's trunk! *Natural History* 106 (10): 37–43.

Shoshani, J. 1998. Understanding proboscidean evolution: a formidable task. *Trends in Ecology & Evolution* 13 (12): 480–487.

Shoshani, J. (consult. ed.). 2000. *Elephants* (Revised Edition). Checkmark Books (an imprint of Facts on File, Inc.), New York, 240 pp.

Shoshani, J. 2005. Proboscidea. In: *Mammal Species of the World: A Taxonomic and Geographic Reference* (eds D. E. Wilson & D. A. Reeder). (3rd edn). Johns Hopkins University Press, Baltimore, pp. 90–91.

Shoshani, J. 2006. Taxonomy, classification, history, and evolution of elephants. In: *Biology, Medicine, and Surgery of Elephant* (eds M. E. Fowler & S. K. Mikota). Blackwell Publishing, Ames, IA, pp. 3–14.

Shoshani, J. & Eisenberg, J. F. 1982. *Elephas maximus*. *Mammalian Species* 182: 1–8.

Shoshani, J. & Tassy, P. (eds) 1996. *The Proboscidea: Evolution and Palaeoecology of Elephants and Their Relatives*. Oxford University Press, Oxford, 472 pp.

Shoshani, J. & Tassy, P. 2005. Advances in proboscideans taxonomy & classification, anatomy & physiology, and ecology & behavior. *Quaternary International* 126–128: 5–20.

Shoshani, J. *et al.* [76 co-authors]. 1982. On the dissection of a female Asian elephant (*Elephas maximus maximus* Linnaeus, 1758) and data from other elephants. *Elephant* 2 (1): 3–93.

Shoshani, J., Dalen, A., Watson, G., Marchant, G. H. & Marsac, E. 1997. The pharyngeal pouch: a unique receptacle in the throat of an elephant. In: *Proceedings of the 23rd National Conference of the American Association of Zoo Keepers*. American Association of Zoo Keepers. pp. 14–24.

Shoshani, J., Golenberg, G. M. & Yang, H. 1998. Elephantidae phylogeny: morphological versus molecular results. *Acta Theriologica* Suppl. 5: 89–122. [Erratum: on page 99, subfamily Elephantinae should include *Primelephas* as shown in Table 1, page 93.]

Shoshani, J., Sanders, W. J. & Tassy, P. 2001. Elephants and other proboscideans: a summary of recent findings and new taxonomic suggestions. In: *Proceedings of the 1st International Congress of 'La Terra degli Elefanti', The World of Elephants* (eds G. Cavarretta, P. Gioia, M. Mussi & M. R. Palombo). Consiglio Nazionale delle Ricerche, Rome, pp. 676–679.

Shoshani, J., Hagos, Y., Yacob, Y., Ghebrehiwet, M. & Kebrom, E. 2004. The elephants (*Loxodonta africana*) of Zoba Gash Barka, Eritrea: Part 2. Numbers and distribution, ecology and behaviour, and fauna and flora in their ecosystem. *Pachyderm* 36: 52–68.

Shoshani, J., Kupsky, W. J. & Marchant, G. H. 2006a. Elephant brain. Part I: gross morphology, functions, comparative anatomy, and evolution. *Brain Research Bulletin* 70: 124–157.

Shoshani, J., Walter, R. C., Abraha, M., Berhe, S., Tassy, P., Sanders, W. L., Marchant, G. H., Libsekal, Y., Ghirmai, T. & Zinner, D. 2006b. A new proboscidean from the late Oligocene of Eritrea, a 'missing link' between Mammutida and Elephantida. *Proceedings of the National Academy of Sciences of the United States of America* 103 (46): 17296–17301. [Errata appeared in *PNAS*, January 16, 2007, vol. 104 (3): 1103.]

Shoshani, J., Ferretti, M. P., Lister, A. M., Agenbroad, L. D., Saegusa, H., Mol, D. & Takahashi, K. 2007. Relationships within the Elephantinae using hyoid characters. *Quaternary International* 169: 174–185.

Shrader, A. M., Ferreira, S. M., McElveen, M. E., Lee, P. C., Moss, C. J. & Van Aarde, R. J. 2006. Growth and age determination of African savanna elephants. *Journal of Zoology, London* 270: 40–48.

Sikes, S. K. 1966. The African elephant, *Loxodonta africana*: a field method for the estimation of age. *Journal of Zoology, London* 150: 279–295.

Sikes, S. K. 1967. The African elephant, *Loxodonta africana*: a field method for the estimation of age. *Journal of Zoology, London* 154: 235–248.

Sikes, S. K. 1971. *The Natural History of the African Elephant*. Weidenfeld and Nicolson, London, xxv, 397 pp.

Slotow, R. & van Dijk, G. 2001. Role of delinquent young 'orphan' male elephants in high mortality of white rhinoceros in Pilanesberg National Park, South Africa. *Koedoe* 44: 85–94.

Slotow, R., Balfour, D. & Howison, O. 2001. Killing of black and white rhinoceroses by African elephants in Hluhluwe–Umfolozi Park, South Africa. *Pachyderm* 31: 14–20.

Smuts, G. L. 1975. Reproduction and population characteristics of elephants in the Kruger National Park. *Journal of the Southern African Wildlife Management Association* 5: 1–10.

Soltis, J., Leong, K. & Savage, A. 2005a. African elephant vocal communication I: antiphonal calling behaviour among affiliated females. *Animal Behaviour* 70: 579–587.

Soltis, J., Leong, K. & Savage, A. 2005b. African elephant vocal communication. II: rumble variation reflects the individual identity and emotional state of callers. *Animal Behaviour* 70: 589–599.

Spinage, C. A. 1994. *Elephants*. T & A. D. Poyser, London, 319 pp.

Stalmans, M. & Anderson, J. 1992. The forest elephants of Togo. *African Wildlife* 46: 71–75.

Tassy, P. 1986. *Nouveaux Elephantoidea (Mammalia) dans le Miocène du Kenya*. Editions du CNRS Cahiers de Paléontologie, Paris, 135 pp.

Tassy, P. 1994. Les proboscidiens (Mammalia) fossiles du Rift occidental, Ouganda. In: *Geology and Palaeobiology of the Albertine Rift Valley, Uganda–Zaire*. Vol. 2: *Palaeobiology/Paléobiologie* (eds B. Senut & M. Pickford). Centre International pour la Formation et les Echanges Géologiques, Orléans, pp. 1–423.

Tassy, P. 1996. Who is who among the Proboscidea? In: *The Proboscidea: Evolution and Palaeoecology of Elephants and their Relatives* (eds J. Shoshani & P. Tassy). Oxford University Press, Oxford, pp. 39–48.

Tassy, P. 1998. Et la trompe vint aux éléphants. *La Recherche* 305: 54–57.

Tassy, P. 2003. Elephantoidea from Lothagam. In: *Lothagam: The Dawn of Humanity in Eastern Africa* (eds M. G. Leakey & J. M. Harris). Columbia University Press, New York, pp. 331–358.

Tassy, P. & Darlu, P. 1987. Les Elephantidae: nouveau regard sur les analyses de parcimonie. *Géobios* 20 (4): 487–494.

Temminck, C. J. 1847. *Coup-d'oeil général sur les possessions néerlandaises dans l'Inde archipélagique*, Vol. 2. A. Arnz and Company, Leide.

Theuerkauf, J., Ellenberg, H. & Guiro, Y. 2000a. Group structure of forest elephants in the Bossematie Forest Reserve, Ivory Coast. *African Journal of Ecology* 38: 262–264.

Theuerkauf, J., Waitkuwait, W. E., Guiro, Y., Ellenberg, H. & Porembski, S. 2000b. Diet of forest elephants and their role in seed dispersal in the Bossematie Forest Reserve, Ivory Coast. *Mammalia* 64: 447–459.

Thomas, L., Beyers, R., Hart, J. A. & Buckland, S. 2001. Recommendations for a survey design for the Central African region. *Technical Report No. 1*. MIKE – Monitoring of the Illegal Killing of Elephants – Central African Pilot Project. CITES, Nairobi, 27 pp.

Thomas, M. G., Hegelberg, E., Jones, H. B., Yang, Z. & Lister, A. M. 2000. Molecular and morphological evidence on the phylogeny of the Elephantidae. *Proceedings of the Royal Society of London B* 267: 2493–2525.

Thouless, C. R. 1996. Home ranges and social organization of female elephants in northern Kenya. *African Journal of Ecology* 34: 284–297.

Todd, N. E. & Roth, V. L. 1996. Origin and radiation of the Elephantidae. In: *The Proboscidea: Evolution and Palaeoecology of Elephants and Their Relatives* (eds J. Shoshani & P. Tassy). Oxford University Press, Oxford, pp. 193–202.

Todd, T. W. 1913. Notes on the respiratory system of the elephant. *Anatomische Anzeiger* 44: 175–183.

Turkalo, A. 1996. Studying forest elephants by direct observation in the Dzanga clearing: an update. *Pachyderm* 22: 59–60.

Turkalo, A. & Fay, J. M. 1995. Studying forest elephants by direct observation: preliminary results from the Dzanga clearing, Central African Republic. *Pachyderm* 20: 45–54.

Van der Merwe, N. J., Bezuidenhout, A. J. & Seegers, C. D. 1995. The skull and mandible of the African elephant (*Loxodonta africana*). *Onderstepoort Journal of Veterinary Research* 62: 245–260.

Vanleeuwe, H. & Gautier-Hion, A. 1998. Forest elephant paths and movements at the Odzala National Park, Congo: the role of clearings and Marantaceace forests. *African Journal of Ecology* 36: 174–182.

Verlinden, A. & Gavor, I. K. N. 1998. Satellite tracking of elephants in northern Botswana. *African Journal of Ecology* 36: 105–116.

Vigne, L. & Martin, E. 2011. Consumption of elephant and mammoth ivory increases in southern China. *Pachyderm* 49: 79–89.

Viljoen, P. J. 1987. Status and past and present distribution of elephants in Kaokoveld, South West/Namibia. *South African Journal of Zoology* 22: 247–257.

Viljoen, P. J. 1988. The ecology of the desert dwelling elephants *Loxodonta africana* (Blumenbach, 1797) of western Damaraland and Kaokoland. PhD thesis, University of Pretoria, South Africa.

Viljoen, P. J. 1989. Spatial distribution and movements of elephants (*Loxodonta africana*) in the northern Namib Desert region of the Kaokoveld, South West Africa/Namibia. *Journal of the Zoological Society of London* 219: 1–19.

Viljoen, P. J. & Bothma, J. du P. 1990. Daily movements of desert-dwelling elephants in the northern Namib desert. *South African Journal of Wildlife Research* 20: 69–72.

Ville, J.-L. 1995. Man and elephant in the Tsavo area of Kenya: an anthropological perspective. *Pachyderm* 20: 69–72.

Vogel, J. C., Talma, A. S., Hall-Martin, A. J. & Viljoen, P. J. 1990. Carbon and nitrogen isotopes in elephants. *South African Journal of Science* 86: 147–150.

Wallace, C. 1978. Chromosome analysis in the Kruger national Park: mitotic and meiotic studies in the African Elephant *Loxodonta africana*. *Koedoe* 21: 197–200.

Walsh, P. D. & White, L. J. T. 1999. What it will take to monitor forest elephant populations. *Conservation Biology* 13: 1194–1202.

Weigl, R. 2005. *Longevity of Mammals in Captivity: From the Living Collections of the World*. Kleine Senckenberg-Reihe 48, Stuttgart, 214 pp.

Weir, J. S. 1972. Spatial distribution of elephants in an African National Park in relation to environmental sodium. *Oikos* 23: 1–13.

Western, D. 1986. The pygmy elephant: a myth and a mystery. *Pachyderm* 7: 4–5.

Western, D. 1989. The ecological role of elephants in Africa. *Pachyderm* 12: 42–45.

Western, D., Moss, C. J. & Georgiadis, N. 1983. Age estimation and population age structure of elephants from footprint dimensions. *Journal of Wildlife Management* 47: 1192–1197.

White, L., Tutin, C. & Fernandez, M. 1993. Group composition and diet of forest elephants, *Loxodonta africana cyclotis* Matschie 1900, in the Lope Reserve, Gabon. *African Journal of Ecology* 31: 181–199.

White, L. J. T. 1994. *Sacoglottis gabonensis* fruiting and the seasonal movements of elephants in the Lope Reserve, Gabon. *Journal of Tropical Ecology* 10: 121–125.

Whitehouse, A. M. 2002. Tusklessness in the elephant population of the Adoo Elephant National Park, South Africa. *Journal of Zoology, London* 257: 249–254.

Whitehouse, A. M. & Hall-Martin, A. J. 2000. Elephants in Addo Elephant National Park, South Africa: reconstruction of the population's history. *Oryx* 34: 46–55.

Whyte, I. J. 2001a. Conservation Management of the Kruger National Park Elephant Population. PhD thesis, University of Pretoria, South Africa.

Whyte, I. J. 2001b. Headaches and heartaches – the elephant management dilemma. In: *Environmental Ethics: What Really Matters, What Really Works* (eds D. Schmidtz & E. Willot). Oxford University Press, New York, pp. 293–305.

Whyte, I. J. & Fayrer-Hosken, R. 2008. Playing elephant god: ethics of managing wild elephant populations. In: *Elephants and Ethics: The Morality of Coexistence* (eds C. Wemmer & K. Christen). Johns Hopkins University Press, Baltimore, Maryland. pp. 399–417.

Whyte, I. J., Van Aarde, R. J. & Pimm, S. 2003. Kruger National Park's elephant population: its size and consequences for ecosystem heterogeneity. In: *The Kruger Experience – Ecology and Management of Savanna Heterogeneity* (eds J.

T. du Toit, H. C. Biggs & K. H. Rogers). Island Press, Washington, DC, pp. 332–348.

Wickler, W. & Seibt, U. 1997. Aimed object-throwing by a wild African elephant in an interspecific encounter. *Ethology* 103: 365–368.

Williamson, B. R. 1975. The condition and nutrition of elephant in Wankie National Park. *Arnoldia Rhodesia* 7 (12): 1–20.

Wittemyer, G., Douglas-Hamilton, I. & Getz, W. M. 2005. The socioecology of elephants: analysis of the processes creating multitiered social structures. *Animal Behaviour* 69: 1357–1371.

Wright, P. G. & Luck, C. P. 1984. Do elephants need to sweat? *South African Journal of Zoology* 19: 270–274.

Yang, H., Golenberg, E. M. & Shoshani, J. 1996. Phylogenetic resolution within the Elephantidae using fossil DNA sequence from the American mastodon (*Mammut americanum*), as an outgroup. *Proceedings of the National Academy of Sciences USA* 93 (3): 1190–1194.

Zumpt, F. & Wetzel, H. 1970. Fly parasites (*Diptera:* Oestridae and Gasterophilidae) of the African Elephant *Loxodonta africana* (Blumenbach) and their problems. *Koedoe* 13: 109–121.

Order Sirenia

Amrine-Madsen, H., Koepfli, K.-P., Wayne, R. K. & Springer, M. S. 2003. A new phylogenetic marker, apolipoprotein B, provides compelling evidence for eutherian relationships. *Molecular Phylogenetics and Evolution* 28: 225–240.

Anderson, P. K. 1981. The behaviour of the dugong (*Dugong dugon*) in relation to conservation and management. *Bulletin of Marine Science* 31: 640–7.

Anderson, P. K. 1982. Studies of dugongs at Shark Bay, Western Australia. II. Surface and subsurface observations. *Australian Wildlife Research* 9: 85–99.

Anderson, P. K. 1997. Shark Bay dugongs in summer. I: lek mating. *Behaviour* 134: 433–462.

Anderson, P. K & Barclay, R. M. K. 1995. Acoustic signals of solitary dugongs: physical characteristics and behavioural correlates. *Journal of Mammalogy* 76: 1226–1237.

Anderson, P. K. & Birtles, A. 1978. Behaviour and ecology of the dugong, *Dugong dugon* (Sirenia): observations in Shoalwater and Cleveland Bays, Queensland. *Australian Wildlife Research* 5: 1–23.

Anderson, P. K & Prince, R. I. T. 1985. Predation on dugongs: attacks by killer whales. *Journal of Mammalogy* 66: 554–556.

Baylis, H. A. 1936. Some parasitic worms from the British Cameroons. *Annals and Magazine of Natural History* 17 (10): 257–272.

Beck, C. & Forrester, D. J. 1988. Helminths of the Florida manatee, *Trichechus manatus latirosiris,* with a discussion and summary of the parasites of sirenians. *Journal of Parasitology* 74: 628–637.

Blair, D. 1981. Helminth parasites of the dugong, their collection and preservation. In: *The Dugong. Proceedings of a Seminar/Workshop Held at James Cook University of North Queensland, 8–13 May 1979* (ed. H. Marsh). Department of Zoology, James Cook University of North Queensland, Townsville, Australia, pp. 275–285.

Boyd, I. L., Lockyer, C. & Marsh, H. 1999. Reproduction in marine mammals. In: *Biology of Marine Mammals* (eds J. E. Reynolds III & S. A. Rommel). Smithsonian Institution Press, Washington, DC. pp. 218–286.

Bryden, M., Marsh, H. & Shaughnessy, P. 1998. *Dugongs, Whales, Dolphins and Seals. A Guide to the Sea Mammals of Australasia.* Allen and Unwin, St Leonards, 176 pp.

Cadenat, J. 1957. Observations de cétaces, siréniens, cheloniens et sauriens en 1955–1956. *Bulletin de l'Institut Française d'Afrique Noire, Série A* 19: 1358–1383.

Campbell, R. S. F. & Ladds, P. W. 1981. Disease of the dugong in northeastern Australia: a preliminary report. In: *The Dugong. Proceedings of a Seminar/Workshop Held at James Cook University of North Queensland, 8–13 May 1979* (ed. H. Marsh). Department of Zoology, James Cook University of North Queensland, Townsville, Australia, pp. 167–181.

Chilvers, B. L., Delean, S., Gales, N. J., Holley, D. K., Lawler, I. R., Marsh, H. & Preen, A. R. 2004. Diving behaviour of dugongs, *Dugong dugon*. *Journal of Experimental Marine Biology and Ecology* 304: 203–224.

Cockcroft, V. G. & Young, D. D. 1998. An investigation of the status of coastal marine resources along the west coast of Madagascar. Unpublished report to the World Wide Fund for Nature (WWF), Gland, Switzerland, 121 pp.

Cockcroft, V. G., Salm, R. V. & Dutton, T. P. 1994. The status of dugongs in the western Indian Ocean. *First International Manatee and Dugong Research Conference, March 11–13*. Gainesville, Florida.

Cole, A. & Okera, W. 1976. Records of the West African manatee, *Trichechus senegalensis*, in Sierra Leone. *Bulletin of the Institute of Marine Biology & Oceanography (Fourah Bay College, University of Sierra Leone)* 1: 42–43.

Dekeyser, P. L. 1952. Note sommaire sur la temperature rectal de Lamatin (*T. senegalensis* Link). *Bulletin du Muséum national d'Histoire naturelle*, 2ème série, 2: 243–246.

Diop, M. 1984. *Operation Lamantins (*Trichechus senegalensis*).* Report to the conservator of Djoudj National Park. Republic of Senegal, Ministry of Nature Protection, 4 pp.

Dollman, G. 1933. Dugongs from Mafia Island and a manatee from Nigeria. *Natural History Magazine* 9 (28): 117–128.

Domning, D. P. 1995. What do we know about the evolution of the dugong? (Abstract) *Mermaid Symposium: First International Symposium on Dugong and Manatees, November 15–17, 1995,* Toba, Mie, Japan. Toba Aquarium, Toba, pp. 23–24.

Domning, D. P. 1996. Bibliography and Index of the Sirenia and Desmostylia. *Smithsonian Contributions to Paleobiology* 80: iii + 611.

Domning, D. P. 2001a. The earliest known fully quadrupedal sirenian. *Nature* 413: 625–627.

Domning, D. P. 2001b. Evolution of the Sirenia and Desmostylia. In: *Secondary Adaptation of Tetrapods to Life in Water. Proceedings of the International Meeting, Poitiers, 1996* (eds J.-M. Mazin & V. de Buffrénil). Verlag Dr Friedrich Pfeil, Munich, pp. 151–168.

Domning, D. P. 2002. Sirenian evolution. In: *Encyclopedia of Marine Mammals* (eds W. F. Perrin, B. Würsig & H. Thewissen). Academic Press, San Diego, California, pp. 1083–1086.

Domning, D. P. 2005. Fossil Sirenia of the West Atlantic and Caribbean region. VII. Pleistocene *Trichechus manatus* Linnaeus, 1758. *Journal of Vertebrate Paleontology* 25: 685–701.

Domning, D. P. & de Buffrénil, V. 1991. Hydrostasis in the Sirenia: quantitative data and functional interpretations. *Marine Mammal Science* 7: 331–368.

Domning, D. P. & Hayek, L.-A. 1986. Interspecific and intraspecific morphological variation in manatees (Sirenia: *Trichechus*). *Marine Mammal Science* 2: 87–144.

Domning, D. P., Zalmont, I. & Gingerich, P. D. 2010: Sirenia. In: *Cenozoic Mammals of Africa* (eds L. Werdelin & W. J. Sanders). University of California Press, Berkeley. pp. 147–160.

Dupuy, A. R. & Maigret, J. 1978. Les mammifères marins des côtes du Sènégal. *Bulletin IFAN* 40: 457–465.

Dutton, P. 1994. *Past and Present Status of Dugong* Dugong dugon *in the Bazaruto Archipelago and other Known Habitats on the Mozambique Coast*. Paper presented at the First International Manatee and Dugong Research Conference, Gainesville, Florida.

Falcon-Matos, L., Mignucci-Giannoni, A. A., Toyos-González, G. M., Bossart, G. D., Meisner, R. A. & Varela, R. A. 2003. Evidence of a shark attack on a West Indian manatee (*Trichechus manatus*) in Puerto Rico. *Journal of Neotropical Mammalogy* 10: 161–166.

Garcia-Rodriguez, A. I., Bowen, B. W., Domning, D., Mignucci-Giannoni, A. A., Marmontel, M., Montoya-Ospina, R. A., Morales-Vèla, B., Rudon, M., Bonde, R. K. & McGuire, P. M. 1998. Phylogeography of the West Indian manatee (*Trichechus manatus*): how many populations and how many taxa? *Molecular Evolution* 7: 1137–1149.

Gheerbrant, E., Domning, D. P. & Tassy, P. 2005. Paenungulata (Sirenia, Proboscidea, Hyracoidea, and relatives). Chapter. 7. In: *The Rise of Placental*

Mammals: Origin and Relationships of the Major Extant Clades (eds K. D. Rose & J. D. Archibald). Johns Hopkins University Press, Baltimore, pp. 84–105.

Giffin, E. B. & Gillett, M. 1996. Neurological and osteological definitions of cervical vertebrae in mammals. *Brain Behavior Evolution* 47: 214–218.

Glaser, K. S. (photographs) and J. E. Reynolds, III (text). 2003. *Mysterious Manatees*. University Press of Florida, Gainesville, 187 pp.

Gohar, H. A. F. 1957. *The Red Sea dugong*. Publications of the Marine Biological Station, Al-Ghardaqa, Egypt 9: 3–50.

Grigione, M. M. 1996. Observations on the status and distribution of the West African Manatee in Cameroon. *African Journal of Ecology* 34: 189–195.

Grubb, P., Jones, T. S., Davies, A. G., Edberg, E., Starin, E. D. & Hill, J. E. 1998. *Mammals of Ghana, Sierra Leone and the Gambia*. Trendrine Press, Zennor, Cornwall.

Guissamulo, A. T. & Cockcroft, V. G. 1997. Dolphin and dugong occurrence and distribution and fisheries interactions in Maputo and Bazaruto Bays, Mozambique. *Proceedings of the 49th International Whaling Commission Meeting*. Scientific Committee Annual Meeting, UK. International Whaling Commission, Cambridge.

Hanafy, M., Gheny, M. A., Rouphael, A. B., Salim, A. & Fouda, M. 2006. The Dugong, *Dugong dugon*, in Egyptian waters: distribution, relative abundance and threats. *Zoology in the Middle East* 39: 17–24.

Hatt, R. 1934. A manatee collected by the American Museum Congo Expedition, with observations on the recent manatees. *Bulletin of the American Museum of Natural History* 66: 533–536.

Heinsohn, G. E., Wake, J., Marsh, H. & Spain, A. V. 1977. The dugong, *Dugong dugon* (Müller) in the seagrass system. *Aquaculture* 12: 235–248.

Heithaus, M. R. 2001. The biology of tiger sharks, *Galeocerdo cuvier* in Shark Bay, Western Australia: sex ratio, size distribution, diet, and seasonal changes in catch rates. *Environmental Biology of Fishes* 61: 25–36.

Hines, E., Reynolds, J. E., III, Mignucci-Giannoni, A. A., Aragones, L. V. & Marmontel, M. (eds). in press. *Sirenian Conservation: Issues and Strategies in Developing Countries*. University Press of Florida, Gainesville. 348 pp.

Howell, J. H. 1968. The Borgu Game Reserve of northern Nigeria. *Nigerian Field* 33: 147–165.

Hughes, G. R. & Oxley Oxland, R. 1971. A survey of dugong (*Dugong dugon*) in and around Antonio Enes, Northern Mozambique. *Biological Conservation* 3: 299–301.

Husar, S. L. 1978. *Trichechus senegalensis*. *Mammalian Species* 89: 1–3.

Ita, E. O. 1994. *Aquatic Plants and Wetland Wildlife Resources of Nigeria*. CIFA Occasional Paper. No. 21. FAO, Rome, 52 pp.

Itoh, S. & Tsuyuki, H. 1974. Fatty acid component of Senegal manatee fats. *Scientific Reports of the Whales Research Institute* 26: 307–311.

Kingdon, J. 1971. *East African Mammals: An Atlas of Evolution in Africa*. Vol. I: *Primates, Hyraxes, Pangolins, Protoungulates, Sirenians*. Academic Press, London, 446 pp.

Lanyon, J. M. & Marsh, H. 1995. Digesta passage times in the dugong. *Australian Journal of Zoology* 43: 119–127.

Lanyon, J. M., Sneath, H. L., Long, T. & Bonde, R. K. 2010. Physiological response of wild dugongs (*Dugong dugon*) to out-of-water sampling for health assessment. *Aquatic Mammals* 36: 46–58.

Lee Long, W. J. & Coles, R. G. 1997. Status of seagrasses in the Great Barrier Reef region. In: *State of the Great Barrier Reef World Heritage Area Workshop: Proceedings of a Technical Workshop held in Townsville, Queensland, Australia, 27–29 November 1995* (eds D. Wachenfeld, J. Oliver & K. Davis). Great Barrier Reef Marine Park Authority, Townsville, pp. 185–193.

Marsh, H. 1980. Age determination of the dugong, *Dugong dugon* (Müller) in northern Australia and its biological implications. *Reports of the International Whaling Commission, Special Issue* 3: 181–201.

Marsh, H. 1995. The life history, pattern of breeding and population dynamics of the dugong. In: *Population Biology of the Florida Manatee* (eds T. J. O'Shea, B. B. Ackermann & H. F. Percival). US Department of the Interior, National Biological Service, Information and Technology Report No. 1, pp. 75–83.

Marsh, H. 2002. Dugong. In: *Encyclopedia of Marine Mammals* (eds W. F. Perrin, B. Würsig & H. Thewissen). Academic Press, San Diego, pp. 344–347.

Marsh, H., Heinsohn, G. E. & Spain, A. V. 1977. The stomach and duodenal diverticula of the dugong (*Dugong dugon*). In: *Functional Anatomy of Marine Mammals*, Vol. 3 (ed. R. J. Harrison). Academic Press, London, pp. 271–295.

Marsh, H., Channells, P. W., Heinsohn, G. E. & Morissey, J. 1982. Analysis of stomach contents of dugongs from Queensland. *Australian Wildlife Research* 9: 55–67.

Marsh, H., Heinsohn, G. E. & Glover, T. D. 1984a. Changes in the male reproductive organs of the dugong, *Dugong dugon* (Sirenia: Dugongidae) with age and reproductive activity. *Australian Journal of Zoology* 32: 743–766.

Marsh, H., Heinsohn, G. E. & Marsh, L. M. 1984b. Breeding cycle, life history and population dynamics of the dugong, *Dugong dugon* (Sirenia: Dugongidae). *Australian Journal of Zoology* 32: 767–788.

Marsh, H., Harris, A. N. M. & Lawler, I. R. 1997. The sustainability of the indigenous dugong fishery in Torres Strait, Australia/Papua New Guinea. *Conservation Biology* 11: 1375–1386.

Marsh, H., Eros, C. & Webb, R. 2000. Dugongs in health and disease. In: *Marine Wildlife: The Fabian Faye Course for Veterinarians*. Proceedings 335, 4–8 September 2000, Sea World Nara Resort, Gold Coast. Post Graduate Foundation in Veterinary Science, University of Sydney, Australia, pp. 301–317.

Marsh, H., Eros, C., Penrose, H. & Hugues, J. 2002. *Dugong: Status Report and Action Plans for Countries and Territories*. UNEP/DEWA/RS. 02–1.

Marsh, H., O'Shea, T. J. & Reynolds, J. E., III. 2011. *Ecology and Conservation of the Sirenia: Dugongs and Manatees*. Cambridge University Press, New York. 544 pp.

Marshall, C. D. 2002. Morphology, Functional. In: *Encyclopedia of Marine Mammals* (eds W. F. Perrin, B. Würsig & H. Thewissen). Academic Press, San Diego, pp. 759–774.

Marshall, C. D., Clark, L. A. & Reep, R. L. 1998. The muscular hydrostat of the Florida manatee (*Trichechus manatus latirostris*): a functional morphological model of perioral bristle use. *Marine Mammal Science* 14: 290–303.

Marshall, C. D., Maeda, H., Iwata, M., Furuta, M., Asano, S., Rosas, F. & Reep, R. L. 2003. Orofacial morphology and feeding behaviour of the dugong, Amazonian, West African and Antillean manatees (Mammalia: Sirenia): functional morphology of the muscular-vibrissal complex. *Journal of Zoology, London* 259: 245–260.

McKenna, M. C. 1975. Toward a phylogenetic classification of the Mammalia. In: *Phylogeny of the Primates* (eds W. P. Luckett & F. S. Szalay). Plenum, New York and London, pp. 21–46.

Meredith, R. W., Janečka, J. E., Gatesy, J., Ryder, O. A., Fisher, C. A., Teeling, E. C., Goodbla, A., Eizirik, E., Simão, T. L. L., Stadler, T., Rabosky, D. L., Honeycutt, R. L., Flynn, J. J., Ingram, C. M., Steiner, C., Williams, T. L., Robinson, T. J., Burk-Herrick, A., Westerman, M., Ayoub, N. A., Springer, M. S. & Murphy, W. J. 2011. Impacts of the Cretaceous Terrestrial Revolution and KPg extinction on mammal diversification. *Science* 334: 521–524.

Morais, M. 2006. Availação da Condição e Distribuição do Manatim Africana (*Trichechus senegalensis*) ao Longo do Rio Cuanza. Unpublished report. Luanda.

Müller, P. L. S. 1776. *Des Ritters Carl von Linné … vollständigen Natursystems Supplements – und Register-Band über alle sechs Theile oder classens des Thierreichs. Mit einer ausführlichen Erklürung* … Gabriel Nicolaus Raspe, Nuremberg, 384 pp.

Nishiwaki, M. 1984. Current status of the African manatee. *Acta Zoologica Fennica* 172: 135–136.

Nishiwaki, M. & Marsh, H. 1985. Dugong *Dugong dugon* (Müller, 1776). In: *Handbook of Marine Mammals*, Vol. 3 (eds S. H. Ridgway & R. Harrison). Academic Press, London, pp. 1–31.

Pabst, D. A., Rommel, S. A. & McLellan, W. A. 1999. The functional morphology of marine mammals. In: *Biology of Marine Mammals* (eds J. E. Reynolds, III, & S. A. Rommel). Smithsonian Institution Press, Washington, DC, pp. 15–72.

Parr, L. A. & Duffield, D. A. 2002. Interspecific comparison of mitochondrial DNA variation between extant species of manatee. In: *Cell and Molecular*

Biology of Marine Mammals (ed. C. J. Pfeiffer). Krieger Pulishing Co., Malabar, Florida, pp. 152–158.

Powell, J. A. 1978. Evidence of carnivory in manatees (*Trichechus manatus*). *Journal of Mammalogy* 59: 442.

Powell, J. A. 1996. *The Distribution and Biology of the West African Manatee (*Trichechus senegalensis* Link, 1795).* United Nations Environment Programme, Regional Seas Programme, Oceans and Coastal Areas. Nairobi, Kenya, 68 pp.

Preen, A. R. 1989a. *Technical Report, Dugongs.* Vol. 1: *The Status and Conservation of Dugongs in the Arabian Region.* MEPA Coastal and Marine Management Series, Saudi Arabia, 200 pp.

Preen, A. R. 1989b. Observations of mating behaviour in dugongs (*Dugong dugon*). *Marine Mammal Science* 5: 382–387.

Preen, A. R. 1995a. Impacts of dugong foraging on seagrass habitats: observations and experimental evidence for cultivation grazing. *Marine Ecological Progress Series* 124: 201–213.

Preen, A. R. 1995b. Diet of dugongs: are they omnivores? *Journal of Mammalogy* 76: 163–171.

Preen, A. R. 2004. Distribution, abundance and conservation status of dugongs and dolphins in the southern and western Arabian Gulf. *Biological Conservation* 118: 205–218.

Reep, R. L., Marshall, C. D. & Stoll, M. L. 2002. Tactile hairs on the postcranial body in Florida manatees: a mammalian lateral line? *Brain Behavior Evolution* 59: 141–154.

Reeves, R. R., Tuboku-Metzger, D. & Kapindi, R. A. 1988. Distribution and exploitation of manatees in Sierra Leone. *Oryx* 22: 75–84.

Reynolds, J. E., III. 1999. Efforts to Conserve the Manatees. In: *Conservation and Management of Marine Mammals* (eds J. R. Twiss, Jr & R. R. Reeves). Smithsonian Institution Press, Washington, DC, pp. 267–295.

Reynolds, J. E., III & Odell, D. K. 1991. *Manatees and Dugongs.* Facts on File, New York.

Reynolds, J. E., III & Powell, J. A., Jr. 2002. The manatees – Family Trichechidae (*Trichechus manatus*, *T. senegalensis*, and *T. inunguis*). In: *Encyclopedia of Marine Mammals* (eds W. F. Perrin, B. Würsig & H. Thewissen). Academic Press, San Diego, pp. 709–720.

Reynolds, J. E., III & Rommel, S. A. 1996. Structure and function of the gastrointestinal tract of the Florida manatee, *Trichechus manatus latirostris*. *Anatomical Record* 245: 539–558.

Reynolds, J. E., III, Rommel, S. A. & Bolen, M. E. 2001. Anatomical dissection – thorax and abdomen. In: *Encyclopedia of Marine Mammals* (eds W. F. Perrin, B. Würsig & H. Thewissen). Academic Press. San Diego, pp. 21–30.

Reynolds, J. E., III, Rommel, S. A. & Pitchford, M. E. 2004. The likelihood of sperm competition in manatees – explaining an apparent paradox. *Marine Mammal Science* 20: 464–476.

Rice, D. W. 1998. *Marine Mammals of the World. Systematics and Distribution.* Special Publication Number 4. The Society for Marine Mammalogy, 231 pp.

Robineau, D. & Rose, J. M. 1982. Le dugong (Dugong dugon Muller, 1776) Sirenia Dugongidae en République de Djibouti. *Biological Conservation* 24: 233–238.

Rommel, S. A. & Reynolds, J. E., III. 2000. Diaphragm structure and function in the Florida manatee (*Trichechus manatus latirostris*). *Anatomical Record* 259: 41–51.

Rommel, S. A. & Reynolds, J. E., III. 2002. Postcranial skeleton of marine mammals. In: *Encyclopedia of Marine Mammals* (eds W. F. Perrin, B. Würsig & H. Thewissen). Academic Press, San Diego, pp. 1089–1103.

Savage, R. J. G., Domning, D. P. & Thewissen, J. G. M. 1994. Fossil Sirenia of the West Atlantic and Caribbean region. V. The most primitive known sirenian, *Prorastomus sirenoides* Owen, 1855. *Journal of Vertebrate Paleontology* 14: 427–449.

Sheppard, J. K., Preen, A. R., Marsh, H., Lawler, I. R., Whiting, S. & Jones, R. E. 2006. Movement heterogeneity of dugongs, *Dugong dugon* (Müller) over large spatial scales. *Journal of Experimental Marine Biology and Ecology* 334: 64–83.

Shoshani, J. 1993. Hyracoidea-Tethytheria affinity based on myological data. In: *Mammal Phylogeny: Placentals* (eds F. S. Szalay, M. J. Novacek & M. C. McKenna). Springer Verlag, New York, pp. 235–256.

Shoshani, J. 2005. Order Sirenia. In: *Mammal Species of the World: A Taxonomic and Geographic Reference* (eds D. E. Wilson & D. M. Reeder). Johns Hopkins University Press, Baltimore, pp. 92–93.

Sikes, S. 1974. How to save the mermaids. *Oryx* 12: 465–470.

Silva, M. A. & Araújo, A. 2001. Distribution and current status of the West African manatee (*Trichechus senegalensis*) in Guinea-Bissau. *Marine Mammal Science* 17 (2): 418–424.

Skinner, J. D. & Chimimba, C. T. 2005. *The Mammals of the Southern African Subregion* (3rd edn). Cambridge University Press, Cambridge, UK.

Spain, A. V. & Heinsohn, G. E. 1975. Size and weight allometry in a North Queensland population of *Dugong dugon* (Müller) (Mammalia: Sirenia). *Australian Journal of Zoology* 23: 159–68.

Sprent, J. F. A. 1983. Ascaridoid nematodes of sirenians – a new species in the Senegal manatee. *Journal of Helminthology* (Suppl.) 57: 69–76.

Thewissen, J. G. M. & Domning, D. P. 1992. The role of phenacodontids in the origin of the modern orders of ungulate mammals. *Journal of Vertebrate Paleontology* 12: 494–504.

Travis, W. 1967. *The Voice of the Turtle.* Allen and Unwin Ltd, UK.

Vianna, J. A., Bonde, R. K., Caballer, S., Giraldo, J. P., Lima, R. P., Clark, A., Marmontel, M., Morales-Vela, B., Souza, M. J., Parr, L., Rodriquez-Lopez, M. A., Mignucci, A. A. & Powell, J. A. 2006. Phylogeography, phylogeny and hybridization in trichechid sirenians: implications for manatee conservation. *Molecular Ecology* 15: 433–47.

Vincent, T. 1996. Occurrence of a Dugong, *Dugong dugon* (Muller, 1776) (Mammalia, Sirenia, Dugongidae) in October 1994, near Hurghada (Egypt) in the Red Sea. *Annales de l'Institut Océanographique* 72: 179–183.

Wamukoya, G. M., Mirangi, J. & Ottichilo, W. K. 1995. *Marine Aerial Survey: Sea Turtle and Marine Mammals.* Kenya Wildlife Service Technical Series Report No. 1, Kenya Wildlife Service, Kenya, 65 pp.

Wamukoya, G. M., Mirangi, J. M. & Ottichilo, W. K. 1996a. *Marine Aerial Survey: Marine Mammals, Sea Turtles, Sharks and Rays.* Kenya Wildlife Service Technical Series Report No. 1, Kenya Wildlife Service, Kenya.

Wamukoya, G. M., Ottichilo, W. K. & Salm, R. V. 1996b. *Aerial survey of dugongs (*Dugong dugon* (Müller)) in Ungwana Bay and the Lamu Archipelago.* Kenya Wildlife Service Technical Series Report No. 2, Kenya Wildlife Service, Kenya.

Wells, R. S., Boness, D. J. & Rathbun, G. B. 1999. Behavior. In: *Biology of Marine Mammals* (eds J. E. Reynolds, III & S. A. Rommel). Smithsonian Institution Press, Washington, DC, pp. 324–422.

WWF Eastern African Marine Ecoregion. 2004. *Towards a Western Indian Ocean Dugong Strategy: The Status of Dugongs in the Western Indian Ocean Region and Priority Conservation Actions.* WWF, Dar es Salaam, 68 pp.

Cohort Afroinsectiphillia

Amrine-Madsen, H., Koepfli, K-P., Wayne, R. K. & Springer, M. S. 2003. A new phylogenetic marker, apolipoprotein B, provides compelling evidence for eutherian relationships. *Molecular Phylogenetics and Evolution* 28: 225–240.

Goswami, A., Prasad, G. V. R., Upchurch, P., Boyer, D. M., Seiffert, E. R., Verma, O., Gheerbrant, E. & Flynn, J. J. 2011. A radiation of arboreal basal eutherian mammals beginning in the Late Cretaceous of India. *Proceedings of the National Academy of Sciences of the United States of America* 108: 16333–16338.

Meredith, R. W., Janečka, J. E., Gatesy, J., Ryder, O. A., Fisher, C. A., Teeling, E. C., Goodbla, A., Eizirik, E., Simão, T. L. L., Stadler, T., Rabosky, D. L., Honeycutt, R. L., Flynn, J. J., Ingram, C. M., Steiner, C., Williams, T. L., Robinson, T. J., Burk-Herrick, A., Westerman, M., Ayoub, N. A., Springer, M. S. & Murphy, W. J. 2011. Impacts of the Cretaceous Terrestrial Revolution and KPg extinction on mammal diversification. *Science* 334: 521–524.

Murata, Y., Nikaido, M., Sasaki, T., Cao, Y., Fukumoto, Y., Hasegawa, M. & Okada, N. 2003. Afrotherian phylogeny inferred from complete mitochondrial genomes. *Molecular Phylogenetics and Evolution* 28: 253–260.

Nishihara, H., Satta, Y., Nikaido, M., Thewissen, J. G. M., Stanhope, M. J. & Okada, N. 2005. A retroposon analysis of afrotherian phylogeny. *Molecular Biology and Evolution* 22: 1823–1833.

Poulakakis, N. & Stamatakis, A. 2010. Recapitulating the evolution of Afrotheria: 57 genes and rare genomic changes (RGCs) consolidate their history. *Systematics and Biodiversity* 8: 395–408.

Robinson, T. J., Fu, B., Ferguson-Smith, M. A. & Yang, F. 2004. Cross-species chromosome painting in the golden mole and elephant-shrew: support for the mammalian clades Afrotheria and Afroinsectiphillia but not Afroinsectivora. *Proceedings of the Royal Society of London B* 271: 1477–1484.

Seiffert, E. R. 2007. A new estimate of afrotherian phylogeny based on simultaneous analysis of genomic, morphological, and fossil evidence. *BMC Evolutionary Biology* 7: 224.

Seiffert, E. R. 2010. The oldest and youngest records of afrosoricid placentals from the Fayum Depression of northern Egypt. *Acta Palaeontologica Polonica* 55: 599–616.

Orders Afrosoricida and Macroscelidea

Agnelli, P., Azzaroli, M. L. & Simonetta, A. M. 1988. Some remarks on the mammals of Somalia. *Biogeographia* 14: 499–512.

Allen, G. M. 1939. A checklist of African mammals. *Bulletin of the Museum of Comparative Zoology* 83: 1–763.

Allen, G. M. & Loveridge, A. 1927. Mammals from the Uluguru and Usambara Mountains, Tanganyika Territory. *Proceedings of the Boston Society of Natural History* 38: 413–441.

Allen, G. M. & Loveridge, A. 1933. Reports on the scientific results of an expedition to the south-western highlands of Tanganyika Territory. II Mammals. *Bulletin of the Museum of Comparative Zoology at Harvard University* 75: 47–140.

Allen, J. A. 1922. The American Museum of Natural History Expedition Collection of Insectivora. *Bulletin of the American Museum of Natural History* 47: 1–38.

Amrine-Madsen, H., Koepfli, K-P., Wayne, R.K. & Springer, M.S. 2003. A new phylogenetic marker, apolipoprotein B, provides compelling evidence for eutherian relationships. *Molecular Phylogenetics and Evolution* 28: 225–240.

Ansell, W. F. H. 1960. Mammals of Northern Rhodesia. Government Printer, Lusaka, 155 pp + 24 pp maps & figs.

Ansell, W. F. H. & Ansell, P. D. H. 1969. *Petrodromus tetradactylus* at Ngoma. *Puku* 5: 211–213.

Ansell, W. F. H. & Ansell, P. D. H. 1973. Mammals of the north-eastern montane areas of Zambia. *Puku* 7: 21–69.

Arthur, D. R. 1965. *Ticks of the genus Ixodes* in Africa. The Athlone Press, London, 348 pp.

Asher, R. J. 1999. A morphological basis of assessing the phylogeny of the 'Tenrecoidea' (Mammalia, Lipotyphla). *Cladistics* 15: 231–252.

Asher, R. J. 2000. Phylogenetic history of Tenrecs and other insectivoran mammals. PhD dissertation, State University of New York at Stony Brook, USA, 223 pp.

Asher, R. J., Novacek, M. J. & Geisler, J. H. 2003. Relationships of endemic African mammals and their fossil relatives based on morphological and molecular evidence. *Journal of Mammalian Evolution* 10: 131–194.

Aulagnier, S. & Thévenot, M. 1986. Catalogue des Mammifères sauvages du Maroc. *Travaux de l'Institut Scientifique Rabat, Series Zoologie* 41: 1–164.

Baker, A. J., Lengel, K., McCafferty, K. & Hellmuth, H. 2005. Black-and-rufous sengi (*Rhynchocyon petersi*) at the Philadelphia Zoo. *Afrotherian Conservation – Newsletter of the IUCN–SSC Afrotheria Specialist Group* 3: 6–7.

Bennett, N. C. & Spinks, A. C. 1995. Thermoregulation and metabolism in the Cape golden mole (Insectivora: *Chrysochloris asiatica*). *Journal of Zoology, London* 236: 521–529.

Bernard, R. T. F., Bronner, G. N., Taylor, P. J., Bojarski, C. & Tsita, J. N. 1994. Aseasonal reproduction in the Hottentot golden mole, *Amblysomus hottentotus*, from the summer rainfall region of South Africa. *South African Journal of Science* 90: 547–549.

Bernard, R. T. F., Kerley, G. I. H., Doubell, T. & Davison, A. 1996. Reproduction in the round-eared elephant shrew (*Macroscelides proboscideus*) in the southern Karoo, South Africa. *Journal of Zoology, London* 240: 233–243.

Bredenkamp, G. & van Rooyen, N. 1996. Moist Sandy Highveld Grassland. In: *Vegetation of South Africa, Lesotho and Swaziland*. (eds. A. B. Low & A. G. Rebelo). Department of Environmental Affairs and Tourism, Pretoria. 85 pp.

Bronner, G. N. 1992. Notes on the early post-natal development of a giant golden mole *Chrysospalax trevelyani* (Günther, 1875) born in captivity (Mammalia: Insectivora; Chrysochloridae). *Koedoe* 35: 57–58.

Bronner, G. N. 1995a. Cytogenetic properties of nine species of golden moles (Insectivora: Chrysochloridae). *Journal of Mammalogy* 76: 957–971.

Bronner, G. N. 1995b. Systematic revision of the golden mole genera *Amblysomus*, *Chlorotalpa* and *Calcochloris* (Insectivora: Chrysochloromorpha; Chrysochloridae). PhD thesis, University of Natal, Durban, South Africa.

Bronner, G. N. 1996. Geographic patterns of morphometric variation in the Hottentot golden mole, *Amblysomus hottentotus* (Insectivora: Chrysochloridae). A multivariate analysis. *Mammalia* 60: 729–751.

Bronner, G. N. 2000. New species and subspecies of golden mole (Chrysochloridae: *Amblysomus*) from Mpumalanga, South Africa. *Mammalia* 64: 41–54.

Bronner, G. N. & Jenkins, P. D. 2005. Order Afrosoricida. In: *Mammal Species of the World: A Taxonomic and Geographic Reference* (3rd edn) (eds D. E. Wilson & D. M. Reeder). Johns Hopkins University Press, Baltimore, pp. 71–81.

Bronner, G. N., Hoffman, M., Taylor, P. J., Chimimba, C. T., Best, P. B., Mathee, C. A. & Robinson, T. J. 2003. A revised systematic checklist of the extant mammals of the southern African subregion. *Durban Museum Novitates* 28: 56–106.

Broom, R. 1907a. On some new species of *Chrysochloris*. *Annals and Magazine of Natural History* 19: 262–268.

Broom, R. 1907b. A contribution to the knowledge of the Cape golden moles. *Transactions of the South African Philosophical Society* 18: 283–311.

Broom, R. 1908. Further observations on the Chrysochloridae. *Annals of the Transvaal Museum* 1: 14–16.

Broom, R. 1916. Some observations on the dentition of *Chrysochloris*, and on the tritubercular theory. *Annals of the Natal Museum* 2: 129–140.

Broom, R. 1946. Some new and some rare golden moles. *Annals of the Transvaal Museum* 20: 329–335.

Broom, R. 1950. Some further advances in our knowledge of the Cape golden moles. *Annals of the Transvaal Museum* 21: 234–241.

Brown, J. C. 1964. Observations on the elephant shrews (Macroscelididae) of equatorial Africa. *Proceedings of the Zoological Society of London* 143: 103–119.

Butler, P. M. 1956. The skull of *Ictops* and the classification of the Insectivora. *Proceedings of the Zoological Society of London* 126: 453–481.

Butler, P. M. 1978. 4: Insectivora and Chiroptera. In: *Evolution of African Mammals* (eds V. J. Maglio & H. B. C. Cook). Harvard University Press, Cambridge, Massachusetts, pp. 56–68.

Butler, P. M. 1988. Phylogeny of the insectivores. In: *Phylogeny and Classification of the Tetrapods*, Vol. 2 (ed. M. J. Benton). Clarendon Press, Oxford, pp. 117–141.

Butler, P. M. 1995. Fossil Macroscelidea. *Mammal Review* 25: 3–14.

Cabrera, A. 1932. Los mamiferos de Marruecos. *Trabajos del Madrid Museo Nacional de Ciencias Naturales* 57: 1–361.

Carter, A. M., Enders, A. C., Kunzle, H., Odour-Okelo, D. & Vogel, P. 2004. Placentation in species of phylogenetic importance: the Afrotheria. *Animal Reproduction Science* 82–83: 35–48.

Chabaud, A.-G., Bain, O. & Puylaret, F. 1966. Description de trois nouveau Nématodes Molineinae et considérations sur la systématiques et le caractère archaïque de cette sous-famille. *Bulletin du Muséum national d'Histoire naturelle*, 2ème série 38: 904–920.

Chabaud, A.-G., Bain, O. & Tcheprakoff, R. 1967. Redescription de *Galeiceps cucullus* (Linstow, 1899) et remarques sur l'osmo-régulation des Nématodes Aniskides. *Annales de Parasitologie (Paris)* 42: 321–326.

Corbet, G. B. 1974a. Subfamily Potamogalinae. In: *The Mammals of Africa: An Identification Manual* (eds J. Meester & H. W. Setzer). Part 1.2. Smithsonian Institution Press, Washington, DC, 2 pp.

Corbet, G. B. 1974b. Family Macroscelididae. In: *The Mammals of Africa: An Identification Manual* (eds J. Meester & H. W. Setzer). Part 1.5. Smithsonian Institution Press, Washington, DC, 6 pp.

Corbet, G. B. 1995. A cladistic look at classification within the subfamily Macroscelidinae based upon morphology. *Mammal Review* 25: 15–17.

Corbet, G. B. & Hanks, J. 1968. A revision of the elephant shrews, family Macroscelididae. *Bulletin of the British Museum of Natural History (Zoology)* 16(2): 47–111 + plate.

Coster, S. & Ribble, D. O. 2005. Density and cover preferences of black-and-rufous elephant-shrews (*Rhynchocyon petersi*) in Chome Forest Reserve, Tanzania. *Belgian Journal of Zoology* 135(Suppl.): 175–177.

Cutler, J. B., Dumas, A. & Wynn, R. M. 1998. Immunohistochemical analysis of the interhemal membranes of the elephant shrew. *Gynecologic and Obstetric Investigation* 46: 46–48.

Cuzin, F. & Séguignes, M. 1990. Capture d'*Elephantulus rozeti* (Insectivora, Macroscelididae) dans le Haut-Atlas marocain au-dessus de 2700 m. *Mammalia* 54: 164–165.

Delanoe, P. 1931. Le reservoir de virus du spirochete marocain, *Sp. hispanicum* S. de Buen, var. *marocanum* Ch. Nicolle et Ch. Anderson 1928. *Archives de l'Institut Pasteur de Tunis* 20: 286–314.

Dempster, E. R., Perrin, M. R. & Nuttall, R. J. 1992. Postnatal development of three sympatric small mammal species of southern Africa. *Zeitschrift für Säugetierkunde* 57: 03–111.

Douady, C. J., Catzeflis, F., Kao, D. J., Springer, M. S. & Stanhope M. J. 2002. Molecular evidence for the monophyly of Tenrecidae (Mammalia) and the timing of colonization of Madagascar by Malagasy tenrecs. *Molecular Phylogenetics and Evolution* 22: 357–363.

Douady, C. J., Catzeflis, F., Raman, J., Springer, M. S. & Stanhope, M. J. 2003. The Sahara as a vicariant agent, and the role of Miocene climatic events, in the diversification of the mammalian order Macroscelidea (elephant shrews). *Proceedings of the National Academy of Sciences of the United States of America* 100: 8325–8330.

Downs, C. T. 1996. Renal structure, and the effect of an insectivorous diet on urine composition of southern African elephant-shrew species (Macroscelidea). *Mammalia* 60: 577–589.

Du Toit, S. 1993. Ecophysiology and host status of the rock elephant shrew, *Elephantulus myurus* (Thomas & Schwann, 1906). MSc thesis, University of the Orange Free State, South Africa.

Dubost, G. 1965. Quelques renseignements biologiques sur *Potamogale velox*. *Biologia Gabonica* 1: 257–272.

Duckworth, A. C. & Hickman, G. C. 1985. Defence strategies in coexisting species of oniscomorph and juliform millipedes against predation by the giant golden mole (*Chrysospalax trevelyani*). *South African Journal of Science* 81: 700–701.

Duncan, P. & Wrangham, R. W. 1971. On the ecology and distribution of subterranean insectivores in Kenya. *Journal of Zoology, London* 164: 149–163.

Durrell, G. 1953. Giant water shrew. *Zoo Life, London* 8: 145–147.

Eccard, J. A., Walther, R. B. & Milton, S. J. 2000. How livestock grazing affects vegetation structures and small mammal distribution in the semi-arid Karoo. *Journal of Arid Environments* 46: 103–106.

Ellerman, J. R., Morrison-Scott, T. C. S. & Hayman, R. W. 1953. *Southern African Mammals 1758 to 1951: A Reclassification*. British Museum (Natural History), London, 363 pp.

Emerson, G. L., Kilpatrick, C. W., McNiff, B. E., Ottenwalder, J. & Allard, M. W. 1999. Phylogenetic relationships of the order Insectivora based on complete 12S rRNA sequences from mitochondria. *Cladistics* 15: 221–230.

Evans, F. G. 1942. The osteology and relationships of the elephant shrews (Macroscelididae). *Bulletin of the American Museum of Natural History* 80: 85–125.

Fain, A. & Lukoschus, F. S. 1976. Observations sur la Myobiidae d'Insectivores avec description de taxa nouveaux (Acarina: Prostigmates). *Acta Zoologica et Pathologica Antwerpiensia* 66: 121–188.

Fain, A., Lukoschus, F. S. & Rack, G. 1974. Notes on parasitic mites from some small mammals in Liberia. *Mitteilungen aus dem Hamburgischen Zoologischen Museum und Institut* 71: 165–174.

Faurie, A. S. 1996. A comparative study of communication in six taxa of southern African elephant-shrews (Macroscelididae). PhD thesis, University of Natal, South Africa.

Faurie, A. S., Dempster, E. R. & Perrin, M. R. 1996. Foot drumming patterns of southern African elephant-shrews. *Mammalia* 60: 567–576.

Fielden, L. J., Perrin, M. R. & Hickman, G. C. 1990a. Water metabolism in the Namib Desert golden mole, *Eremitalpa granti namibensis* (Chrysochloridae). *Comparative Biochemistry and Physiology* 96A: 227–234.

Fielden, L. J., Perrin, M. R. & Hickman, G. C. 1990b. Feeding ecology and foraging behaviour of the Namib Desert golden mole, *Eremitalpa granti namibensis* (Chrysochloridae). *Journal of Zoology, London* 220: 367–389.

Fielden, L. J., Hickman, G. C. & Perrin, M. R. 1992. Locomotory activity in the Namib Desert golden mole *Eremitalpa granti namibensis* (Chrysochloridae). *Journal of Zoology, London* 226: 329–344.

Filipucci, M. G., Hickman, G. C., Capanna, E. & Nevo, E. 1991. Genetic diversity and differentiation of the endemic subterranean golden moles of South Africa. *Biochemical Systematics and Ecology* 19: 461–466.

FitzGibbon, C. D. 1994. The distribution and abundance of the golden-rumped elephant shrew *Rhynchocyon chrysopygus* in Kenyan coastal forests. *Biological Conservation* 67: 153–160.

FitzGibbon, C. D. 1995. Comparative ecology of two elephant-shrew species in a Kenyan coastal forest. *Mammal Review* 25: 19–30.

FitzGibbon, C. D. 1997. The adaptive significance of monogamy in the golden-rumped elephant-shrew. *Journal of Zoology, London* 242: 167–177.

FitzGibbon, C. D. & Rathbun, G. B. 1994. Surveying *Rhynchocyon* elephant-shrews in tropical forest. *African Journal of Ecology* 32: 50–57.

FitzGibbon, C. D., Mogaka, H. & Fanshawe, J. H. 1995. Subsistence hunting in Arabuko-Sokoke Forest, Kenya, and its effects on mammal populations. *Conservation Biology* 9: 1116–1126.

Fleming, P. A. & Nicolson, S. W. 2002. How important is the relationship between *Protea humiflora* (Proteaceae) and its non-flying mammal pollinators? *Oecologia* 132: 361–368.

Ford, C. E. & Hamerton, J. L. 1956. Chromosomes of five rodent species. *Nature* 177: 140–141.

Fourie, L. J., du Toit, J. S., Kok, D. J. & Horak, I. G. 1995. Arthropod parasites of elephant-shrews, with particular reference to ticks. *Mammal Review* 25: 31–37.

Freitag, S. & Van Jaarsveld, A. S. 1997. Relative occupancy, endemism, taxonomic distinctiveness and vulnerability: prioritising regional conservation actions. *Biodiversity & Conservation* 6: 211–232.

Gargett, V. & Grobler, J. H. 1976. Prey of the cape eagle owl, *Bubo capensis mackinderi* Sharpe 1899, in the Matopos, Rhodesia. *Arnoldia (Rhodesia)* 8: 1–7.

Guth, C., Heim de Balsac, H. & Lamotte, M. 1959. Recherches sur la morphologie de *Micropotamogale lamottei* et l'evolution des Potamogalinae. I. Ecologie, denture, anatomie crânienne. *Mammalia* 23: 423–447.

Guth, C., Heim de Balsac, H. & Lamotte, M. 1960. Recherche sur la morphologie de *Micropotamogale lamottei* et l'évolution des Potamogalinae. II. – Rachis, viscères, position systematique. *Mammalia* 24: 190–217.

Hanna, N. & Anderson, J. 1994. Njule 92. Final Report, assessing the status and distribution of the black-and-rufous elephant-shrew. Unpublished report. Oxford University expedition to Tanzania, Oxford, UK, 67 pp.

Heim de Balsac, H. 1954. Un genre inédit et inattendu de Mammifère (Insectivore Tenrecidae) d'Afrique Occidentale. *Comptes Rendus de l'Académie des Sciences, Paris* 239: 102–104.

Heim de Balsac, H. 1956. Morphologie divergente des Potamogalinae (Mammifères Insectivores) en milieu aquatique. *Comptes Rendus de l'Académie des Sciences, Paris*: 242: 2257–2258.

Heim de Balsac, H. & Bourlière, F. 1955. Ordre des Insectivores. Systématique. In: *Traité de Zoologie* (ed. P.-P. Grassé), 17: 1653–1697.

Helgen, K. M. & Wilson, D. E. 2001. Additional material of the enigmatic golden mole *Cryptochloris zyli*, with notes on the genus *Cryptochloris* (Mammalia: Chrysochloridae). *African Zoology* 36: 10–112.

Hickman, G. C. 1990. The Chrysochloridae: studies toward a broader perspective of adaptation in subterranean mammals. In: *Evolution of Subterranean Mammals at the Organismal and Molecular Levels* (eds E. Nevo & O. A. Reig). Wiley–Liss, New York. pp. 23–48.

Hollister, N. 1918. East African mammals in the United States National Museum. Part I. Insectivora, Chiroptera, and Carnivora. *United States National Museum Bulletin* 99: 1–194.

Holm, E. 1969. Contribution to the knowledge of the biology of the Namib Desert golden mole *Eremitalpa granti namibensis* Bauer & Niethammer 1959. *Scientific Papers of the Namib Desert Research Station* 41: 37–42.

Holroyd, P. A. & Mussell, J. C. 2005. Macroscelidea and Tubulidentata. In: *The Rise of Placental Mammals* (eds K. D. Rose and J. D. Archibald). Johns Hopkins University Press, Baltimore, pp. 71–83.

Hoogstraal, H., Huff, C. G. & Lawless, D. K. 1950. A malarial parasite of the African elephant shrew, *Elephantulus rufescens dundasi* Dollman. *Journal of the National Malaria Society* 9: 293–306.

Hugot, J.-P. 1979. Description de cinq nouveaux Nématodes d'un Tenrecoidea africain: *Potamogale velox* du Chaillu. *Bulletin du Muséum national d'Histoire naturelle, 4ème série, section A* 4, *Zoologie, Biologie et Écologie animale* 1: 1057–1073.

Hutterer, R. 1993. Order Insectivora. In: *Mammal Species of the World: A Taxonomic and Geographic Reference* (2nd edn) (eds D. E. Wilson & D. M. Reeder). Smithsonian Institution Press, Washington, DC, pp. 69–130.

Hutterer, R. 2005. Order Soricomorpha. In: *Mammal Species of the World: A Taxonomic and Geographic Reference* (eds D. E. Wilson & D. M. Reeder) (3rd edn). Johns Hopkins University Press, Baltimore, pp. 220–311.

Jarvis, J. U. M. 1974. Notes on the golden mole, *Chrysochloris stuhlmanni* Matschie, from the Ruwenzori Mountains, Uganda. *East African Wildlife Journal* 12: 163–166.

Jennings, M. R. & Rathbun, G. B. 2001. *Petrodromus tetradactylus*. *Mammalian Species* 682: 1–6.

Kerley, G. I. H. 1995. The round-eared elephant-shrew *Macroscelides proboscideus* (Macroscelidea) as an omnivore. *Mammal Review* 25: 39–44.

Keymer, I. F. 1971. Blood protozoa of insectivores, bats and primates in Central Africa. *Journal of Zoology, London* 163: 421–441.

Kingdon, J. 1974. *East African Mammals: An Atlas of Evolution in Africa.* Vol. II, Part A: *Insectivores and Bats.* Academic Press, New York, 341 pp.

Kolebinova, M. 1979. *Leptotrombium (Ericotrombium) rheinwaldi* n. sp., a new larval Trombuliculidae (Acarina) from Morocco. *Bonner Zoologische Beitrage* 30: 487–490.

Kolebinova, M. 1981. *Neotrombicula (Neotrombiculoides) elegantissima* n. sp., an elephant shrew parasite from Tanzania. *Senckenbergiana Biologica* 61: 293–296.

Koontz, F. W. 1984. Sternal gland scent communication in the rufous elephant-shrew, *Elephantulus rufescens* Peters, with additional observations on behavior and reproduction in captivity. PhD thesis, University of Maryland.

Koontz, F. W. & Roeper, N. J. 1983. *Elephantulus rufescens*. *Mammalian Species* 204, 5 pp.

Kowalski, K. & Rzebik-Kowalska, B. 1991. *Mammals of Algeria*. Polish Academy of Sciences/Ossolineum Publishing House, Warsaw, 370 pp.

Kuhn, H.-J. 1964. Zur Kenntnis von *Micropotamogale lamottei* (Heim de Balsac). *Zeitschrift für Säugetierkunde* 29: 152–173.

Kuhn, H.-J. 1971. An adult female *Micropotamogale lamottei*. *Journal of Mammalogy* 52: 477–478.

Kuyper, M. A. 1979. A biological study of the golden mole *Amblysomus hottentotus*. MSc dissertation, University of Natal, Pietermaritzburg, South Africa.

Kuyper, M. A. 1985. The ecology of the golden mole *Amblysomus hottentotus*. *Mammal Review* 15: 3–12.

Lamotte, M. & Petter, F. 1981. Une taupe dorée nouvelle du Cameroun (Mt Oku, 06.15N, 10.26E): *Chrysochloris stuhlmanni balsaci* ssp. nov. *Mammalia* 45: 43–48.

Laurie, W. A. 1970. The food of the barn owl in the Serengeti National Park, Tanzania. *Journal of the East African Natural History Society* 28: 1–4.

Lawrence, B. & Loveridge, A. 1953. Mammals from Nyasaland and Tete. *Bulletin of the Museum of Comparative Zoology at Harvard University* 110: 3–80.

Lawrence, R. F. 1944. A new parasitic mite from the golden mole. *Proceedings of the Zoological Society of London* 114: 302–306.

Leirs, H., Verhagen, R., Verheyen, W. & Perrin, M. R. 1995. The biology of *Elephantulus brachyrhynchus* in natural miombo woodland in Tanzania. *Mammal Review* 25: 45–49.

Leon, B., Shkolnik, A. & Shkolnik, T. 1983. Temperature regulation and water metabolism in the elephant shrew *Elephantulus edwardi*. *Comparative Biochemistry and Physiology* 74A: 399–407.

Liu, F.-G. R., Miyamoto, M. M., Freire, N. P., Ong, P. Q., Tennant, M. R., Young, T. S. & Gugel, K. F. 2001. Molecular and morphological supertrees for eutherian (placental) mammals. *Nature* 291: 1786–1789.

Lovegrove, B. G., Lawes, M. J. & Roxburgh, L. 1999. Confirmation of pleisiomorphic daily torpor in mammals: the round-eared elephant shrew *Macroscelides proboscideus* (Macroscelidea). *Journal of Comparative Physiology B* 169: 453–460.

Lovegrove, B. G., Raman, J. & Perrin, M. R. 2000a. Heterothermy in elephant-shrews *Elephantulus* sp. (Macroscelidea): daily torpor or hibernation? *Journal of Comparative Physiology B* 171: 1–10.

Lovegrove, B. G., Raman, J. & Perrin, M. R. 2000b. Daily torpor in elephant shrews (Macroscelidea: *Elephantulus* spp.) in response to food deprivation. *Journal of Comparative Physiology B* 171: 11–21.

Loveridge, A. 1922. Notes on East African Mammalia – other than horned Ungulates – collected and kept in captivity 1915–1919. Part 2. *Journal of the East Africa and Uganda Natural History Society* 17: 39–69.

Lumpkin, S. & Koontz, F. W. 1986. Social and sexual behaviour of the rufous elephant-shrew (*Elephantulus rufescens*) in captivity. *Journal of Mammalogy* 67: 112–119.

Lumpkin, S., Koontz, F. & Howard, J. G. 1982. The oestrous cycle of the rufous elephant-shrew, *Elephantulus rufescens*. *Journal of Reproduction and Fertility* 66: 671–673.

Lundholm, B. G. 1955. Description of new mammals. *Annals of the Transvaal Museum* 22: 279–304.

MacPhee, R. D. E. & Novacek, M. J. 1993. Definition and relationships of Lipotyphla. In: *Mammal Phylogeny: Placentals*, Vol. 2 (eds F. S. Szalay, M. J. Novacek & M. C. McKenna). Springer Verlag, New York, pp. 13–31.

Maddock, A. H. 1986. *Chrysospalax trevelyani*: an unknown and rare mammal endemic to southern Africa. *Cimbebasia, Series A* 8: 87–90.

Maddock, A. H. & Hickman, G. C. 1985. A preliminary report on locomotory activity in wild and captive *Chrysospalax trevelyani* (Mammalia: Chrysochloridae). *South African Journal of Zoology* 20: 271–273.

Madsen, O., Scally, M., Douady, C. J., Kao, D., Debry, R. W., Adkins, R., Amrine, H. M., Stanhope, M. J., de Jong, W. W. & Springer, M. S. 2001. Parallel adaptive radiations in two major clades of placental mammals. *Nature* 409: 610–614.

Malia, M. J., Adkins, R. M. & Allard, M. W. 2002. Molecular support for Afrotheria and the polyphyly of Lipotyphla based on analyses of the growth hormone receptor gene. *Molecular Phylogenetics and Evolution* 24: 91–101.

Mason, M. J. & Narins, P. M. 2001. Seismic signal use by fossorial mammals. *American Zoologist* 41: 1171–1184.

Matson, J. O. & Blood, B. R. 1997. Morphological variability and species limits in elephant shrews (*Elephantulus intufi* and *E. rupestris*) from Namibia. *Mammalia* 62: 77–93.

Matson, J. O., Courtright, M. & Lester, L. A. 1984. Nongeographic variation in the rufous elephant shrew, *Elephantulus rufescens* (Peters, 1878), from Kenya. *Mammalia* 48: 593–598.

Matthey, R. 1954. Les chromosomes de *Macroscelides rozeti* Duvernoy (*Mammalia–Insectivora*) existe-t-il une serie polypoide chez les Macroscelidae? *Revue Suisse de Zoologie* 61: 669–677.

Mayer, A. von, O'Brien, G. & Sarmiento, E. 1995. Functional and systematic implications of the ear in golden moles (Chrysochloridae). *Journal of Zoology, London* 236: 417–430.

McConnell, C. S. 1986. Coexistence of the golden mole *Amblysomus hottentotus* (Chrysochloridae) and the mole-rat *Cryptomys hottentotus* (Bathyergidae). MSc thesis, University of Natal (Pietermaritzburg), South Africa.

McDowell, S. B. 1958. The Greater Antillean insectivores. *Bulletin of the American Museum of Natural History* 115: 115–213.

McKenna, M. C. 1975. Towards a phylogenetic classification of the Mammalia. In: *Phylogeny of the Primates, a multidisciplinary approach* (eds W. P. Luckett & S. Szaley). Plenum Publishing, New York, pp. 21–46.

Meester, J. 1953. A new golden mole from Spitzkop, Sabie. *South African Journal of Science* 49: 207–208.

Meester, J. 1964. Revision of the Chrysochloridae. I. The desert golden mole *Eremitalpa* Roberts. *Scientific Papers of the Namib Desert Research Station* 26: 1–7.

Meester, J. 1974. Family Chrysochloridae. In: *The Mammals of Africa: An Identification Manual* (eds J. Meester & H. W. Setzer). Smithsonian Institution Press, Washington, DC, pp. 1–7.

Meester, J. A., Rautenbach, I. C., Dippenaar, N. J. & Baker, C. M. 1986. Classification of Southern African Mammals. *Transvaal Museum Monograph* 5: 1–359.

Monadjem, A. 1998. Distributional patterns and conservation status of mammals of Swaziland, southern Africa. *Koedoe* 41: 45–59.

Moreau, R. E., Hopkins, G. H. E. & Hayman, R. W. 1946. The type localities of some African mammals. *Proceedings of the Zoological Society of London* 115: 387–447.

Murphy, W. J., Eizirik, E., Johnson, W. E., Ya Ping Zhang, Ryder, O. A. & O'Brien, S. 2001. Molecular phylogenetics and the origins of placental mammals. *Nature* 409: 614–618.

Murphy, W. J., Eizirik, E., O'Brien, S. J., Madsen, O., Scally, M., Douady, C. J., Teeling, E., Ryder, O. A., Stanhope, M. J., de Jong, W. W. & Springer, M. S. 2001. Resolution of the early placental mammal radiation using Baysian phylogenetics. *Science* 294: 2348–2351.

Narins, P. M., Lewis, E. R., Jarvis, J. & O'Riain, J. 1998. The use of seismic signals by fossorial southern African mammals: a neuroethological gold mine. *Brain Research Bulletin* 44: 641–646.

Neal, B. R. 1982. Reproductive biology of the Rufous elephant-shrew, *Elephantulus brachyrhynchus* (Macroscelididae) in Kenya. *Zeitschrift für Säugetierkunde* 47: 65–71.

Neal, B. R. 1984a. Relationship between feeding habits, climate and reproduction of small mammals in Meru National Park. *African Journal of Ecology* 22: 195–205.

Neal, B. R. 1984b. Seasonal feeding habits of small mammals in Kenya. *Zeitschrift für Säugetierkunde* 49: 226–234.

Neal, B. R. 1995. The ecology and reproduction of the Short-snouted Elephant-Shrew, *Elephantulus brachyrhynchus*, in Zimbabwe with a review of the reproductive ecology of the genus *Elephantulus*. *Mammal Review* 25, 51–60.

Nicoll, M. E. 1985. The biology of the Giant otter-shrew *Potamogale velox*. *National Geographical Society Research Report* 21: 331–337.

Nicoll, M. E. & Rathbun, G. B. 1990. *African Insectivora and Elephant-Shrews. An Action Plan for their Conservation.* International Conservation of Nature and Natural Resources, Gland, Switzerland, 53 pp.

Niethammer, G. 1970. Beobachtungen am Pyrenäen-Desman, *Galemys pyrenaica*. *Bonner Zoologische Beiträge* 21: 157–182.

Novacek, M. 1982. Information for molecular studies from anatomical and fossil evidence of higher phylogeny. In: *Macromolecular Sequences in Systematic and Evolutionary Biology* (ed. M. Goodman). Plenum, New York, pp. 3–41.

Novacek, M. 1984. Evolutionary stasis in the elephant-shrew, *Rhynchocyon*. In: *Living Fossils* (eds N. Eldredge & S. M. Stanley). Springer Verlag, New York, pp. 4–22.

Novacek, M. J. & Wyss, A. R. 1986. Higher level relationships of the recent eutherian orders: morphological evidence. *Cladistics* 2: 257–287.

Nowak, R. M. 1999. *Walker's Mammals of the World* (6th edn). Johns Hopkins University Press, Baltimore, 2 vols, 1936 pp.

Oduor-Okelo, D., Gombe, S. & Amoroso, E. C. 1980. The placenta and fetal membranes of the short-nosed elephant shrew, *Elephantulus rufescens* (Peters, 1878). *Saugetierkundliche Mitteilungen* 28: 293–301.

Olson, L. E. & Goodman, S. M. 2003. Phylogeny and biogeography of tenrecs. In: *Natural History of Madagscar* (eds S. M. Goodman & J. P. Benstead). University of Chicago Press, Chicago and London, pp. 1235–1242.

Patterson, B. 1965. The fossil elephant shrews (Family Macroscelididae). *Bulletin of the Museum of Comparative Zoology at Harvard University* 133: 297–336.

Perrin, M. R. 1995. Comparative aspects of the metabolism and thermal biology of elephant shrews (Macroscelidea). *Mammal Review* 25: 67–78.

Petter, F. 1981. Remarques sur la systématique des Chrysochloridés. *Mammalia* 45: 49–53.

Quentin, J.-C. 1978. *Pterygodermatites* (*Neopaucipectines*) *zaiedi* n. sp. nouveau nematode rictulaire parasite du macroscelide en Tunisie. *Bulletin du Muséum national d'Histoire naturelle, 3ème série* 510, *Zoologie* 351: 37–41.

Quentin, J.-C. 1979. Place zoologique de *Spirura rothschildi* Seurat, 1915, nematode parasite de Macroscelide, dans l'evolution du genre *Spirura*. *Bulletin du Muséum national d'Histoire naturelle, 4ème série, section A* 4: *Zoologie, Biologie et Écologie animales* 1: 1051–1056.

Quentin, J.-C. & Seureau, C. 1978. Identification et biologie de gongyloème parasite du macroscelide en Tunisie. *Annales de Parasitologie Humaine et Comparée* 55: 631–640.

Rahm, U. 1960. Note sur les spécimens actuellement connus de *Micropotamogale* (*Mesopotamogale*) *ruwenzorii* et leur répartition. *Mammalia* 24: 511–515.

Rahm, U. 1961. Beobachtungen an der ersten in Gefangenschaft gehaltenen *Mesopotamogale ruwenzorii* (Mammalia-Insectivora). *Revue Suisse de Zoologie* 68: 73–90.

Rahm, U. 1966. Les mammifères de la forêt équatoriale de l'Est du Congo. *Annales du Musée Royal de l'Afrique Centrale, Sciences Zoologiques* 149: 35–121.

Raman, J. & Perrin, M. R. 1997. Allozyme and isozyme variation in seven southern African elephant shrew species. *Zeitschrift für Säugetierkunde* 62: 108–116.

Ranger, G. 1927. Notes on some small mammals found near Kei Road. *Records of the Albany Museum* 3: 487–490.

Rathbun, G. B. 1978. Evolution of the rump region in the golden-rumped elephant-shrew. *Bulletin of Carnegie Museum of Natural History* 6: 11–19.

Rathbun, G. B. 1979. The social structure and ecology of elephant-shrews. *Zeitschrift für Tierpsychologie* (Suppl.) 20: 1–77.

Rathbun, G. B. & Kingdon, J. 2006. The etymology of 'sengi'. *Afrotherian Conservation – Newsletter of the IUCN/SSC Afrotheria Specialist Group* 4: 14–15.

Rathbun, G. B. & Kyalo, S. N. 2000. Golden-rumped Elephant-shrew. In: *Endangered Animals – a Reference Guide to Conflicting Issues* (eds R. P. Reading & B. J. Miller). Greenwood Press, Westport, pp. 125–129, 340–341.

Rathbun, G. B. & Rathbun, C. D. 2006. Social structure of the bushveld sengi (*Elephantulus intufi*) in Namibia and the evolution of monogamy in the Macroscelidea. *Journal of Zoology, London* 269: 391–399.

Rathbun, G. B. & Redford, K. 1981. Pedal scent-marking in the rufous elephant-shrew, *Elephantulus rufescens*. *Journal of Mammalogy* 62: 635–637.

Rathbun, G. B. & Woodall, P. F. 2002. A bibliography of elephant-shrews or sengis (Macroscelidea). *Mammal Review* 32: 66–70.

Rathbun, G. B., Beaman, P. & Maliniak, E. 1981. Capture, husbandry and breeding of rufous elephant-shrews, *Elephantulus rufescens*. *International Zoo Yearbook* 21: 176–184.

Rautenbach, I.L. 1978. A numerical reappraisal of the southern African biotic zones. *Bulletin of the Carnegie Museum of Natural History* 6: 175-187.

Rautenbach, I. L. & Nel, J. A. J. 1980. Mammal diversity and ecology in the Cedarberg Wilderness Area, Cape Province. *Annals of the Transvaal Museum* 32: 101–124.

Rautenbach, I. L. & Schlitter, D. A. 1977. Non-geographic variation in elephant shrews (Genus *Elephantulus* Thomas and Schwann, 1906) of southern Africa. *Annals of the Carnegie Museum of Natural History* 46: 223–243

Ribble, D. O. & Perrin, M. R. 2005. Social organization of the eastern rock elephant-shrew (*Elephantulus myurus*): the evidence for mate guarding. *Belgian Journal of Zoology* 135 (Suppl): 167–173.

Robbins, L. W. & Baker, R. J. 1978. Karyotypic data for African mammals with a description of an *in vivo* bone marrow technique. *Bulletin of the Carnegie Museum of Natural History* 6: 188–210.

Roberts, A. 1924. Some additions to the list of South African mammals. *Annals of the Transvaal Museum* 10: 59–76.

Roberts, A. 1931. New forms of South African mammals. *Annals of the Transvaal Museum* 14: 221–236.

Roberts, A. 1935. Scientific results of the Verney–Lang Kalahari expedition, March to September, 1930. *Annals of the Transvaal Museum* 16: 187–249.

Roberts, A. 1936. Report on a survey of the higher vertebrates of north-eastern Zululand. *Annals of the Transvaal Museum* 18: 163–251.

Roberts, A. 1951. *The Mammals of South Africa*. Trustees of the 'Mammals of South Africa' Book Fund, Pretoria, 700 pp.

Roeper, N. 1981. Foot drumming in the elephant shrew, *Elephantulus rufescens*. MSc thesis, University of Maryland, USA.

Rovero, F. & Rathbun, G. B. 2006. A potentially new giant sengi (elephant-shrew) from the Udzungwa Mountains, Tanzania. *Journal of East African Natural History* 95: 111–115.

Rovero, F., Rathbun, G. B., Perkin, A., Jones, T., Ribble, D. O., Leonard, C., Mwakisoma, R. R. & Doggart, N. 2008. A new species of giant sengi or elephant-shrew (genus *Rhynchocyon*) highlights the exceptional biodiversity of the Udzungwa Mountains of Tanzania. *Journal of Zoology, London* 274: 126–133.

Roxburgh, L. & Perrin, M. R. 1994. Temperature regulation and activity pattern of the round-eared elephant shrew *Macroscelides proboscideus* (Shaw, 1800). *Journal of Thermal Biology* 19: 13–20.

Sauer, E. G. F. 1973. Zum sozialverhalten der kurzohrigen elefantspitzmaus, *Macroscelides proboscideus*. *Zeitshrift für Säugetierkunde* 38: 65–97.

Schlitter, D. A. 1993. Order Macroscelidea. In: *Mammal Species of the World: A Taxonomic and Geographic Reference* (2nd edn) (eds D. E. Wilson & D. M. Reeder). Smithsonian Institution Press, Washington, DC, pp. 829–830.

Schlitter, D. A. 2005. Order Macroscelidea. In: *Mammal Species of the World: A Taxonomic and Geographic Reference* (3rd edn) (eds D. E. Wilson & D. M. Reeder). Johns Hopkins University Press, Baltimore, pp. 82–85.

Séguignes, M. 1983. La torpeur chez *Elephantulus rozeti* (Insectivora, Macroscelididae). *Mammalia* 47: 87–91.

Séguignes, M. 1988. Facteurs de repartition d'*Elephantulus rozeti* (Macroscelidea, Macroscdelididae) en Afrique du nord. Unpublished PhD thesis, Université des Sciences et Techniques du Languedoc, Montpellier, France.

Séguignes, M. 1989. Contribution à l'étude de la reproduction d'*Elephantulus rozeti* (Insectivora, Macroscelididae). *Mammalia* 53: 377–386.

Seurat, L. G. 1913. Sur quelques nematodes du Sud Tunisien. *Bulletin de la Société d'Histoire naturelle de l'Afrique du Nord* 4: 126–130.

Seurat, L. G. 1914. Sur une filare peritoneale du Macroscelide. *Comptes Rendus de Séances de la Société de Biologie* 77: 524–526.

Seymour, R. S., Withers, P. C. & Weathers, W. W. 1998. Energetics of burrowing, running and free-living in the Namib Desert golden mole (*Eremitalpa namibensis*). *Journal of Zoology, London* 244: 107–117.

Shortridge, G. C. 1934. *The Mammals of South West Africa.* 2 vols. William Heinemann, London, 779 pp.

Shortridge, G. C. 1942. Field notes on the first and second expeditions of the Cape Museums' mammal survey of the Cape Province; and descriptions of some new subgenera and subspecies. *Annals of the South African Museum* 36: 27–100.

Simonetta, A. M. 1968. A new golden mole from Somalia with an appendix on the taxonomy of the family Chrysochloridae (Mammalia: Insectivora). *Monitore Zoologico Italiano* (Suppl.) 2: 27–55.

Simpson, G. G. 1945. The principles of classification and a classification of mammals. *Bulletin of the American Museum of Natural History* 85: 1–350.

Skinner, J. D. & Smithers, R. H. N. 1990. *The Mammals of the Southern African Subregion* (2nd edn). University of Pretoria Press, Pretoria, 771 pp.

Smit, H. A., Robinson, T. J., Watson, J. & Jansen Van Vuuren, B. 2008. A new species of elephant-shrew (Afrotheria: Macroscelidea: *Elephantulus*) from South Africa. *Journal of Mammalogy* 89: 1257–1269.

Smithers, R. H. N. 1971. The Mammals of Botswana. *Museum Memoir, National Museums and Monuments of Rhodesia* 4: 1–340.

Smithers, R. H. N. 1983. *The Mammals of the Southern African Subregion*. University of Pretoria Press, Pretoria, 734 pp.

Smithers, R. H. N. 1986. South African Red Data Book – Terrestrial Mammals. *South African National Scientific Programmes Report* 125: 1–216.

Smithers, R. H. N. & Lobão Tello, J. L. P. 1976. Checklist and Atlas of the Mammals of Moçambique. *Museum Memoir, National Museums and Monuments of Rhodesia* 8: 1–184.

Smithers, R. H. N. & Wilson, V. J. 1979. Checklist and Atlas of the mammals of Zimbabwe Rhodesia. *Museum Memoir, National Museums and Monuments of Zimbabwe Rhodesia* 9: 1–193.

Spinks, A C. & Perrin, M. R. 1995. The digestive tract of *Macroscelides proboscideus* and the effects of diet quality on gut dimensions. *South African Journal of Zoology* 30: 33–36.

St Leger, J. 1931. A new golden mole (*Chlorotalpa*) from Mt Elgon, Uganda. *Annals and Magazine of Natural History* ser. 10, 8: 605–607.

Stanhope, M. J., Waddell, V. G., Madsen, O., de Jong, W., Blair Hedges, S., Cleven, G. C., Kao, D. & Springer, M. 1998. Molecular evidence for multiple origins of Insectivora and for a new order of endemic African insectivore mammals. *Proceedings of the National Academy of Sciences of the United States of America* 95: 9967–9972.

Stimson, C. & Goodman, M. 1966. Chromosomes of the elephant-shrew, *Nasilio brachyrhynchus*. *Mammal Chromosome Newsletter* 22: 188.

Stuart, C., Stuart, T. & Pereboom, V. 2003. Aspects of the biology of the Cape Sengi, *Elephantulus edwardii*, from the Western Escarpment, South Africa. *Afrotherian Conservation – Newsletter of the IUCN/SSC Afrotheria Specialist Group* 2: 2–4.

Taylor, P. J., Richardson, E. J., Meester, J. & Wingate, L. 1994. New distribution records for six small mammal species in Natal, with notes on their taxonomy and ecology. *Durban Museum Novitates* 19: 59–66.

Thomas, O. & Schwann, H. 1906. The Rudd explotation of South Africa. V. List of mammals obtained by Mr. Grant in the North East Transvaal. *Proceedings of the Zoological Society of London* 1906: 575–591.

Thompson, S. D. & Nicoll, M. E. 1986. Basal metabolic rate and energetics of reproduction in therian mammals. *Nature* 321: 690–693.

Tolliver, D. K., Robbins, L. W., Rautenbach, I. L., Schlitter, D. A. & Coetzee C. G. 1989. Biochemical systematics of elephant shrews from southern Africa. *Biochemical Systematics and Ecology* 17: 345–355.

Trautmann, K. & Carbone, C. 1991. The exhibition and management of the short-eared elephant shrew *Macroscelides proboscideus* at the Philadelphia Zoo. *AAZPA Regional Conference Proceedings* 1991: 663–670.

Tripp, H. R. H. 1971. Reproduction in elephant-shrews (Macroscelididae) with special reference to ovulation and implantation. *Journal of Reproduction and Fertility* 26: 149–159.

Unger, R. 2000. Anmerkungen sur handaufzucht von kurzohr-russelspringern, *Macroscelides proboscideus* (Shaw, 1800). *Zeitschrift der Kolner Zoo* 43: 121–126.

Unger, R. & Schratter, D. 2000. Nahrungspraferenzen von Kurzohr-Russelspringern, *Macroscelides proboscideus* (Shaw, 1800) in Tiergarten Schonbrunn. *Zoologische Garten* 70: 60–69.

Van der Horst, C. J. 1946. Some remarks on the biology of reproduction in the female of *Elephantulus*, the holy animal of Set. *Transactions of the Royal Society of South Africa* 31: 181–199.

Van der Horst, C. J. 1954. *Elephantulus* going into anoestrus; menstruation and abortion. *Philosophical Transactions of the Royal Society of London* 238B: 27–61.

Van der Horst, C. J. & Gillman, J. 1941. Extreme polyovulation and the factors determining the survival of a single embryo in each uterine horn in *Elephantulus*. *South African Journal of Science* 37: 249.

Van Dijk, A. M., Madsen, O., Catzeflis, F., Stanhope, M. J., de Jong W. W. & Pagel, M. 2001. Protein sequence signatures support the African clade of Mammals. *Proceedings of the National Academy of Sciences of the United States of America* 98: 188–193.

Van Wyk, J. H. 1988. The diet of the rhombic skaapsteker, *Psammophylax rhombeatus rhombeatusi* (Serpentes: Colubridae). *Navorsinge van die Nasionale Museum Bloemfontein* 5: 455–471.

Vaucher, C., Durrette-Desset, M.-C. & Hugot, J.-P. 1979. *Molineus eburnensis* n. sp. (Nematoda: Trichostrongyloidea), parasite de *Micropotamogale lamottei* Heim de Balsac. *Bulletin de la Société neuchâteloise des Sciences naturelles* 102: 49–53.

Verheyen, W. N. 1961a. Recherches anatomiques sur *Micropotamogale ruwenzorii*. 1. La morphologie externe, les viscères et l'organe génital mâle. *Bulletins de la Société Royale de Zoologie d'Anvers* 21: 1–16.

Verheyen, W. N. 1961b. Recherches anatomiques sur *Micropotamogale ruwenzorii*. 2. La myologie de la tête, du cou et de la patte antérieure. *Bulletins de la Société Royale de Zoologie d'Anvers* 21: 1–27.

Verheyen, W. N. 1961c. Recherches anatomiques sur *Micropotamogale ruwenzorii*. 3. La myologie de la patte posterieure, de la région sacrale et de la queue. *Bulletins de la Société Royale de Zoologie d'Anvers* 22: 1–28.

Verheyen, W. N. 1961d. Recherches anatomiques sur *Micropotamogale ruwenzorii*. 4. Observations ostéologiques et considérations générales. *Bulletins de la Société Royale de Zoologie d'Anvers* 22: 1–7.

Vernon, C. J. 1972. An analysis of owl pellets collected in southern Africa. *Ostrich* 43: 109–123.

Vogel, P. 1983. Contribution à l'écologie et à la zoogéographie de *Micropotamogale lamottei* (Mammalia, Tenrecidae). *Revue d'Ecologie (La Terre et la Vie)* 38: 37–49.

Vogel, P., Odartchenko, N. & Graf, J.-D. 1977. Formule chromosomique de *Micropotamogale lamottei* (Mammalia, Tenrecidae). *Mammalia* 41: 81–84.

Waddell, P.J., Kishino, H. & Ota, R. 2001. A phylogenetic foundation for comparative mammalian genomics. *Genome Informatics* 12: 141-154.

Wenhold, B. A. & Robinson, T. J. 1987. Comparative karyology of three species of elephant shrews (Insectivora: Macroscelididae). *Zeitschrift für Säugetierkunde* 52: 1–9.

Withers, P. C. 1978. Bioenergetics of a 'primitive' mammal, the Cape golden mole. *South African Journal of Science* 74: 347–348.

Withers, P. C. 1979. Ecology of a small mammal community on a rocky outcrop in the Namib Desert. *Madoqua* 2: 229–246.

Woodall, P. F. 1987. Digestive tract dimensions and body mass of elephant shrews (Macroscelididae) and the effects of season and habitat. *Mammalia* 51: 537–545.

Woodall, P. F. 1995a. The male reproductive system and the phylogeny of elephant-shrews (Macroscelidea). *Mammal Review* 25: 87–93.

Woodall, P. F. 1995b. The penis of elephant shrews (Mammalia: Macroscelididae). *Journal of Zoology, London* 237: 399–410.

Woodall, P. F. & Currie, G. J. 1989. Food consumption, assimilation and rate of food passage in the Cape Rock Elephant Shrew, *Elephantulus edwardii* (Macroscelidea, Macroscelidinae). *Comparative Biochemistry and Physiology A* 92: 75–79.

Woodall, P. F. & FitzGibbon, C. 1995. Ultrastructure of spermatozoa of the yellow-rumped elephant shrew *Rhynchocyon chrysopygus* (Mammalia: Macroscelidea) and the phylogeny of elephant shrews. *Acta Zoologica (Stockholm)* 76: 19–23.

Woodall, P. F. & Mackie, R. I. 1987. Caecal size and function in the rock elephant shrew *Elephantulus myurus* (Insectivora, Macroscelididae) and the Namaqua rock mouse *Aethomys namaquensis* (Rodentia, Muridae). *Comparative Biochemistry and Physiology A* 87: 311–314.

Woodall, P. F. & Skinner, J. D. 1989. Seasonality of reproduction in male rock elephant shrews, *Elephantulus myurus*. *Journal of Zoology, London* 217: 203–212.

Woodall, P. F., Woodall, L. B. & Bodero, D. A. V. 1989. Daily activity patterns in captive elephant shrews (Macroscelididae). *African Journal of Ecology* 27: 63–76.

Zack, P. S., Penkrot, T. A., Bloch, J. I. & Rose, K. D. 2005. Affinities of 'hyopsodontids' to elephant shrews and a Holarctic origin of Afrotheria. *Nature (London)* 434: 497–501.

Order Tubulidentata

Asher, R., Novacek, M. J. & Geisler, J. H. 2003. Relationships of endemic African mammals and their fossil relatives based on morphological and molecular evidence. *Journal of Mammalian Evolution* 10: 131–194.

Benirschke, K., Wurster, D. H., Low, R. J. & Atkin, N. B. 1970. The chromosome complement of the aardvark, *Orycteropus afer*. *Chromosoma* 31: 68–78.

Carpaneto, G. M. & Germi, F. P. 1989. The mammals in the zoological culture of the Mbuti pygmies in north-eastern Zaire. *Hystrix – Italian Journal of Mammalogy* (n.s.) 1: 1–83.

Clark, W. E. Le Gros & Sonntag, C. F. 1926. A monograph of *Orycteropus afer*. III: The skull. The skeleton of the trunk and limbs. *Proceedings of the Zoological Society of London* 30: 445–485.

Franz, F. 1908. Das Auge von *Orycteropus afer*. *Zoologischer Anzeiger* 32: 148–150.

Frechkop, S. 1946. L'Oryctérope, ne serait-t-il pas le prototype de l'incarnation de Seth-Typhon? *Chroniques d'Egypte* 41: 91–92.

Grubb, P., Jones, T. S., Davies, A. G., Edberg, E., Starin, E. D. & Hill, J. E. 1998. *Mammals of Ghana, Sierra Leone and the Gambia*. Trendrine Press, Zennor, Cornwall.

Hatt, R. T. 1934. The pangolins and aardvarks collected by the American Museum Congo Expedition. *Bulletin of the American Museum of Natural History* 66: 643–672.

Hunt, R. M. & Korth, W. W. 1980. The auditory region of Dermoptera: morphology and function relative to other mammals. *Journal of Morphology* 164: 167–211.

Keimer, L. 1944. L'oryctérope dans l'Egypte Ancienne. *Etude d'Egyptologie* 6: 1–20.

Kingdon, J. 1971. *East African Mammals: An Atlas of Evolution in Africa*, Volume 1. Academic Press, London. 446 pp.

Lehmann, T. 2004. Les Orycteropodidae (Mammalia, Tubulidentata) du Mio-Plio-Quaternaire d'Afrique. Ph.D. thesis, University of Poitiers, France.

Lehmann, T. 2007. Amended taxonomy of the order Tubulidentata (Mammalia, Eutheria). *Annals of the Transvaal Museum* 44: 179–196.

Lehmann, T. 2008. Plio-Pleistocene aardvarks (Mammalia, Tubulidentata) from East Africa. *Fossil Records* 11: 67–81.

Lehmann, T. 2009. Phylogeny and systematics of the Orycteropodidae (Mammalia, Tubulidentata). *Zoological Journal of the Linnean Society* 155: 649–702.

Lehmann, T., Vignaud, P., Likius, A. & Brunet, M., 2004. A fossil aardvark (Mammalia, Tubulidentata) from the lower Pliocene of Chad. *African Journal of Earth Science* 40: 201–217.

Lehmann, T., Vignaud, P., Mackaye, H. T. & Brunet, M., 2005. A new species of Orycteropodidae (Mammalia, Tubulidentata) in the Mio-Pliocene of Northern Chad. *Zoological Journal of the Linnean Society* 143: 109–131.

Lynch, C. D. 1994. The mammals of Lesotho. *Navorsinge van die Nasionale Museum, Bloemfontein* 10 (4): 177–241.

MacInnes, D. G. 1956. Fossil Tubulidentata from East Africa. *Fossil Mammals of Africa*. British Museum (Natural History), London. 10: 1–38.

MacPhee, R. D. E. 1994. Morphology, adaptations, and relationships of *Plesiorycteropus*, and a diagnosis of a new order of eutherian Mammals. *Bulletin of the American Museum of Natural History* 220: 1–214.

Malbrant, R. & Maclatchy, A. 1949. *Faune de l'Equateur Africain Français*. Tome II: *Mammifères*. Paul Lechevalier, Paris. 323 pp.

Manlius, N. 2002. Aardvarks to otters: ancient Egypt's unusual fauna. *Ancient Egypt* 3: 24–31.

Meester, J. 1971. Order Tubulidentata. Part 10. In: *The Mammals of Africa: An Identification Manual* (eds. J. Meester & H.W. Setzer). Smithsonian Institution Press, Washington, DC. pp. 1–2.

Meeuse, A. D. J. 1958. A possible case of interdependence between a mammal and a higher plant. *Archives Néerlandaises de Zoologie* 13: 314–318.

Melton, D. A. 1976. The biology of the aardvark (Tubulidentata–Orycteropodidae). *Mammal Review* 6: 75–88.

Nel, P. J., Taylor, A., Meltzer, D. G. A. & Haupt, M. A. 2000. Capture and immobilisation of aardvark (*Orycteropus afer*) using different drug combinations. *Journal of the South African Veterinary Association* 71: 58–63.

Novacek, M. J. 1989. Higher mammal phylogeny: the morphological-molecular synthesis. In: *The Hierarchy of Life* (eds. B. Fernholm, K. Bremer & H. Jornvall). Elsevier, Amsterdam. pp. 421–435.

Pagès, E. 1970. Sur l'écologie et les adaptations de l'oryctérope et des pangolins sympatriques d'Afrique. *Biologia Gabonica*. 6: 27–92.

Patterson, B. 1975. The fossil aardvarks (Mammalia: Tubulidentata). *Bulletin of the Museum of Comparative Zoology at Harvard College* 147: 185–237.

Patterson, B. 1978. Pholidota and Tubulidentata. In: *Evolution of African Mammals* (eds. V. J. Maglio & H. B. S. Cooke). Harvard University Press, Cambridge, MA. pp. 268–278.

Pickford, M. 1975. New fossil Orycteropodidae (Mammalia, Tubulidentata) from East Africa. *Netherlands Journal of Zoologie* 25: 57–88.

Pickford, M. 2004. Southern Africa: a cradle of evolution. *South African Journal of Science* 100: 205–214.

Pickford, M. 2005. *Orycteropus* (Tubulidentata, Mammalia) from Langebaanweg and Baard's Quarry, Early Pliocene of South Africa. *Comptes Rendus Palevol* 4: 715–726.

Pocock, R. I. 1924. The external characters of *Orycteropus afer*. *Proceedings of the Zoological Society of London* 1924: 697–706.

Rahm, U. 1966. Les mammifères de la forêt équatoriale de l'est du Congo. *Annales du Musée Royal de l'Afrique Centrale, Sciences Zoologiques* 149: 39–121.

Robinson, R. J. & Seiffert E. R. 2004. Afrotherian origins and interrelationships: new views and future prospects. *Current Topics in Developmental Biology* 63: 37–60.

Rögl, F. 1999. Circum-Mediterranean Miocene paleogeography. In: *Land Mammals of Europe* (ed. G. Rössner & K. Heissig). Verlag Dr. Friedrich Pfeil, Munich. pp. 39–48.

Romer, A. S. 1938. Mammalian remains from some Paleolithic stations in Algeria. *Logan Museum Bulletin* 5: 165–184.

Rovero, F. & De Luca, D. W. 2007. Checklist of mammals of the Udzungwa Mountains of Tanzania. *Mammalia* 71: 47–55.

Schouteden, H. 1948. Faune de Congo Belge et du Ruanda–Urundi. I. Mammifères. *Annales du Musée du Congo Belge, Zoologie* 8 (1): 1–331.

Shoshani, J. & McKenna, M. C. 1998. Higher taxonomic relationships among extant mammals based on morphology, with selected comparisons of results from molecular data. *Molecular Phylogenetics and Evolution* 9: 572–584.

Shoshani, J., Goldman, C. A. & Thewissen, J. G. M. 1988. *Orycteropus afer*. *Mammalian Species* 300: 1–8.

Skinner, J. D. & Chimimba, C.T. (eds.) 2005. *The Mammals of the Southern African Subregion*. (Third Edition). Cambridge University Press, Cambridge, UK. 814 pp.

Smithers, R. H. N. 1971. *The Mammals of Botswana*. Museum Memoirs National Museums and Monuments of Rhodesia 4. 340 pp.

Smithers, R. H. N. 1983. *The Mammals of the Southern African Subregion. University of Pretoria, Pretoria, 736 pp.*

Smithers, R. H. N. & Wilson, V. J. 1979. *Check List and Atlas of the Mammals of Zimbabwe Rhodesia*. Museum Memoirs National Museums and Monuments of Rhodesia 9. 147 pp.

Sokolov, V.E., Chernova, O.F., Van Hoven, V. & Van Hoven, S. 1995. Skin structure of *Orycteropus afer* (Tubulidentata). In: *Theriological Investigations in Ethiopia* (ed. V. E. Sokolov). Nauka, Moscow. pp. 297–315 [in Russian].

Sonntag, C. F. 1925. A monograph of *Orycteropus afer*. I: Anatomy except the nervous system, skin, and skeleton. *Proceedings of the Zoological Society of London* 23: 331–437.

Sonntag, C. F. & Woollard, H. H. 1925. A monograph of *Orycteropus afer*. II: Nervous system, sense organs and hairs. *Proceedings of the Zoological Society of London* 1925: 1185–1235.

Springer, M. S., Cleven, G. C., Madsen, O., de Jong, W. W., Waddell, V. G., Amrine, H. M. & Stanhope, M. J. 1997. Endemic African mammals shake the phylogenetic tree. *Nature* 388: 61–64.

Stanhope, M. J., Waddell, V. G., Madsen, O., de Jong, W. W., Hedges, S. B., Cleven, G. C., Kao, D. & Springer, M. S. 1998. Molecular evidence for multiple origins of Insectivora and for a new order of endemic African insectivore mammals. *Proceeding of the National Academy of Sciences of the United States of America* 95: 9967–9972.

Taylor, W. A. 1998. The ecology of the aardvark, *Orycteropus afer* (Tubulidentata – Orycteropodidae). M.Sc. thesis, University of Pretoria, South Africa.

Taylor, W. A. & Skinner, J. D. 2000. Associative feeding between Aardwolves (*Proteles cristatus*) and Aardvarks (*Orycteropus afer*). *Mammal Review* 30: 141–143.

Taylor, W. A. & Skinner, J. D. 2001. Associative feeding between Ant-eating chats, *Myrmecocichla formicivora*, and Aardvarks, *Orycteropus afer*. *Ostrich* 72: 199–200.

Taylor, W. A. & Skinner, J. D. 2003. Activity patterns, home ranges and burrow utilisation of aardvark (*Orycteropus afer*) in the Karoo. *Journal of Zoology, London* 261: 291–297.

Taylor, W. A. & Skinner, J. D. 2004. Adaptations of the aardvark for survival in the Karoo: a review. *Transactions of the Royal Society of South Africa* 59: 105–108.

Taylor, W. A., Lindsey, P. A. & Skinner, J. D. 2002. The feeding ecology of the aardvark *Orycteropus afer*. *Journal of Arid Environments* 50: 135–152.

Thewissen, J. G. M. 1985. Cephalic evidence for the affinities of Tubulidentata. *Mammalia* 49: 257–284.

Thewissen, J. G. M. & Badoux, D. M. 1986. The descriptive and functional myology of the fore-limb of the Aardvark (*Orycteropus afer*, Pallas 1766). *Anatomische Anzeiger* 162: 109–123.

Van der Made, J. 2003. The aardvark from the Miocene hominoid locality Çandir, Turkey. *Courier Forschungs-Institut Senckenberg*, Frankfurt. 240: 133–147.

Vernon, C. J. & Dean, W. R. J. 1988. Further African bird-mammal feeding associations. *Ostrich* 59: 38–39.

Weigl, R. 2005. *Longevity of Mammals in Captivity: From the Living Collections of the World*. Kleine Senckenberg-Reihe 48. pp. 214.

Yalden, D.W., Largen, M. J., Kock, D. & Hillman, J. C. 1996. Catalogue of the mammals of Ethiopia and Eritrea. 7. Revised checklist, zoogeography and conservation. *Tropical Zoology* 9: 73–164.

Yang, F., Alkalaeva, E. Z., Perelman, P. L., Pardini, A. T., Harrison, W. R., O'Brien, P. C., Fu, B., Graphodatsky, A. S., Ferguson-Smith, M. A. & Robinson, T. J. 2003. Reciprocal chromosome painting among human, aardvark, and elephant (superorder Afrotheria) reveals the likely eutherian ancestral karyotype. *Proceeding of the National Academy of Sciences of the United States of America* 100: 1062–1066.

Glossary

Emlen, S.T. & Oring, L.W. 1977. Ecology, sexual selection, and the evolution of mating systems. *Science* 197: 215–223.

Mayr, E., Linsley, E. G. & Usinger, R. L. 1953. *Methods and Principles of Systematic Zoology*. McGraw-Hill, New York, x + 328 pp.

Rosevear, D. R. 1953. *Checklist and Atlas of Nigerian Mammals, with a Foreword on Vegetation*. The Government Printer, Lagos, Nigeria, 131 pp. + maps.

Rosevear, D. R. 1965. *The Bats of West Africa*. Trustees of the British Museum (Natural History), London, xvii + 418 pp., 1 map.

Smithers, R. H. N. 1983. *The Mammals of the Southern African Subregion*. University of Pretoria Press, Pretoria, 734 pp.

White, F. 1983. *The Vegetation of Africa: A Descriptive Memoir to Accompany the UNESCO/AETFAT/UNSC Vegetation Map of Africa*. UNESCO, Paris.

Authors of Volume I

Richard F.W. Barnes
Ecology, Behavior & Evolution
Division of Biological Sciences 0116
University of California at San Diego
La Jolla, CA 92093-0116
USA
email: rfwbarnes@znet.com

Ron E. Barry
Biology Department
Bates College
44 Campus Avenue
Lewiston, ME 04240
USA
email: rbarry@bates.edu

Paulette Bloomer
Genetics Laboratory
University of Pretoria
Pretoria Campus
Gauteng Province 0002
South Africa
email: p.bloomer@up.ac.za

Gary N. Bronner
Department of Zoology
University of Cape Town
Private Bag, Rondebosch 7701
South Africa
email: gary.bronner@uct.ac.za

Thomas M. Butynski
King Khalid Wildlife Research Centre, Thaumama
Thaumama Research Centre
PO Box 61681
Riyadh 11575
Kingdom of Saudi Arabia
email: tbutynski@aol.com

Daryl P. Domning
Department of Anatomy
Howard University
Laboratory of Evolutionary Biology
520 W. St. NW
Washington, DC 20059
USA
email: ddomning@howard.edu

Paul Dutton
Department of Fisheries, Western Australia
WIOME Research Project
3 Baron-Hay Court
South Perth WA 6151
Australia

Angela Gaylard
South African National Parks (SANParks)
PO Box 3542
Knysna 6570
South Africa
email: angela.gaylard@sanparks.org

Colin P. Groves
School of Archeology and Anthropology
Australian National University
Canberra, ACT 0200
Australia
email: colin.groves@anu.edu.au

David C. D. Happold
Research School of Biology
Australian National University
Canberra, ACT 0200
Australia
email: David.Happold@anu.edu.au

S. Blair Hedges
Pennsylvania State University
Biology Building
208 Mueller, University Park
PA 16802
USA
email: sbh1@psu.edu

Hendrik N. Hoeck
formerly Max-Planck Institute, now retired
Correspondence address:
Sonnenbergstrasse 3
8280 Kreuzlingen
Switzerland
email: h.hoeck@gmx.ch

Michael Hoffmann
IUCN Species Survival Commission
c/o United Nations Environment Programme – World Conservation Monitoring Centre
219c Huntingdon Road
Cambridge CB3 0DL
UK
email: mike.hoffmann@iucn.org

Paula Kahumbu
WildlifeDirect
Africa Conservation Fund (KENYA)
PO Box 24467
Karen 00502
Nairobi
Kenya
email: paula@wildlifedirect.org

Jonathan Kingdon
Department of Zoology
University of Oxford
WildCRU, Tubney House
Abingdon Road
Tubney OX13 5QL
UK

Thomas Lehmann
Abteilung Paläoanthropologie und Messelforschung Sektion Säugetierphylogenie
Senckenberg-Forschungsinstitut und Naturmuseum Senckenberganlage 25
D-60325 Frankfurt am Main
Germany
email: Thomas.Lehmann@senckenberg.de

Daniel Livingstone
Department of Geology
Biological Sciences Building
Duke University
2671 Davis Street, Raleigh
NC 27608
USA
email: livingst@duke.edu

J. Michael Lock
(formerly Royal Botanic Gardens Kew, Richmond, Surrey, UK)
Glen Fern
Whitford Road, Musbury
Axminster EX13 7AP
UK
email: mike@glenfern.eclipse.co.uk

Helene Marsh
Earth & Environmental Sciences
James Cook University
Sir G. Fisher Building
Townsville Campus
Townsville, N. Queensland
Australia
email: Helene.Marsh@jcu.edu.au

Jos Milner
Balhallach Cottage
Girnoc, Ballater
Aberdeenshire AB35 5SR
UK
email: jos.milner@gmail.com

Robert J. Morley
Department of Earth Sciences
Royal Holloway, University of London
Queens Building, Egham
Surrey TW20 0EX
UK
email: palynova@btinternet.com

David Moyer
Tanzania WCS Flight Program
PO Box 936, Iringa
Tanzania
email: daudi.simoya@gmail.com

Mike Perrin
School of Botany and Zoology
University of KwaZulu-Natal
P/Bag X01
Scottsville 3209
South Africa

Joyce Poole
Buskhellinga 3
Sandefjord
Norway
email: jpoole@elephantvoices.org

James A. Powell
Sea2Shore Alliance
4411 Bee Ridge Road
Sarasota
Florida 34233
USA
email: jpowell@sea2shore.org

Galen B. Rathbun
Department of Ornithology and Mammalogy
California Academy of Sciences (San Francisco)
c/o PO Box 202
Cambria, CA 93428-0202
USA
email: grathbun@calacademy.org

John E. Reynolds III
Mote Marine Laboratory
Laboratory of Marine Life
1600 Ken Thompson Parkway
Sarasota, FL 34236
USA
email: reynolds@mote.org

Diana Roberts
Department of Zoology
University of Oxford
WildCRU, Tubney House
Abingdon Road
Tubney OX13 5QL
UK
email: diana.roberts@zoo.ox.ac.uk

Susanne Shultz
Institute of Cognitive and Evolutionary Anthropology
University of Oxford
64 Banbury Road
Oxford OX2 6PN
UK
email: susanne.shultz@anthro.ox.ac.uk

Erik R. Seiffert
Department of Anatomical Sciences
Stony Brook University
Health Sciences Center T-8, Stony Brook
New York, NY 11794-8081
USA
email: erik.seiffert@stonybrook.edu

Jehesky Shoshani
Deceased

Pascal Tassy
Histoire de la Terre
Muséum national d'Histoire naturelle
Centre de recherche sur la paléobiodiversité et les paléoenvironnements (CP38)
57 rue Cuvier
Paris, 75231 Cedex 05
France
email: ptassy@mnhn.fr

William A. Taylor
Suite 4A
PO Box 71664
Bryanston
2021
South Africa
email: taylor.wa@gmail.com

Elmer Topp-Jørgensen
Insitute of Bioscience
Aarhus University
Denmark
email: elmer_topp@hotmail.com

Andrea Turkalo
International Programs
Wildlife Conservation Society
2300 Southern Blvd
Bronx, NY 10460
USA
email: dzanga2000@yahoo.com

Peter Vogel
Institut d'Ecologie – Zoologie et Ecologie Animale, Bâtiment de Biologie
Université de Lausanne
CH-1015 Lausanne
Switzerland
email: Peter.Vogel@unil.ch

Fritz Vollrath
Department of Zoology
University of Oxford
South Parks Road
Oxford OX1 3PS
email: fritz.vollrath@zoo.ox.ac.uk

Ian Whyte
PO Box 814
Graskop 1270
Mpumalanga
South Africa
email: ian@mpu.co.za

Indexes

Subject index for chapters 1–8

Index of species profiles

French names

German names

English names

Scientific names